The Bloomsbury Handbook of Sonic Methodologies

The Bloomsbury Handbook of Sonic Methodologies

Edited by Michael Bull
and Marcel Cobussen

BLOOMSBURY ACADEMIC
NEW YORK • LONDON • OXFORD • NEW DELHI • SYDNEY

BLOOMSBURY ACADEMIC
Bloomsbury Publishing Inc
1385 Broadway, New York, NY 10018, USA
50 Bedford Square, London, WC1B 3DP, UK
29 Earlsfort Terrace, Dublin 2, Ireland

BLOOMSBURY, BLOOMSBURY ACADEMIC and the Diana logo
are trademarks of Bloomsbury Publishing Plc

First published in the United States of America 2021
This paperback edition published 2023

Copyright © by Michael Bull and Marcel Cobussen, 2021

Each chapter © of Contributor

Cover design: Louise Dugdale
Cover image © SJG / Joost Grootens, Clémence Guillemot

All rights reserved. No part of this publication may be reproduced or transmitted in any form or by any means, electronic or mechanical, including photocopying, recording, or any information storage or retrieval system, without prior permission in writing from the publishers.

Bloomsbury Publishing Inc does not have any control over, or responsibility for, any third-party websites referred to or in this book. All internet addresses given in this book were correct at the time of going to press. The author and publisher regret any inconvenience caused if addresses have changed or sites have ceased to exist, but can accept no responsibility for any such changes.

Library of Congress Cataloging-in-Publication Data
Names: Bull, Michael, 1952- editor. | Cobussen, Marcel, 1962- editor.
Title: The Bloomsbury handbook of sonic methodologies /
edited by Michael Bull and Marcel Cobussen.
Description: New York City : Bloomsbury Academic, 2020. |
Includes bibliographical references and index. |
Summary: "An interdisciplinary overview of the variety of sonic methodologies used by sound scholars and artists based on contemporary theories and empirical analyses"– Provided by publisher.
Identifiers: LCCN 2020036716 (print) | LCCN 2020036717 (ebook) |
ISBN 9781501338755 (hardcover) | ISBN 9781501338762 (epub) |
ISBN 9781501338779 (pdf)
Subjects: LCSH: Sound–Research. | Interdisciplinary research.
Classification: LCC QC226 .B46 2020 (print) | LCC QC226 (ebook) | DDC 534–dc23
LC record available at https://lccn.loc.gov/2020036716
LC ebook record available at https://lccn.loc.gov/2020036717

ISBN:	HB:	978-1-5013-3875-5
	PB:	978-1-5013-9350-1
	ePDF:	978-1-5013-3877-9
	eBook:	978-1-5013-3876-2

Typeset by Integra Software Services Pvt. Ltd

To find out more about our authors and books visit www.bloomsbury.com
and sign up for our newsletters.

Contents

List of Illustrations xi
List of Contributors xvii

Introduction
Michael Bull and Marcel Cobussen 1

Part I Disciplines, Methodologies, Epistemologies

1. **Introduction to Part I: Sounds Inscribed onto the Face – Rethinking Sonic Connections through Time, Space, and Cognition**
   Michael Bull 17

2. **Sonic Methodologies in Anthropology**
   Alexandrine Boudreault-Fournier 35

3. **Sonic Methodologies by Way of Deconstruction**
   Naomi Waltham-Smith 57

4. **Nature's Music: Sonic Methodologies in the Study of Environmental Biology**
   Wouter Halfwerk 75

5. **Hearing With: Researching the Histories of Sonic Encounter**
   James G. Mansell 93

6. **Sonic Methodologies in Urban Studies**
   Christabel Stirling 115

7. **Sound and Pedagogy: Taking Podcasting into the Classroom**
   Neil Verma 141

8   **Sonic Methodologies in Literature**
    Justin St. Clair 155

9   **Sonic Materialism and/as Method**
    Tyler Shoemaker 169

10 **Sonic Methodology in Philosophy**
    Elvira Di Bona 187

11 **Sonic Methodologies in Science and Technology Studies**
    Joeri Bruyninckx and Alexandra Supper 201

12 **The Sonic Environment in Urban Planning, Environmental Assessment and Management**
    A. Lex Brown 217

13 **Sonic Methodologies in Medicine**
    Jos J. Eggermont 235

14 **Soundscape as Methodology in Psychoacoustics and Noise Management**
    André Fiebig and Brigitte Schulte-Fortkamp 253

15 **Sonic Methodologies of Sound**
    Salomé Voegelin 269

Part II Sound Arts, Musics, Spaces

16 **Introduction to Part II: Art – Research – Method**
    Marcel Cobussen 283

17 **Ambulatory Sound-Making: Rewriting, Reappropriating, 'Presencing' Auditory Spaces**
    Elena Biserna 297

18 **Sound Installations for the Production of Atmosphere as a Limited Field of Sounds**
    Jordan Lacey 315

19 **Fragile Devices: Improvisation as an Interdisciplinary Research Methodology**
Rebecca Caines 325

20 **'The Music Comes from Me': Sound as Auto-Ethnography**
Darla Crispin 341

21 **Sound beyond Representation: Experimental Performance Practices in Music**
Lucia D'Errico 357

22 **Performing Centrifugal Sound**
G Douglas Barrett 369

23 **How to Cut Up a Record?**
Paul Nataraj 383

24 **Directing Listening: Sound Design Methods from Film to Site-Responsive Sonic Art**
Ben Byrne 397

25 **Sound, Space, and Pneumatic Valves: Using Pneumatic Valves as Sound Sources to Create Spatial Environments**
Edwin van der Heide 407

26 *The Overheard*: **An Attuning Approach to Sound Art and Design in Public Spaces**
Marie Højlund, Jonas R. Kirkegaard, Michael Sonne Kristensen, and Morten Riis 423

27 **Sound on Sound: Considerations for the Use of Sonic Methods in Ethnographic Fieldwork inside the Recording Studio**
Paul Thompson 437

28 **Ecological Sound Art**
Jonathan Gilmurray 449

29 **Hydrophonic Fields**
Jana Winderen interviewed by Stefan Helmreich 459

30 Melt Me into the Ocean: Sounds from Submarine Spaces
Yolande Harris 469

31 Attentive Listening in Lo-Fi Soundscapes: Some Notes on the Development of Sound Art Methodologies in Vietnam
Stefan Östersjö and Nguyễn Thanh Thủy 481

Part III Geographies, Politics, Histories

32 Introduction to Part III: Listening as Method
Marcel Cobussen 499

33 Auditory Diagramming: A Research/Design Practice
Alex Arteaga 511

34 Close Listening: Approaches to Research on Colonial Sound Archives
Anette Hoffmann 529

35 Sonic Feminisms: Doing Gender in Neoliberal Times
Marie Thompson 543

36 Sound as City Maker: Developing a Participatory-Collaborative Process to Work with Sound as an Urban Resource; the Case of Mr. Visserplein (Amsterdam, the Netherlands)
Edda Bild, Michiel Huijsman, and Renate Zentschnig 557

37 Dropping Down Low: Online Soundmaps, Critique, Genealogies, Alternatives
Angus Carlyle 581

38 Listening as Methodological Tool: Sounding Soundwalking Methods
John L. Drever 599

39 **Sounding Wild Spaces: Inclusive Map-Making through Multispecies Listening across Scales**
    Alice Eldridge, Jonathan Carruthers-Jones, and Roger Norum 615

40 **The Emergence of Voices in an Indian Bus Stand: An Ethnographic and Acoustic Approach**
    Christine Guillebaud 633

41 **Historical Sounds: A Case Study**
    Aimée Boutin 647

42 **Sonic Writing**
    Holger Schulze 659

43 **Silence of Mauá: An Atmospheric Ethnography of Urban Sounds**
    Jean-Paul Thibaud 671

44 **Sound Design Methodologies: Between Artistic Inspiration and Academic Perspiration**
    Nicolas Misdariis and Daniel Hug 685

45 **Listening to the 2001 Argentine Crisis: Soundscapes of Protest, Music, and Sound Art**
    Violeta Nigro Giunta 705

46 **The Sound System of the State: Critical Listening as Performative Resistance**
    Tom Tlalim 719

47 **Sonifications Sometimes Behave So Strangely**
    Paul Vickers 733

48 **The Conflicting Sounds of Urban Regeneration in Liverpool**
    Jacqueline Waldock 745

49 **Ethnographies Sounded on What? Methodologies, Sounds and Experiences in Cairo**
    Vincent Battesti 755

50 **Podcast Preservation and the Noise of Saved Sounds**
Jeremy Wade Morris 779

51 **The Earview as a Border Epistemology: An Analytical and Pedagogical Proposition for Design**
Pedro J. S. Vieira de Oliveira 795

52 **Hacking Composition: Dialogues with Musical Machines**
Ezra J. Teboul 807

Index 821

Illustrations

Figures

4.1 Sonograms of different species. A cricket, a frog, and two birds, one with a simple song and the other with a complex song 77

4.2 Sonogram of an Andean solitaire (*M. ralloides*) that demonstates the two-voice phenomenon 83

4.3 Sonograms of a recording made with a microphone and a laser-Doppler vibrometer 83

5.1 Photograph of an RAC 'Quiet' road sign in Darlington 95

5.2 Front cover of a booklet of positive thoughts to be recited silently in time with the chimes of Big Ben at 9.00 p.m. 97

5.3 Section of an advertisement for the Underwood Noiseless Typewriter 101

5.4 Cadbury's 'Silent Theatre Box' editorial 102

5.5 Cover of leaflet announcing the Darlington Quiet Town Experiment 107

5.6 Logo of the Darlington Quiet Town Experiment 108

5.7 Detail from 'Do you have a noisy gnome in your home?' leaflet 110

5.8 Image entitled 'Noise reading being taken on a building site' 111

6.1	Chris's 'musical London' map, 2014 132
6.2	Ali's 'musical London' map, 2014 133
6.3	Martin's 'musical London' map, 2014 134
13.1	The sound conduction pathway in the human ear 245
13.2	Cross section of the cochlea showing the location of the organ of Corti containing the outer and inner hair cells from which the spiral ganglion cells leave 246
13.3	Frequency-specific click-evoked otoacoustic emission waveforms obtained from the human ear 246
13.4	Auditory brainstem response sources in the brainstem 247
13.5	Three examples of a comparison between behavioural audiograms and audiograms based on auditory brainstem response – threshold responses 248
13.6	Auditory brainstem responses (ABR) and obligatory auditory evoked potentials (AEP) on a logarithmic timescale 248
14.1	The perceptual construct of a soundscape according to ISO 12913–1 255
14.2	Types of soundscape studies and their main actors 256
14.3	Methods and instruments frequently applied in soundscape studies 257
14.4	Assessments of loudness of eight sites repeatedly visited by different soundwalk groups over several years 263
14.5	Assessments of unpleasantness of eight sites repeatedly visited by different soundwalk groups over several years 263

14.6	Unpleasantness group judgements over measured loudness values according to ISO 532–1 (left) and over LAeq-values in decibels (dB)(A) (right)	264
17.1	Scratch Orchestra, *Richmond Journey*, 1969, programme	306
17.2	Scratch Orchestra, *Richmond Journey*, 1969, map with the itinerary	307
17.3	Elana Mann, Take a Stand Marching Band, documentation of the Los Angeles May Day march, 1st May 2017	309
17.4	Elana Mann, Take a Stand Marching Band, documentation of the Los Angeles May Day march, 1st May 2017	310
23.1	'Popcorn', La Strana Società – inscribed record by Paul Nataraj	390
25.1	The pneumatic valve used in *Pneumatic Sound Field*	410
25.2	*Pneumatic Sound Field* during DEAF07, Museum Boijmans van Beuningen, Rotterdam, 2007	412
25.3	*Schwingungen – Schwebungen*, bonn hoeren, Bonn, 2015	417
26.1	Screenshot from *The Overheard* website	427
26.2	*Forest Megaphones* by Birgit Öigus	429
26.3	The memorial monument at Mindeparken, Aarhus	430
36.1	Visualization of the process of 'Crowdsourcing Mr. Visserplein' with its three core parts	561
36.2	Historical photograph of Mr. Visserplein, 1983	563
36.3	Mr. Visserplein, aerial photograph, February 2016	563
36.4	Self-guided soundwalk: trajectory (left) and walk (right)	565

36.5	Space-use-sound model	566
36.6	Workshop 1 on characterizing the space, use, and sound of Mr. Visserplein	568
36.7	Square redesign proposals	569
36.8	Workshop 2 on visualizations and auralizations of square redesign proposals	570
36.9	'Geluid als stadmaker' event attendees on the augmented soundwalk	570
36.10	Project exhibition: banners	571
36.11	'Geluid als stadmaker' – event brochure	572
39.1	Research study site at Abisko National Park showing the walking transect (black) and waypoints (numbered crosses) and the river (white) running into lake Torneträsk	622
39.2	Schematic of proposed conceptual framework detailing co-design of mixed methods approach to inclusive wilderness mapping	624
40.1	Buses at Saktan Tampuran, 2015–2016	638
40.2a	Sonogram of the main vendor making the utterance 'Palakkad'	639
40.2b	Sonogram focused on utterance B	640
40.3	Sonogram of the main vendor making the utterance 'Peecheedam', while the two secondary vendors call 'Kuntakulam' and 'Palakkad'	642
40.4	A multiple-configuration sonogram with seven vendors	643
43.1	A view of Condominío Barão de Mauá	672
44.1	Image of the minute-repeater device and its physical model	691

44.2	Functional scheme of the prototyped sound synthesis engine: a wavetable synthesis with four parallel buffers whose frequency and gain are driven by the vehicle's speed	694
44.3	(a) Two 'wizards', performing their interaction mock-up. (b) Live try-out and exchange with some participants	697
45.1	Score of Luciano Azzigotti's International Errorista	713
45.2	Buenos Aires Sonora, *Mayo, los sonidos de la Plaza* (2003), press release	714
49.1	Ahmed Wahdan, Giza, Cairo, Egypt p.m.	756
49.2	Workshop open on the street, al-Gamaliyya, Cairo, Egypt, 28 November 2016, 3.30 p.m.	758
49.3	In the street of Gamaliyya, Cairo, Egypt, 28 November 2016, 3.00 p.m.	761
49.4	Part of a loud sound system unpacked for a birth celebration (subu'), Bashtīil, Cairo, Egypt, 18 November 2016, 4.30 p.m.	765
49.5	Promenade on a bridge over the Nile, Downtown, Cairo, Egypt, 3 November 2016, 5.00 p.m.	771
50.1	Two sound waves from different podcasts, indicating different levels of production, editing, and mastering for each	788
52.1	The inside of the oscillator box, containing one Hex Schmitt Trigger, 74C14 chip	812
52.2	The oscillator box connected to the mixer circuit components using a ribbon connector attached to fishing weights for *The Royal Touch* set-up	813

Tables

14.1 Methodical Aspects of a Soundwalk Method 260

36.1 Characterization of Mr. Visserplein in Terms of Space, Use, and Sound 564

39.1 Data Types Associated With Each of the Surveys Carried Out in Abisko National Park 622

Contributors

G Douglas Barrett is Assistant Professor of Communication at Salisbury University.

Vincent Battesti is a researcher in social anthropology at the CNRS. Website: https://vbat.org.

Edda Bild is a postdoctoral fellow at the School of Information Studies, McGill University.

Elena Biserna is associate researcher at PRISM (AMU /CNRS) and TEAMeD (Université Paris 8).

Aimée Boutin is Professor of French at Florida State University.

Lex Brown is Professor Emeritus in environmental planning, Griffith University.

Joeri Bruyninckx is Assistant Professor of Science and Technology Studies at Maastricht University.

Michael Bull is Professor of Sound Studies at the University of Sussex.

Ben Byrne is a Senior Lecturer, Digital Media at RMIT University.

Rebecca Caines is an Associate Professor in Interdisciplinary Programs, University of Regina.

Angus Carlyle is Professor of Sound and Landscape at the University of the Arts, London.

Marcel Cobussen is Full Professor of Auditory Culture and Music Philosophy at Leiden University.

Darla Crispin is Vice Rector for Research & Artistic Development at the Norwegian Academy of Music (NMH), Oslo.

Jonathan Curruther-Jones is a Marie Skłodowska Curie doctoral research fellow at the University of Leeds.

Lucia D'Errico is a postdoc fellow at the Orpheus Institute (Ghent, Belgium).

Pedro J S Vieira de Oliveira is a researcher, sound artist and educator.

Elvira Di Bona is Assistant Professor of Philosophy, University of Turin.

John Drever is Professor of Acoustic Ecology and Sound Art at Goldsmiths, London.

Jos J. Eggermont is Emeritus Professor at the University of Calgary.

Contributors

Alice Eldridge is a Lecturer in Music and Music Technology at the University of Sussex.

André Fiebig is a visiting professor at the Technical University of Berlin.

Jonathan Gilmurray is a sound artist, writer and lecturer.

Violeta Nigro Giunta is a PhD candidate at the CRAL-EHESS, Paris.

Christine Guillebaud is a research fellow at the French National Centre for Scientific Research (CNRS).

Wouter Halfwerk is an Assistant Professor at the Vrije Universiteit.

Yolande Harris teaches at the University of California, Santa Cruz.

Edwin van der Heide is an artist who runs his own studio and is part-time lecturer and researcher at Leiden University.

Stefan Helmreich is Professor of Anthropology at MIT.

Anette Hoffmann is Lise Meitner fellow at the Academy of Fine Arts in Vienna, Austria.

Marie Højlund is an Assistant Professor in Sound Studies at Aarhus University.

Daniel Hug lectures at the University of Applied Sciences Northwestern Switzerland.

Michiel Huijsman is an artist, researcher and independent curator based in Amsterdam.

Jordan Lacey is Senior Lecturer at RMIT University, Melbourne.

Jonas R. Kirkegaard is an Associate Lecturer at the Sonic College, Denmark.

Michael Sonne Kristensen lectures at the Sonic College, Denmark.

James G. Mansell is an Associate Professor at the University of Nottingham.

Nicolas Misdariis is head of the Ircam STMS Lab / Sound Perception & Design team.

Jeremy Morris is an Associate Professor of Media and Cultural Studies at the University of Wisconsin-Madison.

Paul Nataraj is a sound artist, writer, podcast producer and music researcher.

Roger Norum is a social anthropologist who works among communities in the Arctic and Asia.

Stefan Östersjö is Professor of Musical Performance and Head at Luleå University of Technology.

Morten Riis is an Associate Professor in electronic music composition at the Royal Academy of Music, Aarhus, Denmark.

Justin St. Clair is Associate Professor of English at the University of South Alabama.

Brigitte Schulte-Fortkamp is a Professor of Psychoacoustics and Noise Effects recently retired from the Technische Universität Berlin, Germany.

Holger Schulze is Full Professor in musicology at the University of Copenhagen.

Tyler Shoemaker is a PhD student at the University of California, Santa Barbara.

Christabel Stirling is a musical ethnographer and sound studies researcher and visiting lecturer at the University of Westminster.

Alexandra Supper is an Assistant Professor in the Department of Society Studies at Maastricht University.

Ezra J. Teboul is an artist and researcher. More information is available at redthunderaudio.com.

Jean-Paul Thibaud, sociologist, is CNRS senior researcher at Cresson (https://aau.archi.fr/cresson).

Tom Tlalim is a Senior Lecturer at the University of Winchester.

Marie Thompson is a Lecturer in music at The Open University, UK.

Paul Thompson has worked as a professional recording engineer for over ten years, in and around Liverpool.

Nguyễn Thanh Thủy is a leading Vietnamese zither player and holds a teaching position at the Vietnam National Academy of Music.

Neil Verma is Assistant Professor of sound studies at Northwestern University.

Paul Vickers is Associate Professor of Computer Science and Computational Perceptualisation at Northumbria University in Newcastle.

Salomé Voegelin is a Professor of Sound at the London College of Communication, University of the Arts London. www.salomevoegelin.net

Jacqueline Waldock is researcher at the University of Liverpool.

Jana Winderen is an artist who currently lives and works in Oslo, Norway.

Renate Zentschnig is currently working on Urban Sound Lab, a long-term participatory sound project in Amsterdam Zuid.

Introduction
Michael Bull and Marcel Cobussen

Like many such endeavours, this general introduction is written after all else has been completed – the beginning comes at the end. The editors had not foreseen how a certain contingency might arise to completely upend daily life, such as the development of a global pandemic that has now surrounded their working and living environments. So this introduction to The Bloomsbury Handbook of Sonic Methodologies *is carried out as a virtual dialogue between the two editors – one in the UK, the other in the Netherlands. In addition to this general introduction there are introductions for the three parts that comprise this book, the first written by Michael Bull and the second and third written by Marcel Cobussen. The editors had already decided that these three part introductions were to function as 'interventions' rather than as traditional overviews of contents, a common format that so often goes unread by those browsing through volumes such as this.*

The two editors – joined by, among others, their interests in philosophy and football (not the same teams) – come from differing trajectories and backgrounds. Marcel was educated at the Rotterdam Conservatory and worked for over fifteen years as a professional (jazz) pianist and teacher before he changed his focus to cultural studies and (Continental) philosophy and became a Professor at Leiden University, primarily supervising artistic researchers, thereby combining his experiences in both the academic and the art world. Michael, whose first degree was in philosophy and sociology, initially had a career in adult education before branching out to establish and run a jazz club in Central London. At the age of forty he took a risky 'leap of faith,' re-entering the academy to write a doctorate in the Sociology Department of Goldsmiths College on the use of Walkmans. The rest, they say, is history!

This book is somewhat larger than the editors had initially anticipated, with over fifty chapters deriving from scholars working in the arts, humanities, social sciences, and sciences. Both editors agreed in not wishing the volume to represent a descriptive catalogue of a wide range of methods used by all of the contributors. Rather, they actively encouraged authors to critically reflect upon their own use of methods within their own research, thereby explicating their own theoretical assumptions whilst also describing 'how' they carried out their research. This provides the volume – uniquely, when it comes to great tomes on methodology – with a theoretical and empirical precision that will be both instructive and interesting to read.

Marcel

The history of this book goes back to November 2016. Michael and I were meeting for breakfast in the restaurant of a hotel in Leiden, The Netherlands, as I had organized a conference there and he was one of the invited speakers. At a certain moment Michael introduced his idea to begin collecting a bunch of essays on the relationship between sound studies and methodology or on the various methods that are used among sound scholars and sound artists. At first this seemed like a rather limited topic, only interesting for die-hard academics. However, taking into consideration that there is perhaps not one single academic discipline that is *not* dealing with sound in one way or another and realizing that all these disciplines have their own ways of organizing their research strategies, it soon became clear that presenting a rather broad overview of all the different methods in which sound is either the subject of research, or indeed the methodical medium itself, seemed like a challenging idea that could result in exciting and valuable material for almost everyone interested in and dealing with sound. Hence, our joint project started off as an endeavour to map the many and various practices and methodical strategies through which corpuses of knowledge, experience, intuition, prehension, and engagement are established, disestablished, and reestablished, thereby tracing the constitutive role of methodical processes in the construction of knowledge related to or based on sound.

Reading through the book now, I am amazed by the breadth of research being done on and through sound, the range of sonic and physical spaces explored, the amount of methodological diversity, and the variety of disciplines in which sound has somehow found its place, either as a research object or as a research method. What this Handbook makes clear is that methods function in thoroughly heterogeneous assemblages, consisting of discourses, institutions, regulatory decisions, and scientific traditions (Barad 2007: 63). Nevertheless, one criticism of sound studies has recently come from ethnomusicologist Gavin Steingo, who has argued that the southern hemisphere – perhaps with the exception of Australia – has often been neglected in sonic research. Do you think we have addressed this issue fully in the present volume?

Michael

Whilst I am in broad agreement with Gavin Steingo and Jim Sykes, the editors of *Remapping Sound Studies* (2019), over their critique of sound studies as being overly Eurocentric, I think the reasons for this are both institutional and historically situated. That is not to say that these issues are in any way solved. I am pleased that the contents of our Handbook were already decided by the time I had read Steingo and Sykes's critique as I think we have gone at least as far as they have gone in their volume in relation to the sounds of the South. This leaves open the question as to how one might go further than this. To begin by summarizing their critique, they argue, quite convincingly, that, historically, sound studies (let's leave aside for the moment what this demarcation of the subject might mean) has not sufficiently covered the sounds of the South with all of the conceptual, theoretical, and methodological limitations and consequences that this might imply. They argue for the need of a 'new cartography of global modernity for sound studies'. In pursuit of this they provide excellent alternative contents for an imaginary southern sound studies reader together with twelve chapters investigating various facets of the sounds of the South.

So, to the historical nature and meaning of the claim of southern exclusion. This is where I am most in tune with their critique although this agreement also depends upon what we might mean by the term sound studies. Let me explain. I was introduced to the grandly termed World Forum of Acoustic Ecology just over twenty years ago after being invited to give a talk at their conference in Dartmoor in the UK. The setting was impressive and I would wake at the crack of dawn in order to attend listening exercises with the wonderful Pauline Oliveros. Whilst at the conference, I skimmed through the literature relating to the World Soundscape Project, perplexed that it appeared to refer to work carried out only in Canada (Vancouver) and Finland. A start for sure. I don't want to go into the theoretical limitations of the movement (I allude to them in my introduction to section one of this book) but merely to point out that the intellectual pioneers of the movement such as R. Murray Schafer, Barry Truax, Hildegard Westerkamp, and others, despite their wonderful work, conceptual and methodological, came with their attendant Western intellectual baggage, like many of us. My own take on the sonic was largely urbanist and one of the issues in the urban West was postcolonialism in all of its forms. I worked in a department with Paul Gilroy and attended lectures given by Stuart Hall, Edward Said, James Baldwin, and many others on race and colonialism. This was miles away intellectually from the interests of the World Soundscape Project at that time. But since then it has been precisely in the interdisciplinary coming together of sonic research that this volume attests to, that a more global approach to the sonic is being realized – with the political, economic, and institutional challenges duly noted by many of the contributors to this Handbook.

Steingo and Sykes themselves recognized their critique as a partial one – that there exists a paradox at the heart of their book, namely that all of the contributors to their volume, whilst researching the South, worked in music and ethnomusicology departments of North American universities. There were no voices of those working and living in the South, although Steingo is himself South African. This is not meant as a criticism – merely as a statement of where we are. Whilst our volume is wonderfully 'global' in reach, most of the work is written within institutions of the wealthy West, where the money and power largely resides. It is difficult to redress the imbalance of southern voices as it is difficult to decide in advance what they will want to write about and research or how they will theorize their own voices.

Marcel

Let me just add a few sentences to yours, Michael. First of all, I am happy to see that our Handbook does, at least, include studies from South Africa, South America, and Vietnam and ethnographic research done in Egypt, India, and Brazil, although I readily admit that more geographical areas and different cultural approaches could have been covered, especially within Africa and Asia.

On the other hand, in my capacity as editor-in-chief of *The Journal of Sonic Studies* I can say that we have published special issues on sound studies, soundscapes, and sound art in Southeast Asia as well as Latin America (and are planning future issues on Africa and Russia). What became clear from these two issues is that the contributors – mainly based and working in these areas – did not employ significantly different theoretical frameworks or methodical tools as compared to their Western colleagues. We had eagerly hoped for new approaches, for new concepts, or to discover and become acquainted with important authors unknown in the West (or North), but this happened only very sparsely.[1] Of course this principally says a lot about the dominance of Western scholarly discourses, perhaps most clearly evident in the hardly questioned and pervasive use of the English language.

With regard to this Handbook, however, the main issue for us, of course, was not so much geographical coverage (or, for that matter, gender, age, or ethnicity) as to present and critically reflect on a wide variety of methodical tools used in various disciplines. In other words, perhaps it is time to be a bit more precise about the question of what this volume set out to achieve.

Michael

Yes, I agree with your point here, Marcel, about the slow rate of change within the writings of the South. I experienced something very similar when I founded, with David Howes, our journal *Senses and Society* in 2003. We had expected radical changes in methodologies, theoretical overviews, and subject matter, but these came very gradually over the years. So perhaps we simply have to be more patient.

So, the aim of the book was to bring together theoreticians and practitioners who either work in sound or are interested in the sonic to reflect upon both the role of sound in their chosen discipline(s) and indeed to reflect methodologically upon their sonic practices. We had understood that in the widely proliferating subject of sound studies there had been no work trying to draw in the issues and problems surrounding *how* we actually research the sonic – whether that be in the hard sciences, the social sciences, the humanities, or the arts.

Marcel

I think it is immediately necessary to add something here, in order to avoid any misunderstanding, and that is that researching the sonic is of course only one side of the coin. Several chapters in this book present, as their central topic, how through recording, processing, and listening to sounds or our sonic environments, new knowledge can be gained, new experiences are possible, unfamiliar worlds can be discovered – knowledge, experiences, and worlds that in themselves are not necessarily audible. Although not included in our book, I am thinking of the work done by, for example, Bernie Krause, Andrea Polli, David Dunn, Åsa Stjerna, and so many others, who through their sound artworks and projects, make people aware of climate change, ecological issues, or specific natural features. Take, for example, the Electrical Walks by the German sound artist Christina Kubisch: she equips her audience with specially designed headphones that pick up the infra- and ultrasounds of magnetic fields present in our environment and transposes them into frequencies that can be registered by our ears, thereby making people aware of the continuous exposure to frequencies that, although in principle inaudible, do affect their functioning, their neurological system, and (therefore) their well-being.

Perhaps more so than visual or textual information, sound can expose its audience to (gradual) transformations taking place over longer periods of time, as it is itself time based. Besides – as Jon Gilmurray touches upon in his contribution to this volume – applying the sonic as a methodical tool can give us an 'enhanced understanding' of certain topics or events, as sound appeals not only to the cognitive and the rational but also gives space for associations, imaginations, and speculation as alternative forms of knowing. Or think of 'sonic journalism', a term coined by sound artist Peter Cusack and defined by him as 'the idea that all sound, including non–speech, gives information about places and events and that listening provides valuable insights different from, but complimentary to, visual images and language' (Cusack 2011). Sonic journalism, Cusack continues, can, for example, transmit 'a powerful sense of spatiality, atmosphere, and timing', thus adding substantially to our understanding of events and issues (Cusack 2011).

What might become clear from these examples is that sound is not a passive element waiting to be investigated by researchers; sound has its own, specific possibilities for agency.[2] A performative approach to sound and an embodied engagement with sonic matter is an important and even necessary supplement to already established scholarly methods: knowledge can also be gained from a methodology that has a direct material engagement with sound, with sound as a matter that matters. In short, I think the volume somehow responds to the methodological dualism implied in the following quote:

> We identify two broad methodological strands: sonic ethnographies, which rely on both conventionally written and more-than-textual representations of sonic qualities: and soundscape studies, which encompass a wide range of methods, including field recording, sound mapping, and sound walks.
>
> (Gallagher and Prior 2014: 272)

Doing and thinking, diffracting[3] and reflecting, experimenting and theorizing are dynamic practices that play a constitutive role in our relation to sound or in our relation to the world through the sonic; they are material-discursive practices and methodological tools through which the interactions between sound and human as well as nonhuman beings can be explored and produced. (What I try to achieve by connecting the material and the discursive with this hyphen, is to disrupt the alleged ontological difference between theory and practice: as much as theorizing is a [material] practice, practice is always permeated by theory, by conceptualizations, by reflection.)

So, providing a space where people coming from the hard sciences, social sciences, humanities, and the arts can encounter one another; providing a space where philosophical reflections, analyses of concrete artistic interventions, historical overviews, and attention for human and nonhuman agents intersect; providing a space where the discursive and the material, the theoretical and the practical are no longer regarded as antagonisms – if these were some of the objectives we also had in mind, does this mean that we are permitted to be pleased with this volume?

Michael

Yes, Marcel, I like your description of the way in which the methodological tools in the book provide a platform for an 'enhanced understanding' of the role of sound filtered through a set of interdisciplinary 'encounters'. Too often, the academy pays lip service to interdisciplinarity whilst locking us into our intellectual silos – this issue confronts us every day when we have to decide where to publish our work, be it an anthropological, a history, or a music journal rather than, say, a sound studies journal. If you permit me a bit of a personal anecdote here that sheds a little light on my interest in 'interdisciplinarity' but also importantly into the 'multisensory' which I think our volume also succeeds in. I first came across these barriers when at school. As a teenager I didn't know very much about university other than wanting to attend one. Nobody in my family had been to university and I didn't know that my teachers had to write a reference supporting my application. During my A level studies I had chopped and changed subjects – from mathematics to art for example – and had not been a particularly obedient student. My personal statement was full of my interest in art and twentieth-century French philosophical thought as well as my love of football and so on. I remember my art teacher who had to compile my school reference telling me, 'Michael, I don't think you'll get an offer from a university. You see, I think you're a bit of an intellectual dabbler.' This struck me then as both a complement and criticism – breadth but not enough depth! As it turned out I did get a university offer, but only one – after all how many do you need! But I recount this experience to highlight the deep suspicion of interdisciplinarity within areas of the academy. This suspicion of interdisciplinarity of Enlightenment 'cultivation' (in German *Bildung*) runs deep. This is why my discovery of the Frankfurt School and their archetypal interdisciplinary project has been so influential to me as indeed was their embrace of the multisensory investigation of experience as early as the 1930s, as this quote from Horkheimer illustrates:

> The objects we perceive in our surroundings – cities, villages, fields and woods – bear the mark of having been worked upon by man. It is not only in clothing and appearance, in outward form and emotional make up that men are the products of history. Even the way they see and hear is inseparable from the social life process, as it has evolved over the millennia. The facts, which our senses present to us, are socially pre-formed in two ways: through the historical character of the object perceived and through the historical character of the perceiving organ.
>
> (Horkheimer 1972: 200)

So this sensory sensibility is significant methodologically; indeed it is a challenge to our understanding of experience and the contribution that the sonic makes towards it, both historically and in terms of the present. I will give just one illustrative example. Five years ago I wrote a large research proposal submitted to a UK Research Council on Sensory Borders within Europe. This was before the refugee crisis really took hold in Europe. The idea was to research multiple European entry points using a range of sensory methodologies as migrants went through the various entry procedures, both legal and social. The aim was holistic – incorporating sound, vision, taste, smell, and touch. The idea behind this was that researching any one sense would be inadequate: how are we to understand the present repressive visual regime in France with reference to the forbidding of the wearing of the Burka within a context of sensory colonialism/orientalism? Of course this prohibition not only has a visual component but also is connected to the sounds of Islam (speech, prayers, and music) and to its supposed touch and smells (foods and social rituals); taken together they orientate attitudes towards the Muslim 'other' in European culture. So, reading through our edited volume I am pleased with both its interdisciplinary scope and in its engagement with the multisensory nature of experience.

As for myself whom I would describe as an 'urbanist' I have been amazed, and indeed have learnt a lot, by the breadth of sonic spaces and places covered in this book. Reading through it one gets a feeling of sonic vertigo – or at least I do. Sound in space is covered throughout the three sections and from the sciences, arts, and social sciences – a veritable cornucopia of coming together, from the oceans of the world to the music clubs of London, from the northern climes of Norway to the bustling sounds of Cairo, from hacking as a creative musical act to the role sound plays in medical practice and research, from the drongo warning a meerkat of a nearby predator by making a specific call to podcasts as a relatively new way of providing knowledge.

Now that we are all in coronavirus lockdown the soundscapes of many of the cities of the world are transformed. There are already databases recording these sonic (and by extension physical) transformations. What, I wonder, will we make of these transformations in years to come? The other day, in the UK, we were all encouraged to come to the front door of our homes and clap for the efforts of the medical staff in hospitals caring for those who were infected by the virus. As we walked out of our door we could hear the clapping of people in the neighbourhood – clear, as no sound of traffic interrupted the sounds. We, rather than clapping our hands, struck our metal kitchen utensils – metal on metal, much to the embarrassment of our son. These scenes were relayed across the country – a participatory sonic social transmitted nationally and globally, active, yet nevertheless ideological for the majority of those hospital workers in the UK who had not been given the protective clothing required for their own safety. We were – and simultaneously were not – all in the same boat. The UK prime minister and other 'notables' were all tested for the coronavirus whilst the doctors treating them were not: our clapping was simultaneously 'hollow' and 'heartwarming'.

For now I would like to focus upon the heartwarming sonic. We had viewed on our television sets and on the internet many people up and down Italy and then France and Spain singing, clapping, and playing musical instruments from their balconies. Confined to their homes the sonic carried beyond the empty streets into the opened windowed apartments; in this sense all of those making 'noise' felt in the same boat and the sense of community passed through the walls of isolation within which they lived. For those southern European towns and cities in which the bustle of outside life is so significant, the sonic transformation of their cities and towns was all the more apparent than for those of us huddled up in London, Manchester, or Birmingham. On looking at the silent streets of Rome I was reminded of Carlo Levi's description of the festival of San Giovanni in the Piazza Navona in the 1950s:

> Even if you never leave your home, stay shut up in your room, never look out of the window […] a convict, a monk, or an invalid, even the blind, in Rome, cannot help but notice festivals […]. From afar, you can sense a sort of throbbing and shrilling in the air. The closer you get to Piazza Navona, the greater this throbbing becomes, growing little by little, into a vague, thunderous din; and, as if by some piece of magic, as you are swept into the crowd, it seems as if there is a rushing river in the broad lake of the piazza […]. Everyone has a whistle and everyone is blowing into their own, trying to drown out the others […] I grew more and more to be part of the crowd. I realized that the sound issuing from my little instrument was enveloping me like a compact atmosphere, as if within an invisible suit of armor […]. United like some great swarm beneath this cupola of sound and separated from one another by a personal resonating diaphragm […] the crowd flowed around the stalls of toys and candy floss.
>
> (Levy 2004: 29–30)

Now the Roman crowds are dispersed into their own homes – the sonic rhythm of the street interrupted, the street silent. The whistles, singing, and playing reconfigure the physical presence of the street, not alone together as Levi describes but together alone – a reversal or collapsing of the sonic duality of city and countryside.

Maybe we are existing in a temporary sonic limbo where we all live in the quietude redolent of the countryside so loved by the acoustic ecology movement where we can all hear one another clapping our hands.

Marcel

What you make clear here, Michael – at least the way I hear you – is how important sound is as a political instrument, as a medium to express togetherness and solidarity (as well as separation and protest), as a tool to establish and disestablish identity. Sound is always more than sound; it immediately exceeds the sonic, the audible, the sonorous, and oscillates between sense and the sensorial (see Nancy 2007). How does sound do this? How does it work? How does sound matter as a topic and as a method? These are difficult questions for me, first of all because the methodical was never the main point of attention in my research. Somehow, I always felt uncomfortable when I had to react to questions about my research methods, and my standard answer became: reading books, listening to music and sound art, and then engaging in some critical reflection. I had the same reservations when asking my PhD students, for example, about the way they intended to organize their research. Perhaps it has something to do with my rather long and ongoing interest in deconstruction. In 'Letter to a Japanese Friend' Jacques Derrida was very explicit about it: 'Deconstruction is not a method and cannot be transformed into one.' Even stronger: 'It is *not enough* to say that deconstruction could not be reduced to some methodological instrumentality or to a set of rules and transposable procedures' (Derrida 1988: 3, my emphasis). The main reason Derrida so vehemently opposed the reduction of deconstruction to a reading method at the service of in-depth analyses and critical reflections is that 'deconstruction takes place, it is an event that does not await the deliberation, consciousness, or organization of a subject, or even of modernity' (4). As most of my work is still deeply influenced by Derrida's thinking, I have always resisted the idea that I somehow (should) choose a set of rules and procedures to investigate sound and music; instead I try to be as responsive and perceptive as possible to what these sounds and music are telling me.[4] In that sense I always search for a methodology that is responsive, responsible, and attentive to the specificity of the sonic material under investigation – an attitude that Hans-Georg Gadamer would perhaps call 'objective' – dynamically de-framing and reframing the method, aiming more for methodological creativity than methodological pluralism.

However, even if a real dialogue between subject and object, between researcher and research object were possible, a dialogue in which subjects do not control the object, for example by determining the research method, both subject and object are involved in a process amidst many other agents. In my opinion, attention for the sonic can benefit from a post-humanist account in which any practice should not be considered a human-based activity but a (re)configuration of the world through which meanings, differences, and systems are enacted. As historian and literary scholar Hayden White makes clear in *Tropics of Discourse*, no escape is possible from certain predeterminations that exceed the individual or personal level: narrative modes (aesthetics), explanatory models (epistemology), ideological backgrounds (ethics), and the various 'schools' within one discipline (institutionalization) – conflicting as they might be sometimes – affect the way scholars and artists somehow organize their research, from the initial research question, hypothesis, or mere topic, to the methodology, analyses, critical reflections, choice of case studies, and ultimate outcomes (White 1978: 66–72).

To confine myself to the topic of this book: specific methods provide the lenses through which we view or, better, construct something that we call sound, the sonic, sonic ambiance, music, or auditory culture. Sounds do not have an independent existence separate from, for example, the material and theoretical methodical tools that are used. And although these tools are constantly open to rearrangements, rearticulations, reworkings, and recontextualizations, using one specific method necessarily means excluding others, which does affect the outcomes. The nature of the observed phenomenon changes with corresponding changes in the methodical tools being applied. Methods actively contribute to the sonic events that scholars try (in vain) to capture in words – different interactions produce different events, although I would like to stress once more that the materiality of these phenomena is not a given nor simply an effect of human agency.

Before passing the baton again, however, I would like to come back to this concept of 'diffraction' once more, introduced above in the way it is used by Haraway, here in a quote from Karen Barad:

> Unlike methods of reading one text or set or ideas against another, where one set serves as a fixed frame of reference, diffraction involves reading insights through one another in ways that help illuminate differences as they emerge: how different differences get made, what gets excluded, and how those exclusions matter.

(Barad 2007: 30)

The good thing about handbooks like ours is that various ways of approaching a topic and organizing a research strategy obviously, though not deliberately, disclose the pros and cons of each research method. Perhaps Barad's diffraction doesn't always work within one chapter or within the work of one scholar, but by simply reading more than just a few texts from this Handbook, it will soon become clear what the methodical differences are, how one method necessarily excludes certain benefits of another, and how methodological choices directly affect the way a topic is perceived, presented, and produced. The diffraction will thus take place in the act of reading, in the mind and attitude of the reader. Contradictions, counterarguments, antagonisms (or agonisms), discrepancies, etc. are not to be avoided in such a handbook; they constitute its richness.

Michael

Yes Marcel, for me it is both a question of 'anti-foundationalism' and 'mediation'. As you know, for me it is a matter of how we might understand 'dialectics', from which I take Theodor Adorno's recognition of it as 'the consistent consciousness of non-identity' (Adorno 1973: 5). I also realize that many of the books that I might refer to are lying in my university office – out of reach. But luckily, my dog-eared copy of Adorno's *Minima Moralia* is to hand. In his aphorism (what a method!) 'Antithesis' he states,

> He who stands aloof runs the risk of believing himself better than the others and misusing his critique of society as an ideology for his private interest. While he gropingly forms his own life in the frail image of a true existence, he should never forget its frailty, nor how little the image is a substitute of true life. Against such awareness, however, pulls the momentum of the bourgeois within him. The detached observer is as much entangled as the active participant.

(Adorno 1974: 26)

Underlying this observation is the status of the 'non-identical' and the status of the speculative relation between subject and object. Yet of course Adorno's epistemology was materialistic whilst recognizing that no 'object' is merely given – because, as you say, any object is there only in relation to a subject – and that all objects are historical and cultural and thereby provisional.

What has always drawn me to Adorno's dialectics is his base level of human suffering as a material base, 'the need to let suffering speak is a condition of all truth. For suffering is objectivity that weighs upon the subject' (Adorno 1973: 17–18). This negative is articulated rather simply by Barrington Moor Jr who claimed that human misery was more easily understood than human happiness with its vast cultural, historical, and individual variations. The contributors to this volume do a pretty good job of steering away from the Charybdis and Scylla of methodological positivism and idealism. Positivism and idealism are somewhat joined in a dream of unmediated experience. From the work of Bishop Berkeley who believed that the world was there only by virtue of it being perceived; a mimetic fantasy, reality was identical to the retina of the eye. This form of idealism was replicated in the work of Jean Baudrillard, of course, in his treatment of the first Gulf War as real only on the 'screen' of the observer. The notion of pure description also lies at the theoretical-methodological heart of phenomenology deriving from Edmund Husserl who believed you could bracket out the cultural elements of knowledge to look upon (and by extension hear) what you experienced untainted by those artefacts – pure consciousness untethered from the world. The dream of unmediated knowledge thus takes a major place in the pantheon of Western knowledge claims. Something which we grapple with in this book and from which, historically, scholars of the sonic have drifted into like everybody else.

Sonic positivism was embodied in the early use of the phonograph by ethnographers. The attractiveness of being able to fix, transpose, and transport sound arose with the phonograph in the late nineteenth century. Erica Brady (1999) estimated that fourteen thousand cylinder recordings of North American Native Americans were made by ethnologists between 1890 and 1935 – these cylinders kept in many museums and university departments are both a testament to the cultural value attached to 'collecting' history (what could be better than archiving the dying sounds of a culture for future reference in order to gain a clearer understanding of lost worlds?) and to the positivism and colonial mentality hidden within these sonic documents. At the time these recordings were considered to be accurate representations of that which was recorded. Yet for the most part, recorders of these interviews and the sounds of rituals and so on failed to mention how the material was gathered. The recordings we now understand represented a 'fetishization' of the sonic, a 'false objectivity'. These recordings were blind to all forms of nonverbal contextualization embedded in and acting beyond the recorded sound – the physicality of the culture in its ritualistic and material form. Also hidden from 'view' was the asymmetrical power relation embodied in the ethnographic encounter. Lest we think that these concerns merely deal with the 'past' I am reminded of a presentation given just a few years ago at a conference in London concerning the relation between music and the emotions. I was listening to a presentation with an impressive use of visual material that charted the playing of chord sequences to subjects who had been wired up so as to display patterns of physiological response to the changing chord sequences. These researchers had been given a large research grant to do this research. As I listened to this, an uneasy feeling crept over me as I remembered all those terrible experiments carried out on women in the 1950s to measure sexual response – the hidden misogyny lost on the researchers. Towards the end of the presentation the young researcher proclaimed with some pride that the physiological response of the subjects to these changing chord sequences was similar to those who took pleasure in eating a good meal and to those responses of those engaging in sex. My initial response was that any methodology that couldn't tell the difference between listening to music, eating food, or engaging in sexual activity didn't seem particularly useful. The talk was part of a number of research presentations that had the ostensibly noble pursuit of the role of sound in the reduction of suffering of those who were in the final days of life. Yet, they merely posited both the objective meaning of music and the objectification of human responses to it. The philosopher Roger Scruton, now sadly dead himself, looked at me in alarm, saying he would sooner drop down dead than suffer the torture of music imposed upon him in his final moments of life!

So where does this get us methodologically? Returning to Adorno's take on human suffering in relation to the above, I am reminded of the diary of a very brave French woman in the Second World War, Agnes Humbert. An active member of the Parisian Resistance, Humbert was captured and imprisoned. Whilst in prison, amongst many of her sonic experiences – she was in solitary confinement and only heard the sounds of others – she writes; 'Yesterday I heard the screams of a man being tortured. When the screams died down, they were followed by deep, throaty laughter. All day I have been haunted by these two sounds: screams and laughter. I don't know which was more terrible. The laughter, I think' (Humbert 2008: 97). The experience unspeakable – I have merely read these words, it was Humbert who experienced them. Their meaning individual, historical, contextual – the screams real – everything else variable – mediated for our own purposes?

Michael and Marcel

What will you gain or discover by reading this book? And how have we organized the content? First, we would like to state once again that this is not a straightforward methods book in which methodologies are divorced from the theoretical, institutional, disciplinary, and cultural contexts within which they are used. This book is 'situated'. It teaches by example, and *by example* here means with attention to context, attention to research methods and types of knowledge that are specific to a given situation, accounting for the agencies of the researcher, the object of study, and how the research has been executed. Part I provides a theoretical overview as to how researchers have studied or 'used' sound within each discipline. For example, how historians have studied sound, how anthropologists, biologists, and urban scholars have studied sound, but also how sound is used in medicine, in ethnography, in sociology, and so on. The volume, in a unique way, also traverses the historically formed intellectual division of labour that we tend to inhabit by intermittently crossing the divide between the arts, humanities, the social and the hard sciences and their treatment of sound. In Parts II and III the reader is taken through a wide array of sonic research that critically contextualizes the methods that have been employed. Part II consists largely of (descriptions of) artistic research projects, that is, research done by artists in which the methodical tools such as field recordings, soundwalks, improvisation exercises, or musical performances are deployed in order to gain knowledge about, for example, environmental issues, social behaviour, musical and musicological developments, or the well-being of humans and nonhumans. Part III is mainly dedicated to concrete case studies in which sound is the lens through which non-sonic issues are studied. Methods are presented which serve to increase attention for sonic events (for example in public urban spaces but also within the musical domain) and propositions are made for rethinking and transforming the relation between language and sound as well as how sounds can have an impact on design methods.

So we have come to the end of our virtual dialogue, an endeavour which has been as pleasurable as the putting together of this rather massive book. We would like to thank our Bloomsbury editor, Leah Babb-Rosenfeld, who has supported us as this book grew ever larger.

Notes

1. I should add, straight away, that several contributions to these issues did, however, reveal different emphases. The issue on Southeast Asia made clear how extensive the sonic differences are between rural and urban areas in these countries; and within the issue on Latin America many authors accentuated the political significance of making sound. Additionally, both issues introduced their readers to an invaluable variety of field recordings and sound art from those regions.
2. It is, among others, Karen Barad who propagates and defends an emancipation of 'matter' as an independent and important agent: 'What compels the belief that we have a direct access to cultural representations and their content that we lack toward the things represented? How did language come to be more trustworthy than matter? Why are language and culture granted their own agency and historicity, while matter is figured as passive and immutable or at best inherits a potential for change derivatively from language and culture?' (Barad 2007: 132).
3. According to Donna Haraway (1992: 300) diffraction is a 'mapping of interference, not of replication, reflection, or reproduction. A diffraction pattern does not map where differences appear, but rather maps where the effects of differences appear.' The word 'diffraction' was first coined in 1660 by the Italian scientist Francesco Maria Grimaldi and refers to the ways waves – also sound waves – combine, overlap, bend, and spread when they encounter obstacles or slits. As such I think the concept works well in relation to sound studies and sonic methodologies.
4. Of course the word 'try' should be understood here in an actively passive sense. Being susceptible or receptive to 'the other' – in this case sound – requires a certain amount of passivity. Yet it is necessary to prepare for it, and this asks for a deliberate and conscious effort. So, even though this attitude escapes all programming, it is certainly no inertia. Derrida called it 'invention' because one gets ready, one makes a step destined to let the other come, *come in*.

References

Adorno, Theodor (1973). *Negative Dialectics*. New York: Seabury Press.
Adorno, Theodor (1974). *Minima Moralia: Reflections on a Damaged Life*. London: New Left Books.
Barad, Karen (2007). *Meeting the Universe Halfway: Quantum Physics and the Entanglement of Matter and Meaning*. Durham, NC: Duke University Press.
Brady, Erica (1999). *A Spiral Way: How the Phonograph Changed Ethnography*. Jackson: University Press of Mississippi.
Cusack, Peter (2011). 'Sounds from Dangerous Places'. Available online: https://www.sounds-from-dangerous-places.org/ (accessed 28 June 2020).
Derrida, Jacques (1988). 'Letter to a Japanese Friend'. In David Wood and Robert Bernasconi (eds), *Derrida and Différance*, 1–5. Evanston, IL: Northwestern University Press.
Gallagher, Michael and Jonathan Prior (2014). 'Sonic Geographies: Exploring Phonographic Models'. *Progress in Human Geography* 38 (2): 267–284.
Haraway, Donna (1992). 'The Promises of Monsters: A Regenerative Politics for Inappropriate/d Others'. In Lawrence Grossberg, Cary Nelson, and Paula Treichler (eds), *Cultural Studies*, 295–337. New York: Routledge.
Horkheimer, Max (1972). *Critical Theory, Selected Essays*. New York: Herder and Herder.
Humbert, Agnes (2008). *Resistance: Memoirs of Occupied France*. London: Bloomsbury Press.
Levy, Carlo (2004). *Fleeting Rome: In Search of la Dolce Vita*. London: John Wiley.
Nancy, Jean-Luc (2007). *Listening*. Trans. Charlotte Mandell. New York: Fordham University Press.
Steingo, Gavin and Jim Sykes (eds) (2019). *Remapping Sound Studies*. Durham, NC: Duke University Press.
White, Haydn (1978). *Tropics of Discourse: Essays in Cultural Criticism*. Baltimore: Johns Hopkins University Press.

Part I

Disciplines, Methodologies, Epistemologies

1

Introduction to Part I: Sounds Inscribed onto the Face – Rethinking Sonic Connections through Time, Space, and Cognition

Michael Bull

Theodor Adorno once commented that everything was related to everything else. Adorno was approaching connectivity from a Marxist perspective, but the same statement might also refer to the structural functionalism of Talcott Parsons. Of course, the appropriate question might be what is the nature of this connectivity; what cultural, political, and economic values and understandings are embedded in any attempt at connectivity? These are also questions of epistemology and methodology as the fifteen chapters in this section testify to. Scholars investigating the sonic 'abstract out', as do all other scholars – the written testimony, the sound of the aircraft flying above, the beat of the heart that a doctor listens to, the morning chorus of birds that we might be lucky enough to wake to, the sounds within a London music venue or on an urban street, the squelch of mud beneath our feet. This abstracting out is followed by a reassembling, a reconnecting that enables us to situate and extrapolate from that which we have abstracted out from. The sound of the heart beat which the doctor listens to and interprets, reconnects to the history of the technology; this in turn connects the sounds of the hospital ward to the daily rhythms of those inhabiting the ward and contextualizes the silence of those who suffered in those same wards one hundred years ago during the First World War. A listening filtered through the repressive masculine stereotypes enforced by both men and women of the time representing an ideology of a 'stiff upper lip' that masks the internalized screams of patients (Sterne 2003; Rice 2013; Carden-Coyne 2014). The nature of this connecting and reconnecting is frequently informed by the shifting sands of disciplinary interests that we tend to inhabit despite the frequent protestations of multidisciplinarity;

the ethnomusicologist, sociologist, sound artist, urban designer, medical researcher, historian, literary theorist, and so on each listen out within their disciplinary domain. And then there is sound studies itself. Jonathan Sterne recently commented that sound studies should 'be grounded in a sense of its own partiality, its authors' and readers' knowledge that all the key terms we might use to describe and analyze sound belong to multiple traditions, and are under debate' (Sterne 2012: 4). In order to achieve this critical self-evaluation Sterne replicates C. Wright-Mills's invitation to sociologists in the 1960s to exercise their 'sociological imagination' by invoking sound scholars to cultivate their own 'sonic imaginations' in order to 'rework culture through the development of new narratives, new histories, new technologies, and new alternatives' (Sterne 2012: 6). In tune with Sterne's directive the rest of this Introduction will interrogate and question a range of received knowledge within sound studies as to the nature of sonic connections through time, space, and cognition. This will be broken down into a simple question: when does a sound begin and when does it end? In doing so I attempt to shed 'light' upon the possibilities and potential for traversing traditional subject boundaries whilst also questioning a range of historical material used in the analysis of sound and by extension an epistemology of sound which informs much of the work we carry out in sound studies. The analysis will focus primarily upon examples drawn from the First World War – a war in which all of those who experienced it are now dead and for which almost no sounds survive other than the sweet strains of the songs of the time materialized in shellac, living testimony of the complex relationship that exists between entertainment, sentimentality, longing, and propaganda (Brooks, Bashford, and Magee 2019). Writings on the First World War continue to be subject to a 'heterophonia' of competing sonic streams that live on in contemporary accounts. It is also a war interrogated by all of the disciplines noted above, and many more besides. This Introduction will 'trace' a series of sonic connections within the First World War and beyond. In doing so the 'trace' will act as both an epistemological and methodological tool. The connections to be made are not unified or holistic but rather prismatic whereby the sonic is refracted through disparate historical and cultural material producing a range of 'soundways' defined by R. C. Rath as 'paths, trajectories, transformations, mediations, practices and techniques' (Rath 2003: 2). These traces might be understood as a web moving simultaneously in many directions and times. The use of the term is an adaptation from Ernst Bloch's method of writing in *Traces* (2006). In this work Bloch often begins a trace by drawing upon a childhood experience before extending his observations through tracing the initial thought through a wide range of cultural experiences and examples. In reviewing Bloch's work Benjamin Korstvedt describes the method as 'to stimulate an imaginatively critical, questioning, even questing, attitude that can read clues and signs from ordinary lived experience in ways that reveal the mutually determining relationship between existential and social being' (Korstvedt 2007). Adorno, in a nuanced critique of Bloch, describes his use of the method thus: 'These experiences are no more esoteric than whatever it was about the sound of Christmas bells which moved us so profoundly and which we never wholly outgrow: the feeling that this can't be all, that there must be something more than just the here and now' (Adorno 1980: 97).

In the following trace I begin not with Christmas bells but with the sound of boots sinking into mud one hundred years ago, before moving on to connect these sounds, and by extension silences, as embedded in a solitary object whose materiality will be imaginatively listened to so that we might hear both the silences and sounds contained within it. The object is a 100-year-old blue wooden park bench displayed in Sidcup, Kent. Its only distinguishing feature are the words 'FOR WOUNDED SOLDIERS ONLY' written in large white letters on its back. This singular park bench in turn leads to a range of reflections concerning the duration and spaces of sounds, silences, and the very materiality of the sonic, before finally concluding with two sonic commemorations of war, *We're Here Because We're Here* (2016), which commemorated the first day of the Battle of the Somme in which 26,000 young men lost their lives, and a sound installation commissioned by the Imperial War Museum, London entitled *Coda to Coda* (2018), which recreates the moment that the guns of war were silenced on the eleventh hour of the eleventh day of the eleventh month on the Western Front in 1918.

The sounds of mud under feet

Data can begin anywhere, abstracted from the labyrinthine nature of the world. Much of our sonic evidence from the First World War comes in the form of written or otherwise recorded testimonies from those who experienced it. Testimony is a contested concept and is here subjected to a cultural, political, and sonic critique filtered through the present analysis. Who is it that is engaged in the saying and who is interpreting the words? Much has been made of the partiality of many of these testimonies, from the war poetry of Wilfred Owen and Siegfried Sassoon to the novels of Robert Graves, Erich Maria Remarque, David Jones, and many others; partial in terms of the class and educational nature of these writers which, it is suggested, contribute to an overplaying of the despair of war (Hynes 1992; Fussell 2013; Winter 2014). This critique itself can be contested, as indeed I do here, by drawing upon alternative voices of the war in the form of diaries never meant to have been read or published. For example, the diaries of James McCudden, the air ace who shot down fifty-seven aircraft and who was killed in 1918, who left school at fourteen and had never written a word before he put pen to paper whilst fighting. His diaries, like so many others, were published posthumously. McCudden's journals, like many other working-class accounts, do not diverge greatly from those written by their well-known, 'educated', and illustrious counterparts. Furthermore, these 'private' reflections are not subject to the interpersonal mores of a 'stiff upper lip' that resonates through so much existing audiovisual testimonies of those who survived the war and who were interviewed often fifty years after the end of the war. And then there were those who had not even this voice, who, whilst surviving the war, remained silent – those many thousands who suffered facial disfigurement.

Memory and testimony traces shift in time – personally, socially, and politically; middle-aged grandchildren of those who fought in the war are now in turn rewriting and reimagining aspects of that conflict. I have just returned from the premiere of the

Sam Mendes movie *1917* dedicated to his grandfather Alfred Mendes who recounted his wartime experiences to Mendes many years ago; 2018 saw the release of Peter Jackson's documentary film *They Shall Not Grow Old,* dedicated to his serving grandfather Sgt William Jackson, commemorating the 100th anniversary of the end of the war. These films will be analysed fully elsewhere, but Jackson's film is noted for its recreation of black-and-white original film footage into painstaking and effective colour, but also for employing forensic speech specialists to decode the speech of troops and for employing actors to speak these words in his film – thus adding a novel take on giving combatants a voice and endowing the film with a heightened sense of 'realism' (Bull 2019). The present sonic trace follows the example of personal testimony some fifty years distant.

I am sifting through family photographs that my mother had kept and have come across a photograph I remember taking as a teenager. It is a photograph of my Corsican grandfather taken in his home high up in the Corsican mountains, taken in the 1960s. As I look at this photograph I have a Marcel Proust moment: the sounds of the French chanteur Jean Ferrat singing 'La Montaigne' comes flooding back to me. The song is being played on a portable record player outside the house and is accompanying my cousin's morning physical exercise routine. The front door is open and the warm sweet smell of eucalyptus wafts through the door from the huge tree in front of the house. I remember sitting on a chair; my grandfather, Pepe, is recounting his wartime experiences, primarily those at the Battle of Verdun where he was an infantryman (Poilu), to his rapt grandson – myself. His was a story of the hardships of life in the trenches, told by so many; the cold, the interminable rain, the terrible conditions in the trenches that killed as many as the incessant shelling, the stench of decaying bodies, rats, and human excrement, the sound of the squelching of mud as his boots sunk into one, two feet of mud. He looked at me with his large, open, farmers face and I paraphrase, 'Michael, there was one sound worse than the squelching of the mud. It was when there was no sound of mud. We were treading on the dead laying beneath us. Treading on the faces of the dead!' He still had nightmares of this desecration, as indeed did I after he told me this. He never returned to Northern France. Brought up a devout Catholic, he lost his faith during the war and on returning home when the local priest told him that he should carry out confession, to put into words that which he had done, my grandfather punched him to the floor and never set foot in a church again, even on his own death. The memory of this encounter has stayed with me. There are many corroborating accounts of soldiers treading on the decaying bodies of the dead in the trenches along with all of the other atrocities that troops encountered in Northern France and elsewhere. My grandfather lived these experiences fifty years after having survived them and passed them on to me where they now reappear, mediated on these pages, abstracted out from the flow of time and now reconnected, at once personal, intimate, familial, historical, and cultural.

What does this Proustian moment of recollection tell us about the question of when does a sound begin and when does it end? Gavin Steingo and Jim Sykes have recently argued that we need to 'reconceptualise sonic history as a nonlinear process that is fraught with cultural frictions' (Steingo and Sykes 2019: 11). This Proustian moment and, indeed, my grandfather's recollection are both non-linear. Ernst Bloch concurs in arguing for non-synchronous, 'not all people exist in the same now. They do so only externally, by virtue of the fact that they may all be seen today. But that does not mean that they are living at the same time with others' (Bloch 1977: 22). The nature of this non-synchronism might be cognitive, individual, cultural, and historical. Bloch's French compatriot, the historian Marc Bloch, a trooper in the First World War, and a member of the French Resistance in the Second World War where he died after being brutally tortured by the Nazis, points to both the abstracting out of 'memories' and to their persistence, their reoccurrence after the drift of years:

> I shall never forget the 10th of September, 1914. Even so, my recollections of that day are not altogether precise. Above all they are poorly articulated, a discontinuous series of images, vivid in themselves but badly arranged like a reel of a movie film that showed here and there large gaps and the unintended reversal of certain scenes.
>
> (Bloch 1980: 89)

Twenty years later, during the occupation of Paris he wrote:

> Ever since the Argonne in 1914, the buzzing sound of bullets has become stamped on the grey matter of my brain as on the wax of a phonograph record, a melody instantly recalled by simply pushing a button; too, even after twenty-one years, my ear still retains the ability to estimate by its sound the trajectory and probable target of a shell.
>
> (Bloch 1980: 41)

The time and nature of war produces its own configuration of time – sonic and otherwise, both non-synchronic and linear as Walter Benjamin recognized 'a generation that had gone to school on horse-drawn streetcars now stood under the open sky in a landscape where nothing remained unchanged but the clouds and, beneath those clouds, in a force field of destructive torrents and explosions, the tiny, fragile human body' (Benjamin 2002: 144). The linearity lies in the development of the technologies of destruction experienced non-synchronistically. And the cultural frictions – my peasant grandfather possessed no political vote like the majority of those who fought, their bodies owned by the state to do with them as it chose, their bodies 'matter' like the shells that killed them, their voices often reduced to humour, parody, and cynicism as expressed in the reinterpretation of songs of the time and their lyrics. Song has remained an ever present 'we' in theatres of war with troops frequently subverting the lyrics of popular wartime songs, replacing them with their own dystopian or sexually explicit lyrics (Sweeney 2001). In the 1932 French movie *Wooden Crosses*, the troops sing, 'they tell us we'll be getting bronze crosses [medals] but all we get are wooden crosses' to be placed on their graves. No commercial discs were made of these sonic transgressions at the time. The voices of propaganda continued to drown out the voices of wartime experience, silencing them, with their simple, unambiguous, dulcet yet 'patriotic' tones of '*Keep the Home Fires Burning*'.

The sonic connection – non-synchronous, contingent – leads from my grandfather's wartime account to a blue park bench. It moves from the dead soldiers trodden underfoot to the surviving facially disfigured soldier; to their treatment, silencing, their lifelong trauma and its political and technological antecedents that prefigure the split second in which their faces were disfigured and traumatized. What then are the sonic connections between the blue park bench, the injuries, the treatments, the technologies of destruction and personal narratives, memories, testimonies of those who sat on benches like these? These sonically inflected questions are not merely sonic but also epistemological, methodological, historical, economic, political, and aesthetic questions.

The blue park bench and sounds etched onto the faces

My grandfather, like so many others, was witness to the indescribable. Whilst he might have trembled at the thought and experience of treading on the remains of the dead, he would have also witnessed all kinds of injuries, some fatal, some not. This trace now moves from the faces of the dead to those who lived with the consequences of often catastrophic facial injuries. The power of the disfigured face as cultural stigma is well documented culturally and historically yet remains largely absent in accounts of the First World War (Gilman 2014). Yet facial injuries were a common occurrence in the First World War; in the UK alone over 60,000 troops were treated for such injuries, often in specialist medical units. These injuries were caused by a variety of factors: war time terrain, the nature of the conflict, and the destructive power of the weaponry used. The destruction wrought thus produced swift developments in surgical techniques, modes of medical intervention and the development of plastic surgery in efforts to ameliorate the destructive power of the weapons used on all sides. Many victims of head wounds died instantly but their survival rates were higher than those injured in the stomach, 90 per cent of whom never made it alive to a hospital. Whilst the grave nature of many of their injuries were instantaneous, their screams of pain would often last for many hours for all to hear. Their futures often inscribed with the sound of their nightmares, their pain uttered through years of surgery, their speech stifled through the destruction of mouths, tongues, and windpipes, their eating thus sonically transformed as this trooper's written record describes: 'It was several months after leaving hospital before I regained my speech, and not for a couple of years later could I speak plainly or eat solid food' (Bamji 2017: 38). The words fail to convey the trauma and duration of the experience. But what is it that I am describing here? The following two graphic descriptions suffice. The first is of an injury sustained by shelling, the second from a bullet. The speed and sharpness of shell fragments often caused catastrophic disfiguring injuries. Woods Hutchinson, a US war doctor stated that whilst 'a bullet would go completely through the face from side to side, and perhaps break one jaw or put out an eye, a shell splinter will often shear away the whole lower half of the face, leaving the tongue hanging down on the chest, or tear away an eye, all the front of the upper jaw and teeth' (Hutchinson 1918: 227).

Equally Louis Barthas kept a detailed diary through his four years of war, describing an adjacent soldiers' facial injuries in graphic detail:

> Another soldier who was crawling up suddenly leapt up and fell right in the middle of us, but we were frozen with horror. This man had almost no face left. An explosive bullet had blown up in his mouth, blasting out his cheeks, ripping out his tongue (a piece of which hung down), and shattering his jaws, and blood poured copiously from these horrible wounds […] finally we were able to get him to a first-aid station.
>
> (Barthas 2014: 43–4)

How then should we understand the sonic nature of these descriptions? I propose that we think of these soldiers as having the sounds of their trauma etched onto their faces much like music was etched onto the shellac of the discs of the time, thereby interrogating the connections between materialism, sound, and experience. Perhaps we need to move beyond the sonically literal, beyond mere 'vibrations' towards new connections; epistemological, cultural, sensory, technological, and political in order to question when a sound begins and when it ends by using the example of these soldiers? The literalness of vibrations as an epistemological sonic starting point has already been questioned by Mark Grimshaw and Tom Garner (2015), but I do not wish to replace one starting point with another: the sonic imaginary should broaden out our understanding of sonic duration, not replace the literalness of vibrations.

How sound is understood is an issue of sonic duration, but it is also about how we conceive of and draw boundaries around the sonic. The gun discharges, the sound is instantaneous as it impacts the soldiers face – a mere fraction of a second. Artillery shells took longer to arrive and were subject to intense conjecture by troops as to where they might land – troops developed a sonic skillset that sometimes might ameliorate the contingency of their survival (Daughtry 2015). Why separate the immediacy of sonic duration – the split second – from that which went before and from that which comes later? There exists the shifting soundscape of the battlefield, the strategic and policy meetings of those designing and planning the weapons to be used, representing an alternative, hidden range of soundscapes, and there are the munition factories that make these weapons of destruction, indeed a sonic mobilization of a whole country.

> A tobacco factory had turned to making shells, a gramophone works was making shell fuse, and a magneto-maker, a piano factory, and a coach-builder had turned to the manufacture of one or another kind of munitions […]. Similar conversions were happening around the nation, such that the whole country is one seething munition factory.
>
> (Woollacott 1994: 28)

Questions surrounding how this might be discretely conceptualized are ones of personal recollection, military history, and strategic interpretation; of military technology and the development of killing weapons – the long-range rifle, the aeroplane able to drop bombs onto troops below, the machine gun, and so on. Does the sound of a delayed fuse seventeen-inch shell originate in its design, manufacture or the values that underpin it? The momentary sound and impact of a shell on human flesh can be traced back to its sonic design, innovation, and production – a political, economic, and cultural conveyor belt of death. Where might such an analysis lead? If you were Theodor Adorno or Max Horkheimer the question might take you back to the origins of Western culture, to Homer's *Odyssey* and the dialectic of enlightenment, sonically inflected? These issues will be dealt with in greater depth in a later publication but suffice to say at what such an analysis of sonic antecedents and consequences might draw upon. These epistemological questions are intimately connected to a range of methodological concerns and by extension to an imaginative interdisciplinary connectivity.

Just as the written testimony of war veterans might tell us more about the nature of gender stereotypes of the time than the 'actual' experience of war, so written documents might point to the extension of our understanding of sonic design embedded in the individual experience of war and its destructiveness. Written manuals concerning the design, development, and manufacture of a wide array of weaponry was common even during the prosecution of the First World War. Pride of place goes to the writings of Douglas T. Hamilton who became the 'high priest' of the technological manufacture of death through a series of manuals written during the war that describe in minute detail the production techniques of weaponry and the measurement of their destructive power. His books are also distinct for the length of their titles; for example his 1916 treatise on shell manufacture is entitled, *High-Explosive Shell Manufacture: A Comprehensive Treatise on the Forging, Machining and Heat-Treatment of High-Explosive Shells and the Manufacture of Cartridge Cases, Primers, and Fuses, Giving Complete Directions for Tool Equipment and Methods of Setting.* His books are exemplary examples of detailed and functional analysis of how the weapons of destruction are designed so as to successfully achieve optimum results: 'High explosive shrapnel shell combines the principles of both the high explosive shell with the common shrapnel shell […]. It detonates upon impact, causing considerable damage, and is capable of destroying the shield in protecting a field gun' (Hamilton 1916: 46).

Whilst in 1914, 50 per cent of 75mm shells used were loaded with shrapnel, troops didn't have metal helmets to give them any protection until early 1916. Design preceded protection.

> This shell [the British 18 pounder] is set off by what is known as the delay action fuse. This allows the shell to penetrate fortifications or earth works before it is detonated, and, consequently, enables the explosion to have a much more destructive effect than if it took place instantaneously upon impact.
>
> (Hamilton 1916: 7)

Hamilton also produced an inventory of the cost of each bullet, shell, rifle, and cannon. For example, a Lee-Enfield rifle used by the British army cost $20 in 1914 with its bullets a mere few cents each.

Just as it is possible to trace the sound of the shell back in time, so sound does not cease on impact. I suggest that sound is transformed, traduced onto the disfigured face that bears its sonic scars for all to 'see', comment on, hide from, or avoid, thus breaking down the binary between sound and vision. Indeed the sonic is fully embodied. These sonic scars were both embodied and materialized, manifest facially not just on the face but on the psyche, with sonic time disassembled, frozen, returned. Body and cognition both 'in tune' yet out of tune. The cognitive, in the form of shell shock becoming visible in the bodily movement of those who suffered thus. Just as sound moves backwards on the 'production line of death' so it moves forwards into the personal and social narratives of victims, etched into their psyches and faces.

Whilst sonic trauma was both visible (facial disfigurement, bodily movement) and invisible (cognitive impairment), the invisible, in the form of shell shock or other psychological trauma, was often preferred to the overtly visible. Yet the psychological effects of continual exposure to the conditions experienced by troops in the First World War were recognized at the time. Grafton Elliot Smith, a serving doctor, argued as early as 1917 that:

> Before this epoch of trench warfare very few people have been called upon to suppress fear continually for a very long period of time […] harrowing sight and sounds, disgust and nausea at the happenings in the trenches […]. Frequently the assaults made upon him nowadays are impersonal, undiscriminating and unpredictable, as in the case of heavy shelling […]. The noise of the bursting shells, the premonitory sounds of approaching missiles during exciting periods of waiting, and the sight of those injured in his vicinity whom he cannot help, all assail him, while at the same time he may be fighting desperately within himself. Finally, he may collapse when a shell bursts near him though he need not necessarily be injured by actual contact with particles of the bursting missile.
>
> (Elliot Smith 1917: 8)

The lasting implications of trauma were also recognized:

> Loss of memory, insomnia, terrifying dreams, pains, emotional instability, diminution of self-confidence and self-control, attacks of unconscious or changed consciousness sometimes accompanied by convulsive movements resembling those characteristic of epileptic fits, incapacity to understand any but the simplest matters, obsessive thoughts, usually of the gloomiest and most painful kind, even in some cases hallucinations and incipient delusions – make life for some of their victims a veritable hell.
>
> (Elliot Smith 1917: 13)

The consequences of the etching of sound onto the victim's face were experienced in many settings, from the individual's psyche to the everyday, to the social spaces they would move through to the hospital ward. Sound manifests itself in the medical procedures undertaken to repair that which technology had damaged or destroyed together with a range of gendered sonic expectations and performances carried out by both patients and staff. The year 1916 saw the development of inter-tracheal anaesthesia, which replaced chloroform and ether, making it possible to operate in sterile environments for longer and enabling more effective treatment for those whose windpipes had been blown away. For those with facial or throat wounds the silences were prolonged and intensely painful. Whilst medical procedures had made rapid advances in the First World War, Paul Alverdes, a German war veteran, who had part of his throat shot out in the war, described the pre-1916 listening procedures necessary for doctors to sometimes operate on the facially disfigured:

It was next to impossible for a surgeon to operate on a patient's face or throat if a gas mask was in use for controlling the anaesthetic – the surgeon performing the procedure would need to be able to hear the patient's breathing at all times – so no anaesthetic. The patients referred to this as 'the healing torture'.

(Alverdes 2017: 16)

Through these treatments, often lasting years, the hospital wards which treated the wounded were frequently full of 'the groans and screams' which penetrated the minds and bodies of those who treated them. Nevertheless, complex 'codes of silence' within hospital wards were operationalized both by the soldiers who experienced pain and suffering and the female nurses who managed the hospital wards. Patients, including the facially disfigured, were encouraged to display high levels of cheerfulness – 'happy though wounded'. One nurse treating the wounded on the hospital wards commented that she experienced the 'very spirit of suffering, [in hospital wards] the silence is full of it' (Carden-Coyne 2014: 324).

So, in returning to our blue park bench in Sidcup and to the facially disfigured soldiers who sat on this bench, we might ask why these blue benches were inscribed with the words 'FOR WOUNDED SOLDIERS ONLY' painted onto their back? Silence is inscribed into this park bench. All those that once sat on these benches, their face or throats in tatters, are dead of course, silent as indeed many were silent in their lives. Sidcup, like many other towns in the UK, strongly supported the war. The inhabitants had flocked to the local cinema in 1917, the same year as the hospitals opened their doors to those disfigured as a result of it, to watch the first war documentary *The Battle of the Somme*, universally congratulated at the time for its gritty 'realism'. Equally, Sidcup, like so many other towns, had possessed its 'white feather' brigades of women who would plant white feathers on those men they thought were shirking their responsibility to go to war. A silent rebuke embodying a repressive sonic masculinity. We might surmise that the physical effects of war were not as 'attractive' to those citizens as the gritty reality of the silent screen. Indeed 'the town's residents petitioned for the men to be excluded from the town' (Bamji 2017: 142). As a consequence of this, the local council and hospital reached a compromise whereby the words written on the park bench would warn the people of Sidcup to 'look away' at the horrors that would confront them if they were to look at those sitting on the bench. The 'realism' embodied by long-lasting injuries both physical and mental proved problematic for many civilians. The bench was the first stage of their public invisibility and indeed silence. Even those with minor facial disfigurements were not permitted to return to the trenches as it was thought that the uninjured did not want to see what they might become. Their silence manifest on these park benches both personal and interpersonal were mirrored in their experiences on the hospital wards and operating theatres. The gendered expectations and performances of both patients and staff sometimes mirrored those who walked past the blue benches. The silencing of those who suffered in their everyday lives was replicated by their silencing in the hospital wards of the time. On the wards of the Queen Mary Hospital mirrors were not allowed so that the facially injured should not look at the destruction wrought upon their faces – their own 'blue bench'.

The sounds of silence and the silence of sounds

> Silence is not acoustic. It is a change of mind, a turning around.
>
> —John Cage

Since John Cage we have recognized that silence is elusive. Even in anechoic chambers, if we listen intently, we can hear ourselves – our blood pumping, the embodiment of sound within environmental 'silence' – if we choose to or have the skill to listen thus. When audiences first listened intently to 4′33″ they became aware of an array of environmental sounds. They experienced sound as any bourgeois audience might – from a Beethoven concert conducted by Toscanini to that of Varèse conducted by Stokowski – in rapt attentive silence, yet in the process, paradoxically, they became attentive to the world and also to themselves. Music became sound just as sound became music, yet irredeemably bourgeois nevertheless. Yet the Cagean approach to sound, and therefore silence, remains merely one possibility within cultures of listening. Cage's 'change of mind' paradoxically returns us to the dualism so loved within Western culture of mind and body. In choosing to reorientate ourselves to the sounds of the world and, by extension, to ourselves – the blood pumping through our veins – an analytical puritanism creeps in. Yet one more literalism within the study of sound. Cage's anechoic chamber is mind listening to body after all. A personal dualism unlike that of his audience who, after all, were listening primarily to the world. If we were to take a bodily approach to sound and silence, a somatic one, sometimes preconscious but not necessarily so, in which emotion and fear, cognitive to be sure, play their role in the phenomenology of sonic experience, then a range of 'different' silences might be discovered that challenge us methodologically. Returning to the silences of the First World War, many soldiers became deaf due to the sonic destruction of their hearing, the physical result of sonic assault. Yet many also became mute, unable to speak at all despite no physical injury. Others suffered from aphonia and dysphonia, products of what we today would refer to as forms of post-traumatic stress disorder. Their sensory system shut down with modes of speech and hearing crowded out by trauma, the result of many experiences coupled with temperament – sounds, sights, fear, tension, time, and exhaustion. Just as the mind can attend to sounds, so the emotions can block those sounds out. In discussion with those who have experienced the sudden death of loved ones, they frequently mention the silence of death – the waking up in the middle of the night alone, the feeling of deathly silence; they do not hear the clock ticking, the feet shuffling, the heart beating. The push against the resistant door produces no sound with death lying on its other side. The mind crowds out sound in a personal, temporary silence. This silence need not merely be a private and enclosed silence, it can also be a public silence. On visiting Auschwitz, the site of so much human suffering, its historical soundscape different from the fields of Verdun or the domestic bathroom, visitors sometimes (but not always) mention the deathly silence of the site. They do not hear the sounds of the birds flying overhead or the subdued voices of others. Silence works in tandem to sound, follows it, and is frequently out of time with it.

Culturally silence often becomes a form of social invisibility – the Native Indians of America had no voice in the nineteenth century, remaining unrecognized by Murray Schafer in his description of the sonic richness of the settler trains arrival in American Frontier towns that entailed the death of Native American livelihoods, lands, and much more besides.

> By comparison with the sounds of modern transportation, those of the trains were rich and characteristic: the whistle, the bell, the slow chuffing of the engine at the start […]. The sound of travel have deep mysteries […]. The train whistle was the most important sound in the frontier town, the solo announcement of contact with the outside world.
>
> (Schafer 1977: 81)

Equally, the slaves portrayed in Joseph Conrad's *Heart of Darkness* (1899), with the rhythmic clinking of the chains around their feet and necks, brutalized in order to fuel the industrialization of those profiting from their misery, are 'heard' in Britain as 'railway spine', caused by the continuous vibration of the trains they travelled on and complained about (Trower 2012). The sonic draws us in closely to the immediacy of experience, to the here and now: sound juxtaposed with deadly cultural silences produces its own sonic dialectic. The parochial cultural norms of those who listen to the sounds that are merely around them, that do not seek connections beyond the here and now of sonic vibrations, is a dominant and continuing trope of Western sonic culture.

We need not look so far for examples of 'silencing' in the First World War. Whilst browsing through the sonic records in the Imperial War Museum in London and listening to the BBC radio and television archives of interviews with those who fought in the First World War, one might be struck by their similarity. These interviews carried out many years after the soldiers' wartime experiences appear to embody a specific ideology, that of the 'stiff upper lip'. After a period of over fifty years when these accounts were recorded, they remain selective, presented, and conform to an internalized embodiment of an ideologically charged masculinity; one in which the war is described as 'bad' but 'not that bad'. Whilst the ideology of suffering 'in silence' had lost much of its ideological justification by the 1960s and 1970s when these interviews occurred – it was an era of the Vietnam War after all – it nevertheless remained buried in the psyches of these old troopers. Their sonic testimonies become simultaneously 'authentic' and 'misleading'. The words do not speak merely for themselves but are embodied in a set of cultural values existing in and out of their time.

Yet all these words are spoken by those whose faces, if not their psyches, remained 'intact'. Their 'silences', an internalization of pain and suffering, also preclude the literal silencing of others, not so lucky. This is a silencing not only of the dead but of those unfortunate men who sat on our blue park benches in Sidcup and elsewhere. In the search for 'silences', obscured by the words spoken, there is the silence of the facially disfigured. There is no existing testimony given by the disfigured, either on tape or on television, and very rarely in written diaries. Existing accounts are written by doctors and nurses, not the victims themselves. It is as if they never existed, a continuing thorn in the sonic testimonies of the First World War. If we listen we can only 'hear' the silence of the disfigured. The silence greeting the facially disfigured was merely an extreme that faced those who had merely lost arms or legs. These unfortunates received a portion of a disability pension, their poverty resulting in the presence of many soldiers needing to beg in the streets of UK cities. The facially disfigured received 100 per cent disability pensions – the state willing to do almost anything to keep them off the streets.

Beyond the hospital wards of the First World War silence was represented both as a form of 'turning away' and as a respectful commemoration that has remained to the present day. Cultural silences remain inscribed into contemporary wartime memorials, their sonic connections extending into the past, their lineages traced from ideologies of the First World War, embodied in the sonic aesthetics of contemporary commemoration.

We're Here Because We're Here to Coda to Coda: Listening to the silences of commemoration

It is to the sounds and silences of contemporary First World War commemorations that I now turn in order to discuss the persistence of both a sonic turning away coupled with politically sensitive forms of silence embedded within commemoration.

I begin with Jeremy Deller's *We're Here Because We're Here* which commemorated the 100th anniversary of the beginning of the Battle of the Somme on 1 July 2016, a day in which 26,000 troops were killed. The commemoration was nationwide consisting of 1,600 volunteers moving through everyday sites, often transit sites, like those experienced by soldiers of the day – railway stations and so on – dressed as First World War troopers. Each soldier represented one trooper who had been killed on that first day of battle; they carried small cards with the details of that one soldier: their age, family, regiment, and death; and if approached, they would silently hand out the white card with the details of the dead soldier whom they were representing. This moving memorial (many *were* moved) chimes with the directive of Henri Barbusse, the writer of the first novel depicting the horror of the First World War from the viewpoint of a serving soldier, when talking of those who had fallen that, 'the only acceptable attitude, in our view, to bring to their tombs, as one comrade put it, is "impeccable silence". At least, if we do not speak their name, no one else should dare to do so' (Barbusse 2009 quoted in Winter 2014: 182). Deller, in constructing the day's commemoration, had been influenced by stories deriving from the war of relatives believing that they had seen their dead loved ones walking the streets of London and elsewhere. Death was invisible for many of these relatives with no repatriation of bodies or indeed no remaining body at all. Deller's memorial was largely silent. On film taken on the day it is possible to hear the thud of the soldiers boots as they marched and mingled in with the normal city sounds of railway stations, people talking, and information system announcements. This 'silence' was periodically shattered as the 'troops' suddenly broke out into song – 'We're Here Because We're Here', sang to the tune of 'Auld Lang Syne'. This was the song that some of the soldiers sang on that day one hundred years ago, soldiers who had no voice in the geopolitics of the time. In their renditioning of the song we hear a trace of sonic irony and futility unrecognized in the popular songs of the time in which the sonic manufacture of propaganda is all that remains on shellac in the form of 'Keep the Home Fires Burning' and 'Pack up your Troubles'. The singing of 'We're Here Because We're Here' signposts an alternative soundscape to the sonic propaganda of music at the time listened to both by troops and families. No commercial discs were made of these sonic transgressions at the time. The voices of propaganda continued to drown out the voices of wartime experience with their simple, unambiguous, dulcet yet 'patriotic' tones. At the end of the day Deller's 'troops' marched around in opposing concentric circles before whatever audience existed, visually illustrating the meaninglessness of their deaths, before giving out a short, shattering, collective roar of pain, not much longer than the sound that an approaching shell might have taken before it killed them on the battlefield. The technologies of death intertwined literally and philosophically with the flesh of its victims. After this final roar of the voice the participants dispersed silently into the night as if they had never existed.

Deller describes himself as a conceptual artist, not a sound artist, but *We're Here Because We're Here* uses silence, sound, and vision equally and to powerful effect. To the sound recordist measuring the sounds of the urban environment, the piece would go largely unnoticed. The power of the piece resides in its audiovisual presence that remains on multiple YouTube feeds.

Two years later I am sitting in the Imperial War Museum in London, headphones over my ears and hands over the headphones. I am listening to and 'feeling' *Coda to Coda*, a sound piece lasting a little over a minute commissioned by the Imperial War Museum to commemorate the final sounds of the First World War on the Western Front. As I listen there is the 'lifelike' sound of a variety of shells whistling through the air before exploding, then slowly fading away to be replaced by the sounds of bird calls before silence. By placing my hands over my head, I also experience sonic vibrations deriving from the piece. As a sound piece the aesthetics work well. It is 'as if' we are eavesdropping in on a momentous historical occasion. In addition to the aesthetics of the piece, reminiscent of the ending of the 1930s American movie *All Quiet on the Western Front*, the designers were attentive to the 'realism' of the tape through the use of original sound pressure impulses used on the front that assessed the direction and distance from which shells were fired along the front. The equipment measured the time it took for the sound impulses to reach the front.

> The sound ranging equipment used six tuned low frequency microphones arranged in a wide arc behind the allied lines. The microphones were connected to a string galvanometer at a forward listening position. A low frequency signal picked up by one of these microphones would move a thin wire in the galvanometer and cast a shadow onto a piece of moving film […] to create a visual recording of these sound impulses.
>
> ('Making a New World: Armistice Soundwave' 2018)

Recordings of appropriate gunfire were used to interpret the sound and intensity of shelling with the simulation of shock waves through vibrations that the listener within the

Imperial War Museum could experience. Will Worsley, director and principal composer of *Coda to Coda*, stated that 'we hope that our audio interpretation of sound ranging techniques […] enables visitors to project themselves into that moment in history and gain an understanding of what the end of the First World War may have sounded like' ('Making a New World: Armistice Soundwave' 2018). Yet as I listen to the piece I am overcome with a sense of unease. I look around; others, primarily young and eager, are at desks listening intently to what I have just listened to. We are surrounded by the material artefacts of the war housed in the museum dedicated to the memory of that war. The sound of *Coda to Coda* was not that loud – it cannot be, or else our ears would suffer damage. Equally, the sonic vibrations of the shells could be felt, but not that much. Sonic 'realism' remains a myth of course, and so we are left with aesthetics and perhaps the imagination. Yet it is something else about the soundscape that I have listened to that is troubling me. The material inscription of the final sounds of war, transformed from the materiality of the paper into the sound that I have just listened to, appears as too 'literal'. The joining of 'sonic realism' or 'sonic fidelity' to aesthetics is equally impressive and troubling. The sounds that I listen to through the headphones bear little resemblance to the sounds of gunfire on that day. What could replicate the sonic intensity of such an experience as described by an American Red Cross volunteer in 1916 of a shell landing 200 metres away from him:

> There was a sound like the roar of an express train, coming nearer at tremendous speed with a loud singing, wailing noise. It kept on coming and I wondered when it would ever burst. Then when it seemed right on top of us, it did, with a shattering crash that made the earth tremble. It was terrible. The concussion felt like a blow in the face, the stomach and all over; it was like being struck unexpectedly by a huge wave in the ocean.
>
> (Alexander 2010: 2)

Whilst listening to *Coda to Coda* I am reminded of a statement made by a serving trooper in the First World War after he had watched *The Battle of the Somme* at the cinema in 1917. The film had been lauded for its realistic depiction of the war, using only original war footage. The deficit 'realism' of *The Battle of the Somme* was apparent to the trooper. On being asked about the film's realism, he answered: 'Yes … about as like as a silhouette is like a real person, or as a dream is like a waking experience. There is so much left out' (Reeves 1997: 16). The materiality of sonic representations of war, whilst shadow-like in themselves, might nevertheless remain constructive and informative. Yet there is something else troubling me about *Coda to Coda*. Its soundscape is contextual, geographically specific, but more worrying, as we listen to the dying sounds of the war and the rebirth of nature through the sound of birdsong – a common trope of the time and subsequently – is that there is one sound that is omitted: the sounds of the troops who were there, who experienced and suffered. We hear no voices, merely bird song and then silence. In *Coda to Coda* the shadowy realism of the sonic is mingled with an aesthetic silencing of those who experienced the war as the sounds of war die away to be replaced by the sounds of nature. Just as the troops are silenced in *Coda to Coda*, so one hundred years before, the facially disfigured patients in the hospital in Sidcup were silenced in the recording of history and in the turning away of those who walked past.

This Introduction has sought out new ways to conceptualize how sonic connections might be approached, researched, and understood. In doing so it has questioned how we deal with a variety of historical testimonies but also with contemporary sound works that both subvert or confirm ideologies entrenched in past accounts. The graphic descriptions of those who suffered or died in the First World War and who are memorialized in film, print, and sound art are not meant as blanket pronouncements, but rather as an opening of disciplinary and methodological doors. The authors of this rather large first section do likewise. They produce a rich tableau of sonic connectivity amidst a dizzying array of methodologies and methodological issues, from music venues in London to the innovative use of digital audio in the classrooms of America; from the soundscapes of urban cultures to reflections upon the very nature of sonic methodologies that are 'merely' written down; from the innovative use of sonic methodologies to understand the past to the 'oral tradition' rooted in the heart of literature; from the philosophical grasp of auditory experience to the intervention of environmental planning of urban spaces that many of us inhabit; from listening to the sounds of nature to reflecting upon the materiality of those sounds. Materiality and culture surface as topics of discussion and investigation in many of these chapters, as medical, scientific, and technical investigations extending to those embedded in the cultures examined in anthropology, history, and urban culture and beyond.

References

Adorno, Theodor (1980). 'Bloch's "Traces": The Philosophy of Kitsch'. *New Left Review* 1 (121) (May–June): 1–9.

Alexander, Caroline (2010). 'World War 1: 100 Years Later; The Shock of War'. In *Smithsonian. Com*. Special Report. Available online: http://www.smithsonianmag.com/history/the-shock-of-thr-war-55376701/ (accessed 1 January 2020).

Alverdes, Paul (2017). *The Whistlers Room*. London: Casemate UK.

Bamji, Andrew (2017). *Faces from the Front: Harold Gillies, The Queen's Hospital, Sidcup and the Origins of Modern Plastic Surgery*. Exeter: Helion and Company.

Barbusse, Henri (2009). *Under Fire: The Story of a Squad*. New York: Feather Trail Press.

Barthas, Louis (2014). *Poilu: The World War One Notebooks of Corporal Louis Barthas, Barrelmaker, 1914–1918*. New Haven, CT: Yale University Press.

Benjamin, Walter (2002). *Selected Writings*, vol. 3, *1935–1938*, Cambridge, MA: Belkanap and Harvard University Press.

Biernoff, Suzannah (2011). 'The Rhetoric of Disfigurement in First World War Britain'. *Social History of Medicine*, 24 (3) (December): 666–685.

Birdsall, Carolyn (2012). *Nazi Soundscapes: Sound, Technology and Urban Space in Germany, 1933–1945*. Amsterdam: Amsterdam University Press.

Bloch, Ernst (1977). *The Principle of Hope*. Oxford: Oxford University Press.

Bloch, Ernst (2006). *Traces*. Stanford, CA: Stanford University Press.

Bloch, Marc (1980). *Memoirs of War 1914–15*. Ithaca, NY: Cornell University Press.

Brooks, William, Christina Bashford, and Gayle Magee (eds) (2019). *Over Here, Over There*. Champaign, IL: Illinois University Press.

Bull, Michael (2019). 'Into the Sounds of War: Imagination, Media, and Experience'. In Mark Grimshaw-Aagaard, Mads Walther-Hansen, and Martin Knakkergaard (eds), *The Oxford Handbook of Sound and Imagination*, vol. 1. Oxford: Oxford University Press.

Carden-Coyne, Ana (2014). *The Politics of Wounds: Military Patients and Medical Power in the First World War*. Oxford: Oxford University Press.

Daughtry, J. Martin (2015). *Listening to War: Sound, Trauma, and Survival in Wartime Iraq*. Oxford: Oxford University Press.

Dorgelès, Roland (1921). *Wooden Crosses*. London: Aegypan Press.

Elliot Smith, Grafton (1917). *Shell Shock and its Lessons*. Manchester: Manchester University Press.

Fussell, Paul (2013). *The Great War and Modern Memory*. Oxford: Oxford University Press.

Gilman, Sander L. (2014). *Seeing the Insane*. London: Echo Point Books.

Grimshaw, Mark and Tom Garner (2015). *Sonic Virtuality: Sound as Emergent Perception*. Oxford: Oxford University Press.

Hamilton, Douglas T. (1916). *High-Explosive Shell Manufacture: A Comprehensive Treatise on the Forging, Machining and Heat Treatment of High-Explosive Shells and the Manufacture of Cartridge Cases, Primers, and Fuses, Giving Complete Directions for Tool Equipment and Methods of Setting*. London: Ulan Press.

Horkheimer, Max and Theodor Adorno (2007). *Dialectic of Enlightenment*. Palo Alto, CA: Stanford University Press.

Hutchinson, Woods (1918). *The Doctor in the War*. Boston, MA: Houghton Mifflin.

Hynes, Samuel (1992). *A War Imagined: The First World War and English Culture*. London: Pimlico Press.

Keegan, John (1976). *The Face of Battle*. London: Jonathan Cape.

Korstvedt, Benjamin (2007). 'Review of Bloch, Ernst, *Traces*'. H-German, H-Net Reviews, October. Available online: http://www.h-net.org/reviews/showrev.php?id=13778 (accessed 28 June 2020).

Jünger, Ernst (2003). *Storm of Steel*. London: Penguin Books.

Leese, Peter (2014). *Shell Shock: Traumatic Neurosis and the British Soldiers of the First World War*. Basingstoke: Palgrave Macmillan.

'Making a New World: Armistice Soundwave' (2018). [Blog] *Coda to Coda*, 6 November. Available online: https://codatocoda.com/blog/making-a-new-world-armistice-soundwave/ (accessed 3 January 2020).

Rath, Richard Cullen (2003). *How Early America Sounded*. Ithaca, NY: Cornell University Press.

Reeves, Nicholas (1997). 'Cinema, Spectatorship and Propaganda: "Battle of the Somme" (1916) and its Contemporary Audience'. *Historical Journal of Film, Radio and Television*, 17 (1): 5–28.

Rice, Thomas D. (2013). *Hearing the Hospital: Sound, Listening, Knowledge and Experience*. Canon Pyon: Sean Kingston Press.

Schafer, R. Murray (1977). *Tuning the World*. New York: Alfred A. Knopf.

Sterne, Jonathan (2003). *The Audible Past: Cultural Origins of Sound Reproduction*. Durham, NC: Duke University Press.

Sterne, Jonathan (ed) (2012). *The Sound Studies Reader*. London: Routledge.

Steingo, Gavin and Jim Sykes (eds) (2019). *Remapping Sound Studies*. Durham, NC: Duke University Press.

Sweeney, Regina M. (2001). *Singing Our Way to Victory. French Cultural Politics and Music during the Great War*. Middletown, CT: Wesleyan University Press.

Trower, Shelley (2012). *Senses of Vibration: A History of the Pleasure and Pain of Sound*. New York: Continuum Press.

Winter, Jay (2014). *Sites of Memory, Sites of Mourning: The Great War in European Cultural History*. Cambridge: Cambridge University Press.

Woollacott, Angela (1994). *On Her Their Lives Depend: Munitions Workers in the Great War*. Berkeley, CA: University of California Press.

2

Sonic Methodologies in Anthropology

Alexandrine Boudreault-Fournier

The volume *Writing Culture: The Poetics and Politics of Ethnography* (1986), edited by James Clifford and George Marcus launched a series of groundbreaking criticisms about the discipline of anthropology. The critics targeted the ways in which anthropologists traditionally conducted fieldwork and *wrote* ethnographies. More specifically, the volume emphasized the necessity of including a polyphony of voices in ethnographic writing, and encouraged ethnographers to think in terms of fragmentation, incompleteness, non-linearity, and partial truth. Importantly, it also raised the issue of reflexivity, that is, the anthropologist's own subjectivity. This challenged notions of objectivity associated with the ethnographic enterprise. Also, the volume addressed the issues of representation and the temporality of ethnographic accounts, which tended to position 'the other' as not living in the same time frame as the researcher (Fabian 1983). These marked a shift in the 'anthropological imagination' with the world no longer perceived as complete and coherent but rather fragmented and ambiguous (MacDougall 2005: 244).

In *Writing Culture*, Clifford argued that anthropology should give more attention to expressive speech and gesture instead of focusing on a detached 'observing eye'. Following this line of thought, he asked: 'But what about the ethnographic ear?' (Clifford 1986: 12). Similarly, Johannes Fabian who also contributed to the debates surrounding the 'crisis of representation', encouraged anthropologists to think about ethnography – anthropologists' methodological strategy of choice – as primarily about speaking and listening rather than observing. In comparison to visually oriented approaches – observation, participant observation – metaphors that point to sonic methodologies emphasize a temporal process, and demand coelvaness, which refers here to the idea of sharing a common time and space with the research participants (Fabian 1983). Despite this courageous call for sonic ways of engaging and representing ethnographic experiences, the writer's voice still pervades in anthropology (Erlmann 2004: 1). In fact, until the 1990s, anthropologists largely turned a deaf ear to the potential of sounds to broaden their avenues of research (Calzadilla and Marcus 2006; Marcus 2010). Yet, it would not be exaggerated to argue that we are currently witnessing a 'resurgence of the ear' within the discipline (Erlmann 2004), and that anthropology is contributing in meaningful ways to the exploration of the sound world.

It is generally understood that the 'anthropology of sound' considers all sound-related dimensions (noise, music, voice, silence, etc.) as significant elements of research and analysis. This chapter concentrates on the extramusical phenomena and elements that are included in the 'anthropology of sound'. Hence, it does not dedicate much attention to music per se, although many of the methodological strategies mentioned in this chapter could also apply to the discipline of ethnomusicology. Interestingly, and quite contradictorily, the 'anthropology of sound' also shares many connections with visual anthropology as both subdisciplines have similar concerns for recording and editing, and both generate conversations about the issues of representation, creativity, and imagination, among other things (Boudreault-Fournier 2016a).

In investigating sounds through ethnographic fieldwork, anthropologists propose an attentive listening to the everyday in addition to exploring, from an in-depth perspective, how sounds are entangled in the lives and practices of listeners. Furthermore, sounds offer a cultural and historical contextualization of auditory perception, and encourage delving into sonic phenomena, textures, and events as social and cultural constructs. Indeed, anthropology is not alone in giving serious consideration to how sounds relate to social and cultural life, and to considering the context in which listening practices take place. This explains why the 'anthropology of sound', or the 'anthropology *in* sound' (Feld 1996), is fundamentally interdisciplinary.

This chapter sheds light on the methodological strategies developed and adapted by anthropologists who are interested in the sonic dimensions of social and everyday life. In order to provide a general landscape of the methods employed by anthropologists, this chapter is divided in two sections. The first section explores six epistemological approaches – soundscape, relational ontology, the senses, sound and space, arts-based research and collaboration, and ethnographic film-making – that have a deep impact on how sounds became part of the anthropological project and, by extension, how methodologies were developed to engage with emerging questions and challenges that relate to the sound world. These six approaches are not mutually exclusive but complement each other. The second section develops four concrete methodological components of an ethnographic approach attentive to sounds – listening, recording, editing, and representation (following the conversation in Feld and Brenneis 2004: 461).

Ethnographic fieldwork – the anthropological method par excellence – is broadly understood as a research approach that necessitates long-term involvement with a group, collective, or community in order to ultimately 'speak nearby' instead of 'speaking about' as pointed out by Trinh T. Minh-ha in her film Reassemblage (1982), produced during her fieldwork in Senegal. Ethnographic fieldwork is guided by strict ethical concerns and aims, using methods such as simple observation, participant observation, and interviewing (among others), to investigate a research question or interest. It is now generally understood that ethnography refers to 'a reflexive and experiential process through which academic and applied understanding, knowing and knowledge are produced' (Pink 2015: 4). Finally, projects are increasingly developed in collaboration or with the involvement of participants 'in the field', and sound-related projects contribute to anthropology's recent interdisciplinary dynamism.

Epistemological approaches to the sound world in anthropology

Soundscape

Despite a few exceptions (for instance, Feld [1982] 1990; Helmreich 2007; Rice 2008), the concept of soundscape has not had as much impact in anthropology as it has had in other disciplines, particularly in sound studies, but also in arts, cultural studies, urban planning, and communication. R. Murray Schafer defines the concept of soundscape as 'any acoustic field of study' and any 'events *heard*, not objects *seen*' (Schafer [1977] 1994: 8, emphasis in the original). Although from an anthropological point of view, the term soundscape as understood by Schafer offers possibilities of considering the cultural nature of sound, existing methodologies that allow us to think about the sound world, and exploring material spaces of performances (Samuels et al. 2010: 330), the concept has been criticized by anthropologists (among others) for representing a utopian and romantic-ecological-environmentalist vision of sound, especially in relation to the urban space which Schafer most often associates with noise pollution. Among the oft-cited critics, Tim Ingold emphasizes how the concept of soundscape proposes a model that separates hearing from the other senses. Furthermore, Schafer's approach presents sound as a landscape that is admired from a distance without engaging with debates on agency and issues of perception (Ingold 2007; Helmreich 2010; Eisenberg 2015; Feld 2015a). Finally, Schafer's perspective is reminiscent of 'salvage ethnography', associated with the work of famous anthropologist Franz Boas, which focused on preserving, describing, and recording cultural forms threatened with extinction over time. Similarly, Schafer and his team focused on collecting 'non-contaminated' sounds before noise pollution and to collect recordings in a repository of 'endangered sounds'.

In spite of this, Schafer's concept inspired – and still inspires – social scientists to pay attention to sound material in their fields of investigation and to come up with different understandings of the term soundscape. For example, historian Emily Thompson, who explores the emergence of the modern sound through delving into a sonic history of architecture during the post-Fordist era, defines soundscape as 'an auditory or aural landscape [...] simultaneously a physical environment and a way of perceiving that environment; it is both a world and a culture constructed to make sense of that world' (Thompson 2002: 1). Stefan Helmreich (2007) described the experience of sound under water – not immediately perceptible to human ear – as 'soundstate'. And Alain Corbin (1998) deals with the presence – then the disappearance – of bells in the soundscape of the French countryside in the nineteenth century, evoking a rich sensory landscape in transformation. Schafer's work also influenced in one way or another several anthropologists working with sound, including Steven Feld, who made several recordings of bells in Europe and Africa (Feld 2002; 2004a; 2004b; 2005). More recently, Andrew Eisenberg (2013) discusses the politics of sound in Mombasa Old Town in Kenya by engaging with the multiple layers of an 'Islamic soundscape'.

There are two ways of understanding the concept of soundscape, which speak to the methodological approaches adopted by anthropologists. First of all, there are the layers of sounds that surround us in everyday life and that make up the sonic world, for example when we take a walk in a busy market or in a forest. Secondly, there are the soundscapes that we create or alter voluntarily. As Schafer ([1977] 1994: 205) explains, we can all guide the orchestras that perform the soundscapes of this world, and we can also help reduce noise pollution. Thus, we can transform the soundscapes that surround us by creating more harmonic spaces that are receptive to the complexity of the sounds. For example, we may decide not to use air conditioning at home to reduce noise pollution or, as Schafer did, we may compose an original symphony, which is also a strategy, albeit a more substantial one, for transforming the soundscape of our everyday life. According to Schafer, noise is considered the enemy of sound and that is why we must try to eliminate it. Yet, the anthropological perspective on what distinguishes 'sound' from 'noise' is based on cultural, contextual, and even personal criteria rather than universal principles. By attempting to embed noise-related phenomena within cultural and social contexts, it is therefore possible to generate reflections that go beyond mere criticism and the rejection of sounds 'we do not like'.

Also, ethnographic research on noise from an infrastructural and ethnomusicological point of view encourages the emergence of conversations about how sounds circulate (or not) through certain networks, and how this influences creative forms of listening as well as generates unique sonic aesthetics (Larkin 2008; Novak 2013; Eisenberg 2013; Steingo 2016; Boudreault-Fournier, forthcoming). A contextual approach to noise pushes anthropologists to develop multisited ethnographic fieldwork that 'follows the sounds' as they circulate through various networks and infrastructures.

To go back to Schafer's invitation to take part in the composition of the soundscapes of this world, it remains a bold, if not courageous, invitation from the perspective of anthropologists who often situate themselves within a *social sciences* approach (rather than from an arts discipline). Yet, sound processing and editing, as well as arts-based research and collaboration with sound artists, stimulate new ways of imagining the ethnographic enterprise. From this, new conversations emerge about the role and impact of the anthropologists while conducting fieldwork (discussed below). Directly or indirectly, the concept of soundscape as developed by Schafer influenced many anthropologists to consider sound as a significant element of ethnographic research – something they should pay attention to – and to consider sound recording as a serious aspect of ethnographic fieldwork.

Relational ontology

During a conversation with Donald Brenneis, the ethnomusicologist, linguist, and anthropologist Steven Feld (Feld and Brenneis 2004) remembers the first paper he wrote while he was studying under the supervision of Alan P. Merriam at Indiana University. His paper was a response to the book *The Anthropology of Music* (1964), written by Merriam himself. Feld explains that two questions were at the base of the argument developed in his essay at the time, questions that still guide his anthropological approach today: What about an anthropology of sound? What about ethnographies that are tape recordings? (Feld and Brenneis 2004: 463). It was this shift of perspective towards listening and recording that led Feld to delve into the 'anthropology of sound' and into what he later called an 'anthropology *in* sound' (Feld 1996).

During his fieldwork in Papua New Guinea in the 1970s, Feld described the songs performed by the Kaluli people of Bosavi as adaptations of the tropical forest in which they lived, but it was much more than that. By training his ears to the ways in which people produce and perceive sounds, Feld realized that the songs performed by the Kaluli people were vocal cartographies of the rainforest, and that they were sung from the point of view of the birds that represented the Kaluli people's dead ancestors. To immerse himself in the phenomenology of perception, body, place, and voice, he had to explore a different ontological approach. For example, rather than considering trails at ground level, he transposed himself into the skin of birds to understand the poetics of over-river airways (Feld [1982] 1990; 2015a). His reflections, fueled by recordings (Feld 1991; more below on recordings), generated a relational approach to the sound domain that responded to Schafer's fundamental shortcomings (Feld 1994; 1996).

In the same conversation with Brenneis mentioned before, Feld explained that 'an ethnography should include what it is that people hear every day' (Feld and Brenneis 2004: 462), what he came to refer to as 'acoustemology' (Feld 1996), a fusion between acoustics and epistemology, to refer to the primacy of sound and listening as a modality of knowledge and of being in the world. Based on a relational ontology, the concept of acoustemology transmits the idea that we learn not through an acquired form of knowledge but through a cumulative and interactive process based on participatory and interactive experiences (Sterne 2012: 13–14). Feld was inspired by the concept of soundscape, but he also dissociated himself from it. In comparison with soundscape as developed by Schafer, who takes sound as an indicator of how humans live in their environment, acoustemology is relational, situated, fluid, reflexive, and contextual; it is the study of sound as a way of knowing. Acoustemology contrasts with the impression of rigidity, distance, and division of the senses associated with soundscape.

The essence of relationality embedded in acoustemology inspired many anthropologists to pursue how we learn from interacting with a specific sound world, positing that listening involves a form of knowledge that is relational and based on the awareness of our acoustic presence. For instance, Thomas Porcello draws on acoustemology to refer to the impact of technologies in the mediation of sensation and sound experiences. He came to refer to this as 'techoustemology'. The empirical experience of sound consumption through the use of technologies (radio, recorded music, electric musical instruments, television, video games, etc.) is embedded within a cultural and social space and, in a historical period, associated with perceptions of listening (Porcello 2005: 270). Andrew Eisenberg (2013) explores the acoustemology of Muslim citizenship in Kenya, and Tom Rice (2003) delves into the 'soundselves', the acoustemology of the self in relation to the experience of care in the hospital environment.

Senses

The music historian and sound scholar Veit Erlmann argues that until recently, anthropologists have generally approached the senses as texts to be read. As a consequence, few ethnographies have engaged with the domain of sound and 'actual listening practices' (Erlmann 2004: 2). Yet, ethnographies focusing on the senses and on embodied sensorial perceptions, including hearing and listening, contributed to a 'sensory turn' from the 1990s onward (Stoller 1989; Taussig 1993; Howes 2003, 2019a; Rice 2008, 2013). As part of this growing interest for the senses, Lucien Castaing-Taylor founded the Sensory Ethnography Lab at Harvard University in 2006 and David Howes founded the Centre for Sensory Studies at Concordia University, which have contributed to emphasizing the multisensoriality of the ethnographic process (Pink 2015).

In giving attention to the sensorial world, some anthropologists propose to destabilize the hegemony of the visual often associated with Western culture (Stoller 1989; Howes 1991; Bull and Back 2003), while others point out that 'subterranean histories' have always confronted the presumption of ocularcentrism associated with modernity (Hirschkind 2006; Samuels et al. 2010). Indeed, there might be a form of anti-visualism associated with the anthropology of the senses (Rice 2008; Eisenberg 2015). Yet, the objective is not to argue that hearing should take priority over the other senses in proposing a sono-centric approach (Bull and Back 2003; Sterne 2003; Erlmann 2004; Feld 2015a) but to think of sound as part of a set of sensations that remain at the centre of the relational experiences defining the ethnographic project.

Two more or less defined branches have come out with strength from this 'sensory turn' and have encouraged the emergence of new paths in academic debates: an 'anthropology of the senses' branch associated with the work of David Howes and followers; and 'sensory anthropology' aligning with the work of Tim Ingold and Sarah Pink, among others. The two branches division is arbitrary but it still helps to indicate a general idea of how anthropologists tend to approach the senses from a methodological perspective.

Following an anthropology of the senses type of approach, Paul Stoller (1997) encourages anthropologists to develop a 'sensuous scholarship' that confronts the sensual constitutions of local epistemologies. This implies that each culture should be approached on its own sensory terms and that anthropologists should consider from a comparative perspective each culture's own 'ways of sensing' (Howes and Classen 2013). Scholars associated with the 'anthropology of the senses' tend to argue that 'senses are made, not given', that sensory values are embedded within a social and cultural context, and that they are socially conditioned, which implies that sensorial perception is not simply a psychophysical phenomenon (Howes 2019b). Also, following Howes, although the senses should be understood as complementing each other, they may work in conflict. Therefore, 'perception is best understood in terms of performance rather than psychophysics' (Howes 2019b). In contrast, indebted to the ecological psychology of James J. Gibson and the phenomenology of Maurice Merleau-Ponty, Ingold does not engage with a cultural construction or collective representation of the senses. He bases his exploration of sensorial perception on psychophysics and how people listen, touch, taste, etc. and his perceptual approach corresponds to the sensorial anthropology branch mentioned above. Along with Ingold, Pink argues for a 'sensory anthropology' that encourages ethnographers to think about 'how people sense the world', in comparison to 'ways of sensing the world' as characterizing the 'anthropology of the senses' branch. In her book *Doing Sensory Ethnography*, Pink proposes various methodological approaches to develop sensorial ethnographies:

> The experiencing, knowing and emplaced body is [...] central to the idea of a sensory ethnography. Ethnographic practice entails our multisensorial embodied engagements with others (perhaps through participation in activities or exploring their understandings in part verbally) and their social, material, discursive and sensory environments. It moreover requires us to reflect on these engagements, to conceptualise their meanings theoretically and to seek ways to communicate the relatedness of experiential and intellectual meanings to others.
>
> (Pink 2015: 18)

To reflect on our own engagement requires a reflexive approach to our own bodily experiences as social scientists, a call that echoes the debate launched by the edited volume *Writing Culture*, mentioned above. Pink places verbal communication, and more specifically interviewing, as a key 'social, sensorial and affective' method to better understand other people's sensory categories (Pink 2015: 76). To conduct research about the senses, including hearing, however, anthropologists need to think in terms of multisensory ethnographic projects (Feld 1996). This implies that anthropologists should consider sonic ways to share experiences and interpretations emerging from ethnographies conducted in sound.

Sound and space

Sound is inherently spatial. It is also positional, which means that one's location in an environment affects how one perceives sounds. In terms of human experience, 'sound is constitutive of space, just as space is constitutive of sound' (Eisenberg 2015: 194). A space is thus perceived, facilitated and imagined as a site of experience and experiment 'for thinking relations between bodies, concepts, and materials of various kinds' (McCormack 2008: 7). Derek McCormack uses the term 'thinking-space' to refer to an encounter with a site where space is conceived and generated. Echoing Michel de Certeau, McCormack provides examples of actions such as walking, dancing, writing, and we could add listening, as techniques to encourage anthropologists to develop 'thinking-space' as a *process*. In other words, space is not just there to be discovered by the researcher, rather, it is a relational engagement or an acoustemology of place. Through listening and recording practices anthropologists are encouraged to co-constitute a sense of space (Westerkamp 2002; Pink 2015).

Therefore, conducting research 'in the field' includes a sonic awareness of the surrounding environment that speaks to a 'cultured' sonic interpretation of space. For instance, Eisenberg describes the Old City of Mombasa, where vocal performances including sermons and religious songs in multiple languages – Arabic, Swahili, English – are heard as a 'multiaccentual public space' (Eisenberg 2013, 2015). These sermons mark spatial territories and neighbourhoods and contribute to building the Muslim citizenship of the Kenyan coast. Eisenberg adopts a 'participant audition'[1] method to explore the Islamic soundscape, and more specifically the multiple boundaries that characterize Mombasa as a heterogenous city. He argues that participant audition 'calls for attention to sonic spatialities not only as multiple but also as overlapping and mutually mediating' (Eisenberg 2015: 200).

In a different geographic location, the anthropologist Martijn Oosterbaan also pays attention to the charismatic cacophony, or multiple layering of sounds, created by different groups in a favela in Rio de Janeiro, including Pentecostal churches and partygoers listening to funk. Through amplified music that communicates a 'politics of presence', the different groups build the boundaries of their territories. Hence, music reproduces group identities in the favela city-space, making audible the power struggles of the neighbourhood and the positions occupied by its inhabitants (Oosterbaan 2009: 85).

When conducting research in the field, anthropologists navigate inside layers of sounds, at the same time as they reflect on the social and political questions that emerge from the research. When adopting attentive listening practices, sonic perceptions of a space generate ethnographic reflections about boundaries, identities, and politics of presence that shape the city and other social territories (LaBelle 2010) such as the 'sensory home' (Pink 2004), museums (Zisiou 2011; Bubaris 2014), hospitals (Rice 2013), and ancient sites (Devereux 2001; Reznikoff 2006; Blesser and Salter 2007). Ethnographers conducting 'anthropology in sound' may pay attention to controlled sonic spaces, for example recording studios, which are said to inform modern epistemologies of sound (Born 2013; see also Thompson, in this volume), and even mediate social spaces and aesthetics (Meintjes 2003). In adopting a 'fresh ear' and in being receptive to other sonic materialities, alternative layers of meanings and interpretations often emerge. Social spaces such as churches (Blesser and Salter 2007) and shopping centres (Sterne 1997; LaBelle 2010) reveal intentional religious and capitalist architectonics respectively. Ultimately, auditory spatial awareness 'is more than just the ability to detect that space has changed sounds, it includes as well the emotional and behavioral experience of space' (Blesser and Salter 2007: 11). Indeed, 'it is difficult to identify any work of sound studies that does *not* deal in some way with space' (Eisenberg 2015: 195, emphasis in the original). Thus, anthropologists and sound scholars contribute to the particularization of social spaces by delving into broader questions of politics, identity, performance, and aesthetics.

Arts-based research and collaboration

Despite the fact that anthropology has 'largely treated the work of sound artists as tangential to its enterprise' (Samuels et al. 2010: 334), sound art does attract the attention of anthropologists who wish to push the boundaries of more conventional forms of conducting and understanding ethnographic fieldwork and explore novel ways of engaging with the sound world. Anthropologists sometimes borrow from methodologies associated with the arts, often referred to as 'research-creation' or 'arts-based research' (Boudreault-Fournier

2016b). They may also collaborate with sound artists for the realization of joint projects. Collaborations between artists and anthropologists who combine their efforts to design and implement an art work that is also part of a research venture contribute to the exploration of novel approaches to ethnographic fieldwork (Schneider and Wright 2006, 2010, 2013). In this context, the process of recording sounds and editing them into original compositions can be conceived as part of a practice of ethnography and can reveal layers of meanings that are not necessarily attainable through more traditional ethnographic methods (Drever 2002; Boudreault-Fournier 2012, 2016a, b; Boudreault-Fournier and Wees 2017).

The *Displace 1.0* (2011) and *2.0* (2012) performances by artists Chris Salter and TeZ together with anthropologist David Howes are examples of artist-anthropologist initiatives that combine approaches to create an installation and a performance aimed at stimulating sensorial experiments. In an interview about the experiment *Displace 2.0: Performative Sensory Environment*, Salter explains that the 'critical aim is to design large performative environments that hybridize cultures and begin to develop the aesthetic potential of the non-visual senses' in order 'to see if we could start a dialogue between sensory anthropology and the design of new kinds of so-called "multimodal" [...] environments' (Bertolotti n.d.: n.p.). Similarly, Howes (2019a) expresses that the 'performative sensory environments' created through the *Displace 1.0* installation, 'sought to disrupt conventional habits of perception by rearranging the senses and thereby open a crack in the Western sensorium'.

Other examples of collaborations include Steven Feld, who worked with artist Virginia Ryan on a multimedia installation called the *Castaways Project*, inspired by the West Africa coastline (Ryan and Feld 2007). The project redefined the boundaries between art and anthropology, and proposed a multisensorial way of engaging with memories and the legacy of slavery (see Virginia Ryan Virtual Artist n.d.). Another collaboration, between Feld and Ghanaian multi-instrumentalist Nii Otoo Annan, gave birth to an experimental CD called *Bufo Variations* (Annan and Feld 2008; see also Feld 2015b). Felds's methodological approach consisted of a mix of playback techniques (that he borrowed from anthropologist and film-maker Jean Rouch), dialogic editing, and conversations with Annan about 'his listening practices, his sonic knowing, his acoustemological way of hearing' (Feld 2015b: 94). Feld referred to this last methodological strategy as 'listening to histories of listening'. In attempting to explore how listening is embedded into forms of learning and practicing, the dialogic nature of listening emerges and connects with the relational ontology principle of acoustemology.

Ethnographic film-making

Many people forget that cinema is as much about sound as it is about images. Michel Chion (1994) defines the false impression that sound is 'invisible' or unnecessary to the cinematographic experience as 'audiovisual illusion'. Ethnographic films are usually characterized by low budgets, small teams, and handheld cameras, based on fieldwork and guided by ethical considerations. Very few ethnographic or documentary film-makers have paid serious attention to sound, be it in practice – from a methodological or theoretical perspective – or in the considerations of the sonic dimensions as contributing to a film narrative (Jeffrey Ruoff [1992] and Paul Henley [2007] are rare exceptions).

The history of the ethnographic film is tightly woven with that of the discipline of anthropology on the one hand, and technological developments on the other. Most notably, the invention of shoulder cameras and synchronized sound in the 1960s – permitting sound to be recorded simultaneously with the images – allowed for greater flexibility of movement and filmic possibilities. With this innovation, it became more common to see and hear participants in a film dialoguing or responding to the camera (for example *Chronicle of a Summer* [1961] by Jean Rouch and Edgar Morin) rather than having voice-over comments by a 'narrator god' (for example *The Hunters* [1957] by John Marshall and Robert Gardner). Australians David and Judith MacDougall also helped pave the way for an approach that encouraged the use of subtitles rather than voice-overs in order to hear the intonation of the vocal exchanges (Grimshaw 2008).

But essentially, sounds in ethnographic films are still deemed as secondary to images, and music added in the editing suite is often regarded with a suspicious eye (Henley 2007). The French anthropologist-film-maker Jean Rouch went as far as describing music added to images in the editing suite as the 'opium of cinema'. That is because observational film – a genre of ethnographic film – is thought to engage with the everyday, captured by the camera in the most non-obtrusive way possible. Still today, the 'frank admission of the manipulation of the sound-track sits rather uncomfortably with the empirical rhetoric not just of observational cinema but with most contemporary modes of ethnographic documentary-making' (Henley 2007: 57).

Without denying that sound manipulation might not best fit with the aesthetics of observational ethnographic film, Henley recognizes that the design of a soundtrack could contribute to the *thickness* of ethnographic description (Geertz 1973; Samuels et al. 2010). This thickness is in the representation of space and place. In other words, carefully recorded environmental sounds and a soundtrack designed in the editing suite can contribute to better communicating a sense of place and, according to Henley, does so in three main ways: thickening ethnographic description, enhancing the spectator's understanding, and proposing new interpretations of the film's subject matter (Henley 2007: 56).

In using sounds more creatively and effectively in ethnographic film-making, 'visual' anthropologists contribute not only to enhancing the experience of the audience but also provide new forms of engagement with listening, design, aesthetics, composition, and collaboration (for example Boudreault-Fournier and Diamanti 2018). Recently, the *American Anthropologist* journal relaunched its 'Visual Anthropology' section as 'Multimodal Anthropologies', defined as 'an umbrella term which encompasses all the audio-visual affordances of contemporary media' (Collins, Durrington, and Gill 2017). Multimodal anthropology invites ethnographers to work across multiple media and also to develop collaborative work, engaging in public anthropology 'through a field of differentially linked media platforms' (Collins, Durrington, and Gill 2017: 142). Sound (recording) is one element among others that are combined in a composite of media with an emphasis on the 'processes of knowledge production' instead of the final product (a book, a film, etc.). Growing from visual anthropology, 'modal approaches' go further than just the visual, in also using the other senses and media to build innovative ways in which anthropology could be increasingly engaged with the public in the near future.

Ethnographic fieldwork

Conducting ethnographic fieldwork has traditionally been equated with long-term *immersion* in a cultural setting other than the one to which the anthropologist is accustomed. The obligation to conduct empirical research in a *different* cultural setting than the ethnographer's own is not expected anymore. Yet, the idea of immersion is still implicit in how ethnographic fieldwork is imagined and taught in graduate programmes. This immersion ideally involves a long and intensive commitment during which the anthropologist is expected to learn the language (among other things) and acquire a sense of the multiple dimensions entangled in the exploration of an original research question. In describing his first days of fieldwork in Papua New Guinea, Feld expresses feelings of displacement and confusion during his sudden immersion in another cultural setting:

> The first day I was there, within two hours of arriving in the village, we hear sung weeping. Somebody had died. They [Bambi and Buck Schieffelin] said, 'Get your tape recorder.' I didn't understand the language. I didn't know anything! So here I am, wham! With big Bagra [tape recorder] and headphones and microphone sitting among all these people who were weeping. I just sort of closed my eyes and listened and realized that I could easily spend a year trying to figure out the first sounds I was hearing.
>
> (Feld and Brenneis 2004: 464)

Therefore, participant observation is imagined as a form of experience during which the ethnographer is 'soaked in' a cultural context in which he or she will live for a long period of time (one year is usually expected for a PhD student).

Based on his unique fieldwork in a submarine called *Alvin*, which descends deep down to the ocean floor of the Juan the Fuca Ridge located on the Pacific Northwest coast of North America, the anthropologist Stefan Helmreich developed a powerful critique of the idea of cultural immersion. His work shows that 'doing anthropology in sound' (Feld 1996) can generate epistemological and methodological debates about the nature of ethnographic fieldwork – a reflection that can be applied beyond sound related projects. More specifically, Helmreich puts forward the term 'transduction', to refer to both the process of transmuting one medium into another. In the case of the submarine, there is a transfer from the medium of water to the medium of air. He also defines the anthropological project to encompass a 'transduced sensing' that permits cultural immersion (Helmreich 2007: 622, 2009).

As Helmreich dives into the darkness, his attention is captured by the sounds of the deep ocean, perceptible by human ears thanks to a process of transduction afforded by the submarine's apparatuses. This vessel, which both acts like a bubble and a cyborg, allows the anthropologist to experience a sense of immersion within another sensorial environment. As mentioned above, Helmreich draws a fascinating parallel between the electric transduction of the submarine's machinery and the role of the anthropologist. This comparison also allows him to articulate a critique of the concept of immersion. More specifically, Helmreich argues that to 'think transductively is […] also to consider ethnography itself as transduction – and the ethnographer as a kind of transducer' (Helmreich 2007: 633). In comparison with the concept of immersion, which presumes 'dissolving into' or 'becoming one with', transduction is a process that can tune us into disjuncture, resistance, and distortion (Helmreich 2007). Transduction emphasizes the process of transformation from one form of energy (in a broad sense) to another. Similarly, ethnographers act as transducers when they convert their sensorial experiences of fieldwork into digestible ethnographic accounts – which often take the shape of written texts. Helmreich is cautious about arguing that transduction could be applied to all forms of ethnographic enterprise. Yet his call to consider the potential for a transductive ethnography 'in sound' can certainly apply to research on media, infrastructure, sonic controlled spaces such as recording studios (Meintjes 2003; Theberge 2004), and ethnographic epistemologies of the senses (Helmreich 2007).

The work of Helmreich on transduction shows that paying attention to sounds while conducting fieldwork can generate in-depth epistemological and methodological reflections that impact the discipline of anthropology at a theoretical level as well as the nature of its empirical approach. While recognizing the critique of immersion developed by Helmreich, we also need to acknowledge that ethnographic fieldwork remains anthropology's methodological strength. Based on long-term relationships as well as direct and embodied experiences, ethnographic fieldwork is a – mostly – qualitative methodological approach that allows for in-depth engagement with everyday life, especially the particular and the sensorial. In order to contribute to a 'sounded anthropology' (Samuels et al. 2010) and to train our 'ethnographic ear' (Erlmann 2004), the rest of this chapter will delve into the ethnographic enterprise with a focus on sonic practices. In order to do so, I will discuss four dimensions – listening, recording, editing, and representation – identified by Feld as fundamental to thinking about an ethnography of sound (Feld and Brenneis 2004).

Listening

Listening is an art to be cultivated, which implies that it takes time and practice to refine our hearing sensitivities and our sound awareness. Schafer (1994) proposes various techniques to sharpen our ears to the sounds that surround us. Exercises to improve listening include: stop making sounds for a while, not talking for a whole day, meditating by focusing on a specific sound, and/or soundwalking. He explains that a soundwalk is more than just a walk because it involves careful listening to sounds throughout the journey. Schafer proposes the use of a guide and a map on which one can annotate the different sounds encountered while wandering in a territory, but there are many ways one can think about conducting a soundwalk.

Composer and environmentalist Hildegard Westerkamp also firmly believes in the power of soundwalks because 'in opening our ears, we can open our minds' ('How

Opening Our Ears Can Open Our Minds: Hildegard Westerkamp' 2017). As mentioned, soundwalks may take different forms, but for Westerkamp, the essential remains the same: a designated person guides the walk during which participants, in silence, must pay close attention to the sounds that surround them. The journey may include listening stops, instructions, moments of mediation and appreciation, and added audible elements. Westerkamp explains:

> Instructions given before soundwalks are a tool to enable deeper listening. They are always simple and don't change much. The experiences of the participating listeners, however, are always different. The main and only instruction is to walk together without speaking. Everything else is said to make sure that all questions are answered, and everyone feels safe to allow a total immersion into the listening.
>
> (Westerkamp 2017: 151)

Once the soundwalk is over, the guide discusses participants' experiences. Westerkamp notes that the group conversation is

> an essential ingredient of a soundwalk experience […] because it brings to consciousness the real significance of such a listening process. It gives an invaluable opportunity to compare what was heard and noticed, what was significant for participants, what was not and why, which parts of the walk impressed the most and so on. The variety of responses highlights how complex our listening perception really is.
>
> (Westerkamp 2017: 155)

Interestingly, soundwalking stimulates the senses and encourages an increased awareness of sounds. But, as Westerkamp points out, one needs to reflect on how to think and express it. The concept of deep listening, whether in the context of a soundwalk or when an ethnographer pays close attention to the presence of sounds in a specific space, can refer to the process of attuning 'our ears to listen *again* to the multiple layers of meaning potentially embedded in the same sound' (Bull and Back 2003: 4, my emphasis). But more than agile listening, as Westerkamp reminds us, deep listening also involves 'practices of dialogue and procedures for investigation, transposition, and interpretation' (4).

Attuning our ears may serve diverse purposes and we may adopt varied forms of listening in different contexts and situations. For instance, a piano tuner, a bird song enthusiast, a self-taught musician, an employee working in an industry, an ecologist, and an anthropologist, do not necessarily listen attentively for the same reasons. As part of his ethnographic research on sounds in hospital settings, Tom Rice (2013) defines the stethoscope as 'an autobiographical object', and his methodological listening approach as 'stethoscopic' (Rice 2008). More specifically, in using a stethoscope to listen to patients' bodies, Rice experimented with the subtleties involved in medical listening practices. Interestingly, the stethoscope also facilitated his relationships with medical specialists (Harris 2015; Pink 2015: 105). David Novak (2013), who conducted research on noise music in Japan, argues that listening is a form of circulation and that it should be approached as an active and creative practice. In observing how listeners react in particular affective-volitional ways to an Islamic sermon, which he qualifies as an 'ethical performance', Charles Hirschkind (2006) also emphasizes that listening is active and participatory. As suggested by Rice's fieldwork experience in a hospital setting, the technique of listening can be used during ethnographic research as a way to facilitate conversation with participants. Listening to recordings can also trigger memories and stories associated with the past (Harris 2015). Through sonic elicitation, such conversations can further prompt research participants to talk about their ways of sonic knowing and can teach us about 'how to be an ethnographer listener' (Feld and Brenneis 2004). Feld used recording as a form of sonic-elicitation to stimulate conversation among the Kaluli people. While playing back recordings to small groups of people and inviting them to listen, he also stimulated conversations about the interpretations of sounds. In referring to the production of his album *Voices of the Rainforest*, Feld explains how the participation of the Bosavi through sonic elicitation was key to his ethnographic listening approach because he better understood what sounds meant for them. Through listening, he also received their input on how to edit the recording, a technique he refers to as *dialogic editing*. Again, Feld needs to be quoted at length:

> So I had three cassette players in the bush. I would record tracks on the Nagra and then transfer them onto cassettes. And then, I would sit with people and listen to the cassettes and invite my listeners to scroll the knobs on the cassettes. It was an ethnoaesthetic negotiation, trying to work with Bosavi people to understand how they listened, how they heard the dimensionality of forest sound, how they would balance a mix of birds, water, cicadas, voices, and so forth.
>
> (Feld and Brenneis 2004: 467)

Listening is then a method to cultivate, and it can also become a method of ethnographic exploration to deepen our sonic approach as well as reflect on the nature of recordings.

Recording

Jonathan Sterne (2003) argues against the idea that recording devices promote a disorienting effect on our senses. On the contrary, a fundamental feature of sound recording and composition is intensive listening, both directly with our ears and via the microphone, a process that immerses creator and listener in their environment, fostering increased spatial awareness. In soundscape composition, 'knowledge of specific contexts shapes the composer's work and invokes the listener's knowledge of those contexts […] at the intersections of the listener's associations, memories and imaginations related to that place' (Eylul Iscen 2014: 127). Anthropologists who use an audio recorder interact with the environment in a particular way; they move cautiously as the senses awaken. That is because recording involves special attention to where one stands. It is about detecting the sound details that one wants to collect. Anthropologists conducting fieldwork through sound shift their attention to what surrounds them when recording a sound. They become more aware of their presence, the sounds that emanate from their body and the movements they make, more so as it is difficult to eliminate the presence of the anthropologist who records even with the use of a unidirectional microphone. Therefore, the process of recording provides an opportunity to develop a reflexive sensory approach, as it creates routes to multisensorial knowing (Pink 2004). Thanks to the recording device, anthropologists experience a place through their bodies and their senses. Intensive listening practices during the process of recording fully immerse anthropologists in their environment, fostering an increased spatial awareness.

Despite the fact that the publication of sound recordings is not that common among ethnographers, recording is conceived as a fundamental practice for anthropologists. Notably, Feld argued that 'recordings always seemed to me an important alternative as a form of ethnographic practice' (Feld and Brenneis 2004: 462). This implies that sound recording becomes a strategy to connect with research participants, and also considers how the ethnographer can reflect on their own listening practices and sensitivities. Yet as the writer and artist Paul Carter (2004) highlights, recordings do not only mimic or echo the environment in which the participants live, as for example in the rainforest recordings collected by Feld. Recordings are full of noises, disruptions, technological, and historical traces, and the place and context in which they are played often add another layer of sonic disruption and cultural discontinuity. More than just a mnemonic device, recordings offer the possibility to reflect critically and reflexively on the positionality of the anthropologist; they encourage a careful and multisensorial engagement with place, and stimulate a dialogue about what listening means and for whom.

Editing

The process of editing sounds and images into original compositions and clips should be conceived as part of the practice of ethnography, as it is the case with writing (Drever 2002; Feld and Brenneis 2004; Boudreault-Fournier 2012, 2017). Recording and editing soundscapes or sound clips can shape and transform our understanding and representation of a space. In other words, editing a field recording becomes a process through which the ethnographer is making sense of the data of a specific place. The ethnographer is involved in transducing (to echo Helmreich's term) their ethnographic experiences into a narrative, a story, or a sonic representation. The use of the sound recorder and the editing techniques need to be approached as a creative process, 'which requires craft and editing and articulation just like writing', and if not, 'little will happen of an interesting sort in the anthropology of sound' (Feld and Brenneis 2004: 471). As a response, it can be argued that both editing sounds and writing are processes of synthesis and creative production. Of course, and as mentioned in relation to ethnographic films, there is a certain discomfort, still prevalent in anthropology, with the idea of constructing meaning through the manipulation, transformation, and staging of data. This is even more relevant with sound, which is often perceived as a rather abstract medium. Yet, it could be argued that sound compositions are highlighting 'people's sensory experiences and mental worlds in more expressive and relational ways' (Eylul Iscen 2014: 127), and provide greater freedom to explore the issue of imagination (Boudreault-Fournier and Wees 2017). In borrowing from arts-based research approaches, anthropologists can explore the dimension of sound processing and editing, often missing from anthropological accounts, in transparent and productive ways (Boudreault-Fournier and Wees 2017). Sound opens new avenues of sensing, experiencing, and learning that cannot be compared to the visual and/or the written. Sound presents opportunities to anthropologists to other ways of knowing and sharing that include the creative and reflective potential of the editing process.

Representation

Sound recording and editing can take various forms, from oral storytelling to recorded soundwalks, sonic documentaries, and podcasts. Walter Gershon proposed a 'sonic ethnography', which he defines as 'the sounded representation of "data" collected and analyzed over the course of an ethnographic study' (Gershon 2013). Gershon conducted fieldwork with a fifth-grade classroom, and adopted a methodology based on collaborative sound recording and editing to explore how songwriting might alleviate racial and gender gaps in science education. One of his outputs is a piece made of various sonic layers, including a beat track, with students rapping and interacting with the teachers and the ethnographer himself ('ResoundingScience' 2013). In Gershon's piece, sound is both used as a research methodology and as a tool of dissemination and representation.

Like any ethnography, sonic methodologies (including sonic ethnography, sonic elicitation, etc.) are an act of interpretation. They involve selective listening, a specific collection of sound clips, as well as the construction of a narrative. The ethnographer and research participants may be involved in the process of selecting, editing, and creating a sound piece that will represent the voices of the ones who are implicated. The presence of the ethnographer is always embedded in sonic representations (as in any other ethnographic texts) and it is the role of the researcher to adopt a transparent and reflexive position when recording and editing sound clips. Issues of power relations, marginality, and silencing may emerge as present or absent in sonic forms of representation and they should be acknowledged. Writing ethnographies remains the most common practice used by anthropologists to express and translate ethnographic experiences. Yet, conducting anthropology *in* sound should also imply the production of *sonic* outputs such as sound installations and performances, recordings, and podcasts providing multisensorial engagements with ethnographic research. Sonic forms of representation offer novel strategies of learning and sharing ethnographic knowledge at the same time as they stimulate new conversation about the nature of ethnographic fieldwork itself.

Unfortunately, little inquiry has been conducted into the ethics of participatory modes of researching through listening, Gershon being an exception. As mentioned by the historian Anna Harris (2015), most of the research on the ethics of using sonic participatory methodological approaches in ethnographic and historical research has focused on the repatriation of sound archives to indigenous communities (Hennessy 2009; Reddy and Sonneborn 2013). Producing and sharing recordings, soundscapes, and clips (among other media) position the 'anthropology in sound' as an effective methodological approach that generates forms of knowing, teaching, and transmitting that are complementary to, and certainly as valuable as the written text, which is still the dominant medium in anthropological research. It is time for anthropologists to keep their ears open to other forms of knowing and sensing the world.

Conclusion

Following the concept of acoustemology developed by one of the main pioneers of the anthropology of sound, Steven Feld, sonic methodologies in anthropology are characterized by an ontological model that is fundamentally relational and that relies on first-hand sensorial experiences. Active listening is required to generate dialogues and exchanges based on fieldwork and multisensorial encounters. Conducting ethnographic fieldwork in sound involves being present with the research participants and sharing the same temporal and spatial dimensions, to paraphrase Johannes Fabian's work.

Therefore, what defines the discipline's sonic methodologies is its emphasis on ethnographic fieldwork as a method of experiencing the sounds present in the environment, as well as the ethnographer's involvement in transducing their own empirical encounters. As suggested in this chapter, in addition to including active listening practices, ethnographic fieldwork should also involve creative engagements with sounds through recording and editing techniques. James Clifford might not have thought about the opportunities afforded by sounds to reshuffle the ethnographic project when he asked, thirty years ago: 'But what about the ethnographic ear?' Yet, ethnographers are increasingly engaged in refining their ears in creative ways to engage with emerging questions and concerns.

The six epistemological approaches developed here – soundscape, relational ontology, the senses, sound and space, arts-based research and collaboration, and ethnographic film-making – guide the exploration of the sonic world in anthropology. They impact the selection of methodological strategies for specific research questions, and much clearly remains to be said about the sonic methodologies adopted by anthropologists to explore topics such as voice and storytelling, digital media, technological mediation, and innovative pedagogical strategies. The use of sonic methodologies in anthropological research invites interdisciplinarity and draws on other fields of study, including art, music, communication, and history, to build a dynamic line of inquiry that is committed to empirical ways of knowing. In ethnographic fieldwork, these methodologies are typically applied to generating locally embedded explorations of sonic worlds. This highlights the necessity of delving into sound while considering the cultural, social, and historical contexts in which sounds take place. Situated, detailed exploration of soundscapes allow anthropologists to engage with social and cultural issues through taking different vantage points within the nested, co-occurring spaces created by sound at scale, close or expanded. Through careful listening to the world around us, the anthropologists' methodologies are both refined and expanded by researching *in* sound; an empowering response to a fast-moving world.

Note

1. Similar to participant observation but from an auditory perspective, participant audition implies the adoption of in-depth listening practices and careful attention to the multiple layers of sounds that circulate – sometimes in conflict – in an urban landscape.

References

Annan, Nii Otto and Steven Feld (2008). [CD] *Bufo Variations*. VoxLox.

Bertolotti, Silvia (n.d.). 'Displace 2.0: Mediation of Sensations. A Conversation with Chris Salter'. *Digicult*. Available online: http://digicult.it/news/a-conversation-with-chris-salter-on-displace-2-0/(accessed 19 February 2019).

Blesser, Barry and Linda-Ruth Salter (2007). *Spaces Speak, Are you Listening? Experiencing Aural Architecture*. Cambridge, MA: MIT Press.

Born, Georgina (2013). 'Introduction –Music, Sound and Space: Transformations of Public and Private Experience'. In Georgina Born (ed.), *Music, Sound and Space: Transformations of Public and Private Experience*, 1–70. Cambridge: Cambridge University Press.

Boudreault-Fournier, Alexandrine (2012). 'Écho d'une rencontre virtuelle: Vers une ethnographie de la production audio-visuelle'. *Anthropologica* 54(1): 1–12.

Boudreault-Fournier, Alexandrine (2016a). 'Recording and Editing'. In Denielle Elliott and Dara Culhane (eds), *A Different Kind of Ethnography: Practices and Creative Methodologies*, 69–89. Toronto: University of Toronto Press.

Boudreault-Fournier, Alexandrine (2016b). 'Microtopia in Counterpoint: Relational Aesthetics and the Echo Project'. *Cadernos de Arte e Antropologia* 5(1): 135–154.

Boudreault-Fournier, Alexandrine (Forthcoming). 'Street Net and Electronic Music in Cuba'. In Kyle Devine and Alexandrine Boudreault-Fournier (eds), *Organized Sound: Musicologies and Anthropologies of Infrastructure*. Oxford: Oxford University Press.

Boudreault-Fournier, Alexandrine and Nick Wees (2017). 'Creative Engagement with Interstitial Urban Spaces: The Case of the Vancouver's Back Alleys'. In Martha Radice and Alexandrine Boudreault-Fournier (eds), *Urban Encounters: Art and the Public*, 192–211. Culture of Cities Series. Montréal: McGill University Press.

Bubaris, Nikos (2014). 'Sound in Museums –Museums in Sound'. *Museum Management and Curatorship* 29(4): 391–402.

Bull, Michael and Les Back (2003). 'Introduction: Into Sound'. In Michael Bull and Les Back (eds), *The Auditory Culture Reader*, 1–18. Oxford: Berg.

Calzadilla, Fernando and George E. Marcus (2006). 'Artists in the Field: Between Art and Anthropology'. In Arnd Schneider and Christopher Wright (eds), *Contemporary Art and Anthropology*, 95–116. Oxford: Berg.

Carter, Paul (2004). 'Ambiguous Traces, Mishearing, and Auditory Space'. In Veit Erlmann (ed.), *Hearing Cultures: Essays on Sound, Listening, and Modernity*, 43–63. New York: Berg.

Chion, Michel (1994). *Audio-vision: Sound on Screen*. New York: Columbia University Press.

Chronicle of a Summer (1961). [Film] Dir. Jean Rouch and Edgar Morin. France: Argos Films.

Clifford, James (1986). 'Introduction: Partial Truths'. In James Clifford and George E. Marcus (eds), *Writing Culture: The Poetics and Politics of Ethnography*, 1–26. Berkeley, CA: University of California Press.

Clifford, James and George E. Marcus (eds) (1986). *Writing Culture: The Poetics and Politics of Ethnography*. Berkeley, CA: University of California Press.

Collins, Samuel Gerald, Matthew Durrington, and Harjant Gill (2017). 'Multimodality: An Invitation'. *American Anthropologist* 119(1): 142–153.

Corbin, Alain (1998). *Village Bells: Sound and Meaning in the Nineteenth-Century French Countryside*. New York: Columbia University Press.

Devereux, Paul (2001). *Stone Age, Soundtracks: The Acoustic Archaeology of Ancient Sites*. London: Vega.

Drever, John L. (2002). 'Soundscape Composition: The Convergence of Ethnography and Acousmatic Music'. *Organised Sound* 7(1): 21–27.

Eisenberg, Andrew (2013). 'Islam, Sound and Space: Acoustemology and Muslim Citizenship on the Kenyan Coast'. In Georgina Born (ed.), *Music, Sound and Space: Transformations of Public and Private Experience*, 186–202. Cambridge: Cambridge University Press.

Eisenberg, Andrew (2015). 'Space'. In David Novak and Matt Sakakeeny (eds), *Keywords in Sound*, 193–207. Durham, NC: Duke University Press.

Erlmann, Veit (2004). 'But What of the Ethnographic Ear?: Anthropology, Sound, and the Senses'. In Veit Erlmann (ed.), *Hearing Cultures: Essays on Sound, Listening, and Modernity*, 1–20. Oxford: Berg.

Eylul Iscen, Ozgun (2014). 'In-Between Soundscapes of Vancouver: The Newcomer's Acoustic Experience of a City with a Sensory Repertoire of Another Place'. *Organised Sound* 19 (2): 125–135.

Fabian, Johannes (1983). *Time and the Other: How Anthropology Makes Its Object*. New York: Columbia University Press.

Feld, Steven [1982] (1990). *Sound and Sentiment: Birds, Weeping, Poetics, and Song in Kaluli Expression*. Philadelphia: University of Pennsylvania Press.

Feld, Steven (1987). 'Dialogic Editing: Interpreting How Kaluli Read Sound and Sentiment'. *Cultural Anthropology* 2(2): 190–210.

Feld, Steven (1991). [CD] *Voices of the Rainforest: A Day in The Life of The Kaluli People*. Rykodisc.

Feld, Steven (1994). 'From Schizophonia to Schizmogenesis: On the Discourses and Commodification Practices of "World Music" and "World Beat"'. In Charles Keil and Steven Feld (eds), *Music Grooves*, 257–289. Chicago: Chicago University Press.

Feld, Steven (1996). 'Waterfalls of Song: An Acoustemology of Place Resounding in Bosavi, Papua New Guinea'. In Steven Feld and Keith H. Basso (eds), *Senses of Place*, 91–135. Santa Fe, NM: School of American Research Press.

Feld, Steven (2001). [CD] *Rainforest Soundwalks*. EarthEar.

Feld, Steven (2002). [CD] *Bells and Winter Festivals of Greek Macedonia*. Smithsonian Folkways Recordings.

Feld, Steven (2004a). [CD] *The Time of Bells, I: Soundscapes of Italy, Finland, Greece, and France*. VoxLox 104. P/R/A/PH.

Feld, Steven (2004b). [CD] *The Time of Bells, II: Soundscapes of Finland, Norway, Italy and Greece*. VoxLox 204. P/R/A/PH.

Feld, Steven (2005). [CD] *The Time of Bells, III: Musical Bells of Accra, Ghana*. VoxLox 205. P/R/A/PH.

Feld, Steven (2015a). 'Acoustemology'. In David Novak and Matt Sakakeeny (eds), *Keywords in Sounds*, 12–21. Durham, NC: Duke University Press.

Feld, Steven (2015b). 'Listening to Histories of Listening: Collaborative Experiments in Acoustemology with Nii Otoo Annan'. In Gianmario Borio (ed.), *Musical Listening in the Age of Technological Reproduction*, 91–103. London: Routledge.

Feld, Steven and Don Brenneis (2004). 'Doing Anthropology in Sound'. *American Ethnologist* 31(4): 461–474.

Geertz, Clifford (1973). 'Thick Description: Toward an Interpretive Theory of Culture'. In *The Interpretation of Cultures: Selected Essays*, 3–30. New York: Basic Books.

Gershon, Walter (2013). 'Resounding Science: A Sonic Ethnography of an Urban Fifth Grade Classroom'. *Journal of Sonic Studies* 4. Available online: https://www.researchcatalogue.net/view/290395/290396/905/1161 (accessed 8 July 2020).

Goodman, Steve (2009). *Sonic Warfare: Sound, Affect, and the Ecology of Fear*. Cambridge, MA: MIT Press.

Grimshaw, Anna (2008). 'From Observational Cinema to Participatory Cinema – and Back Again? David MacDougall and the Doon School Project'. *Visual Anthropology Review* 18(2): 80–93.

Guardians of the Night (2018). [Film] Dir. Alexandrine Boudreault-Fournier and Eleonora Diamanti. Cuba: Sonoptica.

Harris, Anna (2015). 'Editing Sound Memories'. *Public Historian* 37(4): 14–31.

Helmreich, Stefan (2007). 'An Anthropologist Underwater: Immersive Soundscapes, Submarine Cyborgs, and Transductive Ethnography'. *American Ethnologist* 34(4): 621–641.

Helmreich, Stefan (2009). *Alien Ocean: Anthropological Voyages in Microbial Seas*. Berkeley, CA: University of California Press.

Helmreich, Stefan (2010). 'Listening Against Soundscapes'. *Anthropology News*, December. Available online: http://anthropology.mit.edu/sites/default/files/documents/helmreich_listening_against_soundscapes.pdf (accessed 28 January 2019).

Henley, Paul (2007). 'Seeing, Hearing, Feeling: Sound and the Despotism of the Eye in "Visual" Anthropology'. *Visual Anthropology Review* 23(1): 54–63.

Hennessy, Kate (2009). 'Virtual Repatriation and Digital Cultural Heritage: The Ethics of Managing Online Collections'. *Anthropology News* 50(4): 5–6.

Hirschkind, Charles (2006). *The Ethical Soundscape: Cassette Sermons and Islamic Counterpublics*. New York: Columbia University Press.

Howes, David (1991). *The Varieties of Sensory Experience: A Sourcebook in the Anthropology of the Senses*. Toronto: Toronto University Press.

Howes, David (2003). *Sensual Relations: Engaging the Senses in Culture and Social Theory*. Ann Arbor, MI: University of Michigan Press.

Howes, David (2019a). 'Multisensory Anthropology'. *Annual Review of Anthropology* 48: 17–28.

Howes, David (2019b). 'Embodiment and the Senses'. In Michael Bull (ed.), *The Routledge Companion to Sound Studies*. London: Routledge.

Howes, David and Constance Classen (2013). *Ways of Sensing: Understanding the Senses in Society*. London: Routledge.

'How Opening Our Ears Can Open Our Minds: Hildegard Westerkamp' (2017). CBC radio, 31 August. Available online: https://www.cbc.ca/radio/ideas/how-opening-our-ears-can-open-our-minds-hildegard-westerkamp-1.3962163 (accessed 28 June 2020).

The Hunters (1957). [Film] Dir. John Marshall and Robert Gardner. USA: Documentary Educational Resources.

Ingold, Tim (2000). *The Perception of the Environment*. London: Routledge.

Ingold, Tim (2007). 'Against Soudscapes'. In Angus Carlyle (ed.), *Autumn Leaves: Sound and the Environment in Artistic Practice*, 10–13. Paris: Double Entendre.

Labelle, Brandon (2010). *Acoustic Territories: Sound Culture and Everyday Life*. London: Bloomsbury.

Larkin, Brian (2008). *Signal and Noise: Media, Infrastructure, and Urban Urban Culture in Nigeria*. Durham, NC: Duke University Press.

MacDougall, David (2005). *The Corporeal Image: Film, Ethnography, and the Senses.* Princeton, NJ: Princeton University Press.

Marcus, George E. (2010). 'Contemporary Aesthetics in Art and Anthropology: Experiments in Collaboration and Intervention'. *Visual Anthropology* 23(4): 263–277.

McCormack, Derek P. (2008). 'Thinking-Space and Research-Creation'. *Inflexions* 1 (1). Available online: http://www.inflexions.org/n1_mccormackhtml.html (accessed 4 February 2019).

Meintjes, Louise (2003). *Sound of Africa: Making Music Zulu in a South African Studio.* Durham, NC: Duke University Press.

Merriam, Alan P. (1964). *The Anthropology of Music.* Evanston, IL: Northwestern University Press.

Novak, David (2013). *Japanoise: Music at the Edge of Circulation.* Durham, NC: Duke University Press.

Oosterbaan, Martijn (2009). 'Sonic Supremacy: Sound, Space and Charisma in a Favela in Rio de Janeiro'. *Critique of Anthropology* 29(1): 81–104.

Pink, Sarah (2004). *Home Truths: Gender, Domestic Objects and Everyday Life.* Oxford: Berg.

Pink, Sarah (2015). *Doing Sensory Ethnography.* 2nd edition. Los Angeles: Sage.

Porcello, Thomas (2005). 'Afterwords'. In Paul D. Greene and Thomas Porcello (eds), *Wired for Sound: Engineering and Technologies in Sonic Cultures*, 269–274. Middletown, CT: Wesleyan University Press.

Reassemblage (1982). [Film] Dir. Trinha T. Minh-ha. USA: Women Make Movies.

Reddy, Sita and D. A. Sonneborn (2013). 'Sound Returns: Toward Ethical "Best Practices" at Smithsonian Folkwats Recordins'. *Music Anthropology Review* 7(1–2): 127–139.

'ResoundingScience' (2013). Sound Cloud. Available online: https://soundcloud.com/resoundingscience/lss-jss-final (accessed 28 June 2020).

Reznikoff, Iegor (2006). 'The Evidence of the Use of Sound Resonance from Palaeolithic to Medieval Times'. In Chris Scarre and Graeme Lawson (eds), *Archaeoacoustics*, 77–84. Cambridge: MacDonald Institute for Archaeological Research.

Rice, Tom (2003). 'Soundselves: An Acounstemology of Sound and Self in the Edinburgh Royal Infirmary'. *Anthropology Today* 19(4): 4–9.

Rice, Tom (2008). '"Beautiful Murmurs": Stethoscopic Listening and Acoustic Objectification'. *Senses and Society* 3(3): 293–306.

Rice, Tom (2013). *Hearing and the Hospital: Sound, Listening, Knowledge and Experience.* Canon Pyon: Sean Kingston Publishing.

Rouch, Jean (1995). 'The Camera and Man'. In Paul Hockings (ed.), *Principles of Visual Anthropology*, 79–98. Berlin: Mouton de Gruyter.

Ruoff, Jeffrey (1992). 'Conventions of Sound Documentary', in Robert B. Altman (ed.), *Sound Theory/Sound Practice*, 217–234, New York: Routledge.

Ryan, Virginia and Steven Feld (2007). [CD, DVD] *The Castaways Project.* Santa Fe, NM: VoxLox.

Samuels, David W., Louise Meintjes, Ana Maria Ochoa, and Thomas Porcello (2010). 'Soundscapes: Toward a Sounded Anthropology'. *Annual Review of Anthropology* 39: 329–345.

Schafer, R. Murray [1977] (1994). *The Soundscape: Our Sonic Environment and the Tuning of the World.* Rochester: Destiny Books.

Schneider, Arnd and Christopher Wright (2006). 'The Challenge of Practice'. In A. Schneider and C. Wight (eds), *Contemporary Art and Anthropology*, 1–28. New York: Berg.

Schneider, Arnd and Christopher Wright (2010). 'Between Art and Anthropology'. In Arnd Schneider and Christopher Wright (eds), *Between Art and Anthropology: Contemporary Ethnographic Practice*, 1–21. New York: Berg.

Schneider, Arnd and Christopher Wright (2013). 'Ways of Knowing'. In Arnd Schneider and Christopher Wright (eds), *Anthropology and Art Practice*, 1–23. New York: Bloomsbury.

Steingo, Gavin (2016). *Kwaito's Promise: Music and the Aesthetics of Freedom in in South Africa*. Chicago: University of Chicago Press.

Sterne, Jonathan (1997). 'Sound like the Mall of America: Programmed Music and the Architectonics of Commercial Space'. *Ethnomusicology* 41(1): 22–50.

Sterne, Jonathan (2003). *The Audible Past: Cultural Origins of Sound Reproduction*. Durham, NC: Duke University Press.

Sterne, Jonathan (2012). 'Sonic Imaginations'. In Jonathan Sterne (ed.), *The Sound Studies Reader*, 1–17. London: Routledge.

Stoller, Paul (1989). *The Taste of Ethnographic Things: The Senses in Ethnography*. Philadelphia: University of Pennsylvania Press.

Stoller, Paul (1997). *Sensuous Scholarship*. Philadelphia: University of Pennsylvania Press.

Taussig, Michael (1993). *Mimesis and Alterity: A Particular History of the Senses*. London: Routledge.

Theberge, Paul (2004). 'The Network Studio: Historical and Technological Paths to a New Idea in Music Making'. *Social Studies of Science* 34(5): 759–781.

Thompson, Emily (2002). *The Soundscape of Modernity: Architectural Acoustics and the Culture of Listening in America, 1900–1933*. Cambridge, MA: MIT Press.

Virginia Ryan Virtual Artist (n.d.). 'The "Castaway Project"'. Available online: http://www.virginiaryanart.ifp3.com/page/the-castaways-project-2003-2007/#/page/the-castaways-project-2003-2007/ (accessed 28 June 2020).

Westerkamp, Hildegard (2002). 'Linking Soundscape Composition and Acoustic Ecology'. *Organised Sound* 7(1): 51–56.

Westerkamp, Hildegard (2017). 'The Natural Complexities of Environmental Listening: One Soundwalk – Multiple Responses'. *BC Studies* 194: 149–162.

Zisiou, Michail (2011). 'Towards a Theory of Museological Soundscape Design: Museology as a Listening Path'. *Soundscape* 11(1): 36–38.

3
Sonic Methodologies by Way of Deconstruction

Naomi Waltham-Smith

Methods of invention

Folded and tucked inside Jacques Derrida's own published copy of *Psyché: Inventions de l'autre* in his personal library, now housed at Princeton University, are photocopies of pages from two French music dictionaries. On the two folded sheets, there are lines in the margins alongside parts of the entries on 'Invention' – some straight, some double tram lines, some more of a squiggle. While one cannot say with absolute certainty that Derrida made all of these marks, they appear to be in the same hand, made by the same pen, and they are broadly consistent with the shape of other such marginalia in other volumes in his library. On one of the sheets, there are other annotations for other entries of the pages that have clearly been photocopied, which leads one to conclude that Derrida most likely added the marginalia against the entries on 'Inventions' and 'Inventions Musicales' having copied these particular pages with the specific notion of invention in mind – perhaps while revisiting his own text. Whether these marks represent traces of his thinking in any meaningful sense and at what point is much harder to say.

Taken collectively (though not without a couple of exceptions), the annotations suggest an interest in the structural aspects of Bach's inventions, in the way in which they expose their rhythmic and melodic material according to a harmonic unfolding, and also in their deployment of technical contrapuntal devices. For both entries, Derrida's marginalia point to the typical tonal scheme of a double exposition of thematic material in the tonic and dominant followed by a series of episodes and middle entries culminating in a final entry in the tonic. For one entry, Derrida puts double lines against the final sentence, which reads: 'The writing of these didactic pieces is more often than not canonic because Johann Sebastian Bach's goal [*but*] was to demonstrate to his students all the resources of the learned style, of which canonic imitation forms the essential framework.' In the other entry from *Larousse de la musique* (1957), he underlines the more nuanced interpretation that 'these *Inventions* do not have an exclusively didactic goal [*but*]; they surpass it'. He puts a further line against the end of the entry which develops this point, arguing that Bach's 'genius' lies in 'reus[ing] the most beautiful discoveries of the art of polyphony and demonstrat[ing] the variety of effects that can be spun out of dry and mathematical formulas'.

This chimes with Derrida's paleonymic reuse of the notion of genius in the collection's titular essay, 'Psyche: The Invention of the Other', to describe the intervention of an unexpectedly new invention that comes from the other (paleonymy itself, of course, being one such invention of Derrida): 'There is no invention without the intervention of what was once called genius' (Derrida 2007: 418). But in the same breath he warns that the 'brilliant flash of *Witz*' cannot be accounted for by 'a restricted economy of différance'. Derrida's reinvented genius is concerned not with the possibilities of invention – the finds or discoveries (*trouvailles*) that Bach uncovers in the repository of the learned contrapuntal style – but with 'the invention of the impossible', understood as a double genitive, inventing what is impossible and the invention that comes only from the hand of the impossible – a chance even 'beyond the incalculable as a still possible calculus' (418). And yet, Derrida's annotations suggest that he is also to some extent stopped short by the technical rigour of Bach's inventions which precisely are calculable, teachable even – by what might be described as their mathematical, systematic, even *methodical* elements. Specifically, these draw attention to the way in which Bachian invention is not so much opposed to methodical composition but rather represents its limit, raising its status and credibility to a pinnacle (*ses plus hautes lettres de noblesse*), as one of the dictionaries puts it. These copied pages and their marginalia thus show Derrida turning to the musical inventions of Bach as a way to reflect on a certain abiding tension in his own thought between the unconditional incalculable and conditional calculation. A careful reading of Derrida's writings over the duration of his career shows that, far from advocating a messianic deferral in favour of the absolutely other, he insists that any unconditional 'worthy of the name' (*digne de ce nom*) can only come in the conditioned and conditional acts done in its name.[1] Thus Derrida will say that any decision – a category he associates with the event of invention and chance –

> must follow a law [*loi*] or a prescription, a rule. It must be able to be of the order of what is calculable or programmable, for example as an act of equity. But if the act simply consists in applying a rule, of enacting a program of effecting a calculation, we might say that it is legal […] but we would be wrong to say that the *decision* was just. Quite simply because in that there was no decision.
>
> (Derrida 2001: 251, translation modified, emphasis in the original)

Perhaps it is the case that Bach offers, for Derrida, a glimpse of the more or less tight interlacing of calculation and the incalculable – of method and invention.

Sound in deconstruction

Why do I dwell on these folded pieces of paper from Derrida's library? Why focus on the notion of musical invention underscored there? It might seem as if I were suggesting that musical invention provided a template – a *method* even – for deconstruction. Indeed, in the essay 'What Remains of Music' in *Psyche*, Derrida discovers, via reflections on Roger

Laporte's *Fugue* and *Supplément* of 1970 and 1973 respectively, a certain affinity between music, specifically rhythm, and his notion of *écriture* as an exteriority or relation to the other that compromises any interiority or inner voice from the outset. Commenting on a lengthy quotation from Laporte, Derrida proposes:

> This reinscription of the blank of writing has an essential relation to music and rhythm. Rhythm counts more than all the themes it carries off and relaunches and scans constantly. That is why, instead of an inventory […] of all the 'themes' that *Fugue* fugues and *Supplément* supplements (the fugue and the supplement are at once the title, the form, and the theme of this musical transport of writing), instead of drawing up a false list of themes treated (the signature, the privilege of the psychoanalytic, metaphor […] etc.), I will briefly mark, so as to send one back as quickly as possible to the text itself, if one can still say that, the affinity between the muscle or the 'rhythmic beat of a blank' (these are almost the last words of *Supplément*) and the rest.
>
> (Derrida 2007: 88–89)

Despite certain statements that suggest a privileging of aphonia over the sonorous, it would be mistaken to conclude that Derrida straightforwardly eschews sound for silence, since the effort of what has come to be known as deconstruction is not simply to invert oppositions, such as that between voice and writing, but, moreover, to put oppositionality itself in question. As I argue elsewhere, Derrida's thought entails less a departure from or a destruction of the *phonē* than it does its paleonymic reinvention (Waltham-Smith 2021: ch. 3). If the grammatological project involves a generalization of writing beyond its narrow or vulgar concept, this notion of *archi-écriture* is inextricably caught up with the sounding voice, as he explains in the context of a joint interview with Hélène Cixous:

> Those who do not read me reproach me at times for playing writing against the voice, as if to reduce it to silence. In truth, I proposed a reelaboration and a generalization of the concept of writing, of text or of trace. Orality is also the inscription [*frayage*] of a trace.
>
> (Derrida 2005: 1)

Notwithstanding these remarks – no doubt to some extent encouraged by Cixous's own solicitousness towards aurality – Derrideans have remained suspicious of or/aurality. Exemplary of this tendency is a chapter of David Wills's *Inanimations* devoted to Cixous's habit of inserting unusually large white spaces in the middle of sentences or, alternatively, of compressing words so that they run on together as if without breaths between them (Wills 2016: 11–52). It is probably fair to say that Wills's reading remains scrupulously faithful to a certain Derrida who in *Of Grammatology* links the spacing of *différance* to Mallarmé – faithful to the letter, one might say. His focus is on how these striking blank spaces – Mallarmé's '*les "blancs" frappent*' – wean writing off its subordination to sound, in this case to being nothing more than the dictation of the breath. In an article on Derrida's recently discovered *Geschlecht III*, Rodrigo Therezo assimilates these Mallarmean *blancs* to the silent *Grundton* that Martin Heidegger discerns as the foundation of Trakl's poetry, while also acknowledging that this silence is problematic for Derrida. 'All would be well and good with Heidegger's metaphorical emphasis on silence', writes Therezo (2018: 253), 'except for the fact that this fundamental tone is inextricable from the unity of the German language and hence from a philosophical nationalism. There is enough in this text and in Derrida's writings more generally to suggest a more dispersive, disseminated account of idiom rather than one gathered into a stultifying unity – one that would not so much be 'bleached', as Therezo wonders (2018: 259), as it would sound altogether otherwise. Or maybe it would be a bleached sound, whatever that might be (I will have to come back to this).

In another interview, Derrida gives a striking account of the kind of interweaving of gathering and dissemination, of binding and loosening, he thinks is eclipsed by metaphysical unity – striking because the image he uses is diverted by way of a sonic metaphor, specifically the sound of thousands of voices and their myriad rhythms and intonations as they travel great distances over telephone lines (Derrida 2002: 29–30). The metaphor of the telephone, telephony as metaphoricity – these themes will detain us later. For now, I want to note that this remarkable passage stands out against a more pervasive hesitation in relation to the sonorous. There is no wholehearted embrace of the sonorous in Derrida's work and he is certainly not a writer on sound and music in the way that Jean-Luc Nancy, Philippe Lacoue-Labarthe, Gilles Deleuze, Roland Barthes, or even Alain Badiou have all been. On several occasions, Derrida has spoken of the multiple voices in his texts, especially women's voices (cf. Derrida 1995: 394),[2] but there are also times when he echoes a long tradition going back to Plato of thinking of music and sound as that which lies outside philosophy and thus of according them either condemnation for their irrationality and barbarity or reverence for their sublimity. A little shy of writing expressly about music, when asked in an interview if he was tempted to write on the multiplicity of voices in music, Derrida replied:

> I wonder if philosophy […] has not meant the repression of music or song. Philosophy cannot, as such, let the song resonate in some way … I do not write *about* these voices […] I try to let them speak […]. The music of voices, if there is any, I do not sign it […] first of all I listen to it.
> Interviewer: So, let's listen.
> JD: Let's listen.
>
> (Derrida 1995: 394–395, emphasis in the original)

Cixous maintains that this hospitality of listening that Derrida gives to these multifarious voices, human and animal, is precisely a buttress against the rise of nationalisms and racisms (Cixous 2009b: 53). Metaphysics, by contrast, makes common purpose with exclusionary regimes insofar as it seeks to suppress this multiplicity by appropriating it – by making method out of madness, so to speak. This is the argument that Derrida advances by way of the motif of listening in 'Tympan', in a passage that merits citing at length in order to grasp the full effect of its metaphorics of resonance, vibration, and percussion.

> If philosophy has always intended, from its point of view, to maintain its relation with the nonphilosophical […] if it has constituted itself according to this purposive *entente* with its outside, if it has always intended to hear itself speak [*entendue à parler*], in the same language, of itself and of something else, can one, strictly speaking, determine a nonphilosophical place, a place of exteriority or alterity from which one might still treat [*trait*] of *philosophy*? Is there

any ruse not belonging to reason to prevent philosophy from still speaking of itself, from borrowing its categories from the logos of the other, by affecting itself [*s'affectant*] without delay, on the domestic page of its own tympanum (still the muffled drum, the *tympanon*, the cloth stretched taut in order to take its beating, to amortize impressions, to make the *types* (*typoi*) resonate, to balance the striking pressure of the *typtein*, between the inside and the outside), with heterogeneous percussion? Can one violently penetrate philosophy's field of listening without its immediately […] making the penetration resonate within itself, appropriating the emission for itself, familiarly communicating it to itself between the inner and middle ear […]. In other words, can one puncture the tympanum of a philosopher and still be heard and understood by him?

(Derrida 1990: xii, emphasis in the original)

In this colourful passage, listening becomes a metaphor for the operation of philosophy in its relation to the nonphilosophical, to its outside – an outside here figured as a violently percussive sound. And yet, Derrida wonders whether this sound would have to be so violent as to rupture philosophy's eardrum, rendering it unable to hear, if it is to avoid being made to sing philosophy's own tune? Either way, the sonorous is what goes unheard by philosophy. Derrida makes a very similar point in the discussion of Laporte in 'Psyche':

Fugue musics. The irreducibility of the musical here does not stem from any melocentrism. And I will try later to relate this unheard-of musical effect to a remainder unassimilable by any possible discourse, that is by all philosophical presentation in general.

(Derrida 2007: 88)

The metaphorics of aurality operates in multiple ways in deconstructive thought. In this passage, sound represents inassimilable exteriority and heterogenous difference, and listening functions as a metaphor for philosophy's appropriative desire. Sound figures as the limit to philosophy's drive to mastery and sovereignty, for it threatens its integrity and self-identity by penetration. It is not only its force but, moreover, sound's resistance to containment, its wayward dispersion – its 'promiscuity', as Brandon LaBelle (2010: xxvi) puts it – that disrupts the sovereignty of philosophy. But we ought also to recognize that, especially in appeals to a sonic turn within the humanities, the sonorous is often ontologized as a principle of dispersion and multiplication. I argue that this hypostatization of sonic difference repeats the metaphysical gesture of appropriation and containment. As much as sound is a figure for the unheard outside of philosophy, it is also, as resonant dispersal, a way of metamorphizing philosophy's overflowing its own bounds in the direction of the other. As Derrida never tires of repeating, there is nothing more philosophical than this overflowing (Derrida 2019: 71).

But is sound nothing more, for deconstructive or post structuralist thought, than a figure of the eminently philosophical overflowing of conceptuality and hence of philosophy's own limits? And what would it mean for the question of sonic methodologies to think sound as philosophy's own *de*-limiting? If the sonorous provides rhetorical figures for what spreads out, disperses, vibrates, wanders in multiple directions, and beats against itself – in short, if these figures point to what is anything but systematic or methodological – in what sense can we speak of sonic methodologies? This line of questioning suggests that sonic methodologies would only be methodological to the extent that they transgress any methodologism. Or it might imply a methodology whose rigour and unity are defined only by its stance *against* method.

Derrida, though, will also warn us that this overflowing of limits, for which sound is a metaphor, is at once a matter of constraint: 'Each time there is overflow, it resembles what is overflowed, overflowing remains in *affinity* with what is overflowed, *affined* and I'll even say confined to what is overflowed' (Derrida 2019: 86, emphasis in the original). If sound is what threatens to perforate and overflow (philosophical) method, it also then retains a certain affinity with (philosophical) thought and the methodological character of (its) thinking about method. For this reason, then, this apparent paradox of sonic methodologies poses some fascinating challenges for thinking about method and methodology today – challenges which, I would argue, deconstruction is well suited to address. What I have hoped to indicate is that, notwithstanding the reputation that poststructuralism has for privileging language, textuality, and writing in particular, its deconstructive variants offer just as many, if not more, possibilities for thinking about the sonorous as phenomenology or the new materialisms. If, despite suspecting the voice, deconstruction is unthinkable without a certain solicitation of the sonorous, it is necessary, furthermore, to resist the temptation to reduce the role of sound in the writings of Derrida, Nancy, Lacoue-Labarthe, Cixous, Avital Ronell, and Peter Szendy, among others, to nothing more than a metaphor for the working out of philosophical-methodological questions.

Against this reduction, the sonorous works in these texts to unsettle the very metaphorics to which it appears to fall prey. This is to no small degree because the distinction between figurative metaphor and the (proper) method for which it supposedly substitutes is unsustainable, for method, as Derrida shows, is already metaphorical. Method, and not just philosophy's, turns out to be metaphorical insofar as its generalizability to different situations relies upon an analogical transference or transposition. A metaphor for (philosophical) method is thus a re-marking of method. Sound – as method and as metaphor – is able to get at the metaphoricity of metaphor, at the substitutability that makes metaphor at once possible and slippery. It both promises to replace the metaphorics that is coextensive with metaphysics and at the same time exposes philosophy as something that has been replacing and displacing itself from the outset, like a vibration. To grasp more rigorously (more methodically?) what might be at stake in a deconstructive notion of sonic methodologies where sound is not reducible to a metaphor for philosophy's self-deconstruction, it is therefore necessary first to make a detour via a discussion of Derrida's arguments about method and its relationship with metaphor.

A detour via method

One of the greatest challenges to this task is Derrida's suspicion of method. What could a body of thought have to say about sonic methodologies when, at first blush, it appears to rest upon a rejection of philosophical methodologism – the teleology at work in metaphysics and in the practice of hermeneutics? Derrida famously denied that deconstruction was another method, despite what the anglophone reception of his thought seemed to suggest. And yet it might also with justification be said that he never stopped writing about method. His texts are littered with what can be characterized as *quasi*-methodological remarks to reflect the precautions that he always takes with this notion of method. Derrida devotes considerable time especially to reflections on the method that is reading. There is a recognizable Derridean quality of thought – a *type* of reading, as he describes it in *Geschlecht III* – that lends itself to a certain synthesizing, formalizing, even systematic gesture, as Geoffrey Bennington's work has shown; and yet, in the book that they wrote together, Derrida's contribution remains distrustful of even this cautious formalization. The method of deconstruction, 'if there is any' (*s'il y en a*) – both a way of doing deconstruction and the concept of method that deconstruction produces – consists as much in a deconstruction of method, which is to say in an event or invention of method. As Derrida puts it in 'Psyche',

> For a deconstructive operation, *possibility* is rather the danger, the danger of becoming an available set of rule-governed procedures, methods, accessible approaches. The interest of deconstruction, of such force and desire as it may have, is a certain experience of the impossible: that is [...] of the other – the experience of the other as the invention of the impossible, in other words, as the only possible invention.
>
> (Derrida 2007: 15, emphasis in the original)

In other words, it is the possibilization of method that deconstruction resists. It disrupts the idea of method as something of which one is capable, which one can possess and master, which can be made present. This is why, from a Derridean standpoint, method is something that only ever takes shape in the process of its practice. Deconstruction, if it can be called a method, is something that does not exist outside its taking place in a text and hence in a singular event. It is neither preformed nor separable as something distinct after the fact. Method, however, as Derrida recognizes, is not an entirely one-time thing. By definition, it is repeatable, yet open to unanticipated modification and strictly impossible in(ter)ventions. Derrida develops this sense of method's hospitality to the impossible by noting that method is a *meta-hodos*, that is, the pursuit or following of a way or path. When he says that dissemination admits of 'no method' (*pas de méthode*), he plays on this homonym in French which also means *step* (Derrida 1981: 271). Rejecting any teleological conception of method – of progressing from beginning to end, or from simple to complex, for example Derrida nonetheless continues to think of method by way of the figure of the path (*hodos*), albeit one that is circuitous rather than straightened or narrowed. With another homonym, he quips: 'We here note a point/lack of method [*point de méthode*]: this does not rule out a certain marching order [*une certaine marche à suivre* – literally 'a walk to follow']' (271). Derrida frequently describes this way as a *via rupta* or *voie frayée*, a path that has to be opened or cleared by being broken or beaten, and I would suggest that we hear in this breach an echo of the violent noise that threatens to rupture philosophy's eardrum.

Following a path that Heidegger had already opened, Derrida notes that for the Greeks *hodos* also had the sense of perversion or going astray and that, with regard to every methodology, Heidegger thus allows that any *Denkweg* (path of thought) is liable to be an *Irrweg* (errancy or aberrance). From this emerges Derrida's notion of destinerrance in *The Post Card*, according to which it is necessarily possible that a letter will not arrive at its destination or, thinking now of his exchanges with Cixous, that a telephone call will go unanswered or be misdirected. The recently published *Geschlecht III*, which tackles some quasi-methodological or 'pre- or a-methodological precautions', as Derrida phrases it here (2018: 43),[3] reveals the proximity of Derrida's thinking to Heidegger's problematization of method. Moreover, in this text, unlike the discussions of method alongside René Descartes or Baruch Spinoza in his 1981–1982 seminar, 'La langue et le discours de la méthode', for example, or the famous section on 'The Question of Method' in *Of Grammatology*, sonic metaphors assume a striking significance as a way of pointing to and destabilizing the presuppositions of traditional methodology. To begin the approach towards the question of sonic methodologies – and you can see already that this will have been a roundabout path carried off track by multiple diversions – it would be necessary to follow the trajectory of the detour in Derrida's thinking as it winds its way from the texts of 1967 to *Rogues* and the beastly steps of the final seminar of 2001–2002. This journey can only here be sketched out incompletely, in fits and starts, jumping from one moment to another, but that will allow me to test the hypothesis of method's inventive singularity. Otherwise put, is there a systematic thought, even a methodology, of diversion in Derrida's writings? And is there a methodical role that sound plays on this course?

The figure of wandering or diversion is already present in the passage from *Of Grammatology*, which approaches the question of method via the notion of an ex-*orbitant* reading:

> Starting from this point of exteriority [in relation to the totality of the age of logocentrism], a certain deconstruction of that totality which is also a traced path, of that orb (*orbis*) which is also orbitary (*orbita*), might be broached. The first gesture of this departure and this deconstruction, although subject to a certain historical necessity, cannot be given methodological or logical intraorbitary assurances.
>
> (Derrida 1974: 161–162)

This departure, which appears to be radically empiricist, 'proceeds like a wandering thought on the possibility of itinerary and of method'. And yet it only appears to be empiricist, for this empiricism ends up destroying itself in the process.

> To exceed the metaphysical orb is an attempt to get out of the orbit (*orbita*), to think the entirety of the classical conceptual oppositions, particularly the one within which the value of empiricism is held: the opposition of philosophy and nonphilosophy, another name for empiricism, for this incapability to sustain on one's own and to the limit the coherence of one's own discourse.
>
> (Derrida 1974: 162)

Five years later in 'Tympan', as we have seen, this empirical outside of philosophy will explicitly be figured as sonorous. So, if reading is to be ex-orbitant – or extra-*vagant*, as Derrida will have it in *Geschlecht III* – it will require that it breach the ear of philosophical methodologism. What this means for listening becomes clearer in *Geschlecht III*, where

Derrida will want to distinguish a destinerrant reading from a certain Heideggerian hearing that, insofar as it adheres to a long tradition of phonocentrism, blunts the radicality of his own warnings against method and methodologism. *Geschlecht III* announces its task, at least in part, as a reflection on Heidegger's method of reading Trakl and sets about asking a series of questions 'beyond method' (*d'outre méthode*): about the paths that Heidegger takes which may not yet or no longer be methods, about the rhythm of his step on such paths, and about the relationship his reading has to traditional methodologies and bodies of knowledge. As Heidegger confesses, his approach to reading is wayward. In Derrida's words, it moves via 'abrupt jumps, leaps, and zigzags' such that it is unclear whether these 'singular ruptures' are the result of careful calculation or come by surprise (Derrida 2018: 35). Rather than a systematic investigation or methodical interpretation of Trakl's poetry, Heidegger picks out verses from poems according to a 'metonymic sliding [*glissement métonymique*]' (83).

On the basis of this haphazard reading, Derrida aims to 'generalize' and 'problematize' a 'type' of Heideggerian reading where type does not surrender to the notions of model, procedure, or method, but instead points to typing, to the impression and strike of inscription, that is, to the *Schlag* in *Geschlecht*. Even though Derrida suggests that he is not primarily referring here to the tympan, what he has to say about typing and iterability recalls the argument he advanced in 'Tympan' about the doubling effects of the printing press's hammer. Typing, Derrida proposes, is never a single stroke or hammer but always a matter of over-typing – a double blow – and this is the quasi-palimpsestic effect he seeks to achieve by allowing his reading to re-mark Heidegger's.

It is important to hold onto the sonic metaphor that is indirectly brought into play here with the reference to the tympan, notwithstanding Derrida's desire to distinguish the thought of *Geschlecht* from the sense of typing developed in the earlier essay. I argue this is in no small part because the difference that Derrida will locate between Heidegger's blows and their Derridean over-typing hinges once more on a sonic metaphor. Heidegger's reflections on the arbitrary path that he takes, with its unexpected leaps from one poem to another, are not exactly methodological, Derrida argues, insofar as they warn against method and against the very methodologism that would reproach Heidegger for being capricious and improvisatory. They can only be described therefore as 'pre- or a-methodological' to the extent that they retreat from the discourses and knowledges that ground their authority in method. The question is: what authorizes the leaps and metonymic transitions from one place to another in Trakl's poems, if not some kind of method? Derrida's (over-)reading reveals that Heidegger's is not as radical in its waywardness as it might claim to be. The answer turns out to be a certain kind of listening. Underpinning all of Trakl's different poems, gathering all the different places into a single place, is a singular resonance, a unique consonance or unison (*einzige Einklang*) whose unity stems from what Heidegger calls the silent fundamental or tonic (*Grundton*), which spreads out like the ripple of a wave to all the individual poems. It is the unity of this tone, which gathers difference together in a single place, that permits the various interpretative leaps and metonymic transitions of Heidegger's seemingly haphazard method of reading. It guides the choice of specific poems and verses because Heidegger 'allows himself to be oriented by "the hearing" (*l'entente*) or 'precursive listening' (*précursive écoute*)' to this tone in its unity. The method-without-methodologism of Heidegger's reading is thus revealed to be both far less radical than he claims and also to hinge on a sonic or aural methodology – that is, a path that passes, is diverted, by way of the sonorous and whose very detours are directed by sound. Except that Heidegger's fundamental tone does not sound but remains absolutely unspoken, a soundless sonority. While one might think that this attunement to the silent voice of being would be welcomed by the author of *Voice and Phenomenon*, this is significantly complicated in *Geschlecht III* by the association Heidegger makes between the silent fundamental tone and the German idiom. The linguistic and philosophical nationalism of this unifying silence poses insurmountable difficulties for Derrida who must therefore insist on reading *over* Heidegger, in the sense of printing over but also perhaps in the sense of speaking over or listening past.

This over-reading should, I want to argue, be understood not simply as replacing or displacing but, moreover, as intensifying or exacerbating reading, as reading more and more to the point of exhaustion. We might then speak, instead of Heidegger's precursory aural attunement to the unity of the fundamental, of an *over*-hearing that would splinter and shatter that purported unity. I promised earlier to return to Therezo's suggestion that this unity be bleached. For him, this involves a turn from the aural to the visual via the insertion of blank spaces between the letters, a standard, though now obsolete, German typographical practice (*sperren*) for showing emphasis. Therezo wonders whether this graphic dimension allows one 'to *see* more easily – and not so much hear' the divisibility and spacing that is the condition of (im)possibility of the place and, by extension, of any method (Therezo 2018: 258, emphasis in the original).[4] This emphasis on aphonic textuality and its spacing is closely aligned with Derrida's own thinking. That having been said, the difficulty with this turn *away* from sound is that it overlooks Derrida's own observation in *Rogues* that any turning away from or turning one's back on is always also a turning back or return (Derrida 2005). What I want to suggest, then, is that the unity and abhorrent nationalism of Heidegger's silence (in every sense) is shattered only via a turn back to aurality, which would necessarily involve aurality turning around and about itself to then return to itself, diverting and deviating from itself, taking a bypath, another path to bypass itself, and hence to get past itself, to overtake and 'exceed itself' (*dépasser*), to overstep and thereby to override and overhear itself by analogy with overprinting.

The notion of overhearing that I am proposing here is very close to what Peter Szendy has described as *surécoute*: a power of overhearing that, in striving towards a total surveillance, always comes up against a deaf point. As Szendy acknowledges, this follows from the Derridean understanding of the quasi-transcendental as a condition of (im)possibility that is both inside and outside the field it makes possible (Szendy 2016: xi). I want, though, to extend Szendy's notion of overhearing by associating it explicitly with the idea of an *usure* of the ear, its wearing out and exhaustion through overuse. Impossibility is not simply a structural, positive condition but, moreover, the effect of an intensification and an excess through which hearing destroys itself. This follows Derrida's characterization of *différance* in *Of Grammatology* as something that strives to expand without constraint but which necessarily limits its own expansion precisely in order to preserve itself. Later in his life,

Derrida would describe this self-destructive tendency as autoimmunity – as a power that turns (back) on itself. In this *over*-hearing there is not simply wandering or diversion but also an irreducible, originary perversion of listening. If listening is that which is always already drifting away from the straight and narrow, in what sense can one speak of sonic methodologies? This is the destination away from which we keep twisting and turning.

Telephonic metaphoricity

One final detour, then. And this is where the idea of bleaching also comes into play as a way to think how the question of method is entangled with that of metaphor. In the well-known essay 'White Mythology', Derrida describes the operation of metaphor, according to its traditional philosophical definition, as a process of bleaching: through repeated usage, like the face of a coin, metaphor loses its concrete, particular quality as it enters into linguistic currency, becoming transferrable and fungible (Derrida 1990: 210). As Derrida is at pains to point out in his response to Paul Ricœur's (mis)reading of this essay, 'The *Retrait* of Metaphor', metaphor cannot be reduced to the former sense of *usure* as becoming worn-out but must also be understood as a kind of usury which, through multiplication and transfer to myriad contexts, produces a surplus value of meaning (Derrida 2007: 56). And it is also a metaphor that implies a *continuist presupposition*: the history of a metaphor appears essentially not as a displacement with breaks, as reinscriptions in a heterogenous system, mutations, separations without origin, but rather as a progressive erosion (Derrida 1990: 215). In other words, the path *usure* takes is too direct and unbroken. This wearing out is too methodical. Even though Derrida settles on privileging the figure of economy in 'White Mythology' and leaves the figure of the path – of Heidegger's *Weg zur Sprache* – in the background, 'The *Retrait* of Metaphor' begins with a meditation on the circulation and transport that metaphor shares with method. Shifting modes of transport (from automobile to ship), Derrida speaks metaphorically of metaphor by way of the figures of 'drifting, skidding, or sideslipping [*dérapage*]' before suggesting that metaphor only 'withdraws' (*retrait*) at the moment that it overextends and overflows itself.

Derrida immediately associates this overflowing with re-marking and re-turning, with the *trait* or stroke's 're-tracing' (*re-trait*) itself, and hence we can see that *usure* in its usuriousness chimes with the overprinting of the tympan – and, by extension or metaphorical displacement, we might say, with philosophy's eardrum being over-beaten (almost) by sound. If the path is a figure or metaphor, it is also the case that metaphor is itself a figure of displacement, a *meta-hodos* or circuitous path. As Derrida argues in 'La langue et le discours de la méthode', 'metaphor itself, if one can say that, is a way [*chemin*], a way followed by a displacement of sense or words, of discourse. One even speaks, metaphorically, it is said, of the vehicle of a metaphor' (Derrida 1983: 40). What is more, 'all method presupposes a certain metaphoricity not only insofar as it is itself a certain practice of the way [...] but also to the extent that a method calls upon the analogical transposition of the rules or procedures it puts forward' (40–41). At the same time Derrida resists the reduction of the way to a metaphor, preferring instead to speak of the 'motif' of the path. He warns that doing so would be to believe that one could master a figure by means of rhetorical technique without recognizing that rhetoric only emerged on the basis of a kind of thinking itself determined as a way. What metaphor cannot think is its own metaphoricity – the very drift and errancy that makes metaphorical transport (im)possible.

Such precipitation of rhetoric ultimately fixes the errancy and drift of the path. We see something similar in the burgeoning field of sound studies where there is a tendency to ontologize the difference and dissemination of sound – as vibration or resonance, for instance – and it is precisely this hypostatization of sound, which threatens to stifle all that might be radical about the sonic turn, that deconstruction resists. One way to avoid this destination would perhaps be to multiply and criss-cross the passages of metaphor. A third figure that Derrida deploys in 'The *Retrait* of Metaphor' is suggestive for this purpose. He invokes Heidegger's word *Geflecht* (network or braid) to speak here, as he does in several other places, notably in *The Truth in Painting* and *Negotiations*, of an 'interlacing [*entrelacement*]' of threads, which are neither completely loose nor absolutely tied up but more or less tightly bound.

In the remarkable passage from *Negotiations* mentioned earlier and in his exchanges with Cixous, this motif of writing as interlaced and entangled paths drifts through another series of metaphorical guises: from knots (another nautical metaphor) to the sounds of voices carried over telephone lines under the sea.

> There is a word that keeps coming back to me, and the image of the knot. Negotiation as a knot, as the work of the *knot*. In the *knot* of negotiation there are different rhythms, different forces, different differential vibrations of time and rhythm. The word *knot* came to me, and the image of a rope. A rope with an entanglement, a rope made up of several strands knotted together. The rope exists. One imagines computers with little wires, wires where things pass very quickly, wires where things pass very slowly: negotiation is placed along all of these wires. And things pass, information passes, or it does not pass, as with the telephone. Also, cables that pass under the sea and thousands of voices with intonations, that is, with different and entangled tensions. Negotiation is like a rope and an interminable number of wires moving or quivering with different speeds or intensities.
>
> (Derrida 2009: 29–30)

What, though, would motivate a turn to telephony if it is yet another metaphor for the transport or displacement of which metaphor and method are also figures, albeit one transferred to the sonic sphere? I want by way of conclusion to explore what value the metaphorics of telephony might have for theorizing the relationship between philosophical ways of thinking and specifically sonic methods and methodologies. I also want to suggest that the figure of telephony allows us to see how the type of thinking that goes by the name of deconstruction is itself a sonic methodology. Telephony exhibits in sonorous form the kind of typographical spacing marked by the white spaces between letters. Carrying sounds over long distances, with calls at risk of going unanswered or being misdirected, the telephone spaces and displaces the *phonē* of metaphysics. This deconstructive sense of telephony comes to the fore in Derrida's reflections on Joyce, whose writing often wants to be read aloud and thereby elicits a reading acutely solicitous of sound and hearing.

In 'Ulysses Gramophone' to underscore the originarily telephonic quality of speech as a paradigm for the deconstruction of sovereignty, Derrida quips: 'In the beginning was the telephone' (Derrida 2013: 51). Derrida locates an originary 'telephonic *technē* [...] at work within the voice' long before the invention of such technology that disturbs the Husserlian circuit of hearing-oneself-speak: 'A mental telephony which, inscribing the far, distance, différance, and spacing in the *phōnē*, at the same time institutes, prohibits, and disrupts the so-called monologue' (52).

And yet it is far from trivial that deconstruction, as a way of thinking, has its material conditions in telegraphic and telephonic communications technologies. The postcard and its possibilities were thematized in the book of that title, but Derrida's thinking late in his life on power and sovereignty cannot be divorced from the medium via which it is elaborated. As we learn in Derrida's *H. C. for Life*, together with Cixous's *Insister of Jacques Derrida* and her novel *Hyperdream*, much of their intellectual exchange was conducted *over* the phone in the double sense that they frequently conversed and debated philosophical issues by telephone (Cixous even speaks of the importance for her of merely knowing that she could call him and of those imagined phone calls) and also that they disagreed over telephony and its power (Derrida 2006; Cixous 2007, 2009a).

One of these fault lines has to do with the telephone's function as a lifeline, although it would transpire in the novel written after Derrida's death – itself something of a lifeline to hold onto her friend – that it was more a case of crossed wires. If the telephone line is like an 'umbilical cord of life or death', Derrida points out in the *Death Penalty* seminar that it represents a divine or otherwise sovereign power to decide over life and death (Derrida 2014: 49). The telephone links the death-row prisoner to the governor who can exercise executive sovereign power to grant a last-minute pardon, thus leaving life 'suspended' from the telephone line (139). The telephone is not incidental to Derrida's argument but rather structures the analysis of the death penalty as that which seeks to pin down the unpredictable, chancy, evental – as it were inventive – element of living. By fixing and thus mastering the contingency of death, which is precisely what makes life capable of surprise, it strives to make life more rigorous, more methodical.

If held in one hand, the *tēle* introduces *différance* and spacing in the *phōnē*, in the other hand or at the other end of the line, this receptivity to the other exposes the ear to the sovereign command, to the possibility of being penetrated by and of incorporating the superegoic voice of the parent, the law, the state, or the university. In his reading of Nietzsche's fifth lecture from *On the Future of Our Educational Institutions*, for example, Derrida suggests that 'the umbilical cord of the university [...] has you by the ear' (Derrida 1985: 35). It 'dictates to you what you are writing' (and Derrida has always suspected dictation of subordinating writing to the voice), thereby keeping you on a 'leash', tied to the 'paternal belly of the State [...] like one of those Bic ballpoints attached by a little chain in the post office' (Derrida 2014: 36). In the *Death Penalty* seminar, the motif of telephony also leads to the romantic notion that music has a connection to the divine unrivalled among the arts, rising above sensuousness. Nietzsche, for instance, puts the musician on the telephone with God, who 'speaks metaphysics', making him a 'ventriloquist' and 'mouthpiece' of the sovereign beyond. It makes the distant immediately close. 'As if the telephone then became portable and cellular', remarks Derrida somewhat enigmatically; 'telephony is metaphysics' (Derrida 2014: 146). In this way the metaphor of telephony is associated with the theme of wandering and of the errant step.

In the 1995–1996 seminar on hostility and hospitality, the mother tongue is likened to that 'most mobile of telephones' by which the home and the proper may be carried around with us and whose infinite mobility in this way turns into its opposite, resisting mobilities insofar as it moves around with me (Derrida and Dufourmantelle 2000: 91). Hearing-oneself-speak is

> the most mobile of mobiles, because the most immobile, the zero-point of all mobile telephones, the absolute ground of all displacements; and it is why we think we are carrying it away, as we say, with each step [*pas*], on the soles of our shoes. But always while being separated from oneself like this, while never being quits with that which, leaving oneself, by the same step never stops quitting its place of origin.
>
> (Derrida and Dufourmantelle 2000: 91–93)

The fantasy of telephony as absolute immobility and sovereignty, however, is only possible because language, like the telephone receiver, is a prosthesis and hence a dis/replacement of human or/aurality. The telephone line is what breaches the inviolability of the home. Equally, there is no proper home without the conditional hospitality of my deciding 'to invite whomever I wish to come into my home, *first in my ear*' (Derrida and Dufourmantelle 2000: 51, my emphasis). The increasingly pervasive interception of telecommunications, state surveillance, and censorship only serves to multiply this penetration of the proper – which leads back to the issues at stake in 'Tympan'.

This appeal to the figure of telephony aims to show that, far from being external to philosophical ways of thinking (as irrational noise, the supposed sensible proper before metaphorization) or a metaphor for the continuous, progressive erosion of the edge of the philosophical path (as promiscuous vibration or untrammeled resonance), sound is the prostheticity and technical supplementary that is philosophy's condition of (im)possibility. With its series of substitutions and displacements, transporting sound from one place to another, telephony acts like a concatenation of discontinuous articulations that extend inside philosophy. Following Avital Ronell's extraordinary *Telephone Book* (1989), we can say that sound has a telephonic relation to philosophy, rerouting ways of thinking through its switchboard with all the possibilities for misdirection, mishearing, overhearing, or going unanswered that this entails. It is in this sense that deconstruction might be described as a sonic methodology.

At the same time, the figure of telephony cannot be elevated into a master or meta-metaphor any more than sound can be a meta-method in opposition to speculative hegemony. Foreshadowing 'White Mythology' by almost a decade, the early seminar on Heidegger of 1964–1965 calls for a de-metaphorization or destruction of metaphor that would not so much leave metaphor behind as it would expose the metaphoricity as such of metaphor, which is to say the very force of re- and dis-placement that is its condition of possibility. 'It is not a matter of substituting one metaphor for another, which is the very movement of language and history, but of thinking this movement as such, thinking metaphor in metaphorizing it as such' (Derrida 2016: 190).[5]

This task is not a philosophical one. If there is any method to it, it consists in the destruction of metaphor by another metaphor, of replacing one at such great speed by another, on the spot, one taking the place of the other, such that what is revealed is the power of replacing taking place as such. Derrida finds in Cixous's 'art of replacement' an electric, telegraphic power of the 'might' (*puisse*), which, far from being any hypostatization of the possible, is nothing but the event of the impossible. Again, this replacement is sonic. It is the 'eco-homonymy' by which Cixous at great speed replaces the meanings of or even subtly displaces phonemes (Derrida 2006: 73–74). If the destruction of metaphor happens in Cixous's *é-cri-ture*, it is by no means its only destination. As Derrida explains in the early Heidegger seminar,

> if by another metaphor one calls thinking this vigilance destroying metaphor while knowing what it's doing, there is no need to wonder where there is more thinking, in science, metaphysics, poetry, and so on. There is thinking every time that this gesture occurs, in what is called science, poetry, metaphysics or elsewhere.
>
> (Derrida 2016: 190)

The same should be said of sonic methodologies, which know no proper place, only the movement of drifting and skidding in their constant reinvention.

Notes

1. On this often misunderstood point, see the final chapter of Geoffrey Bennington's *Scatter 1: The Politics of Politics in Foucault, Heidegger, and Derrida* (2016: esp. 269–271), and Michael Naas's *Derrida from Now On* (2008: 24–25), which Bennington cites.
2. See also Derrida's text 'Ants' (2012: 41–42n22) where he is talking about hearing his own voice in that of Cixous's author when she describes the women who have lived their lives in her and with whose tongues she has tasted the world.
3. Derrida is commenting on (and later quoting from) Heidegger's 'Language in the Poem: A Discussion on Georg Trakl's Poetic Work', in *On the Way to Language*.
4. See also David Wills's (2016) appeal to visible white spaces in his discussion of Cixous's idiosyncratic writing practices.
5. On the destruction of metaphor also see Mendoza-de Jesús 2017.

References

Bennington, Geoffrey (2016). *Scatter 1: The Politics of Politics in Foucault, Heidegger, and Derrida*. New York: Fordham University Press.

Cixous, Hélène (2007). *Insister of Jacques Derrida*. Trans. Peggy Kamuf, with original drawings by Ernest Pignon-Ernest. Stanford, CA: Stanford University Press.

Cixous, Hélène (2009a). *Hyperdream*. Trans. Beverley Bie Brahic. Cambridge: Polity.

Cixous, Hélène (2009b). 'Jacques Derrida: Co-responding Voix You'. Trans. Peggy Kamuf. In Pheng Cheah and Suzanne Guerlac (eds), *Derrida and the Time of the Political*, 41–54. Durham, NC: Duke University Press.

Derrida, Jacques (1974). *Of Grammatology*. Trans. Gayatri Chakravorty Spivak. Baltimore: John Hopkins University Press.

Derrida, Jacques (1981). *Dissemination*. Trans. Barbara Johnson. Chicago: University of Chicago Press.

Derrida, Jacques (1983). 'La langue et le discours de la méthode'. *Recherches sur la philosophie du langage: Cahiers du Groupe recherches sur la philosophie et le langage* 3: 35–51.

Derrida, Jacques (1985). *The Ear of the Other: Otobiography, Transference, Translation*. Trans. Peggy Kamuf. New York: Schocken Books.

Derrida, Jacques (1990). *Margins of Philosophy*. Trans. Alan Bass. Chicago: University of Chicago Press.

Derrida, Jacques (2001). 'Force of Law: The "Mystical Foundation" of Authority'. Trans. Mary Quaintance. In Gil Anidjar (ed.), *Acts of Religion*, 230–298. New York: Routledge.

Derrida, Jaques (2002). *Negotiations: Interventions and Interviews*. Ed. and trans. Elizabeth Rottenberg. Stanford, CA: Stanford University Press.

Derrida, Jacques (2005). *Rogues: Two Essays on Reason*. Trans. Pascale-Anne Bault and Michael Naas. Stanford, CA: Stanford University Press.

Derrida, Jacques (2006). *H.C. for Life, That Is to Say …*. Trans. Laurent Milesi and Stefan Herbrechter. Stanford, CA: Stanford University Press.

Derrida, Jacques (2007). *Psyche: Inventions of the Other*, vol. 1. Ed.and trans. Peggy Kamuf and Elizabeth Rottenberg. Stanford, CA: Stanford University Press.

Derrida, Jacques (2009). *Negotiations: Interventions and Interviews*. Ed. and trans. Elizabeth Rottenberg. Stanford, CA: Stanford University Press.

Derrida, Jacques (2012). 'Ants'. Trans. Eric Prenowitz. *Oxford Literary Review* 24 (1): 17–42.

Derrida, Jacques (2013). 'Ulysses Gramophone: Hear Say Yes in Joyce'. Trans. François Raffoul. In Andrew J. Mitchell and San Slote (eds), *Derrida and Joyce: Texts and Contexts*, 41–86. Albany, NY: State University of New York Press.

Derrida, Jacques (2014). *The Death Penalty*, vol. 1. Trans. Peggy Kamuf. Chicago: University of Chicago Press.

Derrida, Jacques (2016). *Heidegger: The Question of Being and History*. Trans. Geoffrey Bennington. Chicago: University of Chicago Press.

Derrida, Jacques (2018). *Geschlecht III*. Ed. Geoffrey Bennington, Katie Chenoweth, and Rodrigo Therezo. Paris: Seuil.

Derrida, Jacques (2019). *Theory and Practice*. Trans. David Wills. Chicago: University of Chicago Press.

Derrida, Jacques and Hélène Cixous (2005). 'From the Word to Life: A Dialogue Between Jacques Derrida and Hélène Cixous with Aliette Armel'. Trans. Ashley Thompson. *New Literary History* 37 (1): 1–13.

Derrida, Jacques and Anne Dufourmantelle (2000). *Of Hospitality: Anne Dufourmantelle Invites Jacques Derrida to Respond*. Trans. Rachel Bowlby. Stanford, CA: Stanford University Press.

Derrida, Jacques and Elisabeth Weber (eds) (1995). *Points: Interviews, 1974–1994*. Stanford, CA: Stanford University Press.

LaBelle, Brandon (2010). *Acoustic Territories: Sound Culture and Everyday Life*. New York: Continuum.
Mendoza-de Jesús, Ronald (2017). 'Historicity as Metaphoricity in Early Derrida: From the History of Being to *Another* Historiography'. *CR: The New Centennial Review* 17 (1): 43–72.
Naas, Michael (2008). *Derrida from Now On*. New York: Fordham University Press.
Ronell, Avital (1989). *The Telephone Book: Technology, Schizophrenia, Electric Speech*. Lincoln, NB: University of Nebraska Press.
Szendy, Peter (2016). *All Ears: The Aesthetics of Espionage*. Trans. Roland Végső. New York: Fordham University Press.
Therezo, Rodrigo (2018). 'When Silence Strikes: Derrida, Heidegger, Mallarmé'. *Oxford Literary Review* 40 (2): 238–262.
Waltham-Smith, Naomi (2021). Shattering Biopolitics: Militant Listening and the Sound of Life New York: Fordham University Press.
Wills, David (2016). 'Living Punctuations'. In David Wills, *Inanimation: Theories of Inorganic Life*, 111–152. Minneapolis, MN: University of Minnesota Press.

4

Nature's Music: Sonic Methodologies in the Study of Environmental Biology

Wouter Halfwerk

Introduction

Being in a tropical rainforest at night can be an overwhelming experience. Hundreds of animals of all kinds are creating a cacophony of different sounds that can be at times deafening to our ears. Frogs call from water puddles formed on the forest floor, crickets chirp from the bushes, and owls hoot from branches. Experiencing early morning in a temperate forest in spring time can be equally impressive. Dozens of birds of ten to twenty different species can sing together to create the so-called dawn chorus (Slabbekoorn 2004). Underwater it is no different: the sounds of a tropical coral reef are as diverse as those of a rainforest (Montgomery et al. 2006).

Studying sounds in an evolutionary framework

All animals make use of sounds to either improve their survival chances or to increase their reproductive success (Bradbury and Vehrencamp 2011). Some use sounds to communicate, others to orient and find food, or to avoid predators (Stevens 2013). Songbirds sing to attract a mate or to let neighbouring rivals know to stay away (Marler and Slabbekoorn 2004). Frogs and crickets also advertise their readiness to mate by singing out loud (Gerhardt and Huber 2002). But communication can have more than just a sexual function. For example, sounds can be used to warn group members of potential danger, such as an approaching predator, or to coordinate a group that is foraging or hunting together (Hollén, Bell, and Radford 2008). Communication can even take place between different species, sometimes in a way that is mutually beneficial, sometimes in a way that mainly benefits the producer of the sound. Fork-tailed drongos (*Dicrurus adsimilis*) are tropical songbirds that can often be found in the vicinity of meerkats (*Suricata suricatta*). The drongo can warn a group of meerkats of a nearby predator by making a specific call, and meerkats respond by fleeing to cover (Flower, Gribble, and Ridley 2014). During the dry season, when food is scarce, the drongo sometimes produces a false alarm, in particular when the meerkats have just caught a prey. The meerkats flee and the drongo gets the prey. The drongo makes sure not to cry wolf too often, so that the meerkats cannot risk ignoring its alarm calls (Flower, Gribble, and Ridley 2014). In the end, both the drongo and meerkat benefit from one another, but the rewards are not equally split.

Producing a sound that carries reliable information from sender to receiver also plays a central role in evolutionary studies on sexual selection (Fitch and Hauser 2003). When singing or calling to attract a potential mate, it is usually males that try to sound as impressive as possible to prospecting females. Consequently, mating sounds are usually loud in intensity or low in frequency. But mating sounds can also provide reliable information on males. Females usually benefit from choosing a partner that can provide valuable resources, such as a large territory or a good food patch. Larger males are usually better at providing these resources as they outcompete smaller males. Females across a wide range of species therefore show a clear preference to mate with a male that makes a low-pitched and loud sound for good reasons: physical constraints keep small males from sounding as low and loud as big ones (Gingras et al. 2013). However, males of some species have found ways to trick females, as they have evolved structures that make them sound bigger than they really are. Some frogs have extremely large lungs and vocal sacs (Halfwerk et al. 2017), and some birds have extremely long trachea that are coiled up in their body cavity (Fitch and Hauser 2003), all in order to produce acoustic cues that makes their audience think they are big.

Conveying acoustic cues that (truthfully or not) represent the physical strength of the sender is one way to impress an audience, however, many animals and in particular songbirds have evolved extremely complex songs to make them more impressive. Songbirds can produce elaborate songs made up of a vast array of particular elements. Additionally, some species can even produce hundreds of distinct songs (see also Figure 4.1). Typically, in species with complex repertoires, females prefer males that sing the most diverse and complex songs. Why females should prefer males with big repertoires over males that have smaller repertoires is still a subject of debate (Kroodsma 2004). One idea is that having a large repertoire requires being in good physical condition during the critical learning phase, which is for many species when they are still in the nest and depend on their parents for food. Therefore, if you sing complex songs, you come from a high-quality family that has been able to provide sufficient resources during development (Catchpole 1980). If there is some genetic variation that determines the quality of raising young, it would benefit females to use information from the song to determine the potential parenting quality of the male. Other hypotheses have focused more on a perceptual explanation, for example that songs that are repeated time after time get boring (as the brain habituates to the same perceptual input), and variable songs are simply more attractive to listen to (Kroodsma 2004).

However, producing sounds does also come at certain costs. Many predators and parasites rely on the sounds produced by their prey or hosts to detect and locate them. For example, these sounds can be intentional, having evolved as a so-called 'signal' to serve a

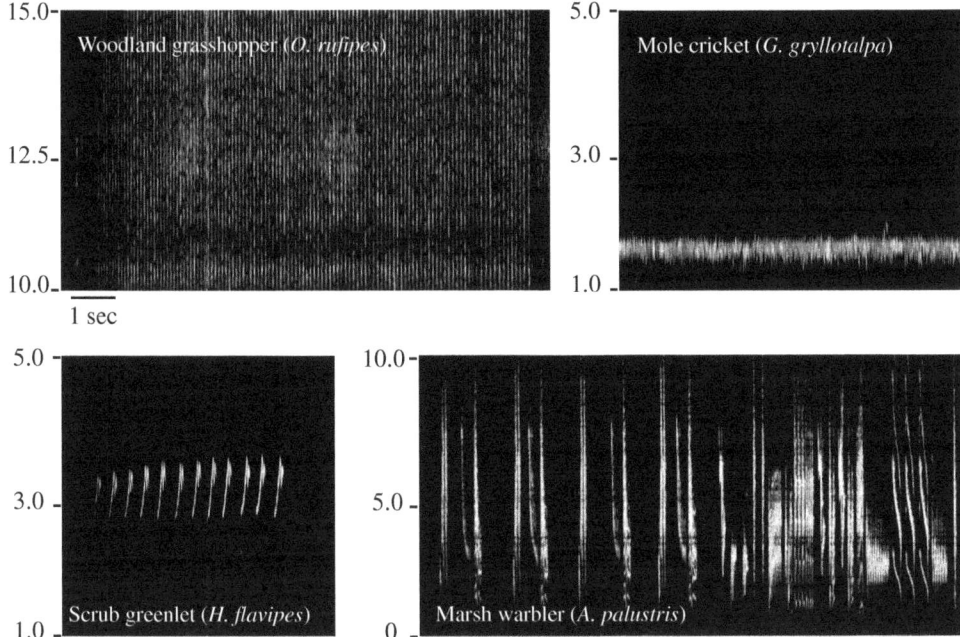

Figure 4.1 Sonograms of different species. A cricket, a frog, and two birds, one with a simple song and the other with a complex song. Note the different frequency scales (in kilohertz) on the *y*-axis. The temporal scale (in seconds) is indicated by the bar in the upper left panel.

specific communication function. Alternatively, sounds can be made unintentionally, as a by-product of certain activities, for example the rustling sounds of insects walking among leaves. Predators and parasites that eavesdrop on mating signals can in particular play an important role in signal evolution. Some frogs suffer from predation of bats that have evolved special hearing sensitivities to detect and localize them using the frog's mating call (Halfwerk et al. 2014). Predation by bats keeps these frogs from calling at high rates, high amplitude or with high complexity, features that would otherwise make them favourable to females. However, these bats are absent from areas with high levels of noise and light pollution, for example urban environments. Not surprisingly, urban frogs that are released from this predation risk create more extreme and thus more attractive calls than their forest counterparts (Halfwerk et al. 2019).

Using sounds to locate and monitor animals and their populations

Besides studying sounds related to ecological and evolutionary processes, they also provide an easy way to locate species and monitor populations. Trained human observers can easily use the acoustic activity of a species to determine its presence or absence in a certain area, and even to estimate its local density. For many species, acoustic monitoring is the main method to estimate their population size over time. These estimates are particularly accurate for species that make sounds to advertise territorial occupancy, such as the majority of song birds. The number of singing individual birds is thought to be a good representation of the total number of individuals in a given area (Ralph, Droege, and Sauer 1995). Nowadays, trained observers are more and more supplemented by automated recordings, which is especially useful when studying rare species or species that produce sounds irregularly (Acevedo and Villanueva-Rivera 2006).

Automated recorders have become particularly useful in the past decade to study animals that are otherwise hard to see, such as nocturnal mammals and birds. For example, bat populations are frequently monitored with the use of ultrasonic automated set-ups that can monitor their activity and identify different species throughout the night (Obrist, Boesch, and Flückiger 2004).

A brief history of technologies used to study natural sounds

The first attempts to capture sounds of nature were put on paper by musicians. Bird song was the easiest to translate into a musical notation system, as most of their tones are relatively pure and often harmonically related. One of the most famous examples of putting bird song into quantifiable units comes from Wolfgang Amadeus Mozart's starling. Presumably, Mozart kept a starling as a pet and was so touched by its song that he incorporated some of the bird's motives in his own music (West and King 1990).

Recording sounds

The scientific field of bioacoustics started when analogue sound recorders became available. Driven by curiosity, the first naturalists who were eager to record animal sounds hauled equipment with them on expeditions that are by today's standards enormous. The phonograph, invented in 1877, was the first equipment readily available to any researcher on animal sounds. Most of the early phonographic setups used by the end of the nineteenth century engraved the sound wave vibrations onto a rotating disc known as the 'record'. These records, or 'vinyls', were made of different sorts of material, such as tinfoil, wax-coated cardboard, rubber, or shellac compound, and produced mixed results under challenging field conditions. Some of these setups could be taken into the field, although it often required researchers to caste their own 'records' using a mixture of ethanol, rubber or shellac. Importantly, the phonographs and their later successors, the gramophones, could also be used to playback sounds. This allowed researchers to start testing for the first time what the biological functions of animal sounds were by broadcasting these sounds back to the animals from which they were recorded (Fischer, Noser, and Hammerschmidt 2013).

Analysing sounds

Although sound recorders were available for some time, the field of bioacoustics really kicked-off in the 1950s and 1960s when machines that could turn a sound wave into an image containing spectral information became available (Marler 1955; Struhsaker 1967). These Sonagraphs (e.g. from Kay Electric) made use of Fourier analysis, which divides a sound into discrete frequency ranges and calculates both the amplitude and phase within them (Marler and Tamura 1962). Fourier analysis could be used on a whole section of sound, for example a frog call or bird song, with a length in the order of seconds. This approach allowed researchers to determine the lowest and highest frequencies of a sound, its peak frequency (defined as the frequency with the highest spectral energy), as well as its harmonic structure. Nowadays, we refer to this way of visualization as a power-spectral-density plot (or power spectrogram). This approach opened up all sorts of questions related to animal communication, such as what ways are sounds produced as well as how are sounds used? Dividing sounds into smaller time units, using discrete or fast Fourier analysis algorithms, and plotting them on paper also allowed researchers to study how an animal call or song changed its spectral properties over time. In the early days, most of these Fourier analyses took hours if not days, depending on the length of the recordings. Nowadays, these so-called spectrograms can be produced using a wide range of acoustic software, such as Audacity or Raven, in milliseconds.

The first insights

Species can often look alike but sound rather different. Until the dawn of the field of bioacoustics, most individuals that looked alike in terms of morphology and colour patterns were considered to belong to the same species. Sound recording and analyses led to the discovery that single species complexes often consisted of multiple species that had previously gone unnoticed. In addition to the use of sounds for improved species identification, sound recorders also led to the discovery of totally new species, in particular nocturnal groups such as owls and frogs (Laiolo 2010).

Another major insight came from observations of the songs of birds that were recorded in different areas. In the 1950s and 1960s, the ethologist Peter Marler compared sonograms and found that birds that share a neighbourhood also share the same type of song (Marler 1955) whereas birds from different neighbourhoods had different songs (Marler and Tamura 1962). These 'bird dialects' reminded him of the subtle differences found between dialects in human language. Subsequent studies revealed that these bird song dialects had very abrupt boundaries. Some individuals that occupied a territory at these dialect boundaries were found to sing the songs of both dialects, just like people at borders can often master more than one language (Baptista 1977). Soon, researchers hypothesized that some bird species have to learn how to sing their song during their lifetime, just as humans need to learn their language. Subsequent isolation experiments revealed that songbirds, such as chaffinches and zebra finches, need an adult tutor to copy their songs (Thorpe 1958).

Our closest relatives, the chimpanzee, bonobo, and gorilla, do not show vocal production learning. Songbirds have therefore become the dominant model to understand language learning, including its neurobiological underpinnings and even grammatical structuring (Nowicki and Searcy 2014). All because some field biologists decided to study songs of wild birds.

Finally, and perhaps most amazingly, new recording technologies allowed for the discovery of a totally new way of perceiving the world, namely echolocation. In the 1930s, a professor of zoology at Harvard, Donald Griffin, hypothesized that bats make use of acoustics for orientation, as they must use some special sense to orient in the pitch-dark environment of a cave (Griffin 1944). He argued that bats produce a highly directional, ultrasonic beam and rely on the returning echoes to know whether they are flying towards an object or not. By using an ultrasonic sensitive condenser microphone coupled to a cathode ray oscillograph, Griffin was able to document for the first time what type of sounds bats produced during hunting and how they changed both the interval between echolocation calls as well as the frequency components during a successful attack on an insect prey (Griffin 1950).

Understanding the production and perception of the sounds of nature

If a tree falls in a forest and no one is around to hear it, does it make a sound? The answer is in the ear of the beholder. If trees would be able to perceive the sounds they make themselves the answer to this philosophical question would be 'yes'.

More biologically relevant are questions relating to how sounds are produced and perceived by individuals of the same or different species, and how this production/perception affects a species' behaviour and ecological interactions. Although trees do make sounds, sometimes even at moderate amplitude levels, like when poplar leaves are moved by the wind, most people would agree that this sound production does not serve a communicative function. However, plants, including trees, are sensitive to the vibrations that can be induced by sounds; just think of a Venus flytrap that closes when sensitive hair cells are touched by its insect prey. Sound waves can induce vibrations in plant structures, for example along leaves and roots. Some of these cues can be used by plants to determine where their root system should develop (Gagliano et al. 2017). When playing a tone of a certain frequency that travels through the soil, pea plant (*Pisum sativum*) roots will grow in the direction of the speaker, although it is still unclear why they do so. Another recent and somewhat surprising finding demonstrates that plants can use vibrations produced by caterpillars that chew on their leaves to induce a chemical defense (Appel and Cocroft 2014). Hypothetically, this defense mechanism could also be induced by broadcasting loud sounds with similar frequency characteristics as the vibrations produced by caterpillars at close range.

Mechanisms of sound perception

Compared to plant acoustical perception, researchers have a far better understanding of the sounds produced and used by animals. For most vertebrates such as mammals, amphibians, and even fish, it is clear that they have evolved pressure-sensitive structures to detect sounds.

Most terrestrial vertebrates have an ear that is similar to ours, consisting of an eardrum that is connected via one or more air or fluid-filled channels to a structure that contains hair cells that function as neuronal receptors. There are however also important differences in the morphological and neuronal structures used by vertebrates to process sounds, which partly determines the range of sounds these animals can hear. For example, most vertebrates have a single cluster of neuronal cells in their inner ear. Furthermore, most of the sounds come in via the ear channels and are transduced via different membranes into surface waves that travel along layers of vibration-sensitive hair cells. Frogs and toads on the other hand have up to seven different nuclei in their inner ear that are sensitive to a wide range of different stimuli, including sound waves that travel via the ground and via their lungs (Lewis et al. 2001). Larger lungs lead to a more stretched body wall, which means that more sound waves can be absorbed, culminating in a higher sensitivity to sounds (Christensen-Dalsgaard 2005). Mammals have evolved inner ear bones to improve the impedance between air and the fluid that fills the cochlea and thereby increase their sensitivity. The morphology of the cochlea also differs substantially between the different animal groups. Frogs lack a clear structure within and among the different groups of hair cells, but mammals and birds have their hair cells organized tonotopically along a long, thin membrane (Manley and Fay 2013). When a sound wave passes through the cochlea, different frequencies induce hair cell movements at different positions along the membrane. A longer membrane means that a wider range of frequencies can be detected or that a better spectral resolution can be achieved. The cochlea in birds is stretched out, whereas in mammals it is curled up, allowing for a longer length and possibly a wider range of frequencies that can be perceived. Not surprisingly, mammals have evolved acoustic perceptual systems that are sensitive over an amazingly broad range of frequencies, ranging from the infrasound range used by elephants for their communication to the ultrasound range used by whales, dolphins, and bats to hunt for prey (Manley and Fay 2013).

Invertebrates, and in particular insects, have evolved a very diverse array of morphological and neuronal structures to process acoustic cues. Some insects, for example katydids, a group of grasshoppers, have pressure sensitive eardrums in their legs that are connected via neurons to their brains (Hoy and Robert 1996). Others have evolved these structures on the sides of their body, for example the ears that have evolved in many nocturnal moths to detect the echolocation calls of hunting bats (Hoy and Robert 1996). The acoustic 'bat-detectors' of moths are very simplistic, consisting in many cases of only two neurons, yet they are very effective, as a moth that hears a bat will dive-bomb out of the air in an unpredictable, spiraling flight path, thereby making it hard for its predator to intercept. Other insects, such as mosquitos or midges that use sounds made by their hosts to find them, have evolved very sensitive antennae on their head that consist of many fine hairs. When a pressure wave passes, these hairs are moved back and forth, allowing the insects not only to determine the type but also the location of the sound source (Nadrowski et al. 2011).

Mechanisms of sound production

Sounds in nature can be categorized as either impulsive or continuous. Impulsive sounds can be produced by using one object to hit another object with considerable force. Chimpanzees for example use sticks to make drumming sounds on logs in the forest; and woodpeckers hammer their heads, or more precisely their bills, into trees to advertise their presence (Bradbury and Vehrencamp 2011). Continuous sounds usually consist of air- or water-borne vibrations and are often of more tonal quality, depending on the physics of the morphological sound source and the resonance properties of the morphological structures surrounding the sound source (Bradbury and Vehrencamp 2011). Continuous sounds are very short, like the calls produced by some bat species (less than 1 millisecond in duration), or very long, like the whistle produced by some whales, and can vary in frequency and amplitude over time, which is why we can find such an extraordinary array of different sounds in nature.

Continuous sounds can either be produced via stridulation, where two solid morphological structures are rubbed together, via the passage of air or water along vibrating morphological structures, or via very fast muscle contractions. Most insects produce sounds via stridulation, the process of rubbing various body parts against one an other. Crickets for example have comb-like structures on one of their wing veins that produce chirping sounds when rubbed against the other wing. Cicadas are one of the few exceptions in insects, as they produce their sometimes deafening sounds via a different process that is called 'buckling'. They possess two membranes (tymbals) covered in a row of ribs at the side of their body that are rapidly contracted and released by a set of muscles. When the ribs collide or jump into their original position, a snapping sound is produced; all of this happens at an astonishing rate of several hundreds of times per second.

Most vertebrates, apart from fish off course, use their lungs to vocalize. Upon singing, talking, or calling, an individual pushes air from its lungs into the trachea where one or more vibrating membranes act as the sound source (Suthers and Zollinger 2004). The sounds produced are subsequently filtered by the transmission properties of the remaining part of the trachea and the oral cavity. Usually there is only one vibrating structure that acts as the sound source, such as the larynx in mammals and frogs. However, some songbirds who possess a syrinx can have two vibrating structures, at both ends of the bronchia that come from the two lungs. This allows them to produce two sounds that are not harmonically related, which is known as the two-voice phenomenon (Suthers and Zollinger 2004) (Figure 4.2).

From an evolutionary perspective there are several interesting tradeoffs related to the sound production mechanisms of vertebrates. Birds, for example, need to open and close their bills in synchrony with the sound they produce, because opening the bill affects

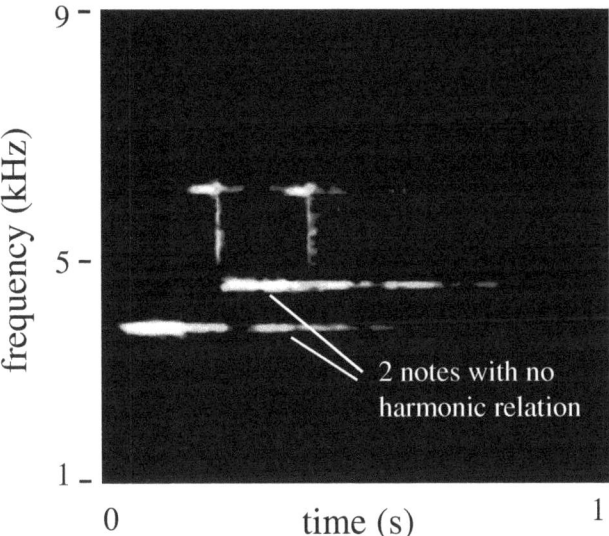

Figure 4.2 Sonogram of an Andean solitaire (*M. ralloides*) that demonstrates the two-voice phenomenon. The two notes are simultaneously produced but have a different fundamental, suggesting that there are two sound sources active simultaneously. Most songbirds have two sound-producing organs (or syrinx) close to where the two bronchea merge into the trachea. Some species can independently control the muscles in their syrinx, which leads to this two-voice phenomenon.

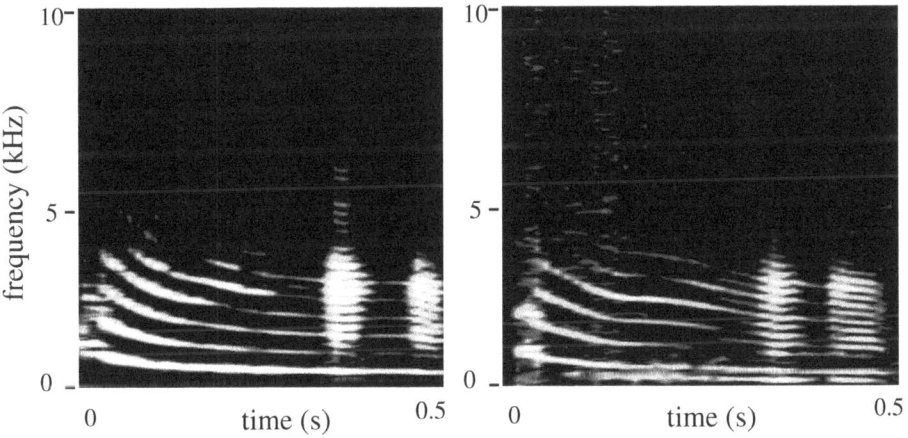

Figure 4.3 Sonograms of a recording made with a microphone and a laser-Doppler vibrometer. The left panel shows the airborne sound made by a calling male tungara frog (*P. postulosus*). The right panel shows the vibrations recorded by pointing the beam of a laser-Doppler vibrometer at a frog's vibrating vocal sac.

the resonance properties of the trachea. Bird species that produce elements with rapid frequency modulations, so-called 'trills', need to open and close their bills fast, which is more difficult with a heavy bill than a thin one. Bird bills vary in shapes and sizes, related to their diet. A heavy and short bill is useful to crack open seeds, but species with heavy bills cannot sing fast trills (Podos 2001). Birds are thus constrained in the range of sounds they can produce by their morphological characteristics (in this case the bill). Likewise, some frogs need to inflate their lungs in order to make a loud and low-frequency sound. Having a large air-filled lung requires that a frog floats on a water surface in order to support its body as well as the physical forces associated with pushing air from their lungs through their larynx. When some frog species that call from water are placed on the ground, their calls sound less loud and are rated as less attractive by females, simply because they cannot inflate their lungs to the maximum size (Halfwerk et al. 2017; Smit et al. 2019).

Current approaches to study mechanism and function of animal sounds

Nowadays there are many ways to study how sounds are produced, how they are transmitted through the environment, and how they are perceived and used by different animals. In recent years a new field has emerged that specifically addresses the diversity of sounds found in nature. This particular field is usually referred to as 'soundscape ecology' and aims to understand the global and local patterns in sounds, as well as their ecological significance. The field of soundscape ecology is mainly driven by technological advancement, in particular the use of remote sensing techniques such as the deployment of a network of automated sound recorders (Pijanowski et al. 2011). Linkage with the more classic field of animal communication is found in particular in studies addressing the role of anthropogenic noise on animal behaviour and reproduction (Halfwerk 2012). Below I will review how current techniques are applied to the different subfields of animal communication, namely the production, transmission, and perception of sound. I will briefly review the technologies used for environmental monitoring and explain how these techniques help us to understand ecological and evolutionary processes.

Measuring production properties

As outlined above, sounds are produced by putting something into motion, be it by hitting an object with a stick or by making something vibrate. Most current measurement techniques focus on the production of vibratory sounds, using laser-Doppler vibrometers, high-speed video cameras, or microphone arrays. Depending on the sampling regime or frame rate, the production of sounds spanning a frequency range of 1 hertz to 100 kilohertz can be studied, either from the outside or the inside of an animal's body. For example, using a high-speed camera that is sensitive to X-rays, the movement of the tongue as well as changes to the vocal tract of a singing bird can be visualized (Ohms et al. 2010). This technique requires the use of small lead beads glued to the animal's tongue and throat that are easy to trace on X-ray images. When a bird is changing its singing posture, for example by changing the position of the tongue inside the bill, the resonance properties of the vocal apparatus are changed and a different sound is produced. Comparing the movements of the lead beads from the video with the acoustic features of the song elements that are simultaneously emitted by the birds thus allows in situ study of the production mechanism. Even the subtle movements of the larynx or syrinx can now be studied in situ, for example with a micro CT-scan. Such approaches can be used to study the oscillations of the different syringeal muscles involved in sound production by different bird species (Elemans et al. 2015). CT-scan videoing allows researchers to test what aspects of sound production are shared by species as diverse as ostriches, pigeons, and zebra finches, and to understand how these complex structures have evolved.

High-speed video imaging has also proved pivotal in revealing some of the more mysterious sound production mechanisms found in nature. For example, some bird species produce loud buzzing or ringing sounds during their courtship displays with their bill closed. Do these birds make these sounds without their syrinx? If so, how can they produce these high-pitched sounds, which require very rapid movements? The club-winged manakin (*Machaeropterus deliciosus*) is a Neotropical passerine bird found only in the western parts of Columbia and Ecuador (Bostwick et al. 2009). Males display in so-called 'leks' and are visited by females that mate with them; they then leave to build nests and raise the young on their own. Male manakins spend most of their adult life courting females and do so by raising their wings high above their heads, while simultaneously making a three-note whistling sound. Using high-speed video cameras, researchers found out that these birds move their wings back and forth at a very high speed, exceeding 100 hertz. However, the sounds are of even higher frequency. Following closer inspection it turns out that the sounds are made by special feathers with multiple spikes on them, creating multiple sound waves that, superimposed on one and another, produce the high-pitched ringing sounds (Bostwick et al. 2009). The high-speed video camera made it possible to uncover a novel evolutionary invention to court females.

Other measurement techniques that have been applied successfully to sound producing animals include the use of laser-Doppler vibrometry. This technique makes use of a monochromatic light beam aimed at vibrating body parts (e.g. the bill, throat, or eardrum of an animal). If the body part is reflective enough, the returning beam can be compared to the outgoing beam to assess the Doppler shift, which is proportional to the velocity at which the body part vibrates. Non-reflective surfaces can be covered with adhesive tape or very small glass beads covered in silver particles to enhance the sensitivity of the laser. Many frogs shuttle air from their lungs into a large vocal sac. After calling, this sac deflates again, pushing the air back into the lungs for the next round of calling. Experiments from the 1990s, during which frogs were placed in helium enriched enclosures, already ruled out that these sacs act as resonators (Rand and Dudley 1993). Using laser-Doppler vibrometry, my colleagues and I have recently found that these vocal sacs do resonate with specific frequencies (Halfwerk et al. 2017; Figure 4.3). Perhaps these sacs function as radiators or amplifiers of specific acoustic features which help frogs to create a more diverse type of call.

The use of microphone arrays has also recently improved our understanding of directional sound production, which is important for research on animals that rely on echolocation.

Bats produce ultrasonic calls with their larynx and emit these sounds either via their nose or their mouth. Depending on the frequency of the sound and the opening of their mouth, the sound beam produced is either very directional, being loud in front and faint to the sides, or more omnidirectional (Jakobsen, Ratcliffe, and Surlykke 2013). Researchers have studied the shape of the bats' sonar beam using two arrays of microphones, orthogonally placed behind a food reward. On the one hand, this approach allows researchers to triangulate the exact position of an approaching bat; on the other hand, the amplitude levels recorded by the different microphones can be used to reconstruct the bats' sound beam morphology (Jakobsen, Ratcliffe, and Surlykke 2013). As it turns out, when bats are hunting in an environment with many different objects they narrow their sonar beam to avoid having to deal with all the echoes that get back to them and interfere with the echoes that only come back from the prey target (Jakobsen, Ratcliffe, and Surlykke 2013). The mechanism is comparable to zooming a flash light in and out when switching from a broad field of view to a narrow beam to look at something of interest in more detail.

Finally, current methods can still be pretty low tech and do not always need state-of-the-art sensing techniques. An easy way to understand how sound is produced is by manipulating the organs that are thought to play a part in it. The spikes and ridges of many a cricket or grasshopper have been filed off by many researchers in an attempt to study their function. Even the larynx of some frog species has been surgically altered in order to get a better understanding of their sound-production mechanisms (Gridi-Papp, Rand, and Ryan 2006). By far the simplest technique is moving an animal to a location with different physical properties to test how this affects its sound production. Floating frogs, for example, call at lower amplitude and with less complexity when placed in very shallow water (Halfwerk et al. 2017).

Measuring transmission properties

I still vividly remember how, years ago, I set out to study the sound-transmission properties in the cloud forests of Ecuador, bringing nothing more than a tape recorder, a microphone, and a bag full of balloons. I studied variation in bird song in relation to differences in habitat, particularly comparing locations higher up the Andean slope to locations at lower elevations (Halfwerk et al. 2016). One of my hypotheses was that the sounds of different bird species were adapted to the specific transmission properties of the two habitat types. At higher elevation, there is much more vegetation close to the ground, which particularly affects the transmission of higher frequency sounds (Dingle, Halfwerk, and Slabbekoorn 2008). I wanted to compare attenuation rates across frequencies of sounds transmitted through these two environments and needed a technique to do so. As it turns out, once inflated to their proper size, popping a balloon produces a consistent broadband sound, that can be recorded with a microphone at various distances. However, it became a struggle to move through the dense forest and to start my recorder at one location and inflate and pop the balloon in a consistent fashion in another location. A year later I returned with a set of speakers and broadcasted an artificial frequency sweep to achieve the same goal.

These sorts of transmission experiments have been carried out since the 1980s and have not changed much since (Richards and Wiley 1980). The main technological advancement is the use of wireless speakers, as this saves the researcher from either laying out hundreds of metres of cable between microphone, recorder, and loudspeaker, or having to run back and forth over difficult terrain as I had to do with the balloons. Nevertheless, these transmission experiments, either over long distances through continuous habitats or in small confined spaces such as nest holes or breeding burrows dug into the soil, have revealed how the environment influences efficient sound propagation (Muñoz and Penna 2016). For example, fast trills are usually more effective in open areas such as meadows and grass plains, as in forested areas they quickly degrade due to the many echoes. Likewise, due to the fact that sound of large wavelengths can bend around objects, high-frequency sound with smaller wavelengths more often bounce off and thereby attenuate faster in areas with many objects (Wiley and Richards 1978).

Measuring perceptual properties

Understanding how sounds are perceived requires first of all knowledge of a species' sensitivity. These sensitivities can be specified in the frequency domain, in the amplitude domain, or in both. Sensitivity measurements provide the lowest and highest possible sound that can be perceived by an individual, as well as the frequency at which optimal hearing is achieved. Typically, the most sensitive frequency tends to match the peak frequency of the most important sounds in an animal's life, be it the sound of its mates, prey, or predators.

Recording the electrophysiological activity directly from the peripheral auditory nerves is the most direct way of assessing an individual's sensitivity of sounds, for example, by measuring the spike activity of individual hair cells in the mammalian cochlea or individual neurons in the auditory centres in the brainstem in relation to the playback of different tones. However, these techniques are very difficult and expensive to use if you happen to study an animal that does not live in zoos. Another approach is using the auditory brainstem response (ABR) while playing sounds of different frequencies and intensities. This technique relies on an electro-encehphalogram (EEG) and records electrical signals by using electrodes placed on the skull. It is a very useful technique to take outdoors and is more and more applied to map the acoustic capacities of wild animals (Brittan-Powell et al. 2010).

Finally, laser-Doppler vibrometry can be applied to study specific resonance properties of the sound receiving morphological structures. The eardrum does not vibrate equally well across frequencies in response to an arriving sound pressure wave and thus determines partly what sounds arrive at the inner ear (Caldwell et al. 2014).

Environmental monitoring

The last two decades, off-the-shelf, stand-alone sound recorders became available. These recorders allow for the monitoring of sound-producing animals in a standardized and synchronized way at a cheap price and for a long period of time. Some of these recorders can be equipped with solar power as well as a satellite connection, and thus allow researchers to study remote areas that usually take days to reach, such as the High Arctic or Antarctica.

These recorders have been also been used to study the soundscape of a specific area in order to know how sounds change over time of day and season (Pijanowski et al. 2011). For example, using an array of twenty different sound recorders, I was able to document how sound produced by traffic transmits through a woodland area. I found that traffic noise transmission changed depending on the temperature, wind direction, and geological morphology of the area (Halfwerk et al. 2011). Furthermore, I could document that traffic noise overlapped with the song of birds throughout most of the breeding season. Finally, I could estimate the noise levels in specific frequency bands at breeding nest boxes occupied by great tits (*Parus major*); I found that noise levels were negatively related to breeding success, particularly those noises that overlapped spectrally with the bird's own song (Halfwerk et al. 2011).

The movement patterns and population size of animals are nowadays also monitored with these automated sound recorders. Specially designed and remote-controlled hydrophones are used to study marine mammals, and recorders with weather-proof ultrasonic microphones are used to study bats. When combining these setups with GPS information one can even pinpoint the location of an animal within a specific range if its sound is recorded simultaneously on two or more recording units.

In areas where animals are difficult to see, for example in the densely vegetated understory of forests or in murky waters, an array or grid of recorders can also be used to track movements. A study on banded wrens used a setup with twelve different microphones to examine how they vocalize in their territory and how they change their song types depending on where they are. The wrens use certain song types specifically when they are in the centre of their territory and other song types when they are close to the boundary of neighbouring territories (Mennill and Vehrencamp 2008).

Conclusion and future perspectives

I have given a short historical overview of the use of different approaches to investigate the ecology and evolution of animal sounds. The various techniques described above are by far not a full and complete overview of all that is currently possible. I have provided some insights into the possibilities that I think have greatly enhanced our understanding of animal sounds, mostly based on personal experience. Likewise, what I think will be important directions for future research and applications is more a matter of opinion than a careful consideration of all possible options.

Currently, many people around the globe possess a powerful sound recorder in the form of their smartphone. More and more people are providing input via citizen science projects, or by uploading their sound and video recordings to dedicated websites for nature observations or recordings, such as xeno-canto.org, observado.org, and fonozoo.

com. These recordings will turn out to be a valuable asset and perhaps allow for a fully automated approach to study the diversity of sounds throughout the world. Comparative analytical tools, perhaps borrowed from the field of genetics, could mine these databases for interesting patterns, which could then be followed up and may lead to the discovery of new species or new communicative functions.

Another major advancement may come from the various tools currently used for speech recognition. These digital tools may help us to identify species as well as individual animals automatically, and perhaps allow us to link their motivational state to specific acoustic features present in their songs and calls. Most of the breakthroughs in bioacoustics came from borrowing already existing techniques from very different fields. The future is not likely to be different.

References

Acevedo, M. A. and L. J. Villanueva-Rivera (2006). 'From the Field: Using Automated Digital Recording Systems as Effective Tools for the Monitoring of Birds and Amphibians'. *Wildlife Society Bulletin* 34 (1): 211–214.

Appel, H. M. and R. Cocroft (2014). 'Plants Respond to Leaf Vibrations Caused by Insect Herbivore Chewing'. *Oecologia* 175 (4): 1257–1266.

Baptista, L. F. (1977). 'Geographic Variation in Song and Dialects of Puget Sound White-Crowned Sparrow'. *Condor* 79 (3): 356–370.

Bostwick, K. S., D. O. Elias, A. Mason, and F. Montealegre-Z (2009). 'Resonating Feathers Produce Courtship Song'. *Proceedings of the Royal Society B: Biological Sciences* 277 (1683): 835–841.

Bradbury, J. W. and S. L. Vehrencamp (2011). *Principles of Animal Communication*. 2nd edition. Sunderland, MA: Sinauer Associates.

Brittan-Powell, E. F., J. Christensen-Dalsgaard, Y. Tang, C. Carr, and R. J. Dooling (2010). 'The Auditory Brainstem Response in Two Lizard Species'. *Journal of the Acoustical Society of America* 128 (2): 787–794.

Caldwell, M. S., N. Lee, K. M. Schrode, A. R. Johns, J. Christensen-Dalsgaard, and M. A. Bee (2014). 'Spatial Hearing in Cope's Gray Treefrog: II. Frequency-Dependent Directionality in the Amplitude and Phase of Tympanum Vibrations'. *Journal of Comparative Physiology A* 200 (4): 285–304.

Catchpole, C. K. (1980). 'Sexual Selection and the Evolution of Complex Songs among European Warblers of the Genus Acrocephalus'. *Behaviour* 74: 149–166.

Christensen-Dalsgaard, J. (2005). 'Directional Hearing in Nonmammalian Tetrapods'. In A. N. Popper and R. R. Fay (eds), *Sound Source Localization*, 67–123. New York: Springer.

Dingle, C., W. Halfwerk, and H. Slabbekoorn (2008). 'Habitat-Dependent Song Divergence at Subspecies Level in the Grey-Breasted Wood-Wren'. *Journal of Evolutionary Biology* 21 (4): 1079–1089. https://doi.org/10.1111/j.1420-9101.2008.01536.x.

Elemans, C., J. H. Rasmussen, C. T. Herbst, D. N. Düring, S. A. Zollinger, H. Brumm, K. Srivastava, N. Svane, M. Ding, and O. N. Larsen (2015). 'Universal Mechanisms of Sound

Production and Control in Birds and Mammals'. *Nature communications* 6 (8978). https://doi.org/10.1038/ncomms9978.

Fischer, J., R. Noser, and K. Hammerschmidt (2013). 'Bioacoustic Field Research: A Primer to Acoustic Analyses and Playback Experiments with Primates'. *American Journal of Primatology* 75 (7): 643–663.

Fitch, W. T. and M. D. Hauser (2003). 'Unpacking "honesty": Vertebrate Vocal Production and the Evolution of Acoustic Signals'. In A. M. Simmons, A. N. Popper, and R. R. Fay (eds), *Acoustic Communication*, 65–137. New York: Springer.

Flower, T. P., M. Gribble, and A. R. Ridley (2014). 'Deception by Flexible Alarm Mimicry in an African Bird'. *Science* 344 (6183): 513–516.

Gagliano, M., M. Grimonprez, M. Depczynski, and M. Renton (2017). 'Tuned In: Plant Roots Use Sound to Locate Water'. *Oecologia* 184 (1): 151–160.

Gerhardt, H. C. and F. Huber (2002). *Acoustic Communication in Insects and Anurans: Common Problems and Diverse Solutions*. Chicago: University of Chicago Press.

Gingras, B., M. Boeckle, C. Herbst, and W. Fitch (2013). 'Call Acoustics Reflect Body Size across Four Clades of Anurans'. *Journal of Zoology* 289 (2): 143–150.

Gridi-Papp, M., A. S. Rand, and M. J. Ryan (2006). 'Animal Communication: Complex Call Production in the Tungara Frog'. *Nature* 441 (7089): 38–38. https://doi.org/10.1038/441038a.

Griffin, D. R. (1944). 'Echolocation by Blind Men, Bats and Radar'. *Science* 100 (2609): 589–590.

Griffin, D. R. (1950). 'Measurements of the Ultrasonic Cries of Bats'. *Journal of the Acoustical Society of America* 22 (2): 247–255.

Halfwerk, W. (2012). 'Tango to Traffic: A Field Study into Consequences of Noisy Urban Conditions for Acoustic Courtship Interactions in Birds'. PhD dissertation, Leiden University, Leiden.

Halfwerk, W., L. J. M. Holleman, C. M. Lessells, and H. Slabbekoorn (2011). 'Negative Impact of Traffic Noise on Avian Reproductive Success'. *Journal of Applied Ecology* 48 (1): 210–219.

Halfwerk, W., P. Jones, R. Taylor, M. Ryan, and R. Page (2014). 'Risky Ripples Allow Bats and Frogs to Eavesdrop on a Multisensory Sexual Display'. *Science* 343 (6169): 413–416.

Halfwerk, W., C. Dingle, D. M. Brinkhuizen, J. W. Poelstra, J. Komdeur, and H. Slabbekoorn (2016). 'Sharp Acoustic Boundaries across an Altitudinal Avian Hybrid Zone Despite Asymmetric Introgression'. *Journal of Evolutionary Biology* 29 (7): 1356–1367.

Halfwerk, W., J. A. Smit, H. Loning, A. M. Lea, I. Geipel, J. Ellers, and M. J. Ryan (2017). 'Environmental Conditions Limit Attractiveness of a Complex Sexual Signal in the Túngara Frog'. *Nature communications* 8 (1): 1891.

Halfwerk, W., M. Blaas, L. Kramer, N. Hijner, P. A. Trillo, X. E. Bernal, R. A. Page, S. Goutte, M. J. Ryan, and J. Ellers (2019). 'Adaptive Changes in Sexual Signalling in Response to Urbanization'. *Nature Ecology and Evolution* 3 (3): 374.

Hollén, L. I., M. B. Bell, and A. N. Radford (2008). 'Cooperative Sentinel Calling? Foragers Gain Increased Biomass Intake'. *Current Biology* 18 (8): 576–579.

Hoy, R. R. and D. Robert (1996). 'Tympanal Hearing in Insects'. *Annual Review of Entomology* 41 (1): 433–450.

Jakobsen, L., J. M. Ratcliffe, and A. Surlykke (2013). 'Convergent Acoustic Field of View in Echolocating Bats'. *Nature* 493 (7430): 93–96. https://doi.org/10.1038/nature11664.

Kroodsma, D. E. (2004). 'Diversity and Plasticity of Bird Song'. In P. Marler and H. Slabbekoorn (eds), *Nature's Music: The Science of Bird Song*, 108–130. London: Elsevier.

Laiolo, P. (2010). 'The Emerging Significance of Bioacoustics in Animal Species Conservation'. *Biological Conservation* 143 (7): 1635–1645.

Lewis, E. R., P. M. Narins, K. A. Cortopassi, W. M. Yamada, E. H. Poinar, S. W. Moore, and X. L. Yu (2001). 'Do Male White-Lipped Frogs Use Seismic Signals for Intraspecific Communication?'. *American Zoologist* 41 (5): 1185–1199. https://doi.org/10.1093/icb/41.5.1185.

Manley, G. A. and R. R. Fay (2013). *Evolution of the Vertebrate Auditory System*, vol. 22. New York: Springer.

Marler, P. (1955). 'Characteristics of Some Animal Calls'. *Nature* 176 (4470): 6.

Marler, P. and M. Tamura (1962). 'Song "Dialects" in Three Populations of White-Crowned Sparrows'. *Condor* 64 (5): 368–377.

Marler, P. and H. Slabbekoorn (2004). *Nature's Music: The Science of Birdsong*. San Diego, CA: Elsevier.

Mennill, D. J. and S. L. Vehrencamp (2008). 'Context-Dependent Functions of Avian Duets Revealed by Microphone-Array Recordings and Multispeaker Playback'. *Current Biology* 18 (17): 1314–1319.

Montgomery, J. C., A. Jeffs, S. D. Simpson, M. Meekan, and C. Tindle (2006). 'Sound as an Orientation Cue for the Pelagic Larvae of Reef Fishes and Decapod Crustaceans'. *Advances in Marine Biology* 51: 143–196.

Muñoz, M. I. and M. Penna (2016). 'Extended Amplification of Acoustic Signals by Amphibian Burrows'. *Journal of Comparative Physiology A* 202 (7): 473–487.

Nadrowski, B., T. Effertz, P. R. Senthilan, and M. C. Göpfert (2011). 'Antennal Hearing in Insects: New Findings, New Questions'. *Hearing Research* 273 (1–2): 7–13.

Nowicki, S. and W. A. Searcy (2014). 'The Evolution of Vocal Learning'. *Current Opinion in Neurobiology* 28: 48–53.

Obrist, M. K., R. Boesch, and P. F. Flückiger (2004). 'Variability in Echolocation Call Design of 26 Swiss Bat Species: Consequences, Limits and Options for Automated Field Identification with a Synergetic Pattern Recognition Approach'. *Mammalia mamm* 68 (4): 307–322.

Ohms, V. R., P. C. Snelderwaard, C. Ten Cate, and G. J. L. Beckers (2010). 'Vocal Tract Articulation in Zebra Finches'. *PLoS ONE* 5 (7). https://doi.org/e1192310.1371/journal.pone.0011923.

Pijanowski, B. C., L. J. Villanueva-Rivera, S. L. Dumyahn, A. Farina, B. L. Krause, B. M. Napoletano, S. H. Gage, and N. Pieretti (2011). 'Soundscape Ecology: The Science of Sound in the Landscape'. *BioScience* 61 (3): 203–216.

Podos, J. (2001). 'Correlated Evolution of Morphology and Vocal Signal Structure in Darwin's Finches'. *Nature* 409 (6817): 185–188.

Ralph, C. J., S. Droege, and J. R. Sauer (1995). 'Managing and Monitoring Birds Using Point Counts: Standards and Applications'. In C. J. Ralph, J. R. Sauer, S. Droege (eds), *Monitoring Bird Populations by Point Counts*, 161–168, 149. Gen. Tech. Rep. PSW-GTR-149. Albany, CA: US Department of Agriculture, Forest Service, Pacific Southwest Research Station.

Rand, A. S. and R. Dudley (1993). 'Frogs in Helium: The Anuran Vocal Sac Is Not a Cavity Resonator'. *Physiological Zoology* 66 (5): 793–806.

Richards, D. G. and R. H. Wiley (1980). 'Reverberations and Amplitude Fluctuations in the Propagation of Sound in a Forest: Implications for Animal Communication'. *American Naturalist* 115 (3): 381–399.

Slabbekoorn, H. (2004). 'Singing in the Wild: The Ecology of Birdsong'. In P. Marler and H. Slabbekoorn (eds), *Nature's Music: The Science of Birdsong*, 178–205. San Diago, CA: Elsevier.

Smit, J. A., H. Loning, M. J. Ryan, and W. Halfwerk (2019). 'Environmental Constraints on Size-Dependent Signaling Affects Mating and Rival Interactions'. *Behavioral Ecology* 30 (3): 724–732.

Stevens, M. (2013). *Sensory Ecology, Behaviour, and Evolution*. 1st edition. Oxford: Oxford University Press.

Struhsaker, T. T. (1967). 'Social Structure among Vervet Monkeys (*Cercopithecus aethiops*)'. *Behaviour* 29 (2–4): 83–121.

Suthers, R. A. and S. A. Zollinger (2004). 'Producing Song: The Vocal Apparatus'. *Annals of the New York Academy of Sciences* 1016 (1): 109–129.

Thorpe, W. (1958). 'Further Studies on the Process of Song Learning in the Chaffinch (*Fringilla coelebs gengleri*)'. *Nature* 182 (4635): 554.

West, M. J. and A. P. King (1990). 'Mozart's Starling'. *American Scientist* 78 (2): 106–114.

Wiley, R. H. and D. G. Richards (1978). 'Physical Constraints on Acoustic Communication in Atmosphere: Implications for Evolution of Animal Vocalizations'. *Behavioral Ecology and Sociobiology* 3 (1): 69–94.

5
Hearing With: Researching the Histories of Sonic Encounter

James G. Mansell

I
sit on
this hill
beneath the
shade of this
gnarled, tall,
wise, old oak
looking out
around me at the
amazing beauty that
surrounds me. As far as
the horizon everything brilliant-
ly magnificent. Suddenly, there is a
hush, is it me? Am I imagining it? No, not
even an undertone, utter silence. The birds are
mute, the animals dumb, the wind has ceased in soft sibil-
ation, the clatter and grind of the tractor has been stifled, the
chuckle of the stream is frozen, the world seems to be quiescent, waiting
with bated breath. In this silent paradise I think of what I have to
return to, compared with this it seems like aural purgatory, the noises that
pollute our envir-
onment, the pandem-
onium and hullabaloo
of our modern word.
The rasping and
grating of our

> labour saving devices
> the instant chatter
> of the television,
> raucous pop music
> played too loud,
> the drone of jets
> as they pass over
> head, the tinny
> sound of car
> engines rushing everywhere
> Even as I think this, heaven is
> pierced by the squeal of children
> at play, the yelp of a dog, the dis-
> tant peal of a church bell and the irritat-
> ing rasp of a saw. The world is revived. Gone are
> the Angels of Silence and back is the
> demon
> NOISE.
> (Katherine Londesbrough, age fourteen years)

This poem was published in a collection entitled *Children on Noise* following a literary competition held in the schools of Darlington, North-East England, in 1978. Prizes of £100, as well as consolation T-shirts, were available to the children who produced the best stories and poems about noise in their town. The competition was part of the Darlington Quiet Town Experiment of 1976–1978 run by the UK government's Noise Advisory Council. The experiment sought to 'determine if noise levels can be reduced by creating an "awareness" of noise by publicity and education' (Darlington Borough Council n.d.). The literary competition was one of many activities undertaken in the town over the two years of the experiment designed to encourage residents – adults and children alike – to listen to their town, and to themselves. The Royal Automobile Club (RAC) erected road signs that read: 'Darlington is a Quiet Town. Please drive quietly' (Figure 5.1). Posters, leaflets, and social activities (such as quiet bingo) were circulated and organized. The people of Darlington were trained to hear noise and to enact quietness. Though unusual in its format, the Darlington Quiet Town Experiment is typical of the strategic work that goes on around and through everyday sound. How we hear is socially shaped. It has a history.

At midday on 21 August 2017, a crowd gathered on the streets of Westminster, central London, to hear to the final bongs of 'Big Ben' – the 13-ton bell atop what is now known as the Elizabeth Tower at the UK Houses of Parliament – before a four-year cessation to allow restoration work to take place. Newspapers reported that some in the crowd fought back tears, including at least one Member of Parliament who had assembled there with colleagues, heads bowed to reverently mark the occasion. A row had earlier erupted in the British media about why it was necessary to silence Big Ben for so long. Critics, including some MPs, pointed out that the bell had tolled through most of the Second World War and that stopping it for an extended period was an indictment of national ingenuity and a threat to Britain's place in the world. The sound of Big Ben was presented by these critics as

Figure 5.1 Photograph of an RAC 'Quiet' road sign in Darlington. *Source:* Noise Advisory Council 1981: 23.

inextricably bound up with Britain, a heartbeat almost. For some it was auditorily symbolic of national self-determination in the context of negotiations to leave the European Union in 2019. Even academic commentators were drawn into the frenzy, with one writing that the silencing of 'an essential component of the landscape of London, and of the pantheon of national icons that present "Britishness" to the rest of the world' was the result of 'a failure of management in the heart of Westminster' (Clapson 2017). The British Broadcasting Corporation (BBC), which ritually broadcasts the live sounds of Big Ben on national radio in the UK, faced a decision about whether to maintain a live relay of an alternative bell (Nottingham's Council House bell was considered) or use a non-live recording of Big Ben. It eventually opted for the latter.

How did the sound of Big Ben become such a powerful symbol of the British nation? Was it by sheer virtue of the bell's geography, placed as it was in 1858 at the heart of imperial political power? The answer is that Big Ben's affective power to bind hearers together is not innate to its stature, geography, or even a straightforward result of its place in broadcasting schedules but was, rather, actively produced more deliberately than that by campaigns to direct how it was heard. Prominent among these was the 'Big Ben Silent Minute', sometimes known just as the 'Big Ben Minute', a campaign launched in 1940 by an organization known as the Big Ben Council with close ties to the Conservative Party and the Church of England, which promoted a daily minute of silence and silent prayer throughout the British Empire during and in time with the chimes of Big Ben broadcast on BBC radio in the run up to the news at 9.00 p.m. The full 9.00 p.m. chimes of Big Ben were broadcast live, taking around a minute from beginning to end (Dakers 1943). The minute was originally intended to provide a daily moment of hopefulness about Britain's prospects in the Second World War and to reconnect separated loved ones in a minute of synchronized remembrance. Books of prayer and positive thoughts, written specifically with the rhythm of Big Ben in mind, were published (such as the 'Golden Thoughts' booklet shown in Figure 5.2). The Big Ben Minute remained a feature of BBC radio until 1960, when the 9.00 p.m. news bulletin was moved to 10.00 p.m. and the minute cut amid angry controversy (Briggs 1995: 325–340).

Like the Darlington Quiet Town Experiment, the Big Ben Silent Minute was designed to direct hearing attention and to produce an affective and meaningful relationship between sound and hearer, in this instance, among other things, providing an auditory focal point for the nation and its empire and the sonic conditions needed for it to be sanctified as such. This is the stuff of sound history. Research in this field seeks to establish not only what was audible in the past but also how and why that audibility was produced: how and why sounds – from the chirp of a bird to the roar of a motorbike, from recorded music to the tone and accent of voice in daily speech – were shaped and given meaning, made valuable or denigrated, and brought to attention or left in the background. Sound historians do this not simply for the sake of adding sensory context to our understanding of the past but because, they argue, what and how we hear shapes subjectivity and community in important ways. Sounds are socially active, producing us as subjects and drawing us together as sensing collectives and, to use Tom Western's phrase, 'securing the aural border' (Western 2015: 77–97).

This chapter sets out a sonic-historical methodology drawing on existing work in the field of historical sound studies that is attentive to the conscious shaping of auditory perception in the past.[1] It proposes two central principles. The first is that historians could think of what they do as *hearing with* rather than *listening to* the past. This is an approach to historical source material which would seek to historicize sound's role in shaping subjects and naturalizing relations of power, remaining alert to a range of auditory subject positions in the past, and acknowledging the listening ear of the historian in the act of hearing with. I use the term hearing rather than listening deliberately to emphasize that, beyond moments of listening and campaigns to direct listening attention in the past, historical ways of hearing have evolved over time in which sounds have gained

Figure 5.2 Front cover of a booklet of positive thoughts to be recited silently in time with the chimes of Big Ben at 9.00 p.m. *Source:* Junior 1941.

common-sense meanings and associations for which listening is no longer consciously required and which have unequal social effects (see also Mansell 2018: 343–352). A way of hearing is more than an act of listening: it takes shape in text, image, and social discourse as much as in sound. It impacts beyond the auditory in the ways that subjects think, feel, and manage their bodies. Ways of hearing produce the possibility of listening. The second methodological principle is a focus on what I term here the production of *sonic encounter*, a socially shaped and culturally specific affective relationship between hearer and heard. The sonic encounter is the meeting point of 'soundscapes' (for my purposes, a useful shorthand for the sounds that surround us in everyday life) and 'soundselves' (as theorized by Tom Rice [2003], listening subjects whose sense of self is shaped by sound): it is the affective field of feeling and sense-making which those who wish to produce ways of hearing seek to shape.[2]

Hearing with

When I tell fellow historians what I research, they usually say something to the effect of: that must be interesting, but aren't you reliant on sound recordings being available, doesn't that mean you can only write about the recent past, there probably *aren't* many historical sound recordings around, *are there*? Because of a deeply ingrained tendency to privilege reading and writing in historical scholarship, when they hear the word 'sound' most historians think, probably *media studies*, definitely *somebody else's* business. The assumption is that listening to the past is a distinctively poor relation to reading from its surviving textual source materials. Matthew Rubery describes a similar response to his research on the history of audio books: listening to rather than reading a book is considered by many to be distinctly second rate (Rubery 2016). These kinds of prejudices help to explain why relatively little of the historical work in sound studies is undertaken by historians working in university history departments. Many, like myself, are based in other disciplines, including media, cultural, and communications studies. This is an unfortunate misunderstanding, since the methodological premise of most scholarship in historical sound studies is textual rather than auditory, and deliberately so.

Mark M. Smith, an early exponent of sound history with his *Listening to Nineteenth-Century America* (2001), argues that by rereading historical source material for evidence of encounters with sound we can understand both the acoustic environments of the past and how those environments were perceived and made meaningful in cultural-historical contexts.[3] Drawing on a sensorially attuned social history tradition with origins in the Annales school and culminating in Alain Corbin's *Village Bells: Sound and Meaning in the Nineteenth-Century French Countryside* (1998), Smith helped to establish a new subfield of sound history.[4] But the 'Listening to' in the title of Smith's book refers primarily to the listening done by historical subjects rather than by the historian. Smith, like others who have attended to past sounds, is rightly sceptical about the historian's ability to gain direct and uncomplicated access to how the past sounded precisely because we do not hear in the same way today, culturally, as those in the past did. Smith's realist-constructivist approach is typical of sound historical scholarship. In her history of architectural acoustics in early twentieth-century America, Emily Thompson defines the 'soundscape of modernity' as 'simultaneously a physical environment and a way of perceiving that environment; it is both a world and a culture constructed to make sense of that world' (Thompson 2004: 1). In textual sources, Smith argues, we find written record of how sounds were experienced and interpreted, as well as the sound environments themselves. Even where we have access to recorded sound in the form of radio broadcasts, field recordings, or film and television soundtracks, Smith and others argue that we need a wider body of source material to understand the historicity of these sounds. This approach is put to good use, for example, in Carolyn Birdsall's (2012) research on Nazi soundscapes, which uses radio archive materials alongside oral history and written archives to understand historical hearers as 'earwitnesses' to fascism. Smith has written of his hope that, rather than a self-contained field of sound history, listening through the ears of historical subjects will become more of a 'habit' for historical scholarship as a whole, a 'methodological, epistemological, and even ontological embeddedness – a way of examining the past that becomes second nature so that evidence is read, consciously and even subconsciously, for tidbits of the acoustic, smatterings of the auditory, gestures of silence, noise, listening, and sound' (Smith 2014: 13–14).

The listening involved in historical sound studies is a reconstructed listening, then, an excavation of how past subjects paid attention to sound. As Daniel Morat has noted, however, 'listening to' is far from a settled historical methodology: 'You can find', as Morat observes, 'different notions of "sound history," "aural history," "auditory history," "history of hearing," and "history of listening"', deployed across the field (Morat 2014: 3). Here, I wish to propose that in order to avoid the misunderstandings to which the 'listening to' label gives rise, as well as some of its limitations, we might more usefully describe what the sound historian does, or could do, as *hearing with*. Listening has been closely associated with the production of knowledge in historical sound studies. Emily Thompson focuses her attention on the expert culture of listening developed by architectural acousticians. Jonathan Sterne's (2003) research on the history of sound recording and reproduction technologies identifies a range of expert listening practices evolving over the course of the nineteenth century as the basis of an 'Ensoniment' (an auditory equivalent of the Enlightenment) in which expert auditory ways of knowing, from telegraphy to phonography, formed the basis for modern knowledge culture. Though not experts in the same sense, Kassandra Hartford argues that soldiers developed 'attentive listening practices typically associated with music and musicians' on the Western Front during the Great War because, 'in a war fought largely in trenches and tunnels, in the dark or in a dugout, visual observation was limited' (Hartford 2017: 98). This focus on the historical listener has tended to privilege a particular kind of audition, that done by a conscious listener, seeking knowledge, with the ability to make their listening matter in the historical record. In other words, it has limited the range of auditory subject positions that historians have attended to. 'Listening to' has been a very particular kind of 'listening with'.

A *hearing-with* approach would focus not so much on records of this conscious listening as a route to insight into the historical soundscape, but rather on the production and sustaining of ways of hearing which operated through atmospheres of everyday sounds and had different kinds of auditory impacts on hearers. These ways of hearing were produced in the past by the organization of audibility and the directing of hearing. The organizing and directing was undertaken consciously, as in the case of the Darlington Quiet Town Experiment and the Big Ben Silent Minute. However, those involved in such campaigns were not necessarily aware of the ways in which their attempts to shape everyday hearing gave audible form to cultures of class and gender (as in the case of quiet) or of an imperial Britain (in the case of the Big Ben Minute) because those cultural dispositions were not always recognized as such by those who intervened in sound. The historian's role, then, is to return to these historical ways of hearing, to understand how they were assembled, and to assess what effects they had in the shaping of social life. The organizing and directing that goes into producing a way of hearing can be found in the historical record, even if it was not consciously listened to in the past.

Western's approach to the analysis of 'audio nationalism' illustrates the hearing-with approach I have in mind. He shows, in the context of the BBC's post-war broadcasting of folk music and ethnographic field recordings, 'how sound was at once nationalized and nationalizing' on the radio. He argues that 'certain sounds were selected to represent national qualities, used to construct national character, and delimit the nation' (Western 2015: 88). There was listening work involved here, since the field recordists who gathered materials for the radio programmes analysed by Western were the ones selecting appropriate music and sound to embody the nation and its bordered distinctiveness. Yet there was also the bringing into being of an aural truth of the nation, the production of a way of hearing music and sound as British that happened in the presentation of these sounds on the radio and their framing in print contexts such as the *Radio Times*: the BBC was 'training people how to listen', according to Western (2015:89). A hearing-with approach hears with this training. Western's analysis can be extended to the Big Ben Silent Minute whose aim was similarly to enter into the daily routines of radio listeners and produce an active auditory engagement with the national community.

A hearing-with approach does not adopt the listening position of the powerful but rather hears through it what only a historian can hear – the historicity of truths as they have taken sonic form. But a hearing-with approach also demands that the historian hears with both those who shaped ways of hearing and with those who were subject to their influence. Take the example of the sonic category of quiet. Historians have now dedicated a good deal of attention to noise, understanding why and how past societies made sense of which sounds to exclude as meaningless or harmful.[5] This work has focused on what Jennifer Lynn Stoever describes as the 'listening ear' of the noise abatement campaigner – a close historical attention to the way in which noise was defined and its eradication justified (Stoever 2016: 7). The same attention has not been given to quiet, even though most anti-noise campaigns, such as the Darlington Quiet Town Experiment, have been designed to produce it. In the 1930s, the Anti-Noise League published a magazine called *Quiet* and in the 1950s the successor Noise Abatement Society published one called *Quiet, Please*, underscoring the kind of sound they wished to produce. In twentieth-century anti-noise campaigns, noise was listened to critically, but quiet was quietly produced as a category of good sonic conduct. Far from the neutral category it may seem, quiet contains and activates social relations of power. In the absence of noise there is not silence but a kind of sound that contains the normative values of the society that produced it. In the 1930s, quiet was described by the Anti-Noise League's leader, Lord Horder, as 'acoustic-civilization'.[6] As the other to noise, 'acoustic-civilization' was assumed to be readily understood; it was comfort, peace, and privacy. It was auditory common sense. Yet upon close inspection of the surviving historical source materials it is clear that this 'civilized' quiet was not the public good it was made out to be: it was based on a middle-class auditory habitus and was specifically *for* those whose work required concentration, mainly professional men. Still, it was promoted *to everyone* as universally good behaviour.

The noise abatement advocates of the Anti-Noise League did not listen to the acoustics of gendered difference, but their way of hearing produced it. The close-up from a 1935 advert for the Underwood Noiseless Typewriter in Figure 5.3, which was advertised in an

Anti-Noise League exhibition handbook, shows quiet in the context of the 1930s office: a male office worker leans forward and listens to his visitor without distraction from his nearby female typist. Quiet here was an atmosphere that reproduced patriarchal 'civilization' in sound. The female typist is identified as a source of noise. The male office worker is an active listener. His female visitor, to whom he listens, conforms to the ideal of the softly spoken woman (otherwise the quiet afforded by the typewriter might not be needed). As Marie Thompson has argued, 'women have often been represented as "naturally" noisy in comparison to their male counterparts; within popular consciousness,' she goes on, 'they are imagined to be more talkative, choosing to discuss the trivialities of life and surrounding themselves with a noisy, meaningless babble' (Thompson 2013: 300). The field of quiet was gendered, with women burdened with greater responsibility for producing quiet than men. Men's capacity to listen was simultaneously reproduced in such contexts. Cadbury's silent theatre box (Figure 5.4), introduced in 1930, was a box of chocolates designed to offer quiet refreshment at the theatre and provides further evidence of this gendering of quiet. Cadbury said that it was intended to 'enable the rapt or bored playgoer to take her nourishment without distracting the actors by rustling the packing material'. A further note beside the editorial added that the silent theater box would also allow the playgoer to avoid 'arousing the indignation of one's neighbours' in the theater. A closer inspection of noise abatement archives therefore reveals the coming into being of a field of quiet produced by anti-noise campaigning or everyday antipathy to noise which shaped everyday behaviours

Figure 5.3 Section of an advertisement for the Underwood Noiseless Typewriter.
Source: Anti-Noise League 1935: 75.

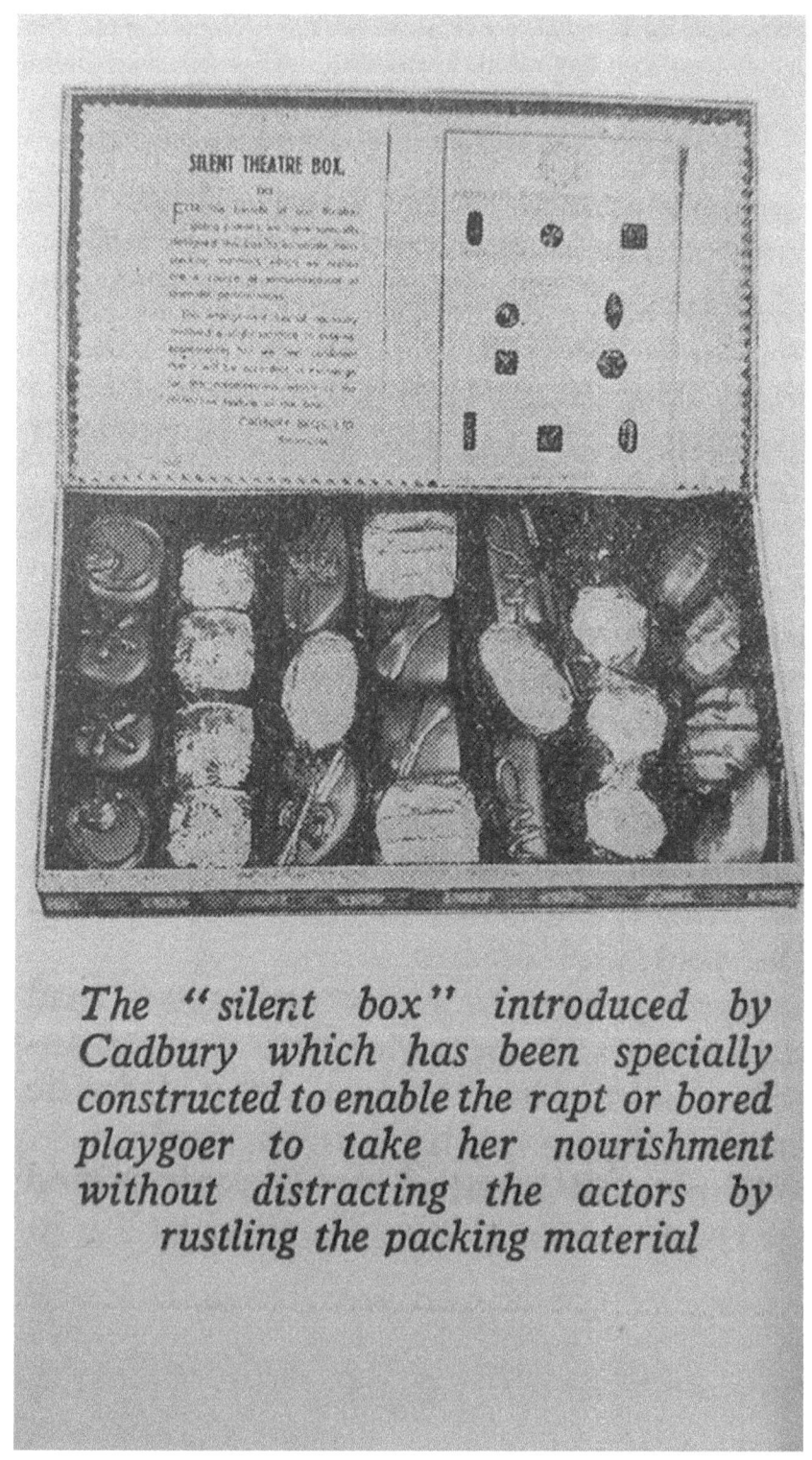

Figure 5.4 Cadbury's 'Silent Theatre Box' editorial. *Source: Advertiser's Weekly*, 12 December 1930: 423.

and gave sonic form to time-bound ideas of 'civilization'. Quietness is a way of hearing. Of course, it produced not only class and gendered subjectivities, but, as other kinds of source material would show more clearly, the sonic contours of race, ethnicity, religion, and sexuality. The colonial archive, for example, might help us to understand the quietness in Figure 5.3 as a dimension of whiteness.[7]

The way we hear quiet also changes historically. From 'acoustic civilization' in the 1930s quiet has become financialized 'natural capital' in the twenty-first century. A 2011 report produced for the UK government explained that:

> Quiet and 'quiet areas' contribute to economic welfare through the generation of human well-being and prevention of illness and ecosystem decline that inflict costs on society. However, many of the benefits of quiet are not directly priced in the market and therefore risk being under-valued with resulting degradation or total loss of 'quiet' or 'relatively quiet' areas.
>
> (URS Scott Wilson 2011)

Promoting its Quiet Mark consumer goods, such as the quiet Magimix kettle and noise-cancelling Sennheiser headphones available at the upmarket department store John Lewis, or even holidays on which hotels guarantee quiet to the discerning holidaymaker in search of creative recuperation, today's Noise Abatement Society explains that 'recognizing that ultimately all wealth derives from natural systems, we need to find practical ways of expressing the value of ecosystem services in all decisions which risk reducing natural capital. This includes soundscape quality' (see Quiet Mark 2016). What might be described as a neoliberal way of hearing is at work here, which will one day be the subject of historical research.

The sonic encounter

Any restatement of sonic-historical methodology must consider its position in relation to the methodological debate currently taking place in sound studies on the question of what has been called the ontological challenge. This challenge has been usefully set out by Brian Kane as a questioning of 'the relevance of research into auditory culture, audile techniques, and the technological mediation of sound in favor of universals concerning the nature of sound, the body, and media' (Kane 2015: 3). Spearheaded by philosopher of music Christoph Cox (2011) and cultural theorist of sound Steve Goodman (2009), the ontological turn in sound studies has sought to shift attention away from meaning and cultural experience and towards the materiality of sound and its precognitive power to affect the human body. In Goodman's case, which is closer than Cox's to the interests of most historians, the argument is that during times of war, in particular, what sound means matters much less than what sound does in terms of its vibrational impact on bodies. Rather than cultural ways of hearing, Goodman argues that we should be concentrating our attention on what he calls the 'politics of frequency' – by which he means specifically the bodily impact of low-frequency sounds in war (Goodman 2009: xv) – or as Cox puts it, ask of our evidence 'not what it means or represents, but what it does, how it operates, what changes it effectuates' (Cox 2011: 157). Connected to a wider turn to materiality, object-oriented ontology, and affective atmospheres in the humanities and social sciences, Cox's and Goodman's critique of auditory culture approaches, such as the one I have set out in the previous section, demands a reply.

It has readily found one in several powerful rebuttals. Marie Thompson and Annie Goh, most notably, have put forward persuasive critiques of the 'origin myth' of sonic nature operating in Cox's work. Thompson argues that, while Cox claims to have found evidence of the nature of sonic flux within the best sound art works, 'his pursuit of the "nature of sound" risks uncritically naturalizing what is ultimately a specific onto-epistemology of sound' (Thompson 2017b: 270). This onto-epistemology draws on a lineage of sound art extending from John Cage which Thompson identifies as being 'entangled with, amongst other things, histories of whiteness and coloniality'. She argues that, 'ontologies bear the traces of their historical moment even when those ontologies "withdraw" from mediation' (266–282). Thompson does not reject sonic ontologies, indeed her book *Beyond Unwanted Sound* (2017a) is an argument in favour of understanding noise as an active, affective force rather than a moral category of bad sound. She suggests, instead, that, 'situating rather than simply dismissing sonic ontologies enables us to ask how "the nature of the sonic" is determined – what grounds the sonic ground – while remaining open to how it might be heard otherwise' (Thompson 2017b: 278). Thompson's is thus an approach that advocates for the necessity of sound history. It is via attention to what here I have called ways of hearing that claims to the 'nature' of sound can be unmasked as culturally specific and invested in politics and society. Goh's argument is even more explicit in its advocacy of an historical approach to sound. Critiquing what she terms a dominant 'sonic naturalism' in sound studies' preoccupation with auditory knowledge, she proposes instead a principle of 'sounding situated knowledges' in the past as a way of interrogating the conditions of sonic knowing (Goh 2017: 283–304). In this section, I intend to develop an approach equal to the challenge of researching what Jim Sykes has described as 'culturally-constituted ontologies of sound and listening that structure social relations' (Sykes 2018: 56).

The ontological challenge of Cox and Goodman raises the question of whether historians of sound can credibly adopt the concept of affect or affective atmosphere to explain what sound did and meant in the past. Above, I have hinted, though not yet explicitly claimed, that ways of hearing are in part affective: they harness sound's power to affect the body and produce feeling as a way of actualizing gender and national belonging. If the affect and the atmospheres that it produces is pre-cultural as Goodman and other affect theorists maintain, can there be a history of sonic affect? If, as Kane explains, 'Goodman discourages accounts of the sonic in terms of conscious hearing or listening in favor of an unconscious, affective, intensive account of sound as material impact', where does that leave the historian of sound? Kane's answer is that studies in auditory culture have never in fact been 'simply studies in "representation" or "signification" without consideration of the body. Rather,' he goes on, 'scholars in auditory culture seek to demonstrate the successions and relays between cognition and affect, or, speaking broadly, between the mind and the body. As listeners acquire new skills,' he argues, 'much of the cognitive effort involved in the initial training is offloaded onto the body. At the same time, bodily capacities constitute both the basis upon which training occurs and the ground for potential future cultivation' (Kane 2015: 8). Kane's theory of auditory training is precisely what is at stake in the production of ways of hearing.

Others have mounted similar defenses of a method for researching sonic affect that retains a place for cultural analysis. Marie Thompson argues in *Beyond Unwanted Sound* that noise should be thought of as characterized by affectivity rather than negativity, 'a perturbing force-relation that, for better or worse, induces a change'. But, she goes on, 'noise (and affect) is frequently entangled with signifying registers' (Thompson 2017a: 42–48). Anahid Kassabian proposes a theory of 'distributed subjectivity' to explain the role that the affective encounter with sound plays in generating identity. Distributed subjectivity is 'a nonindividual subjectivity, a field, but a field over which power is distributed unevenly and unpredictably, over which differences are not only possible but required, and across which information flows, leading to affective responses' (Kassabian 2013: xxv). The kind of information that produces the most powerful affective response, for Kassabian, is sound. She notes that 'identity is one of the formations that are left behind after affect does its work' (xxvii). She goes on to argue that although 'identities seem static and positional, they are anything but, and they are constituted microsecond by microsecond according to affects that are in motion'. She uses the example of the singing of a national anthem to illustrate her point.

> For many people – though certainly not all – their national anthem invokes pride and community, a warm feeling of belonging. Each singing is an affective event, creating a wave of feeling that flows across a group of any size, from one to thousands. Affect like that leaves behind residue that appears to produce a static identity. But the very fact that it needs to be done over and over suggests that something rather different is happening.
>
> (Kassabian 2013: xxviii)

Sound, in the example of national anthem singing, is maintaining individual and collective identity via its ritual performance, according to Kassabian. Sound's power to affect is being deployed in the project of nation-building. Kane, Thompson, and Kassabian's theorization of the affective *and* meaningful power of sound lays the ground for my suggestion that historians might go in search of the production of *sonic encounters* in the past.

Affect can sometimes appear to be a rather abstract notion. Certainly, a lot of historians would perceive it this way. Under other names, however, it has been active in historical understandings of what sound is and does. Late nineteenth- and early twentieth-century discussions of noise's tendency to produce nervousness or neurasthenia was the language of its day to describe the bodily impact of sound and its ability to generate bad feelings. During the Second World War, government health authorities were alert to the extent to which bomb sounds could produce fear responses in civilian populations and embarked on propaganda campaigns to encourage people to hear bombs without fear or not to hear them at all (by using ear plugs).[8] Sound's atmospheric materiality is what makes it such an effective medium for the circulation of values and production of behaviours. The Big Ben Silent Minute is an example of the production of a sonic encounter. The sound of Big Ben tolling is not in any sense intrinsically British but was produced as such by interventions in the affective field, what might otherwise be described as everyday atmospheres of sound. The feeling of national pride or sense of collective endeavour that is associated with Big Ben is an affective response produced culturally. It is produced by broadcasting, pamphlets, and ritual, materials which produce the auditory 'training' that Kane identifies.

The Darlington Quiet Town Experiment is an example of the kind of work that goes into producing a sonic encounter, in this instance, with everyday urban sounds. This two-year experiment was designed to solve the 'human factor' in what was now firmly established as the 'noise problem' in public policy planning. While other areas of the Noise Advisory Council's work in the 1960s and 1970s focused on scientific investigation of the technological sources and medical effects of noise, the Darlington experiment was primarily an educational mission to reshape everyday auditory behaviour in urban space. It aimed to help the residents of the town realize that 'much unnecessary noise was made by people who did not realise that they were causing a nuisance to others' (Noise Advisory Council 1981: 4). Children were targeted for involvement not because they were perceived as the primary creators of noise, but because it was thought that, through education, they could be transformed into a noise-sensitive generation. The posters, pamphlets, and social activities undertaken as part of the Darlington experiment were designed to generate a specific kind of encounter between hearers and specific kinds of sounds, heard as noise.

In the background of the Darlington experiment was an assumption that noise was a symptom of anti-social behaviour and that fixing the noise problem was a route to producing a better-behaved and more socially harmonious town. Although some familiar technological culprits were identified in publicity materials as especially noisy, such as motorbikes, in general noise was defined as time and context dependent, such as playing the radio too loudly in the morning or leaving too noisily from the pub late at night. The experiment was designed to protect domestic, private life in the face of intrusions from the public realm, but what was being produced was very much a privatized quietness, a quiet to be enjoyed from the comforts of one's home. As in the context of the 1930s, gendered labour underpinned the production of this quietness. In promotional materials circulated as part of the Darlington experiment, the creators of noise are largely identified as men who do not appreciate the acoustic needs of women and children for domestic and educational life. In these publicity materials, it was women who urged their fellow townsfolk to be quiet (as in the cover of the leaflet announcing the experiment in Figure 5.5 and the logo used on promotional materials in Figure 5.6). In a leaflet which asked, 'Do you have a noisy gnome in your home?' (Figure 5.7), women were asked to reflect on whether the man of their house 'does odd jobs at odd times', 'likes music too loud', 'never closes doors quietly', or 'leaves his dog uncontrolled'. This leaflet had its desired effect in the story submitted by one fifteen-year-old girl to the literary competition. The girl in her story complained of her father's 'rendering of Pomp and Circumstance […] blasted around the house at full volume […] I don't think he is actually quiet for five minutes each day', she wrote. 'When he wakes up in the morning, he starts his day off by switching on the radio, full blast.' The story concludes:

So if Darlington wants to make a success of its Quiet Town Campaign, I think they had better get rid of my father, while they finish the experiment. He often talks about going back to the places he went to during the war, such as Iceland, Norway and France. I don't mind which country you decide to send him to, just somewhere, where he can be as noisy as he likes.

(Darlington Borough Council n.d.: 39)

Figure 5.5 Cover of leaflet announcing the Darlington Quiet Town Experiment. *Source:* Noise Advisory Council 1981.

Figure 5.6 Logo of the Darlington Quiet Town Experiment. *Source:* Noise Advisory Council 1981.

The Darlington Quiet Town Experiment produced a sonic encounter that naturalized noise as social disruption. It did so primarily for a female hearer who was constructed as an agent of quiet in her town. It did so for sounds that threatened the integrity of the private family home. There *was* an active listener in the experiment: he is pictured frequently in reports wielding noise-measuring equipment (see, for example, Figure 5.8) and represented the knowledge-gathering Noise Advisory Council. By hearing with the female hearer promoted in the experiment's publicity, however, the historian might gain a different perspective on quiet, a perspective framed by gendered ways of hearing. There is more to hear with in historical instances where sonic encounter has been produced. The

THE NOISY GNOME DOES ODD JOBS
AT ODD TIMES . . . .

.... LIKES MUSIC TOO LOUD

Figure 5.7 Detail from 'Do you have a noisy gnome in your home?' leaflet. *Source:* Noise Advisory Council 1981.

example of gendered ways of hearing that I have set out here is intended only to illustrate the necessity of hearing beyond the listening-to and of attending to the effects of ways of hearing in the past.

Sonic encounters also take place out of the immediate context of their production. The Darlington Quiet Town and Big Ben Silent Minute examples used here have been intended only to draw attention to the most obvious forms of auditory attention shaping. The sonic encounter between a Western colonizing ear and, say, the traditional musical tradition of a colonized people, of the kind that finds form in ethnographic field recordings now held in institutions such as the British Library, has not been stage-managed in the sense of the Darlington Quiet Town Experiment, but is nonetheless still the product of a Western, colonial, way of hearing. In hearing with these ways of hearing, we must reflect carefully on the role of the historian's listening ear. There remains further work to be done to theorize the ways in which historians might realize Goh's (2017) aim of sounding situated knowledges, but reflexivity about what makes historical knowledge of the auditory past possible is a necessary first step.

Figure 5.8 Image entitled 'Noise reading being taken on a building site'. *Source:* Noise Advisory Council 1981: 15.

Notes

1. The chapter focuses on modern sound cultures because the methodologies needed to research ancient, medieval, and early modern sound are somewhat different to those needed for the post-1800 period. It should nevertheless be noted that there is a flourishing field of study on pre-1800 sound history. Foundational texts are Richard Cullen Rath's *How Early America Sounded* (2003) and Bruce Smith's *The Acoustic World of Early Modern England: Attending to the O-Factor* (1999). Recent works include Shane Butler and Sarah Nooter's edited volume *Sound and the Ancient Senses* (2019) and Niall Atkinson's *The Noisy Renaissance: Sound, Architecture and Florentine Urban Life* (2016).
2. The concept of the soundscape is most closely associated in sound studies with R. Murray Schafer's *The Soundscape: Our Sonic Environment and the Tuning of the World* (1994).
3. Smith's argument on the necessity of textual sources in historical sound studies can be found in Smith (2015: 55–64).
4. Smith's edited essay collection *Hearing History: A Reader* (2004) helped cement the sense of a field in the making.
5. The key text on the history of noise is Karin Bijsterveld's *Mechanical Sound: Technology, Culture and Public Problems of Noise in the Twentieth Century* (2008).
6. Lord Horder papers, Wellcome Library, GP/31/B.4/4; GP/31/B2/23.
7. Anette Hoffman and Phindezua Mnyaka's (2014) research on the colonial sound archive points to the importance of further work in this area. See also Hoffman, in this volume.
8. For a discussion on noise and neurasthenia and the management of civilian hearing in the Second World War, see Mansell 2017.

References

Anti-Noise League (1935). *Noise Abatement Exhibition: Science Museum, South Kensington, 31st May–30th June 1935*. London: Anti-Noise League.

Atkinson, Niall (2016). *The Noisy Renaissance: Sound, Architecture and Florentine Urban Life*. University Park, PA: Pennsylvania State University Press.

Bijsterveld, Karin (2008). *Mechanical Sound: Technology, Culture and Public Problems of Noise in the Twentieth Century*. Cambridge, MA: MIT Press.

Birdsall, Carolyn (2012). *Nazi Soundscapes: Sound, Technology and Urban Space in Germany, 1933–1945*. Amsterdam: Amsterdam University Press.

Briggs, Asa (1995). *The History of Broadcasting in the United Kingdom*, vol. 5. Oxford: Oxford University Press.

Butler, Shane and Sarah Nooter (eds) (2019). *Sound and the Ancient Senses*. New York: Routledge.

Clapson, Mark (2017). 'Big Ben Silenced: Britain's Bong Furore is a Sign of National Insecurity'. *The Conversation*, 21 August 2017. Available online: https://theconversation.com/big-ben-silenced-britains-bong-furore-is-a-sign-of-national-insecurity-82715 (accessed 12 March 2019).

Corbin, Alain (1998). *Village Bells: Sound and Meaning in the Nineteenth-Century French Countryside*. Trans. Martin Thom. New York: Columbia University Press.

Cox, Christoph (2011). 'Beyond Representation and Signification: Toward a Sonic Materialism'. *Journal of Visual Culture* 10: 145–161.

Dakers, Andrew (1943). *The Big Ben Minute*. London: Andrew Dakers.

Darlington Borough Council (n.d.). *Children on Noise: Darlington Quiet Town Experiment*. Darlington: Darlington Borough Council.

Goh, Annie (2017). 'Sounding Situated Knowledges: Echo in Archaeoacoustics'. *Parallax* 23: 283–304.

Goodman, Steve (2009). *Sonic Warfare: Sound, Affect and the Ecology of Fear*. Cambridge, MA: MIT Press.

Hartford, Kassandra (2017). 'Listening to the Din of the First World War'. *Sound Studies: An Interdisciplinary Journal* 3: 98–114.

Hoffmann, Anette and Phindezua Mnyaka (2014). 'Hearing Voices in the Archive'. *Social Dynamics: A Journal of African Studies* 41: 140–165.

Junior, Allan (1941). *As Big Ben Strikes: Some Thoughts for 'The Big Ben Minute'*. Dundee: Valentine & Sons.

Kane, Brian (2015). 'Sound Studies Without Auditory Culture: A Critique of the Ontological Turn'. *Sound Studies: An Interdisciplinary Journal* 1: 2–21.

Kassabian, Anahid (2013). *Ubiquitous Listening: Affect, Attention, and Distributed Subjectivity*. Berkeley, CA: University of California Press.

Mansell, James G. (2017). *The Age of Noise in Britain: Hearing Modernity*. Urbana, IL: University of Illinois Press.

Mansell, James G. (2018). 'Ways of Hearing: Sound, Culture and History'. In Michael Bull (ed.), *The Routledge Companion to Sound Studies*, 343–352. New York: Routledge.

Morat, Daniel (2014). 'Introduction'. In Daniel Morat (ed.), *Sounds of Modern History: Auditory Cultures in 19th and 20th-Century Europe*, 1–12. Oxford: Berghahn.

Noise Advisory Council (1981). *The Darlington Quiet Town Experiment: September 1976–September 1978*. London: Her Majesty's Stationery Office.

Quiet Mark (2016). 'Home'. Available online: https://www.quietmark.com/ (accessed 23 March 2016).

Rath, Richard Cullen (2003). *How Early America Sounded*. Ithaca, NY: Cornell University Press.

Rice, Tom (2003). 'Soundselves: An Acoustemology of Sound and Self in the Edinburgh Royal Infirmary'. *Anthropology Today* 19: 4–9.

Rubery, Matthew (2016). *The Untold Story of the Talking Book*. Cambridge, MA: Harvard University Press.

Schafer, R. Murray (1994). *The Soundscape: Our Sonic Environment and the Tuning of the World*. Rochester, VT: Destiny Books.

Smith, Bruce R. (1999). *The Acoustic World of Early Modern England: Attending to the O-Factor*. Chicago: University of Chicago Press.

Smith, Mark M. (2001). *Listening to Nineteenth-Century America*. Chapel Hill, NC: University of North Carolina Press.

Smith, Mark M. (2014). 'Futures of Hearing Pasts'. In Daniel Morat (ed.), *Sounds of Modern History: Auditory Cultures in 19th and 20th-Century Europe*, 13–22. Oxford: Berghahn.

Smith, Mark M. (2015). 'Echo'. In David Novak and Matt Sakakeeny (eds), *Keywords in Sound*, 55–64. Durham, NC: Duke University Press.

Smith, Mark. M. (ed.) (2004). *Hearing History: A Reader*. Athens, GA: University of Georgia Press.

Sterne, Jonathan (2003). *The Audible Past: Cultural Origins of Sound Reproduction*. Durham, NC: Duke University Press.
Stoever, Jennifer Lynn (2016). *The Sonic Color Line: Race and the Cultural Politics of Listening*. New York: New York University Press.
Sykes, Jim (2018). 'Ontologies of Acoustic Endurance: Rethinking Wartime Sound and Listening'. *Sound Studies: An Interdisciplinary Journal* 4: 35–60.
Thompson, Emily (2004). *The Soundscape of Modernity: Architectural Acoustics and the Culture of Listening in America, 1900–1933*. Cambridge, MA: MIT Press.
Thompson, Marie (2013). 'Gossips, Sirens, Hi-Fi Wives: Feminizing the Threat of Noise'. In Michael Goddard, Benjamin Halligan, and Nicola Spelman (eds), *Resonances: Noise and Contemporary Music*, 297–311. London: Bloomsbury.
Thompson, Marie (2017a). *Beyond Unwanted Sound: Noise, Affect and Aesthetic Moralism*. London: Bloomsbury.
Thompson, Marie (2017b). 'Whiteness and the Ontological Turn in Sound Studies'. *Parallax* 23: 266–282.
URS Scott Wilson (2011). 'The Economic Value of Quiet Areas'. Available online: http://randd.defra.gov.uk/Document.aspx?Document=TheEconomicValueofQuiet_FinalReport.pdf (accessed 13 March 2019).
Western, Tom (2015). 'Securing the Aural Border: Fieldwork and Interference in Post-War BBC Audio Nationalism'. *Sound Studies: An Interdisciplinary Journal* 1: 77–97.

6

Sonic Methodologies in Urban Studies

Christabel Stirling

Introduction

What does it mean to talk about the music and sound culture of a particular city at a time when the production, circulation, and consumption of music is increasingly trans- or post-urban? As opera festivals are broadcast from the theatres of one city to the cinema screens of another, rapid cultural flows between Accra, Johannesburg, and London culminate in new 'global' genres and cross-cultural modes of musical production, and the 'worldwide crews' of electronic/dance music flit remotely between frenzied dance battles in Southside Chicago gymnasiums and the smooth wooden floors of Manhattan Records in Shibuya, Tokyo. Meanwhile our notions of where and how to locate the urban grow increasingly complex. How can we understand, and research, the relationships between music, sound, and the city in an era of hyper-connectivity and digital mediation? How important are the affective qualities and sociopolitical potentialities of urban locality, spatial proximity, and live musicality in such an era? How should one go about conducting qualitative research of large-scale urban music events where audience numbers are in the tens of thousands? And what methodological demands are placed on researchers engaging with music and sound cultures in monstrously convoluted megacities such as São Paolo, Mumbai, or Manila?

Glancing at the literature on cities, the diverse and even incommensurable approaches towards analysing the post-industrial city seem to announce the difficulty that contemporary urban scholars face in dealing with cities that are increasingly fractured, centrifugal, and enveloped by a vast mediascape of local, regional, and transnational networks. On the one hand, cultural geographers and non-representational theorists celebrate the virtual spatiality of the dematerialized 'information city' with its promise of global interconnectivity and a sociality irreducible to spatial propinquity (Amin and Thrift 2002; Amin 2012). Brimful of seductive metaphors such as 'flow', 'hybridity', 'excess', and 'emergence', this literature emphasizes the radical potentials of wireless infrastructures and the non-anthropocentric public spheres that they make possible. On the other hand, urban anthropologists and architectural theorists critique the notion that virtual space could ever supersede or displace material space, pointing to the paradoxical enhancement of spatial propinquity in the digital age, where power and wealth are reconcentrated in specific places and locales (Sassen 2001; Gandy 2005; Harvey 2006). As these scholars note, it is the global metropolitan elite who are lifted out of the chaos of the concrete city in air-conditioned 'citadels of connectivity' (Gandy 2005: 37). Meanwhile sprawling vistas of congestion, poverty, and infrastructural collapse rage on around and below – vistas that are themselves encased in new media, often operating via parallel or 'pirate' distribution circuits, but that nonetheless remain precarious, subject to continual breakdown.

In accounts of the city where music and sound are prominent, it is, however, the recursive and 'nested' relationships between co-present and mediated space that become especially palpable (Born 2013). Amidst the buzz of Cairo's popular neighbourhoods, Charles Hirschkind describes how Islamic cassette sermons 'spill into the street from loudspeakers in cafés', at once reconfiguring the acoustic architecture of the city as the recorded voices of well-known orators collide with car horns, bustling crowds, and a Michael Jackson bassline in a passing car (Hirschkind 2006: 7). Reaching the ears of sensitive listeners on the street, in shops, and on buses, the cassettes draw individuals into moments of private ethical reflection and shared affective unity. Taxi drivers and shop owners become part of a pious virtual public nested within the private space of their vehicle or establishment, '[exploiting] moments of boredom and labor' as they hone their virtuous selves through visceral modes of appraisal (Hirschkind 2006: 28). Meanwhile, in post-industrial Detroit, Carla Vecchiola traces a different kind of virtual public – one that has evolved from the city's grassroots electronic music community and its capillary global movement. As she notes, transnational networks not only take Detroit and its music 'out across the globe', but also draw streams of 'international techno tourists' to the city from Asia, South America, and Europe, generating a physical coming together of Detroit's global music fan base in ways that strengthen local community building and disrupt images of urban decay that abound in Detroit (Vecchiola 2011: 96). In this context, online communications and mail-order custom initiate new trans-urban socialities that exceed the locality of the city while remaining inextricably tied to it: as a 'social network of friends not yet met and familiar places not yet physically experienced' (108).

At a time when the boundaries between material and immaterial, concrete and virtual, have become so intensely interwoven, what difficulties are posed to scholars engaging with sound and music in heterogeneous urban settings? How can we get to grips with the methodological requirements of cities that are so culturally, politically, and physically different, but that – through ongoing currents of immigration, displacement, digital circulation and exchange – are also intimately connected? And how might we capture the potential fluidity and 'openness' of the networked city while continuing to challenge the spatial exclusions and immobilities that erode public life in the physical city? Considering such questions, this chapter explores possible approaches and methods for dealing with the complexity of the twenty-first-century city. I begin by providing an overview of recent research conducted at the intersection of music, sound, and urban studies, highlighting the methods that those engaging in such work have developed. Next, I reflect upon how methods and techniques from across the musical sub-disciplines might combine to create more critical urban methodologies. Finally, I discuss how I have put some of these methodological strategies into practice in my own urban musical research. In particular, I reflect upon the potentials of using a number of audiovisual and participatory methods alongside more conventional ethnographic techniques and approaches. As I argue, different cities have different methodological needs, and successful ways of working in one urban context are not always transferable to another. Nonetheless, it is my hope that this chapter will offer a set of tools to be taken up, experimented with, and adapted across a range of empirical urban contexts in order to better grasp the complex realities of our time.

Music in the city and the city in music

Increasingly, urban studies scholars working in geography, sociology, and architecture have engaged with music and sound as a major part of their research. With cultural geographers such as Susan Smith (1997), George Revill (2000), and Arun Saldanha (2002) having probed the spaces and places of music since the mid-1990s, more recent work in this field has seen a shift to music and sound's ability to *initiate* spatialities through practices of performance, encounter, and the 'fleshy dynamics of embodiment' (Anderson, Morton, and Revill 2005: 643; Revill 2013; Simpson 2017). Grounded in a conception of urban space not as bounded or preconceived but as dynamic and continually unfolding, such a shift has had methodological implications too, encouraging a participatory and experimental engagement with the 'now' of musical practice and performance – an approach dubbed by Nichola Wood and colleagues as 'doing and being' geographies of music (Wood, Duffy, and Smith 2007). Rosemary Overell's (2012) work on 'brutal belonging' in Australia and Japan's grindcore scenes is a strong example of this approach being taken up. Drawing on Wood et al.'s (2007) notion of 'participant-sensing', Overell uses a digital recorder to capture 'on-the-spot' experiences of grindcore scene members at different gigs and venues, as well as supplying participants with their own digital recorders through which to spontaneously log their thoughts and feelings (Overell 2012: 90–94). While not an entirely 'non-representational' method, these audio diaries, she notes, help to 'close the gap a little' between the affective dimensions of musical urban life and the 'clinical ethnographic interview', generating a livelier, more embodied account of the spaces and atmospheres produced by grindcore (90).

Paralleling this, sociologists such as Les Back have, for a long time, been 'listening' to urban multiculture, attending ethnographically to the musical and cultural dialogues arising between South Asian and African Caribbean immigrants in niches of London and Birmingham, as well as, more recently, examining how the movement of music across borders – the 'trafficking of sampled sounds' (Back 2016: 191) – can generate transnational and trans-urban connections that challenge 'racially inflected nationalism[s]' (Back 1996, 2016). Key to Back's work is his striving towards what he calls a 'sensuous' or 'live' sociology: a sociology that favours a wide range of sensory experiences and multimedia methods, from film-making and soundscape recording to thick situated description of 'social life in process', thus broadening ethnography's reliance on interview (Back 2009: 3). Listening to Deptford market in South East London, for example, Back bears witness to a thriving multiculture characterized by 'rituals of sociality and banter', good-natured haggling, and the convivial sharing of food recipes – a vibrant sonic social scene that contradicts xenophobic claims made by his participants in interview and that emphasizes the need for method triangulation (15). Similarly, in his account of London bus soundscapes, sociologist Richard Bramwell highlights the 'ad hoc' social and technological networks that emerge around the playing and sharing of music on bus journeys – a sociability that disrupts the 'anti-sociality' invoked by the buzzes, beeps, and automated voice-overs of the 'official' bus soundscape, while also subverting the government narrative of London transport as a site of suspicion and mistrust (Bramwell 2015).

Complementing these social scientific studies, musicology and sound studies have also shown a burgeoning interest in urban geography over the past two decades, developing areas of research such as iPod listening and urban experience (Bull 2007); 'gigographies' and the cartographies of live performance (Laing 2009; Lashua, Cohen, and Schofield 2010); music's intertwinement with tourism, travel, and gentrification (Cohen 2007; Holt and Wergin 2013; Garcia 2016); and the role of music in diasporic urban placemaking, particularly as a spatializing or 'homing' device through which to cultivate shared spaces of belonging (Dueck and Toynbee 2011; Henriques and Ferrara 2016). Of this literature, Sara Cohen's ethnographic work on 'popular musicscapes' in Liverpool is particularly useful methodologically, mobilizing critical forms of cartography alongside archival materials, photographs, and interviews to draw out the hidden musical histories of the city. By juxtaposing several different kinds of music city maps – from tourist music heritage maps to participants' hand-drawn maps of their music-making activities in the city – Cohen and her collaborators reveal how particular narratives, musicians, and venues (e.g. the Beatles, the Cavern Club) have taken on a skewed mythological status in Liverpool, coming to symbolize 'entire musical genres and eras' at the expense of the journeys and trajectories of other musicians and styles (Lashua, Cohen, and Schofield 2010: 126; Cohen 2011: 240). In particular, these 'master maps' of music heritage obscure Liverpool's black musical histories and legacies, including the constraints on black musicians' mobilities in the post-war period and the ongoing exclusion of black-originating genres such as grime from urban public spaces. Mapping, in the hands of these scholars, then, becomes a tool through which to draw out the disparities and contradictions between 'official', historical, sociocultural, and personal characterizations of the musical city, and to illuminate a city's musical obstructions and absences as well as flows.

Notably, mapping has also been a key method for sound studies scholars. Primarily associated with the World Soundscape Project and the emergence of acoustic ecology in the 1970s, 'noise maps' have evolved as a way of charting the volume, density, and movement of noise in cities, using both quantitative decibel charts and qualitative pictorial diagrams and graphic notations (cf. Schafer 1970). Meanwhile, 'sound maps' constitute a more playful, artistic engagement with urban sound, less associated with public health and noise as a pollutant, and more with sound as a defining quality of a city's character, and thus as potentially crucial to urban planning and design (Cusack 2017; Lappin,

Ouzounian, and O'Grady 2018). With the explosion of web-based maps in the last decade, sound and noise mapping have largely become crowdsourced activities, generating new kinds of 'participatory' sonic urbanism and communal sound archiving, as well as raising concerns about free labour, access to technology, and acoustic surveillance (Waldock 2011; Ouzounian 2021).

Other kinds of sound mapping, such as soundwalking and field recording, have also become popular among sound studies scholars, particularly those engaging with the social and corporeal dimensions of urban sound and/or sound art. Significant, here, is David Pinder's (2001: 8) auto-ethnographic account of Janet Cardiff's *Missing Voice (Case Study B)*, which unfolds as an aural psychogeography of London's East End mediated by the doubtful, fanciful, subjective listener who walks to excavate 'hidden histories and geographies'; Linda O'Keeffe's (2015) participatory soundwalks with teenagers in Dublin, which expose the 'missing voices' of young people in urban design and the role of the urban soundscape in exacerbating social exclusion; and Tom Hall and colleagues' (2008: 1033) 'touring interviews' – 'interviews as, or nested within, soundwalks' – in which young people in South Wales 'walk' their interviewers through the city, with street noise often emerging as an 'innovative disturbance' that shifts dialogues, sheds light upon urban reconstruction, and highlights disquieting levels of acclimatization to overwhelmingly loud industrial sounds. As Marcel Cobussen, Vincent Meelberg, and Barry Truax have noted, such in situ urban sonic practices expand the sensorial dimensions of listening considerably, generating experiences of sound that are simultaneously tactile, kinaesthetic, olfactory, and gustatory as well as sociocultural and situated (Cobussen, Meelberg, and Truax 2016: 6). Consequently, when taken as a qualitative research method, soundwalking acts as a particularly powerful articulator of the differentiation of urban acoustic experience, illuminating the conflicting sonic atmospheres, (im)mobilities, and histories that permeate the city and rendering the experiences of those who are marked by fixity and marginality as well as choice and fluidity. In this way, soundwalking might be seen to proffer a sensorial counterpart to Cohen's cartographic practice, unsettling 'official' accounts of the spatially open, networked city and revealing instead the diverse, often limited ways in which individuals and social groups navigate urban space in the physical city.

Building on sound's intertwinement with urban social and cultural identities, a further important tributary to emerge from musicology and sound studies pertains to histories of sound in/of the city. Documenting the changes wrought to cities such as Madrid, New York, and Lyon during the nineteenth century, historians of European and American music have noted how urban and economic developments of this era not only altered the acoustics of the street and the trajectories of sound through the city, but also fuelled the emergence of new social class identities, marked, in turn, by conflicting sound cultures that jostled for space in the modern metropolis (Picker 2003; Thompson 2004; Boutin 2015; Balaÿ 2016; Llano 2018). Particularly frequent in this literature are references to the 'silence-seeking' bourgeoisie, whose display of contempt for noisy (often immigrant) street musicians and the 'shrill cries' of peddlers signalled both their legitimacy as part of an elite social-class category, and their desire to control and impose order onto literal neighbourhoods of the city. Sound and music as instruments of power and order are also at the forefront of an emerging body of work on colonial urban music history, attentive to the attempts made to impose European urban values on colonized societies through sonic-sensory regulation and the propagation of European music (Irving 2010; Baker and Knighton 2011; Rotter 2019). Employing different methodologies and consulting a wide range of archival sources – poetry, guidebooks, historic urban plans, paintings, and personal diaries – these studies act as valuable historical forebears to contemporary forms of audio mapping and 'sensuous' sociology in their ability to shed light upon how a city's sounds were perceived by different social and cultural groups at particular historical moments. Moreover, in charting the point at which urban noise started to emerge as a public health issue in the West, such accounts are vital to understanding the historical trajectories of contemporary noise mapping.

A final significant area of research relates to anthropologies of urban sound. In recent years, ethnomusicologists have engaged compellingly with the relations between affect, the social, and the spatial in urban environments, emphasizing sound's ability to implore, repel, and provoke in ways that instigate shifts between public and private experience, reconfiguring or reinforcing socio-spatial relations (Stokes 2010; Born 2013; Hankins and Stevens 2013). Much of this literature has focused on postcolonial and/or post-conflict cities currently undergoing rapid urbanization in Africa, the Middle East, and Asia (Hirschkind 2006; De Witte 2008; Eisenberg 2013). Adopting 'listening', participant observation, and film-making among other ethnographic techniques, these accounts make palpable the deeply encultured nature of city sound. In Beirut, for example, the acoustically magnifying derelict buildings around which the urban soundscape ricochets coupled with the relentless drilling and hammering of an enterprise-driven post-war reconstruction programme amount to a situation in which the grievances of a troubled history literally resonate (Royaards 2019). Urban sound, in this context, thus takes on a profound historicity: imbued with the acoustics of disintegrating architectural shells and yet-to-be-populated towers, traffic noise and muezzin calls carry the sonic trail of ongoing political instability, spatial rem(a)inder and erasure, and an uncertain identity and future. Meanwhile, in Accra, public space is similarly cacophonous but differently contested, here saturated by the sounds of the various religious groups that vie for audible presence in the cityscape. As Marleen De Witte notes, the combination of technological mediation, in the form of powerful PA systems, and open-air architecture due to the hot climate, means that 'private sound easily becomes public and public sound permeates into spaces as private as one's bed', leading to an 'auditory sacred space that is never contained' and that fuels frequent clashes over territory, cultural history, and citizenship (De Witte 2008: 693, 706).

If holding the sounds of these and other cities together exposes their differences, it also allows similarities to come to the fore, particularly regarding the evolving aurality of so-called 'media urbanism'. Defined by Ravi Sundaram (2009: 6) as the convergence of crisis-level urban growth and ubiquitous media, the soundworlds of media urbanism are those promulgated by low-cost mobile telephony, fast-moving electronic music devices, and increasingly 'hackable' technological infrastructures in cities that are themselves expanding at dizzying rates. Under such conditions, the endless sounds of construction work and the perpetual car horn blowing of informal transport services that use 'beeps' to pick up passengers are overlaid with electronically boosted music, political campaigns, religious chants, news, prayer, radio sermons, and jingles, most of which extend far beyond their physical locations (Hirschkind 2006; De Witte 2008; Sundaram 2009). Government and local authorities are thus confronted with a multiplicity of mediated sound cultures, which, due to the escalating movement of peoples, are growing in diversity as well as volume, are often antagonistic to one another, and are increasingly seen as pervasive, 'unmanageable', emerging from the body politic '*as if without limits*' (Sundaram 2009: 24, emphasis in the original). Such exhilarating levels of urban-technological intensity and sonic maelstrom do not, however, obscure sound's potential to act as an ideological force in the city. On the contrary, as Delhi's portable media playing youth are vilified as 'ear contaminators' by civic campaigners seeking to affirm their middle-class identities (24–25), while the Ghanaian government mobilizes a noise abatement discourse to resolve a cultural religious sound clash (De Witte 2008: 707), urban sound's intertwinement with identity formation, social control, cultural-historical friction, and attempts to silence and segregate the 'other' appears as strong as it did in the nineteenth century. Such degrees of difference and similarity across both geography and history bring into articulation the potential gains to be made from studying cities in comparative cross-cultural and temporal perspective, rather than merely as singular-complex entities (Klotz et al. 2018).

Methods and methodologies

This latter point raises the question of methodology, and how it might be distinguished from and brought into a critical relation with questions of method. Indeed, taken together, the above literatures offer a wealth of innovative methods for researching music, sound, and urban matters. Where 'participant-sensing', listening, and soundwalking enable particular proximity to the micro-social and embodied dynamics of urban musical experience, ethnographic and archival approaches to mapping (popular) music expose the higher-level institutional and economic forces that are at work in (re)producing particular versions of the music city. Meanwhile, historical source analysis affords unique levels of insight into the lost auditory worlds of cities undergoing modernization, colonization, and other irrevocable sociocultural and economic changes, while noise and sound mapping, as analytical and artistic tools, have significantly altered how cities are perceived, planned, and designed, and will likely continue to do so as environmental discourses gain force.

Perhaps less common in the literature is a critical interrogation of why particular methodological approaches are deemed more or less suitable for engaging with music/sound and the urban, what specific benefits and limitations they bring, and what the different stances could amount to together, particularly when brought into a relation with theoretical discourses. Auto-ethnography, for example, has numerous advantages for researching the affective propensities of urban sound and/or sound art, enabling one to detect changes in adrenaline levels or heightened sensation in the skin and flesh in conjunction with other aspects of the 'assemblage' – sounds, technologies, personal and other histories, spaces, discourses, and social relations (Born 2010a: 88). Yet, it is also limited to the experiences of the individual researcher. Supplementing this with ethnography, which might involve participating in, observing, filming, recording, and 'listening' to particular field sites or installations over time, as well as talking to and interviewing participants, reveals more about how different people going about their lives experience and respond to city soundscapes and sonic practices, while also facilitating a sensitivity towards what Danilyn Rutherford refers to as 'affect and "affect"': the affects felt by the researcher engaging with the ethnographic field, and the affects experienced by the participants being researched (Rutherford 2016: 289). An important benefit of ethnography, then, is its capacity to expose the existence of multiple, situated perspectives and vantage points, and the propinquity it affords to the embodied socio-spatial relations produced by music and sound.

Nonetheless, without an historical perspective, it is difficult to fully comprehend and diagnose the contemporary. This is true both at the micro-social level, given the way that social and political histories saturate the everyday urban sonic landscapes in which we live in 'intimate, up-close terms' (Back 2016: 1027); and at the macrosocial level, in terms of being able to deduce the 'cumulative outcome' of such everyday processes as 'historical trajectories of variation or transformation, stability or stasis' (Born 2010c: 235). Triangulating history with (auto-)ethnography thus presents numerous advantages. It enables, for example, insight into the continuities and breaks between past- and present-day street music cultures, including how and why certain modes of perception and ideology 'became available' at particular historical moments, what discourses and legislative measures emerged as a consequence, and the extent to which these achieved stability over time. It reveals how the unequal movement of sounds, genres, and people through the contemporary city – exemplified in London by the expansion of classical music and other predominantly white cultural forms into non-traditional urban spaces conterminously with the relentless shutdown of black-run venues and genres such as grime – have long historical precedents, from the sonic-spatial domination of classical music over immigrant street music in Victorian London, to the violent exclusion of black musical expression from urban space via the 'colour bar' in post-war Britain. And it shows how historical forms of embodied 'sensitivity' and white middle-class boundary drawing, including the power to command silence over urban space, not only penetrate through to the present in European cities in the form of noise complaints, racist policing, and revoked venue licenses, but also congeal in new geographical and political spaces, under new media conditions, as the bedrock for new ethnic and class identities – as Sundaram's account of New Delhi's denigrated 'ear contaminators' makes clear. Bringing these diachronic perspectives into dialogue with theory, it becomes apparent that recent work in cultural geography, which wants to see the city as radically emergent through a conceptual emphasis on affect, process, and performativity, poses problems for understanding experiences and events that are characterized more by continuity than change.

The overarching point, following Georgina Born and Will Straw, is thus that we need methods capable of articulating both stability and dynamism in urban musical cultures – ways of working that grasp the 'effervescence' and sensory richness of city sounds and socialities as well as their direction of movement and scale (Straw 2001: 252–4; Born 2005). Combining questions of temporality and history with 'up-close' descriptive and ethnographic work, as Born suggests, allows us to trace 'the historical trajectories of musical assemblages', uncovering the ways in which seemingly unstable, fast-moving urban musical practices expand into larger processes of historical change or continuity, transformation or reproduction (Born 2005: 34, 15). Moreover, working comparatively across geography and topography, as well as history, sheds light upon the often-surprising similarities and differences that emerge between cities, their soundworlds, and their rates of change/stability at particular historical conjunctures. As Straw's (1991) work on 'scenes' demonstrates, the empirical challenges that this kind of work generates include thinking about how 'indigenously' produced sounds can propagate to new urban centres and subsequently evolve at a different rate; how 'native' and 'dispersed' scenes may enter into mutually influential relations and precipitate unintended musical developments and trans-urban connections; and how cities can become host to a vast range of musical practices and publics that diverge from each other 'physically', at the face-to-face level, but coincide and overlap 'virtually', via the shared taste communities that they engender globally.

How, then, one might ask, is it possible to work in all of these different ways at once? How can one design and conduct rigorous ethnographic fieldwork in complex urban settings while also attending rigorously to history? What is gained or compensated by choosing multi-sited over single-sited research, and is the capacity to carry out intensive fieldwork jeopardized in opting for the former? And if digital technologies have transformed the sonic fabric of cities, have they not also transformed the methodological possibilities for researching sound in/of/and the city? In the remainder of this chapter, I discuss how I have grappled with some of these questions in my own urban musical research. Indeed, while combining multi-sited ethnography with history and theory enabled unique insights and perspectives, it still left me with the practical problem of how to conduct qualitative research in a city that spans 610 square miles and has an estimated population of nine million (London). As I describe, such a challenge not only entailed that I 'cast my net' appropriately but also that I think in more experimental ways about methods that might do justice to musical urban sprawl.

Comparison, difference, and diachrony

For the past five years, my ethnographic research has focused on live music audiences in London, drawing insight from classical music, sound art, dub reggae, and electronic/dance music. Specifically, I have been concerned with the social and affective processes by which music and sound generate collectivities, and with how one can or might gain proximity – methodologically and representationally – to the visceral, non-discursive aspects of musical experience. Working comparatively across genres, some of the questions I have sought to answer are: how do music and sound act upon the physical body in ways that potentially shift embodied social boundaries and power relations? What kinds of social spaces do music and sound make possible, and what role do these spaces play in the production of urban public life? Can music and sound catalyze social coalitions that are emergent, and that simultaneously reorder existing social hierarchies and divisions? To what extent could this facilitate a reimagining of the concept of affect for a musical and sonic politics?

Comparison has always been central to this project. One significant reason for this was that, since the project was not 'about' a particular community, institution, or otherwise easily describable entity, but was rather constructed around more open questions about what might or might not be possible (musically, socially, spatially, politically) at a particular historical conjuncture (contemporary London), it was important to draw difference into the ethnographic picture. Comparison, which was built into the research through multiple field sites and 'juxtapositions of locations' (Marcus 1995: 105), seemed an obvious solution, given its ability to situate the present as pluralistic and multifaceted rather than as unitary. By traversing, discovering, and moving between an array of musical spaces – some familiar, some strange, many placed at considerable distances from each other, others adjacent but oblivious to each other – comparison allowed me to channel the close-up, local perspective of ethnography along multiple tributaries. It enabled me to build a map of the urban musical terrain in London that drew a huge amount of diversity into it, generating a richer, more complex, if necessarily partial, ethnographic, and historical understanding of the present. This employment of 'difference' as a methodological principle proved central to my theoretical concerns too: it facilitated what Michel Foucault (1981) refers to as a 'polyhedron' of empirical information through which to understand the workings of musical affect, thus moving away from the theory-driven empiricism of many affect theorists (see Stirling 2019).

Regarding the study of music and sound cultures in such a big city, it also seemed important that the genres and field sites I selected had the propensity to occupy a range of sites and neighbourhoods – not just collectively but in and of themselves too, so as to allow for different levels of comparison. At the time of fieldwork (2013–2015), the migration of classical music out of the concert hall and into unusual urban spaces and venues was gaining particular traction in London, fronted by initiatives such as Nonclassical (est. 2004), the Night Shift (est. 2006), and the London Contemporary Music Festival (est. 2013) (see Nonclassical n.d.; Orchestra of the Age of Enlightenment 2020; London Contemporary Music Festival [LCMF] n.d.). Studying this 'new music' movement alongside classical concerts taking place in traditional concert hall settings thus allowed me to analyse the live performance socialities of classical music across nightclubs, car parks, warehouses, train stations, and Second World War air raid shelters, as well as concert hall auditoria. Similar levels of comparison were made possible by my sound art fieldwork, which drew me to a range of urban spaces: canal towpaths, churches, residential streets, housing estates, galleries, and arts cafes. Electronic/dance music, encompassing various styles and sounds, presented an interesting inversion of the classical music scene in terms of its increasing 'intellectualization' and the prevalent emphasis on certain subgenres as 'art' forms to be consumed in concert halls and galleries as well as nightclubs. During fieldwork, for instance, I saw prominent DJs perform at the Southbank's Festival Hall, the Barbican Centre, and the Tate galleries, and this was paralleled by a growing number of collaborations between DJ/producers and symphony orchestras.[1] Finally, the dub reggae scene was, at the time of fieldwork, very wide ranging, incorporating relatively 'mainstream' events at established inner-London nightclubs, smaller-scale dances in non-gentrified neighbourhoods and community spaces, and large-scale street carnivals such as Notting Hill and Brixton Splash. As a field site, it thus presented a prime opportunity for comparative work between a range of indoor and outdoor sound system sessions.

Working between and across these genres and scenes, then, took me to all kinds of social and musical spaces in all corners of the city – from an outdoor disco festival in Enfield to an historic Caribbean venue in Southall. It demanded that I travel long distances – by train, (night) bus, bicycle, and foot – at all times of the day and night. It generated overlap and similarity as well as difference, as individuals who I had met as part of one scene popped up unexpectedly in another, while a single multipurpose venue hosted a reggae night, an experimental classical concert, and an all-night techno event in the space of a few days. Further, it allowed me to take unexpected trajectories, following the fragmented and dispersed activities of musical and cultural formations across multiple online/offline locations. While the research thus didn't move between cities – though it might productively in the future – it still encompassed multi-sited ways of working, requiring that I negotiate different degrees of familiarity and estrangement in relation to my field sites, moving between 'public and private spheres of activity', and demanding that I constantly recalibrate my positioning in terms of what George Marcus refers to as the multi-sited researcher's 'shifting affinities for […] as well as alienations from, those with whom he or she interacts with at different sites' (Marcus 1995: 112–113).

What did comparison between these four broadly defined field sites allow that single-sited research might not have? Two points are worth drawing attention to here. First, holding these genres together, as contiguous sites of urban musical activity with distinct histories and discourses, enabled both differences and surprising commonalities to come to the fore. For instance, while opposed in many ways, a number of striking similarities emerged between the dub reggae and classical music scenes, particularly with regard to the honing and enclaving of the historical and cultural spaces in which these musics exist in their live forms, and the disciplined forms of embodiment and listening that occur within and help produce these spaces. Parallels surfaced between dub reggae and sound art, too, notably in the experimental aesthetic techniques shared by both genres – montage technique, spatial manipulation, transplanting 'found' sounds – and the creative trajectories and experiences of those who produced and participated in them. At the same time, thinking about the nature of the social relations brought into play by the different field sites, sound art's place-based, participatory, and collaborative potential, which enables artists to work in diverse urban neighbourhoods with various communities, afforded very different forms of social and affective engagement than, say, electronic/dance music, which in turn encompassed a huge amount of difference in itself given its incorporation of multiple subgenres. Only through comparison was I able to trace these links between field sites, translating what in one site was comparable to or divergent from, similar but not necessarily equivalent to, another.

A second important vector of comparison was my ability to map the movement of individuals across different musical collectivities, and in so doing, to understand both the interrelations and disconnections between scenes, and the potential reasons why certain musical performance situations made more sense to certain individuals than others. It became possible to see, for example, why those I'd met at one field site felt unable or unwilling to participate in the co-present spaces of another, in spite of liking and listening to the music of that other field site and feeling part of its 'virtual' community. One way this came to light during fieldwork was when a number of women expressed a strong affinity for dub reggae, drum and bass, and grime but admitted that they wouldn't participate in these musics' live scenes because the masculine atmospheres and protocols of the spaces in which the musics were embedded made them uncomfortable. Not only, then, did comparison allow me to grasp the particularities and differences between the genres themselves and their collective spaces of performance. It also enabled me to trace the musical pathways of individuals distributed across those collective spaces, and thereby to grasp the differing degrees of access and urban mobility that different people harbour in relation to diverse musical genres. Comparison as a methodology thus helped me, in the words of Moira Gatens and Genevieve Lloyd, to 'think in the space between individuals and groups' (Gatens and Lloyd 2002: 72); to realize the collective dimensions of selfhood, and to understand that, as individuals, we are 'inserted into economies of affect and imagination which bind us to others in relations of joy and sadness, love and hate, co-operation and antagonism' (73). I do not believe that these insights would have come to fruition with single-sited research.

The opportunity to grapple with the qualitative complexity of live crowds was partly also attributable to my decision to use ethnography as a primary research method. By virtue of its situated, local perspective, ethnography allowed me to get right up close to the fleeting, sensory, and ephemeral aspects of urban musical experience. It facilitated detailed observation of the movements, gestures, and actions of individuals within musical collectivities; the demographics and social relations (convivial, apathetic, hostile, etc.) brought into play by such collectivities; and the elusive immaterial quality often referred to as 'vibe' or 'energy' that circulates through a musical/sonic body. More than this, though, ethnography allowed me to enact continual shifts in perspective between multiplicity and singularity: to attend qualitatively to the threshold mechanisms that enable people to move between private and public experience in the presence of music and sound, and in so doing, to see how relations of difference and individuality coexist with, and are crossed by, relations of unity and similarity. I was thus able to approach a question that has perplexed social theorists for over a century – that being the question, as Lisa Blackman frames it, of how the many can act as one, and how one can act as many (Blackman 2012) – with a methodological stance that neither reduced the musical public to a unitary totality or entity, nor permitted descent into bifurcating plurality and heterogeneity. Moreover, when triangulated with comparative and diachronic analysis, such an approach brought to light how particular socio-musical formations exhibit far greater degrees of stability and continuity than others, and how relatedly, as Born puts it, certain genres are transmitted through time and space 'much more successfully than others' (Born 2010c: 244).

To give a simplified example of this: as part of my fieldwork, I sought to bring analyses of London's contemporary classical music scenes – both the 'new music' and established concert hall scenes – into dialogue with literature on the social history of concert life in Europe and America. What this approach revealed was an extraordinary degree of continuity between past- and present-day audiences. Customs, postures, and practices that were established among bourgeois concertgoers in the mid-nineteenth century, such as silently submitting to the 'work of art', suppressing outward emotional responses to the music, and policing the manners of fellow concertgoers, endure practically unchanged into the twenty-first century. Further, such practices – as well as the primarily white, middle-class, musically educated publics that enact them – endure *in spite of* contemporary classical music curators' explicit attempts to draw new kinds of audience and alleviate the formalities associated with classical performance by relocating the music to nightclubs and other non-traditional concert spaces and reprogramming it alongside popular and non-Western genres. The picture that emerges is thus one of profound historical longevity and resistance to change. 'New music' initiatives seek to initiate transformation by seemingly returning to a pre-nineteenth century model of concert life, emphasizing 'miscellany' as a programmatic principle, encouraging informal behaviours, and relocating the music to quotidian urban spaces such as parks and public squares, as was common in the eighteenth century; yet audiences not only remain normative to the genre, particularly in terms of race and class, but also struggle to relinquish the listening habits and affective registers of nineteenth-century white, male, heterosexual bourgeois idealism. Classical music's antiquated social and embodied norms are, then, seemingly ingrained to such an extent that changes in spatial location and musical programming tend to be fairly inconsequential.

By contrast, the dub reggae and dubstep assemblages exhibit a much greater degree of contingency, with alterations to the spaces and sites of performance impacting the musics' social identity formations in significant ways. When dubstep crossed over to mainstream in the mid to late 2000s, for example, the genre's migration to new, less 'underground' spaces helped to redraw gatekeeping boundaries, making the scene more accessible to women as well as to white middle-class groups. Unlike the intractability of classical music, changes in venue, promotion, and publicity were thus seen to shift the demographics of dubstep audiences quite dramatically. Similar processes have taken place in dub reggae, specifically in relation to Jah Shaka, who runs one of the UK's oldest sound systems. As my interlocutors reflected, Shaka dances held at cultural centres in the 1970s and 1980s were predominantly black and male – much like the sound system events that take place at Caribbean cultural centres today. Yet, during the 1990s, this changed in a fundamental way. Around 1992, the Black Arts administration service Culture Promotions took over Shaka's management and started promoting to a wider audience. Booking Shaka gigs at venues such as the Rocket on Holloway Road, which was popular with students, as well as the Dome in Tufnell Park, significantly modified Shaka's crowds, bringing in a considerable white middle-class following and many more women in addition to his mixed-class black and Asian crowd. Gradually becoming a staple of Shaka dances, this social heterogeneity again demonstrates the power of promotion, venue, and urban location to shift audience demographics in certain musical assemblages.

The broader observation, however, is that there appears to be a racial and class dimension to these processes. White middle-class audiences are drawn into black-originating or multiracial genres such as dub, dubstep, and more recently, grime, at moments when these genres have crossed over to mainstream or changed their promotional strategies and venues; while conversely, black and/or working-class audiences have not been able or do not wish to move into historically white and/or higher-class musical spaces such as classical music, regardless of the changes made to space, site, and publicity. Such findings suggest, firstly, that certain musical public spheres are much more resistant to change than others; and secondly, that social boundaries – particularly those of (higher) class and race (whiteness) – are being inadvertently recreated by the classical music assemblage itself, even as claims are made for trying to transform them.

These kinds of comparative insights came to fruition, in part, by repeatedly attending, observing, and documenting relevant musical events; building trusting relations with, and interviewing, audience members, musicians, promoters, venue owners, and sound engineers; spending time in record shops, record production houses, venues, cafes, and other neighbourhood spaces; and 'following' the activity of musical initiatives from offline to online spaces. I then sought to read across from this ethnographic work to relevant histories and theories. Nonetheless, I still faced three major challenges in undertaking my fieldwork. The first was the question of how to research and convey the mercury-like qualities of musical affect and atmosphere in ways that didn't simply fall back on discursive methods. The second challenge was how to conduct qualitative crowd research, sometimes in situations where audience numbers were in the thousands, or where my hopes of talking to more than a handful of people during the course of an event were dashed by the rules and taboos of the genre. And the third challenge was how to approach the study of music and sound art in a city as vast and as rapidly changing as London. How, in other words, could I even scratch the surface of this musically saturated, densely populated city, barely recognizable from one year to the next in its high streets, backstreets, nightclubs, and skylines? Responding to these challenges, I developed a toolbox of audiovisual, participant-based, and collaborative methods, which I mobilized alongside conventional ethnographic techniques. In the next and final section, I unpack this toolbox in more detail.

Live methods

A key source of inspiration when designing my fieldwork was Les Back and Nirmal Puwar's *Live Methods* (2013). Writing from a sociological viewpoint, Back and Puwar argue that digital technologies have transformed our ways of apprehending and analysing the social world, creating space for an 'expanded' sociology. With the smartphone having largely eclipsed the notebook as the ethnographer's storage device, digital methods such as photography, video, and audio recording – all of which are embedded in a smartphone – offer new tools for 'real-time' or 'live' investigation and 'inter-corporeal understanding' (Back and Puwar 2013: 7). By making use of such tools, they suggest, we might get closer to 'the fleeting, distributed, multiple, [and] sensory [...] aspects of sociality' through research techniques that are mobile and operate from 'multiple vantage points' (28).

Several of the methods reviewed earlier can be classified as 'live' – from Overell's use of digital recording at grindcore gigs to Hall and colleagues' soundwalking interviews. In my own fieldwork, I also attempted to put a number of 'live methods' into practice. Among the most fruitful was a 'think-out-loud' technique that I adapted from Tia DeNora's (2000) pioneering work on music in everyday life. Similar to Overell's use of 'participant-sensing', this method involved inviting audience members at different musical and sonic events to literally 'think-out-loud' into a recording device – my iPhone – about their real-time social and embodied experiences. Part of the appeal of this method was that it attributed a certain agency to my participants, allowing them to make spontaneous utterances without me intervening or taking notes. But these audio snapshots of dancefloors and concert spaces also proved to be an invaluable way of documenting the minutiae of urban musical experience. At electronic/dance music nights, for example, participants would use 'think-out-loud' to express disgust at the pungent bodily smells that had suddenly interrupted their musical pleasure; comment on the way that an event mutates from one hour to the next, as crowds flood in to see their favourite DJ and then vacate the dancefloor immediately after; and lament the tendency for intense crowdedness to breed sexual harassment. Further, these audio memos were revealing in terms of the (dis)connections they exposed between sonic foreground and background. In one memorable example, a participant can be heard complaining about a high-profile DJ's mixing skills not being up to scratch, just as a distorted but distinctly 'dodgy mix' becomes audible overhead. In conjunction with my own observations of individual-collective relations, as well as informal dialogue with crowd members, this method thus helped me to build a rich sensory-affective picture of music, sound, and sociality in their live forms.

Encouraged by the success of 'think-out-loud', I also pursued the idea of mobilizing a 'team-based auto-ethnography'. Conducted once again through audio-recorded voice memos, I asked a group or 'team' of three or four participants to become 'co-researchers' by accompanying me to a particular event, recording their observations and experiences into their phones, and forwarding them to me at the end of the night. Though this method proved difficult to coordinate, and I only succeeded in making it work a handful of times, the data it generated was illuminating, offering glimpses into the potentials that digital technologies harness for transforming ethnographic crowd research. Indeed, such a technique was an effective way of 're-imagining [participant] observation', producing what Back and Puwar call a 'pluralization of observers' (Back and Puwar 2013: 7): a group of individuals who document the same event from multiple vantage points, as different social-subjective nodes in a complex crowd or public. Not only did such a technique allow me to involve my participants in the research, acknowledging them as peers and listening to their thoughts and concerns; it also illuminated possible new ways of researching 'live' and 'live-streamed' musical events simultaneously, with a group of researchers potentially dispersed across co-present and mediated publics, working collaboratively between different cities and even time zones. Finally, what both the 'team-based' and 'think-out-loud' methods drew attention to was how the affective and the sensory were almost always the first points of reflection for participants in documenting their sonic experiences. This often worked as a complement to my own text-based field notes, which sometimes centred more on larger-scale observations, such as audience demographics, entry fees and dress codes, spatial and material properties, venue capacity, and levels of policing. As such, I was able to amass data that moved constantly between music and sound's micro-socialities and macrosocial conditions and qualities.

A final 'live method' that I put to use was field recording. Initially, I would make recordings of the musical events I attended purely for mnemonic purposes – to help me remember what was going on or what something sounded like. As such, these recordings were often low fidelity and semi-random: sporadic snapshots of a dancefloor or snatches of conversation captured in the smoking area, sometimes no more than a few seconds long. Yet, listening back at home, I was often amazed at the level of sonic detail that my iPhone had managed to capture, rendering audible imperceptible, forgotten moments and affective transitions that would have otherwise passed me by. One could hear, in the form of shouts and cheers, for example, the jubilant collectivizing energy that erupts across a dancefloor when a well-loved tune drops; the mediation of sounds and vibrations through the physical materials of a spatial environment such that those sounds can then tell us something about the textural surfaces of that cultural space; the distortion on the recording and the levels of shouting that pulsate into audibility between bass kicks, often indicating a deliberate cranking up of the volume by venues to encourage people to 'drink more, talk less', as one engineer told me; and the moment when the selector started the record at the wrong speed by accident, and everyone had a good laugh.

I started to see how these soundscapes were imbued with much wider urban political issues and cultural histories. Audible expressions of disgust and exasperation at the overcrowding of a dance club event, for instance, were often a trickle-down effect of intensifying gentrification and social control, with venues forced to 'oversell' their events in order to cover the costs of extortionate commerce-driven DJ fees, soaring rents, and compulsory security measures. By listening, I was able to gain an alternative insight into how these issues manifest audibly and physically on dancefloors: how certain musical public spaces in London are becoming sites of rigorous control permeated by a crushing, individualizing crowd density. On the other hand, capturing the rattling windows and vibrating wooden-panelled toilets of a bass-infused reggae dance was simultaneously to become sensitive to long cultural histories of migration, homemaking, and survival. Indeed, the refraction of sound through the 'homely' surfaces of wood and carpet that have sustained African Caribbean cultural centres since the post-war period speaks back to a time when black British communities were violently excluded from urban public space and compelled to create their own venues. Initially little more than living room dances with the furniture pushed back ('shebeens'), these cultural spaces or 'public homes' today remain invested with sonic histories of resistance and defiance by virtue of their specific material and spatial properties. Field recording as a method, then, revealed sound's potential to impart alternative or additive knowledges about the urban social world and its musical and sonic environments – to do justice to the impassioned and textured qualities of sonic sociality and history in ways that writing, speaking, and vision struggle to. Such a method in turn raises questions about the epistemological work that sound has the capacity to do, and how sound might be incorporated into the research process *as sound*, rather than as transcription or other kinds of discursive translation.[2]

'Draw Your Musical London'

As a final methodological tool, relating particularly to the challenges of urban ethnography, I took inspiration from urban sociologist Emma Jackson (2012), who, in her work on contemporary spaces of homelessness, invites her participants to produce mental maps of the city under the instruction, 'Draw Your London.' Through their creativity and willingness, Jackson is able to chart the trajectories of young homeless people in London: their routes through particular neighbourhoods, their attachment to specific urban places, and the forms of violence and governance they encounter. Moreover, by virtue of the composite maps, Jackson is able to identify similarities and differences between her participants – mutual fears, danger zones, shared spaces of loss, belonging, and opportunity (Jackson 2012).

Repurposing this method, I experimented with asking my interviewees to 'Draw Your Musical London,' inviting them to create a musical mind map of the city that showed the spaces and places that were of musical significance to them.[3] Part of my reasoning for deploying such a method was so that I could better understand how people become implicated in wider socio-spatial, affective, and musical currents, whilst remaining disconnected from, and unable to 'make sense' of, others. And indeed, an important finding to arise from the 'musical mapping' project was how participants perceived themselves to be spatially and musically 'distributed'. Brief descriptions scribbled on the maps often relayed a deep sense of attachment and nostalgia to multiple spaces, people, and sounds, many of which were placed at a temporal as well as spatial distance from each other. One participant, for example, included a color-coded 'Key' to delineate different decades of musical life (1970s, 1980s, 1990s, etc.), while another mourned the loss of bygone life-changing nights experienced in his twenties. In addition to this palimpsestic quality, what the maps also conveyed was a strong sense of the socio-musical circles through which people deemed themselves to move. Of particular interest, here, was how participants' cartographic portrayals of themselves sometimes reflected a merging of 'imagined community' and physical reality, incorporating venues and musical spaces that they'd never actually been to before but still felt they belonged to. Equally, there were times when participants would omit certain musical 'selves' from their maps, wanting to be perceived in a certain way, only for these 'hidden' musical identities to surface in an interview or discussion at a later date. Linked to this, in turn, was the question of people's musical-geographic 'radiuses' and degrees of urban mobility, often detectable from the size of their genre maps and the breadth of the spaces that were accessible to them. Indeed, studying the maps in conjunction with interviews and participant observation became an important way of analysing the eclecticism and scale of people's musical affiliations and participatory horizons – their 'omnivorousness' (Peterson and Kern 1996) – which I often found to be heavily mediated by class, race, gender, sexuality, and age. In this sense, the maps revealed patterns and disconnections between the private musical tastes and listening habits of particular individuals, and the degree to which those individuals were, or were not, able to traverse public musical-spatial boundaries and urban thresholds (see Figures 6.1, 6.2, and 6.3).

When triangulated with ethnographic, historical, and theoretical approaches, 'live' and experimental methods such as these have the potential to significantly enhance our ways of knowing and understanding cities and their complex music and sound cultures, not least by offering ways of overcoming the practical challenges of qualitative crowd research and generating new techniques for exploring the sonic texture of urban nightscapes and the spatial distribution of sonic 'selves'. Moreover, working with critical forms of cartography and field recording that are participant- as well as researcher-based seems to go some way towards allowing the researcher to experience the world beyond their own mind. At the same time, it is clear that such methods also present new challenges, particularly in relation to questions of representation, ethics, and transferability. How, for example, does one go about naming, dissecting, and representing experiences that are felt, sensed, or only half-known, and what evaporates or gets lost in the process? How might a soundwalk or 'on-the-spot' voice memo be incorporated into the research process without recourse to description or text? How should one credit those participants who become central to the research through collaborative methods such as mapping and field recording? And what can be done about the potential non-transferability of digital and 'live' methods – something that I encountered in my own fieldwork upon realizing that 'think-out-loud' was extremely

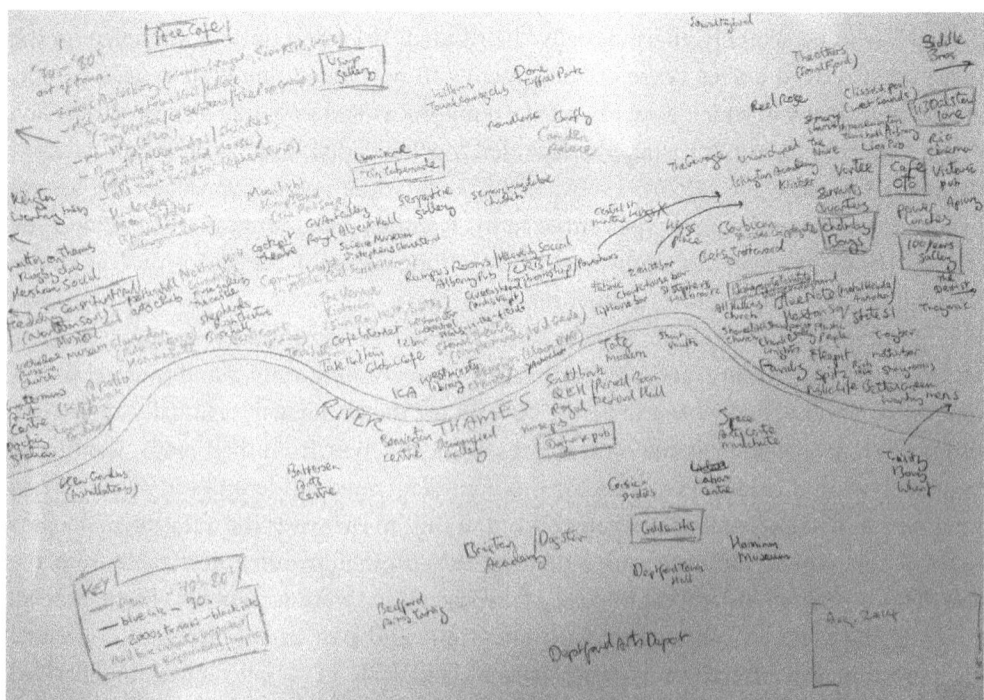

Figure 6.1 Chris's 'musical London' map, 2014.

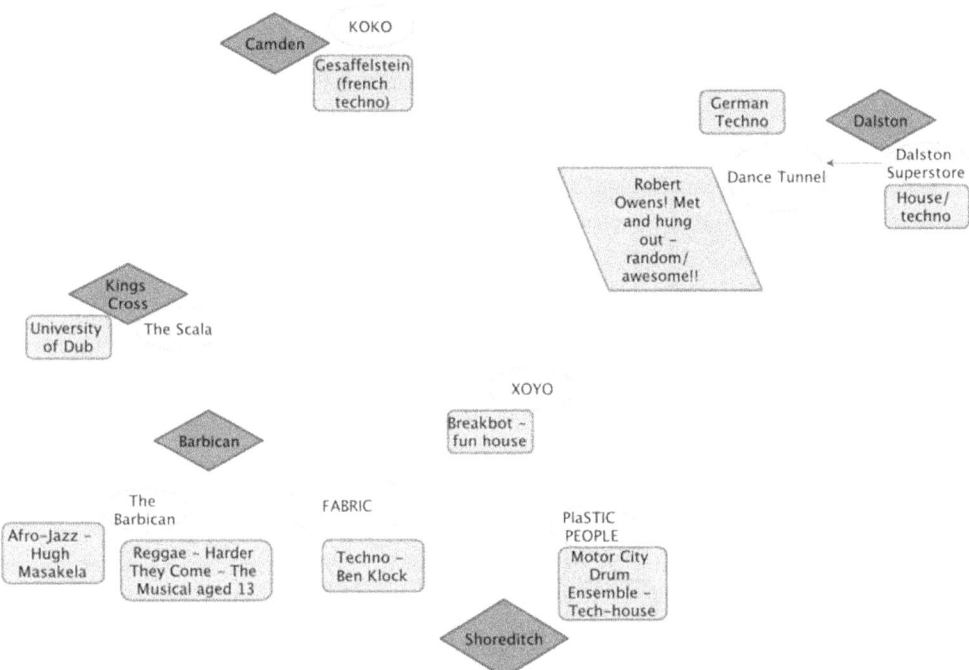

Figure 6.2 Ali's 'musical London' map, 2014.

productive across the electronic/dance music spectrum but not at all feasible during a classical concert, but which is also a conceivable problem for those working comparatively across cities. How might practices of field recording and 'think-out-loud' work in a city such as Beirut, for example, where cameras and sound equipment (particularly in the hands of Westerners) are viewed with intense suspicion and distrust?

Across sound studies and urban sociology, responses to some of these questions have started to emerge in the form of multimedia publication platforms, 'compound' sound-text-image research outputs, and reflections upon what it means to collaborate and co-author with our participants (Back, Shimser, and Bryan 2012; Gandy and Nilsen 2014; Ouzounian and Bingham-Hall 2019). To this I would add that there remains considerable scope for experimenting with 'live' methods across diverse urban musical contexts, and that if different cities have different methodological requirements, some of the methods outlined above might productively be tested, transplanted, and potentially modified according to the particular encultured cities/sites that they seek to reveal and transcribe.

Conclusion

The sheer range of techniques and approaches discussed in this chapter speaks both to the expansive interdisciplinary nature of the research being conducted between music, sound, and urban studies, and the challenges faced by those pursuing such research, as cities

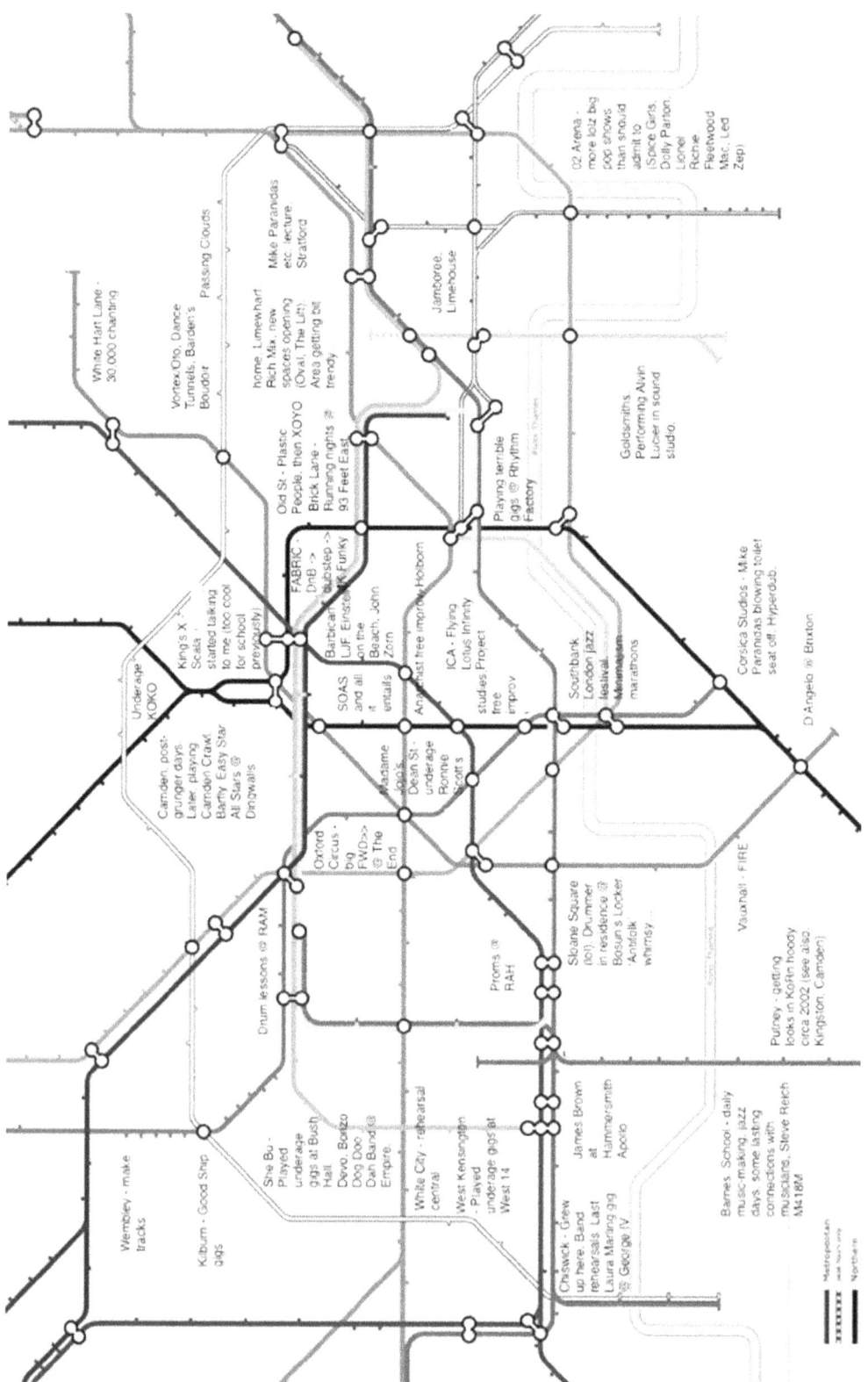

Figure 6.3 Martin's 'musical London' map, 2014.

themselves grow increasingly complex and demanding in terms of the methods and tools required to work effectively within them. Journeying through cultural geography, urban sociology, historical musicology, sound studies, and anthropology, one finds an array of innovative and carefully honed techniques for understanding specific dimensions of the sounding city – from the 'live' methods of listening, soundwalking, and digital recording, which have a particular capacity to render the processual sociality and impressionistic quality of contemporary urban life, to historical and literary depictions of the nineteenth-century acoustic city, which impart a vivid sense of the changing affective and ideological power of urban sound as cities themselves underwent dramatic change. Notwithstanding the specialist capacities of these methods, scholars such as Back (2009) have emphasized the need for method triangulation in grasping the contradictory, multifaceted, and often inconsistent nature of city life. In his own work on racism and multiculture in London's East End, Back moves between interviews with his participants, soundscape recordings of their daily social interactions, and historical analyses of migration, class, and belonging in East London to reveal significant disparities between words, sounds, and actions: interviewees' racist melancholia and historical amnesia around 'whiteness' and community cohesion are undermined by the convivial intercultural exchanges and multiracial friendships that they perform and participate in daily on the streets. As with the contradictions that emerged in my own fieldwork between participants' cartographic and interview-based portrayals of themselves, Back's findings reiterate the importance of traversing different spheres and scales of sociality – from the intimate one-on-one interview through the public social arena to the diachronic 'long' view – in order to grasp the chasms as well as connections that arise between the said and the seen/done, between imagination and reality, biography and history.

    Building upon this notion of method triangulation, I pointed – in the second half of the chapter – towards the potentials of a relational methodology that moves pluralistically and at times agonistically between history, (comparative) ethnography, and theory. Such an approach takes inspiration from Born (2005), Straw (2001), and others such as Lawrence Grossberg (2014), who argue for a closer methodological relationship between the affective, performative dimensions of musical urban sociality and the wider institutional forces and 'weighty histories' that 'give each seemingly fluid surface a secret order' (Straw 2001: 248); but who also – particularly in Born's and Grossberg's case – stress the importance of holding theoretical discourses to account through rigorous historically informed empiricism. In this way, speculative concepts and theories can be treated as 'tools' whose feasibility has to be 'constantly constructed and contested' in relation to specific concrete situations, while the complexity of the empirical, in turn, may be enlivened and potentially reconceived by imaginative conceptual thinking (Grossberg 2014: 13; cf. Born 2010b). Indeed, only by pursuing such a methodology – one that places theory in the teeth of ethnography and history, that refuses, in Deleuzian terms, to choose one 'or' the other – can the limits of conceptual or empirical or historical work alone be deciphered, and the potentials for more radically collaborative and generative ways of working be brought into being. Within this, as I have shown, digitally enabled 'live' methods can take on a critical role in triangulating the empirical, rendering audible the historical, and dramatizing or modifying the conceptual.

Notes

1. For example, techno producer Jeff Mills's collaboration in 2015 with the BBC Symphony Orchestra and dubstep innovator Mala's collaboration in 2018 with the Outlook Orchestra.
2. This is something that I have explored in a short sound piece published as part of the Optophono edition 'Acoustic Cities: London & Beirut' (Ouzounian and Bingham-Hall 2019) and is something that I continue to explore with my friend and collaborator Freya Johnson Ross.
3. With hindsight, I realize that this method in many ways resembles the hand-drawn maps of Lashua and colleagues' (2010) participants in their study of music in Liverpool.

References

Amin, A. (2012). *Land of Strangers*. Cambridge: Polity Press.
Amin, A. and N. Thrift (2002). *Cities: Reimagining the Urban*. Cambridge: Polity Press.
Anderson, B., F. Morton, and G. Revill (2005). 'Editorial: Practices of Music and Sound'. *Social & Cultural Geography* 6 (5): 639–644.
Back, L. (1996). *New Ethnicities and Urban Culture: Racisms and Multiculture in Young Lives*. London: University College London Press.
Back, L. (2009). 'Researching Community and its Moral Projects'. *21st Century Society: Journal of the Academy of Social Sciences* 4 (2): 201–214.
Back, L. (2016). 'Moving Sounds, Controlled Borders: Asylum and the Politics of Culture'. *YOUNG* 24 (3): 185–203.
Back, L., S. Shimser, and C. Bryan (2012). 'New Hierarchies of Belonging'. *European Journal of Cultural Studies* 15 (2): 139–154.
Back, L. and N. Puwar (eds) (2013). *Live Methods*. London: Wiley.
Baker, G. and T. Knighton (eds) (2011). *Music and Urban Society in Colonial Latin America*. Cambridge: Cambridge University Press.
Balaÿ, O. (2016). 'The Soundscape of the City in the Nineteenth Century'. In K. Gibson and I. Biddle (eds), *Cultural Histories of Noise, Sound and Listening in Europe, 1300–1918*, 221-234. London: Routledge.
Blackman, L. (2012). *Immaterial Bodies: Affect, Embodiment, Mediation*. London: SAGE.
Born, G. (2005). 'On Musical Mediation: Ontology, Technology and Creativity'. *Twentieth Century Music* 2 (1): 7–36.
Born, G. (2010a). 'Listening, Mediation, Event: Anthropological and Sociological Perspectives'. *Journal of the Royal Musical Association* 134 (1): 79–89.
Born, G. (2010b). 'The Social and the Aesthetic: For a Post-Bourdieuian Theory of Cultural Production'. *Cultural Sociology* 4 (2): 1–38.
Born, G. (2010c). 'On Tardean Relations: Temporality and Ethnography'. In M. Candea (ed.), *The Social after Gabriel Tarde: Debates and Assessments*, 230–247. Oxford: Routledge.
Born, G. (ed.) (2013). *Music, Sound and Space: Transformations of Public and Private Experience*. Cambridge: Cambridge University Press.

Boutin, A. (2015). *City of Noise: Sound and Nineteenth-Century Paris*. Urbana, IL: University of Illinois Press.
Bramwell, R. (2015). *UK Hip-Hop, Grime and the City: The Aesthetics and Ethics of London's Rap Scenes*. London: Routledge.
Bull, M. (2007). *Sound Moves: iPod Culture and Urban Experience*. London: Routledge.
Cobussen, M., V. Meelberg, and B. Truax (eds) (2016). *The Routledge Companion to Sounding Art*. New York: Routledge.
Cohen, S. (2007). *Decline, Renewal and the City in Popular Music Culture: Beyond the Beatles*. Aldershot: Ashgate.
Cohen, S. (2011). 'Cavern Journeys: Music, Migration and Urban Space'. In B. Dueck and J. Toynbee (eds), *Migrating Music*, 235–250. London: Routledge.
Cusack, P. (2017). *Berlin Sonic Places: A Brief Guide*. Berlin: Wolke Verlag.
DeNora, T. (2000). *Music in Everyday Life*. Cambridge: Cambridge University Press.
De Witte, M. (2008). 'Accra's Sounds and Sacred Spaces'. *International Journal of Urban and Regional Research* 32 (3): 690–709.
Dueck, B. and J. Toynbee (eds) (2011). *Migrating Music*. London: Routledge.
Eisenberg, A. (2013). 'Islam, Sound and Space: Acoustemology and Muslim Citizenship on the Kenyan Coast'. In G. Born (ed.), *Music, Sound and Space: Transformations of Public and Private Experience*, 186–202. Cambridge: Cambridge University Press.
Foucault, M. ([1981] 1991). 'Questions of Method'. In G. Burchell, C. Gordon, and P. Miller (eds), *The Foucault Effect: Studies in Governmentality*, 73–86. Chicago: University of Chicago Press.
Gandy, M. (2005). 'Cyborg Urbanization: Complexity and Monstrosity in the Contemporary City'. *International Journal of Urban and Regional Research* 29 (1): 26–49.
Gandy, M. and B. Nilsen (eds) (2014). *The Acoustic City*. Berlin: Jovis.
Garcia, L-M. (2016). 'Techno-Tourism and Post-Industrial Neo-Romanticism in Berlin's Electronic Dance Music Scenes'. *Tourist Studies* 16 (3): 276–295.
Gatens, M. and G. Lloyd. (2002). *Collective Imaginings: Spinoza, Past and Present*. London: Routledge.
Grossberg, L. (2014). 'Cultural Studies and Deleuze-Guattari, Part 1'. *Cultural Studies* 28 (1): 1–28.
Hall. T., B. Lashua, and A. Coffey (2008). 'Sound and the Everyday in Qualitative Research'. *Qualitative Inquiry* 14 (6): 1019–1040.
Hankins, J. D. and C. Stevens (eds) (2013). *Sound, Space and Sociality in Modern Japan*. Oxford: Routledge.
Harvey, D. (2006). *Spaces of Global Capitalism: A Theory of Uneven Geographical Development*. London: Verso.
Henriques, J. and B. Ferrara (2016). 'The Sounding of the Notting Hill Carnival: Music as Space, Place and Territory'. In J. Stratton and N. Zuberi (eds), *Black Popular Music in Britain Since 1945*, 131–152. Oxford: Routledge.
Hirschkind, C. (2006). *The Ethical Soundscape: Cassette Sermons and Islamic Counterpublics*. New York: Columbia University Press.
Holt, F. and C. Wergin (eds) (2013). *Musical Performance and the Changing City: Post-industrial Contexts in Europe and the United States*. Oxford: Routledge.
Irving, D. R. M. (2010). *Colonial Counterpoint: Music in Early Modern Manila*. Oxford: Oxford University Press.

Jackson, E. (2012). 'Fixed in Mobility: Young Homeless People and the City'. *International Journal of Urban and Regional Research* 36 (4): 725–741.

Klotz, S., P. Bohlman, and L-C. Koch (eds) (2018). *Sounding Cities: Auditory Transformations in Berlin, Chicago, and Kolkata*. Zurich: Lit Verlag.

Laing, D. (2009). 'Gigographies: Where Popular Musicians Play'. *Popular Music History* 4 (2): 196–219.

Lappin, S., G. Ouzounian, and R. O'Grady (2018). 'The Sound-Considered City: A Guide for Decision Makers'. *Recomposing the City*. Available online: https://static1.squarespace.com/static/520167f1e4b0b5e68d9b07cd/t/5b3a0bff8a922dbccfbab57c/1530531021679/The+Sound-Considered+City+Low+Resolution+Web+Copy.pdf (accessed 29 June 2020).

Lashua, B., S. Cohen, and J. Schofield (2010). 'Popular Music, Mapping, and the Characterization of Liverpool'. *Popular Music History* 4 (2): 126–144.

Llano, S. (2018). *Discordant Notes: Marginality and Social Disorder in Madrid, 1850–1930*. Oxford: Oxford University Press.

London Contemporary Music Festival (LCMF) (n.d.). 'About'. Available online: http://www.lcmf.co.uk/About (accessed 29 June 2020).

Marcus, G. E. (1995). 'Ethnography in/of the World System: The Emergence of Multi-Sited Ethnography'. *Annual Review of Anthropology* 24: 95–117.

Nonclassical (n.d.). 'About Us'. Available online: http://www.nonclassical.co.uk/about-us/ (accessed 29 June 2020).

O'Keeffe, L. (2015). 'Thinking Through New Methodologies. Sounding Out the City with Teenagers'. *Qualitative Sociology Review* 11 (1): 6–32.

Orchestra of the Age of Enlightenment (2020). 'The Night Shift: 2020–21'. Available online: https://oae.co.uk/season/the-night-shift/ (accessed 29 June 2020).

Ouzounian, G. (2021). *Stereophonica: Sound and Space in Science, Technology, and the Arts*. Cambridge, MA: MIT Press.

Ouzounian, G. and J. Bingham-Hall (eds) (2019). *Acoustic Cities: London & Beirut*. Optophono edition 4. Oxford: Optophono. Available online: https://www.optophono.com/acousticcitieslondonbeirut (accessed 29 June 2020).

Overell, R. T. (2012). 'Brutal Belonging: Affective Intensities in, and between, Australia's and Japan's Grindcore Scenes'. PhD thesis, University of Melbourne.

Peterson, R. and R. Kern (1996). 'Changing Highbrow Taste: From Snob to Omnivore'. *American Sociological Review* 61 (5): 900–907.

Picker, J. (2003). *Victorian Soundscapes*. Oxford: Oxford University Press.

Pinder, D. (2001). 'Ghostly Footsteps: Voices, Memories and Walks in the City'. *Cultural Geographies* 8 (1): 1–19.

Revill, G. (2000). 'English Pastoral: Music, Landscape, History and Politics'. In I. Cook, D. Crouch, S. Naylor, and J. Ryan (eds), *Cultural Turns/Geographical Turns*, 140–158. London: Longman.

Revill, G. (2013). 'Points of Departure: Listening to Rhythm in the Sonoric Spaces of the Railway Station'. *Sociological Review* 61 (1): 51–68.

Rotter, A. (2019). *Empires of the Senses: Bodily Encounters in Imperial India and the Philippines*. Oxford: Oxford University Press.

Royaards, M. (2019). 'City of Impulse'. In G. Ouzounian and J. Bingham-Hall (eds), *Acoustic Cities: London & Beirut*. Oxford: OPTOPHONO – OP4.

Rutherford, D. (2016). 'Affect Theory and the Empirical'. *Annual Review of Anthropology* 45: 285–300.

Saldanha, A. (2002). 'Music, Space, Identity: Geographies of Youth Culture in Bangalore'. *Cultural Studies* 16: 337–350.

Sassen, S. (2001). *Global City: London, New York, Tokyo*. Princeton, NJ: Princeton University Press.

Schafer, R. M. (1970). *The Book of Noise*. Vancouver: Price Print.

Simpson, P. (2017). 'Sonic Affects and the Production of Space: "Music by handle" and the Politics of Street Music in Victorian London'. *Cultural Geographies* 24 (1): 89–109.

Smith, S. (1997). 'Beyond Geography's Visible Worlds: A Cultural Politics of Music'. *Progress in Human Geography* 21: 502–529.

Stirling, C. (2019). 'Orbital Transmissions: Affect and Musical Public-Making in London'. DPhil thesis, University of Oxford.

Stokes, M. (2010). *The Republic of Love: Cultural Intimacy in Turkish Popular Music*. Chicago: University of Chicago Press.

Straw, W. (1991). 'Systems of Articulation, Logics of Change: Communities and Scenes in Popular Music'. *Cultural Studies* 5 (3): 368–388.

Straw, W. (2001). 'Scenes and Sensibilities'. *Public* 22/23: 245–257.

Sundaram, R. (2009). *Pirate Modernity: Delhi's Media Urbanism*. Oxford: Routledge.

Thompson, E. (2004). *The Soundscape of Modernity: Architectural Acoustics and the Culture of Listening in America, 1900-1933*. Cambridge, MA: MIT Press.

Vecchiola, C. (2011). 'Submerge in Detroit: Techno's Creative Response to Urban Crisis'. *Journal of American Studies* 45 (2011): 95–111.

Waldock, J. (2011). 'SOUNDMAPPING: Critiques and Reflections on this New Publicly Engaging Medium'. *Journal of Sonic Studies* 1. Available online: https://www.researchcatalogue.net/view/214583/214584 (accessed 29 June 2020).

Wood, N., M. Duffy, and S. J. Smith (2007). 'The Art of Doing (Geographies of) Music'. *Environment and Planning D: Society and Space* 25: 867–889.

7

Sound and Pedagogy: Taking Podcasting into the Classroom

Neil Verma

There is no obvious point from which to methodically explore the integration of sound into learning settings. The best-known critical lineages, from Plato's colloquy on speech in *Phaedrus* and R. Murray Schafer's classes on noise in 1970s British Columbia to John Cage's Black Mountain College seminars and Pauline Oliveros's Deep Listening Institute, fail to capture the long and elaborate mutual imbrication across cultures of instruction with sound-making, a category of human activity that could arguably include any learning that involves vocalization as its minimal condition. 'Listening is important in all educational experiences', Schafer has written, 'whenever verbal or aural messages are exchanged' (Schafer 1992: 7). Rather than trying to relate all that – but as a way to give a much more practical perspective on 'sound teaching' and the myriad issues that attend it – I focus in this chapter on a recent case of a much more diffuse phenomenon: how post-secondary instructors have been taking up podcasting as a methodology for student work, something being developed in an improvised way across several institutions and disciplines in the United States. Although it draws on traditions – most proximately, radio documentary and sound art – podcasting is perceived to be a 'new' sound practice, maybe the first sound-pedagogical methodology to emerge in dialogue with sound studies itself, what Sterne calls the 'interdisciplinary ferment' (Sterne 2012: 2) evident across the humanities and social sciences over the past decades. Musical instruction and appreciation, rhetoric and public speaking, radio and recording, soundwalking: all these existed at least as semi-institutionalized training practices before 'sound studies' consolidated as a critical mode. Podcasting, by contrast, did not. It is a sonic pedagogy that emerged at the same time as sound studies, one that is especially self-aware of its irreducible sonorousness.

We can follow the emergence of this teaching practice precisely. Podcasting first came to post-secondary classrooms just after the iPod launched in 2001, as educators across settings began to distribute lectures on iTunes, a movement that used the neologism 'podagogy' (Campbell 2005; Rosell-Aguilar 2007; Bell 2008). More recently, in part as a response to the explosion in narrative podcasting since the popularity of *Serial* in 2014, an overlapping approach has begun to emerge as instructors ask students to produce podcasts for classes rather than just distributing their own lectures digitally. The goal of this chapter is to gain insight on this newer 'Podagogy 2.0' moment from instructors who ask students to learn 'by' podcasting rather than 'through' the podcasts of others. To this end, I interviewed nineteen faculty members from around the United States, learning about their motives, methods, and outcomes. While the interviews suggested a feeling of success, three areas of ambivalence emerged. The first had to do with finding a balanced approach to teaching technology. The second had to do with writing skills, and how podcast teaching contributes to or neglects them. Finally, there was a question about how podcasting troubles boundaries of what is 'inside' and what is 'outside' the class. Together, these three areas of uncertainty reflect perennial difficulties and affordances of sound-based pedagogical practices, while at the same time showing some of the particular issues that instruction through digital technologies tend to entail.

Podagogy 1.0/Podagogy 2.0

There is a long-standing lack of qualitative, experience-driven advice on podcast creation by students, although there is a reasonable body of research on podcast creation by instructors (Campbell 2005; King and Gura 2007; Bell 2008). Educators were among the earliest to make use of the 'culture of mobile listening' that emerged in the early years of the twenty-first century associated with the rise of the iPod, sending out lectures and assignments as audio (Bull 2005). In 2004 a well-publicized project at Duke University saw all freshman given free iPods; two years later philosopher Susan Stuart's lectures on Immanuel Kant unexpectedly became among the most downloaded tracks in the education category of iTunes (Macleod 2006). Scholarly literature that followed these and other developments focused on a few key areas, such as the use of podcasts in language learning (Rosell-Aguilar 2007), along with a number of pilot studies of techniques in specific fields (Dale and Pymm 2009; Ng'ambi and Lombe 2012). Some argued that podagogy promoted flexibility and motivation among students, increasing their level of engagement (Salmon and Nie 2008; Dale and Pymm 2009), while others were sceptical about mere lecture capture as a meaningful change from existing practices or good use of time (King and Gura 2007; Kazlauskas and Robinson 2012).

For the purposes of this chapter, I want to step away from that literature a little to focus on what I will call 'Podagogy 2.0', which emphasizes the integration of podcast creation on the part of students, rather than faculty, into coursework. If the 'pod' in podagogy in the early 2000s referred to putting podcasting in the hands of educators, today it increasingly refers to putting podcasting in the hands of students. I will focus especially on the classroom setting, although that is not the only place where podcasts are appearing on campuses. For five years or so there have been high-profile examples of student-driven and student-produced podcasts at universities across the United States, such as Stanford's *State of the*

Human podcast and *Serendipity* at Sarah Lawrence College, as well as a wide variety of clubs, but in this chapter I will focus on everyday uses of podcasting in classrooms, where student work primarily faces peers and instructors in a situation where it is intended to contribute to curricular outcomes.

When it comes to student perceptions and assessments of outcomes of learning through podcasting, there is rich literature on what can or might be accomplished (Salmon and Nie 2008; Forbes 2015; Galloway 2017). But it has been hard to come by reflection by educators about why they teach in this way, what they have learned about it, and how they prepare syllabi and conduct classes. Without this perspective, collective wisdom about the practice, as well as details about the day-to-day work of assignments and assessments could be lost. My question for the project was an open-ended one: what issues face instructors who are a part of the sonic methodology of 'Podagogy 2.0'? The study has pointed to three answers: uncertainties in approaching technology, ambivalence about writing, and problems and opportunities of audience and context.

Project background and methodology

I began asking students to do a podcast in lieu of writing weekly responses in my class 'Film Among the Sound Arts' in 2014. The concept was to look at film sound in conjunction with sound art of other kinds: radio, pop music, installation media, tape art, etc. Rather than require students to write essays about the films we watched, I asked them to create experimental sound pieces, which we called 'sound responses' or 'podcasts', interchangeably. For a class on *The Red Shoes*, for instance, one student imagined a key sequence of the film in stereo, recording the sound of ballet dancers in rehearsal forming geometric circles. In another assignment, a student created a reading of Poe's 'The Tell-tale Heart' using an effect she borrowed from *The Exorcist*. In both cases, students learned to 'make things' with listening and became more intimately aware of sound in the process. Satisfied with the experience, in 2015 I started a class in podcast studies, 'Podcasting and New Audio', which I have repeated at undergraduate and graduate levels. As an industry seemed to be growing around podcasting, I wanted to unsettle student habits of listening. I had long observed that while students could be adept at unpacking the aesthetics, ethics, and politics behind images, TV programmes, and films, they tended to listen to audio works lazily, unaware of how editing, mixing, and storytelling produce rhetorical effects; my podcasting class became a venue to experiment with ways to improve media literacy in this area. In the class, I explain some processes that go in to creating podcasts, and I coach students in listening as I have them create pieces each week in groups – students have prompts that ask them to create a sonic 'manifesto', create the shortest piece of audio that tells a complete 'story', and record something 'fake', as a way of identifying the ways in which audio establishes the meaning of each of these terms.

In 2015, I wrote a reflection on these matters for peers on the *Antenna* website (Verma 2015) hosted by the University of Wisconsin, and the current project grew out of how I might learn from my peers in order to expand that article. In consultation with colleagues I devised an interview-based methodology. In December of 2017, I used two Facebook groups – Teaching Media, and the Sound Studies Scholarly Interest Group – to solicit peers who do this kind of work and would be willing to talk about it. I also recruited interviewees at two professional meetings, the Radio Preservation Task Force meeting (a national research organization meeting on Capitol Hill in Washington, DC, in November 2017), and the Great Lakes Association for Sound Studies (GLASS, a regional group that met at the University of Chicago in November 2017 and at the University of Wisconsin in April 2018). This yielded a set of nineteen willing respondents who were interviewed between January and June 2018. Interviews were anonymous, so untenured or contingent faculty could speak frankly about their experience.

Respondents worked at a reasonably wide variety of institutions, from R1 universities and large state schools to liberal arts colleges and vocational schools. They taught in departments from communications and media studies to English, history, theatre, creative writing, music, and journalism. Some classes for which student-created podcasts were used included: New Media, Digital Storytelling, Public History, History of Radio, Environmental History, Pop Music Journalism, Digital Nonfiction, and Audio Storytelling, as well as several classes on podcasting itself. Courses could be as broad as an introduction to mass media or a survey of modern theatre, or as specific as a class on the fiction of James Joyce. The respondents skewed towards early-career faculty, including mostly tenure-track assistant professors, with only four at the associate level and one full professor, along with a few grad students and adjunct instructors.

After asking each interviewee what classes they used podcasts for and when they began to do so, I raised four main subject areas, which we discussed on tape for a period of between 25 minutes and 1 hour:

- *Motivation.* I asked what made the interviewee interested in pursuing this kind of sonic methodology, what learning goals were set, and what sort of institutional support there had been for the teaching of podcast creation.
- *Implementation.* I asked about what types of technology were employed and how, what the classroom listening experience was like, what prompts seemed effective, whether group work or individual exercises were foregrounded, and what problems came up.
- *Integration.* I asked where podcasting fit in the larger aims of the class, the curriculum, and the discipline. Crucially, I wanted to know what instructors felt the students actually learn, particularly if the interviewee thought that it could not otherwise be taught.
- *Assessment.* To conclude, I asked what sort of feedback the interviewee had received on her or his work in this area, either from students, peers or colleagues. I also asked about personal reflections, as well as what remains to be tried and what needs are yet unmet.

Project findings

Motivation

Nearly all interviewees cited a personal or research interest in podcasting media as a key reason for pursuing its use in the classroom. Many cited celebrated podcast programmes such as *This American Life*, *Radiolab*, and *Serial* and competitions, for instance the Third Coast Audio Festival, as inspirations. Almost all instructors were interested in narrative-driven podcasts rather than talk, comedy, or music formats. Some noted that students were already avid podcast listeners, while others said the opposite – that their students had never even heard of podcasting. Many engaged in this type of teaching as a requirement of their work, and at least four felt that they were hired because they presented themselves as able to teach in this area. Two were advisors to campus radio stations, but another just happened to come across a design for a mobile podcasting cart with a mixer, and stumbled upon institutional funding to build it. A few were developing certificate or degree programmes for the making of creative audio. Several instructors, particularly in communication and media studies, used making a podcast as an experimental unit within an existing core course, while others looked at it as a chance to give 'hard skills' in digital editing to students who may not have a specific career interest in the actual topic area of the class. Instructors in the latter category often found themselves in a conundrum, trying to balance actual readings along with skill-building tutorials, sometimes on two or three editing platforms.

Institutional support for podagogy proved varied. Some instructors felt that they were treated as if their work was simply a gimmick, but others found strong support, even earning internal publicity and clout in their institutions eager to seem that their offerings were cutting edge. It seems clear that the term 'podcast' itself was important in this negotiation, particularly as many in the wider world were still becoming familiar with the term. All instructors used the term in our discussions, even though many realized that some of the assignments they devised could easily be classified as 'field recording' exercises. Two instructors, for example, asked students to record the sounds of their daily lives, to explore 'what kind of story can you communicate using just sounds', in a modern version of Pauline Oliveros's idea of a 'listening journal' (Oliveros 2005). Others could be radio projects. One instructor took students to a basketball game and asked them to interview attendees inside and outside the venue. Oral history is another concept – several instructors included the task to interview a loved one. There are new terms, too. Musicologist Kate Galloway refers to the creations from a maker practice in her classes on soundwalking and ecology as 'digital ecomusicology objects' and 'digital audio assignments' (Galloway 2017: 54, 56). Others have found that calling these and other types of recording 'podcasts' made a difference for recruiting students and was a useful shorthand when convincing department chairs outside the sound studies idiom to list a course in the area.

Several stated that podcast-creation was a way to 'work in' a unit of sound studies into a larger architecture, enhancing and complimenting work on critical listening. One interviewee characterized podcasting as 'a medium for [students] to reproduce the kind of teaching I do in class where we do careful listening together'. Another insisted that by podcasting, students could learn how to read audio for implicit and explicit messages. Goals tended to focus on using podcasts as a way of getting students more engaged and to provide them with a new creative outlet that let students introduce more personality into their work, sharpening their sense of audience. 'You can get lockjaw when writing', one instructor pointed out. 'Podcasts open that and get them thinking critically in real time.' Like me, many instructors also wanted to make their students better critical listeners and thought that doing so required making them more creative recorders and editors. One interviewee emphasized that students 'realize they've come to take for granted how they process their environment through their ears […] focusing on podcasts has taught them to really maximize what they can do as listeners'. Nearly all faculty found it to be a relief both for students and instructors to create something without manifesting it in the written word, to 'think conversationally' as one put it. Podagogy 2.0 also seemed to meet other existing priorities, from enhancing student pitching skills to teaching community engagement. Loftier objectives were never separable from production technique. Instructors said that podcasting could teach creativity, structure, innovation, research, but they also said it could teach how to mix. One hallmark of teaching in this way is how class conversations could deal with the most delicate questions of structure, ethics, representation, and art, but then bounce right away to how to set gain levels to avoid clipping.

Not all instructors asked students to work in groups, but those who did so considered it to be a very valuable element of the pedagogy. It was felt that by forcing students to divide tasks and obey a production schedule as a unit, these courses more closely mimicked the kinds of jobs they would likely have outside of the college setting. Almost uniformly, faculty discovered that they had a similar surprise as did their students when it came to workload: students thought producing a podcast would be easier than writing an essay, but ended up spending much more time than they would have with writing; similarly, instructors had a rude awakening when thinking that grading podcasts might take less time than weekly response papers. The need for better time management was a frequent topic of discussion, but students didn't seem to mind – one interviewee said that after learning that it takes more time to produce audio than they expected, 80 to 90 per cent of students in one class reported that they would still rather do production. No one who was interviewed suggested they would abandon podcasting, and many had ideas for new iterations, suggesting that motivation remained high.

Implementation

One of the broadest topics of conversation had to do with the tools instructors employed in their classes. Some instructors favoured in-studio production, while others favoured field-based recordings, and many taught students how to sound-treat their own closets and offices. There was a wide variety of technical approaches to recording itself. For instance, many instructors took a do-it-yourself (DIY) approach to recording and editing, letting students use whatever was easiest to hand, which in practice meant using smartphones as recorders and editing with simple, widely available or free programs such as Garageband and Audacity. Several instructors used Adobe Audition because their institutions already subscribed to the Adobe suite of software, which made it easy for students to access in a lab. These solutions seemed to appeal most to instructors who were learning the technology themselves, and for situations in which there was a low level of recording and editing skills to begin with and no mandate to improve those skills. In the literature on podcasting, one advantage of podagogy is the need of students to acquire technical skills and problem-solve on their own (Forbes 2015). This was a philosophical choice, for some; podcast culture has a DIY ethic, and having that as a constraint in the creative classroom space is one way of exploring the meaning of that culture. Several instructors felt that teaching skills is less important than community involvement and social justice, that there is something important about inculcating the idea of making media with technologies and skills the students already have. Another felt proud students could make media out of 'spit and duct tape'. A third emphasized that the main thing students at their institution need to be convinced of is that 'they can dive in to a new thing that they've never done before and just make something […] that they can be a part of something new'. That said, one or two instructors pointed out that an emphasis on low-level skills tends to increase a 'slacker factor', giving students the impression that it was all right to throw something together at the last minute.

And there is another side to the coin. Several instructors took it as their role to help teach portable skills for professionally oriented students in creative writing and journalism, but also in theatre, history, and media studies, fields where marketing oneself today involves a broad set of media tools. Instructors in this camp tended to use professional or prosumer-grade Tascam or Zoom recorders, particularly the Zoom H5, occasionally with external dedicated microphones (Rode and Shure were preferred brands), and to use more ambitious Digital Audio Workstations (DAWs), such as ProTools, Hindenberg, or Reaper. Some relied on extra staff or teaching assistants to teach advanced tutorials in DAWs, while others directed their students to YouTube instructional videos or the training website Lynda.com. This approach had a philosophy too. Instructors who came from the radio or podcasting industries often spoke of the difference in the physical experience of interviewing with an external microphone and of responsibly teaching industry-standard software to students. Particularly in educational settings that emphasize portfolios – whether that is an art school context, or a two-year college whose students need a body of work to transfer to a four-year college – skills and outputs are at a premium. A surprising number of instructors were able to raise funds for hardware from partners in their departments, scholarly divisions, or at their libraries. Others discovered that the microphones and recording facilities of legacy radio stations already on campus could be borrowed – good microphones can last decades with little maintenance and are interoperable between systems over time. Some also found useful recording facilities through audiology or clinical facilities. Instructors starting out would do well to begin by inventorying their own institutions, looking for existing library resources and legacy radio stations, where excellent equipment is often available, and to scour their campuses for acoustic environments suitable for recording in.

Whatever the instructor's philosophy, students do expect to learn skills in these classes that they can export to other contexts, along with social skills around pitching or workflow, experience using a mic and working in a DAW remain key skills. It can also be hard to establish uniform rules for skill building, since many classes have students with variations in interest in acquiring skills as well as in existing skills, so it is hard to find a way of accommodating all members of the group. And the more complex the DAW, the more instruction time could be taken up with tutorials. One instructor let students use more than one DAW, which meant toggling between two very different programs in tutorials on common practices such as normalization and cross-fading. Decisions about tools also necessarily had ramifications on assessment. Those who put a low premium on equipment tended to assess students based on their concepts as represented in accompanying statements, rather than the professionalism of their work, while others insisted that a story should be as close to 'ready-to-air' as possible. There could even be a discrepancy in 'what' was being assessed in the first place – some would listen to a final work as an mp3 file, while others would also evaluate the ProTools session that went in to creating it.

As far as assignments go, most of the instructors for whom podcasting was just one unit among others tended to generate relatively broad prompts, letting students make whatever they felt would engage with the written material in a survey course. In other cases, assignments emulated well-known programmes such as *Song Exploder* or *This American Life*, or had very well-defined parameters, such as the creation of a 'sonic postcard'. But for the most part instructors tended to conceive of assignments with the whole course in mind. One student collected the various kinds of 'silences' he heard around campus; another told the shortest possible 'story' by recording two seconds of the snap of a mousetrap. In many cases, the work was expressive as well as intellectual: a musician wrote a fugue to imitate Joyce, a historian produced dramatic readings of primary documents, and cinema students were asked to look at paintings or photographs and compose soundscapes that comment upon them. In other cases, assignments were more social in nature. For instance, students in a gender and women's studies class used a podcast assignment to profile workers at a sexual assault bystander prevention programme. Some work was site-specific. In one case, students were asked to investigate the history of an abandoned African American migrant camp from the last century, in another students did a series of audio projects about a river that runs through campus, highlighting ecological perspectives. Several instructors opted for a cumulative approach, inviting students to make projects that follow a brand or style over the course of a semester, while others asked students to curate pieces of music on a specific theme (say, the music of Belize, or musical responses to dictatorship), and another tried a three-part structure that began with a *Moth*-style recorded storytelling event, followed by journalistic audio documentary, and finally a work of audio fiction.

I had expected instructors to have more to say about issues of content, specifically as it concerns the use of copyright material in recordings, but it turned out that a more difficult matter was privacy. Many found that students used podcast platforms to express highly personal stories for which propositional writing seldom affords the opportunity. 'I've been really impressed with the students and the way they will do incredibly personal stories,' one instructor mentioned, also citing examples of material too sensitive to even discuss with me. Themes of trauma, loss, and abuse were common in submitted materials, which made some instructors feel put in the position of a therapist – it could be thrilling, but at the same time many felt suddenly unqualified. In many cases, listening together as a group in class could become a little risky, something exacerbated by the solitary way we normally consume podcasts. As Kate Lacey has pointed out, there is a paradox to the fact that we seem to listen more 'socially' than ever, sharing links to works through social media; at the same time we listen more individually than ever, alone in the micro-airspace of a personal device (Lacey 2013a). As one of the instructors who devotes time to listening to student work in class, these issues had come up for me – for instance, I wasn't sure about playing stories about personal struggles with illness in class, or accepting recordings made covertly by students of others. Elsewhere, instructors have struggled with a wide variety of issues, including what to do, for instance, about releasing audio online of student interviews with undocumented immigrants, or others who might be in jeopardy if their voices were heard in a digital space outside the classroom context (Holmstrom 2017).

Among my interviewees there was interest in establishing clearer ideas about how podcasts should or should not reach out beyond the confines of a classroom. Some instructors asked students to submit their podcasts through the online platform SoundCloud, a common distribution method used by professional podcasters. Of the instructors who used SoundCloud, several asked students to use password protection, but others left it open on purpose and were surprised at the number of plays that student works received. Many instructors encouraged students to develop their podcasts for actual release, or to submit work to competitions. A few instructors selected pieces to air on college radio stations, or held live events at the end of their semester, during which students could exhibit their work, sometimes with a professional from the podcasting world giving critiques. Those who had no capstone event expressed a desire to be able to do one. Like the issue of technology, the extent to which podcasts were created for extramural purposes tended to have a lot of ramifications in terms of the kind of prompts that were chosen, the extent and type of feedback given and the overall rhythm of the course.

Integration

One question that brought out a great deal of reflection was: what do you feel students actually learn? Some of the answers became entangled with how instructors felt about their own disciplines. A historian noted that while few of their students might become actual historians, many more might use media in their careers, and so teaching students how to tell stories in the medium was worthwhile. Others noted that their respective disciplines were increasingly becoming broader, incorporating multimodal ways of expression and argumentation, in part because of perceptions about the job market. Many skills podcasters need – pitching, evaluating story structure, selecting tape, working in teams – would therefore have broader application. One referred to students as 'media creators' rather than writers, and another as 'content creators' rather than radio broadcasters, terms which suggest a breakdown in the silos that had once separated media. Other skills cited were more specific – creative writers who took on podcasting learned about how to write better dialogue in prose, historians learned about how literary voice plays in to questions of representation and ethics, print journalists learned interviewing skills, theatre historians developed modes of performance analysis as well as how to market themselves. There was a clear impression among instructors that as they enter the workforce, students will have to work in many media, some of which have yet to emerge, and podcast creation could help develop the habit of learning new media.

Curiously, we spent a lot of time talking about student writing. Many of my interviewees admitted that they felt relieved allowing students to do something other than write in order to receive credit. One commented that the thinking you can do in audio is different than it is in writing, in that the former encodes the time of thought itself; 'you can hear them working things out' which provides a kind of analytical encounter 'above and beyond anything I could get from them from a writing prompt'. It also feels good. Instructors frankly felt that their students find argumentative writing frustrating, and few have a passion for it; it is not surprising that creating podcasts instead was a tonic both to students frustrated by writing and to instructors exasperated by that frustration. Of course, a form of writing remained part of the process, in part because of the emphasis on narrative-driven podcasting rather than interview-based podcasting, as students and instructors realized that writing and rewriting scripts was integral to the creative process, and it is certainly true that editing recorded audio through software is a kind of 'writing by other means'. One instructor spoke of using research-based podcasting to 'trick them into writing a research paper and not knowing it', and another of overcoming the bad rap that writing suffers, explaining to students 'you already know how to be creative, you already know how to be analytical, *you just don't think you know how to be analytical*'. For some, the whole project of teaching through podcasting made a deeper intervention – 'trying to break the humanities from the fetish of the book', as one instructor put it. Several instructors insisted, moreover, that becoming a better podcaster oddly made their students better writers, since they learned to write for voice in a way that traditional essay and excursive formats did not generally allow. One instructor noticed that poets tended to take to the form well, because they are already used to thinking about economy of words, as well as how they sound out loud. This is something that is reflected in the scholarly literature (Porter 2018), which views the development of voice as one of the great boons of the new medium.

Be that as it may, the majority of interviewees were ambivalent about using podcasting over writing. By embracing 'writing by other means', some asked, are we letting our students (and ourselves) off the hook? In expanding our curriculum – and to a certain extent our disciplines – a balance needs to be struck with core disciplinary competencies. The more things we do in classrooms, the less sharp some of the goals of classes become, and this was a clear concern for podcast teachers who balance readings with critical listening, tutorials in ProTools, student presentations, narrative theory, and radio history all at the same time.

Assessment

Although this qualitative study did not collect student evaluation data, anecdotally it seems that podcast-based teaching classes and units receive strong reviews. Interviewees describe these courses as some of the most successful classes they teach in terms of student satisfaction. Given the landscape of higher education, this suggests these courses and units have a chance of surviving, even when the medium is not as fresh as it seems today. How did instructors self-assess? Many felt that they were still new to the area, and issues of improving the workflow seemed to be top of their minds. Much of the discussion in my interviews involved stories about how certain assignments were fine-tuned from one year to another. Often this involved introducing constraints that checked simplistic approaches (when creating a podcast out of your daily sounds, no toilet or personal grooming sounds, please) or that brought checks to the process of creation – nearly all of my interviewees discussed their pursuit of further use of rough drafts for critiques.

The field still has many needs. There are only a couple of effective texts that teach production strategy, interpretation, or editing in a way that is designed to fit humanistic higher learning (Abel 2015; Biewen and Dilworth 2017). There are many online resources, however. Some key resources from which instructors in this study drew included: interviews on transom.org, the Third Coast International Audio Festival, The Sarah Awards, Louisa Lim's *Masterclass* podcast series at the University of Melbourne, the sound studies site SoundingOut! as well as the *Aca-Media* podcast, freesound.org, and the *State of the Human* podcast at Stanford. Podcasting has the problem of lacking a canon of common examples, while at the same time moving very swiftly, which means that scholars often have to rewrite their syllabi and cannot rely on students coming to class with some key pieces in mind. Finally, there is no obvious forum in which to share pedagogical strategies. One direct outcome of this study has been the idea for a shared document folder in which all the instructors who contributed to the study might pool assignment ideas and syllabi.

Conclusions for sound pedagogy

What does this research project tell us about a kind of sound pedagogy that develops in light of sound studies? First, sound pedagogy – like sound studies – has a deeply ambivalent relationship to technology. On the one hand, there is an impulse to embrace it, to speak the language of technical affordance, and to emphasize creative possibilities, letting these become the guiding principles in the learning setting. In this sense, the relationship between sound studies and sound pedagogy is similar to the relation between sound studies and media archaeology. On the other hand, the emphasis on technology tends to lead to the de-emphasis of lived social relations and community-based mission. Many instructors feel they must balance the choice of spending time on devices and the choice of spending time on stories, although all believe these two things are entwined. Second, sound pedagogy points towards the acute and fraught relation between writing and sound. While recent authors have spoken of a 'sonic turn' in the humanistic sciences, the institutions of modern instruction remain deeply tethered to the written word as the key outcome of thought and discourse. It is telling that many of those who feel podcasting is a better way of teaching justify this belief on the basis that the method makes students in to better writers, thereby tacitly conceding the supremacy of the written word over the utterance. Finally, the case of Podagogy 2.0 (particularly because it is a *digital* methodology) illustrates the exciting but also dangerous propensity for sound to travel beyond contexts initially intended. Everywhere in this study, traditional classroom relationships seemed to alter as a result of the methodology: students have more personal relationships with one another and with their instructor as a result of the propensity of confessional storytelling as a mode; works produced for a class could end up on the web in public, or be presented for communities; students do their work moving around in the community rather than sitting in libraries or work carols. In many cases, taking podcasting into the classroom entails taking the class out of the classroom, embeds it more clearly into communities and audiences, with all the risks and rewards that such a shift affords.

And what is true of Podagogy today has been true of other settings of sound pedagogies throughout history, from Socrates leaving the city for his famous discourses in Plato to Schafer's seminars exploring the built environment as composition. In so many cases, when it comes to sound pedagogies, we see the same three attributes – the ambivalent relation to technology, the persistent and sometimes nagging primacy of textual media, and the tendency to move out of traditional scenes of instruction, indeed to remake them – but perhaps it is only in the context of the growth of sound studies that the resonance of these elements become truly audible.

Acknowledgements

I would like to thank my interviewees for their time and expertise, as well as the many students with whom I have worked over the years on these issues. Research for this project was made possible by the Searle Center for Teaching and Learning at Northwestern University, where I was a fellow in 2017/2018. I want to thank members of my cohort of fellows, particularly David Boyk, Jolie Matthews, and Asma Ben Romdhane. I also want to thank Searle leadership Susie Calkins and Bennett Goldberg for their advice and wisdom, as well as Sharisse Grannan for her invaluable help designing this project.

References

Abel, J. (2015). *Out on the Wire: Storytelling Secrets of the New Masters of Radio*. New York: Broadway Books.

Bell, D. (2008). 'The University in Your Pocket'. In G. Salmon and P. Edirisingha (eds), *Podcasting for Learning in Universities*, 178–179. New York: Open University Press.

Biewen, J. and A. Dilworth (eds) (2017). *Reality Radio: Telling True Stories in Sound*. 2nd edition. Chapel Hill, NC: University of North Carolina Press.

Bull, M. (2005). 'No Dead Air! The iPod and the Culture of Mobile Listening'. *Leisure Studies* 24 (4): 343–355.

Campbell, G. (2005). 'There's Something in the Air: Podcasting in Education'. *EDUCAUSE Review* 40(6): 32–47.

Dale, C. and J. Pymm (2009). 'Podagogy: The iPod as Learning Technology'. *Active Learning in Higher Education* 10 (1): 84–96.

Forbes, D. (2015). 'Beyond Lecture Capture: Student Generated Podcasts in Teacher Education'. *Waikato Journal of Education* 16 (1): 195–205.

Galloway, K. (2017). 'Making and Learning with Environmental Sound: Maker Culture, Ecomusicology and the Digital Humanities in Music History Pedagogy'. *Journal of Music History Pedagogy* 8 (1): 45–71.

Holmstrom, B. (2017). 'Podcasting in the Composition Classroom: Writing, Research and Activism'. *IMPACT: The Journal of the Center for Interdisciplinary Teaching and Learning* 6 (2): 12–16.

Kazlauskas, A. and K. Robinson (2012). 'Podcasts Are Not for Everyone'. *British Journal of Educational Technology* 43 (2): 321–330.

King, K. and M. Gura (2007). *Podcasting for Teachers: Using a New Technology to Revolutionize Teaching and Learning*. Charlotte, NC: Information Age Publishing.

Lacey, K. (2013a). 'Listening in the Digital Age'. In J. Loviglio and M. Hilmes (eds), *Radio's New Wave: Global sound in the Digital Era*, 9–23. New York: Routledge.

Lacey, K. (2013b). *Listening Publics: The Politics and Experience of Listening in the Media Age*. New York: Polity.

Macleod, D. (2006). 'Kant Takes My iTunes Off You'. *Guardian*, 14 December. Available online: https://www.theguardian.com/education/2006/dec/14/highereducation.uk1 (accessed 29 June 2020).

Ng'ambi, D. and A. Lombe (2012). 'Using Podcasting to Facilitate Student Learning: A Constructivist Perspective'. *Educational Technology & Society* 15 (4): 181–192.

Oliveros, P. (2005). *Deep Listening: A Composer's Sound Practice*. Lincoln, NE: Deep Listening Publications.

Porter, J. (2018). 'The Pleasure of the Voice: Speakerly Writing in the Digital Age'. In Laura Gray-Rosendale (ed.), *Getting Personal: Teaching Personal Writing in a Digital Age*, 239–254. Albany: SUNY University Press.

Rosell-Aguilar, F. (2007). 'Top of the Pods—In Search of a Podcasting "Podagogy" for Language Learning'. *Computer Assisted Language Learning* 20 (5): 471–492.

Salmon, G. and M. Nie (2008). 'Doubling the Life of iPods'. In G. Salmon and P. Edirisingha (eds), *Podcasting for Learning in Universities*, 1–12. New York: Open University Press.

Schafer, R. (1992). *A Sound Education: 100 Exercises in Listening and Sound Making*. Indian River, ONT: Arcana Editions.

Sterne, J. (2012). 'Sonic Imaginations'. In J. Sterne (ed.), *The Sound Studies Reader*, 1–18. London: Routledge.

Verma, N. K. H. (2015). 'Podagogy, a Word I Didn't Make Up'. *Antenna*, 25 June. Available online: http://blog.commarts.wisc.edu/2015/06/25/podagogy-a-word-i-didnt-make-up/ (accessed 13 July 2020).

8

Sonic Methodologies in Literature
Justin St. Clair

> There are ways you can use a studio. Things you can do that open up impossible spaces in the mind. You can put the listener in a room that doesn't exist, that couldn't exist. You can put them in an impossible room.
>
> —Hari Kunzru, *White Tears* (2017: 26)

An impossible room

In the digital age, it is easy to forget that literature, from its very inception, has been one of our most significant audio technologies. The pen is the ur-stylus, the originary and prototypical instrument of audio transcription, and the literary record predates its electroacoustic complements by millennia. Consequently, R. Murray Schafer insists, 'while we may utilize the techniques of modern recording and analysis to study contemporary soundscapes, for the foundation of historical perspective, we will have to turn to earwitness accounts from literature and mythology' (Schafer 1993: 8).

In recent decades, numerous scholars have followed Schafer's suggestion, lending their ears to literary texts in search of lost sound.[1] Nevertheless, this enterprise is, as many of its own practitioners admit, necessarily fraught. At a minimum, the literary record refracts the audible past doubly, bending bygone sound through authorial perception and the opacity of language alike. While literature might provide some tantalizing, complementary data on the soundscapes of antiquity, it is at best approximative and conjectural. What is more, if we overemphasize its role as a pre-phonographic transcription technology, we risk diminishing literature as merely artefactual. The protestations of McLuhanites notwithstanding, print is neither antiquated nor obsolete. Nor, for that matter, are the sonic methodologies of literary studies limited to scholarly exercises in audio archaeology. Contemporary literature has an extensive and profound engagement with sound, and literary art and scholarship continue to be – as in epistemes past – an intrinsic part of sound culture.

Rather than detailing the methodologies by which scholars have attempted to extrude historical soundscapes from literary texts, this chapter will instead consider literature as a sounded site within contemporary media culture, and one, moreover, that can play a significant role in processing the very notion of 'soundscape' itself. In 'The Stereophonic Spaces of Soundscape', Jonathan Sterne examines the roots of what has become a ubiquitous (and elastic) term within the field of sound studies. He situates the concept historically, arguing that 'soundscape is very much a creature of mid-century sound media culture' and best understood as 'part of an electroacoustic moment in sound history' (Sterne 2015: 67). Rejecting conventional presuppositions, Sterne argues that 'the essence of the soundscape [...] *is not physical space or the relation between physical space and its representation*' (emphasis in the original). Instead, he contends, 'its essence is a stable audioposition, one from which the entire world is available to be heard'. 'The soundscape concept is attractive,' Sterne concludes, 'because it simultaneously invokes a unified auditory perspective, a stable audioposition, and then hides the work of shaping perspective' (79–80). This self-effacement is inherently ideological: what seems but 'audio-technical discourse' obscures not only the means by which it can 'figure and amplify modes of subjectivity', but also how our 'ways of hearing the world [are] rooted in the post-war consumerist structure of listening' (80).

This pivot from the spatial to the perspectival – from an artefactual or environmental understanding of soundscape to one that emphasizes the audioposition of the listening subject – is particularly resonant within literary studies. The reading experience is quite literally scripted: it is, to appropriate and repurpose one of Sterne's formulations, a kind of 'situated omniscience' in which the very 'notion of sonic space [...] is created through the act of comprehension' (Sterne 2015: 73).[2] Contemporary literature, in other words, simulates a stable audioposition, and one from which anything that can be imagined is available to be heard. Likewise, the soundscapes of literature should also be understood in the context of contemporary media culture. They are embedded within the same consumerist structures of listening and part of the same electroacoustic moment in sound history that gave rise to our critical formulations. As I have argued elsewhere, contemporary writers 'have an aural fixation, and their obsession with sound echoes larger anxieties regarding the supposedly diminished position of print fiction in the contemporary media pantheon' (St. Clair 2013: 2).

Like all media, print has obvious limitations, some so apparent they hardly warrant mention. If one is after a high-fidelity recording of a live orchestral performance, for example, print is certainly not the most suitable of options. Nevertheless, literature has several remarkable qualities that make its contributions to audio culture particularly unique. First, as suggested above, a literary soundscape is unconstrained by audibility: on the printed page, anything that can be imagined can be heard. There is an inherent paradox here, of course, but if we set aside the physiology of listening, momentarily, and substitute sensory comprehension, the possibilities become apparent. Second, rather than hiding its perspective-shaping work, contemporary literature is deeply engaged in metadiscursive practice. Literary fiction, in particular, is far more than mere transmission. It is continuously appraising its own cultural relevance, assessing itself in relation to ascendant media forms, and evaluating the role of language within audiovisual mass media. This reflexivity provides the basis for literature's media engagement and allows it to offer a minority report on aural culture – side-channel feedback, if you will, on the social implications of sound and listening practice. As Sam Halliday argues in *Sonic Modernity*, 'literature […] is especially well suited for revealing sound's "configured" quality', which he defines as 'sound's imbrication in the non- or trans-acoustic'. Literature is inherently a metalinguistic medium and is therefore 'especially well-suited for revealing such para-sonic factors as sound's social connotations, its relationship with other senses, and – perhaps most importantly of all – the qualitative dimension that means certain sounds are actually of interest to people, things they actively seek out or shun' (Halliday 2013: 12).

Hari Kunzru's novel *White Tears*, from which the epigraph to this chapter is lifted, is an excellent example of how literary fiction functions as a sounded site within contemporary media culture. In short, Kunzru constructs an 'impossible room' that is uniquely literary, one that no amount of studio magic could ever quite conjure. At its core, Kunzru's novel is a meditation on audibility – on whose voices get heard, on what gets recorded, and on how the historical record, ever palimpsestuous, has more gaps and erasures than readable tape. The novel's narrator, Seth, is an audio-flâneur who wanders the city, eavesdropping as he goes. 'No one ever noticed,' he reports in the opening paragraph: 'I had a binaural setup, two little mics in my ears that looked like headphones, [and] a portable recorder clipped to my belt' (Kunzru 2017: 3). When he plays the recordings back in the studio, Seth invariably discovers 'phenomena [he] hadn't registered, pockets of sound [he]'d moved through without knowing' (3). For Seth, listening becomes an obsession. 'I was trying to hear something in particular,' he admits, 'a phenomenon I was sure existed: a hidden sound that lay underneath the everyday sounds I could hear without trying' (7). 'Marconi', Seth explains, 'believed that sound waves never completely die away, that they persist, fainter and fainter, masked by the day-to-day noise of the world' (43).

This fantasy of everlasting sound becomes emblematic within the novel. As Seth tells the story, 'Marconi thought that if he could only invent a microphone powerful enough, he would be able to listen to the sound of ancient times' (Kunzru 2017: 43). It is – perhaps unsurprisingly, given Marconi's role in the development of radio – a broadcaster's fantasy. If audio recording is a hedge against the ephemerality of sound, Marconi's imaginary acoustics obviate the need for recording technology altogether: just tune in to the great cosmic radio and bygone soundscapes are always already available. Kunzru adopts Marconi's microphone as a kind of figure for literature itself, and then sets out to expand its possibilities. Being able to hear 'lost' sound (e.g. listening in on jazz pioneer Buddy Bolden, of whom no known recordings survive) is one kind of thought experiment, but what about sounds that were thwarted entirely, sounds that could never have been recorded because they did not occur? This is what the literary record can offer: the irretrievable b/w the irreal.

On his rambles around New York City, Seth begins to capture snippets of this impossible sound, floating 'like ghosts at the edges of American consciousness' (Kunzru 2017: 130). Back in the studio, his friend Carter stitches bits of the field recordings together, and finds, to his amazement, that an a cappella blues vocal fits perfectly atop a separate guitar track, one that Seth had not even realized he had recorded. It was 'as if they were two halves of a single performance' (57). With some prodding, Seth manipulates the composite until 'it sounded like a worn 78, the kind of recording that only exists in one poor copy, a thread on which time and memory hang' (58). Carter invents an artist (Charlie Shaw) and a label (Key & Gate), and posts 'Graveyard Blues' online. And then all hell breaks loose. The recording they have constructed apparently already exists, but impossibly so. In the world of the novel, Charlie Shaw was an actual musician, but one who never managed to record his signature 'Graveyard Blues'. In 1929, the legendary H. C. Speir had booked him for a session in Jackson, Mississippi, but Shaw was subsequently arrested for vagrancy. Unable to pay the $100 fine, he was sentenced – on the very day he was scheduled to record – to one year on a chain gang, building a levee for the judge's brother. The label Key & Gate, thus, is a dark pun: for Charlie Shaw, incarceration serves much the same function as a noise gate in a mixing console, attenuating his signal and preventing 'unwanted sound' from appearing on record.

Much like the history it depicts, the novel is haunted by a record that simultaneously is and is not. It is a meditation on authenticity and appropriation, on repetition, race, and the fetishization of sound. If, metaphorically speaking, literature is Marconi's microphone, then it is certainly an augmented version thereof – one that offers impossible sound alongside the inaccessible. And, indeed, this is precisely how Kunzru develops his sociocultural critique. As the story unfolds, we learn that the record that Seth and Carter have ostensibly invented had already been discovered some fifty years earlier by a pair of mid-century blues obsessives on a collecting expedition in the Deep South. That acquisition ended badly, as does Seth and Carter's episode half a century later, both pairs pursued by the vengeful ghost of Charlie Shaw. 'My lips move but there is no sound,' Shaw laments late in the novel, 'because no one remembers me and no one living will ever hear my music' (Kunzru 2017: 258). The novel's key repetitions – Shaw's unrecorded song and generations of collectors who fetishize black sound – underscore how institutional racism reverberates throughout history, how voices on the margins are muted and silenced even as they are exoticized. Shaw's original incarceration comes at the hands of the Wallace brothers, leveraging the laws and institutions of Jim Crow in Mississippi to extend slavery's promise of free labour. It is the brothers' twenty-first-century progeny who provide the reprise on Shaw's ghostly return: he is locked away in a facility owned by the Wallace Magnolia Group, a private prison provider in an America where mass incarceration has become the new Jim Crow law. If it sounds all too familiar, it is: *White Tears* gives us history as a broken record and asks us to attend to the repetitions. 'If Marconi was right and certain phenomena persist through time,' Seth concludes, 'then secrets are being told continuously at the edge of perception. All secrets, always being told' (77).

In the sections that follow, I will sample the sonics of contemporary print fiction, and describe several ways that literature utilizes sound and figures auditory perception from its increasingly marginal media position. Much of what follows places emphasis on form over content, but we cannot ignore the fact that when it comes to literature, the two channels are never entirely discrete. Formal soundings so often echo their thematic counterparts that in *An Essay on Criticism* Alexander Pope famously made it a prescription: 'The *Sound* must seem an *Eccho* to the *Sense*' (Pope 1711: 22, emphasis in the original). Nevertheless, in an effort to underscore the sonic methodologies we find expressed within print fiction, its formal features – from techniques of characterization to the referential evocation of music – must necessarily be placed at the fore.

Sound identities

Many take print fiction to be a character-driven medium. Such expectations are the unfortunate (if persistent) result of realism's century-plus reign as the dominant mode of popular fiction. The predilections of the reading public (and the merits of those preferences) notwithstanding, character is nonetheless an inescapable element of contemporary fiction, and one with acoustic significance. In *The Noises of American Literature, 1890–1985: Toward a History of Literary Acoustics* (2006), Philipp Schweighauser describes how the process of characterization hinges on what we might call the sonics of embodiment. Fictional characters, particularly in works of realism, stand in for real-world subjects, and as lived experience is inherently sounded, so too must literary lives be articulated within aural matrices. 'Like human beings, fictional characters are not merely passive receivers of acoustic phenomena,' Schweighauser observes, 'but actively participate in the making of the soundscapes they live in.' The noises these characters contribute to their fictional environments, he emphasizes, are limited neither to verbal utterances 'nor to the sounds of the human voice', but encompass a wide range of embodied sound. 'Characters in novels snore, sigh, snarl, and scream; they grunt, gabble, gossip, and grumble; they clamor, cough, cry, and curse. At a more basic level, human bodies, fictional or not, continually produce noises whether they walk, stand still, sleep, get up, or sit down' (Schweighauser 2006: 70).

For Schweighauser, what is of central importance is not simply the sound of lived experience, but rather the extrapolations and inferences that invariably occur when we encounter the sounds of others. 'If observed,' in other words, 'the noises people make will be subject to the value judgments of observers.' Or to put it more bluntly, audition is one of the perceptive senses by which humans stereotype and generalize. Novelists are 'very much aware of such mechanisms', Schweighauser argues, and often craft, as a consequence, 'acoustic profiles for their characters' (Schweighauser 2006: 70). As *The Noises of American Literature* appeared before profiling (i.e. racial or ethnic) became a political buzzword, Schweighauser uses the term 'acoustic profiling' to describe a creative methodology: that is, 'a characterization technique that endows fictional bodies with a set of distinctive acoustic properties designed to position characters with regard to the ensemble of social facts and practices that constitute the fictional world they inhabit'. 'These acoustic properties', he continues, 'may range from characters' accents, dialects, or intonation patterns to the sounds produced by their laughter, snoring, or the acoustic impact of their footsteps' (71). On occasion, a character's acoustics might be situational, limited to a specific scene or episode. Often, however, an acoustic profile is consistently rendered throughout a work – a distinctive, constituent element of a character's totality. For such cases, Schweighauser proposes the neologism *audiograph* – a character's soundmark, if you will, and a term which not only can be literalized as 'sound writing' but also carries connotations of individuality and idiosyncrasy by punning on the word autograph. Much like Schafer's notion of a soundmark – 'a community sound which is unique or possesses qualities which make it specially regarded or noticed by people in that community' (Schafer 1993: 274) – Schweighauser's audiograph incorporates reaction and reception into its very essence. In other words, a soundmark is not so designated on the basis of its own qualities alone, but rather as a result of how it is regarded and noticed. Similarly, an audiograph is not simply the aural qualities of a fictional character but rather a sonic methodology that generates a specific kind of literary audioposition. In short, acoustic profiles 'serve both to position characters on the social scale and to direct readers' judgments of them' (Schweighauser 2006: 70). Thus, while Schweighauser may have intended the expression 'acoustic profiling' to represent an authorial process, the purpose of the device itself, as he explains it, is consonant with contemporary understanding of profiling: it generates extrapolative prejudice. 'The positioning accomplished via audiographs', he notes, 'may involve value judgments on the part of other characters, narrators, and implied authors as well as implied and empirical readers' (71).

While the terms 'acoustic profiling' and 'audiograph' have not been broadly adopted within the profession, the methodologies they describe are ubiquitous. Critics, moreover, have attuned themselves to literature's long history of aural characterization, whatever the chosen terminology. In 'Defining Habits: Dickens and the Psychology of Repetition', for example, Athena Vrettos examines how 'peculiar habits abound in Dickens's characters' (Vrettos 2000: 416). Unsurprisingly, a great many of these behavioural quirks have a kind of sonic signature. In *Dombey and Son* (1848), for instance, we encounter specific examples of aural repetition ranging 'from Mr. Dombey's habitual jingling of his gold watch chain [...] to Mr. Chick's "peculiar little monosyllabic cough; a sort of primer, or easy introduction to the art of coughing"' (Vrettos 2000: 416).[3] 'Nineteenth-century authors often seem obsessed with insignificant details,' Schweighauser observes, but it is a mistake to characterize audiographs, in particular, as simply verisimilitudinous inclusions 'with no apparent narrative function' (Schweighauser 2006: 73). 'Rather,' as Vrettos insists, 'in a balanced economy of behavioral exchange, one person's mannerisms produce corresponding mannerisms in others' (Vrettos 2000: 416). 'Thus,' she contends, 'Mr. Chick's habit of [...] whistling and humming tunes, leads, in turn, to Mrs. Chick's custom of criticizing Mr. Chick's whistling' (416). Mr Chick's acoustic profile, in other words, serves to position him within the novel, inviting not only intradiegetic evaluation (from characters within the world of the story) but also extradiegetic evaluation (from readers in the 'real' world).

Instances of acoustic profiling can be found throughout the literary canon. Even those casually acquainted with the character Sherlock Holmes, for example, know how often the great detective is 'engaged in his favourite occupation of scraping upon his violin' (Doyle 2001: 45). As with other audiographs, his acoustic profile functions as a positioning device, opening a space for Watson's commentary and providing readers with sonic evidence of Sherlock Holmes's social position. This famous example also serves as a reminder that 'embodied sound' is not limited to the somatic sounds that human bodies *emit* but also includes the extracorporeal sounds that bodies *occasion*. It is not only Victorian literature, moreover, that insistently represents identity through sound; contemporary fiction – even while rejecting some of the tenets of realism – has continued this long tradition. David Foster Wallace's *Infinite Jest* (1996), for example, is awash in sound. Among the most memorable of the novel's exercises in acoustic profiling is the sonic signature given to Les Assassins des Fauteuils Rollents, a radical organization largely comprised, it would seem, of legless Québécois separatists. These Wheelchair Assassins, as they are also known, are 'masters of stealth, striking terror into prominent, Canadian hearts, affording no warning excepting the ominous squeak of slow wheels' (Wallace 1997: 1056). In the world of the novel, 'to hear the squeak' becomes a kind of refrain, 'an understood euphemismic locution […] for instant, terrifying, and violent death' (Wallace 1997: 1057). Wallace's darkly comic take on identity not only puns on the concept of separatism (the secession of a province and the amputation of a limb) but it also satirizes conventions of embodiment and embodied sound. By situating the acoustic marker of identity outside the body and mapping it, moreover, onto a symbol of physical impairment, the novel underscores the value-loaded process by which profiling occurs even as it elicits the same from readers.

In the case of Les Assassins des Fauteuils Rollents, Wallace also foregrounds the linguistic aspects of sounded identity. Even the name of the organization itself is significant: Québécois is famously idiosyncratic and colloquial, but *fauteuil rollent* is invented patois, a nonce bastardization of the French *fauteuil roulant* (wheelchair). Wallace, here, is satirizing what Susan Gingell terms 'print textualized orality: that is, writing that brings to the paper or digital page a non-prestige lect or a colloquial or otherwise clearly oral version of a language' (Gingell 2010: 127). In her essay, 'Negotiating Sound Identities in Canadian Literature', Gingell proposes that the study of minoritarian orality in literature is best approached via notions of 'sound identity'.[4] Much in the same way that 'music-making with others helps to constitute a musical, "We"', she argues, so too does a written record of oral practice contribute to 'the formation of identity and the development of a sense of place and social context' (Gingell 2010: 127). Of course, what Gingell is suggesting is not entirely a new idea. In print fiction, identity has long been developed through both direct and indirect speech, using the sound of localized linguistic particulars (i.e. accent, pronunciation, and even syntax) to individualize characters, to position characters socially, and to orchestrate – internally and externally – the judgement of those characters. What may be somewhat different in the case of contemporary fiction, however, is a 'postmodern understanding that subjectivity is constituted in and through language' (127). Furthermore, the realization that human subjectivity is a linguistic construction has opened additional possibilities for literary characterization, particularly in light of two other postmodern turns: a conviction that all speech acts are inherently political and a theoretical valorization of multicultural perspectives.

Media history is littered with examples of aural characterization that amount to oral caricature, the ostensible purpose of which is to debase or deride the other. (See, for instance, American minstrelsy and its Hollywood legacy.) While by no means immune to the racism inherent in popular media forms, contemporary fiction not only engages issues of race in a productive fashion but, given its self-reflexivity, also reflects upon this engagement, thereby providing an important forum for cultural deliberation. From a critical perspective, one of the central questions to be asked of texts that engage in idiolectic characterization (that is, the sonic individuation of characters by non-standard lexical and linguistic markers) is simply this: what are the effects of the simulated audioposition? In other words, when a character's acoustic profile includes elements of marginalized oralities, how do those markers function to position the character for judgement, both within the narrative and for its readers? In Mark Twain's *The Adventures of Huckleberry Finn* (1885), for example, the representation of Jim's dialect repeatedly positions him for ridicule within the world of the novel, but the ways in which other characters react to Jim serve, in turn, as part of a large satirization of bigotry and intolerance. If we compare Jim's sounded identity to that of Uncle Remus, we find – in purely representational terms – little daylight between the two. The effect, however, is far different. Regardless of how charitably the tales of Joel Chandler Harris may have been received in the 1880s, today's readers cannot help but hear 'one of the most powerful and pervasive racist stereotypes in American culture' (Silk and Silk 1990: 15). Simply put, Uncle Remus's dialect does not serve to satirize racism but rather reinforces stereotypes regarding black intellect and capability, while simultaneously promoting white fantasies of the antebellum American South. 'As Harris seems to have understood, sound could trigger white southerners' memories,' however spurious and counterfactual those memories may have been (Ritterhouse 2003: 611). 'Harris reinforced a historical theory of slavery that began with the premise, widespread in his generation, that the human relationships of the peculiar institution had been close and mutually supporting,' writes Robert Hemenway, and 'Remus's dialect especially supports this fantasy' (Hemenway 1982: 21).

When it comes to print textualized orality, context matters. Thus, while accent, pronunciation, and syntax are often deployed as prejudicial operators, they can also serve as constructive positioning devices. When incorporated into a character's acoustic profile, oral representations in literature can even help to further a sense of community identity through the 'signaling of kinship or counter-hegemonic group relations' (Gingell 2010: 130). In fact, print is a particularly useful medium for recording minoritarian oral practice because it necessarily foregrounds difference. In the case of a conventional audio recording, accent, for example, often remains 'invisible' to in-group listeners. Not so in the case of literature. When accents and other elements of oral practice are transcribed, the orthographic trace of so-called normative usage hangs just off the page, an almost-present reminder of linguistic variance. This juxtaposition, in short, allows otherwise 'ear blind' in-group readers to confront their own linguistic diversity, which can, in turn, serve to reinforce group identity.

Imagined music

Many of the sonic methodologies we find employed within print fiction involve the application and evocation of music. As language is better positioned to activate musical memory than it is to present music directly, however, these inclusions often depend upon the reader's imagination. If melodies are to sound in the reader's mind, in other words, the text can do little more than plant notational cues. These, in turn, trigger the memory of past musical experience or exposure, initiating a mental reprise of something already embedded within the cerebrum. It is little wonder, then, that throughout the literary canon we repeatedly encounter examples of musical earworms: songs, melodies, refrains, or other musical snippets that get stuck on repeat, replayed involuntarily in a seemingly endless mental loop. As with literary music, an earworm is both present and absent – there and not there simultaneously.

Edgar Allan Poe's 'The Imp of the Perverse' (1845) contains perhaps the earliest consideration of the 'stuck song' phenomenon in print fiction. 'It is quite a common thing', Poe writes, 'to be thus annoyed with the ringing in our ears, or rather in our memories, of the burthen of some ordinary song, or some unimpressive snatches from an opera' (Poe 1887: 174). Mark Twain's 'A Literary Nightmare' (1876) is another. In this short sketch, Twain's eponymous narrator encounters 'jingling rhymes' that take 'instant and entire possession' of him. 'All through breakfast they went waltzing through my brain,' he writes. 'I fought hard for an hour, but it was useless. My head kept humming.' It is not until he teaches a friend the rhyme that the 'torturing jingle departed out of [his] brain, and a grateful sense of rest and peace descended' (Twain 1996: 16–18). Contemporary fiction continues to employ the trope. In Richard Powers's 'Modulation', for example, one of the central characters wakes up 'with a tune in her head' – or, as the narrator immediately revises, 'not a tune, exactly: more like a motif'. 'She couldn't altogether sing it, but she couldn't shake it either,' he explains. 'She had contracted what the Germans called an *Uhrwurm*, what Brazilian Portuguese called *chiclete de ouvido*: a gum tune stuck in her relentlessly chewing brain' (Powers 2008: 94).

In part, the earworm is attractive as a literary motif because it represents the power of musical recall. Literary practitioners, limited as they are to conjuring music through the use of words, are envious, even as they try to emulate the effect. From a methodological perspective, the primary way that novelists make readers imagine music is through librettization: that is, the technique of embedding song lyrics within the narrative prose. This is common practice within contemporary fiction and is, without a doubt, the result of film and television's ascendance as the primary modes of narrative media over the course of the twentieth century. Both film and television make extensive use of musical soundtracks, and print, in an effort to emphasize its currency, has borrowed liberally from these audiovisual forms.

While by no means the first work of literary fiction to make use of song lyrics, John Dos Passos's *U.S.A.* (1938) trilogy set the tone for novelistic soundtracking in the twentieth century. Scattered throughout the three novels – *The 42nd Parallel* (1930), *1919* (1932), and *The Big Money* (1936) – are sixty-eight 'Newsreels': modernist collages comprised of news clippings, headlines, and popular song lyrics. From the first song that appears ('There's Many a Man Been Murdered in Luzon') to the last ('Yankee Doodle Blues'), *U.S.A.* uses music to index culture. 'Popular songs were manifestations of the "spirit of the time" that John Dos Passos captured,' a barometer on politics and popular sentiment in the opening decades of the so-called American Century (Trombold 1995: 289). But it does more than this. Given both the experimental nature of the sections in which they appear as well as the sociocultural overtones of the songs themselves, the lyrical snippets also serve as musical metacommentary, evaluating both the propagandistic function and the subversive potential of sound media. While much of *U.S.A.*'s music may be unfamiliar to an audience today, contemporaneous readers would have instantly recognized most of the songs and, as a consequence, involuntarily 'heard' the accompanying melodies. Whether these earworms would continue to gnaw at readers after the book was set aside is an open question, but there can be little argument that the melodies extend the fiction's reach, resonating not only within the immediate context of the 'Newsreels' but also harmonizing with nearby episodes (see Seed 1984 and Trombold 1995).

In contemporary fiction, there is no novelist more closely associated with imagined music than Thomas Pynchon. Unlike Dos Passos, however, Pynchon mostly includes lyrics of his own invention. All eight of the novels he has published over the past half century, in fact, have featured his often irreverent songs – from sea shanties to snatches of comic opera and just about everything in between. Pynchon has a fondness for all types of enthymematic constructions: that is to say, literary devices that leave their 'conclusion[s] unexpressed – to be drawn by the reader or the listener' (Hollander 1997). As Charles Hollander notes, 'the enthymeme is the rhetorical technique of choice for a comic who wants his audience to infer the withheld punchline', as in the old ribald one-liner: 'What's the difference between a rooster and a lawyer? The rooster clucks defiance.' Improvisational musicians engage their audiences in much the same manner: 'a great soloist […] flirts with a melody, plays around it, transposes it into various keys or rhythms, offers a fragment or a phrase something like the written melody but not quite', and while 'never quoting the tune directly […] still gets the audience to sing the lyric in their minds' (Hollander 1997). With his love for jazz and a deep commitment to the comedic arts, Pynchon not only replicates both of these techniques, but his lyrics often reverse the process, making the melody the implicit, unexpressed term of the enthymeme.

Among Pynchon's obsessive fandom, determining the tunes to which his songs have been set has become something of a parlour game, for the 'persistent rumor holds that all the songs in Pynchon have actual melodies' (Moody 2011). On occasion, Pynchon reveals his musical sources directly, introducing his parodic lyrics with sing-along instructions – for example, 'sung to a march called Colonel Bogie' (Pynchon 1999: 348) or 'to the tune of the *William Tell* Overture' (Pynchon 2006a: 232). Usually, however, he plays it a bit coy, coaxing the reader towards a particular melody without offering the source title directly. Some of these unnamed tunes, to be sure, are easier to identify than others. In *V.* (1961), for example, the radio plays 'a song about Davy Crockett'. 'This was '56', the narrator tells us: 'the song invited parody' (Pynchon 1999: 65). Even readers who have but a cursory acquaintance with twentieth-century pop music hear Disney's 'The Ballad of Davy Crockett' behind the nine verses that follow. Other times, however, Pynchon's hints are a bit more obscure. 'The tune is known universally among American fraternity boys' (Pynchon 2006b: 310) or 'a sort of medium-tempo Cuban Rhythm' (Pynchon 2006c: 90) don't give the reader much with which to work. In such cases, Pynchon often embeds auxiliary clues within the lyrics themselves. For example, a reader who thinks she hears the standard 'Bye Bye Blackbird' playing behind 'Snap to, Slothrop' in *Gravity's Rainbow*

(1973) has her suspicions confirmed when she arrives at the line 'No one here can love or comprehend me,' which, unlike the rest of Pynchon's bawdy parody, barely revises the original lyric (Pynchon 2006b: 63).

In the contemporary era, the quotation of popular song lyrics is far more difficult an enterprise than it was in John Dos Passos's day for one reason in particular: increasingly aggressive copyright protectionism, which has curtailed literary sampling much in the same way it has altered the practice in pop music. Literary writers – and their publishers – often avoid extensive quotation simply to keep costs under control. As a result, we find paraphrase and other referential techniques deployed more often than lengthy, direct forms of musical quotation. Pynchon provides ample examples of these methodologies as well. One trick of which he is particularly fond is lifting short phrases from popular songs and inserting them into his prose, either as part of the narrative exposition or placed within the dialogue. By limiting the reference to a few scant words, such inclusions typically qualify as fair use and – perhaps more importantly – function as another type of musical enthymeme. Often times, Pynchon will attribute the borrowed phrase with a colloquial 'always sez', as in 'sad but true, as Dion always sez' (Pynchon 2009:11) or 'hey if that's the way it must be, okay, as Roy Orbison always sez' (69). This type of referential soundtracking induces activity on the part of the reader, who must mentally recite a few lines before arriving at the withheld title of the song in question – 'Runaround Sue' (1961) and 'Pretty Woman' (1964), in these particular examples. On other occasions, it is the song title itself that Pynchon includes, as in: 'the car radio, tuned to KFWB, was playing the Doors' "People Are Strange (When You're a Stranger)"' (Pynchon 1990: 133).

Ultimately, techniques such as these serve as a referential matrix within contemporary fiction. With a few well-placed cues, writers trigger musical memories and make melodies resound in the minds of the reading audience. Much like the sounds that comprise a character's audiograph, imagined music is not just superfluous detail used to augment the fiction's illusion of reality. Rather, it is yet another example of literary audiopositioning: the use of sonic devices to provide an auditory perspective, to situate readers and characters in relation to the sounded phenomena of lived experience. In some cases, imagined music serves to supplement a character's acoustic profile; in others, it provides a sonic topography for the fictional environs, positioning its listening subjects along spatiotemporal axes. The early 1960s surf music we find in Pynchon's *Inherent Vice* (2009), for example, serves both functions simultaneously: on the one hand, the songs explicitly figure as part of the protagonist's audiograph ('souvenirs out of a childhood Doc had never much felt he wanted to escape from'), while on the other, they provide a specific sense of time and place, indexing the death of the 1960s with a nostalgic ear to the near past (Pynchon 2009: 125).

Closing notes

The sonic methodologies discussed in this chapter represent only a small sample of literature's rich engagement with aurality and sound culture. For readers interested in further explorations, I would propose two starting points. First, as I suggest in the opening section, racial sonics has emerged as an important topic within literary criticism. Recent book-length studies that provide excellent points of entry include Carter Mathes's *Imagine the Sound: Experimental African American Literature after Civil Rights* (2015) and Jennifer Lynn Stoever's *The Sonic Color Line: Race and the Cultural Politics of Listening* (2016). Second, over the past half century, literature has responded to the ascendance of other media forms with 'a cluster of anxieties about being displaced from some possibly imagined position of centrality in contemporary cultural life' (Fitzpatrick 2006: 201). The result of these anxieties? 'A frenzy of remediation', in which print 'attempts to eat all the other media' (Hayles 2002: 781). In fact, remediation – that is, the way in which one media form replicates and repurposes the tropes, techniques, and devices of another – might be one of the governing logics of literary fiction today. Studies that examine these tendencies with an ear towards media sonics include Mikko Keskinen's collection *Audio Book: Essays on Sound Technologies in Narrative Fiction* (2008) and my own *Sound and Aural Media in Postmodern Literature: Novel Listening* (2013). As both make clear, one evident result of contemporary fiction's media fixation is a superabundance of audible retransmissions, from the loom-echoing, mechanistic triads that reverberate throughout William Gaddis's novella *Agapē Agape* (2002) ('it's Babbage Babbage Babbage but he got his idea from Jacquard's loom so that's all you ever hear, Jacquard's loom Jacquard's loom Jacquard's loom') to the snippets of television audio that punctuate the domestic conversations in Don DeLillo's *White Noise* (1985). Literature is one of our most important fora for cultural deliberation, and the sonic output of our media-saturated world percolates throughout contemporary fiction, providing, in equal measure, both devices that enrich the form and content ripe for critique.

Notes

1. See, for example, Smith 1999; Picker 2003; Cazelles 2006; Boutin 2015; and Pye 2017.
2. Here, Sterne is actually discussing Schafer (1967). My misapplication is intentional.
3. The embedded quotation is from Dickens (1983: 489).
4. Her reference here is to Hudak (1999: 447–474).

References

Boutin, Aimée (2015). *City of Noise: Sound and Nineteenth-Century Paris*. Champaign, IL: University of Illinois Press.
Cazelles, Brigitte (2006). *Soundscape in Early French Literature*. Tempe, AZ: Medieval and Renaissance Texts and Studies.
DeLillo, Don (2016). *White Noise*. New York: Penguin.
Dickens, Charles (1983). *Dombey and Son*. New York, Penguin.

Dos Passos, John (1996). *U.S.A.* New York: Library of America.
Doyle, Arthur Conan (2001). *A Study in Scarlet.* New York: Penguin Classics.
Fitzpatrick, Kathleen (2006). *The Anxiety of Obsolescence: The American Novel in the Age of Television.* Nashville, TN: Vanderbilt University Press.
Gaddis, William (2002). *Agapē Agape.* New York: Viking.
Gingell, Susan (2010). 'Negotiating Sound Identities in Canadian Literature'. *Canadian Literature* 204 (Spring): 127–130.
Halliday, Sam (2013). *Sonic Modernity: Representing Sound in Literature, Culture, and the Arts.* Edinburgh: Edinburgh University Press.
Hayles, N. Katherine (2002). 'Saving the Subject: Remediation in *House of Leaves*'. *American Literature* 74: 779–806.
Hemenway, Robert (1982). 'Introduction: Author, Teller, and Hero'. In Joel Chandler Harris, *Uncle Remus: His Songs and His Sayings*, 7–32. New York: Penguin Classics.
Hollander, Charles (1997). 'Pynchon, JFK and the CIA: Magic Eye Views of *The Crying of Lot 49*'. *Pynchon Notes* 40–41 (Spring–Fall): 61–106.
Hudak, Glenn M. (1999). 'The "Sound" Identity: Music-Making and Schooling'. In Cameron McCarthy, Glenn Hudak, Shawn Miklaucic, and Paula Saukko (eds), *Sound Identities: Popular Music and the Cultural Politics of Education*, 447–474. New York: Lang.
Keskinen, Mikko (2008). *Audio Book: Essays on Sound Technologies in Narrative Fiction.* New York: Lexington Books.
Kunzru, Hari (2017). *White Tears.* New York: Knopf.
Mathes, Carter (2015). *Imagine the Sound: Experimental African American Literature after Civil Rights.* Minneapolis, MN: University of Minnesota Press.
Moody, Rick (2011). 'Serge and the Paranoids: On Literature and Popular Song'. *Post45*, 1 July. Available online: http://post45.research.yale.edu/2011/07/serge-and-the-paranoids-on-literature-and-popular-song/ (accessed 29 June 2020).
Picker, John M. (2003). *Victorian Soundscapes.* Oxford: Oxford University Press.
Poe, Edgar Allan (1887). 'The Imp of the Perverse'. In *The Murders in Rue Morgue and Other Tales*, 170–175. New York: Worthington Co.
Pope, Alexander (1711). *An Essay on Criticism.* London: W. Lewis.
Powers, Richard (2008). 'Modulation'. *Conjunctions* 50: 87–103.
Pye, Patricia (2017). *Sound and Modernity in the Literature of London, 1880–1918.* London: Palgrave.
Pynchon, Thomas (1990). *Vineland.* New York: Little, Brown and Company.
Pynchon, Thomas (1999). *V.* New York: Harper Perennial.
Pynchon, Thomas (2006a). *Against the Day.* New York: Penguin.
Pynchon, Thomas (2006b). *Gravity's Rainbow.* New York: Penguin.
Pynchon, Thomas (2006c). *Mason & Dixon.* New York: Penguin.
Pynchon, Thomas (2009). *Inherent Vice.* New York: Penguin.
Ritterhouse, Jennifer (2003). 'Reading, Intimacy, and the Role of Uncle Remus in White Southern Social Memory'. *Journal of Southern History* 69 (3) (August): 585–622.
Schafer, R. Murray (1967). *Ear Cleaning: Notes for an Experimental Music Course.* Toronto: Clark and Cruichshank.
Schafer, R. Murray (1993). *The Soundscape: Our Sonic Environment and the Tuning of the World.* Rochester, VT: Destiny Books.

Schweighauser, Philipp (2006). *The Noises of American Literature, 1890–1985: Toward a History of Literary Acoustics*. Tallahassee, FL: University Press of Florida.

Seed, David (1984). 'Media and Newsreels in Dos Passos' *U.S.A*'. *Journal of Narrative Technique* 14 (3) (Fall): 182–192.

Silk, Catherine and John Silk (1990). *Racism and Anti-Racism in American Popular Culture: Portrayals of African-Americans in Fiction and Film*. Manchester: Manchester University Press.

Smith, Bruce R. (1999). *The Acoustic World of Early Modern England: Attending to the O-Factor*. Chicago: University of Chicago Press.

St. Clair, Justin (2013). *Sound and Aural Media in Postmodern Literature: Novel Listening*. New York: Routledge.

Sterne, Jonathan (2015). 'The Stereophonic Spaces of Soundscape'. In Paul Théberge, Kyle Devine, and Tom Everrett (eds), *Living Stereo: Histories and Cultures of Multichannel Sound*, 65–83. London: Bloomsbury.

Stoever, Jennifer Lynn (2016). *The Sonic Color Line: Race and the Cultural Politics of Listening*. New York: New York University Press.

Trombold, John (1995). 'Popular Songs as Revolutionary Culture in John Dos Passos' *U.S.A.* and Other Early Works'. *Journal of Modern Literature* 19 (2) (Fall): 289–316.

Twain, Mark (1885). *The Adventures of Huckleberry Finn*. New York: Charles L. Webster and Company.

Twain, Mark (1996). 'A Literary Nightmare'. In Philip Smith (ed.), *Humorous Stories and Sketches*, 16–21. New York: Dover.

Vrettos, Athena (1999/2000). 'Defining Habits: Dickens and the Psychology of Repetition.' *Victorian Studies* 43 (3): 399–426.

Wallace, David Foster (1997). *Infinite Jest*. New York: Little, Brown and Company.

9
Sonic Materialism and/as Method
Tyler Shoemaker

> It's hard to see but think of a sea
> Condensed into a speck.
> And there are waves—
> Frequencies of light,
> Others that may be heard.
> The one is one sea, the other a second.
>
> —Louis Zukofsky

The homophonic slippage that ties together Zukofsky's sea and see, ocean tide and apprehended light, gives voice to the problem of sonic materialism. During anything other than its vocalized activation, sound, here, is not there but only implied, a felt phantom resonating as a residual effect from the two mute things Zukofsky strikes together. Occupying at varying intervals the same phonic space, sea and see rely on a voiced middle C to articulate the parallactic flip that disentangles, briefly, one thing from the other, which then 'may be heard'. 'The one is one sea, the other a second', the one sea is seen a split second later by a second seeing – or heard, rather, since C charges that timed interval. And yet, though this note manages that interchange it remains the unwritten spectre of this poem, unseen. Here sound appears unnamable because it is so closely bound up with other things, not only as one of their felt effects but as a resonate frame in which their qualities are transmitted and perceived. Indeed, the generative ambiguity driving Zukofsky's middle C derives from an inability to discern whether the note in question is the sound of an object (the sea), some secondary quality attending but not constitutive of that object's essential traits, or a sound not sourced, not an effect, but the object itself (middle C), isolated and available for direct contemplation.[1] In the first track, we hear objects. In the second, sound *is* the object we hear. In both, however, a phenomenological analysis poses but cannot alone resolve the ambiguities of distinct components at present commingled in phonic overdetermination.

Present, but perhaps not a thing per se, perceived, but indiscernibly sourced from objects with more well-defined outlines, sound in Zukofsky's poem persists as an after-effect that nevertheless affects and enlivens material interplay. In it, the possibility of teasing out which sound object makes sound remains an open question. That sonic thing in itself (if in fact it is a thing) cannot be cleanly shorn from the things accompanying it (if at all), making it difficult to measure the extent to which sound participates in the materials of Zukofsky's world. It does, that is clear, but how much – or by what means – is anything but.

This is a problem, not *for* sonic materialism, as if this mode of inquiry is the right tool for the job of disentangling Zukofsky's poem, but *of* it. Prepositions matter here, for when sound is possessed by matter (or the other way around), or participates in matter, and when materialism attempts to explore this 'palpable effect on, and affection by the materials through and against which it [sound] is transmitted' (Cox 2011: 148), the tangled terms to which his poem points come alive. As it attempts to delimit the coordinates of those terms materialism exerts a strangely grammatical force on sound's interactions with the world, positioning sound athwart the spaces and things that accompany it. The sound *of* matter puts sound *with* matter, not by it, not on it, nor strictly from it. But how that *with* works – and on what side of that *with* sound sits – remains unclear. If sound is to be the object of a materialist inquiry, and if that inquiry is to understand sound as something that interacts with matter – and in fact is matter itself – that inquiry needs to contend with this problem.

Vibratory energy

In part this problem stems from materialism's search for common denominators. Shelly Trower has argued that physicalists throughout the nineteenth century – among them, Thomas Young, William Herschel, and Hermann von Helmholtz – situated sound as a special case of vibration, wherein its own range of waveforms would merely occupy a small portion of a broader energetic spectrum, one featuring light, heat, and, in the more farfetched theories of Gustav Fechner, supernatural spirits. Different from these other phenomena not in quality but in quantity, the mechanics of sonic waves were to be found in that of vibratory activity more generally; in this framework, energy took on many forms, with sound being only one such instantiation of this force among others. But sound retained one significant feature that set it apart from those other forms: its effects as that of vibration were palpably present to conscious awareness, whereas, say, the vibratory character of heat or light was not. Trower argues that nineteenth-century physicists leveraged sound's felt vibrations to conceptualize the existence and mechanics of those other insensate energies (infrared light in Herschel's case; waveform spirits for Fechner), ones whose qualities, effects, and limits could not be easily ascertained by either phenomenological verification or an experimental setup. Sound 'became a model for energy in general, which physicists described as vibratory, imperceptible, and infinite' (Trower 2012: 39). To attune oneself to that sensible phenomenon was to acquaint oneself, albeit at some remove, with a wider set of energetic activities always surging through and around the natural environment, mechanical artefacts such as heat engines, one's own body, and, according to some, one's mind. When and where conscious awareness could not register the particular effects of that broader spectrum, it fell back on sound as metaphor and model to provide a framework that suffused the matter of the world with energy of varying degrees. Sound, present, made space for a latent potentiality of matter, vibration, rarely felt if yet always there.

My brief account of Trower's argument should make it clear not only that sound has been historically entangled with materialist inquiry, but that an abstract and general concept of vibration (which sound best exemplifies) has served as an underlying model for thinking through the effects and the power of matter. This is no more apparent than in the recent new materialism. In one of this method's most popular formulations, Jane Bennett argues for an understanding of matter's vitality. Things, she writes, have power, and this imbues them with an agential force, a 'thing-power' that remains partially exterior to human agency and knowledge, sometimes slowing, diverting, or altogether blocking our intentions. Things have efficacy in excess of their designs, their uses. They have inertia, a vitalism Bennett equates with affect, a capacity of matter to do and to be and to tug and to pull on anthropomorphic agency; at times things obdurately persist along paths running crosswise to those of human actors and intents. For Bennett, a vitalist take on the material world sets out to asses those frictional junctures, attempting, 'impossibly, to name the moment of independence (from subjectivity) possessed by things, a moment that must be there, since things do in fact affect other bodies, enhancing or weakening their power' (Bennett 2010: 3). And more, at times this force leads things to associate. Things stick and they stick together (Ahmed 2006: 39–40). They self-organize and this gives them agency – or rather, enhances what agency they already have. The vitalist bent of Bennett's new materialism theorizes an inertial force inherent in things that not only hinders effective, anthropomorphic action but also brings those things together, whereupon their collective presence produces a phase shift that moves them from individual thing-power to the impetus of the assemblage as a whole. The larger and more diverse the assemblage, the more it gains influence, and this both impels and compromises the agencies of its constituent elements.

When it comes time to demonstrate this concept of thing-power, Bennett tellingly deploys language that resonates with the epistemologies Trower describes. Note the following analogies the new materialist uses to explain how the capacities of human actors run up against competing energies in an assemblage: 'an intention is like a pebble thrown into a pond, or an electrical current sent through a wire or a neural network: it vibrates' – that word again – 'and merges with other currents, to affect and be affected' (Bennett 2010: 32). Bennett resorts here to the very same semantic, maybe even subatomic field Zukofsky himself thematizes. Her vitalism is his vibration, and both of these are first and foremost energetic. As 'much force as entity, as much energy as matter' (20), the 'elusive recalcitrance' of things Bennett seeks to 'impossibly' name in her investigations finds its articulation in channelled waveforms. Inasmuch as her analysis rests on this general affective capacity it remains in the tradition of a universal energy first sensed in sound, analogically tied to the insensate waves physicalists thought sound could exemplify.

But because it remains to the side of intent and outside of knowability, a thing's 'elusive recalcitrance' appears to prevent Bennett from formulating the precise nature of the participants in her ontology. Her baroque lists – a rhetorical tic in new materialism – simply give no stable sense of what a thing is. Even as she uses vibratory energy as her general model, she takes the existence of its discrete manifestations for granted and does not devote time to ontologically disentangle one material mode from another, sometimes to a deleterious result, as in the abrupt category errors that stem from equating coal and sweat with economic theory (Bennett 2010: 25). If vibratory energy is somehow to be the model for these three elements, or to serve as a common denominator to be found among them all, Bennett skirts that explanation in her eagerness to theorize collective, cross-actor affects embedded in assemblages within and beyond human frames. While one might overlook such category errors in the service of that broader and important point about this kind of connectedness, without a more fully elaborated ontological framework focused directly on the thing (in what way, for example, is economic theory a thing?), it remains difficult to see where and how these connections might work at all. Bennett's vibrancy vibrates to such a fever pitch that its own anchoring points in the matters of the world are lost along the way.

These difficulties serve as something of a cautionary tale for sonic analysis, particularly one like the chapter at hand, which attempts, as thoroughly as possible, to audit sound alongside the world of things. There is no doubt that Bennett's analysis has considerable purchase on my own sonic-materialist dispositions, or of those Christoph Cox and Salomé Voegelin have both elaborated. Indeed, running her waveform analogies through a sonic-materialist channel demonstrates that sound, too, has agential force, its own ability to act upon the elements of its attendant assemblage. Examples of this are close at hand: beyond canonical instances in Western philosophy, such as Martin Heidegger's suggestive comment in the epilogue to 'Das Ding' that we are to 'hear an appeal of Being' (Heidegger 2013: 181–182), or Louis Althusser's infamous scene of interpellation achieved through auditory means (Althusser [1970] 2014: 190–191), there are the rings, squeals, hisses, and clicks of tinnitus, which result from the ear's exposure to prolonged and loud noise; there is what Martin Daughtry (2014: 29) calls the 'weight' of sound, thumping into the cranial and pericardial spaces of clubgoers or, well outside that scene, those of victims suffering from improvised explosive devices in war zones – these, like those earlier philosophic frameworks, evince the world-making (and at times world-destroying) power of sonic activity. And as all of these examples' shared psychological bent indicates, this kind of activity is susceptible to being multisited and multi-mattered, even intersubjective.

But emphasizing these crossovers may too quickly lose sight of the matter at hand; objects affected by sounds are not sounds themselves. And though these objects are susceptible to such affection, surely each resonates in ways particular to its own situation. Rendering this kind of specificity is something Bennett's own analyses do not quite achieve. There is a risk here of reifying sound into a common denominator, one that infiltrates all things but that nevertheless remains unidentifiable as it does so. Mutual resonance may be a powerful category for thinking through sound's involvement with matter, but it may also fail to arrive at a thoroughgoing account of sound as such.

A closer investigation into vibratory activity would seem well equipped to asses these matters, and contemporary theorists of sound have keyed into this approach, well after Helmholtz and others laid it out some two centuries ago. 'Sound is vibration that is perceived and becomes known through its materiality' (Novak and Sakakeeny 2015: 1). Similarly, Nina Sun Eidsheim has suggested that sonic phenomena be thought of as a series of 'intermaterial vibrational practices', with only a small section of that field – or a slice of the spectrum, to recall the epistemologies Trower tracks – designating cochlear activity (Eidsheim 2017: 3). Because sound depends on its materials, and because those materials act upon sound as much as it acts upon them, Eidsheim's own method posits that, 'for all practical purposes, the material that vibrates' alongside sound serves as a candidate for sonic analysis (161). Sonics are thus necessarily 'multisensory' phenomena. Like things, sound is sticky – matters of sound more so. And they range wide. But herein lies the sticking point, one very much in line with the one troubling the vibrancy Bennett attempts to name: is sound an object at all? If sound is inter-material and multisensory, does calling sound 'sound' make sense? What kind of vibration do David Novak and Matt Sakakeeny mean? Which materiality does it become known through? Follow out a physicalist monism to its furthest extent and where along that spectrum is the meaningful difference between one vibrating thing and another? Zukofsky wants to know.[2] It informs the difference between his seeing and seafaring. And these questions are not merely a poet's playthings, for as Trower demonstrates, in empirical investigations sonic things also tend to blur. If, in the wake of quantum mechanics (a subject in which Zukofsky himself was well versed), all matter is reducible to periodic disturbances otherwise called vibration, in what sense does sound, apparently the most obvious of all waveform phenomena, stand out? If waves are heard but also seen and are also water, wherein lies a sound wave's particular materiality, and what epistemological payoff would an analysis of that materiality grant?

This is the problem of sonic materialism. Insisting on the materiality of sound, letting that materiality be felt, does not automatically formulate a clear, conceptual framework for and of the sound object. Nor does it guarantee that there will be any object there to frame at all, among materials, passing through or around them. In fact, by virtue of this paradigm shift it becomes even more difficult to audit the sonic thing in itself.

Sonic in-distinction

The title of a recent essay by John Mowitt multitracks the phenomenal effect of an English onomatopoeia with noumenal and German overtones to frame the problem of sonic materialism this way: 'The Ding in Itself.' John Cage's *Water Walk* (1959) serves as its subject. Performed in January of 1960 for audience members attending the CBS gameshow *I've Got a Secret*, the composition features three minutes of Cage moving about an assemblage of ordinary things, ranging from two mechanical fish, a few ice cubes in a blender, a bathtub, a pressure cooker heated to its steaming point, five radios, a grand piano, and more. At timed intervals (Cage carries with him a stopwatch and looks at it fastidiously) he 'plays' these things, slamming them, squeezing them, blowing on them, switching them on and off – whatever the thing, Cage uses it in a manner that makes sound. But not only sound. Exempting the piano, on which he at times plays a few dense, close-fingered chords, the sonic signature of the things in *Water Walk* lie among what appear to be non-musical, non-aesthetic categories, and yet Cage explicitly frames them in his opening remarks as components of music, integral in fact to it; it is music he is after and it is music he therefore creates, if only nominally in the opinions of those among his chuckling audience. That Cage asserts this musicality walks up to a central question in sound art, Mowitt explains. His claim is that Cage's composition presents to listeners an occasion with which to sound out the limits of the frame of intelligibility that demarcates music from that of sonic phenomena more generally. Posed, suspended, and then posed again, the question, 'are the sounds these things make music?' proliferates throughout all of *Water Walk*.

For Mowitt, if a sound is to count as music, to sound like music, another question will need its own answer: 'what is a thing?' (Mowitt, forthcoming: 2). The disciplinary distinctions *Water Walk* introduces, if only to thwart, stand within the vicinity of a broader ontological problem opened in the rift between perceptual content and aesthetic prejudice. Immanuel Kant, and Heidegger reading Kant, loom large here. By asking after the musicality of its sounds, Cage's composition by proxy demarcates the things of music from the mere sound objects that remain peripheral to it, for as Mowitt explains, the extent to which those sounds are or are not music is also the extent to which this distinction relies upon a listener's framing of those sounds in a manner that distinguishes their sources as either the inert stuff of the world or as things charged with aesthetic potential. But as Kant argued, these distinctions are intuited, pre-made, prejudged, and thus situated so that their precise borders tend to elude the terms of their very investigation. A thing appears to be a thing before being recognized as a thing, for in the Kantian paradigm 'perception is never unmediated by knowing' (Mowitt, forthcoming: 8). Matter is situated in judgement before it matters. Sound, then, is a thing before it is heard.

Though these questions are situated in aesthetic terms, they nevertheless provide a highly useful frame around which to draw a materialist approach to sound. Replace *Water Walk* with the absorptive, defamiliarizing strolls and tours of Christina Kubisch's *Electrical Walks* and Mowitt's track, and that of Bennett's, Eidsheim's, and my own will easily converge. In this later series, which serves as a kind of audial revamp to the city symphony films of Dziga Vertov and Walter Ruttman, urban environments, electrified, are transformed into musical events, sonified. Starting with a first stroll through the city centre of Cologne in 2004, Kubisch has since threaded participants through such places as Riga, Chicago, and London, giving each walker a set of specially designed wireless headphones to wear during their explorations. These pick up electromagnetic waves. ATMs, fluorescent lights, underground cables: in a return to the vibratory activities of Helmholtz, Herschel, and others, the fields these objects produce as they run are transformed into a perceptual field of auditory experience. Each element therein has its own sonic signature; that signal meshes with others; together, they build into a silent soundtrack lying latent in urban environments, environments whose rhythms, timbres, and volumes vary across continents and municipalities (Cox 2006).[3] To hear such differences, one needs only tune into the right frequency – wireless tech wires in listeners to electrical currents running through miles of wires coiled in urban infrastructure.

Kubisch explains in her artist's statement that the background – or underground – radiation of city dwelling has its own surprises: 'Nothing looks the way it sounds. And nothing sounds the way it looks' (Kubisch n.d.). Hers is an art practice of the habituated thing made sonically unfamiliar. But while matter and its manufactured sounding may fail to mimetically correspond, their perceptual disparities share in the work of framing yet another (perhaps happily) subverted assumption: a city's 'sounds are much more varied and musical than you might expect' (Kubisch n.d.). Grey, winding, and bleak miles of infrastructural guts are above all aesthetic: this discovery puts *Electrical Walks* quite clearly in line with the ordinary objects Cage puts to use. And more, it puts Kubisch's series in a position where it is available to the very same line of questioning Mowitt explores: musicality is as much at issue here as it is at the CBS studios in 1960. Kubisch, however, adds a techno-materialist layer to the identification of a sonic occurrence within its disciplinary justification as music (or not), since strictly speaking the things of her *Walks* are non-cochlear. They must be made to make audible sound – or more precisely, they make audible sound through a form of translation. Though Cage 'plays' things, he assumes that, in whatever way he plays them, this will be enough to produce sound, whatever the kind. Sound as such thus goes unquestioned in *Water Walk*.[4] *Electrical Walks* also positions urban infrastructure under a similar aesthetic judgement, but Kubisch's method of electromagnetic listening can only pose, not answer, the question of whether the things that populate the auditory terrain this judgement delimits are themselves sound. That is, though she assumes fluorescent lights can be framed by sonic activity, this assumption cannot definitively state if what one hears as a result of that framing is 'actually' sound – perhaps it is just electricity. In the *Electrical Walks*, matters of sound, not just aesthetic judgement, quickly enter what Mowitt calls a 'zone of in-distinction' (Mowitt, forthcoming: 2). There, sound blends with electricity, with things. Sound happens, this is clear. It can be perceived. Yet a thing that was not sound now is. Or a sound is not sound but simply an electrified thing otherwise silently vibrating.

Whether we parse this zone of in-distinction into stable and set participants with explanations grounded in physicalist vibrations, perceptual content, or ontological first principles, Mowitt ultimately claims that sound's thingness puts disciplinary frames into question. The continuities of sonic phenomena audited in a materialist method figure sound as something available to such a method, but because of this, those continuities thwart the definitional constraints upon which materialism relies. Trower, for example, is careful to note that however well it may frame sound among things, vibration 'is not itself a material object at all', even as it is 'bound up with materiality; vibration moves material, and moves through material' (Trower 2012: 6). It situates sonic matters, resonates with them, and yet stays separate from them; and so, like other matters of sound, vibration is paradigmatic. These matters 'teeter' – Mowitt's phrase – between a thing (translated but perhaps not 'actual' sound in *Electrical Walks*) and a thing in itself (whatever sound 'actually' is). For Mowitt, a sonic thing 'is not something one perceives, whether cochlearly or non-cochlearly, or measures'; rather, it is that which 'surrounds us in the ambience of questioning. It may, in effect, not even be itself' (Mowitt, forthcoming: 8).

In effect: a quiet pun, splitting these matters into sonic practice and the results thereof (phenomenal, empirical, or, to return to Daughtry, violent). The quiddity that sound is, whatever it is, relies on its disciplinary articulations. And crucially, those in science and technology studies would point out that such articulations are always embedded in material-discursive conditions. They work by way of what Karen Barad calls 'agential cuts', wherein technological apparatuses resolve 'the ontological inseparability' of materiality so that it may become available for observation and description (Barad 2007: 348). I will have more to say about these operations later, but for the moment I want to call attention to the way Barad's cuts charge the passive voice in Novak and Sakakeeny's general definition. 'Sound', remember, 'is vibration that is perceived and becomes known through its materiality.' Perceived, known through materiality: in their formulation vibration establishes a paradigm wherein what sound is hinges on these two open variables; recourse to the passive voice of a general definition is necessary when only particulars activate or create the phenomenon in question. The effects of the agents that may come to occupy those placeholders – effects both semiotic and material – will cut matters of sound out of phenomenal processes and determine their particular character. Such a paradigm opens avenues for non-cochlear analyses of deafness and deafening, for experiencing sound without hearing it, as demonstrated by the work of Mara Mills (2011, 2015c), that of Michele Friedner and Stefan Helmreich (2012), and Christine Sun Kim's art practice.[5] And it can explain, too, the sonic quality of non-sonic things in *Electrical Walks*, inasmuch as Kubisch's wireless headphones put electromagnetic waves towards auditory ends.

There is something crucial, then, about how, in spite of its own definitional failures, a sonic materialism of mutual vibration is well primed to examine the precise junctures at which sound resonates, even silently, with and throughout a place. And there is something crucial too in the fact that technological apparatuses can make these junctures known or felt, as if they transpose, however briefly, the intuitions of aesthetic judgement Mowitt discusses into a field in which they may be available for more direct (because felt) contemplation. Whatever vibration is at a given moment, it is only so because material-discursive conditions have made it so. Ontological matters of sonic things notwithstanding, wireless headphones articulating things as sound is something to which materialist analysis can attune itself.

Listening agents

At times this attunement may be as simple as using sound to forensically investigate times and places. This enables a historical materialist approach to sonic matters. As part of a multisited installation he calls the *Hummingbird Clock* (2016–), 'private ear' – not eye – Lawrence Abu Hamdan has installed three binoculars outside and opposite to the law courts of Liverpool. Meant to act like a 'public time piece' (Hamdan 2016), the *Clock*'s lenses are trained to point in a close, scrutinizing focus towards the clock set in the Liverpool Town Hall, sitting just under the brim of its dome, which arches up to a statue of Minerva, seated and looking down. Below each camera is a placard. On them Hamdan explains how the very same power grids supplying energy to the devices Kubisch audits produce a hum on their own, without the help of electromagnetic translation. Though it often sits just below the threshold of normative audibility, this 'mains hum' is a constant, found everywhere on digital audio and video recordings – much to the annoyance of sound engineers seeking to gain clean signals in their studios. As viewers in Liverpool look on, this hum's frequencies surge all around them, silently saturating their urban environment. Small electronic screens on Hamdan's placards track those rates. While the hum is always there, nominally operating at 50 hertz per second, there are small changes in the electrical current of the power grid, and these changes produce corresponding changes in the noise of the mains hum. These minute modulations imprint their own, unique sonic signatures on recording devices across the UK, and as Hamdan explains, they are used by the UK government as a means of state surveillance. Because the material conditions of digital recordings come pre-packaged with sonic 'fingerprints' (Hamdan 2016), forensic investigators can extract that information, scan it, and map the time and place of its making – ubiquitous humming pinpoints its material-discursive conditions.

As a countermeasure to these investigations, Hamdan directs viewers to the second location of his piece, a website: hummingbirdclock.info. There, he has amplified the UK's mains hum well into the decibel range of normative audibility and set the fluctuations of its frequency in his own clock face, complete with a hand each for hours, minutes, and seconds, the last of which visualizes those fluctuations 'like a seismograph', climbing and dipping whenever the mains' nominal 50 hertz cycle wavers as much as a hertz off from its intended mark, up or down. Like the agents of state surveillance, Hamdan does not just livestream these rates but clocks and archives them; starting from 7 July 2016 onward, anything in the UK government's records should also be in his. And whereas state surveillance remains wrapped in secrecy and silent black boxes, Hamdan's website makes the gathering of this data plain plus it offers to check the hum of any digital recording against the artist's files. He provides a submission page for those needing to corroborate a claim. 'Perhaps you are caught on a recording that is being used as evidence against you, but know that it has been edited to make it sound as if you said things you didn't, or that certain material has been edited out' (Hamdan 2016). Fill out a form, send it in, and, with additional help from members of the Forensic Architecture research agency at Goldsmiths College, Hamdan will produce a full report cataloguing the conditions of a recording's sonic making.

Hamdan's *Clock* is one such example of how the arrival of a thing quite literally 'takes time, and the time that it takes shapes "what" it is that arrives'. Sara Ahmed writes that 'what arrives not only depends on time, but is shaped by the conditions of its arrival, by how it came to get here' (Ahmed 2006: 40). That digital recordings carry with them such conditions; that they arrive with clock time reified in and by their data, sequenced, made material, even sensible; that they are shaped by this time; that they interact with other times; that Hamdan detects these times, corroborates them, and disseminates them as a counterstrategy to state surveillance – all this comes together as a practice of sonic materialism. His is a method of transposing into sensuousness sonic things so they may arrive at, and then give some sense of, what Karl Marx and Friedrich Engels (1998: 45) once referred to as the 'historical product' of a 'definite' social system (see also Ahmed 2006: 41–44).[6] He enables audiences to listen for what Douglas Kahn calls the 'long sound', a sound that acquires its particular character from an extended encounter with the materials and spaces across which it has propagated (Kahn 2013: 162). Those, along with that signal's source, make matters of sound matter. Social development, industry, commercial intercourse: the *Hummingbird Clock* audits the very junctures of historical forces Marx and Engels sought to accent, inspecting them, attending to them, hearing them. Here, something like a sonic-materialist method begins to take shape.

What remains crucial in this method is how and what the 'private ear' listens for. Kahn describes an audial attunement to world forces and energies as a 'transperception', which functions like an 'apperception, a consciousness or intrinsic awareness of an energy that includes what has been traversed' (Kahn 2013: 162). In a return to the Kantian tradition in which Mowitt places his own argument, Kahn's prefixed morpheme formulates a posture that puts sound into place before and as it arrives as cochlear activity. This awareness primes listeners to attune themselves to the presence of past auditory phenomena at play during a present moment of audition. It is a vitalism in line with that of Bennet's new materialist analyses; it highlights channels of transmission, accenting the hum of the world that elsewhere vexes audio engineers; it audits sound as the material assemblage that it is, for a sound, Kahn writes, 'is always sounds' (170).

Hamdan's amplifications take a similar approach. And because they do so with an aesthetic that traverses the thresholds of normative audibility, they indicate the lasting usefulness of a phenomenological approach that comports itself to sound's affective weight – and wait times, as in the not-yet but soon-to-be-heard sonic event lying latent in the electrical wiring of cities. Though I wrote above that no such approach would be able to parse the terms of Zukofsky's poem into stable categories and set them in place for further contemplation, sensate sound – though not always cochlear sound – still matters. It most often opens the question that a materialist inquiry takes up. Like Kubisch, Hamdan provides listeners with a technological *aesthesis* that apprehends and articulates the material-discursive conditions of sonic matter, reifying those conditions into felt registers. His *Clock* makes sound representable in and to phenomenal awareness and, with it, makes tangible the things of that sound. What matters, for sonic materialism, is how we ultimately attend to the constructed nature of those felt effects.

Another way to put this is to say that I am not claiming that any and all sensory effects should be anthropomorphic or made to fit only those registers. Even as Hamdan and Kubisch make worlds that conform to human sense-certainty they also indicate a range of sensations and sensibilities that remain exterior to experience. Without the inclusion or acknowledgement of this exteriority they would simply fall into what Mark Hansen has called the phenomenological error of recent media studies, which 'involves the transposition of *our* modes of experience into the heart of *other* modes of experience – specifically, *technical* modes of experience – that are not constituted on the model of what matters to us' (Hansen 2015: 121–122, emphasis in the original). Both artists avoid this error by providing forms of sonic immersion that are tinged with unfamiliarity, slightly off, twanging with the technical conditions of their aesthetic productions in such a way that those conditions remain well within the range of hearing. Kubisch and Hamdan acknowledge the frames of sonic mediation. In doing so they follow on the 'phantasmagoric practice' of Salomé Voegelin's *Listening to Noise and Silence*, manufacturing forms of sonification that treat any-thing as a 'dynamic locale of the agency of perception' (Voegelin 2010: 103), where, in that locale, listening agents – *not* things, *not* objects, *not* assemblages – quite literally make sound for and by themselves.[7] This practice puts these agents' elusive recalcitrance (as Bennett would have it) into direct interaction with the act of audition. Otherwise put, I am, in effect, arguing that sonic materialism should not galvanize but rather mediate the gap between a sonic thing's arrival and a listening agent, and it should do so by asserting that the agents involved therein are open variables. Whether an agent be human or nonhuman, whether its mode of listening aligns with cochlear activity or takes on other sensibilities (technical, material, whatever else), there is a place for the listener in and of sonic materialism. In that place, matter matters most where it engenders the act of whatever it is we decide to call listening. Hence the *trans-* of Kahn's transperception.

This mode of perception gains its prefix not only because it listens across the materials that stick with sound – in Voegelin's account a sonic thing is a 'honeyed thing' (Voegelin 2010: 20) – for their times, their energies, their modulatory powers, but also because this act is almost always bound up with multiple auditing agents. Transperception transforms sound into sounds and the auditor into auditors; for this kind of listening the sonic thing arrives shaped by multiple sites of sensation. It is here where Eidsheim's work on the 'multisensory' phenomena of sound is most relevant – multisensory, it should now be clear, both in terms of involving more than one sense and in that of involving multiple sensors, some of which may be specifically equipped to process sound, some not. Having been worked over by these things, Kahn's long sounds gain and change character. And a key part of this process is that the media technologies involved in those changes do not merely modify a signal and pass it on, as if all perceptual effects of waveform phenomena were due to a single, selfsame signal vibrating every-thing in the world. This is the rhetorical trick of the nineteenth-century physicalists in Trower's account: they relied on sound, typically cochlear, to uncover some prior vibratory connectivity enlivening all elements in the universe. And this too is the potential threat, so Hansen would say, that the works of Zukofsky, Kubisch, and Hamdan pose to a concept of sonic materiality if this concept fails to highlight how these pieces work through a multitude of signals instantiated and actively shaped by the control logics of an experimental setup or aesthetic practice. A focus on the multisensory materials of sound sets us up to audit things along that second track because this focus cannot take for granted the existence of sound as such, even as it perceives (potential) sounds everywhere. And yet, by staying with this uncertainty, this focus may yet carve out space for sonic specificity, even when matters of sound seem to stretch wide and stick to so many things.

Boundary music

It is perhaps for this reason that transduction has stuck so well in sound studies. This process will put these present matters of sound to an end. Appearing in work ranging from media and cultural studies (Henriques 2001, 2011; Sterne 2003; Helmreich and Friedner 2012) to science and technology studies (Barad 2007; Helmreich 2007, 2009, 2015) and certain strains of affect theory and post-phenomenology (Deleuze and Guattari 1987; Goodman 2010; Eidsheim 2017), transduction centres matters of sound at the crossover points between sensate bodies, technological substrates, and cultural techniques – each one an open variable for auditing agents. It is an act of conversion and transcoding, an agential cut whereby some technology pools, condenses, and then modulates world energies from one form into another. Crucially, however, it does not transubstantiate those energies. Transducers produce energies anew, and differently, each device manufacturing them after the manner of its own 'representational recipe' (Helmreich 2009: 212–249, 2015). Though these representations often give off a sense of presence and total immersion for those on their receiving end (hence vibratory physicalism), they are nevertheless available for material-discursive inquiry, much like the one at hand. Along these lines, Kahn's own thinking about transduction leads him to comment, quite suggestively, and in a way that deserves much more elaboration than I can do here, that transduction often adjudicates distinctions between technological objects and those of the natural world (Kahn 2013: 55). While the process can appear to do this quite easily and instantly, even naturally, it does so by engendering distinctions that are inevitably caught up within cultural logics. Whatever sound is, transduction reminds us that sound is only, as Marie Thompson (2017: 274) writes, 'heard as', and so a materialist take on transduced sound can never quite leave the semiotic arena.[8]

Emphasizing how transduction situates audition emphasizes how media technologies actively remake world energies – or better put, make new energies that nevertheless remain in some proximal, analogous, or otherwise roughly translatable position to that of the one a given technology first received. These energies are separate entities enframed by a thoroughgoing logic of technological mediation that an analysis attuned to transduction is best equipped to draw out. Transductive inquiry, writes Stefan Helmreich, 'should remind auditors of the physical, infrastructural conditions that support the texture and temper of sounds we take to be meaningful' (Helmreich 2015: 225). In the context of these conditions sonic matter comes to matter, doing so with a reliance on discursive practices that, as Barad explains, are themselves 'specific material (re)configurings of the world through which the determination of boundaries, properties, and meanings are differentially enacted' (Barad 2007: 148). Transduction is continuous and highly situated among these conditions. So it is perhaps not so much that a sonic thing arrives by way of this process, at least in the sense of 'having travelled from'; a sonic thing stays with the devices entailed in its making, and when those devices interact (intra-act, in Barad's vocabulary), new sonic things appear along with them.[9] What an auditor eventually senses as transmitted sound is first of all the event of material-discursive manufacture.

Because it in part constitutes the very borders of world energies, realizing them after the manner of Barad's agential cut, transduction strangely bounds signals and subjects them to further modulation. The process puts the phenomenon of sound into place, even as it appears to make or let that sound leak into a far wider field (see Eidsheim 2017: 17, 157–178). More, transduction's boundary making often happens with one and the same action, and this action has much to do with the difficulties sonic materialism must traverse – if, in fact, it is not one of the primary causes of those very difficulties. Indeed, bounding sound as a thing is both the end result of transduction and the first thing this process seems to dissolve because this process is the boundary itself – the boundary making – sonic materialism largely works within, sometimes seeks, but cannot itself verify.[10] *Electrical Walks* and the *Hummingbird Clock* make this situation obviously apparent, inasmuch as both use transduction to maintain a gap between things and our experiences of them, all while also making room for further nonhuman auditors. Similarly, in Zukofsky's poem sea and see are condensed into a single graphic capacitor, where they await their transductive articulation by phonic intervention, an intervention that, by activating these resonances, cannot disentangle them. But this blurring works only by virtue of the materials involved. Though a methodological elaboration of matters of sound cannot quite solidify those borders, it can work with the things temporarily bounded therein.

This, finally, is what sonic materialism must audit: bounded things, things and their boundaries, things as boundary making. Amid their silences, their soundings, their vibrations, their elusive recalcitrance, and their enigmatic histories, sonic materialism treats things as impossible brackets cast around the situation of sound. Though this situation may not be wholly reducible to those things, these two phenomena co-occur in such a way that one makes the other available to inquiry. Simply put, sonic materialism puts things and sound in mutual resonance. To repeat the grammatical positioning above: listening for the sound *of* matter puts sound *with* matter, not to equate them, not to establish some sort of causal relation between them, but simply to hear them in and as the thoroughgoing entanglement with which we always find them.

In this way sonic materialism audits, and perhaps even makes a kind of boundary music. To listen for this music, follow the directives Mieko Shiomi ([1963] 2002) has drawn up into the Fluxus event score below. It encourages a form of performance that cuts close to the matters of sound tangled together at least since the power of Zukofsky's middle C and vibrating everywhere. It makes sound sound, even amid silence. Do what she has to say generally, and when it comes time to put these matters to work for sonic-materialist ends, know that Shiomi's first verb applies equally well to production and reception. Each are acts of making. As you keep that in mind, act on this as you see fit:

> <boundary music>
> Make your sound faintest possible to a boundary condition whether the sound is given birth to as a sound or not. At the performance, instruments, human bodies, electronic apparatuses and all other things may be used.

Notes

1. John Locke first makes a distinction between primary and secondary qualities, the former being those qualities that are 'utterly inseparable from the body', such as solidity, number, extension, and figure; the latter, epiphenomenal, are qualities 'which in truth are nothing in the objects themselves but powers to produce various sensations in us' (Locke 1998: 135). For a rebuttal of Locke's theory as it pertains to sound, see O'Callaghan 2007: 15–17.
2. For a discussion of Zukofsky's engagement with quantum mechanics, see Quartermain 1992: 44–58, 70–89.

3. Wolfgang Ernst would say these *Walks* confirm digital media's implicit sonicity. All such media have a time-ordering logic Ernst equates with sonic phenomena: 'Sound in its generalized sense as a temporal enunciation refers to continuous ("analog") and discrete ("digital") vibrational and frequential dynamics of all kinds […]. Sonicity is where time and technology meet' (Ernst 2016: 21).
4. Framed in this way, *Water Walk* poses an encounter between sound and music while leaving unexplored its underlying 'sound-space'. Other outlets of Cage's practice (his work with indeterminacy) and pedagogy (his Experimental Composition course at the New School) do more fully engage with that space, working from out of what Deleuze, adapting Kant, calls the 'problematic Idea', where the idea of sound is 'not an essence defined in abstraction from everyday practices, but is rather seen as a process of learning, creation and experimentation' (Campbell 2017: 369).
5. Many of Kim's pieces engage sonic materiality through notation, displaying sound information without cochlear activity to draw out how inter-material assemblages create and shape sonic events. See, for example, her series 'The Sound Of … ' (2017), which features score notations for events such as 'The Sound of Passing Time' and 'The Sound of Obsessing'.
6. Tina Campt's cross-medial method of listening to the 'felt sound' of images powerfully draws out the temporal dimensions of sonic arrival even further. She superposes haptic and affective frequencies, infrasound and figurative sound to tune into the multiple temporalities of black futurity intimated in passport photos, ethnographic photos of Africans in the Eastern Cape, and criminal identification photos – times past, present, future, and virtual.
7. In an echo of the vibratory physicalists of Trower's account, Hansen argues that something similar to what Voegelin describes happens with all twenty-first-century media. Geo-sensors, structures of dataveillance, high-frequency trading, and the like all work in the 'operational present of sensibility', a register of micro-temporal experiences to which consciousness 'has no natural access'. Like world energies in the nineteenth century, the kind of sense data these media produce simply 'cannot be directly lived by consciousness' (Hansen 2015: 52–53). Perhaps it will fall to sonic aesthetics to once more make this condition more 'real' to us.
8. If there is merit in the recent turn towards sonic ontology, argues Thompson, it lies in the way 'sonic ontologies enable us to ask how "the nature of the sonic" is determined – what grounds the sonic ground – while remaining open to how it might be heard otherwise' (Thompson 2017: 278). An analysis that does not engage these determinants threatens to reify what she calls a masculine and Eurocentric 'white aurality', which claims privileged access to sound beyond culture and representation 'while invizibilizing its own constitutive presence in hearing the ontological conditions of sound-itself' (274). See also Brian Kane's discussion of exemplification and embodiment in the 'onto-aesthetics' of Christoph Cox, Steve Goodman, and others.
9. Casey O'Callaghan has gone so far as to suggest that sound itself may not travel 'through' a medium but rather stays put in a location (O'Callaghan 2007: 46). However, he does not leverage this suggestion to understand what sound has to teach us about discourse's entanglement with matter. But Annie Goh does. Her proposed method of 'sounding situated knowledges' reassesses the nature-culture division in sonic realism by claiming, 'sound studies' central positioning of the body and embodiedness positions sounding as

predisposed to the political-philosophical project of situated knowledges' elaborated in Donna Haraway's feminist epistemology (Goh 2017: 289).
10. In this sense, one of transduction's cardinal functions in a techno-materialist analysis of sound may be that of the parergon (see Derrida 1987: 54–64; and Moten 2003: 247–251).

References

Ahmed, Sara (2006). *Queer Phenomenology: Orientations, Objects, Others*. Durham, NC: Duke University Press.

Althusser, Louis ([1970] 2014). 'Ideology and Ideological State Apparatuses'. In *On the Reproduction of Capitalism: Ideology and Ideological State Apparatuses*, trans. Ben Brewster, ed. G.M. Goshgarian, 232–272. New York: Verso.

Barad, Karen (2007). *Meeting the Universe Halfway: Quantum Physics and the Entanglement of Matter and Meaning*. Durham, NC: Duke University Press.

Bennett, Jane (2010). *Vibrant Matter: A Political Ecology of Things*. Durham, NC: Duke University Press.

Campbell, Iain (2017). 'John Cage, Gilles Deleuze, and the Idea of Sound'. *Parallax* 23 (3): 361–378.

Campt, Tina M. (2017). *Listening to Images*. Durham, NC: Duke University Press.

Cox, Christoph (2006). 'Invisible Cities: An Interview with Christina Kubisch'. *Cabinet* 21: 93–96.

Cox, Christoph (2011). 'Beyond Representation and Signification: Toward a Sonic Materialism'. *Journal of Visual Culture* 10 (2): 145–161.

Daughtry, J. Martin (2014). 'Thanatosonics: Ontologies of Acoustic Violence'. *Social Text* 119, 32 (2): 25–51.

Deleuze, Gilles and Félix Guattari ([1980] 1987). *A Thousand Plateaus: Capitalism and Schizophrenia*. Trans. Brian Massumi. Minneapolis, MN: University of Minnesota Press.

Derrida, Jacques ([1978] 1987). *The Truth in Painting*. Trans. G. Bennington and I. McLeod. Chicago: University of Chicago Press.

Eidsheim, Nina Sun (2017). *Sensing Sound: Singing and Listening as Vibrational Practice*. Durham, NC: Duke University Press.

Ernst, Wolfgang (2016). *Sonic Time Machines: Explicit Sound, Sirenic Voices, and Implicit Sonicity*. Amsterdam: Amsterdam University Press.

Friedner, Michele and Stefan Helmreich (2012). 'Sound Studies Meets Deaf Studies'. *Senses & Society* 7 (1): 72–86.

Goh, Annie (2017). 'Sounding Situated Knowledges: Echo in Archaeoacoustics'. *Parallax* 23 (3): 283–304.

Goodman, Steve (2010). *Sonic Warfare: Sound, Affect, and the Ecology of Fear*. Cambridge, MA: MIT Press.

Hamdan, Lawrence Abu (2016). 'What is the Hummingbird Clock?'. Available online: http://www.hummingbirdclock.info/about/what-is-the-hummingbird-clock (accessed 17 April 2018).

Hansen, Mark B. N. (2015). *Feed-Forward: On the Future of Twenty-First-Century Media*. Chicago, IL: University of Chicago Press.

Heidegger, Martin ([1950] 2013). 'The Thing'. *Poetry, Language, Thought*. Trans. Albert Hofstadter, 161–184. New York: Harper Perennial.

Helmreich, Stefan (2007). 'An Anthropologist Under Water: Immersive Soundscapes, Submarine Cyborgs, and Transductive Ethnography'. *American Ethnologist* 34 (4): 621–641.

Helmreich, Stefan (2009). *Alien Ocean: Anthropological Voyages in Microbial Seas*. Berkeley, CA: University of California Press.

Helmreich, Stefan (2015). 'Transduction'. In David Novak and Matt Sakakeeny (eds), *Keywords in Sound*, 222–231. Durham, NC: Duke University Press.

Henriques, Julian F. (2001). 'Sonic Dominance and Reggae Sound System Sessions'. In Karin Bijsterveld and Trevor Pinch (eds), *The Oxford Handbook of Sound Studies*, 151–175. Oxford: Oxford University Press.

Henriques, Julian F. (2011). *Sonic Bodies: Reggae Sound Systems, Performance Techniques, and Ways of Knowing*. London: Bloomsbury.

Kahn, Douglas (2013). *Earth Sound Earth Signal: Energies and Earth Magnitude in the Arts*. Berkeley, CA: University of California Press.

Kane, Brian (2015). 'Sound Studies without Auditory Culture: A Critique of the Ontological Turn'. *Sound Studies* 1 (1): 2–21.

Kim, Christine Sun (2017). "The Sound Of …," http://christinesunkim.com/work/the-sound-of-non-sounds/ (accessed 17 April, 2019).

Kubisch, Christina (n.d.). 'Electrical Walks: 2004–; Electromagnetic Investigations in the City'. Available online: http://www.christinakubisch.de/en/works/electrical_walks (accessed 14 April 2018).

Locke, John ([1689] 1998). *An Essay Concerning Human Understanding*. Revised and ed. Roger Woolhouse. London: Penguin.

Marx, Karl and Friedrich Engels ([1845] 1998). *The German Ideology*. Amherst, NY: Prometheus Books.

Mills, Mara (2011). 'Deafening: Noise and the Engineering of Communication in the Telephone System'. *Grey Room* 43: 118–143.

Mills, Mara (2015). 'Deafness'. In David Novak and Matt Sakakeeny (eds), *Keywords in Sound*, 45–54. Durham, NC: Duke University Press.

Moten, Fred (2003). *In the Break: The Aesthetics of the Black Radical Tradition*. Minneapolis, MN: Minnesota University Press.

Mowitt, John (forthcoming). 'The Ding In Itself'. In Jane Grant, John Matthias, and David Prior (eds), *The Oxford Handbook of Sound Art*. Oxford: Oxford University Press.

Novak, David and Matt Sakakeeny (2015). 'Introduction'. In David Novak and Matt Sakakeeny (eds), *Keywords in Sound*, 1–11. Durham, NC: Duke University Press.

O'Callaghan, Casey (2007). *Sounds: A Philosophical Theory*. Oxford: Oxford University Press.

Quartermain, Peter (1992). *Disjunctive Poetics: From Gertrude Stein and Louis Zukofsky to Gertrude Stein*. Cambridge: Cambridge University Press.

Shiomi, Mieko ([1963] 2002). 'Boundary Music'. In Ken Friedman, Owen Smith, and Lauren Sawchyn (eds), *The Fluxus Performance Workbook*, 96. Performance Research e-publications.

Sterne, Jonathan (2003). *The Audible Past: The Cultural Origins of Sound Reproduction*. Durham, NC: Duke University Press.

Thompson, Marie (2017). 'Whiteness and the Ontological Turn in Sound Studies'. *Parallax* 23 (3): 266–282.

Trower, Shelley (2012). *Senses of Vibration: A History of the Pleasure and Pain of Sound*. London: Bloomsbury.

Voegelin, Salomé (2010). *Listening to Noise and Silence: Towards a Philosophy of Sound Art*. London: Bloomsbury.

Zukofsky, Louis ([1946] 2006). 'It's Hard to See but Think of a Sea'. In *Selected Poems*, ed. Charles Bernstein. New York: Library of America.

10
Sonic Methodology in Philosophy
Elvira Di Bona

Introduction

Investigating the nature of sound and how we experience it is important not only for the comprehension of the functioning of audition but also because, on the one hand, it helps us understand the role of sensory perception in general within the architecture of the mind and, on the other hand, it tests commonsensical claims on the functioning of perception which are usually based on vision. That is, the investigation of hearing and sounds can be seen as a 'sonic' methodology which serves to understand perception with regard to its peculiar object and its relationship with space and time.

Firstly, focusing on the object of auditory perception and characterizing it as a stream of happenings or events (sounds) have challenged the claim, based on the analysis of visual perception, according to which the object of perception has to be a material object, such as, say, tables or chairs. That is, the mere analysis of what is the object of auditory perception might help to rethink what we perceive when we are in touch with the world, by raising concerns, at least, with respect to the usual theoretical framework of the theories of perception that see vision as a reference point for all sense modalities. However, the same analysis might also lead to the conclusion that auditory objects overlap with material objects (sound sources), allowing us to confirm what is already stated in vision. Therefore, the study of the auditory object has a very important function: on the one hand, it might reinforce the commonsensical idea that we perceive material objects – this will be shown by embracing a specific conception of the auditory object which claims that we hear sound sources – on the other hand, it may call into question this commonsensical idea and thus strengthen the disanalogy between vision and audition – by suggesting that the auditory object is constituted only by sound and its properties. Moreover, the focus on hearing and sound, together with the investigation of what we usually characterize as auditory experience, acts as a useful methodology in order to deepen our understanding of spatial and temporal experiences. That is, if the spatial properties we recover through audition are about the location of sound sources, we can claim that, analogously to vision, audition is informative about the location of material objects. In addition, analysing the experience of sound contributes to cast light on some of the most representative examples of temporal experiences, namely, temporal phases, persistency, and temporal contour. The discussion of these temporal experiences will be pursued by comparing the experience of sound and the experience of colour with relation to time.

To sum up, I argue that the study of sound and auditory experience can be seen as a very successful methodology in philosophy in order to develop and enhance the understanding of perception as a whole and to enrich our comprehension of the experience of space and time.

Sonic methodology and the object of perception

When answering the general question of what we perceive when we are in touch with the world, we mainly provide an answer which is based on the analysis of the sense modality of vision. That is, before answering this question, we usually 'take a look' at what is around us and describe what we immediately 'see'. When we ask what we perceive to be around us, we answer by mentioning things that we are visually acquainted to (tables, chairs, trees, animals, and their qualities). We tend to identify such things with material objects and their attributes.

In the history of the philosophy of perception, the majority of philosophers, with few exceptions, have adopted this way of proceeding when elaborating their theories of perception. The philosopher Casey O'Callaghan suggests that this tendency, which has established a paradigm often referred to as 'visuocentrism', has strongly shaped our understanding of perception for a long time (O'Callaghan 2007: 4). Taking into account sense modalities that are different from vision is a relatively recent development in the philosophy of perception. This development indicates a distancing from the idea of relaying upon vision in the discussion of the object of perception and its relationship to space and time. Therefore, when answering the question of what we perceive when we are in touch with the world, due to the increasing study of sense modalities which are different from vision, it is not obvious that, from the perspective of these sense modalities, we would provide the same answer we would give when focusing on vision.

If we take audition into account and ask what we perceive when we are in touch with the world, we usually answer that we hear sounds, melodies, or rhythms which are things that we do not usually characterize as material objects. As the philosopher Matthew Nudds points out: 'The idea that our experience of sounds is of things which are distinct from the world of material objects can seem compelling. All you have to do to confirm it is close your eyes and reflect on the character of your auditory experience' (Nudds 2001: 210). Therefore, focusing on audition challenges the granitic tendency of claiming that we perceive material objects and their attributes; it constitutes a valid method to put this commonsensical idea into question. At the same time, though, this method of investigating hearing and sound could also reinforce the commonsensical idea that we perceive material objects just by showing that, as in vision, we perceive material sources in audition as well.

Sound theories employ their own notion of an auditory object which basically depends on how the theory answers the following interrelated questions: What do we hear? Are sounds the unique objects of auditory perception, or is there something else that we hear when we have an auditory experience? Does it make sense to say that along with sounds we hear also sound sources? When answering these questions, we come up with two possibilities: either we hear only sounds and their audible qualities, which are pitch, loudness, and timbre, or we hear the causes of these sounds, namely, sound sources (Di Bona and Santarcangelo 2018: 43). If we merely reflect upon our auditory experience when listening to, say, a high-pitched sound, we tend to describe it as either a bold sound with its audible qualities, or we might go deeper and say we hear something richer, something like the cry of a baby or the siren of an ambulance which comes from far away. These two possibilities generate a strong disagreement in the field of auditory perception and determine two different philosophical positions: on the one hand, there is the position according to which the only things we hear are sounds, which are taken to be items that 'exhaust' the content of auditory experience; on the other hand, there is the position for which we hear the objects that produce sounds, which means that we are in touch with the sound sources. Both views are legitimate and grounded on different but equally acceptable intuitions: the intuition for which our auditory system tracks sound sources in order to avoid obstacles and dangers, allowing us to navigate the environment (this intuition justifies the view for which we auditorily perceive sound sources), and the intuition suggesting that sounds are items which can be 'separated' from their causes and are somehow 'disembodied' entities. This intuition warrants the claim that we perceive only sounds (43). Developing the first intuition would challenge the commonsensical idea that we perceive material objects; developing the second intuition would reinforce it.

Two opposite views on auditory perception

Let me analyse in greater detail these two different views and start from the discussion of philosopher D. L. C. Maclachlan's position according to which we merely hear sounds. This is what he asks us to do:

> Suppose that there is a car passing the window. I am asked how I know that there is a car passing the window and I answer: 'Because I can hear it.' Again, suppose that there is a burglar moving about downstairs. I am asked how I know that there is a burglar downstairs and I answer: 'Because I can hear him.' Usually we would be perfectly happy to accept these answers as quite satisfactory, unless, for example, we had reason to believe that the burglar downstairs was only the cat. But do we really hear the car passing the window and the burglar downstairs? In general, do we ever hear anything except sounds and noises of various kinds? […] All I really hear are certain suspicious noises, and I say I hear a burglar only because I assume that a burglar is responsible for the suspicious noises in question.
>
> (Maclachlan 1989: 8)

Maclachlan claims that we hear nothing but sound. He distinguishes between knowledge by perception and knowledge by inference, recalling the Russellian distinction between knowledge by acquaintance and knowledge by description. According to Maclachlan, through the sense of hearing we perceive only sounds and noises and then, only indirectly, we perceive the things responsible for these sounds and noises. The things we indirectly perceive are merely inferred from what is actually given to the experience. His view echoes George Berkeley's, who introduced the difference between the proper object of audition, which is sound, and what we perceive only indirectly, which are sound sources, in the following passage:

> For instance, when I hear a coach drive along the streets, all that I immediately perceive is the sound; but from my past experience that such a sound is connected with a coach, I am said to 'hear the coach'. Still, it is obvious that in truth and strictness nothing can be heard but sound; and the coach in that example is not properly perceived by sense but only suggested from experience.
>
> (Berkeley 1975: 54)

Berkeley suggests that nothing can be heard but sound. Then, sound sources, which in his example are exemplified by a coach, are not 'properly perceived by sense but only suggested from experience' (Berkeley 1975: 54). In the Berkeleyan view, sounds are the unique objects of auditory experience and, since we have already had the experience of perceiving a coach, we are able to associate the sound we currently hear with the passage of a coach. With regard to the idea that sounds are the immediate objects of auditory perception, Berkeley adds:

> PHILONOUS. This point then is agreed between us, that *sensible things are those only which are immediately perceived by sense*. You will farther inform me, whether we immediately perceive by sight anything beside light, and colours, and figures: or by hearing, anything but sounds: by the palate, anything beside tastes: by the smell, beside odours: or by the touch, more than tangible qualities.
> HYLAS. We do not.
>
> (Berkeley 1975: 8, emphasis in the original)

In this passage of the 'Three Dialogues Between Hylas and Philonous', Berkeley claims that just as light, colours, and figures in vision, tastes in taste, odours in smell, and tangible qualities in touch, sounds are the immediate object of hearing and are sensible things.

I have briefly presented the view for which what we perceive in auditory perception is only sounds. Let me now introduce the position for which we literally hear sound sources. In order to defend this view I could start by individuating the characteristics of sound sources that we can have an access to auditorily. By 'sound sources' I mean all the objects which are capable of emitting a sound if correctly stimulated. It seems to be intuitive that there are some attributes of sound sources which we cannot perceive by audition. The redness of a ripe tomato or the heat of a piece of bread just taken out of the oven are among them. There are, however, attributes of objects which, even if prima facie do not seem to be 'audible', turn out to be auditorily perceivable. For example, studies in ecological psychology show that some material objects' properties, such as hardness, texture, or the length of an object, which are commonly perceived by vision or touch, might be perceived also by hearing (Lederman 1979; Freed 1990; Carello, Anderson, and Peck 1998). Hardness, texture, and object length are not only characteristics of ordinary material objects but can be taken also as specific features of sound sources which are audible. All objects capable of making sounds have hardness, texture, and a certain length. The wood of a violin or the surface of a table have a specific hardness; the metal which constitutes a gong might have a hardness as does a pan made of cast iron; and both a flute and rods dropped to the floor have a certain length. Therefore, by auditory perception, we are able to capture some of the sound sources' properties. Among them we have also the gender properties expressed by human voices. I have supported this claim by investigating auditory adaptational effects on gender properties and contrasting auditory experiences before and after the adaptational effects take place (Di Bona 2017b). In light of this investigation, I concluded that auditory experience is not limited to sounds' audible properties, such as pitch, loudness, and timbre (which are also usually labelled as the 'low-level properties' of auditory perception). The existence of adaptation effects on a property is taken to be good evidence that that property is part of the content of perception (Fish 2013; Block 2014). The logic behind this idea is that in vision, for example, all the perceivable properties uncontroversially agreed upon – such as luminance, contrast, or motion – are susceptible to adaptation (Antal et al. 2004; Chen et al. 2005). Therefore, if other properties are susceptible to adaptation it suggests that they might at least be part of the content of perception as well (Di Bona 2017b: 2632).

Claiming that we perceive sound sources when having an auditory experience does not mean to affirm that we can always recognize exactly the object which produced the sound we listened to, since recognizing the source of the sound is not always possible. To argue in favour of the auditory perception of sound sources is enough to account for the perceptual experience of the audible properties of sound sources. This is why we focus on specific attributes such as hardness, texture, the length of an object, or the gender properties of human voices. The fact that, by audition, we can be acquainted with different objects' features shows that perception appears to be much more informative about the auditory surrounding than it is usually believed.

The richness of auditory perception is further confirmed by a number of studies which show that we are able to recognize the activity in which material sources are involved while producing sounds. An experiment demonstrates that listeners who were asked to identify different recorded sounds of jars and bottles of different size falling to the ground either bouncing or breaking, were almost always accurate (Warren and Verbrugge 1984). In another experiment, when listeners were asked to identify thirty common natural sounds (common natural sounds are those generated by clapping, tearing paper, or footsteps), they were able to recognize their sources very reliably. Nudds mentions that, in a similar experiment, seventeen sounds were played and listeners were asked to identify what they were listening to (Nudds 2010: 111). They nearly always described the sounds in terms of their sources, and their descriptions were quite accurate. Several perceivers could 'distinguish the sounds made by someone running upstairs from those of someone running downstairs, others were correct about the size of objects dropped into water' (111). Another study analysed whether a perceived walk was performed by a female or by a male (Li, Logan, and Pastore 1991). Xiaofend Li, Robert J. Logan, and Richard E. Pastore asked eight males and eight females to walk across a hardwood surface. Listeners simply had to listen to a recording of four steps and say whether the walker was a male or female. The probability of identifying a male walker correctly was 69 per cent and the probability of identifying a female correctly was 74 per cent. Another experiment was conducted by the psychologist Bruno Repp and was on clapping. Repp's basic question was whether when we hear clapping, we can also get information about the configuration of hands.[1] In Repp's experiment, perceivers listened to eight different ways of clapping. The clapping was ranked from a flat, parallel mode in which two hands perfectly overlapped, to a mode in which the clapping was made by the contact between fingers and palm. The task was to identify palm-to-palm clapping, finger-to-palm clapping, and one intermediate type of clapping. Even if they were not able to distinguish among the ways of clapping produced by a group of different subjects, (1987: 1104) perceivers were quite able to identify ordinary claps.

Agreeing with the claim that we hear sound sources will confirm the analogy with vision, according to which, as the objects of visual perception are material objects, also the objects of auditory perception are sound sources, namely, material objects producing sounds, some of their properties, and the activities in which they are involved. On the contrary, agreeing with the claim for which we hear only sounds will support the disanalogy with vision, according to which, as the object of visual perception are material objects, the object of auditory perception are sounds with their loudness, timbre, and pitch. As previously shown, there is empirical evidence demonstrating that when having an auditory experience, we mostly go beyond the mere perception of sounds and their properties, and tend to be acquainted with the actual material sound sources. Therefore, the sonic methodology applied to the problem of the object of perception has ultimately demonstrated that the analogy with vision might work so that we can start providing a unitary answer to what is a perceptual object. Nevertheless, sounds are the intermediaries between the perceiver and sound sources; they allow us to get in touch with sound sources, so that we can say that we hear sound sources by hearing sounds. Therefore, when claiming that the auditory object is constituted by sound sources, we still need to tell a story of what is the relationship between sound and sound sources and how sounds allow us to be acquainted with sound sources. Answering this question is the demanding task of future research.

Sonic methodology and space

If what we hear are sound sources or, at least, some of their attributes, we can investigate to what extent we get spatial information about the location of sound sources. As I said, the focus on hearing and sound is a useful methodology not only to understand perceptual experience in general but also to understand the functioning of spatial experience. The sonic methodology applied within issues on the experience of space goes from the particular way of recovering spatial information in audition, to an account of spatial experience in general. That is, if the spatial properties we recover through audition are about the location of sound sources, I can claim that analogously to vision, audition is also informative on the location where material objects are. This is the starting point to provide a unitary account on spatial experience that is based on the study of the common spatial features of material objects we recover through audition and vision.

Moreover, if I can explain how we auditorily perceive spatial properties of sound sources, I will contribute to enrich the list of sound sources properties we can get through audition in addition to the properties I already mentioned as to be part of the content of auditory experience, namely, hardness, texture, object length, and the gender properties of human voices.

In everyday listening, sounds are often perceived as located in the surrounding environment at some distance from the perceiver and coming from a specific direction. We hear the sound of the knocking on the near door; we hear someone crying down the street; we hear the baritone singer rehearsing in the apartment adjacent to ours. By audition we not only get information on the kind of objects in the environment which produce these sounds but also we are acquainted with sounds which seem to have a location in the surrounding space. We turn to look towards a sudden thud or a loud explosion because sounds seem to have an indeterminate location outside our head. Simple reflections on the phenomenology of auditory experience seem to tell us that we hear sounds as being somewhere.

Empirical evidence supports the thesis that, despite the fact that the spatial characteristics we detect by audition are neither as accurate nor as precise as the spatial characteristics we get through vision, audition gives us directional information:

> Research has shown that the region of most precise spatial hearing lies in, or close to, the forward direction and that, within this region, a lateral displacement of the sound source most easily leads to a change in the position of the auditory event [...]. The spatial resolution limit of the auditory system [about 1 degree of arc] is, then, about two orders of magnitude less than that of the visual system, which is capable of distinguishing changes of angle of less than one minute of arc.
>
> (Blauert 1997: 38–39)

The scientist Jens Blauert claims that even though the spatial resolution of the auditory system is less accurate than the spatial resolution of the visual system, in audition we do not perceive only spatial characteristics related to direction, but we experience also what he calls 'distance hearing', according to which 'for familiar signals such as human speech at its normal loudness, the distance of the auditory event corresponds quite well to that of the sound source' (Blauert 1997: 45–46). Sound engineers use researches about the localization of sound to shape the experience of music in different environments (concert halls, theatres), while contemporary composers, too, are often familiar with such researches which they exploit in order to create special effects in their compositions (Di Bona 2017a). (I am referring to composers such as Karlheinz Stockhausen, Luigi Nono, Pierre Boulez, and Alvin Lucier.) If we combine empirical research on spatial hearing with reflections on the phenomenology of auditory experience related to spatiality, we can conclude that the spatiality of audition is essentially about the location of sound sources and the sound that comes from them.

In the philosophical literature, a taxonomy of sound theories has been already proposed, based on the location we assign to sounds and sound sources when reflecting upon hearing. The philosophers Roberto Casati and Jérôme Dokic distinguish four different groups: The *distal* theories group, which argues that sound is located where sound sources are; the *medial* theories group, according to which sound is in the transmitted medium, which separates the listener from the source; the *proximal* theories group, which claims that sound is either in the ears of the listener or in the space in the vicinity of the perceiver; and the *a-spatial* theory group, corresponding to Strawson's position,[2] which claims that sound is an a-spatial item which we hear as occupying no location whatsoever (Di Bona 2019). Each of these groups of theories assigns a specific location to sound and, consequently, accounts for a different relation between sound and its source, and argues for a different spatiality of sound sources.

Let us see in detail how each group of theories elaborates on the relationship between sound and sound source. Medial and proximal theories consider sounds as caused by their sources and as somehow distinct from them. If sounds are sound waves coming from an object and propagating in the environment, or if they are proximal stimuli, they are distinct from the source. These theories claim that, by audition, we get spatial information that is about sounds and do not explicitly say whether this information is also about sound sources. Which is why they face some worries. Both positions assign a location to sound which happens to be different from the location where the sound sources are; therefore, they need to justify why, very often, when asked about where sound is, we mention the place where a sound is coming from, which corresponds to the place where that sound has been produced, namely, its source. Moreover, even if the proximal and the medial views might claim that we can still get spatial info on sound sources which happen to be different from spatial info on sound, they need to justify how, when having an auditory experience, we can distinguish between the spatial info on sound and the spatial info on sound sources.

The distal position (Casati and Dokic 1994; Pasnau 1999; O'Callaghan 2007; Kulvicki 2008, 2014), according to which we hear sounds as items located at the object that produces them, is less controversial than the proximal and the medial positions with regard to the spatiality of audition. This is because, given that it claims that sounds are heard to be where sound sources are, it can easily account that when we get spatial information on sounds, this is information *tout court* also on sound sources, due to the co-location of sound and sound sources. They justify co-location by virtue of metaphysical considerations on the nature of sound, namely, by saying that sound is either an event-like individual located at the source and identical to the event source (Casati, Dokic, Di Bona 2005; Casati, Di Bona, and Dokic 2013), or that sound is a relational event that is a medium disturbance at the interface between the vibrating object and the surrounding air (O'Callaghan 2007). Within the distal view, co-location can be warranted also by classifying sound as a categorical property of the source (Pasnau 1999) or as a dispositional property of the sounding object (Kulvicki 2008, 2014).[3]

All these different metaphysical options on sound reveal a tight relationship between sound and sound sources. Since sound can be identical to the event source, it can be the medium disturbance originated at the interface between the vibrating object and the surrounding air, a categorical or a relational property of the source. The tight relationship between sound and sound sources revealed by the metaphysical status that the distal view attributes to sound acts as a picklock in order to justify why, when we get spatial information on sound, we get spatial information *tout court* on sound sources. It seems clear that among the different groups of views, only the distal view of sound conforms with the phenomenology of spatial experience and the empirical data suggesting that the spatiality of audition is essentially about the location of sound sources and the sounds coming from them.

I did not discuss the a-spatial view of sound for which sound is an a-spatial item which we hear as occupying no location whatsoever since it seems to be strongly counterintuitive and in disagreement with the basic phenomenological considerations and the empirical data on the spatial experience I have already mentioned.

As in vision we can easily say that the material objects we perceive have a location in space and a specific spatial dimension, we can also affirm that audition is informative on the spatiality of the material objects which produce sounds, by telling us at least at what distance and in which direction they are located with respect to us. Therefore, the analogy between vision and audition works not only to the extent that in both cases material objects are the objects of perception, but also since their spatial location can be considered as visually and auditorily perceivable. The analogy works, though, only within the framework of the distal view of sound.

Sonic methodology and time

If sonic methodology has helped to test the comparison between vision and audition concerning the object of perception and the experience of space, we need to test whether it works also when talking about temporal features. It will be clear that sonic methodology will be useful also in this respect. This time, I will be more specific and compare sounds and colours.

There seems to be different temporal experiences that interest sounds. We can hear the different temporal elements or phases that shape the temporal evolution of sound, we can hear sound as something which persists in time, and we can also individuate the temporal contour of a sound with the aim of distinguishing it from a simultaneous one (Di Bona and Santarcangelo 2018: ch. 4). Sounds begin, last, and come to an end. This is evident all the time that we hear a dog barking, a siren wailing, or listen to a melody in a concert hall. In acoustics, the envelope describes the evolution of sound in time. This is typically segmented in four phases (attack, peak, sustain, and decay). It refers to a single sound, such as a brief note on a piano, but it can also be used to describe an entire stream of sounds, such as the slamming of a door or footsteps on a street. The ability of detecting the different temporal phases of an unfolding sound or stream of sounds is considered to be at the basis of the difference with the way in which we experience colours.

On this regard, Casati, Dokic and Di Bona write: 'Sounds take up time. They start and cease. They are intrinsically temporal entities. Their temporal profile is essential to individuating them, in a way which has no analogue in the case of colors and shapes' (Casati, Dokic and Di Bona 2005: 20). I should specify that the disanalogy Casati, Dokic and Di Bona mention applies when we take still colours as an example but when we see colours in motion, like patches intermittently coloured in red, we could experience them as items which start, last, and come to an end. At the same time, let us imagine the situation in which we are in a car, waiting at a traffic light. Let us imagine that the green light starts and then its intensity changes before smoothly becoming yellow. We can describe the evolution of the green light by virtue of the envelope whereby we can detect four phases in the evolution of green, as it happens with sound. Therefore, we can conclude that the experience of this first form of auditory temporality does not justify the disanalogy with colour perception (Di Bona and Santarcangelo 2018: 122–123).

Another form of temporality which seems to characterize the experience of sound is that sound resists qualitative changes. O'Callaghan claims that:

> Sounds survive changes to their properties and qualities. A sound that begins high-pitched and loud may continue to exist though it changes to being low-pitched and soft. An object does not lose its sound and gain a new one when it goes from being high-pitched to low-pitched, as with an emergency siren's wail. […] Determinate perceptible or sensible qualities, however, do not survive change in this way. The red colour of the fence does not survive the whitewashing. The dank smell of the dog does not survive the perfuming.
>
> (O'Callaghan 2008: 4)

O'Callaghan suggests that while sounds resist all kinds of qualitative changes, colours do not survive qualitative changes. It seems quite convincing, indeed, to claim that we do experience the persistency of colour and the persistency of sound in a quite disanalogous way.

A last form of temporal experience in audition is the temporal contour, which is at the core of our ability to individuate sounds (Di Bona and Santarcangelo 2018: 126–128). This form of temporal experience is often analysed by scientists when working on the identification of auditory objecthood (Bregman 1990; Kubovy and Van Valkenburg 2001; Griffiths and Warren 2004; Denham and Winkler 2015). Being able to detect the temporal edges of sound is crucial in order to differentiate auditory streams from a chaotic background. According to Albert S. Bregman (1990), the auditory scene analysis is grounded on two basic groupings: the *primitive* grouping and the *schema-driven* grouping. The primitive grouping takes place *sequentially* and *simultaneously*. Both mechanisms of the primitive grouping lie on our ability to have a temporal auditory experience. Sequential integration originates streams segregation allowing us to distinguish the different sensory elements which come from the auditory environment, such as loudness, pitch, and timbre, and to attribute them to the sound or the stream to which they actually belong. We attribute those features, whilst they change over time, to the appropriate streams and we segregate one stream from another. As for the simultaneous grouping, among different factors, a factor that tends to group components that come from the same source is the synchrony of onsets and offsets of components. This factor that is helpful because parts of a single sound usually start at the same time. The grouping of simultaneous components influence auditory perception, including the number of sounds that are perceived, their pitch, timbre, loudness, and location.

Is there a form of temporal contour which is at the core of our ability to individuate colours which justifies the analogy between vision and audition concerning temporal experience? Small colour differences can be discriminated when the coloured areas are large and adjacent to each other. These conditions occur, for example, when looking at continuous coloured data as in maps of weather or temperature. Larger colour differences are needed if the conditions change from this ideal. The adjacent stripes are distinctly discriminable.

As the region of the coloured areas to differentiate is reduced, larger colour differences are difficult to detect. The ideal condition for colour distinction to take place is when a clear limit separates the colours, as in the case of a badge against the background colour of a dress. Instead, when a gradual limit separates two colours, even the smallest detectable difference in colour appears to be quite visible. The experience of temporal contour seems to be crucial for segregating streams of sounds. On the contrary, when we have to discriminate colours, which is an analogous activity to segregating streams in vision, spatial cues seem to be much more relevant than temporal cues.

Therefore, if there is a significant disanalogy between the experience of sound and the experience of colour with regard to time, temporal contour and temporal persistency mark this disanalogy. On the contrary, when focusing on the experience of temporal phases, the way in which we experience colours and the way in which we experience sounds are analogous.

Sonic methodology helps to show that we can justify the commonsensical idea of perceiving material objects only if we embrace the distal view of sound. This idea is further reinforced by empirical and phenomenological considerations on the spatiality of auditory experience. Furthermore, the commonsensical idea for which we experience time disanalogously when looking at colours and when hearing sounds is justified by the way in which we experience temporal contour and temporal persistency. Conversely, the experience of temporal phases seems to tell us that we experience time analogously in vision and audition.

I conclude that sonic methodology turns out to be a useful procedure in order to test some commonsensical ideas about the object of perception and the experience of space and time, especially when comparing visual experience to auditory experience.

Notes

1. Actually, the experiment was meant to show that we can also recognize the gender of the person clapping but the results were not considered convincing.

2. The philosopher Peter Strawson was the initiator of the discussion on the spatiality of audition within analytic philosophy with the second chapter of his book *Individuals* (1959) titled 'Sounds'.
3. Kulvicki changed his view and embraced a position according to which, given the complexity of the auditory world, instead of focusing on mere sounds, philosophers should focus on the different aspects of what we hear, namely, events, individuals, and spaces.

References

Antal, A., E. T. Varga, M. A. Nitsche, Z. Chadaide, W. Paulus, G. Kovacs, and Z. Vidnyánszky (2004). 'Direct Current Stimulation Over MT?/V5 Modulates Motion Aftereffect in Humans'. *NeuroReport* 15: 2491–2494.

Berkeley, G. ([1713] 1975). 'Three Dialogues Between Hylas and Philonous'. In *Philosophical Works, Including the Works on Vision*, ed. M. R. Ayers. London: Dent.

Blauert, J. (1997). *Spatial Hearing: The Psychophysics of Human Sound Localization*. Cambridge, MA: MIT Press.

Block, N. (2014). 'Seeing-as in the Light of Vision Science'. *Philosophy and Phenomenological Research* 89 (1): 560–572.

Bregman, A. S. (1990). *Auditory Scene Analysis: The Perceptual Organization of Sound*. Cambridge, MA: MIT Press.

Carello, C., K. L. Anderson, A. Kunkler-Peck (1998). 'Perception of Object Length by Sound'. *Psychological Science* 9: 211–214.

Casati, R. and J. Dokic (1994). *La philosophie du Son*. Nîmes: Chambon.

Casati, R., J. Dokic, and E. Di Bona (2005). 'Sounds'. In Edward N. Zalta (ed.), *The Stanford Encyclopedia of Philosophy* (Summer 2020 Edition). Available online: http://plato.stanford.edu/archives/win2012/entries/sounds/ (accessed 1 July 2020).

Casati, R., E. Di Bona, and J. Dokic (2013). 'The Ockhamization of the Event Sources of Sounds'. *Analysis*, 73 (3), 462–466.

Chen, A. H., Y. Zhou, H. Q. Gong, and P. J. Liang (2005). 'Luminance Adaptation Increased the Contrast Sensitivity of Retinal Ganglion Cells'. *NeuroReport* 16: 371–375.

Denham, S. L. and I. Winkler (2015). 'Auditory Perceptual Organization'. In J. Wagenmans (ed.), *The Oxford Handbook of Perceptual Organization*, 601–620. Oxford: Oxford University Press.

Di Bona, E. (2017a). 'Listening to the Space of Music'. In E. Di Bona and V. Santarcangelo (eds), 'The Auditory Object'. Special Issue of *Rivista di Estetica* 66 (3): 93–105.

Di Bona, E. (2017b). 'Towards a Rich View of Auditory Experience'. *Philosophical Studies* 174 (11): 2629–2643.

Di Bona, E. (2019). 'Why Space Matters to an Understanding of Sound'. In T. Cheng, O. Deroy, and C. Spence (eds), *Spatial Senses: Philosophy of Perception in an Age of Science*, ch. 6. New York: Routledge.

Di Bona, E. and V. Santarcangelo (2018). *Il suono: L'esperienza uditiva e i suoi oggetti*. Milan: Raffaello Cortina.

Fish, W. (2013). 'High-level Properties and Visual Experience'. *Philosophical Studies* 162: 43–55.

Freed, D. J. (1990). 'Auditory Correlates of Perceived Mallet Hardness for a Set of Recorded Percussive Events'. *Journal of the Acoustical Society of America* 87: 311–322.

Griffiths, T. D. and J. D. Warren (2004). 'What Is an Auditory Object?'. *Nature Review Neuroscience* 5: 887–892.

Kubovy, M. and D. Van Valkenburg (2001). 'Auditory and Visual Objects'. *Cognition* 80: 97–126.

Kulvicki, J. (2008). 'The Nature of Noise'. *Philosophers' Imprint* 8 (11): 1–16.

Kulvicki, J. (2014). 'Sound Stimulants'. In D. Stokes, M. Matthen, S. Biggs (eds), *Perception and Its Modalities*, 205–221. New York: Oxford University Press.

Kulvicki, J. (2016). 'Auditory Perspectives'. In B. Nanay (ed.), *Current Controversies in Philosophy of Perception*, 83–94. Oxford: Routledge, Taylor & Francis.

Lederman, S. J. (1979). 'Auditory Texture Perception'. *Perception* 8: 93–103.

Li, X., R. J. Logan, and R. E. Pastore (1991). 'Perception of Acoustics Source Characteristics: Walking Sounds'. *Journal of Acoustical Society of America* 90: 3036–3049.

Maclachlan, D. L. C. (1989). *Philosophy of Perception*. Englewood Cliffs, NJ: Prentice Hall.

Nudds, M. (2001). 'Experiencing the Production of Sounds'. *European Journal of Philosophy* 9: 210–229.

Nudds, M. (2010). 'What Are Auditory Objects?'. *Review of Philosophy and Psychology* 1: 105–122.

O'Callaghan, C. (2007). *Sounds*. New York: Oxford University Press.

O'Callaghan, C. (2008). 'Object Perception: Vision and Audition'. *Philosophy Compass* 3 (4): 803–829.

O'Callaghan, C. (2009). 'Sounds and Events'. In M. Nudds and C., O'Callaghan (eds), *Sounds and Perception: New Philosophical Essays*, 26–49. New York: Oxford University Press.

Pasnau, R. (1999). 'What Is Sound?'. *Philosophical Quarterly* 49: 309–324.

Repp, B. H. (1987). 'The Sound of Two Hands Clapping: An Exploratory Study'. *Journal of the Acoustical Society of America* 36: 2021–2028.

Russell, B. (1912). *The Problems of Philosophy*. Oxford: Oxford University Press.

Warren, W. H. and R. R. Verbrugge (1984). 'Auditory Perception of Breaking and Bouncing Events: A Case Study in Ecological Acoustics'. *Journal of Experimental Psychology: Human Perception and Performance* 10: 704–712.

11

Sonic Methodologies in Science and Technology Studies

Joeri Bruyninckx and Alexandra Supper

Introduction

In this chapter, we provide a survey of work which engages with sound in relation to knowledge production and technological practice, and which resonates with broader concerns in the academic field of science and technology studies (STS). STS is itself an interdisciplinary amalgam of perspectives, drawing upon concepts, theoretical approaches, and methodologies from a wide range of disciplines and inter-disciplines. Since the 1970s and 1980s, various interdisciplinary programmes and networks have emerged at the intersection of academia and activism. At the core of these networks lies a shared interest in studying the social contexts of scientific and technological practice from historical, philosophical, or sociological angles, but often also an activist effort to sensitize scientists, engineers, or policymakers to the societal stakes of scientific and technological development. Over time, the field has incorporated methodological and theoretical approaches from long-established mother-disciplines such as sociology, history, philosophy, anthropology, or political science, but intermittently also sought association with more recently established domains, such as social geography, gender studies, media studies, or indeed, sound studies. That diversity typifies STS. But although the field lacks a unified, cohesive methodological programme, its various strands share a social constructivist perspective, which attends to categories, ideas, objects, and structures as the product of social choice, negotiation, and convention.

In this chapter, we trace how, since the early 2000s, attention to sound has emerged as a subject within STS. We trace this interest back to two separate, if interrelated, concerns within STS. On the one hand, we consider it as a product of attention to science and technology as a set of material practices, including the bodily skills that practitioners bring to bear on them. On the other hand, we see interest in sound technologies as an expression of a wider interest in tracing the complex interactions of social, scientific, economic, and aesthetic contexts in technological development and use. After providing this two-part genealogy of sonic preoccupations within STS, we turn our attention to a specific case – the relationship between science and music – that exemplifies the diversity of epistemic and methodological approaches through which STS scholars study sound. In the second half of this chapter, we finally outline a set of methodological principles that run through the diverse body of literature at the intersection of STS and sound studies.

Origins of sound studies in STS

Beyond the laboratory: Science and technology as a set of material practices

The origin story of the field of STS has been told in different ways. Some of these stories take academic traditions as a starting point – for instance, by tracing an intellectual lineage from traditional philosophies of science via Thomas Kuhn to the strong programme and actor-network theory (Sismondo 2010). Others place stronger emphasis on developments outside the academy – for instance by highlighting the formative influence of the protest movements of the 1970s and their interrogations of the social impact of science and technology (Guggenheim and Nowotny 2003). There are nonetheless some aspects that unite these seemingly disparate accounts; some recurring themes include the importance of cultural and social *context* for understanding the development of science and technology, or the rejection of *essentialist positions* in favour of *constructivist approaches*. Conspicuously, the *sound* of science and technology is not usually considered in these intellectual histories. While overviews of scholarship on sound routinely highlight the role of science and technology as 'keys to unlock these new worlds of sound' (Pinch and Bijsterveld 2012: 5), the reverse has rarely been true. Yet although many of the classics of STS scholarship have been mute and devoid of references to the sonic environments of science and technology, we argue here that the study of sound is closely intertwined with some of the core interests and concerns of STS. To understand these intertwinements, it makes sense to start looking in (and proceed to continually hark back to) a place that holds special significance in the manifold origin stories of STS: the laboratory.

Indeed, the laboratory has been widely regarded as a key site in the development of STS. As Park Doing argues, its symbolic power as 'the hardest of hard places' (Doing 2008: 277) has made the laboratory an especially welcoming place for STS researchers to forcefully demonstrate the social construction of what may have been regarded as 'pure knowledge'. Michael Guggenheim and Helga Nowotny poignantly recount the heroic story of STS researchers who 'fearlessly enter the laboratories whose threshold no social scientist had ever dared to cross before' and reveal science as 'a conglomerate of (cultural) practices, just like other fields of activities or other kinds of work' (Guggenheim and Nowotny 2003: 235). One of the most influential epistemological and methodological principles at the foundation of STS research has been the so-called 'principle of symmetry', formulated as part of the 'strong programme' in the sociology of scientific knowledge. This principle holds that both the success and the failure of knowledge claims need to be accounted for and explained in the same manner; in other words, 'truth' and 'falsity' can never be explanations for why certain beliefs or claims displace others, but are themselves in need of explanation (Sismondo 2010). Such constructivist approaches to understanding scientific development became vital to the field of STS, and the laboratory has become a key site to study how and why some claims become accepted as scientific and objective, while others are pushed aside as unscientific and biased.

Sound, however, did not typically factor in these explanations. It was not until 2005 that Cyrus Mody explicitly drew attention to the 'sounds and noises, wanted and unwanted' that permeate laboratory walls (Mody 2005: 176). Classic laboratory studies did prepare the ground, however, by highlighting the importance of local, embodied, tacit skills and material practices that are needed to make scientific experiments work. While STS researchers have long acknowledged these elements in relation to the production and consumption of images, models, and visual 'inscriptions' (Lynch and Woolgar 1990; Knorr Cetina 1999), a more recent strand of scholarship has extended this concern to other sensory dimensions of scientific work – the aural in particular. Such work has revealed, for instance, how sound helps to structure the routines and practices of scientists' experimental work (Schmidgen 2003; Mody 2005; Kursell 2008; Bruyninckx 2017) and to articulate expectations and conceptions of their objects of study (Roosth 2009; Roosth 2010; Supper 2015; Helmreich 2016; Stephens and Lewis 2017). Following a long-standing concern with the tacit dimensions of scientific and medical practice (Polanyi 1966; Collins 2010), the embodied nature of auditory knowledge has come into view in relation to various other domains of technoscientific practice, such as medicine (Van Drie 2013; Harris 2016), audio engineering (Horning 2014), and music (Waksman 2004; Atkinson, Watermeyer and Delamont 2013). So, too, in underscoring the importance of scientific instruments' material qualities and affordances (Hankins and Silverman 1995), authors have shown many such instruments to derive from domains of music or acoustics (Jackson 2006; Pesic 2014). Similarly, attention to the material qualities of devices that produce or measure sound has also been extended to non-scientific domains, such as police work (Kim 2016) or the music industry (Devine 2015).

Such attention to the micro-practices of laboratory work has often been combined with tracing processes of translation between these local contexts and the wider world. This concern has frequently tied into questions about the construction of scientific credibility and the circulation of scientific knowledge outside of the laboratory. Indeed, the geographical and local specificities of 'places of knowledge' (Livingstone 1995) and the fluctuating relationship between the laboratory and other places of scientific work, such as the field site or conference, has been an important concern in STS research (Gieryn 2006; de Bont 2009). Similar concerns resonate in studies of the sonic dimensions of scientific practice (Höhler 2002; Bruyninckx 2012; Supper 2015), which trace how knowledge about sound has taken shape in specific cultural and geographic spaces (Lachmund 1999; Nelson 2015). The circulation of auditory knowledge beyond the laboratory, too, has become an important concern at the intersection of STS and sound studies (Hui, Kursell, and Jackson 2013). How and why, for instance, have sound-related knowledge claims and listening habits gained, or failed to gain, credibility and authority (Perlman 2004; Volmar 2013; Supper and Bijsterveld 2015)? And how are the dynamics of these processes involved in the emergence and stabilization of professional and amateur communities (Porcello 2004; Krebs 2012; Bruyninckx 2015)? As a phenomenon associated with art (music) as much as with science (acoustics), one that is sensual, immediate, and ephemeral yet also traced, inscribed, and conceptualized in a myriad of ways, sound has been well suited to explore the ramifications of key STS principles, by showing that how sound is heard and understood is deeply dependent on its social, cultural, and material context.

New technologies within cultural contexts

In the 1980s, meanwhile, sociologists and historians of technology too began adopting the constructivist agenda of science studies and its principles of symmetry. In doing so, they extended their viewfinder from processes of invention, development, and innovation – primarily within contexts of manufacturing and large-scale technical systems such as power grids – to consider categories of everyday, domestic and consumer technologies and the complex interplays between producers, regulators, and users through which they take and shift shape (Staudenmaier 1985). Such work has come to regard technologies as reflective of producers' social context and cultural assumptions, while at the same time showing technologies to be consumed, modified, domesticated, redesigned, and resisted by users in often unforeseen ways (Oudshoorn and Pinch 2003). For scholars in STS, then, understanding how technologies came to dwindle or proliferate has required attending not just to design and development, but also to how they are being appropriated in local contexts and become entangled with social identities and cultural practices. Following on the heels of these developments, sound has come to serve as a valuable index for such complex interactions between a technology and its social, scientific, economic, or aesthetic context. Indeed, in establishing the mutual shaping of sonic and material, scientific, and technological cultures, STS-inspired scholars have shown sound to be not merely an accidental by-product but often also a consequential driver of technological change.

These dynamics have come into view most conspicuously in relation to musical culture. Inspired by classic STS concepts, a long-standing strand of work has demonstrated that musical technologies are amenable to the same sorts of analytical categories as, say, bicycles or kitchens (Pinch and Bijsterveld 2003), and has found musical culture rich with technologically mediated ways of producing, manipulating, recording, and controlling sounds. 'Following the instruments' (Pinch and Bijsterveld 2004), this work has revealed that their histories are often shaped by similar kinds of 'interpretative flexibility', forks, diversions, dead-ends, and local reinterpretations as they are transported across different geographical, social, and professional boundaries. In doing so, this work has consistently foregrounded instrument builders, artists, musicians, and broadcasters as (often unsuspected) agents of scientific and technological change in different capacities (Jackson 2006; Pantalony 2009) and geographies (Zimmermann 2015); for instance, by bending and reconfiguring media technologies into musical instruments (Flood 2016) or by reinterpreting musical devices (such as the metronome, tuning fork, or siren) as scientific instruments and conceptual tools (Hui, Kursell, and Jackson 2013). In turn, inventors, engineers, and manufacturers have influenced broader musical and cultural changes. Advances in electro-acoustics and electronics, for instance, have yielded new instruments, techniques, and concepts that transformed both popular and experimental twentieth-century music (Braun 1994; Dunbar-Hester 2010); in fact, these technologies have also often purposefully been shepherded by their developers to influence new and emerging musical cultures (Pinch and Trocco 2002; Nelson 2015). Moreover, in shifting the boundary between machines and musical instruments, these technologies have been shown to persistently interrogate the boundaries of musical culture, by not just affecting its acoustic, compositional, and conceptual qualities but also questioning what counts as authorship, performance, virtuosity, and eventually, art (Pinch and Bijsterveld 2003).

Technologies have not only influenced musical culture and its aesthetic sensibilities but have also transformed broader sonic and aural cultures. Such cultures have rarely been determined solely by single momentous technological breakthroughs. Rather, particular listening cultures have been shown to sediment through a complex assemblage of long-standing and shifting values, practices, discourses, and techniques (Sterne 2003). One strand in STS-inspired sound studies has demonstrated, for instance, how – rather than an objective measure for sound quality – 'high fidelity' has been a shifting entity, achieved by insidious marketing (Thompson 1995), enforced through changing performance styles and audile techniques (Sterne 2003; Katz 2010), fostered through peculiar practices of consumer appreciation and validation (Perlman 2004; Downes 2010), and resisted through the cultivation of various counter-aesthetics (Supper 2018). Studies such as these illustrate the importance of attending to the micro-dynamics of user practices when considering technological change; for instance, by tracing how new sound technologies become, or fail to become, embedded in existing cultural discourses and practices (Taylor 2001; Bijsterveld 2004; Bijsterveld and Van Dijck 2009; Morris 2015). As such, they underscore the importance of considering sound's symbolic, cultural, and phenomenal value in accounting for the peculiar trajectories of a technology's use and non-use, its acceptance and resistance.

A related strand of STS-inspired scholarship has examined how sound technologies have been implied in the constitution – sonically as well as socially – of new subjectivities (Theberge, Devine, and Everett 2015). These are created, for instance, through sensory engineering and product design, among others in the car industry (Bijsterveld et al. 2013), or through engineering of psycho-acoustic models in new media technologies such as the telephone (Mills, forthcoming) or the mp3 (Sterne 2012). Such media technologies not only offer new possibilities for communicating through sound in specific settings and contexts; they are also part and parcel of broader engineering cultures and form the infrastructure of a trans-local sonic culture. Accordingly, the twentieth-century development of acoustic and electro-acoustic technologies has often been deeply political; whether because such technologies allowed controversial 'noises' to be measured, objectified, and compared, and thus be made the subject of political interventions or industrial struggles in campaigns for noise abatement (Bijsterveld 2008), or because they emerged as a product or an instrument of a military-industrial complex that arose against the background of two world wars and one cold war (Volmar 2013; Ritts and Shiga 2016; Camprubí 2017).

Zooming out to an even larger scale, some scholars have explored these developments as part of a series of broad-based shifts that consolidated a new, distinctly modern sonic culture (or 'soundscape'). These shifts have been understood as originating from early twentieth-century advances in electro-acoustic engineering, which have changed how sound was conceptualized, stored, reproduced, modulated, and transformed, resulting in a complex interplay between techno-scientific advances and wider developments across the domains of politics, industry, medicine, and entertainment (Thompson 2002; Wurtzler 2007; Mansell 2017; Wittje 2016).

STS and sound exemplified: Science and music

In this chapter so far, we have traced a genealogy of curiosity about sound within the field of STS scholarship, following two main paths from their respective points of origin. Yet as we have also hinted in the previous sections, these paths have intersected and often even merged in more recent scholarship. The multifaceted relationship between science and music is one topic that illustrates recent productivity at this intersection and can at the same time exemplify the methodological approaches taken by STS (and fields in close intellectual proximity to it), whose guiding principles we seek to disentangle in the next section.

One of the most fundamental questions about the relationship between science and music concerns the boundaries between these two domains. Indeed, the categories of 'science' and 'music' are themselves not stable, immutable entities; rather, as a product of historically and culturally variable practices, their relation has been continually renegotiated over the past centuries. In Greek natural philosophy and Renaissance mathematics and astronomy, musical performance and scientific experiment were often hard to distinguish (Johnson 1996; Pesic 2014). If music at first served to investigate natural order in the universe or disprove claims to natural magic (Hankins and Silverman 1995; Gouk 1999), by the late eighteenth century, musical culture and its instruments helped to consolidate acoustics as a science of sound. That privileged position of music shifted between the mid-nineteenth and early twentieth century, as scientists and engineers began to broaden their scope of investigation to include less harmonic acoustical sources, such as noise (Hui, Kursell, and Jackson 2013).

Scholars of science, technology, and sound have linked these epistemological shifts to changes in the social, cultural, material, and discursive make-up of both domains. For instance, just as the separation of music and acoustics has been interpreted as a product of social processes of disciplinary specialization and technical innovation – most notably the electrification of sound (Thompson 2002; Wittje 2016) – so a more recent prioritization of interdisciplinarity on policy agendas and the ubiquity of digital instruments has been shown to create new possibilities for convergence (Mody 2013). The effects of these developments are visible in institutionalized experimental music research centres (Born 1995; Nelson 2015), but also in more ad hoc and project-based formations, such as in approaches for sonifying scientific data, in which the boundaries between science and art are continually renegotiated (Supper 2014).

Music and science have not only changed themselves but also mutually redefined each other. In examining the boundaries between science and music, STS scholars have traced substantial traffic across its porous borders. Some of that traffic has been of a primarily rhetorical nature, as musical metaphors have long helped scientists to conceptualize and communicate about their research subjects (Roosth 2009; Supper 2014; Helmreich 2016). The metaphors used to describe sounds themselves have also undergone transformation; for instance, the understanding of electronic sounds as individuals with specific properties co-emerged not only with scientific epistemologies but also with cultural differences and social hierarchies (Rodgers 2011). Indeed, developments in sound technology and engineering have changed how sound/music has been conceptualized, the instruments it has been produced with, and the listening habits it has generated (Peters and Cressman 2016). Such contingencies complicate existing music historical narratives, as seemingly stable aesthetic objects such as musical pitch have turned out to be the product of standardization by musicians, scientists, and instrument makers, as well as international diplomacy by industrial actors and trade unions (Jackson 2006; Gribenski 2018).

Conversely, sonic practices and skills have been shown to be fundamental to the work of professions across science, engineering, and medicine (Supper and Bijsterveld 2015) and 'to structure emergent disciplinary knowledge' and social configurations (Davies and Lockhart 2017: 2). New musical aesthetics such as those of nineteenth-century bourgeois culture, for instance, have been revealed to shape scientific conceptions of sound and hearing in peculiar ways (Hui 2013). Just so, bourgeois pedagogical networks and simultaneous professional occupations such as inventors, performers, lecturers, artisans, and scholars have yielded influential collaborations across the domains of science, music, engineering, and craft work (Jackson 2006; Pantalony 2009).

Such work has designated material conditions, including such unlikely spaces as the workshop, the parlour, the performance stage, and later the sound studio, as sites of both scientific and artistic innovation. It has also involved taking seriously the role of technology and their materiality (Tresch and Dolan 2013): as musical and acoustic instruments travelled between the realms of science, music, arts, entertainment, and natural magic, their meaning as well as that of the contexts in which they were used have transformed. Musical notations have shaped and been reconsidered in such fields as ethnomusicology or even ornithology, before they were reinvented again by the musical avant-garde (Bruyninckx 2018). Likewise, instruments such as the tuning fork, the siren, or electronically generated sounds have revolutionized the investigation of hearing and influenced musical theory – indeed, musical instruments have often helped researchers to come to terms with 'epistemic things', those sources of distinct but as yet unknown knowledge (Rehding 2014). Yet their sounds have also often entered and changed the very definitions of music and sound.

Methodological principles

The above sections have demonstrated the diverse ways in which scholars drawing on STS perspectives have investigated the sonic environment. Yet regardless of this diversity, such work shares several fundamental methodological and theoretical premises – if not as a strict list of fixed requirements, then at least as a web of family resemblances, of which any given piece of STS scholarship on sound will display at least some elements. What, then, does it mean to study sound from a perspective of science and technology studies?

STS approaches avoid essentialist explanations of events, phenomena, and artefacts

As the discussion above suggests, for scholars in STS, there is no such thing as a single definition of sound, nor does sound exist in any 'real', 'natural', or 'metaphysical' state of being. Rather, in line with a social constructivist attitude, STS approaches attend to the processes, theories, methods, practices, relationships, material features, institutional arrangements, and ideologies that modulate how sound is made, heard, known, and understood.

STS approaches may apply a variety of research methods

One implication of this focus is that STS scholars interested in sound rarely attend solely to its sonic qualities; rather, they tend to study sound as it is embedded in specific discourses, narratives, practices, and contexts. In doing so, they may avail themselves of any number of methods that are also used in other domains within the social sciences and humanities. This includes, for instance, ethnographic methods of interviewing and observing individuals and groups who are involved in making, listening to, making sense of, or otherwise being affected by sound; tracing networks of sound-related knowledge and communities through network analyses and citation analyses; or studying historical documents, media discourses, and other (predominantly written) sources for traces of how sound is given meaning and implicated in particular processes, practices, narratives, and institutional arrangements. They often take the form of case study approaches, weaving together theory, method, and empirical data to study knowledge practices in concrete, local settings.

STS approaches attend to situated practices

Rather than pinpointing universal and essentialist foundations of knowledge (about or through the sonic world), they study how ways of understanding and knowing the world are made, performed, and enacted in practice, through a mixture of theories, methods, perceptions, experiments, tools, and institutional arrangements. The classic 'principle of symmetry' of the sociology of scientific knowledge, which holds that true and false beliefs should be explained in the same way, opens up the path for understanding why particular knowledge claims gain credibility and traction rather than taking their truth-value as self-evident. Similarly, against the suggested possibility of a 'god trick of seeing everything from nowhere' (Haraway 1988), the notion of 'situated knowledge' suggests that knowledge, methods, and indeed, even perception are always situated in specific contexts, implicating the position of both the producer and the objects of knowledge. These notions, albeit implicitly, resonate today in work – within and beyond STS – that traces the rise of acoustics, not as a result of inescapable and universal truth about the functioning of sound, but as a particularly Western, scientific epistemology of the acoustic that emerged in the eighteenth and nineteenth century, under specific social and material conditions, which may interact with other acoustemologies or technoustemologies (Greene and Porcello 2005; Feld 2012; Ochoa Gautier 2014).

STS approaches attend to contingencies

The emphasis on situated practices already hints at the importance of local and historical contingencies. Indeed, the claim that 'it could have been otherwise' is a slogan that captures the spirit of STS like no other, referring not only to (scientific) knowledge claims but also the development of technology. The principle of 'interpretive flexibility' extends the notion of symmetry to the success (or failure) of specific technologies, drawing attention to the historical and social contingencies that shaped technological development and to the possibility of alternative interpretations and designs. The fruitfulness of this approach for the study of sound is demonstrated, for instance, in research on the emergence of technologies such as radio broadcasting (Wurtzler 2007) or analogue synthesizers (Pinch and Trocco 2002), or of knowledge practices such as auscultation (Lachmund 1999); but it also, more implicitly, resonates in work that traces the possibility of sound recording not so much in terms of its societal effects as its cultural origins (Sterne 2003).

STS approaches attend to unspoken assumptions

Showcasing the interpretive flexibility of technologies is not an end goal for STS analysts; rather, it is one of several strategies to make visible the unspoken and taken-for-granted assumptions that are built into technologies and knowledge claims. These assumptions relate not only to the technologies and knowledge claims themselves, but also to the broader cultures that they are inserted into; and while uncovering them is the responsibility of the STS analyst, this responsibility is facilitated by moments of disruption that are sparked by the introduction of new technologies or techniques. A new technique for making audible sound recordings of the 1860s, which were originally developed to be seen rather than heard, for instance, may challenge our 'assumptions regarding the givenness of a particular domain called "sound," a process calling "hearing," or a listening subject' (Sterne and Akiyama 2012: 556). The notion of a 'breaching experiment', adopted from ethnomethodology, draws attention to how new technologies may 'make visible norms and values concerning the art of music and music making that are usually taken for granted' (Pinch and Bijsterveld 2003: 543) – whether the technologies in question are player pianos (Pinch and Bijsterveld 2003) or auto-tune (Marshall 2017).

STS approaches attend to and juxtapose context-specific paradigms

Not only does the focus on situated practices imply an interest in the local settings in which knowledge and technologies are made; it also necessitates strategies for making visible what is at stake in those settings. Following in the footsteps of a long tradition of 'controversy studies' within STS (Sismondo 2010), the juxtaposition of two or more different paradigms which coexist and compete for epistemic authority is a helpful tactic for articulating what is specific to, but also shared among, different approaches. For instance, to understand the specific ways in which audiophiles make meaning around sound, it is useful to consider how they resist the scientifically grounded knowledge claims of audio engineers; but in turn, the juxtaposition of these two competing approaches also makes explicit the grounds on which audio engineers themselves claim epistemic authority (Perlman 2004). In some cases, it is precisely on such epistemically contested and controversial terrain that locally stable configurations of knowledge can be found; as the case of research on automatic speech recognition, with its controversy between 'engineering' and 'auditory' paradigms, demonstrates, controversies and competing paradigms can also be fruitful sites to investigate the creation of stability and order (Voskuhl 2004).

STS approaches attend to processes of translation and circulation

While much STS research is concerned with tracing how knowledge is made and performed in local settings, another recurring concern is with recognizing how such ways of understanding and acting upon the world travel: how they become accepted, routine, standardized, distributed, circulated, and appropriated, and how in doing so, they also shape the worlds through which they travel. Hence, STS approaches pay attention to processes of translation and transfer, for instance by studying how processes of the global circulation of music have been implicated by digital transformations (Taylor 2001) or by tracing how artefacts and knowledge, such as recordings of birdsongs (Bruynickx 2018), travel between different cultural domains and communities of practice.

STS approaches attend to a variety of phenomena and entities

They approach the phenomena they study by understanding them in a variety of different contexts – social, cultural, organizational, political, material – and in a variety of different phases of development, circulation, and appropriation. They study the centres and hubs of innovation, but also consider peripheries, for, as Timothy Taylor argues, 'the margins often have much to say about the centers that those in the centers might not be aware of' (Taylor 2001: 9). Furthermore, the entities that are studied are themselves manifold. They range from discursive formations such as the trope of noise (Wittje 2016) to material arrangements such as audio labs (Klett 2014), from norms and standards such as musical pitch (Gribenski 2018) to bodily dispositions such as self-percussion (Harris 2016), from human producers/users and their intermediaries, such as instrument salespeople (Pinch 2003) to nonhuman actants, such as the musical instruments themselves (Bates 2012).

STS approaches are reflexive

Not only do STS approaches study how knowledge about and through sound is produced, circulated, applied, and interpreted, but they also reflect on the foundations of their own knowledge and interpretations. The notion of reflexivity has a long-standing place within STS, for instance as one of the core principles of the sociology of scientific knowledge prescribing that any sociological theories of scientific development should also be applicable to sociology itself. More recently, for instance in the work of Stefan Helmreich, the notion of reflexivity and its underlying visual rhetoric of individual self-reflection has itself been subject to reflection; inspired by auditory conceptions, Helmreich proposes the notion of 'transduction' as an alternative form of inquiry, 'animated by an auditorily inspired attention to the modulating relations that produce insides and outsides, subjects and objects, sensation and sense data' (Helmreich 2007: 622).

The reflexive and transductive nature of STS approaches presents challenges when it comes to producing an overview of the sonic methodologies of this interdisciplinary and diverse field of study; after all, the insides and outsides, the subjects and objects, of the field of STS are themselves fuzzy, unstable, and difficult to pinpoint conclusively. While our chapter has made an attempt to provide some reference points for what it might mean to take an STS approach to sonic materials, its contours and emphasis have also been shaped by our own position in the field.

References

Atkinson, P., R. Watermeyer, and S. Delamont (2013). 'Expertise, Authority and Embodied Pedagogy: Operatic Masterclasses'. *British Journal of Sociology of Education* 34 (4): 487–503.
Bates, E. (2012). 'The Social Life of Musical Instruments'. *Ethnomusicology* 56 (3): 363–395.
Bijsterveld, K. (2004). '"What Do I Do with My Tape Recorder ... ?": Sound Hunting and the Sounds of Everyday Dutch Life in the 1950s and 1960s'. *Historical Journal of Film, Radio and Television*, 24 (4): 613–634.
Bijsterveld, K. (2008). *Mechanical Sound. Technology, Culture, and Public Problems of Noise in the Twentieth Century*. Cambridge, MA: MIT Press.

Bijsterveld, K. and J. van Dijck (2009). *Sound Souvenirs: Audio Technologies, Memory and Cultural Practices*. Amsterdam: Amsterdam University Press.

Bijsterveld, K., E. Cleophas, S. Krebs, and G. Mom (2013). *Sound and Safe: A History of Listening Behind the Wheel*. Oxford: Oxford University Press.

Born, G. (1995). *Rationalizing Culture: IRCAM, Boulez, and the Institutionalization of the Musical Avant-Garde*. Berkeley, CA: University of California Press.

Braun, H.-J. (ed.) (1994). '"I sing the body electric": Der Einfluss von Elektroakustik und Elektronik auf das Musikschaffen im 20. Jahrhundert'. *Technikgeschichte* 61: 353–373.

Bruyninckx, J. (2012). 'Sound Sterile: Making Scientific Field Recordings in Ornithology'. In Trevor Pinch and Karin Bijsterveld (eds), *The Oxford Handbook of Sound Studies*, 127–150. Oxford: Oxford University Press.

Bruyninckx, J. (2015). 'Trading Twitter: Amateur Recorders and Economies of Scientific Exchange at the Cornell Library of Natural Sounds'. *Social Studies of Science* 45 (3): 344–370.

Bruyninckx, J. (2017). 'Synchronicity: Time, Technicians, Instruments, and Invisible Repair'. *Science, Technology & Human Values* 42 (5): 822–847.

Bruyninckx, J. (2018). *Listening in the Field: Recording and the Science of Birdsong*. Cambridge, MA: MIT Press.

Camprubí, L. (2017). 'The Sonic Construction of the Ocean as the Navy's Operating Environment'. In N. V. Dijk, K. Ergenzinger, C. Kassung, and S. Schwesinger (eds), *Navigating Noise*, 219–245. Cologne: Walther König.

Collins, H. (2010). *Tacit and Explicit Knowledge*. Chicago, IL: University of Chicago Press.

Davies, J. Q. and E. Lockhart (2017). *Sound Knowledge: Music and Science in London, 1789-1851*. Chicago, IL: University of Chicago Press.

De Bont, R. (2009). 'Between the Laboratory and the Deep Blue Sea: Space Issues in the Marine Stations of Naples and Wimereux'. *Social Studies of Science* 39 (2): 199–227.

Devine, K. (2015). 'Decomposed: A Political Ecology of Music'. *Popular Music* 34 (3): 367–389.

Doing, P. (2008). 'Give Me a Laboratory and I Will Raise a Discipline: The Past, Present, and Future Politics of Laboratory Studies in STS'. In E. J. Hackett, O. Amsterdamska, M. Lynch, and J. Wajcman (eds), *The Handbook of Science and Technology Studies*. 3rd edition, 279–295. Cambridge, MA: MIT Press.

Downes, K. (2010). '"Perfect Sound Forever": Innovation, Aesthetics, and the Re-Making of Compact Disc Playback'. *Technology and Culture* 51 (2): 305–331.

Dunbar-Hester, C. (2010). 'Listening to Cybernetics Music, Machines, and Nervous Systems, 1950-1980'. *Science, Technology & Human Values* 35 (1): 113–139.

Feld, S. (2012). *Sound and Sentiment: Birds, Weeping, Poetics, and Song in Kaluli Expression*, Third edition. Durham: Duke University Press.

Flood, L. (2016). 'Building and Becoming: DIY Music Technology in New York and Berlin'. PhD thesis, Columbia University.

Gieryn, T. F. (2006). 'City as Truth-Spot: Laboratories and Field-Sites in Urban Studies'. *Social Studies of Science* 36 (1): 5–38.

Greene, P. D. and T. Porcello (eds) (2005). *Wired for Sound: Engineering and Technologies in Sonic Cultures*. Middletown, CT: Wesleyan University Press.

Gouk, P. (1999). *Music, Science, and Natural Magic in Seventeenth-century England*. New Haven, CT: Yale University Press.

Gribenski, F. (2018). 'Negotiating the Pitch: For a Diplomatic History of *A*, At the Crossroads of Politics, Music, Science, and Industry'. In F. Ramel and C. Prévost-Thomas (eds), *International Relations, Music and Diplomacy. Sounds and Voices on the International Stage*, 173–192. Cham: Palgrave Macmillan.

Guggenheim, M. and H. Nowotny (2003). 'Joy in Repetition Makes the Future Disappear: A Critical Assessment of the Present State of STS'. In B. Joerges and H. Nowotny (eds), *Social Studies of Science and Technology: Looking Back, Ahead*, 229–258. Dordrecht: Kluwer.

Hankins, T. L. and R. J. Silverman (1995), *Instruments and the Imagination*. Princeton, NJ: Princeton University Press.

Haraway, D. (1988). 'Situated Knowledges: The Science Question in Feminism and the Privilege of Partial Perspective'. *Feminist Studies* 14 (3): 575–599.

Harris, A. (2016). 'Listening-touch, Affect and the Crafting of Medical Bodies through Percussion'. *Body & Society* 22 (1): 31–61.

Helmreich, S. (2007). 'An Anthropologist Underwater: Immersive Soundscapes, Submarine Cyborgs, and Transductive Ethnography'. *American Ethnologist* 34 (4): 621–641.

Helmreich, S. (2016). 'Gravity's Reverb: Listening to Space-Time, or Articulating the Sounds of Gravitational-Wave Detection'. *Cultural Anthropology* 31 (4): 464–492.

Höhler, S. (2002). 'Depth Records and Ocean Volumes: Ocean Profiling by Sounding Technology, 1850–1930'. *History and Technology* 18 (2): 119–154.

Horning, S. S. (2014). *Chasing Sound: Technology, Culture, and the Art of Studio Recording from Edison to the LP*. Baltimore, MD: Johns Hopkins University Press.

Hui, A. E. (2013). *The Psychophysical Ear: Musical Experiments, Experimental Sounds, 1840–1910*. Cambridge, MA: MIT Press.

Hui, A. E., J. Kursell, and M. E. Jackson (2013). *Music, Sound, and the Laboratory during the Nineteenth and Twentieth Centuries*. Osiris, vol. 28. Chicago, IL: University of Chicago Press.

Jackson, M. E. (2006). *Harmonious Triads: Physicists, Musicians, and Instrument Makers in Nineteenth-Century Germany*. Cambridge, MA: MIT Press.

Johnson, J. H. (1996). *Listening in Paris: A Cultural History*. Berkeley, CA: University of California Press.

Katz, M. (2010). *Capturing Sound: How Technology Has Changed Music*. Revised edition. Berkeley, CA: University of California Press.

Kim, E. (2016). 'The Sensory Power of Cameras and Noise Meters for Protest Surveillance in South Korea'. *Social Studies of Science* 46 (3): 396–416.

Klett, J. (2014), 'Sound on Sound: Situating Interaction in Sonic Object Settings'. *Sociological Theory* 32 (2): 147–161.

Knorr Cetina, K. (1999). *Epistemic Cultures: How the Sciences Make Knowledge*. Cambridge, MA: Harvard University Press.

Krebs, S. (2012). 'Sobbing, Whining, Rumbling: Listening to Automobiles as Social Practice'. In Trevor Pinch and Karin Bijsterveld (eds), *The Oxford Handbook of Sound Studies*, 79–101. Oxford: Oxford University Press.

Kursell, J. (ed.) (2008). *Sounds of Science – Schall Im Labor*. Preprint. Berlin: Max Planck Institute for the History of Science.

Lachmund, J. (1999). 'Making Sense of Sound: Auscultation and Lung Sound Codification in Nineteenth-Century French and German Medicine'. *Science, Technology, & Human Values* 24 (4): 419–450.

Livingstone, D. (1995). 'The Spaces of Knowledge: Contributions towards a Historical Geography of Science'. *Environment and Planning D: Society and Space* 13 (1): 5–34.

Lynch, M. and S. Woolgar (1990). *Representation in Scientific Practice*. Cambridge, MA: MIT Press.

Mansell, J. G. (2017). *The Age of Noise in Britain: Hearing Modernity*. Champaign, IL: University of Illinois Press.

Marshall, O. (2017). 'Tuning in Situ: Articulations of Voice, Affect, and Artifact in the Recording Studio'. PhD thesis, Cornell University, Ithaca, NY.

Mills, M. (2011). 'Deafening: Noise and the Engineering of Communication in the Telephone System'. *Grey Room* 43: 118–143.

Mills, M. (forthcoming). *On the Phone: Hearing Loss and Communication Engineering*. Durham, NC: Duke University Press.

Mody, C. C. M. (2005). 'The Sounds of Science: Listening to Laboratory Practice'. *Science, Technology, & Human Values* 30 (2): 175–198.

Mody, C. C. M. and A. J. Nelson (2013). '"A Towering Virtue of Necessity": Computer Music at Vietnam-Era Stanford'. *Osiris* 28: 254–77.

Morris, J. W. (2015). *Selling Digital Music, Formatting Culture*. Berkeley, CA: University of California Press.

Nelson, A. J. (2015). *The Sound of Innovation: Stanford and the Computer Music Revolution*. Cambridge, MA: MIT Press.

Ochoa Gautier, A. M. (2014). *Aurality: Listening and Knowledge in Nineteenth-Century Colombia*. Durham, NC: Duke University Press.

Oudshoorn, N. and T. Pinch (2003). *How Users Matter: The Co-construction of Users and Technology*. Cambridge, MA: MIT Press.

Pantalony, D. (2009). *Altered Sensations. Rudolph Koenig's Acoustical Workshop in Nineteenth-Century Paris*. Dordrecht: Springer.

Perlman, M. (2004). 'Golden Ears and Meter Readers. The Contest for Epistemic Authority in Audiophilia'. *Social Studies of Science* 34 (5): 783–807.

Pesic, P. (2014). *Music and The Making of Modern Science*. Cambridge, MA: MIT Press.

Peters, P. and D. Cressman (2016), 'A Sounding Monument: How a New Organ became Old'. *Sound Studies* 2 (1): 21–35.

Pinch, T. (2003). 'Giving Birth to New Users: How the Minimoog Was Sold to Rock and Roll'. In Nelly Oudshoorn and Trevor Pinch (eds), *How Users Matter: The Co-construction of Users and Technology*, 247–270. Cambridge, MA: MIT Press.

Pinch, T. and K. Bijsterveld (2003). '"Should One Applaud?" – Breaches and Boundaries in the Reception of New Technology in Music'. *Technology and Culture* 44 (3): 536–559.

Pinch, T. and K. Bijsterveld (2004). 'Sound Studies: New Technologies and Music'. *Social Studies of Science* 34 (5): 635–648.

Pinch, T. and K. Bijsterveld (2012). *The Oxford Handbook of Sound Studies*. Oxford: Oxford University Press.

Pinch, T. and F. Trocco (2002). *Analog Days: The Invention and Impact of the Moog Synthesizer*. Cambridge, MA: Harvard University Press.

Polanyi, M. L. (1966). *The Tacit Dimension*. Chicago, IL: University of Chicago Press.

Porcello, T. (2004), 'Speaking of Sound. Language and the Professionalization of Sound-Recording Engineers'. *Social Studies of Science* 34 (5): 733–758.

Rehding, A. (2014). 'Of Sirens Old and New'. In S. Gopinath and J. Stanyek (eds), *The Oxford Handbook of Mobile Music Studies*, vol. 2, 77–108. New York: Oxford University Press.

Ritts, M. and J. Shiga. (2016). 'Military Cetology'. *Environmental Humanities* 8 (2): 196–214.

Rodgers, T. (2011). '"What, for me, constitutes life in a sound?": Electronic Sounds as Lively and Differentiated Individuals'. *American Quarterly* 63 (3): 509–530.

Roosth, S. (2009). 'Screaming Yeast: Sonocytology, Cytoplascmic Milieus, and Cellular Subjectivities'. *Critical Inquiry* 35 (2): 332–350.

Roosth, S. (2010). 'Crafting Life: A Sensory Ethnography of Fabricated Biologies'. PhD thesis, Massachusetts Institute of Technology, Cambridge, MA.

Schmidgen, H. (2003). 'Time and Noise: The Stable Surroundings of Reaction Experiments, 1860–1890'. *Studies in History and Philosophy of Biological and Biomedical Sciences* 34 (2): 237–275.

Sismondo, S. (2010). *An Introduction to Science and Technology Studies*. 2nd edition. Chichester: Wiley.

Staudenmaier, J. M. (1985). *Technology's Storytellers. Reweaving the Human Fabric*. Cambridge, MA: MIT Press.

Stephens, N. and J. Lewis (2017). 'Doing Laboratory Ethnography: Reflections on Method in Scientific Workplaces'. *Qualitative Research* 17 (2): 202–216.

Sterne, J. (2003). *The Audible Past: Cultural Origins of Sound Reproduction*. Durham, NC: Duke University Press.

Sterne, J. (2012). *MP3: The Meaning of a Format*. Durham, NC: Duke University Press.

Sterne, J. and M. Akiyama (2012). 'The Recording that Never Wanted to Be Heard and Other Stories of Sonification'. In Trevor Pinch and Karin Bijsterveld (eds), *The Oxford Handbook of Sound Studies*, 544–560. Oxford: Oxford University Press.

Supper, A. (2014). 'Sublime Frequencies: The Construction of Sublime Listening Experiences in the Sonification of Scientific Data'. *Social Studies of Science* 44 (1): 34–58.

Supper, A. (2015). 'Data Karaoke: Sensory and Bodily Skills in Conference Presentations'. *Science as Culture* 24 (4): 436–457.

Supper, A. (2018). 'Listening for the Hiss: Lo-fi Liner Notes as Curatorial Practices'. *Popular Music* 37 (2): 253–270.

Supper, A. and K. Bijsterveld (2015). 'Sounds Convincing: Modes of Listening and Sonic Skills in Knowledge Making'. *Interdisciplinary Science Reviews* 40 (2): 124–144.

Taylor, T. (2001). *Strange Sounds: Music, Technology and Culture*. New York: Routledge.

Théberge, P., K. Devine, and T. Everrett (2015). *Living Stereo: Histories and Cultures of Multichannel Sound*. New York: Bloomsbury.

Thompson, E. (1995). 'Machines, Music, and the Quest for Fidelity: Marketing the Edison Phonograph in America, 1877–1925'. *Musical Quarterly* 79 (1): 131–171.

Thompson, E. (2002). *The Soundscape of Modernity: Architectural Acoustics and the Culture of Listening in America, 1900–1933*. Cambridge, MA: MIT Press.

Tresch, J. and E.I. Dolan (2013). 'Toward a New Organology: Instruments of Music and Science'. *Osiris* 28 (1): 278–298.

Van Drie, M. (2013). 'Training the Auscultative Ear: Medical Textbooks and Teaching Tapes (1950–2010)'. *Senses & Society* 8 (2): 165–191.

Volmar, A. (2013). 'Listening to the Cold War: The Nuclear Test Ban Negotiations, Seismology, and Psychoacoustics 1958–1963'. *Osiris* 28 (1): 80–102.

Voskuhl, A. (2004). 'Humans, Machines, and Conversations: An Ethnographic Study of the Making of Automatic Speech Recognition Technologies'. *Social Studies of Science* 34 (3): 393–421.

Waksman, S. (2004). 'California Noise. Tinkering with Hardcore and heavy Metal in Southern California'. *Social Studies of Science* 34 (5): 675–702.

Wittje, R. (2016). *The Age of Electroacoustics: Transforming Science and Sound*. Cambridge, MA: MIT Press.

Wurtzler, S. J. (2007). *Electric Sounds: Technological Change and the Rise of Corporate Mass Media*. New York: Columbia University Press.

Zimmerman, B. (2015). *Waves and Forms. Electronic Music Devices and Computer Encodings in China*. Cambridge, MA: MIT Press.

12

The Sonic Environment in Urban Planning, Environmental Assessment and Management

A. Lex Brown

Introduction

This chapter provides a broad overview of methodologies used in the study and control of the sonic environment of cities. This is not a straightforward task, as there are multiple disciplines that play some role in this and whose many and varied approaches tend to be applied to only a limited facet of the sonic environment – primarily where there are high levels of unwanted sound or noise. Until recently there has been little attention to those parts of the sonic environment of cities that are not 'high noise'. Furthermore, the sonic methodologies that have the widest application are bound up within other processes such as city planning or the assessment of transport infrastructure developments. This chapter is as much an examination of how consideration of the sonic environment is incorporated into the methodologies of these other processes, as it is an exposition of urban sonic methodologies per se.

The sonic environment of cities is introduced briefly below by enumerating the common and dominant sound sources present in most urban areas. It is the outdoor environment which is of interest here or, more precisely, the outdoor sonic environment as heard indoors. The chapter then identifies the primary disciplines that play some role in study or management of the sonic environment generated by these sources. It introduces approaches to monitoring and mapping this environment and to measurement of human responses to sound exposure. These are prerequisite to the formulation of limit criteria that can be applied in acoustic management based on empirically derived exposure-response relationships.

The chapter then overviews various practices of management of the sonic environment that include engineering noise control and specific noise-focused methodologies within regulated environmental assessment. The role of urban planning in the management of the sonic environment of cities is also touched upon. Finally, the chapter moves beyond the current focus on the 'high-noise' part of the urban sonic environment and postulates various loci where the methodology of soundscape planning could be appropriate.

The sonic environment of cities

The *sonic* (or *acoustic*) environment of *urban outdoor spaces*, consists of all the sound generated by sources heard by a person in those spaces, as modified by the propagation of the sounds from the sources to the receiver.

In most cities, road traffic is the most extensive, and dominant, source heard over most outdoor spaces (European Environment Agency 2014b) because in motorized cities the roadway network permeates the urban form, providing not only through routes for traffic between different areas but also extensive access into the different land uses of the city. As populations and motorization grow, the intensity of traffic increases as does its spread into more areas and into more of the diurnal cycle. Taken together, the moving point sources on the roadways can effectively be considered as line sources, and sound propagates hemicylindrically from them to urban receivers, affected by shielding effects of intervening structures and by ground surface effects and atmospheric absorption during propagation. Other transport noise sources of aircraft and railway traffic can also be pervasive, depending on the propinquity of airports and railway lines to where people live. Propagation of sound from these sources is largely spherical and, in the case of aircraft, not affected in their transmission path by the urban form.

Stationary mechanical sound sources of industry and construction also contribute to the outdoor acoustic environment, but the spatial extent of their intrusion into noise sensitive areas, such as residential precincts of a city, is limited. Construction sounds, by their nature, tend to be of fixed and limited duration, though construction activities in dense inner cities may be becoming perennial and occurring in closer proximity to increasing numbers of people now resident within central business districts. Small-scale mechanical sources such as household air-conditioning plants or swimming pool filters form part of this mix but are highly localized because they have much lower emission levels than the transport sources.

Human sounds and the sounds of nature are also components of urban environments. These include the sounds of voices, footsteps, amplified music, and domesticated animals, associated with both residential and commercial areas of cities. The sounds of nature are generated by wind, by water, and by wildlife. In urban areas, the wildlife is insects, frogs, and birdlife, and dependent, more or less, on the extent of urban vegetation and other habitat available in the streets, footpaths, and yards of urban properties, and in the parks and open spaces of cities. Sounds from the latter occur across a city but are far more likely to be associated with locations that are not proximal to the more dominant transport noise sources where they are masked, or at least partly so, by transport sounds (Pablo Kogan et al. [2018] use a Green Soundscape Index across urban areas to describe the ratio of natural sound to traffic noise perception). These sounds of people and of nature have only recently become of interest in the management of the urban acoustic environment of cities because of new attention to the city soundscape.

The outdoor urban acoustic environment is transmitted into dwellings, schools, offices, and other buildings, and to date it is the adverse effects of noise from external sources on people (Jean-François Augoyard's [1998] 'sounds of discomfort') as heard indoors that has tended to be the motivation for most urban noise studies and efforts at management and control of the urban acoustic environment.

For completeness, it should be noted that people in cities also hear sounds whose sources are not located outdoors, where sounds are generated, propagated, and heard inside buildings. This is the province of architectural, room or building acoustics – the science and engineering of how sound behaves within enclosed spaces and usually with a focus on achieving good sound environments for people within buildings. Architectural acoustics has its own specialized methodology for study and design which is outside the scope of this chapter – apart from the penetration of outside sound into buildings that is considered below.

Multiple disciplines and approaches

Various disciplines have defined, but overlapping, involvement in the sonic environment of cities, which further confounds exposition of a straightforward urban sonic methodology. As could be expected, acousticians play a significant role, but primarily in measurement of sound and in both modelling and mapping of sound levels. Psychologists and public health specialists measure human response to acoustic environments and generally link this to exposures, establishing relationships between exposure to sound and human reaction – so-called exposure-response, or dose-response, relationships. Epidemiologists and other health professionals extend this approach to noise effects such as sleep disturbance and cardiovascular effects. The same methodologies also apply where the human outcome of interest is quality of life and well-being.

Engineers largely carry responsibility for noise control of mechanical sources in cities and for control interventions along the propagation path; environmental and public health managers also set limits and control mechanisms for some of these sources, and for some non-mechanical sources found in cities such as domesticated animals and entertainment noise. Urban planners influence the sonic environment of cities, but rarely as a primary focus, more often as a by-product of planning methodologies for other purposes. Transport specialists predict the future traffic flows on planned transport infrastructure which in turn are used in the prediction of future transport noise levels and exposures. Architects and other building professionals are also involved through design requirements on the acoustic insulation provided by the envelopes of buildings.

The chapter briefly elaborates on the approaches followed by some of these disciplines, with an emphasis on noise from transport sources. The emphasis is appropriate given the predominance of transport noise in the sonic environment of cities.

Exposure assessment: Noise monitoring and mapping

Measurement has always been an underpinning of all investigations and management of the acoustic environment of cities. The first urban noise survey, a 138-site study of noise levels across New York City, was undertaken by its Noise Abatement Commission (Brown et al. 1930) nine decades ago. Since then, many cities have embarked on physical measurement programmes of their levels of environmental noise. These have variously been termed ambient noise surveys, background surveys, or community noise surveys. They are generally based on point noise-level measurements, made over at least 24 hours to capture diurnal variation, and spatially distributed in some predetermined way over a city – often at points on a grid overlain on the city, or perhaps by using a stratified sampling across the land use categories of a city. While these surveys have been major data collection exercises, providing early knowledge and experience of sources and sound levels in urban areas, they have generally failed to provide information on populations of urban noise levels that allow comparison of levels between cities and between different urban populations, or the dependence of noise levels on urban form (Morillas et al. 2018). This is primarily because the spatial variability of sound across a city is finely grained as a result of multiple sources and the complex shielding effects provided by buildings and other structures of the built form. The grid sizes used in many surveys have been orders of magnitude larger than that which would be required to capture the real spatial variation in the urban sonic environment.

For this reason, estimates of the sound levels of a city based on measurement methodologies have largely been replaced by empirical modelling techniques. These combine emission models from transport and other sources with detailed forecasts of future traffic flows on transport networks and utilize complex sound-propagation models (Licitra 2012). Emission modelling requires an understanding of the variation in emission strengths of vehicle sources under different loads; the effects of barriers to propagation paths, including refraction of sound around them; and the absorption and reflection by all surfaces in the urban area, including the ground and the facades of buildings. These models are continuously being improved to provide useful estimates, usually as maps, of the exposure of dwellings or other sensitive land uses to urban noise.[1] Most models are 2D, showing the distribution of sound levels spatially across cities, but 3D models, which provide the distribution of noise levels on vertical facades of tall buildings, have also been developed. These are essential in high-rise cities such as Hong Kong (Stoter et al. 2008; Law et al. 2011).

The combination of such modelled noise maps with residential densities in geographical information systems allows estimates to be made of the exposures of populations to noise (European Commission 2007). Detailed knowledge of the exposures of populations at different levels of environmental noise forms the cornerstone of a formalized methodology for strategic management of the sonic environment of cities. This provides estimates of the number of dwellings, schools, hospitals, and other noise-sensitive uses that are exposed to unsatisfactory levels of noise (Fiedler and Zannin 2015), and the estimated numbers of people with exposures above limit levels. This is essential information for the development of policies and action plans to manage the 'high-noise' part of the sonic environment of cities.

There is also increasing utilization, though still experimental, of mobile devices and citizen-science-based initiatives – enabled by inexpensive microphones and sound level devices in mobile telephones – for obtaining much finer grain knowledge of the exact exposures of people to sound as they move about their daily activities (Guillaume et al. 2016; Murphy and King 2016; Shim et al. 2016).

Measurement of human-response and exposure-response relationships

Noise has a range of effects on people. This includes interference with speech communication; sleep disturbance; adverse effects on human performance, for example in learning situations; and physiological impacts such as cardiovascular problems (World Health Organization [WHO] 2018). There are also effects on wildlife, domesticated animals, and property values. There are evolving trends in methods for the assessment of noise effects in a community (as described in the section on strategic-level assessments below) but to date, the primary measure of effects of noise on people in urban areas has been their level of annoyance or dissatisfaction with the acoustic environment – mostly the annoyance caused to people when they are indoors, at home. Thus, most current management of the urban acoustic environment has been based on annoyance as the primary effect outcome of noise, most often expressed as the percentage of affected people in some area who are highly annoyed. The latter tends to be the effect variable used in correlational studies with noise exposure levels in the derivation of exposure-response relationships for environmental noise. The fact that different exposure-response relationships are found for different noise sources adds to the complexity. Limit values for noise exposures beyond which noise control measures may be warranted (such as roadside traffic noise barriers, or dwelling insulation schemes under aircraft flight paths) are currently derived from exposure-response relationships based on subjective annoyance responses.

Methods are well established for measuring self-assessments of annoyance with the acoustic environment as part of community surveys. These are generally conducted in people's homes using face-to-face interviews, or by mail, by telephone, or online. A questionnaire protocol, and specific questions to ascertain annoyance reactions, has been recognized as an international technical specification (ISO/TS 15666), ensuring consistency and comparability across languages and cultures. Surveys of annoyance with different noise sources in exposed populations have been conducted to derive exposure-response relationships in many different cities. Rainer Guski, Dirk Schreckenber, and Rudolf Schuemer (2017) have recently provided a new systematic review and meta-analysis on effects of environmental noise on annoyance for aircraft, road and rail transport, as well as wind turbines. The conclusion of their review was that more recent exposure-response studies indicate some increase in annoyance responses at comparable exposure levels to those found in earlier meta-analyses. This suggests that people have become less tolerant towards sound exposure from these sources over time, or it may be some effect of the changing nature of some of these sources themselves. The authors suggest that limit recommendations need to be adjusted accordingly.

Meta-analyses of exposure-response studies for human outcomes other than annoyance have also been produced. Elise Van Kempen and colleagues (2018) undertook a systematic review of the effects of environmental noise exposure on people's cardio-metabolic systems, showing, amongst other findings, that there was a significant association between exposure to road traffic noise and the incidence of ischemic heart effects. Mathias Basner and Sarah McGuire (2018) reported a systematic review of sleep-disturbance effects. They found a significant increase in the percentage of people who were highly sleep disturbed with increasing exposure to each of aircraft noise, road traffic noise, and rail noise. The method of collection of both cardiovascular and sleep outcomes, as for annoyance, was self-reported accounts. In addition to these subjectively assessed responses to noise, sleep research has also utilized objectively measured outcomes. For example, Basner and McGuire (2018) reported polysomnographic studies on the acute effects of transportation noise on sleep. They found that the probability of cortical awakenings increased with the maximum level of transport noise events, for each of air, road, and rail traffic. For policymaking and mitigation decisions, objective measures of human response would appear superior to self-reported measures, but the objective measures of cortical awakening are of an acute response only, with as yet, no clear link to long-term health effects of noise exposure.

Noise control and urban planning

The general approach to noise control is three-pronged: reduction of the sound level at the source, reduction of the sound along the transmission path (as in noise barriers), or shielding of sensitive receptors through increasing the attenuation of the envelopes of buildings such as dwellings and schools. This broad methodology applies in different ways for different noise sources, and with responsibilities spread over many authorities and disciplines.

Source control of transport vehicles is a primary technique of transport noise management, but this occurs largely at the fleet level and is achieved through vehicle-type certification originating at international (e.g. aircraft) and national (e.g. road vehicle) levels. By contrast, source levels from motor vehicles may also be reduced, for example, by locally installing special asphalt surfaces which significantly reduce noise from tyre–roadway interaction. Other local examples include regulation of heavy vehicle access on certain streets or traffic-planning activities such as defining minimum noise routes for take off and landing aircraft flight paths.

Engineers largely carry responsibility for noise control of mechanical sources and for control interventions along the propagation path, but building professionals and architects have the responsibility for the reduction achieved by transmission sound loss by building envelopes. The latter depends particularly on the windows of buildings, both in terms of achieved transmission loss (for example, single glazing vs double glazing) and the behaviour of residents in terms of their management of ventilation to their dwellings by the extent of window opening, particularly during sleep.

Urban and city planners affect the sonic environment of cities primarily through land-use planning practices, which have had some success in achieving environmental objectives (de Roo 2017), and by development controls over building form and construction. However, management of the acoustic environment of a city is rarely the direct focus of planners' activities – mostly it tends to be a by-product of planning methodologies utilized primarily for other purposes. These include:

- Separation of incompatible land uses such as polluting industry and residential uses. Such separation is one of the primary tools of spatial planning, achieving a wide range of objectives. As a result, noise from industry tends not to impact noise-sensitive land uses. However, planners are encouraging more mixed land uses for purposes such as minimizing transport demand. Recreational and domestic noise within such developments may also be a problem for residential neighbours.
- Planners control urban densities through the permitted height of buildings, amount of floor space, and building bulk. It is the number of people living in areas of different noise levels that determines the exposure of a city's population to noise, and choices about the location of higher density residential use determine this. The bulk of the building form also has significant effects on the propagation of sound through a city.
- Spatial planning of the location of transport infrastructure in the proximity of dwellings, and the location of new urban development relative to existing transport noise sources have a strong influence on exposures (also see the section on Environmental Impact Assessment [EIA] below).
- Planning development permits may be required for buildings that have noise sensitive uses such as schools, dwellings, or hospitals but that are to be located within areas with adverse acoustic environments such as near roadways or under airport flight paths. This may result in development conditions that include requirements on the building envelope to reduce noise.

Jean Miguel Morillas and colleagues (2018) provide a succinct review of the small number of studies available that have examined the influence of various urban design and planning variable decisions on, particularly road traffic noise, exposures.

Environmental assessment methodologies and the urban acoustic environment

Noise and project-based environmental assessment

Incremental yet broadly continuous growth in transport infrastructure tends to be the dominant policy response to population and traffic growth in most countries. Transport-planning tools are used to predict future traffic demand, whether that be surface or air transport modes. Additional transport infrastructure may be provided to meet deficiencies in capacity. As part of this planning process, new infrastructure projects are proposed. Examples include new rail lines, modified flight paths or roadway realignments. Invariably, associated with these types of proposals, is a regulatory requirement for the project to be subject to formal environmental assessment procedures. Environmental Impact Assessment (EIA) is now firmly established as an important part of the planning and development of road, rail, and air transport projects. The same applies to many new non-transport-related projects such as wind farms, industrial or processing facilities, and large developments such as shopping centres or tourism resorts. They too may be subject to project-based EIA, depending on the regulatory screening process, which varies across different jurisdictions.

The methodology of project-based EIA is well recognized and there is wide experience in such assessments. If the screening has triggered a requirement for assessment, the process includes the steps of scoping which impacts and issues should be considered; a prediction of the likely magnitude of the impacts; an evaluation and assessment of the significance of the predicted impacts; the consideration of alternatives; and the identification of mitigation strategies for impacts that have been identified as significant.

The relevance for this chapter is that noise is nearly *always* an issue included in the scoping of a project's potential impacts (Burgess and Finegold 2008) either because the development itself generates noise – as do most road, rail, airport, or industrial developments – or because it may contain noise-sensitive uses that will become exposed to existing noise sources from outside the development site. A major land development proposed under airport flight paths is an example of the latter. There is an extensive literature on the inclusion of noise considerations in project-based EIA (e.g. Canter 1996; Therivel and Wood 2017). This includes extensive development and application of modelling techniques. Modelling involves the prediction of noise emission levels for all source types, and the subsequent estimation of levels at receptors after attenuation by the distance, by the atmosphere, and by the barriers and reflections along the propagation path from the noise sources (Wood 1999; Garg and Maji 2014). Regulations, guidelines, and criterion levels are available to evaluate the significance of future exposure levels predicted within the EIA process. Mitigation strategies to reduce predicted high levels of exposure by noise-control techniques are widely practiced within EIA. They also often include planning approaches such as increasing the separation distance between the sources and the receivers.

Noise and strategic environmental assessment

EIA methodology has been applied primarily at an individual project level. Over the last two decades, however, this has been extended to the environmental assessment of upstream strategic planning and policy instruments such as urban and regional land-use plans, or transport, water management, or energy policies. This is termed Strategic Environmental Assessment (SEA). It is possible, within SEA, to assess broad acoustic impacts of policies and plans early within planning processes, rather than relying on EIA of individual projects to assess noise impacts later in the development process. The advantage is that various alternatives and options that may be foreclosed once the policy or plan is adopted are still on the table. For example, it would be appropriate to assess noise effects arising from urban consolidation policies, the introduction of road pricing, major land-use changes, or modal subsidies. Policies such as these have the potential to change transport flows, the disposition of noise sources relative to sensitive land uses, and the density of receptors exposed to particular sources – all of which affect the sonic environment of a city and people's exposure to it. However, despite the ubiquitous application of environmental assessment and mitigation tools at the project level, to date there has been little equivalent consideration of noise at these broader strategic plan and policy levels. A.E.M. De Hollander and colleagues (1999) confirm that attention to specific environmental health outcomes when alternative policies and plans are being considered tends to be limited.

While they have had little application to date, examples of two quite different strategic-level approaches are described below.

Environmental modelling incorporated in travel demand models

It is possible to incorporate the techniques of project noise modelling directly into the transport network planning process. This effectively provides an SEA, quantifying the noise effects of different transport scenarios simultaneously with the preparation of transport plans. Transport specialists predict the future traffic flow on the transport infrastructure which can then be used to estimate future transport noise exposures. If these exceed limits, mitigation strategies may be introduced by the transport specialists, or they may revise transport strategies. In transport infrastructure design, future levels of exposure to noise may effectively become a limiting factor to traffic load, and thus a warrant for noise reduction strategies. An example is the determination of flight paths at an airport based on minimum noise routings that reduce aircraft noise exposure across the community.

A range of systems for modelling environmental impacts of transport have been developed, many of which have a transport model integrated with the environmental modelling for considering the environmental effects of road transport at the network level. In one example of this methodology, Lex Brown and Joseph Affum (2002) describe a modelling system in a geographic information system intended for use by transport planners as an add-on module to existing transport planning models. The module uses, as its prime input, the output data from travel demand models used in transport planning, overlaid with land-use information in the immediate vicinity of the modelled road traffic network. The system is designed to provide rapid information to the transport planner on the noise effects of any transport proposal being considered, including comparisons of the levels of exposures to road traffic noise under different planning scenarios, and thereby the ability to aid the selection of a preferred transport scenario based on environmental, not just transport-related, outcomes. In this way, the consideration of noise can be brought up front in the planning process.

Burden of disease: Health Impact Assessment

Methods have now been developed for inclusion of transportation noise metrics in quantitative Heath Impact Assessments at aggregated strategic levels in planning. In recent years, evidence has accumulated regarding the health effects of environmental noise – beyond measurement of the annoyance it causes. In order to inform future policy, and to develop management strategies and action plans for its control, national and local governments can now consider this new evidence on health impacts of environmental noise, and utilize it in the application of SEA to plans, policies, and programmes that have an effect on the exposure of the population to noise. It is possible to quantitatively estimate the burden of disease due to environmental noise by a risk assessment approach – combining the identification of hazards, the assessment of population exposure, and the utilization of appropriate exposure-response relationships. The environmental burden of disease (EBD) is expressed in the disability-adjusted life years (DALYs) metric, which sums the potential years of life lost due to premature death and the equivalent years of 'healthy' life lost by virtue of being in states of poor health or disability. The World Health Organization (WHO) has previously estimated the global burden of disease and the EBD of disease due to environmental factors such as outdoor and indoor air pollution, poor water supply, and sanitation (Murray and Lopez 1996; Prüss-Üstün et al. 2003). It has now extended this EBD methodology to environmental noise.

The WHO provides guidance for estimations of the burden of disease for various health endpoints caused by environmental noise (Fritschi et al. 2011). The DALYs lost through environmental noise exposure are calculated for cardiovascular diseases in adults, cognitive impairment in children, sleep disturbance, annoyance, and tinnitus. The EBD process, as applied by the WHO, is one way of synthesizing this evidence in a standardized manner that provides a useful starting point in providing policymakers with quantitative estimates of the health risk of noise in urban areas.

Given the nature of the evidence on which the estimation of the health effects of environmental noise are based (large-scale data sources, multi-study and multi-country estimates of exposure-response) examination of the health effects of noise in this way is unlikely to be suitable for project-based EIA. However, the availability of quantitative assessments of the burden of disease from environmental noise means that noise can now be appropriately considered as one of the consequences within the planning of strategic level activities. These could include consideration of options within regional/national transport plans and policies, and the development of policy settings such as a preferred form of urban development or transport management options such as congestion pricing.

The magnitudes of the EBD for road transport will likely rank, in many health impact assessments, alongside estimates of the EBD of factors such as road vehicle accidents and atmospheric pollution. Further, there is some evidence that, while the EBD for other factors may be dropping over time, that for environmental noise may be increasing. These approaches could be used to incorporate the sonic environment in decision making with respect to option choice in a range of policy and plan-making activities.

Soundscape planning

Beyond the adverse effect of noise in urban areas, there is a growing interest in the urban sonic environment as a resource and its utilization to achieve human well-being objectives (see Cobussen 2016). The resource is primarily that part of the sonic environment of cities that is not 'high noise' and which have thus been rarely considered by most of the planning and management methodologies for the sonic environment already described in this chapter.

Methods for analysis (e.g. Engel et al. 2018) and management of the soundscape of the urban acoustic environment do not yet have widespread acceptance in practice, and there is limited experience as to where soundscape approaches could be applicable (Xiao, Lavia, and Kang 2018). Planners and designers need guidance on how to identify spaces and places for potential soundscape management to achieve positive human outcomes (Cerwén, Wingren, and Qviström 2017). Further, they need to be able to identify specific objectives for the soundscape design of any particular place and translate these into acoustical design criteria that support the beneficial uses of that place (Cerwén, Kreutzfeldt, and Wingren 2017). What is required is analogous to what the designers of indoor spaces already have available in terms of acoustic objectives and criteria – say for facilitating learning in classrooms or for enjoying speech or music in auditoria. Set out below is a conceptual framework within which to identify potential loci for soundscape planning and design in the outdoor spaces of cities – a much needed contribution to soundscape planning methodology for the urban sonic environment.

It is possible to identify generic loci for soundscape planning and design. Some of these are specific urban spaces/places; others apply more broadly across urban residential areas. Developments of new mapping models to aid soundscape planning (Magaritas and Kang 2017) are required to assist in this identification. Four examples of opportunities for the application of soundscape planning are discussed below.

Some specific loci for soundscape planning
Managing quiet areas

In Europe, quiet areas have been recognized as a target for management of the urban sonic environment, and a *Good Practice Guide on Quiet Areas* has been published by the European Environment Agency (2014a). There has been some mapping of quiet areas in urban and non-urban environments, though generally this has been through the portrayal of the inverse of maps of predicted high levels of transportation noise rather than on any specific identification of characteristics of high acoustic quality. This is because, for the most part, identification of quiet areas has been based on low levels of integrated sound, with no distinction between sound sources. This is inadequate. Whilst a low level of sound may be a characteristic of some areas that are of high acoustic quality, quiet is not the antithesis of noisy, and areas that have low levels of sound may not necessarily be ones in which the acoustic environment is assessed by people as being of high quality. For example, there are many places in urban areas where sounds are at low levels but the source of these sounds is traffic on distant roadways. Most people would be unlikely to prefer the sonic environments of these places. Similarly, there are many urban places where sound levels may be high but that would generally be perceived to be tranquil or of high acoustic quality. Locations near to fountains or where there is extensive birdlife, are examples. There is increasing evidence that it is the congruence of the type of sound heard in a particular environment and people's expectations of that place that determines its acoustic quality (Bild et al. 2018). In other words, identification of areas with a low sound level in urban areas is potentially useful as it could highlight areas that may have high acoustic quality – but this is not a sufficient condition. What is also needed is a parallel investigation as to whether the sounds in these areas meet peoples' expectations/preferences (Vogiatzis and Remy 2014; Rey Gozalo et al. 2018). Manon Raimbault and Danièle Dubois (2005) largely reject physical acoustical parameters as measures of preferred soundscapes.

Managing/Making areas of high acoustic quality

Managing and protecting the acoustic environment of particular 'quiet areas' is useful if they are of high acoustic quality. But there are many outdoor areas that are of high acoustic quality that are far from quiet: a forested urban park with wind in the trees; a fountain with loud splashing water; loud singing of birds or insects in urban parks and residential gardens; buskers in a mall or subway tunnel; church bells in a town square; the sounds of children playing, or of sports fans cheering; the hum of a marketplace (with people sounds as the dominant source and free from mechanical or recorded sounds). Within appropriate contexts, people are likely to enjoy, even cherish these sounds. These areas of high acoustic quality contribute to the richness of urban life and can be included in conscious management and planning of the urban sonic environment. Clearly, design criteria for them cannot be based on sound level. Lex Brown and Andreas Muhar (2004) provide an approach, for any particular human activity and specific context, based on establishing appropriate human objectives. Examples for particular places are: moving water should be the dominant sound heard; hear mostly the non-mechanical, non-amplified sounds made by people; or good for hearing unamplified speech or music.

Management or acoustic design can ensure the wanted sounds in such areas are not masked by the unwanted sounds. The inverse approach of increasing the wanted sound can also be utilized through, for example, design of reverberant space in a mall specifically for the purposes of creating a lively space for busking. Simulation and virtual reality methods are being applied, experimentally, for evaluation and design of urban sound environments (e.g. Jiang et al. 2017) and Sonia Alves et al. (2015) provide several examples of the design and management of urban public spaces using soundscape planning.

Protecting iconic or place-defining sounds

These types of sounds are highly specific to particular localities: bells, clocks, chimes, waves on beaches of seaside cities, the sound of particular local transport (San Francisco's cable cars, for example), perhaps even sounds from agricultural or industrial processes which define the economic base of a town. These can be essential components of the identities of specific urban areas, and could also have much wider values through management of cultural heritage and the attraction of tourism (Maffei, Brambilla, and Di Gabriele 2015). Iconic sound events, such as coordinated multiple church bell happenings, could have dramatic acoustic and sociocultural impact.

Design for sound installations

'Sound installations' is used here as a generic term for public works of art that include some acoustic dimension (e.g. Lacey 2016). Examples include those which react to their environment – either driven by natural forces of wind or water, or responding to human interaction such as drums, chimes, or voice trumpets. Others may incorporate recorded sound, of music, voice, or natural sounds, or fed-back amplification of sound from the immediate, or some remote, environment. The soundscaping issue here is twofold – firstly the appropriateness of the introduced sound to the particular locality (is it a wanted sound by most of those who will hear it) and secondly if the sound generated by the work of art will be audible over the area intended, or whether this may be masked by unwanted sounds. Experiences of various acoustic art installations in different cities show that many are either not adequately supported by local stakeholders or are rendered ineffective through their being masked by traffic noise or other mechanical sounds at the site of the installation.

Broader application within residential areas

Ensuring diversity in the acoustic environment

Another concept associated with the acoustic environment as a resource is that of diversity. Diversity in genes, species, and ecosystems underpins the management of systems of biological resources. Maintenance of natural diversity (and cultural diversity) is also a principle adopted in the spatial planning of regions, natural areas, the countryside, and urban centres. The same diversity principle can find application in management of the acoustic environment. For example, matters such as the characteristic of local sounds and tranquillity are important elements of the spatial quality of rural and urban areas.

In an early soundscape study in Boston, Michael Southworth (1969) hypothesized that changes in the soundscape are needed to increase (1) the identity of the soundscape, and (2) the number of opportunities for delight in sounds and to provide responsive settings which contain novel sounds. The study noted the grey blurring of the acoustic environment that was occurring in cities, in terms of transport noise sources becoming the dominant background everywhere, masking natural sounds and local community sounds. Soundscapes studies have the potential to articulate the extent, or absence, of diversity in the urban acoustic environment.

Encouraging attention to sub-criteria exposures and to restoration of human well-being

Studies of the burden of disease for environmental noise show that there is a contribution from noise even at exposures below what might be set as criteria, or cut-off points, for noise-abatement action. This is because, while the risk of any particular outcome response is lower at lower exposure levels, the numbers of people within an urban area exposed to these levels is high. The consequence is that *any* action to reduce exposures across the dwellings of a community will intrinsically have health benefits. The relevance of this for soundscape planning is that it is unlikely that traditional noise-control approaches will ever set noise limits for particular sources lower than criterion limits. However, lowering of levels of sound over parts of a residential area may be the outcome of some broader soundscaping plan and it is important to recognize that this can result in tangible health benefits as the burden of disease from sub-criterion exposures is significant.

Somewhat more speculative is the potential benefit to people of creating availability, even knowledge, of a better-quality acoustic environment somewhere in their neighbourhood. Francesco Aletta, Tin Oberman, and Jian Kang (2018) have documented the limited evidence available of the positive health-related effects of supportive soundscapes and Van Kamp et al. (2015) note that the restorative benefits of good soundscapes elsewhere may also accrue to a person who otherwise is subject to adverse effects of noise at home. A soundscape planning scheme that introduces quiet sides to dwellings, or increases the prevalence of high-quality acoustic environments elsewhere in a neighbourhood, may provide health benefits through this mediating mechanism.

Summary

This chapter has provided an overview of the study and management of the outdoor sonic environment of cities, predominantly its high-noise components. Acousticians and others apply a range of methods to measure or model the noise levels in cities, and exposure of the population, and the nature and extent of the discomfort and disease it causes them. Exposure-response relationships, and the specification of limit criteria are derived from these. Management actions are largely based on limiting human annoyance from noise – even though there is a well-documented range of health effects of transport noise in addition to annoyance outcomes.

Most methods for control and management of the sonic environment are bound within engineering, design, and the planning paradigms of infrastructure and urban development. These include source reduction, changes in the propagation path, changes to transmission loss of the envelopes of buildings, spatial separation of sources and receivers, and changes in transport and other infrastructure. These involve engineering, planning, transport, architectural, and other design disciplines.

One of the primary ways that noise is assessed and managed is within project-based EIA of new and changed infrastructure. EIA utilizes well-tested methodologies in prediction, assessment, and mitigation of noise impacts. However, despite the ubiquity of assessment and control tools at the project level, there has to date been little consideration of noise at the broader strategic planning levels. This is now changing, with methodology developed for inclusion of noise in quantitative Heath Impact Assessments at aggregated strategic levels of planning.

Concepts of soundscape planning and management also have a role within the methodological toolbox of management of the acoustic environment of urban areas. Soundscape approaches focus on human perception of the acoustic environment and have been applied to places such as urban parks and gardens, city malls, and historical and cultural locations. It is complementary to, not a substitute for, the dominant management paradigm of environmental noise control found in environmental assessment and urban planning approaches. The loci of application of noise control methodologies is where predefined noise limits are exceeded – conventionally adjacent to high noise-level sources or at sensitive receptors where adverse impacts and effects arise. By contrast, the range of application of soundscape management methodologies is much less well developed, apart from a basic awareness of protection of existing 'quiet areas'. This chapter redressed this imbalance by providing methodology for identifying a spectrum of opportunities, and potential design criteria, for urban soundscape interventions.

Soundscape approaches will be applicable for managing quiet areas – but not exclusively so. Soundscape design and management may include creative acoustic design to achieve places of high acoustic quality; ensuring the potential for humans to experience diversity in the acoustic environment throughout urban areas; encouraging attention to sub-criterion exposures; providing restorative access for human health; protecting iconic or place-defining sounds; and providing spaces for public acoustic installations.

Note

1. Stylianos Kephalopoulos and colleagues (2014) describes current modelling of road traffic, railway traffic, aircraft, and industrial sources.

References

Aletta, F., T. Oberman, and J. Kang (2018). 'Associations between Positive Health-related Effects and Soundscapes Perceptual Constructs: A Systematic Review'. *International Journal of Environmental Research and Public Health* 15: 2392.

Alves, S., L. Estévez-Mauriz, F. Aletta, G. M. Echevarria-Sanchez, and V. Puyana Romero (2015). 'Towards the Integration of Urban Sound Planning in Urban Development Processes: The Study of Four Test Sites within the SONORUS Project'. *Noise Mapping* 2: 57–85.

Augoyard, J.-F. (1998). 'The Cricket Effect: Which Tools for the Research on Sonic Urban Ambiences?'. Paper presented at the *Stockholm, Hey Listen!* conference, The Royal Swedish Academy of Music, 9–13 June. pp. 1–7.

Basner, M. and S. McGuire (2018). 'WHO Environmental Noise Guidelines for the European Region: A Systematic Review on Environmental Noise and Effects on Sleep'. *International Journal of Environmental Research and Public Health* 15: 519.

Bild, E., K. Pfeffer, M. Coler, O. Rubin, and L. Bertolini (2018). 'Public Space Users' Soundscape Evaluations in Relation to Their Activities: An Amsterdam-Based Study'. *Frontiers of Psychology* 9: 1593. https://doi.org/10.3389/fpsyg.2018.01593.

Brown, A. L. and J. K. Affum (2002). 'A GIS-based Environmental Modelling System for Transport Planners'. *Computers, Environment & Urban Systems* 26 (6): 577–590.

Brown, A. L. and A. Muhar (2004). 'An Approach to the Acoustic Design of Outdoor Space'. *Journal of Environmental Planning and Management* 47 (6): 827–842.

Brown, E. F., E. B. Dennis, J. Henry, and F. E. Pendray (eds) (1930). *City Noises*. New York: Academy Press.

Burgess, M. A. and L. S. Finegold (2008). 'Environmental Noise Impact Assessment'. In M. J. Crocker (ed.), *Handbook of Noise and Vibration Control*, 1501. Hoboken, NJ: John Wiley & Sons.

Canter, L. W. (1996). *Environmental Impact Assessment*. New York: McGraw Hill.

Cerwén, G., J. Kreutzfeldt, and C. Wingren (2017). 'Soundscape Actions: A Tool for Noise Treatment Based on Three Workshops in Landscape Architecture'. *Frontiers of Architectural Research* 6: 504–518.

Cerwén, G., C. Wingren, and M. Qviström (2017). 'Evaluating Soundscape Intentions in Landscape Architecture: A Study of Competition Entries for a New Cemetery in Järva, Stockholm'. *Journal of Environmental Planning and Management* 60 (7): 1253–1275.

Cobussen, M. (2016). 'Towards a "New" Sonic Ecology', Inaugural Lecture, Professor of Auditory Culture, University of Leiden, 28 November 2016. Available online: http://sonicfield.org/2016/12/towards-a-new-sonic-ecology-lecture-by-marcel-cobussen/ (accessed 30 June 2020).

de Hollander, A. E. M., J. M. Melse, E. Lebret, and P. G. Kramers (1999). 'An Aggregate Public Health Indicator to Represent the Impact of Multiple Environmental Exposures'. *Epidemiology* 10: 606–617.

De Roo, G. (2017). *Urban Environmental Planning: Policies, Instruments and Methods in an International Perspective*. London: Routledge.

Engel, M. S., A. Fiebig, C. Pfaffenbach, and J. Fels (2018). 'A Review of Socio-Acoustic Surveys for Soundscape Studies'. *Current Pollution Reports*, September.

European Commission Working Group Assessment of Exposure to Noise (2007). 'Position Paper: Good Practice Guide for Strategic Noise Mapping and the Production of Associated Data on Noise Exposure'. Available online: https://www.lfu.bayern.de/laerm/eg_umgebungslaermrichtlinie/doc/good_practice_guide_2007.pdf (accessed 13 July 2020).

European Environment Agency (2014a). Good Practice Guide on Quiet Areas. EEA Technical Report No. 4/2014. Luxembourg.

European Environment Agency (2014b). Noise in Europe 2014. EAA Report No 10/2014. Luxembourg: Publications Office of the EU.

Fiedler, P. and P. Zannin (2015). 'Evaluation of Noise Pollution in Urban Traffic Hubs: Noise Maps and Measurements'. *Environmental Impact Assessment Review* 51: 1–9.

Fritschi, L., A. L. Brown, R. Kim, D. Schwela, and S. Kephalopoulos (eds) (2011). *Burden of Disease from Environmental Noise: Quantification of Healthy Life Years Lost in Europe*. Bonn: World Health Organization, Regional Office for Europe, and European Commission Joint Research Centre.

Garg, N. and S. Maji (2014). 'A Critical Review of Principal Traffic Noise Models: Strategies and Implications'. *Environmental Impact Assessment Review* 46: 68–81.

Guillaume, G., A. Can, G. Petit, N. Fortin, S. Palominos, B. Gauvreau, E. Bocher, and J. Picaut (2016). 'Noise Mapping Based on Participative Measurements'. *Noise Mapping* 3: 140–156.

Guski, R., D. Schreckenber, and R. Schuemer (2017). 'WHO Environmental Noise Guidelines for the European Region: A Systematic Review on Environmental Noise and Annoyance'. *International Journal of Environmental Research and Public Health* 14.

ISO/TS 15666 (2003). 'Acoustics – Assessment of Noise Annoyance by Means of Social and Socio-Acoustic Surveys'. Geneva: International Standards Organization.

Jiang, L., M. Masullo, L. Maffei, F. Meng, and M. Vorländer (2017). 'A Demonstrator Tool of Web-Based Virtual Reality for Participatory Evaluation of Urban Sound Environment'. *Landscape and Urban Planning* 170: 276–282.

Kephalopoulos, S., M. Paviotti, F. Anfosso-Lédée, D. Van Maercke, S. Shilton, and N. Jones (2014). 'Advances in the Development of Common Noise Assessment Methods in Europe: The CNOSSOS-EU Framework for Strategic Environmental Noise Mapping'. *Science of the Total Environment* 482-483: 400–410.

Kogan, P., J. P. Arena, F. Bernejo, M. Hinalaf, and B. Turra (2018). 'A Green Soundscape Index (GSI): The Potential of Assessing the Perceived Balance between Natural Sound and Traffic Noise'. *Science of the Total Environment* 642: 463–472.

Lacey, J. (2016). 'Sonic Placemaking: Three Approaches and Ten Attributes for the Creation of Enduring Urban Sound Art Installations'. *Organised Sound* 21 (2): 147–159.

Law, C. W., C. K. Lee, A. S. W. Lui, M. K. L. Yeung, and K. C. Lam (2011). 'Advancement of Three-Dimensional Noise Mapping in Hong Kong'. *Applied Acoustics* 72: 534–543.

Licitra, G. (ed.) (2012). *Noise Mapping in the EU: Models and Procedures*. Boca Raton, FL: CRC Press.

Maffei, L., G. Brambilla, and M. Di Gabriele (2015). 'Soundscape as Part of the Cultural Heritage'. In J. Kang and B. Schulte-Fortkamp (eds), *Soundscape and the Built Environment*, 215–242. Boca Raton, FL: CRC Press.

Magaritas, E. and J. Kang (2017). 'Soundscape Mapping in Environmental Noise Management and Urban Planning: Case Studies in Two UK Cities'. *Noise Mapping* 4, 87–103.

Morillas, J. M. B., G. R. Gozalo, D. M. González, P. A. Moraga, and R. Vilchez-Gómez (2018). 'Noise Pollution and Urban Planning'. *Current Pollution Reports* 4: 208. https://doi.org/10.1007/s40726-018-0095-7.

Murphy, E. and E. A. King (2016). 'Smartphone-based Noise Mapping: Integrating Sound Level Meter App Data into the Strategic Noise Mapping Process'. *Science of the Total Environment* 562: 852–859.

Murray, C. J. L. and A. D. Lopez (eds) (1996). 'The Global Burden of Disease: A Comprehensive Assessment of Mortality and Disability from Disease, Injury, and Risk Factors in 1990 and Projected to 2020'. Global burden of disease and injury series, vol. 1. Cambridge, MA: Harvard University Press.

Prüss-Üstün, A., C. Mathers, C. Corvalán, and A. Woodward (eds) (2003). 'Introduction and Methods: Assessing the Environmental Burden of Disease at National and Local Levels'. Geneva: World Health Organization.

Raimbault M. and D. Dubois (2005). 'Urban Soundscapes: Experiences and Knowledge'. *Cities* 22 (5): 339–350.

Rey Gozalo, G., J. M. Barrigón Morillas, D. Montes González, and P. Atanasio Moraga (2018). 'Relationships among Satisfaction, Noise Perception, and Use of Urban Green Spaces'. *Science of the Total Environment* 624: 438–450.

Shim, E., D. Kim, H. Woo, and Y. Cho (2016). 'Designing a Sustainable Noise Mapping System Based on Citizen Scientists Smartphone Sensor Data'. *PLoS ONE* 11 (9). https://doi.org/10.1371/journal.pone.0161835.

Southworth, M. (1969). 'The Sonic Environment of Cities'. *Environment and Behaviour* 1: 49–70.

Stoter, J. E., H. de Kluijver, and K. Kurakula (2008). '3D Noise Mapping in Urban Areas'. *International Journal of Geographical Information Science* 8: 907–924.

Therivel, R. and G. Wood (2017). *Methods of Environmental and Social Impact Assessment*. 4th edition. New York: Routledge.

Van Kamp, I., R. Klaeboe, A. L. Brown, and P. Lercher (2015). 'Soundscapes, Human Restoration and Quality of Life'. In J. Kang and B. Schulte-Fortkamp (eds), *Soundscape and the Built Environment*, 43–68. Boca Raton, FL: CRC Press.

Van Kempen, E., M. Casas, G. Pershagen, and M. Foraster (2018). 'WHO Environmental Noise Guidelines for the European Region: A Systematic Review on Environmental Noise and Cardiovascular and Metabolic Effects: A Summary'. *International Journal of Environmental Research and Public Health* 15: 379.

Vogiatzis, K. and N. Remy (2014). 'From Environmental Noise Abatement to Soundscape Creation through Strategic Noise Mapping in Medium Urban Agglomerations in South Europe'. *Science of the Total Environment* 482–483: 420–431.

Wood, G. (1999). 'Assessing Techniques of Assessment: Post-Development Auditing of Noise Predictive Schemas in Environmental Impact Assessment'. *Impact Assessment and Project Appraisal* 17 (3): 217–226.

World Health Organization (WHO) (2018). 'Environmental Noise Guidelines for the European Region'. Copenhagen: World Health Organization Regional Office for Europe.

Xiao, J., L. Lavia, and J. Kang (2018). 'Towards an Agile Participatory Urban Soundscape Planning Framework'. *Journal of Environmental Planning and Management* 61 (4): 677–698. https:doi.org/10.1080/09640568.2017.1331843.

13

Sonic Methodologies in Medicine

Jos J. Eggermont

Introduction and overview

Sound plays a prominent role in medical practice and research, in diagnostics as well as in therapy. Historically, one distinguishes three types of sound based on their frequency content. Audible sound occupies the frequency range from 20 hertz to 20 kilohertz. Ultrasound consists of frequencies greater than 20 kilohertz, approximately the upper limit of hearing in young normal hearing humans. Similarly, 'infrasound' consists of frequencies less than 20 hertz, the lower limit of human hearing. Audible sound produced or reflected from inside the body was used early on for methodical investigations and diagnosis using stethoscopes, which were invented in 1816 by René Laennec in Paris, but made practical in 1852 by Georg Philip Camman (Wade and Deutsch 2008).

Audible sound in the form of standardized pure tones is used in the determination of frequency-dependent hearing loss, commonly illustrated by the audiogram. The audiogram graphs the audible sound levels in decibels as a function of octave-separated frequencies, typically from 125 hertz to 8,000 hertz, and refers to the normal hearing-level standard. In the nineteenth century, the use of tones in audiology was based on the Galton whistle, tunable to various high frequencies and specifically useful to test the deteriorating high frequencies in age-related hearing loss (Zwaardemaker 1891). The whistle, invented by Sir Frances Galton (1822–1911), can be adjusted to produce high-frequency sounds between 5 kilohertz and 42 kilohertz, and was used by its inventor to test the limits of hearing in the dogs he saw on his walks in London's Hyde Park. Other instruments in audiology use around the end of the nineteenth century included the monochord (Mollison 1917), a single string tunable instrument, and tuning forks. The tuning fork, invented in 1711 by British musician John Shore, was introduced in audiology to test low- and mid-frequency hearing at the end of the nineteenth century (Feldman 1997). In the 1920s the first electronic audiometer came into use (Bunch 1929). Audiograms form the basis for locating the sources and degree of hearing loss, and this information is used to prescribe hearing aids or surgery. Audible sounds are also used in combination with neuroimaging techniques such as functional magnetic resonance imaging (fMRI), and electroencephalographic (EEG) techniques to localize brain areas involved in, for example age-related hearing impairment, mild cognitive impairment, and Alzheimer's disease (Eggermont 2019).

Because of the increasing evidence for a link between hearing loss and early cognitive decline, timely hearing aid fitting is advised. A hearing aid amplifies external sound and allows easier speech understanding, facilitates communication, and may prevent social isolation. Another category of therapeutic use of audible sound is music that is widely advocated in managing among others, behavioural-emotional disorders, types of dementia, and Parkinson's disease where it is known to improve the ability to initiate and control gait. There is also some evidence that music therapy improves social functioning and quality of life for people with psychiatric disorders.

Another important branch of medical diagnostics uses ultrasound to visualize internal body structures in the abdomen such as the pancreas, liver, gall bladder, kidneys, and spleen, including prenatal testing, in order to find the source of a disease or to exclude pathology. The creation of an image from ultrasound, a sonogram, is done in three steps: (1) a series of focused sound pulses are produced by a piezoelectric transducer, (2) the pulses echo off tissue, and (3) the electric pulses from the echoes vibrate the transducer and are used to build up a digital image based on the delay and strength of the sound echo. An example is echocardiography used to diagnose the dilatation of parts of the heart and function of heart ventricles and valves. By using Doppler ultrasound analysis one can also assess the blood flow.

Therapeutic use of ultrasound is also varied and includes breaking up kidney stones and gallstones by using high-energy pulses; cleaning teeth; treating cataract by using focused ultrasound sources; and non-invasive ablating tumours and other tissues. Here high-intensity focused ultrasound is used and often guided by MRI.

Audible sound as a diagnostic tool

Types of hearing loss

Imagine that your hearing sensitivity for pure tones is exquisite – not affected by frequent exposure to loud music or other noises – but that you have problems in understanding speech even in a quiet environment. This occurs if you have a temporal processing disorder. Although hearing loss is in the ear, hearing problems such as those caused by deficits in temporal processing originate in the brain. We often take hearing for granted; not realizing what good hearing allows us to do. Without hearing, communication with our fellow humans largely disappears. Substitutes for total loss of hearing are sign language, which replaces hearing with vision, and cochlear implants, which restore hearing to a large extent. For hard of hearing persons, amplification with hearing aids restores the sense of sound but does not generally result in normal perception, except when the hearing loss is of the conductive type.

Hearing loss comes in two broad types, conductive and sensorineural. Conductive hearing loss results from deficits in the sound-conducting apparatus of the outer and middle ear. Problems such as fluid in the middle ear and immobility of the middle ear bones are the main causes of a conductive hearing loss (Figure 13.1). Sensory hearing losses result from damage of structures in the cochlea, namely, to the hair cells – the microphones and amplifiers—and the stria vascularis – the battery charger in the ear (Figure 13.2). Neural loss occurs when damage to the auditory nerve is involved. The latter often follows hair cell loss – called secondary degeneration and then constitutes sensorineural loss – but can also occur in isolation, as primary degeneration.

Use of sound in audiology for site of lesion testing

Air- or bone-conduction audiograms

Conductive hearing losses are diagnosed by measuring the audiometric difference in the threshold for air- and bone-conducted sound called the air-bone gap. Bone conducted sound, produced by a vibrator on the scalp, bypasses the external and middle ear in stimulation of the cochlea, but is much less effective than air conduction in normal hearing people. By comparing the air and bone conduction audiograms, both relative to those in a normal ear, middle ear problems can be detected. If only the air-conduction threshold is elevated a conductive hearing loss is concluded; if both are similarly elevated the loss is of the sensorineural type. Mixed types occur as well.

Speech discrimination testing

Speech audiometry is a fundamental tool in hearing loss assessment. Together with pure-tone audiometry, it can aid in determining the degree and type of hearing loss. Speech audiometry provides information on word recognition and about discomfort or tolerance to speech stimuli. Speech audiometry outcomes help also in setting the gain and maximum output of hearing aids for patients with moderate to severe hearing losses. An adaptation of speech audiometry is the Hearing-in-Noise Test, in which the stimuli are presented by a loudspeaker in a frontal position and the patient is required to repeat sentences both in a quiet environment and with competing noise being presented by loudspeakers, coming from different directions.

Otoacoustic emission testing

The ear does not only receive sound, normal ears also emit sounds, which are called spontaneous otoacoustic emissions. Otoacoustic emissions were discovered by David Kemp (1979), and can be recorded with a sensitive microphone in the ear canal and provide a noninvasive measure of the working of the cochlea. There are also two main types of sound-evoked otoacoustic emissions in clinical use. The first, transient-evoked otoacoustic emissions are evoked using a click stimulus. The otoacoustic emissions to a click comprise frequencies up to around 4 kilohertz; frequency selective masking allows identification of individual bands (Figure 13.3). The second, distortion product otoacoustic emissions are evoked using a pair of primary tones with frequencies f1 and f2 with a frequency ratio of f2/f1<1.4. The most commonly measured distortion product otoacoustic emission is at the frequency 2f1–f2. Recording of sound-evoked otoacoustic emissions has become the main method for newborn and infant hearing screening (see below).

Auditory brainstem response testing

The auditory brainstem response is a short-latency auditory evoked potential, with amplitudes below 0.5 microvolt obtained by signal averaging from ongoing electrical activity in the brain and recorded via electrodes placed on the scalp. The resulting recording is a series of scalp-vertex positive waves labelled I through V (Figure 13.4). These waves occur in the first 10 milliseconds after onset of an auditory stimulus, typically a click or short tone pips. Wave I is generated in the cochlear part, wave II in the intra-brain part of the auditory nerve, and wave III in the lower brainstem. Waves IV and V are generated in the upper brainstem (Figure 13.4; Ponton, Moore, and Eggermont 1996).

Amplitude and latency of waves I, III, and V are the basic measures for quantifying the auditory brainstem response. Amplitude is dependent on the number of neurons firing action potentials and above all on their synchrony in firing. Latency depends on hearing loss and again on neural synchrony; interpeak latency (the time between peaks) depends on conduction velocity of action potentials along the brainstem, and interaural latency (the difference in wave V latency between ears) is sometimes used in auditory nerve tumour diagnosis (Eggermont, Don, and Brackmann 1980). The auditory brainstem response is used for newborn hearing screening, auditory threshold estimation (Figure 13.5), determining hearing loss type, and detection of auditory nerve and brainstem lesions.

Newborn hearing screening

In 1995, a US 'Joint Committee on Infant Hearing' (JCIH) had 'endorsed the goal of universal detection of infants with hearing loss and encourages continuing research and development to improve techniques for detection of and intervention for hearing loss as early as possible' (JCIH 1995). Following this statement, a feasibility study sponsored by the National Institutes of Health in the United States was set up to determine the accuracy of three measures of peripheral auditory system status (otoacoustic emissions, auditory brainstem response thresholds, and behavioural visual reinforcement audiometry) applied in the perinatal period (Norton et al. 2000). In this study, both babies who had been in a neonatal intensive care unit (ICU) and healthy babies with one or more risk factors for hearing loss were targeted for follow-up testing using visual reinforcement audiometry at eight to twelve months of age. Automated auditory brainstem response audiometry was implemented using a click stimulus of 30 decibels above its threshold in normal ears, which appeared to be reliable for the rapid assessment of hearing in newborns. More than 99 per cent of infants could complete the auditory brainstem response protocol. More than 90 per cent of neonatal ICU babies and well babies at nursery infant age 'passed' given the strict criteria for response, whereas 86 per cent of those with high-risk factors met the criterion for auditory brainstem response detection. This method is now used in a large majority of developed countries around the world. Details can be found in Eggermont, *Hearing Loss: Causes, Prevention and Treatment* (2017).

Auditory evoked potentials

Hallowell Davis (1896–1992) is often called the 'father of the evoked response audiometry', as he was the first to use long-latency auditory evoked potentials to estimate hearing thresholds and obtain so-called objective audiograms. However, we should consider his first wife Pauline as the 'mother of auditory evoked potentials', because she spotted the repetitive changes in the ongoing electroencephalogram when loud sounds were presented (Davis 1939). These electroencephalogram changes were not easy to quantify, and it was only with the introduction of signal averagers (Dawson 1954) that the recording of auditory evoked potentials and its use in audiometry became a practical venture. The recording of auditory evoked potentials at that time was restricted to long-latency potentials because they have about ten times larger amplitudes (~ 5 microvolt) than those of shorter latency (see below for a classification; Figure 13.6). After his retirement from the Directorship of the Institute for the Deaf in St. Louis in 1965, Davis further developed the long-latency auditory evoked potential audiometry (Davis 1976). His first collaborator in that endeavour was Atze Spoor, then chief audiologist at the Leiden University ENT clinic, who introduced me in 1968 to long-latency auditory evoked potential audiometry at the start of my career in audiology. It became obvious that in children, who were instructed to sit still and stay awake (this was before the days of silent video or Game Boys), this procedure was not very reliable and was soon abandoned in favour of short-latency evoked potentials such as the auditory brainstem response, which can even be recorded in sleep or under anesthesia.

An early classification of auditory evoked potentials is based on peak-response latency, and distinguishes short-latency, middle-latency, and long-latency auditory evoked potentials (Figure 13.6). The auditory brainstem response peaks are indicated by Roman numerals (cf. Figure 13.4). The middle latency (10–50 milliseconds) response components are typically indicated by Na, Pa, Nb, and Pb, and the long latency (above 50 milliseconds) components by P1, N1, P2, and N2. The magnetic fields associated with middle- and long-latency auditory evoked potentials follow the same pattern.

Auditory evoked potentials were initially recorded with only a few electrodes, typically placed at the scalp vertex, forehead, and both mastoids—the hard bone behind the ears. Currently, the use of multiple recording sites (32–256) is standard practice. The use of such high-density electrode recordings allows under specific assumptions the calculation of the strength and location of the clusters of neurons contributing to the signal picked up at the scalp (Ponton et al. 1993). These estimated source localizations have been validated by simultaneous recording of multi-electrode auditory evoked potentials and fMRI in young adults (Scarff et al. 2004). The peaks (or valleys) in the auditory evoked potentials have of course served as objects of interest and for most of these components we have a good idea what neural structures they represent, although in most cases (e.g. P1 and N1) there is likely more than one structure that contributes to a given auditory evoked potential component. It is also clear that a given structure (e.g. primary auditory cortex on Heschl's gyrus) can contribute more than one auditory evoked potential component. In some cases there are also reasonable guesses at what information the individual auditory evoked potential components might represent, other than just the presence or absence of a sound (for details, see Ponton et al. 2000; Eggermont 2007).

Age-related declines of neural synchrony in the human auditory system, which determines the amplitude of the electroencephalogram waves, have been inferred through the use of long-latency auditory evoked potentials. These potentials represent sound-locked electroencephalogram activity. One frequently used auditory evoked potential response is the P1-N1-P2 complex (Figure 13.6), and age-related differences affecting its latency and amplitude have been reported. Although the presence of a long-latency auditory evoked potential response indicates that a stimulus was physiologically discriminated, its presence neither quantifies the quality of stimulus encoding nor indicates if subcortical age-related changes, such as reduced neural synchrony, contribute to cortical responses in older adults. Inconsistencies between behavioural and long-latency auditory evoked potential data have been found, in particular age-related differences in the neurophysiological pattern without corresponding effects in the behavioural task accuracy. Typically, ageing has a higher influence on late cognitive processes than on the perceptual (N1) and pre-attentive (mismatch negativity) ones. The presence of the mismatch negativity – basically the difference between the P1-N1-P2 complex recorded for stimuli consisting of a high probability (standard) and an interspersed low probability (deviant) – reflects pre-attentive coding of this stimulus difference. The mismatch negativity is also used to evaluate the integrity of the working-memory system. In particular, a reduced mismatch negativity response is a robust feature among the elderly (Eggermont 2019).

Music therapy

Music therapy is used in medicine with the goal to assist in the physical recovery and health maintenance of patients. I briefly introduce here its use in coronary heart disease, serious mental disorders, tinnitus, and Parkinson's disease, for which evidence-based evaluations are available. However, there is a wide spectrum of other potentially beneficial uses for which stringent evaluations have not yet been done.

Coronary heart disease

Music is often used to reduce anxiety and distress and improve physiological functioning in patients with myocardial infarctions. A Cochrane systematic review covering twenty-six clinical trials with a total of 1,369 participants found that:

> Listening to music also appears to be effective in reducing anxiety in people with myocardial infarction, especially when they are given a choice of which music to listen to. Listening to music may also reduce pain and respiratory rate. However the size of the effects on pain and

respiratory rate is small. Therefore, its clinical importance is unclear. Finally, listening to music appears to improve patients' quality of sleep following a cardiac procedure or surgery. We found no evidence of effect for depression or heart rate variability, and inconsistent results for mood.

(Bradt, Dileo, and Potvin 2013)

Serious mental disorders

For people with serious mental disorders such as schizophrenia, music therapy may help people improve their emotional and relational competencies. A Cochrane systematic review covering eighteen studies with 1,215 participants suffering from schizophrenia found a positive effect of music therapy compared to standard care (Geretsegger et al. 2017). Their overall conclusion was that 'music therapy seems to help people with schizophrenia but further research is needed to confirm the positive effects found in this review' (Geretsegger et al. 2017).

Tinnitus

Tinnitus or ringing in the ears frequently accompanies hearing loss. About 15 per cent of the population suffers from it. Albeit the cause is often hearing loss, tinnitus is a result of maladaptive changes in brain connectivity. Namely, the brain fills in the missing frequencies resulting from the hearing loss, akin to phantom pain in the case of missing limbs (Eggermont 2012). Sound in various forms is used, from simple noise to mask the tinnitus to music therapy aimed at changing the maladaptive changes in the brain that underlie tinnitus. There is no conclusive evidence on the effectiveness of masking, largely due to lack of quality research (Hobson, Chisholm, and El Refaie 2012). Combining hearing aid amplification alone or with a sound generator also is currently insufficiently investigated to draw firm conclusions (Tutaj, Hoare, and Sereda 2018). A potentially effective method appears to be the use of notched music for tonal tinnitus, where, after the frequency of the tinnitus is established, music selected by the patient is filtered such that energy at and around the tinnitus frequency is absent. A clinical trial of this technique, however, found no effect on tinnitus distress, albeit that the loudness of the tinnitus was significantly reduced (Stein et al. 2016). Christoph Krick and colleagues (2017) developed a therapeutic intervention, namely the Heidelberg Neuro-Music Therapy, which showed an effective reduction of tinnitus-related distress following a one-week short-term treatment. Using fMRI, they found that the default mode network in the brain showed increased activity that correlated with the improvement in tinnitus distress.

Parkinson's disease

People with Parkinson's disease (PD) often show gait abnormalities, such as shuffling steps, start hesitation, and freezing. Rhythmic auditory stimulation, such as marching music and dance therapy, appears to be an effective method in improving the gait in PD patients (Ashori et al. 2015). In a systematic review, Shuai Zhang and colleagues found 'evidence of a positive effect of music-based movement therapy in PD supporting its use for the treatment of motor dysfunction. There was neutral effect evidence to support the use of music-based movement therapy for the treatment of cognitive function and quality of life' (Zhang et al. 2017: 1635).

Ultrasound

Ultrasound frequencies commonly used in biological and imaging applications range from about 500 kilohertz to more than 50 megahertz (Halliwell 2010). After X-radiography, ultrasound technologies are currently the most common of all medical imaging. Emerging clinical applications are found among others in breast disease, cardiology, gastroenterology, gynecology, minimally invasive surgery, musculoskeletal studies, urology, and cardiovascular disease (Wells and Liang 2011).

Diagnostic use of ultrasound

An ultrasonic imaging system typically is able to resolve structures of around 1 millimetre in size at depths of up to around 15 centimetres in the body. Ultrasound travels at a speed of about 1,500 metres per second in soft tissues; this means that the frequency needs to be in the low megahertz range because the wavelength, which is one of the factors that determine the spatial resolution, for example is 0.5 millimetre at 3 megahertz. Pulse-echo ultrasonic imaging is based on the principle that a beam of pulsed ultrasound causes echoes that are delayed in time according to their depths. If the ultrasound is reflected by targets with a component of motion along the axis of the ultrasonic beam (e.g. by pulsatile blood flow), the echoes that are received are shifted in frequency by the Doppler effect (Wells and Liang 2011). Here, we only review three common applications in some detail.

The echocardiogram

An echocardiogram is a sonogram of the heart, and is routinely used in the diagnosis, management, and follow-up of patients with any suspected or known heart diseases. It is one of the most widely used diagnostic tests in cardiology. It can provide information on the size and shape of the heart, pumping capacity, and the location and extent of any tissue damage. Not only can an echocardiogram create ultrasound images of heart structures but it can also produce an accurate assessment of the blood flowing through the heart by Doppler echocardiography. This allows assessment of both normal and abnormal blood flow through the heart (for details, see Moran et al. 2013).

Prenatal diagnostics

A prenatal ultrasound test uses ultrasound that is transmitted through the abdomen via a transducer. The echoes are recorded and transformed into video or photographic images of the foetus. The ultrasound can also be used during pregnancy to show images of the amniotic sac, placenta, and ovaries. Major anatomical abnormalities or birth defects are visible through ultrasound. Most prenatal ultrasound procedures are performed with the transducer on the surface of the skin, using a gel as a conductive medium to aid in the image quality (for recent advances, see Gardiner 2018).

Abdominal sonography

In abdominal sonography, the solid organs of the abdomen such as the pancreas, aorta, inferior vena cava, liver, gall bladder, bile ducts, kidneys, and spleen are imaged (Wikipedia 2020). Sound waves are blocked by gas in the bowel and attenuated in different degree by fat, therefore there are limited diagnostic capabilities in this area. The appendix can sometimes be seen when inflamed (as in, for example, appendicitis). It is also used on the abdominal aorta to detect or exclude abdominal aortic aneurysm (Rafailidis et al. 2018).

Ultrasound as a therapeutic tool

In therapy, continuous wave ultrasound is used in applications such as physiotherapy and surgery. Relatively high-power ultrasound can break up stony deposits or tissue, accelerate the effect of drugs in a targeted area, and assist in the measurement of the elastic properties of tissue. In physiotherapy, claims have been made that ultrasound may alleviate muscle pain and improve tissue healing. However, in a systematic review, Valma J. Robertson and Kerry G. Baker (2001) found 'little evidence that active ultrasound is more effective than placebo treatment for treating patients with pain or a range of musculoskeletal injuries, or for promoting soft tissue healing. The few studies judged to have adequate methodology examined a diverse range of medical conditions and the dose of ultrasound varied, often for no discernable reason.' Here, I introduce four typical therapeutic applications of ultrasound.

Kidney stones

Focused high-energy ultrasound pulses can be used to break kidney stones and gallstones into fragments small enough to be passed from the body without undue difficulty. The procedure for kidney stones appears to be safe and effective (May, Bailey, and Harper 2016; for a historical review, see Tzou et al. 2017).

Tumour ablation

Ultrasound can ablate tumours or other tissue non-invasively. This is accomplished using a technique known as focused ultrasound surgery. This procedure uses generally lower frequencies than medical diagnostic ultrasound but significantly higher time-averaged intensities (Zhou 2011).

Topical drug delivery

Delivering chemotherapy to brain cancer cells and various drugs to other tissues is called acoustic targeted drug delivery. These procedures generally use frequencies between 1 and 10 megahertz and intensities up to 20 Watt/centimetres squared. The acoustic energy is focused on the tissue of interest to make it more permeable for therapeutic drugs (Lewis and Olbricht 2008).

Neuromodulation in the brain

Electric and magnetic signals strongly influence cells near the brain's surface but can only penetrate 1 or 2 centimetres. Ultrasound can target brain structures with high precision and penetrates deep regions in the brain selectively. However, precisely how the technique works remains unclear (Landhuis 2017). Recently, two studies have unequivocally established that low-level ultrasound applied to the brain activates the auditory midbrain and cortex in a similar way as audible sound, but at much lower levels than needed for the desired neuromodulation of non-auditory structures (Sato, Shapiro, and Tsao 2018; Guo et al. 2018). Hongsun Guo and colleagues (2018) found that stimulation with ultrasound of different auditory cortex locations elicited extensive activity throughout the auditory midbrain (central nucleus of the inferior colliculus, ICC) with no indication of localized effects. The effect disappears when the cochlea was destroyed, indicating that the inner ear itself is activated by the ultrasound. Auditory activation of the brain may activate other brain areas by cross-modal interaction. Tomokazu Sato, Mikhail G. Shapiro, and Doris Y. Tsao (2018) noted that the mechanisms that underlie activation of the auditory system are unclear but may,

> include mode conversion between primary compressive ultrasound waves and shear waves within bone and the brain's soft tissue, leading to mechanical activation of ear structures. The ultrasound pressure waves themselves may contain power at frequencies in the audible range, including broadband power due to the onset and offset of each pulse, as well as at harmonics of the pulse repetition frequency, which get propagated to the cochlea.
> (Sato, Shapiro, and Tsao 2018).

Conclusion

Sound in medicine is used as a diagnostic and therapeutic tool. Audible sound is used to test the auditory system sensitivity, perceptual qualities, and cognitive patency. In this way it allows topographic diagnostics including neurological abnormalities impinging on the auditory pathway. Combining sound stimulation with imaging techniques allows differentiation between peripheral and central mechanisms underlying age-related hearing impairment. Therapeutic use of audible sound, in particular music, is widespread but the evidence-based assessments still are sceptical for some of its claimed benefits.

Ultrasound is used as an imaging technique, the second most used after X-rays, based on delayed reflection by internal structures, with very good spatial resolution, and it allows the solid organs of the abdomen such as the pancreas, liver, gall bladder, kidneys and spleen to be imaged. Combined with Doppler techniques, echocardiography is used to diagnose the dilatation of parts of the heart and function of heart ventricles and valves. By using Doppler ultrasound analysis one can also assess the blood flow. Therapeutic use of ultrasound is abundant with notable successes in ablation of tumours and crushing kidney stones. Caution must be used in the interpretation of its neuromodulation capabilities.

More recently, Davide Folloni et al. (2019) have shown the effects of transcranial focused ultrasound (tFUS) in the macaque anterior cingulate cortex (ACC) and amygdala. They measured neuromodulatory effects by examining relationships between activity in these areas and the rest of the brain using fMRI. In control conditions without tFUS, activity in a given area was related to activity in interconnected regions, but such relationships were reduced after sonication, specifically for the targeted areas. Dissociable and focal effects on neural activity could not be explained by auditory confounds, because the auditory stimulation associated with the tFUS application ceased after the sonication, but the neural activity measurements were initiated tens of minutes later. Further, the tFUS of each area, ACC and amygdala, had specific effects that were distinct from one another. To date, tFUS is the most promising neuromodulatory technique to reach areas deep below the dorsolateral surface of the brain in a minimally invasive and focal manner, thereby

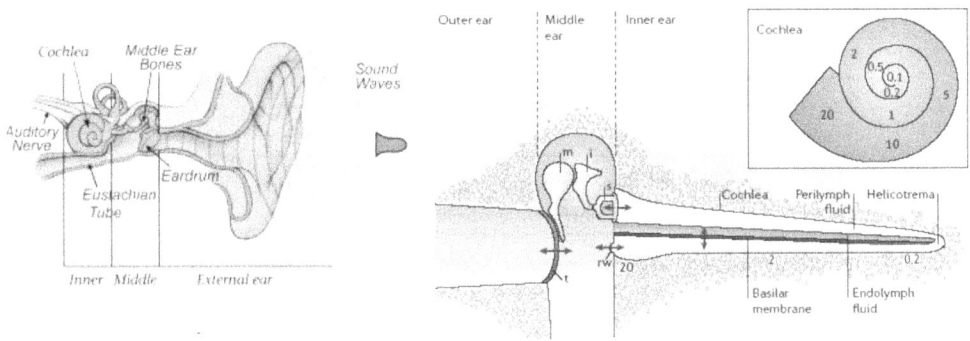

Figure 13.1 The sound conduction pathway in the human ear. Left: Schematic of the outer, middle, and inner ear. Right: A section through the temporal bone showing the sound conduction pathway in the ear. Vibrations of the eardrum or tympanum (t) are relayed via three middle ear bones, the malleus (m), incus (i) and stapes (s), and initiate pressure waves in the cochlear fluids, the pressure being relieved at the round window (rw). The pressure waves set in motion the basilar membrane, on which the organ of Corti and the hair cells are located (cf. Figure 13.2). The cochlea is shown as straight to illustrate its internal structure, but it is normally coiled as in the inset. Different sound frequencies excite different regions of the cochlea, the specific locations being given in kilohertz: from 0.1 to 20 kilohertz in humans. Note that the frequency map is logarithmic, so that each decade occupies an equivalent distance on the basilar membrane. The components are drawn roughly to scale for the human ear, in which the cochlea is 35 millimetres in length. *Source:* Fettiplace and Hackney 2006.

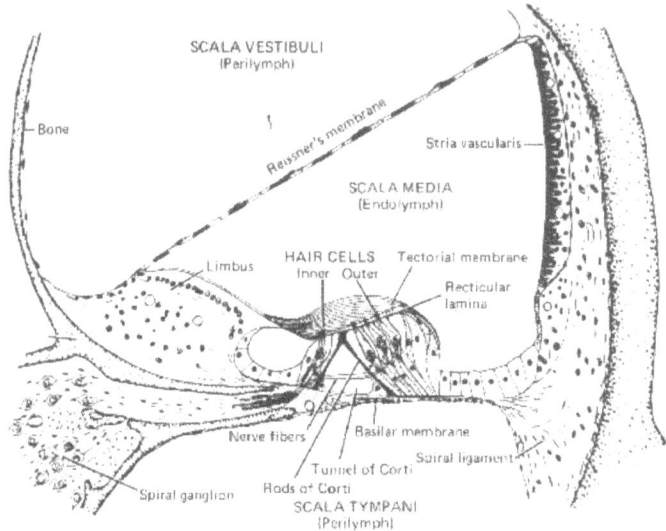

Figure 13.2 Cross section of the cochlea showing the location of the organ of Corti containing the outer and inner hair cells from which the spiral ganglion cells leave. The cochlea is divided into compartments filled with fluids of distinct ionic composition. Perilymph is similar to extracellular fluid with a high Na+ concentration, i.e. close to seawater, a reminder of our evolutionary past. Endolymph, which is found in the central compartment above the tops of the hair cells, contains a high K+ concentration. *Source:* Davis et al. 1953.

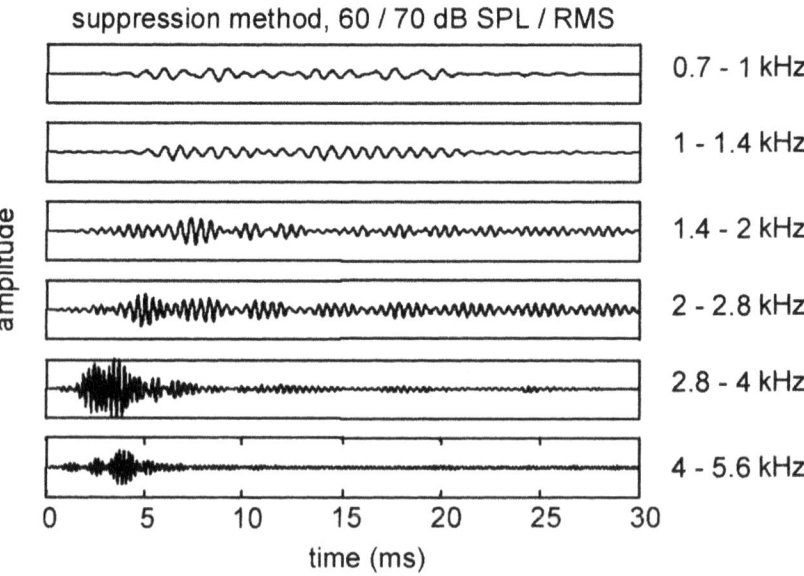

Figure 13.3 Frequency-specific click-evoked otoacoustic emission waveforms obtained from the human ear. Decomposition of wide-band otoacoustic emissions is obtained by frequency-selective masking. From top to bottom, the spectral bands are 0.7–1.0, 1.0–1.4, 1.4–2.0, 2.0–2.8, 2.8–4.0, and 4.0–5.6 kilohertz. *Source:* Molenaar, Shaw, and Eggermont 2000.

providing it with the potential for causally mapping brain functions within and across species. A study in mice demonstrated that sharp edges in a tFUS rectangular envelope stimulus activated the peripheral afferent auditory pathway, and that smoothing these edges eliminated the auditory responses without affecting the motor responses in normal hearing mice (Mohammadjavadi et al. 2019).

Studies such as Sato, Shapiro, and Tsao (2018) and Guo et al. (2018) are important for assessing the fundamentals of the tFUS field; the size of the animals may have some effect, but acoustic edges need to be considered as a major confound.

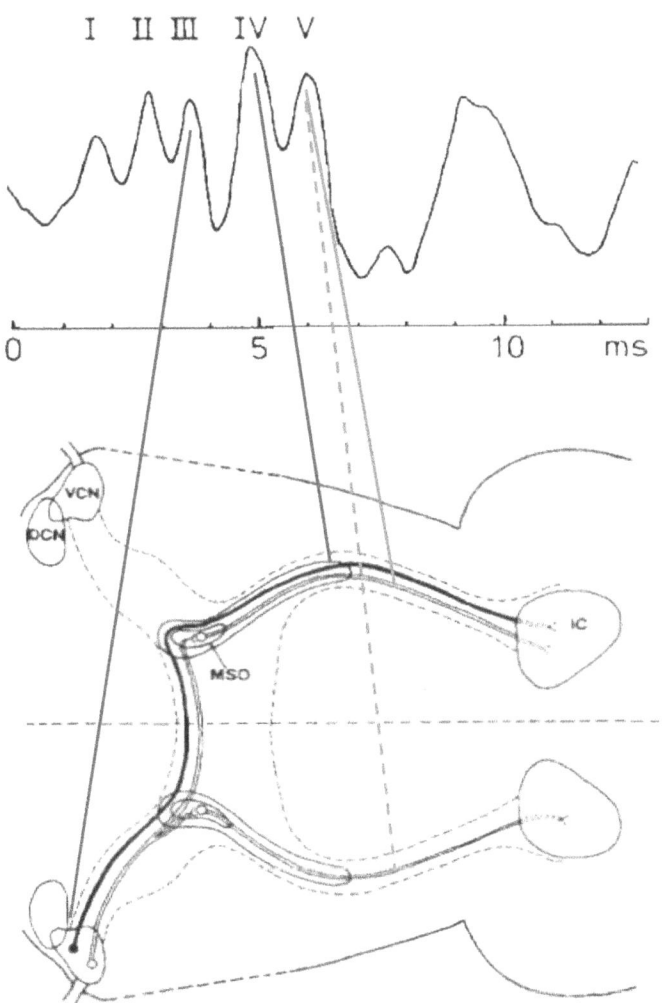

Figure 13.4 Auditory brainstem response sources in the brainstem. Wave III cochlear nucleus; wave IV dominantly from the purely contralateral axonal pathway at the bend near, or ending in, the lateral lemniscus. Wave V from the pathway synapsing in the medial superior olive (MSO) and continuing to the lateral lemniscus, dominantly contralateral to the cochlear nucleus (CN) (full line) but also a minor contribution from the ipsilateral lemniscus (dashed line). *Source:* Ponton, Moore, and Eggermont 1996.

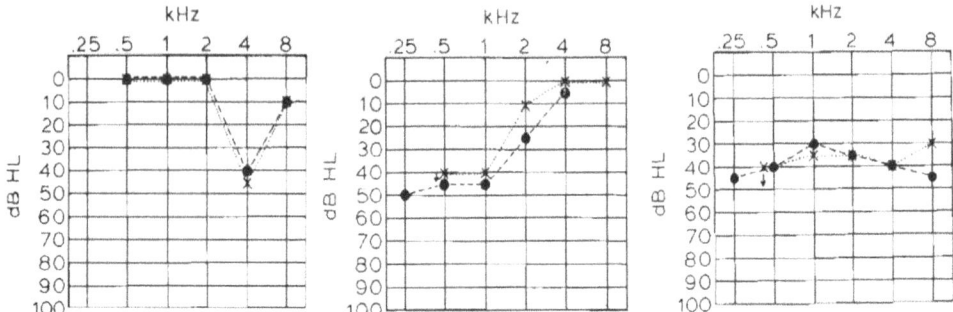

Figure 13.5 Three examples of a comparison between behavioural audiograms (large-dashed line) and audiograms based on auditory brainstem response – threshold responses (small-dashed line). The x-axes represent tone frequency (log scale), the y-axes represent sound level in decibels (dB) relative to normal hearing. *Source:* Based on Don, Eggermont, and Brackmann 1979.

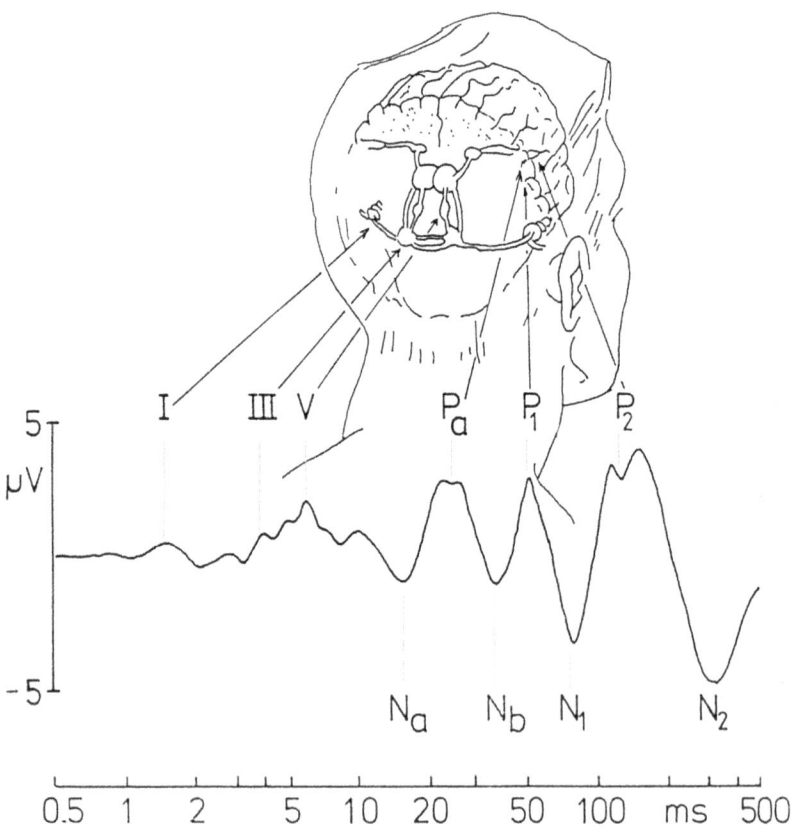

Figure 13.6 Auditory brainstem responses (ABR) and obligatory auditory evoked potentials (AEP) on a logarithmic timescale. The ABR components ('waves') are labelled I, III, and V. The middle-latency components are indicated with Na, Pa, Nb, Pb (P1). The long-latency components are indicated with P1, N1, P2, and N2. Note that Pb typically overlaps with P1. *Source:* Eggermont 2014.

References

Ashoori A., D.M. Eagleman, and J. Jankovic (2015). 'Effects of Auditory Rhythm and Music on Gait Disturbances in Parkinson's Disease'. *Frontiers in Neurology* 6: 234. https://doi.org/10.3389/fneur.2015.00234.

Bradt, J., C. Dileo, and N. Potvin (2013). 'Music for Stress and Anxiety Reduction in Coronary Heart Disease Patients'. *Cochrane Database of Systematic Reviews* 12: CD006577. https:doi.org/10.1002/14651858.CD006577.pub3.

Bunch, C. C. (1929). 'Age Variations in Auditory Acuity'. *Archives of Otolaryngology* 9: 625–636.

Davis, H. (1953). 'Acoustic Trauma in the Guinea Pig'. *Journal of the Acoustic Society of America* 25: 1180–1189.

Davis, H. (1976). 'Principles of Electric Response Audiometry'. *Annals of Otology, Rhinology and Laryngology* (Suppl. 28) 85: 4–96.

Davis, P. A. (1939). 'Effects of Acoustic Stimuli on the Waking Human Brain'. *Journal of Neurophysiology* 2: 494–499.

Dawson, G. D. (1954). 'A Summation Technique for the Detection of Small Evoked Potentials'. *Electroencephalography and Clinical Neurophysiology* 6: 65–84.

Don, M., J. J. Eggermont, and D. E. Brackmann (1979). 'Reconstruction of the Audiogram Using Brain Stem Responses and High-pass Noise Masking'. *Annals of Otology, Rhinology and Laryngology* (Suppl. 57) 88: 1–20.

Eggermont, J. J. (2007). 'Electric and Magnetic Fields of Synchronous Neural Activity Propagated to the Surface of the Head: Peripheral and Central Origins of AEPs'. In R. R. Burkard, M. Don and J. J. Eggermont (eds), *Auditory Evoked Potentials*, 2–21. Baltimore, MD: Lippincott Williams & Wilkins.

Eggermont, J. J. (2012). *The Neuroscience of Tinnitus*. Oxford: Oxford University Press.

Eggermont, J. J. (2014). *Noise and the Brain. Experience Dependent Developmental and Adult Plasticity*. London: Academic Press.

Eggermont, J. J. (2017). *Hearing Loss: Causes, Prevention and Treatment*. London: Academic Press.

Eggermont, J. J. (2019). *The Auditory Brain and Age-related Hearing Impairment*. London: Academic Press.

Eggermont, J. J., M. Don, and D. E. Brackmann (1980). 'Electrocochleography and Auditory Brainstem Electric Response in Patients with Pontine Angle Tumors'. *Annals of Otology, Rhinology and Laryngology* (Suppl. 75) 89: 1–19.

Feldmann, H. (1997). 'History of the Tuning Fork: I; Invention of the Tuning Fork, Its Course in Music and Natural Sciences'. *Laryngorhinootologie* 76 (2): 116–122.

Fettiplace, R. and C. M. Hackney (2006), 'The Sensory and Motor Roles of Auditory Hair Cells'. *Nature Reviews Neuroscience* 7 (1): 19–29.

Folloni, D., L. Verhagen, R. B. Mars, E. Fouragnan, C. Constans, J.-F. Aubry, M. F. S. Rushworth, and J. Sallet (2019). 'Manipulation of Subcortical and Deep Cortical Activity in the Primate Brain Using Transcranial Focused Ultrasound Stimulation'. *Neuron* 101: 1–8.

Gardiner, H. M. (2018). 'Advances in Fetal Echocardiography'. *Seminars in Fetal & Neonatal Medicine* 23: 112–118.

Geretsegger, M., K. A. Mössler, U. Bieleninik, X.-J. Chen, T. O. Heldal, and C. Gold (2017). 'Music Therapy for People with Schizophrenia and Schizophrenia-like Disorders'. *Cochrane Database of Systematic Reviews* 5: CD004025. https://doi.org/10.1002/14651858.CD004025.pub4.

Guo, H.,M. Hamilton II, S. J. Offutt, C. D. Gloeckner, T. Li, Y. Kim, W. Legon, J. K. Alford, and H. H. Lim (2018). 'Ultrasound Produces Extensive Brain Activation via a Cochlear Pathway'. *Neuron* 98 (5): 1020–1030.

Halliwell, M. (2010). 'A Tutorial on Ultrasonic Physics and Imaging Techniques'. *Journal of Engineering in Medicine* 224 (2): 127–142.

Hobson, J., E. Chisholm, and A. El Refaie (2012). 'Sound Therapy (Masking) in the Management of Tinnitus in Adults'. *Cochrane Database of Systematic Reviews* 11: CD006371. https://doi.org/10.1002/14651858.CD006371.pub3.

Joint Committee on Infant Hearing (JCIH) (1995). 'Joint Committee on Infant Hearing 1994 Position Statement'. *International Journal of Pediatric Otorhinolaryngology* 32 (3): 265–374.

Kemp, D. T. (1978). 'Stimulated Acoustic Emissions from within the Human Auditory System'. *Journal of the Acoustic Society of America* 64: 1386–1391.

Krick, C. M., H. Argstatter, M. Grapp, P. K. Plinkert, and W. Reith (2017). 'Heidelberg Neuro-Music Therapy Enhances Task-negative Activity in Tinnitus Patients'. *Frontiers in Neurology* 11: 384. https://doi.org/10.3389/fnins.2017.00384.

Landhuis, E. (2017). 'Ultrasound for the Brain'. *Nature* 551: 257–259.

Lewis, G. K. and W. Olbricht (2008). 'A Phantom Feasibility Study of Acoustic Enhanced Drug Perfusion in Neurological Tissue'. *Proceedings of IEEE/NIH Life Sciences and Applications Workshop*, 67–70.

May, P. C., M. R. Bailey, and J. D. Harper (2016). 'Ultrasonic Propulsion of Kidney Stones'. *Current Opinion in Urology* 26: 264–270.

Mohammadjavadi, M., P. P. Ye, A. Xia, J. Brown, G. Popelka, and K. B. Pauly (2019). 'Elimination of Peripheral Auditory Pathway Activation Does Not Affect Motor Responses from Ultrasound Neuromodulation'. *Brain Stimulation* 12 (4): 901–910. https://doi.org/10.1016/j.brs.2019.03.005.

Molenaar, D. G., G. Shaw, and J. J. Eggermont (2000). 'Recording of Transient-evoked Otoacoustic Emissions Using Noise Suppression: A Comparison with the Non-Linear Method'. *Hearing Research* 143: 197–207.

Mollison, W. M. (1917). 'A Note on the Monochord, with some Illustrative Figures'. *Proceedings of the Royal Society of Medicine (Otology Section)* 10: 39–40.

Moran, C. M., A. J. W. Thomson, E. Rog-Zielinska, and G. A. Gray (2013). 'High-resolution Echocardiography in the Assessment of Cardiac Physiology and Disease in Preclinical Models'. *Experimental Physiology* 98 (3): 629–644.

Norton, S. J., M. P. Gorga, J. E. Widen, R. C. Folsom, Y. Sininger, B. Cone-Wesson, B. R. Vohr, and K. A. Fletcher (2000). 'Identification of Neonatal Hearing Impairment: A Multicenter Investigation'. *Ear and Hearing* 21: 348–356.

Ponton, C. W., M. Don, M. D. Waring, J. J. Eggermont, and A. Masuda (1993). 'Spatio-Temporal Source Modeling of Evoked Potentials to Acoustic and Cochlear Implant Stimulation'. *Electroencephalography and Clinical Neurophysiology* 88: 478–493.

Ponton, C. W., J. J. Eggermont, B. Kwong, and M. Don (2000). 'Maturation of Human Central Auditory System Activity: Evidence from Multi-Channel Evoked Potentials'. *Clinical Neurophysiology* 111: 220–236.

Ponton, C. W., J. K. Moore, and J. J. Eggermont (1996). 'Auditory Brainstem Response Generation by Parallel Pathways: Differential Maturation of Axonal Conduction Time and Synaptic Transmission'. *Ear and Hearing* 17: 402–410,

Rafailidis, V., S. Partovi, A. Dikkes, D. A. Nakamoto, N. Azar, and D. Staub (2018). 'Evolving Clinical Applications of Contrast-enhanced Ultrasound (CEUS) in the Abdominal Aorta'. *Cardiovascular Diagnosis and Therapy* 8 (Suppl. 1): S118-S130.

Robertson V. J. and K. G. Baker (2001). 'A Review of Therapeutic Ultrasound: Effectiveness Studies'. *Physical Therapy* 81 (7): 1339–1350.

Sato, T., M. G. Shapiro, and D. Y. Tsao (2018). 'Ultrasonic Neuromodulation Causes Widespread Cortical Activation via an Indirect Auditory Mechanism'. *Neuron* 98 (5): 1031–1041.

Scarff, C. J., A. Reynolds, B. G. Goodyear, C. W. Ponton, J. C. Dort, and J. J. Eggermont (2004). 'Simultaneous 3T fMRI and High Density Recording of Human Auditory Evoked Potentials'. *NeuroImage* 23: 1129–1142.

Stein, A., R. Wunderlich, P. Lau, A. Engell, A. Wollbrink, A. Shaykevich, J-T. Kuhn, H. Holling, C. Rudack, and C. Pantev (2016). 'Clinical Trial on Tonal Tinnitus with Tailor-made Notched Music Training'. *BMC Neurology* 16: 38. https://doi.org/10.1186/s12883-016-0558-7.

Tutaj, L., D. J. Hoare, and M. Sereda (2018). 'Combined Amplification and Sound Generation for Tinnitus: A Scoping Review'. *Ear and Hearing* 39 (3): 412–422.

Tzou, D. T., M. Usawachintachit, K. Taguchi, and T. Chi (2017). 'Ultrasound Use in Urinary Stones: Adapting Old Technology for a Modern-day Disease'. *Journal of Endourology* 31 (Suppl. 1). https://doi.org/10.1089/end.2016.0584.

Wade, N. J. and D. Deutsch (2008). 'Binaural Hearing – Before and after the Stethophone'. *Acoustics Today* 7: 16–27.

Wells, P. N. T. and H-D. Liang (2011), 'Medical Ultrasound: Imaging of Soft Tissue Strain and Elasticity'. *Journal of the Royal Society Interface* 8: 1521–1549.

Wikipedia (2020). 'Abdominal Ultrasonography'. Last updated I January. Available online: https://en.wikipedia.org/wiki/Abdominal_ultrasonography (accessed 2 July 2020).

Zhang, S., D. Liu, D. Ye, H. Li, and F. Chen (2017). 'Can Music-based Movement Therapy Improve Motor Dysfunction in Patients with Parkinson's Disease? Systematic Review and Meta-Analysis'. *Neurological Science* 38: 1629–1636.

Zhou, Y. F. (2011). 'High Intensity Focused Ultrasound in Clinical Tumor Ablation'. *World Journal of Clinical Oncology* 2 (1): 8–27. https://doi.org/10.5306/wjco.v2.i1.8.

Zwaardemaker, H. (1891). 'Der Verlust an hohen Tönen mit zunehmendum Alter: Ein neues Gesetz'. *Archiv für Ohren, Nasen und Kehlkopf Heilkunde* 32: 53–56.

14

Soundscape as Methodology in Psychoacoustics and Noise Management

André Fiebig and
Brigitte Schulte-Fortkamp

Introduction

The soundscape concept, which was first introduced by R. Murray Schafer in the late 1960s, conceived of environmental noise as a musical composition (Schafer 1969). Schafer's work prefigured soundscape studies as the middle ground between science, society, and the arts: from acoustics and psychoacoustics we learn how sound is interpreted by the human brain; from society we learn how sound affects and changes behaviour; and from music we learn how to create ideal soundscapes. These three perspectives together laid the foundation of a new interdisciplinary field: *acoustic design* (Schafer 1977). This diverse footing led to a heterogeneous development of methods for investigating soundscapes. In the 1990s, the soundscape concept was increasingly used in the context of community noise and environmental noise assessment. It received more attention from noise consultants and researchers due to the notion that the perception of sound is a 'multi-stage process' and cannot be understood without studying the context and meaning of sound (Schulte-Fortkamp and Nitsch 1999). During this time, the first sessions were organized at acoustical conferences and congresses, presenting soundscape concepts and early studies to an audience of acousticians, noise consultants, and noise-control engineers, and the benefit of the soundscape concept for noise-control engineering was discussed. The first special session at an acoustic conference that focused on soundscape, titled 'Sound Amenity and Soundscape', took place in Yokohama at the International Congress on Noise Control Engineering (1994). Today, special sessions dedicated to soundscape take place regularly in renowned acoustic conferences, with presented papers ranging from theory to applications across diverse fields. Special issues about soundscape research have been routinely published in various journals, steadily increasing the number of publications in the scientific literature of the field (Kang et al. 2016). Today, over three hundred papers on soundscape, mostly non-open access, are published annually and are expanding by the year (To et al. 2018).

The concept of soundscape was adopted to provide a holistic approach to the acoustic environment beyond considerations of noise and its effect on the quality of life. The concept assesses all sounds perceived in any particular environment without simplification. To do this, soundscape studies use a variety of data collection and measurement methods related to human perception, the acoustic environment, and the interrelationships between person, activity, and place, in space and in time – defined as 'context'. The context influences the soundscape through auditory sensation, its initial interpretation, and one's subsequent responses to the acoustic environment.

When perception is 'measured' by gathering data about individual responses to the acoustic environment, different data collection methods are applied simultaneously. This increases the validity and reduces the uncertainty of the measurements. Importantly, studying soundscapes relies primarily upon human perception and only turns to physical measurements as supplemental information. Different techniques of perception and physical measurement will be discussed in this chapter.

Soundscape study draws together these different assessment tools to understand the perception of the complex sonic environment. Due to the context-related procedure in the study of the environment, soundscape studies reach a high ecological validity (Schulte-Fortkamp and Genuit 2011).

The increasing use of the soundscape approach in practical applications and systematic studies led to the development of scientific methodologies that improved the credibility of the soundscape approach. These were further accompanied by standardization efforts across the discipline, which resulted in the series of the international soundscape standard ISO 12913–1 being adopted in 2014 and its 2nd and 3rd parts in 2018 and 2019 respectively.

Development of soundscape standards

Starting in 2009, an ISO working group was formed to develop and establish standards in the field of soundscape investigations. This ran parallel with the COST TUD Action TD0804 project 'Soundscape of European Cities and Landscapes', which brought diverse soundscape researchers from all over the world together. According to the working group, it became clear that progress in soundscape research was being impeded by a lack of a clear, shared understanding of what was meant by the term soundscape, and the group decided that a clear definition was a requirement for further progress (Schomer et al. 2010). In 2014, the first international soundscape standard, the ISO 12913–1, was published, providing definitions of soundscape and related terms, as well as a conceptual framework (ISO 12913–1 2014). The standard offers a definition of soundscape reflecting the common use and concept determined by the working group consisting of several experts from different

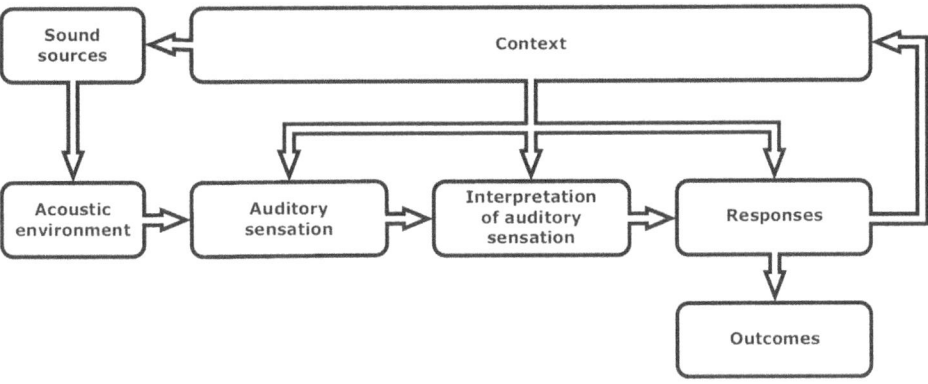

Figure 14.1 The perceptual construct of a soundscape according to ISO 12913–1 (2014).

disciplines. The standard describes soundscape as 'an acoustic environment as perceived or experienced and/or understood by a person or people in context' (ISO12913–1 2014). The conceptual framework describes the process of perceiving an acoustic environment and indicates its basic elements: *context, sound sources, acoustic environment, auditory sensation, interpretation of auditory sensation, responses*, and *outcomes*, as shown in Figure 14.1.

These elements and their specific interrelations contribute to the soundscape – the perceptual construct of an acoustic environment. The standard provides the basic elements for supporting a common basis for investigations and studies. The strong impact of this first standard on soundscape investigations and research can be estimated by considering the frequent citations of this standard in soundscape publications. For example, at the Internoise conference 2017, held in Hong Kong, over a third of the papers of the special session 'Soundscape in Architecture, Urban Planning and Landscape' quoted ISO 12913–1 and used its definitions.

The second part of the ISO series, ISO/TS 12913–2:2018, is a technical specification – which indicates that it addresses work still under technical development. This second part deals with data collection, reporting requirements, and proposing methods and tools for soundscape investigations. The proposed explorative methods are mainly based on the work done in the COST Action TD0804. The third part of the ISO series ISO/TS 12913–3:2019 dealing with the analysis of soundscape data is a further technical specification.

Soundscape methods

Due to the holistic theoretical basis of the soundscape concept, numerous methods from various disciplines have been proposed and applied in research: 'The field of soundscape investigations and projects has evolved differently across disciplines leading to multiple definitions of soundscape and its aims' (ISO 12913–1 2014). Thus, a broad diversity of methods and tools are applied in soundscape investigations according to the respective objective.

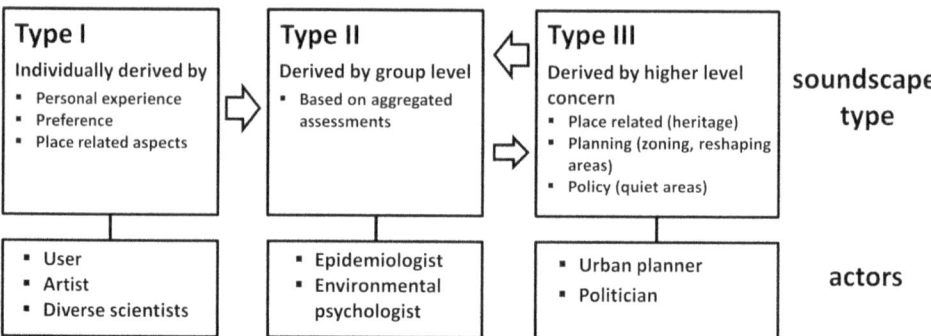

Figure 14.2 Types of soundscape studies and their main actors. *Source:* Lercher and Schulte-Fortkamp 2013.

According to Peter Lercher and Brigitte Schulte-Fortkamp (2013) generally three different types of soundscape studies can be distinguished: type I focuses on the individual, type II deals with larger groups of individuals, and type III deals with higher level concerns by studying soundscape on a large scale (see Figure 14.2).[1]

Aletta and colleagues (2016) delineate soundscape studies as varying in terms of applied methods, such as soundwalks, laboratory experiments, narrative interviews, behavioural observations, and the tools and instruments used such as questionnaires, semantic scales, interview protocols, physiological measurements, and observation protocols. In fact, diverse qualitative and quantitative methods as well as various combinations of those methods find their way into soundscape investigations. Some methods focus on the collection of perceptual data whereas others intend to measure mostly physical data. The range of applied methods is shown in Figure 14.3, illustrating the variety of different approaches available for investigating soundscapes.

The different investigative approaches apply varying measurement instruments or instrument combinations – generally, most methods apply several instruments for investigating human perception of sound. Yet measurements of perception alone do not constitute a real soundscape study because acoustic measurements do not include the most important requirement derived from the definition of the term soundscape, namely the focus on the perceptual construct of an acoustic environment (Brown et al. 2016). The integration of psychoacoustic principles provides a quantitative link between physical stimuli and their evoked auditory sensations (Fastl and Zwicker 2007). However, more parameters beyond classical psychoacoustic metrics are needed to derive information about perception related categories such as, for example, *pleasantness* and *acceptability* of sounds (Genuit and Fiebig 2016).

Currently, the most common methods for data collection applied in soundscape investigations are the soundwalk and the interview. Unfortunately, researchers often claim to investigate soundscapes when they are relying on physical measurements alone. The frequent occurrence of this misunderstanding underscores the need for the standard. Such research that only uses physical data (see Figure 14.1), with its elementary understanding

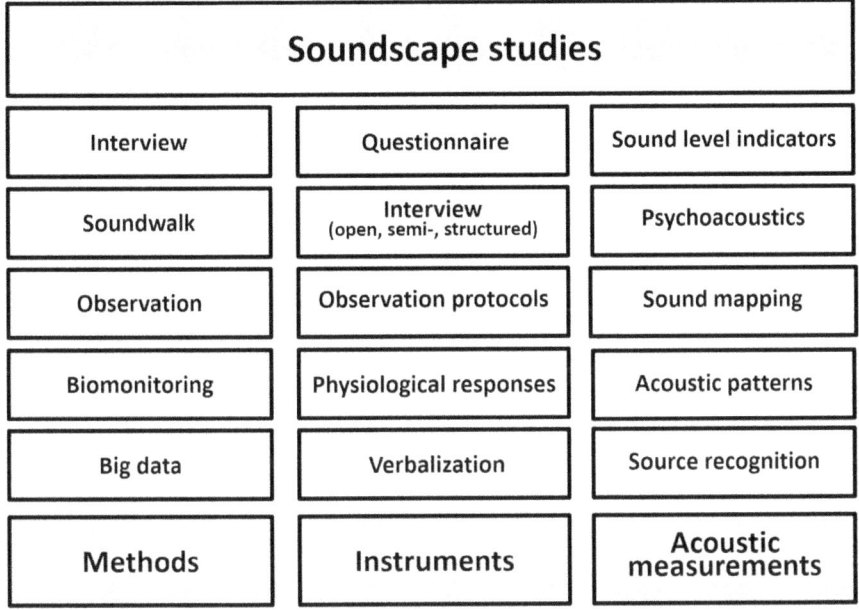

Figure 14.3 Methods and instruments frequently applied in soundscape studies.

of human perception of an acoustic environment as the leading criteria, is not sufficient to study soundscapes appropriately. The ISO standardization is designed to be a practical and useful tool in soundscape ecology (Picker 2018).

Recently, big data approaches in soundscape investigations have been proposed. For example, Aiello et al. (2016) applied a new methodology that relies on tagging information of geo-referenced pictures to the cities of London and Barcelona. They determined sound-related words from the most popular crowdsourced online sound repository, related those to social media data, and observed that the words matched most of the picture tags and offered the widest geographical coverage. Simply said, social media data was mapped onto the streets. Eoin A. King and colleagues (2017) intend to harness the potential of big data to better assess public sentiments towards soundscapes by analysing NYC311 complaints and geo-localized data mined from Twitter. A third, more direct approach is to offer online platforms where users can directly provide soundscape related data. For example, Antonella Radicchi (2018) developed a free mobile application with a focus on quiet areas to crowdsource, evaluate, and map soundscapes by collecting audio recordings, pictures, and user feedback on the location where the sounds are recorded. This data collection method allows researchers to explore correlations between a soundscape and emotional responses, semantic descriptors, perceived quietness, social communication, or personal data.

Another methodological direction is the use of biomonitoring techniques investigating and assessing human responses to (acoustic) environments by means of physiological data. According to Dick Botteldooren et al. (2016) these techniques are more objective than questionnaires, even when it comes to aesthetics or pleasure. Pyoung Jik Lee and colleagues (2018) observed in a laboratory setting that a rural area with water sounds showed the greatest psychological restoration leading to the most physiological restoration. A method that avoids redirecting the natural attention of humans specifically towards acoustic environments is non-participatory observation. For example, changes in human behaviour can be studied by observing specific sonic interventions, such as installing a fountain or broadcasting music or natural sounds in certain areas. It is generally known that humans behave more naturally if they are not informed about being observed (Fischer et al. 1984). Of course, in this context ethical issues must always be addressed. Although observational methods offer advantages regarding ecological validity, they are resource-demanding and established, robust protocols are needed to make non-participatory soundscape studies comparable for particular cases (Lavia et al. 2018).

It is expected that the detailed description of questionnaires in ISO/TS 12913–2 will lead to more harmonized data collection across soundscape studies, which in turn will improve the comparability of studies and allow the application of meta-analysis. However, since no bias-free method exists, researchers or investigators must be aware of both the advantages and limitations of the different approaches (Aletta et al. 2016) and must choose the method or combination of methods according to the research objective. The use of different methods is recommended to achieve convergent validity and overcome the shortcomings of one single method. For example, Daniel Steele at al. (2016) carried out a mixed methods study based on observation, questionnaires, and recordings, where multiple types of soundscape assessments were collected and analysed. The general idea is that one can be more confident with a result if different methods confirm prior analysis by identical results (Schulte-Fortkamp and Fiebig 2016). In particular, the combination of qualitative and quantitative methods can be productive. Empirical social research pursues different approaches to combine qualitative and quantitative data analyses. A *triangulation* approach is frequently suggested; in contrast to other approaches such as preparatory, elaboration, or generalization models (Mayring 2001), the triangulation concept treats qualitative and quantitative data equally (Fiebig et al. 2006). Accordingly, Norman K. Denzin (1978) explains that the combined use of micro- and macro-level studies, qualitative and quantitative approaches, and theoretical ideas should complement and verify each other in order to achieve robust results. Triangulation for soundscape measurement appears to be a powerful technique that facilitates the validation of data through cross-verification of three components: people, context, and acoustic environment (Schulte-Fortkamp and Fiebig 2016). However, according to Lawrence E. Marks and Daniel Algom (1998) differences between scales might be related to the nature of sensory reality and not on the vagaries of responses, underlining the difficulty in interpreting apparently contradictory results obtained by different methods. Considering the diversity of methods used in different studies, the soundscape standard needs to become known worldwide and practically applied.

Soundwalks

The most popular method in soundscape investigations is the soundwalk, where participants visit certain locations, listen to the ambient noise, and report their feelings and emotions via interviews or questionnaires. The soundwalk is an instrument to explore urban areas by 'lending ears and minds of the local experts' (Voigt and Schulte-Fortkamp 2012), providing meaningful information for appropriately interpreting any numbers and values derived from recordings. To collect such data, many researchers started over the past decade to utilize the soundwalk in diverse ways as tools for investigating soundscapes (Jeon et al. 2013), but methodological differences between various soundwalk designs are significant.

Sampling methods of participants vary from random samples (e.g. random visitors) to systematic samples (e.g. residents). The selection of specific soundwalk sites is carried out in many ways: either it is defined a priori based on explorative pre-studies, specific criteria, or the respective object of investigation; or the participants freely chose relevant sites themselves, which makes the data analysis more difficult.

Soundwalks have been performed individually as well as in groups; the latter accelerate the data collection process and are more popular. The sampling methods of the participants (random vs systematic samples [Berglund and Nilsson 2006]), the instruction of the participants (attention directed towards sounds [Adams et al. 2008] vs emphasis on multi-modality [Fiebig 2015a]), the use of open or closed questions and rating scales (varying in numerous aspects, like bipolar [Axelsson et al. 2009] vs unipolar [Jeon et al. 2018], or discrete [Lindborg 2012] vs analogue [Fiebig et al. 2010]) differ significantly. Most important, the attributes to be judged by the soundwalk participants vary as well. Based on a principal component analysis, Östen Axelsson and colleagues (2010, 2015) identified two main orthogonal components, *pleasantness* and *eventfulness*, which can be evaluated by means of a set of eight affective attributes. Other identified main factors were *comfort*, *quietness*, and *weakness* (Jeon et al. 2010). Some researchers focus on the level of *restorativeness* by collecting in-situ soundscape assessments (Payne 2013). Problems arise when the results of studies using pregiven attributes may lead to the same results. This contradicts the preferred emphasis of foregrounding people's expertise in order to meet the needs of the study, for example, an urban planning investigation.

Table 14.1 summarizes the different methodical aspects used in soundscape investigations. As a measurement method, soundwalks appear to be suitable for exploring urban areas through the minds of local experts and thus opening a field of data for *triangulation* (Schulte-Fortkamp 2013).

Interviews

Interviews are most important for investigating personal and social soundscapes. In most cases, so-called open interviews are conducted. The nature of the soundscapes can be determined effectively by examining the detailed results of open interviews, particularly when these interviews focus on the expertise of people who are familiar with the

Table 14.1 Methodical Aspects of a Soundwalk Method

Aspects	Condition
Acoustic measurements	Monaural-binaural
	Duration
	Measurement position (stationary, mobile)
Way of collecting responses	Interview type (open, structured, semi-structured)
	Ratings on scales
Sampling of participants and sample size	Visitors vs locals
	Ad hoc sample, random sample, systematic sampleNumber of participants
Duration of soundwalk	Snapshot vs 'long-term' measurement interval
Instruction	Tasks (level of attention directed towards sounds, emphasis on multimodality, etc.)
Collection of visual information	Pictures, videos

terrain under study. According to Betina Hollstein, the strong influence of sociocultural backgrounds on perception requires a heterogeneous field of research for measuring soundscape perception:

> This includes different forms of observation, interviewing techniques with a low level of standardization (such as open-ended, unstructured interviews, partially or semi-structured interviews, guided or narrative interviews), and the collection of documents or archival records (e.g., from libraries or public repositories). Despite their differences, such approaches all share common ground, as advocates of the 'interpretive paradigm' agree on certain ideas about the nature of social reality. Social reality is always a 'meaningful' reality, and by representing meaning, it refers to a context of action in which actors organize actions.
> (Hollstein 2011)

Therefore, even though fully structured interviews can be easily conducted, they should only be an addition to open interviews in much the same way as physical data. The second part of the soundscape series, ISO/TS 12913–2, provides a guideline for conducting open interviews.

An established text analysis technique is needed to analyse the data gathering in open interviews. A technique such as Grounded Theory[2] allows the analysis of data through a systematic process and resists possible criticism from strict advocates of quantitative research (Fiebig and Schulte-Fortkamp 2004). In one recent applied example, Fangfang Liu and Jian Kang (2016) used a multistep analysis technique based on Grounded Theory to study factors that affect individuals' preferences and understanding of urban soundscape based on fifty-three in-depth interviews. Using Grounded Theory for analysing interview data also enables the development of integrative diagrams where moderating factors are identified and integrated into a location-related evaluation model (Schulte-Fortkamp and Fiebig 2006).

Interview data allows an exploration of the interviewees' reality and the individual construction of the environment implicitly related to social contexts. Advocates of open interview methods claim that, in contrast to questionnaires, they can be performed without the need for detailed research hypotheses. Moreover, such data shines a light on the knowledge of involved experts and therefore provides needed information for the project, for example for succeeding in the planning phases.

The benefit of soundwalks: An example of a soundwalk study

The soundwalk method is a popular participatory method of obtaining human sensations and responses to a sonic environment (ISO/TS 12913–2 2018). In order to underline the explanatory power and reliability of in-situ assessments collected by the soundwalk method, soundwalks performed in Aachen (Germany) and their respective results are discussed in the following section. Predefined assessment sites were visited and evaluated by means of different tools and questionnaires. At the same time, psychoacoustic measurements were taken to link the in-situ assessments to the properties of the acoustic environments. The advantage of these soundwalk measurements was that, by means of the same measurement procedure, the sites investigated through soundwalking were visited several times over several years, allowing for the consideration of basic quality criteria of empirical research such as reliability. The repeated soundwalks were performed within the framework of the COST Action TD 0804 over the course of several years. The results were analysed using quantitative as well as qualitative analysis methods. The major question was whether specific sites can elicit comparable feelings and emotions over years while the visitors spent only a few minutes there.

Methods

Participants

In total fifty-seven participants (thirty-seven male, twenty female) took part in the repeated measurements. The soundwalk participants were young researchers taking part in soundscape seminars and could be considered to be non-familiar with the investigated sites.

Stimuli

Eight sites in Aachen were visited and evaluated by soundwalk group sizes ranging from four to nine participants. In total, eight soundwalks were conducted, taking place in 2010, 2011, 2012, and 2015, two per year. The sites were deliberately selected with respect to diversity and representativeness (Fiebig 2015a) and included an urban park and central square, among others. Four groups walked from site 1 to site 8, and four groups walked from site 8 to 1.

Procedure

The participants walked together with the moderator. After arriving at a study location, the moderator indicated the direction to face and the start of the active listening period to be considered for assessment. The participants were requested to listen in silence for the indicated period and to use all of their senses to experience the respective sites. After 3 minutes of silent listening to each site the participants provided assessments on an evaluation paper sheet. The participants rated their experience on a provided five-point analogue rating scale with respect to 'loudness' (*How loud is it here?*) and 'unpleasantness' (*How unpleasant is it here?*), ranging from 'not at all' to 'extremely'. Moreover, the participants could note any thoughts and feelings running through their mind after the listening period.

The soundwalk moderator performed psychoacoustic measurements with a binaural measurement system in order to record the environmental noise.

Results

The data gained was then subjected to analysis. Figures 14.4 and 14.5 display the in-situ assessments of *loudness* and *unpleasantness* in terms of average values and the 95 per cent confidence intervals for each year. The relatively large confidence intervals are due to the relatively small samples. Only rarely statistically significant differences occur between the years (Fiebig and Herweg 2017). At site 6, a significantly different assessment occurred in 2012 due to a construction site and its noise located nearby.

When comparing the assessments of a particular site judged by different groups over the years, the majority of ratings by the groups were in fact quite similar; it seems as if the sound tended to evoke similar perceptions and assessments through time. When comparing the assessments of the different sites, statistically significant differences between the sites were observed for both loudness and unpleasantness.

Statistically significant differences were observed when examining the influence of the visiting order to the different study locations. In particular, the first and last site of the soundwalk route were judged differently in the scaling task. This observation illuminates the role that a frame of reference or previous experience plays for participants, particularly at the beginning of the study because there is no predecessor to relate one's judgement to. Regarding the criterion *unpleasantness,* site 1 was judged as less unpleasant ($p<0.05$) at the first visit compared to the ratings provided at the end of the soundwalk, whereas site 8 was assessed as more unpleasant ($p<0.05$) by the end of the soundwalk compared to the ratings of the groups starting at site 8. The same tendencies were observed for the loudness judgements. It must be mentioned that the different groups did not listen to exactly the same sounds, which means that the observed differences between the groups starting at different sites cannot be fully attributed to the factor of order alone.

Figure 14.6 illustrates the relationship between loudness values determined by means of ISO 532–1 (2017) (binaural channels averaged) and the L_{Aeq}[3] with the judged

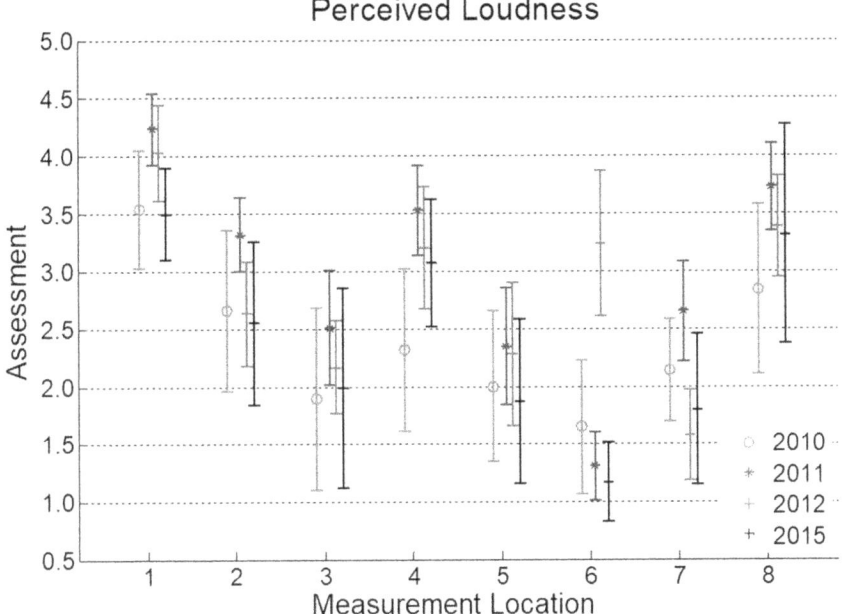

Figure 14.4 Assessments of loudness of eight sites repeatedly visited by different soundwalk groups over several years. Arithmetic mean values and 95 per cent confidence intervals are shown.

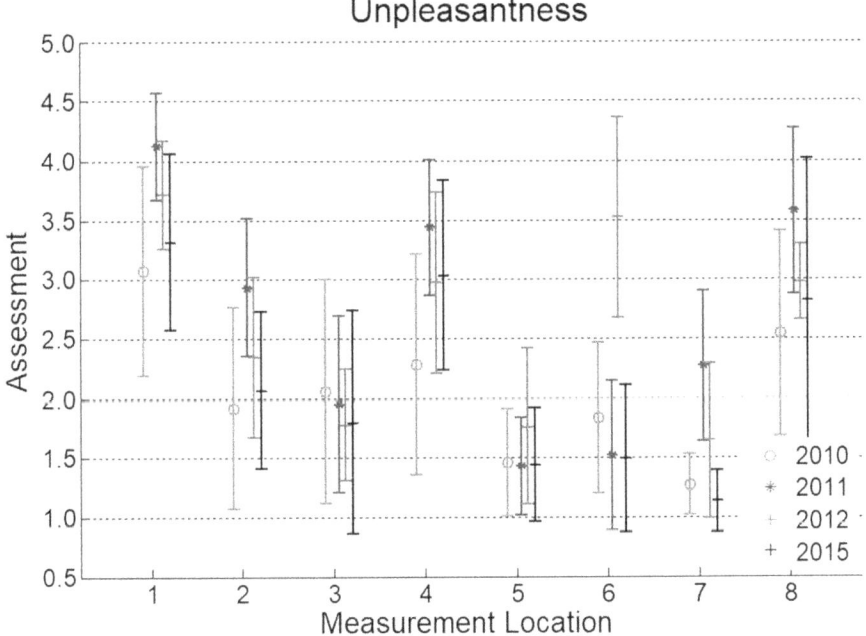

Figure 14.5 Assessments of unpleasantness of eight sites repeatedly visited by different soundwalk groups over several years. Arithmetic mean values and 95 per cent confidence intervals are shown.

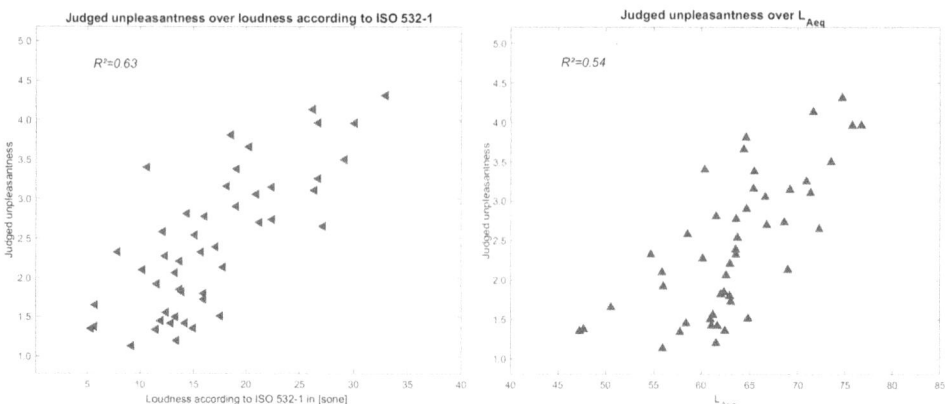

Figure 14.6 Unpleasantness group judgements over measured loudness values according to ISO 532–1 (left) and over L_{Aeq}-values in dB(A) (right).

unpleasantness (respective means of the soundwalk groups). The correlation coefficient clearly decreases compared to the prediction of loudness judgements when hedonic judgements (unpleasantness) are considered, which cannot be predicted by the magnitude of sound (loudness) alone. Psychoacoustic loudness performs better in predicting the in-situ unpleasantness data than the L_{Aeq}. If all single judgements of the *unpleasantness* data are considered, the explained variance by the loudness according to ISO 532–1 drops further down to 42 per cent, resembling the amount of explained variance in classical annoyance surveys (Guski 1999).

Discussion and conclusion

The results of the repeated soundwalks underline the general level of reliability of soundscape investigations performed in-situ under non-laboratory conditions. Since soundscape measurements take place in-situ in uncontrolled conditions and a soundscape experiment (e.g. a soundwalk) cannot be replicated in a strict sense to study objectivity or reliability aspects, the repeatability of results cannot be taken for granted. Altogether, the results of the presented repeated soundwalks suggest a reasonable reliability. It was observed that the soundscape assessments of the considered sites converge over the years and that different groups of participants provided mostly similar assessments (Fiebig and Herweg 2017). Detailed investigations on the basis of the acoustic measurements and the assessments showed a high level of reliability (Fiebig 2016). However, it should be noted that certain sites caused more variance in the assessment data over the different soundwalk measurements and were more unstable acoustically. Thus, the level of reliability and validity of soundwalk measurements cannot be universally defined but depend on the nature of the investigated site. As the order of the visited sites played a role, a variation of the order is recommended (Brambilla et al. 2017).

Future prospects

Nowadays, soundscape research has a rich research tradition and can provide tools to enhance the acoustic quality of an area. There is common consent that the perception and assessment of acoustic environments is not based on the loudness level alone, but also on the meaning of the sound and whether sounds act as carriers of information (Schulte-Fortkamp and Fiebig 2016). Still, solutions must be found for data collection methods and for identifying which kind of analysis is the best for soundscape data. Frequently researchers have sought correlations between perceptual and physical data to achieve a paradigm shift and achieve better results regarding people's needs for a good acoustic environment. But what kind of procedures will be needed when it comes to developing predictive models relevant for soundscape planning and design (Kang et al. 2016)? It is well known that 'any human action, decision making, choice, and prediction about the future is motivated by meaning' (Fiebig 2015b). This leaves us with two big steps to take in the near future: first, guaranteeing the use of platforms that will be known and available for participation, and second, improving the necessity for big data analyses. It will be important to use the soundscape approach based on the ISO standard to ensure that new findings related to any acoustic environment lead to a contextualized, common understanding of the assessment and are based on the participatory expertise of stakeholders.

Notes

1. Considering the majority of soundscape publications, type I studies are performed most. So far studies on higher type levels rarely have been done, preventing generalizations at a group level (Lercher and Schulte-Fortkamp 2013).
2. Grounded Theory is a socio-scientific analysis method with specific systematic procedures often applied to textual analysis (Strauss 1990). It is necessary to enhance the level of abstraction to improve its generalizability without any interpretation. The text must be categorized and conceptualized despite needing to look for simple paraphrases because of their lack of analytical depth. Core categories must be discovered, which then detect and explain ties and dependencies of different categories. Tentative integrative diagrams must be prepared to explore missing interfaces and inadequate knowledge about the links between the identified categories and concepts. The analysis process of coding and developing diagrams is continued until the model is saturated, i.e. new categories or concepts could not be detected in spite of coding new material (Fiebig and Schulte-Fortkamp 2004).
3. L_{Aeq} is an A-weighted equivalent continuous sound pressure level, which means that the sound pressure level is averaged over time. The L_{Aeq} metric is usually applied in all kinds of noise regulations.

References

Adams, M., N. Bruce, R. Cain, P. Jennings, P. Cusack, K. Hume, and C. Plack (2008). 'Soundwalking as Methodology for Understanding Soundscapes'. *Proceedings of the Institute of Acoustics*, vol. 30.

Aiello, L. M., R. Schifanella, D. Quercia, and F. Aletta (2016). 'Chatty Maps: Constructing Sound Maps of Urban Areas from Social Media Data'. *Royal Society Open Science* 3 (3): 150690.

Aletta, F., J. Kang, and Ö. Axelsson (2016). 'Soundscape Descriptors and a Conceptual Framework for Developing Predictive Soundscape Models'. *Landscape and Urban Planning* 149: 65–74.

Axelsson, Ö., M. E. Nilsson, and B. Berglund (2009). 'Swedish Instrument for Measuring Soundscape Quality'. *Proceedings of Euronoise 2009*, Edinburgh, Scotland.

Axelsson, Ö., M. A. Nilsson, and B. Berglund (2010). 'A Principal Components Model of Soundscape Perception'. *J. Acoust. Soc. Am.* 128 (5): 2836–2846.

Axelsson, Ö. (2015). 'How to Measure Soundscape Quality'. *Proceedings of Euronoise 2015*, Maastricht, Netherlands.

Berglund, B. and M. A. Nilsson (2006). 'On a Tool for Measuring Soundscape Quality in Urban Residential Areas'. *Acta Acustica united with Acustica* 92 (6): 938–944.

Botteldooren, D., T. Andringa, I. Aspuru, A. L. Brown, D. Dubois, C. Guastavino, J. Kang, C. Lavandier, M. Nilsson, A. Preis, and B. Schulte-Fortkamp (2016). 'From Sonic Environment to Soundscape'. In J. Kang and B. Schulte-Fortkamp (eds), *Soundscape and the Built Environment*, 17–41. Boca Raton, FL: CRC Press.

Brambilla, G., F. Pedrielli, and M. Masullo (2017). 'Soundscape Characterization and Classification: A Case Study'. *Proceedings of ICSV 24*, London.

Brown, A. L., T. Gjestland, and D. Dubois (2016). 'Acoustic Environments and Soundscapes'. In J. Kang and B. Schulte-Fortkamp (eds), *Soundscape and the Built Environment*, 1–16. Boca Raton, FL: CRC Press.

COST Action TD 0804 (2009). 'About Soundscape of European Cities and Landscapes'. Soundscape of European Cities and Landscapes Network, 2 November. Available online: http://soundscape-cost.org (accessed 1 July 2020).

Denzin, N. K. (1978). *The Research Act*. New York: McGraw-Hill.

Fastl, H. and E. Zwicker (2007). *Psychoacoustics, Facts and Models*. Heidelberg: Springer Verlag.

Fiebig, A. (2015a). 'Acoustic Environments and their Perception Measured by the Soundwalk Method'. *Proceedings of Internoise 2015*, San Francisco, CA.

Fiebig, A. (2015b). 'Cognitive Stimulus Integration in the Context of Auditory Sensations and Sound Perceptions'. PhD Dissertation, TU Berlin, Berlin, Germany.

Fiebig, A. (2016). 'Reliability of In-Situ Measurements of Acoustic Environments'. *Proceedings of DAGA 2016*, Aachen, Germany.

Fiebig, A. and A. Herweg (2017). 'The Measurement of Soundscapes: A Study of Methods and their Implications'. *Proceedings of Internoise 2017*, Hong Kong, China.

Fiebig, A. and B. Schulte-Fortkamp (2004). 'The Importance of the Grounded Theory to Soundscape Evaluation'. *Proceedings of CFA, DAGA 2004*, Strasbourg, France.

Fiebig, A., B. Schulte-Fortkamp, and K. Genuit (2006). 'New Options for the Determination of Environmental Noise Quality'. *Proceedings of Internoise 2006*, Honolulu, HI.

Fiebig, A., V. Acloque, S. Basturk, M. Di Gabriele, M.Horvat, M. Masullo, R. Pieren, K. S. Voigt, M. Yang, K. Genuit, and B. Schulte-Fortkamp (2010). 'Education in Soundscape: A Seminar with Young Scientists in the COST Short Term Scientific Mission "Soundscape - Measurement, Analysis, Evaluation"'. *Proceedings of ICA 2010*, Sydney, Australia.

Fisher, J. D., P. A. Bell, and A. Baum (1984). *Environmental Psychology*. New York: CBS College Publishing.

Genuit, K. and A. Fiebig (2016). 'Human Hearing-related Measurement and Analysis of Acoustic Environments'. In J. Kang and B. Schulte-Fortkamp (eds), *Soundscape and the Built Environment*, 133–160. Boca Raton, FL: CRC Press.

Guski, R. (1999). 'Personal and Social Variables as Co-determinants of Noise Annoyance'. *Noise and Health* 3: 45–56.

Hollstein, B. (2011). 'Qualitative Approaches to Social Reality: The Search for Meaning'. In J. Scott and P. J. Carrington (eds.), *The SAGE Handbook of Social Network Analysis*, 404–416. London: SAGE.

ISO 12913–1 (2014). 'Acoustics – Soundscape, Part 1: Definition and Conceptual Framework'. Geneva: International Standards Organization.

ISO/TS 12913–2 (2018). 'Acoustics – Soundscape, Part 2: Data Collection and Reporting Requirements'. Geneva: International Standards Organization.

ISO/TS 12913–3 (2019). 'Acoustics – Soundscape, Part 3: Data Analysis'. Geneva: International Standards Organization.

ISO 532–1 (2017). Acoustics – Methods for calculating loudness – Part 1: Zwicker method. Geneva: International Standards Organization.

Jeon, J. Y., J. Y. Hong, and P. J. Lee (2013). 'Soundwalk Approach to Identify Urban Soundscapes Individually'. *Journal of the Acoustic Society of America* 134 (1): 803–812.

Jeon, J. Y., P. J. Lee, J. You, and J. Kang (2010). 'Perceptual Assessment of Quality of Urban Soundscapes with Combined Noise Sources and Water Sounds'. *Journal of the Acoustic Society of America* 127 (3): 1357–1366.

Jeon, J. Y., J. Y. Hong, C. Lavandier, J. Lafon, Ö. Axelsson, M. Hurting (2018). 'A Cross-national Comparison in Assessment of Urban Park Soundscapes in France, Korea, and Sweden through Laboratory Experiments'. *Applied Acoustics* 133: 107–117.

Kang, J., F. Aletta, T. T. Gjestland, L. A. Brown, D. Botteldooren, B. Schulte-Fortkamp, P. Lercher, I. van Kamp, K. Genuit, A. Fiebig, L. Bento Coelho, L. Maffei, and L. Lavia (2016). 'Ten Questions on the Soundscapes of the Built Environment'. *Building and Environment* 108 (1): 284–294.

King, E. A., S. Punla-Green, and S. Genovese (2017). 'Soundscapes, Social Media, and Big Data: The Next Step in Strategic Noise Mapping'. *Journal of the Acoustical Society of America* 141 (5): 3622.

Lavia, L., H. J. Witchel, F. Aletta, J. Steffens, A. Fiebig, J. Kang, C. Howes, and P. G. T. Healey (2018). 'Non-participant Observation Methods for Soundscape Design and Urban Planning'. In F. Aletta and J. Xiao (eds), *Handbook of Research on Perception-driven Approaches to Urban Assessment and Design*, 73–99. Hershey, PA: IGI global.

Lee, P. J., S. H. Park, T. Jung, and A. Swenson (2018). 'Effects of Exposure to Rural Soundscape on Psychological Restoration'. *Proceedings of Euronoise 2018*, Crete, Greece.

Lercher, P. and B. Schulte-Fortkamp (2013). 'Harmonising'. In J. Kang, K. Chourmouziadou, K. Sakantamis, B. Wang, and Y. Hao (eds), *Soundscape of European Cities and Landscapes*, 118–127. Available online: http://soundscape-cost.org/documents/COST_TD0804_E-book_2013.pdf (accessed 1 July 2020).

Lindborg, O. M. (2012). 'Correlations between Acoustic Features, Personality Traits and Perception of Soundscapes'. *Proceedings of the International Congress on Music Perception and Cognition*, Thessaloniki, Greece.

Liu, F. and J. Kang (2016). 'A Grounded Theory Approach to the Subjective Understanding of Urban Soundscape in Sheffield'. *Cities* 50: 28–39.

Marks, L. E. and D. Algom (1998). 'Psychophysical Scaling'. In M. H. Birnbaum (ed.), *Measurement, Judgment and Decision Making*, 81–178. San Diego, CA: Academic Press.

Mayring, P. (2001). 'Kombination and Integration Qualitativer and Quantitativer Analysis'. *Forum Qualitative Social Research* 2 (1): Art. 6.

Payne, S. R. (2013). 'The Production of a Perceived Restorativeness Soundscape Scale'. *Applied Acoustics* 74: 255–263.

Picker, J. M. (2018). 'Soundscape(S): The Turning of the Word'. In M. Bull (ed.), *The Routledge Companion to Sound Studies*, 147–157. London: Routledge.

Radicchi, A. (2018). 'The Use of Mobile Applications in Soundscape Research: Open Questions in Standardization'. *Proceedings of Euronoise 2018*, Crete, Greece.

Schafer, R. M. (1969). *The New Soundscape: A Handbook for the Modern Music Teacher*. Scarborough: Berandol Music Limited.

Schafer, R. M. (1977). *The Soundscape: Our Sonic Environment and the Tuning of the World*. Rochester: Destiny Books.

Schomer, P., A. Lex Brown, B. de Coensel, K. Genuit, T. Gjestland, J. Y. Jeon, J. Kang, P. Newman, B. Schulte-Fortkamp, and G. R. Watt (2010). 'On Efforts to Standardize a Graphical Description of the Soundscape Concept'. *Proceedings of Internoise 2010*, Lisbon, Portugal.

Schulte-Fortkamp, B. (2013). 'Soundscape – A Matter of Human Resources'. *Proceedings of Internoise 2013*, Innsbruck, Austria.

Schulte-Fortkamp, B. and A. Fiebig (2006). 'Soundscape Analysis in a Residential Area: An Evaluation Combining Noise and People's Mind'. 'Soundscapes', special issue of *Acta Acustica* 96 (6): 875–880.

Schulte-Fortkamp, B. and A. Fiebig (2016). 'The Impact of Soundscape in Terms of Perception'. In J. Kang and B. Schulte-Fortkamp (eds), *Soundscape and the Built Environment*, 69–88. Boca Raton, FL: CRC Press.

Schulte-Fortkamp, B. and K. Genuit (2011). 'Soundscape Design and its Procedure'. *Proceedings of the ASJ Conference 2011*, Tokyo, Japan.

Schulte-Fortkamp, B. and W. Nitsch (1999). 'On Soundscapes and their Meaning Regarding Noise Annoyance Measurements'. *Proceedings of Internoise 1999*, Fort Lauderdale, FL.

Steele, D., E. Bild, C. Tarlao, I. Luque Martín, J. Izquierdo-Cubero, and C. Guastavino (2016). 'A Comparison of Soundscape Evaluation Methods in a Large Urban Park in Montreal'. *Proceedings of the International Congress on Acoustics 2016*, Buenes Aires, Argentina.

Strauss, A. L. (1990). *Qualitative Analysis for Social Scientists*. Cambridge: Cambridge University Press.

To, W.M., A. Chung, I. Vong, and A. Ip (2018). 'Opportunities for Soundscape Appraisal in Asia'. *Proceedings of Euronoise 2018*, Crete, Greece.

Voigt, K. S. and B. Schulte-Fortkamp (2012). 'Quality of Life – Why Does the Soundscape Approach Provide the Correct Measures?'. *Proceedings of Internoise 2012*, New York, USA.

15

Sonic Methodologies of Sound

Salomé Voegelin

THE TAPERECORDERS are on the floor in front of the TV, their counters turning from one to two. It is four hours later. 'We found the tapes,' Marcia told Joan when they went to see her after supper. 'We are going to play them as soon as we get back. We are going to have a concert.' Marcia claims that Joan clicked her tongue at that. Nobody else heard, but they are all feeling encouraged.

—Barbara Gowdy (1997: 245)

The scope of this chapter as it is indicated by the title might seem self-evident. The expectation being that sound practices and studies by necessity and logic realize themselves through sonic methodologies; that their method of investigation, doing, and interpretation refer themselves to sonic vocabularies and theoretical tools that draw on sonic initiatives and experiences. However, this impression soon gives way to a different insight which shows that sonic practices, from music to audiology, more often than not do not trust their own 'immaterial' base, but seek a visual framework and language to develop their tools of investigation and interpretation in order to confer reliability, repeatability, and consensus.

From an emphasis on the visual score, maps, and spectograms to the philosophical integration of the heard into visual epistemologies, sound cannot take for granted that its material, experience, and expression will be studied and discussed on its own terms. Instead it has to accept that due to historical cultural and disciplinary boundaries and expectations it is the knowledge and thinking of a visual logic that will more often than not frame what it does.[1] This chapter engages in this apparent contradiction and investigates the scope of a sonic understanding of sound: making suggestions about what such a methodology might be and how it might contribute to a current scheme of what we know and how we know it.

Therefore, this text deliberates the tendency to conceptualize and catalogue the sonic in visual terms, and the propensity to ventriloquize sonic practices and sound with a visual voice. Subsequently and in response it will look to sound art and its discourse to propose a different approach and different thinking. In particular Annea Lockwood's composition *Amazonia Dreaming* (1987) will serve as the immaterial ground and practice-based framework for the development of these suggestions.

The motivation for this writing lies not in a critique of the visual and visual methodologies, however, but in the curiosity as to why theory and research eschew sound; an interest to promote its scope; and an intrigue with where the sonic, taken as a legitimate, reliable, and shareable methodology might take us in our understanding of sound and the world heard. What new knowledge we might find, and how we might be able to question the capacity and partiality of our knowledge base by developing tools from the ephemeral itself.

Grasping the ephemeral

'We are all in it,' Doris whispers amazed. She has picked out Gordon saying *orange* and *peanuts*, Sonja saying *nostrils, father* and *jeepers*. Gordon has heard *nostrils* and *jeepers*. *Peanuts*, to him was *penis* but he instantly decided it must be *Venus*. He is concentrating on the second tape, which is playing a short passage of murmured words whose rhythm is syncopated to the 'Mister Sandman' rhythm.

(Gowdy 1997: 246)

The recent turn towards sound in art, theory, cultural studies, philosophy, anthropology, etc., as well as in technology and science fields, has developed on at least two parallel tracks. These tracks are split by methodological and ideological distinctions, between those who seek to understand sound on its own terms, to reach the knowledge of its invisible flow, and those who accommodate sound within conventional (visual) knowledge frames. Those, in other words, that pursue an epistemological assimilation of the invisible into the seen, and those that try to grasp sound despite and against the visual bias and vocabularies. Examples of this distinction are most clearly articulated between theorists who follow a non-cochlear/conceptual path for sound studies or who seek to establish an auditory cultural studies within which sonic artefacts, occurrences, and their contexts and technologies are subject to a cultural rather than a material study (for example, Seth Kim-Cohen, Jonathan Sterne, Michael Bull) and those whose theories and ideas pertain to listening and hearing the material itself, in relation to its virtual flow or its affective power (for example, Christoph Cox, Marie Thompson, Steven Goodman). However, in a possibly involuntary and maybe unavoidable confliction these different paths cross and embrace each other on the pinhead of theory, which ultimately brings their aim level, because as Michael Eng reminds us: this 'desire for sound is still a theoretical desire – a desire of and for Theory' (Eng 2017: 317). And as Adriana Cavarero points out, theory is logocentric. Therefore, theory's access to sound is always a takeover of its sounding, an assimilation of ungraspable noise into devocalized lexical thought. It is a visual and mute pursuit: 'Freed from the acoustic materiality of speech, this pure semantic – which is the privileged object of *theoria* – occupies the place of origin and rules over the phonetic' (Cavarero 2005: 57). Theory then, even of the sonic, promotes a visual and mute thinking of the world. In its hold things are 'this' or 'that', visible and boundaried, framed by referents and distance, rather than sounding speechlessly their invisible and indivisible in-between. Nobody and nothing finds a voice in theory but theory, as it reverberates with its own visuocentric origin and scope. Therefore it is not in the battle between the non-cochlear-conceptual and the material-affective-ontological, nor in the establishment of one certain scholarly territory for the study of sound that the sonic will find its methodology. Rather, it is in their mobile in-between and on the fluid invisible terrain of what their approaches reveal about theory, theoretical thinking, its scope, hierarchy, and authorship that a sonic methodology of sound might find its attitude and where it might practice its own rhythm.

A focus on sound is a focus on the unseen and mobile in-between of things. It stages the world and thought, history and geography as an indivisible volume. This volume is not a measure of decibels but the viscosity of an expanding and connecting sphere: a quasi-aquatic cosmos. In this viscous sphere space and place are not designed by boundaries and differences, walls and doors, but as an indivisible expanse in which we live together as interbeings: as beings with and of each other and other things, whose meaning and sense derive from this lived co-laboration[2] and whose investigative methodology has to co-habit this cosmos.[3]

In sound the world is not made of 'this' or 'that', but is a dimensionality in which things inter-are and where the practice of contingent connecting rather than the shape of separate objects, things, and subjects grants access to knowledge. In this invisible dimension theory is not this or that either, but inter-is as a practice of connecting and expanding what are seemingly oppositional stances from their mobile in-between and in its indivisible sphere in whose viscosity their thoughts co-laborate.

Therefore it is not about the difference between those that seek to write a history or geography of the world in which sound provides new insights and serves as an additional signifier to a visual event or location, and those that seek the (affective-) material of the world to theorize its virtual condition. Instead, it is through the practice of the indivisible volume of sound as a sonic possible world that we reach beyond locations on a map or dates in a chronology, and that we get further than an epistemology of the invisible or an ontology for a materialist articulation of the unseen. In this way we step into the dark simultaneity of interactions and interagencies of all there is; into the world's invisible dimensionality, where history and geography as well as philosophy are the doing and digging of practical thought that yields unlived times and unknown lands[4]: the seemingly impossible possibilities of that which we did not think we could think, include, consider, do, or talk about because it cannot be theorized as 'this' or 'that', but moves unseen in-between.

An issue in this task of considering a sonic methodology of sound is then not only sound itself but the practical engagement its indivisible invisibility encourages us into: to practice a perceptual attitude and sensibility that works on the viscous interbeing of the ungraspable prior to its organization, without evoking mysticism and without grasping it, but by participating in its co-production. Sound as perceptual attitude and sensibility demands a creative movement towards the world. And its invisible vagueness asks that we strain towards what might be left unheard even, where *Penis* might mean *Venus,* and where the patriarchal referent gets swallowed between *nostrils* and *jeepers*.

What is needed is a practical methodology that defies the quietism of theory, and that as a practice of the unseen and mobile in-between – singing, dancing, gardening, and digging – might achieve a generative sense of things. However, a sonic methodology is not a theory as access to, but is access as possibility of. Therefore, the knowledge generated in this practice does not grant access to a sonic world as a certain place, it does not grasp its territory but opens the possibility for vague, mobile, and contingent understandings that challenge the privilege and certainty of existing concepts and categories of the real.

According to Rosi Braidotti this uncertain activity happens at the margins: 'the center is void; all the action is on the margins' (Braidotti 2011b: 42). The margins are where listening becomes a generative field, and where gardening rather than theory produces the reality of a lived world.[5] From these outskirts the actual looks like a construct of power and authority rather than a lived actuality. And even though I have still to gain a voice, and my singing and dancing at the margins remains as yet without impact, at least the overwhelming muteness at the centre has lost its persuasion, and has revealed itself as a naturalized rather than a natural position.

We can try such a sonic sight by squinting our eyes and refocusing the now vague outline of things from the viscosity of their mobile in-between. In this way we gain a sonic sensibility and concept that starts from our interbeing with things and that does not seek distance or absence to get a clear view, but practices a generative doing and closeness to understand what might be there.

In terms of critical thought, 'squinting' provides knowledge that is not based on the distance of a (mathematical) new materialist sense: that does not rely on the absolute truth of the unthought and of calculability, or on a virtual flow, and that does not seek to categorize and structure from the hyper-invisible centre of a visuocentric logic.[6] Instead, it acts from fluid margins with the responsibility to hear difference without distance in what so far remained inaudible because unrepresentable, without simply representing it but by questioning the status and logic of representation and practising another tune: tuning in to the reach and tools of logocentric knowledge, representation and theory, and opening it, through the creative diffraction of its knowledge base, to reveal what remains without impact and consequence.[7] In this respect it is acknowledged, with Jonathan Sterne, that there is no pure hearing and listening that could achieve and take account of a sonic methodology of sound: 'Both listening and technology are prior to hearing and investigating the scene of audibility always reveals power relations that subtend its most basic sonic possibilities' (Sterne 2015: 72).

Rather than taking this cultural and technological always and already of aural (and incidentally also visual) perception as a stumbling block to a sonic methodology of sound, it is exactly because sound is the invisible site of power played out in-between people and things that it compels a critical approach from its own voice. And it is because the auditory is conventionally reached and communicated through technology, scores, spectrograms, and maps, made into a state of quantifiable objectivity, that a sonic methodology is needed to hear and address the mediation and make 'visible' and 'sensible' the power relations that span its possibilities. In this way what remains impossible because ephemeral, indivisible, and incalculable, outside the dominant flow, can enter scientific discourse and practice and show us an expanded possibility of the actual.[8]

The access to this impossible lies not in the spectrogram or the score, it does not lie in theory, neither structuralist or materialist, but in the activities on the margins and in sound art, as a practice that owes nothing to knowledge and delivers so much in relation to how we could understand the world.

The knowledge of sound art

> What is the voice saying? He is about to get up and fiddle with the dials when the voice says distinctly and at such a high volume that it sounds shouted – 'YOU CAN KEEP A SECRET, CAN'T YOU?'
>
> (Gowdy 1997: 246)

Jean-Luc Nancy's often quoted question about the secret that is at stake when we truly listen, when we focus on the sonority rather than the message of a sound, excites possibilities for a practice that can have words without erasing what it talks about; that can keep a secret while telling it (Nancy 2007: 5). This has to be a language that points at its own limits and deformities, that practices a 'cut in the un-sensed [in-sensée]', where we do not hear the source as a quasi-visual and complete appearance or sign but hear the scars and intersections that make a fragile form: 'A friction, the pinch or grate of something produced in the throat, a borborygmus, a crackle, a stridency, where a weighty, murmuring matter breathes, opened into the division of its resonance' (27). What we hear in such a listening is our deviance from norms and expectations, our solitude and our desire for intimacy and sharing, our relationship to other humans and nonhumans, and our failure to be accurate and objective. Listening at the cut in the unsensed is anti-foundational, in limbo, unreliable, and thus apparently it has to be deformed if it is to enter into scientific discourse: represented as scores, spectrographs, according to a referent and grasped by theory, which provide an objective ear and signal our 'dis-ability' in the face of their clear and communicable intelligence. However, as a practical knowledge the indivisible breath of murmuring matter describes another openness to meaning, where our dis-ability is our scope for plural stories, for the inclusion of the fantastical, dreams, and impossibilities into the register of the real. It offers the possibility of a practical sense that reveals itself in the in-between, in the viscosity of an indivisible world, where the spectrogram inhabits the same volume as my formless ears and the two can negotiate and generate a terrain between actuality, possibility and the apparently impossible.

Sound art undertakes this negotiation. It tells a secret while keeping it. It pursues a practical knowledge that enrols scores, drawings, photographs, texts, as well as maps, instrumentation, temporality, etc. as elements of its methodology without insisting on their scientific solidity or unambiguous communication. In the context of the audible work these maps and scores do not turn into visual objects but keep an invisible shape. Mobile and elastic, they bend into possibilities of what they could be. And we too can remain formless in its listening, certain of our uncertainty, participating in the interbeing of ears, body, culture, technology language, and sense.

> Experimenting with a borrowed snare drum, I became aware of sonorities available from almost every part of the instrument – a much wider range than I'd associated with it previously. They suggested delicacy, close focus, a sensual world reminiscent of certain vocal sounds, thus, a duet for hands and voice. Using a plastic jar lid, chopsticks, marbles and various mallets, the player sounds every part of the instrument in a variety of ways without using any of the traditional gestures; the snares are only engaged at the end.
>
> (Lockwood 2017)

Annea Lockwood's composition *Amazonia Dreaming* from 1987 is one such engagement with the knowledge of secrets. The voice on her recording does not say anything but is in duet with other things: 'plastic jar lid, chopsticks, marbles and various mallets'. They interare and co-laborate on the production of an event that is not on a score or a timeline but between things and subjects as things, performing the instrument without tradition but in its sonority. The work's sonic methodology is one of exploration and experimentation. It does not focus on the voice or the instruments in their separate purpose and virtuosity, but on what they propose to be. And in the viscous space of sound they perform their relationship as a contingent knowledge of what it is they suggest.

Listening, I can participate in this experimentation. I can explore the invisible space between myself and the record player, its turning motion, the slight crackle, where 'murmuring matter breathes, opened into the division of its resonance' (Nancy 2007: 27). Where nothing is by itself but everything is known through its interbeing: the being of myself in this room of sound as a volume of things together, human and nonhuman, whose sense of things is contingent: the rhythm of their interaction is a co-laboration that creates the reality of the moment as an interactuality. This listening is not an uncritical indulgence in a solipsistic trance. Instead it is a critical and social listening, aware of the power relations revealed by audibility and conscious of my interbeing in its volume and thus of my accountability to what I hear and what I ignore. However, this critical attitude does not mute the heard, it does not replace hearing with a referent neither musical nor linguistic. I do not think about it in theoretical terms but remain together with it, part of its rhythm and viscosity, and glean a sense of things from this interactuality.

The piece sounds rhythms from touching, from the friction of the in-between and the grating movement of things against each other. Body and instrument are produced not as this or that but in their interbeing as a being with and of each other. Furthermore, they do not perform the centre of their discipline and actuality, understood as conventional instrumentation or according to linguistic meaning, but as marginal activities of the voice and the instrument as sound, playing their formless form. Their movements do not make music but experiment with the vagueness of being sound without a referent. They expand what might be heard through their 'dis-ability' in view of expectations of virtuosity and the recognizable, providing a practical sense and demanding a practical engagement and attitude to the possibilities opened in their unkept sonority.

I get to this understanding of the work by performing its movement: hearing the potential of friction, touching, and the grating of a circular turn in the voluminous sphere of my listening, and in the tiger's eyes pictured on the album cover. In sound, its eyes as well as the image of the Amazon rainforest on the other side cease to be representations and become part of the heard attaining a sonic sensibility and a blurred focus. They are parts but not components of the whole. The composition is not a *Gesamtkunstwerk*, the album not a concept album. The voice, the chopsticks, the snare drum, my body, the record player, the tiger's eyes and the Amazonian trees are autonomous elements of the sound artwork that agitate the invisible viscosity of the in-between and open the work's critical audibility to practical experimentation: to digging, dancing, singing, and gardening.

These autonomous elements move on the margins without unifying into one whole. Instead they stick out, open and break through what a whole might be. And so listening

I do not sit in a homogenous room defined by walls, ceilings, doors, and windows, but in the volume of the work and the architecture of my location as a heterogeneous space and possible world of complex and even contradictory interaction and interbeing. Here reality is not centred but marginal; the contingent interactuality of what there is: an experimentation of the real that has no reference and whose theorization would render it mute, but whose practice resounds with the possibilities of an unseen and unmapped world.

It is in narrating this experimentation rather than theorizing it that I reach the knowledge of the work as its continuation. I have to become a storyteller to retain the ambiguity and the elasticity the work produces, and to participate in rather than grasp what the work does. This storytelling is not literature, it does not produce the fiction of parallel worlds. Literary worlds are created from elements of the primary world they relate to, but they always remain autonomous from the actual world and its ontology, its causes and consequences: 'fictional worlds are based on a logic of parallelism that guarantees their autonomy in relation to the actual world' (Ronen 1994: 8) and that guarantees the autonomy of the actual world in relation to fictional worlds. They remain a proposition rather than an action, and while they can fictionally thematize and discuss real events, their interests and ideologies, they are unable to intervene in their construction.

By contrast, the telling of a sonic story is the telling of sonic possible worlds that do not remain autonomous from the actual world, merely a parallel fiction, but that show the limits of its concepts and categories, and that augment its thinking and knowledge by generating and gesturing towards an invisible real. Sonic fictions are not limited to an actual ontology or to the ontology of literary worlds, but produce the non-ontology of an unseen world as a sonic science fiction that yields the insight of unseen lands as unknown lands that once heard can become part of our present.

To narrate these sonic science fictions and make their knowledge count, we might use Luce Irigaray's gesture-words and 'appeal to language as a path towards sharing the mystery of the other rather than mute its voice' (Irigaray 2001: 20). Through her notion of words as caress, language regains the physical, its resonance on the body, and therefore also its sound: 'This *touching upon* needs attentiveness to the sensible qualities of speech, to voice tone, to the modulation and rhythm of discourse, to the semantic and phonic choice of words' (Irigaray 1996: 125, emphasis in the original). Such gesture-words produce, through their reverberation on the body and on things, the imagination and experience of the invisible volume of sound. And from within this volume they do not act as descriptive references but produce the reality of the indivisible interactuality of interbeing bodies, things, sounds, and voices.

To amplify these gestures we might sing scholarly texts and revocalize what is theorized already, re-sounding mute categorizations and realities to their own rhythm. We might revocalize history, geography, and philosophy by adding our own voice tone into the volume of concepts and ideas that are its theorizations. Singing 'a short passage of murmured words whose rhythm is syncopated to the "Mister Sandman" rhythm' (Gowdy 1997: 246) we could create an interactuality that would not sound as inferior to the clarity of the spectrogram, theory, or the score, but signal the plural scope of our dis-ability to render perfectly. Since, our dis-ability to produce a communicative intelligence enables the possibility for the creative 'actualization of multiple ecologies of belonging' (Braidotti 2011b: 41) and of multiple ecologies of perception and production.

A sonic methodology of sound is thus not a theory but a plural practice of touching and singing: the practice of writing as a making of gestures that sound the contact between theory, work, and experience in an indivisible sphere; and the practice of sounding the rhythm and tone rather than the outline of the work, as a creative actualization of the composition. In this way sound, the composition, is not heard as a sum of components, grasped by theory, but as autonomous elements continued in words and in song that show fissures, contradictions, and disagreement, where things keep on sticking out, denying a certain shape but produce a work nevertheless. Thus we could promote as a sonic methodology of sound the creative movement of the voice and of words, understood as a diffractive mechanism that hears differences without making them and includes secrets without divulging them.

Instead of hoping we will come to a sonic understanding of the world by coercing listening and hearing into (visual/logocentric) theory we could turn to the work of Irigaray, Barad, and Braidotti, in its emphasis on gesture-words, diffraction, and creativity, to find tools to approach the invisible and for it to gain its own voice. And we could enrol literature and song in Cavarero's call for the revocalization of theory to secure a different voice for the heard from the actual fictions of sonic possible worlds. I am thinking here particularly of Barbara Gowdy's writing, whose book *Mister Sandman* draws from the invisible, the fantastical, and the imaginary the possibility of a different knowledge, and which accompanies and narrates this essay through citation. And I am also thinking of Steven Feld whose anthropological narrations produce a knowing as acoustemology: as 'the experience and agency of listening (hi)stories, understood as relational and contingent, situated and reflexive' (Feld quoted in Novak and Sakakeeny 2015: 15).[9]

Accessing all the stories

However fruitful and exciting both singing and writing fiction could be in bringing an artistic and auditory imagination to knowledge and to promote a sound studies based on sonic methodologies of sound I don't know if they will entirely succeed. I am not sure that they will be accepted into discourse to solve the contradiction this chapter started with: the expectation that sonic phenomena and practices are investigated via sonic methodologies, when in fact more often than not they are approached and interpreted within visual thinking, languages, and referents.

The problem is that they are both and each easily dismissed as pertaining to the mythical, the feminine; to a pre-enlightenment fantasy of the non-representational that defies scientific logic and a calculable world, extravagant in its refusal of consensus and repetition; or that they are plain goofy.[10] Sound, a sonic sensibility, is the sorceress in the room of masculine, mute thinking, and the elephant in the footnote of academic discourse. As long as we can grasp the sonic with theory, graphs, notations, and frequencies that are the bricks of a conventional knowledge base which stay mute themselves while categorizing what sounds; and as long as we achieve academic expectations of clarity and communicability, we can and will ignore sound's unwieldy power to think things differently. But in doing so we also ignore its plural songs and what Doreen Massey terms the 'simultaneity of stories-so-far' (Massey 2005: 9–10) that resonate within its knowledge. In other words, by dismissing the unrepeatable vagueness of sound from the register of scholarly discussion, we not only ignore its knowledge but also its authorships and contexts. And by failing to engage in its indivisibility and insisting on the theorization of 'this' or 'that', we do not reach its possibility, which is not its theorization but the access to different ways things can be thought: since the point is not to read sonic fictions, in the process of which they might become theory, but to write them and to sing them.

Once a sonic methodology of sound acquires, from the ambiguity of sound art, from caresses, gesture-words, and the contingency of singing scholarly texts, a methodology that can persuade in its invisible formlessness, we are offered the secret of a different sense. This is a sense that does not need to compete with the visual lexicon and register, but can make it move to reveal the cut at the unsensed from where a plurality of senses might emerge.

At this cut the voices of those that so far have not been heard might become audible. Disciplines and historical givens might open into the indivisible sphere of a sonic sensibility and borders might be diffracted to hear with Barad not difference but 'inventive provocations' that illuminate 'the indefinite nature of boundaries': the lack of clear lines and outlines that allows disciplines and territories to be read through one another, and whose indivisibility invites a reimagination of their crossovers and interbeing (Barad 2003: 80). And to sense with Braidotti the need for a 'transdisciplinary approach that cuts across established methods and conventions of many disciplines' (Braidotti 2011a: 7) and that, as I would like to suggest, produces the continuity, co-laboration, and tuning of a cosmic world, where the audible and the unheard can find an ear and a methodology of their own expression. Because, if we sing together and at the same time and there is only our own echo we know that we have to catch our breath and make space for another sound to come back to us.

> In unison the two tapes click off their reels. 'That's the end of side one,' Gordon says, slapping his knees and coming to his feet. He is feeling fine now. More than fine – fired up. As far as he is concerned, Joan's rhythmic variations are as sophisticated as anything he ever heard on a David Rayne recording. 'This is extraordinary,' he says as he turns the tapes over. 'Disquieting in places, there's no question about that. But once you accept that her intention is to provoke there are levels within levels.'
>
> (Gowdy 1997: 251)

Notes

1. The notion of the visual in this context does not stand for what can be seen but how we look: how our cultural engagement in the world as a place to be seen and to be written about is constructed through a historical and geographical viscuocentrism and a conceptual logic that holds investments and consequences on what we take as

real, truthful, and reliable, and that restricts how else we might think of the world and ourselves. The aim is not to simply critique the dominance of the visual and its impact on scholarly enquiries of sound, however, but ultimately via a sonic methodology to augment and pluralize its reality and values.

2. I deliberately replace the expected word collaboration with the word *co-laboration* to emphasize the experimental, laboratory, nature of this collective production as well as its transdisciplinary approach. This is not a collaborative effort towards a greater aim or outcome, but a co-laborative exploration of things together.
3. 'To be is to inter-be. You cannot just be by yourself alone. You have to inter-be with every other thing. This sheet of paper is because everything else is' (Nhat Hanh 2012: 57). While not adhering to the Buddhist context of Thich Nhat Hanh's philosophy, his notion of interbeing, developed in relation to sound and listening, is a useful term to acknowledge the sociality of a sonic consciousness and thinking and to pursue its consequences and impact into the potential of a sonic methodology of sound.
4. I am borrowing the expression of unknown lands from Nigel Thrift who in his essay 'Performance and Performativity: A Geography of Unknown Lands' promotes the performance of geography to challenge its abstract knowledge, to aid the articulation of a geography of the unknown, and to create a different territoriality and a different sense of boundaries and participation.
5. Braidotti encourages 'a sort of intellectual landscape gardening' (Braidotti 2011b: 46) of an embodied mind as a way to draw a shifting landscape. Pursuing her thoughts into sound, the soundscape and sonic atmosphere, we need an actual gardening, digging, and shifting of earth to revocalize and rephysicalize theory.
6. With the notion of a hyper-invisible centre, and hyper-invisibility in general, I am referring to the norms and conventions at the centre of our sociocultural lives, which, while ideologically and culturally constructed, are so omnipresent and accepted as to have become entirely invisible. They present a naturalized reality that pretends a convenient actuality that is hyper-invisible: its construction and the investments and norms embedded therein are unseen and invisible while its form and the responses and attitudes it demands present the only visibility possible.
7. I borrow the term diffraction from Karen Barad, who via Donna Haraway and quantum physics, proposes the practice of reading the world diffractively rather than reflectively: as a reading of difference and detail rather than a looking for sameness and outlines. In this sense, diffraction is a performative reading of the world from its interactions and interferences. It is a practice rather than a theory that allows us to see the world as a heterogeneous entanglement of plural patterns (Barad 2003: 803).
8. At this point Sterne asks: 'What would sound studies become if we started without the automatic assumptions that we have direct, full access to our own hearing, or through our hearing, direct access to the sonic world?' (Sterne 2015: 74). He suggests that such a project is a little difficult to imagine. In response, I would like to suggest that it is not particularly difficult to imagine at all, but that it might not realize itself through theory and a theoretical thinking alone. Since theory, as mute thinking and a logocentric pursuit, while fulfilling the demands of academic clarity as the expectations of a (seemingly) transparent exchange and the aim of actual knowledge, works along horizontal lines, on points of reference, and thus confirms the possibility of universal access rather than generating the access to possibility. In that sense it can point at the cultural construction

of listening: the impact of technology, gender, race, class, and historical contexts, etc. on how we listen and what we hear, showing their influence on the perception of a current and 'immediate ear'. But it cannot include what misaligns with its line of articulation, what has no words and falls outside its visual lexicon and history. Thus it cannot produce 'another ear'. Instead it runs the danger of reaffirming the reality of the historical prejudice, of what the audible subject and object of a sonic culture are in a theoretical actuality, unable to hear and make count what is outside its comprehension in a practical possibility. Sound's immediacy does not produce imminent intelligibility but presents misalignments, errors, and awkward perspectives in the between-of-things. Its ephemeral instability and flow remind us of its interbeing: its being with and of other things. This directness is material and sensorial rather than intellectual, and provides a material sense, which is always just now, but not therefore natural, exempt from enculturation. Since, while it is about the now, it is not about the instantaneous. Instead, the now of sound is thick with cultural memory and prejudice. In practice these prejudices are not deciphered and restaged in theoretical language, but are made to sound. We do not seek to grasp this now in words that outlast its immediacy, but engage in the gap between what sounds and what is heard, the misalignments that make language and interpretation difficult, but that reveal a practical knowledge that reminds us to hear sounding as well as listening. Consequently a cultural study that aims to explore the construction of what appears immediate, directly accessible, might want to co-laborate with practical sonic methodologies of sound, to include the outside of (theoretical) language. Not in order to pretend a direct access but to engage in the inaccessible of culture through the contiguity of its temporal and invisible materiality.

9. In anthropology and ethnography more than in philosophy, visual theory, and science discourses, stories have a legitimate place amongst the methodological tools available, particularly when they are backed up and accompanied with more quantifiable data sets, or embedded in theoretical writing, to which they are a parallel and expanding stream. There are also other fields that develop sonic fictions as a way to discuss sound in its own audibility. Most notably Kodwo Eshun's seminal *More Brilliant than the Sun: Adventures in Sonic Fiction*, whose Afrofuturist science fiction critiques and expands how music is written about and thus how it could be listened to.

10. 'Goofy' is the term used by Manuel De Landa when describing Irigaray's work in a conversation with Christoph Cox. In answer to a question about Gilles Deleuze and his circle of friends, that is, like-minded philosophers, De Landa suggests that 'Deleuze was close to Foucault and Lyotard, but not to Derrida, and certainly not to Irigaray and her goofy notion of a "masculinist epistemology"' (De Landa quoted in Cox, Jaskey, and Malik 2015: 87). This statement is not only embarrassing but also paradoxically demonstrative of a masculinist viewpoint, deliciously unaware of the dominance of its logic and the suppression of the other, of what is unfamiliar and unknown, while searching for the unthought.

References

Barad, Karen (2003). 'Posthumanist Performativity: Toward an Understanding of How Matter Comes to Matter'. *Signs: Journal of Women in Culture and Society* 28 (3): 801–831.

Braidotti, Rosi (2011a). *Nomadic Subjects, Embodiment and Sexual Difference in Contemporary Feminist Theory*. 2nd edition. New York: Columbia University Press.

Braidotti, Rosi (2011b). *Nomadic Theory, The Portable Rosi Braidotti*. New York: Columbia University Press.

Cavarero, Adriana (2005). *For More than One Voice, toward a Philosophy of Vocal Expression*. Stanford, CA: Stanford University Press.

Cox, Christoph and Manuel De Landa (2015). 'Possibility Spaces: Manuel De Landa in Conversation with Christoph Cox'. In Christoph Cox, Jenny Jaskey, and Suhail Malik (eds), *Realism Materialism Art*, 87–94. Berlin: Sternberg.

Eng, Michael (2017). 'The Sonic Turn and Theory's Affective Call'. *Parallax* 32 (3): 316–329.

Eshun, Kodwo (1998). *More Brilliant than the Sun: Adventures in Sonic Fiction*. London: Quartet Books.

Feld, Steven (2015). 'Acoustemology'. In David Novak and Matt Sakakeeny (eds), *Keywords in Sound*, 15–21. Durham, NC: Duke University Press.

Gershon, Walter (2013). 'Resounding Science: A Sonic Ethnography of an Urban Fifth Grade Class Room'. *Journal of Sonic Studies* 4 (1) (May). Available online: https://www.researchcatalogue.net/view/290395/290396 (accessed 13 July 2020).

Gowdy, Barbara (1997). *Mister Sandman*. London: Flamingo.

Irigaray, Luce (1996). *I Love to You: Sketch of a Possible Felicity in History*. London: Routledge.

Irigaray, Luce (2001). *To Be Two*. New York: Routledge.

Lockwood, Annea (2017). *Tiger Balm/Amazonia Dreaming/Immersion*. Black Truffle Records, BT028, liner notes.

Massey, Doreen (2005). *For Space*. London: Sage.

Nancy, Jean-Luc (2007). *Listening*. New York: Fordham University Press.

Nhat Hanh, Thich (2012). *The Pocket Thich Nhat Hanh*. Boston, MA: Shambhala Pocket Classics.

Ronen, Ruth (1994). *Possible Worlds in Literary Theory*. Cambridge: Cambridge University Press.

Sterne, Jonathan (2015). 'Hearing'. In David Novak and Matt Sakakeeny (eds), *Keywords in Sound*, 65–77. Durham, NC: Duke University Press.

Thrift, Nigel (2007). 'Performance and Performativity: A Geography of Unknown Lands'. In James S. Duncan, Nuala C. Johnson, and Richard H. Schein (eds), *A Companion to Cultural Geography*, 121–136. London: Blackwell.

Part II

Sound Arts, Musics, Spaces

16

Introduction to Part II: Art – Research – Method

Marcel Cobussen

In the beginning of the 1990s some intriguing and provocative music theatre performances took place in The Hague and Amsterdam. The equally famous and notorious Dutch composer, theatre maker, and theoretician Dick Raaijmakers (1930–2013) presented a new series of works, called *Intona* (after Luigi Russolo's sound or noise machines *Intonarumuri* from the 1920s), in which the microphone played the leading role. In *Intona* the microphone no longer performed a reproductive function but was treated as a music instrument or, perhaps more accurately, as a patient: the audience witnessed its groaning, wailing, and even its dissolution while attempting to resist the way the musicians treated it. Indeed, *Intona* was created to investigate whether the microphone has a voice of its own, whether it can talk, sing, play, or communicate, outside of our consent. In order to achieve this, Raaijmakers searched for and experimented with several 'rapprochement techniques': the microphones could be sawn, milled, drilled, wrenched, or swung around; they could be burnt by gas burners, drenched, and cooked or boiled in water; they could be irradiated by compressed air, so that their membranes cannot receive sound waves anymore; they could be treated with chemicals so that they exploded or dissolved, etc. (Mulder and Brouwer 2007: 316–319).

Dick Raaijmakers was an artist, and he was a researcher. As an artist-researcher he had a rather specific research method. Perhaps this method was (slightly) less systematic than his colleague-researchers in medical labs investigating stem cells or trying to refine cancer treatments. Raaijmakers's research was certainly less teleological, less focused on obtaining a predefined objective or clear-cut end result. Perhaps he was less disappointed when things went wrong, simply because he could not predict what would be a good or, conversely, an undesirable outcome; perhaps he was also more interested in the process than in the final product, an emergent process unfolding outside of the confines of a controlled experimental context.[1] In the words of Tim Ingold, Raaijmakers's method was not 'a set of regulated steps to be taken towards the realization of some predetermined end. It is a means, rather, of carrying one and of being carried […], speculatively open to the possibilities of the future' (Ingold 2015: vii). Speculatively open to all the possible sonic reactions of a tortured mic – I assume that Raaijmakers could have recognized his own approach in this description; his systematicity emerged from flexibility and adaptation, rather than from mechanistic predispositions.

However, like his fellow researchers in the academic world or at more commercial research institutes, he was experimenting on the basis of existing knowledge, in this case knowledge as to how microphones register vibrations in the air by way of their ultra-thin diaphragms, their magnet, and their coil as well as the embodied cognition and situated knowledge he had gathered as an artist. *Intona* was, therefore, not simply the result of an artistic brainwave or the outcome of some genius's aesthetic intuition; *Intona* was (in)formed by Raaijmakers's many years of working for the Philips NatLab, the research institute of this multinational. However, his research was not undertaken only for the benefit of art; it mostly took place in and through art. New art works were the *outcomes* of his research, but artistic experimentation, improvisation, and responsiveness to the materials he was working with were also the *methods* he used to develop new performances, installations, and compositions.

The majority of the contributors to this second section of the Handbook are, like Raaijmakers, artists and/or artist-researchers. Most of them do not have the opportunities and means Raaijmakers had at his disposal in the laboratories or studios in which he worked. Nevertheless, they also often deploy their artistic practice as a method, either to add something to already existing knowledge, raise awareness, and make new experiences possible or to intervene in, transform, and affect already existing situations, events, and sensitivities.

Usually opening sentences such as these are followed by a brief overview of the content of each chapter. I will refrain from following this convention here, nor will I attempt to cluster or group the chapters other than having them all included in Part II of this book. The main reason for deviating from the tradition is that summaries invariably do injustice to the richness of a text. Besides, each reader can and should decide for her- or himself which trajectories and threads to follow, in which order to read the contributions, and which parts of the texts are the most interesting. Instead, this Introduction will concentrate on and offer a critical reflection on the triangle art-research-method, while certainly deriving inspiration from the chapters that follow.

(Sound) art and research

In her contribution to this volume, sound artist-researcher Yolande Harris starts by asking if attentive listening can be a method to change us as human beings and our relationship to the environment. What can we learn about our environment through sounds, and how can they help to transform the way we affect and are affected by our milieu? Through her performances, multimedia installations, workshops, soundwalks, field recordings, lectures, and writings Harris attempts not only to increase our awareness of the environment we inhabit but also to influence our attitude towards it. Although she is well aware that (ecological) science has more or less the same goals, she regards the input of sound artists as essential in establishing such a renewed relatedness. Art works appeal to more than just comprehension and rational argumentation: empathy, imagination, and embodied knowledge, activated by and through sounds, are human faculties that can support, inform, and change a primarily scientific relatedness.

Harris's words resonate and return in somewhat modified forms in several other contributions: Jana Winderen, for example, explains in her interview with Stefan Helmreich how her art works may augment a certain active connectedness to the environment. For Jonathan Gilmurray it is imperative that we also learn to engage with our environment through other means than just science and/or politics. Especially sound and sound art are important in this respect, as they can – in contrast to the more static and stable image that visual information provides – make us aware of the dynamic processes that are constantly operative in our environment. Marie Højlund and colleagues term it *perceptual attunement*, referring to a particular sensitivity to movement and change that does not result in representational knowledge, while Jordan Lacey speaks of artistic ruptures to create new experiences, to transform, and to establish new encounters with and within public urban spaces, a strategy that becomes very concrete in the contribution of Edwin van der Heide as well.

All these scholars and artists attribute to art the ability to act as a medium or method through which new/other knowledge, new/other experiences, new/other awareness, new/other sensibilities, or new/other affects become accessible. However, the questions that become pertinent are what kind of methods can be deployed by research in and through the arts, whether this type of research fundamentally differs from other research traditions, and whether there is an unbridgeable gap between the arts and the sciences in terms of methods and methodologies. And, within this field of artistic research, is there a specific role to be played for sound and sound art? These questions lead us into the next paragraphs.

Philosophical objections against methods and methodologies

Almost any definition of 'method' foregrounds terms such as 'systematic procedures', 'modes of inquiry employed by a proper plan', 'tools through which we can collect empirical material', 'a body of principles and techniques', and/or 'a regular and orderliness of thought, action, etc.' (see, for example, Gray and Malins 1993; Merriam-Webster.com Dictionary n.d.; Vannini 2015: 10). Methodology, then, is the system of methods and principles used in a particular discipline or the science of method; it is the larger body of knowledge – consisting of practical applications, abstract reflections, and epistemological foundations – on which choices of methods are based.

However, especially in and through contemporary Continental philosophy, the history of science, and artistic research, the idea – rightly or wrongly – that one can latch on to a clear and strict method has been under attack. In *Practice as Research in the Arts*, Robin Nelson, Professor of Theatre and Intermedial Performance in the UK, writes with regard to artistic research that it is 'no longer tenable to take the methodologies of the sciences as the gold standard of knowledge' (Nelson 2013: 48), suggesting the erosion of the self-evidence of their applicability. Echoing the post-structuralist criticisms on the fundaments of Western thinking, he continues by stating that, for several decades, the methodological paradigms of positivism, rationalism, and empiricism – research methods such as observation, data-gathering, testability, and falsifiability, deployed by objective and neutral researchers and leading to general truths and knowledge of an independent reality – have already been supplemented or even replaced by methodical principles based on the interrelatedness of subject researcher and object as well as by the notions of an interminable potential for new discoveries, the ineluctable situatedness of both research and knowledge, and reality as a linguistic construct (49–53). What is important for Nelson is that these changes or adaptations should be regarded as 'a recognition that knowing is processual and a matter of multiple perspectives', with the consequence that current methods and methodical rigour need to be rethought (53–55).[2]

A fundamental criticism of the rather rigid models, methods, and research strategies of the sciences has also been offered by the Swiss historian of science Hans-Jörg Rheinberger. In *Toward a History of Epistemic Things* he questions the hegemony of theory in our post-Kuhnian era and replaces it by emphasizing the dynamics of research and its related methods of experimentation. Rheinberger explicitly connects this new scientific and methodical paradigm to the concepts of invention and change that are also operative in the art world: the function of experimentation is no longer to provide empirical proofs for theoretical propositions but, rather, to stress the importance of a conceptual indeterminacy during the journey into the unknown in order to produce knowledge that is not yet at the scientist's disposal. Rheinberger here seems to propose a methodical move that shifts from focusing on *known unknowns* (what we know we don't know) to *unknown unknowns* (what we don't know we don't know): 'I perceive thinking as remaining a constitutive part of experimental reasoning, conceived as an embodied disclosing activity that transcends its technical conditions and creates an open reading frame for the emergence of unprecedented events' (Rheinberger 1997: 31).[3]

It is telling that the Canadian philosopher Erin Manning begins her essay 'Against Method' with a quote from Alfred North Whitehead: 'Some of the major disasters of mankind have been produced by the narrowness of men with a good methodology' (Manning 2015: 52). Manning's text is a frontal attack on the idea of method as a way of organizing knowledge in pre-established categories and frames, thereby becoming a 'safeguard against the ineffable' (Manning 2015: 58). In other words, Manning warns (academic) scholars that methods all too often contribute to the creation of new orthodoxies in relation to – or actually in contrast to – our experiences and thinking processes. What is at stake is a micropolitical disciplining of a model modelling the researcher.

Although Manning does not mention him, her arguments against rigid methods and determined methodologies echo the objections formulated by Austrian philosopher of science Paul Feyerabend in his book bearing the same title as Manning's essay: *Against Method*. Feyerabend developed arguments against the endeavours of other philosophers of science – such as Karl Popper, Imre Lakatos, and the logical positivists – to establish a fixed, general, and universal scientific method: it is *unrealistic* (ignoring human developments), *pernicious* (enforcing rules is inhuman), and *detrimental to science* (neglecting the complex physical and historical conditions which influence scientific change). He argues that scientific progress or growth of knowledge is often only possible when researchers are able and willing to break laws; 'proliferation of theories is beneficial for science, while uniformity impairs its critical power' (Feyerabend 1993: 5). Therefore, Feyerabend claims, the only 'method' which will not inhibit scientific progress is an epistemological anarchism in which 'everything' is permissible, for example contradicting well-confirmed theories and working counter-inductively (20).

Methodological criticism in music studies

Also within the discourses around music, objections have been raised against certain methodical and methodological habits. Already in 1991, music theorist and pianist Lawrence Ferrara in his *Philosophy and the Analysis of Music* argued against the tendency of researchers in music to often establish their method first and then start examining a musical work. The trap is that they become predominantly concerned with executing the chosen method instead of giving precedence to the music:

> Methods have developed or evolved in ways that do not fully respond to the multiplicity of levels of musical significance. Methods define the tasks and scope of inquiry into musical significance. As the method replaces the immediacy of the analyst, music comes to mean only what methods allow it to mean.
>
> (Ferrara 1991: xvi)

According to Ferrara – less inspired by post-structuralism but rather by the hermeneutics of Hans-Georg Gadamer – each method that objectifies musical data, for example through precise language and predesignated tasks, subjugates the music and 'controls music by asking questions that grow from a preconstructed schematic of what music can mean' (Ferrara 1991: 38). The strengths and weaknesses of any pre-established method determine to a great extent the analytical judgements and outcomes. In other words, it is not the researcher who can be accused of subjectivism; the subjective role is transferred from the analyst to the method, as it is the method which 'interrogates' and thereby restrains the music.[4]

Although Ferrara's alternative, a methodological pluralism or eclecticism, is still in the service of systematically investigating and mapping the various levels of musical significance – thereby closely connected to a functionality that, for example, Rheinberger tries to overcome – his criticism of the prescriptive power of methods traditionally used in music analysis offers possibilities to reformulate and renew the epistemological tools with which music can be approached by music scholars. The key terms with which Ferrara presents his eclectic method are openness, engagement, and objectivity, the latter understood in an hermeneutical sense as letting the object speak of and for itself:

> A responsive subject who openly engages a freer object is on the way to the classical notion of objectivity. Within this view, the object is given the freedom to show itself. In order to allow a piece of music to be a free object, the analyst must release his methodological will to dominate it. Real objectivity, to the degree that it is attainable, occurs when analytical tasks support the freedom of the music object to show itself in its multi-dimensional polyphony of sound, form, and reference.
>
> (Ferrara 1991: 46)

Philosopher Gary Peters criticizes methodology through musical improvisation in his book *The Philosophy of Improvisation* (2009). While an absence of methodology in the academic world would equate with an absence of credibility, he claims that improvisation cannot bloom within this type of rigour (Peters 2009: 148). Peters creates a fundamental opposition between methodology and method, strongly deviating from the hierarchical difference presented above. In Peters's view methodology is first of all teleological, aiming for the straightest line possible, leading to a clearly formulated and predetermined goal. Aberration is allowed but only temporarily, briefly, and strictly limited: 'Such a curvature of thought is always measured against the teleological straightness that the methodology provides' (162). Improvisation has an uneasy and problematic relation to such a methodological straitjacket because it has no clear end point to which it should converge and because it is almost always built on the principle of trial and error. Of course, Peters is quick to clarify, this does not mean that improvisation is lacking rigour; only, this rigour – which shows itself for example in all kinds of experientially rooted conventions and (implicit) rules – is of a different order: 'methodical rather than methodological', that is, developed 'from work to work and from moment to moment' (148–149). Therefore, Peters states, improvisation has no (need of) methodology, but it does have a method, as a method doesn't necessarily need to rely on the dogmas of progress and teleology. Improvisation 'depends upon error and the failure to reach a goal' (162).

Whereas Ferrara questions the role of methods and methodologies in research on music, Peters criticizes the use of methodologies within the musical practice itself. Methodologies obstruct a necessary amount of creativity and freedom in the production of art. Art benefits from the right to fail, the right to stray from a straight and narrow path, the right to err. Interesting for the rest of this text is how Peters ends his short reflection on the ontological difference between methodology and method: 'Blanchot describes such erring as "research," an endless "turning" and returning that resists the desire to terminate the "fascination" of error in the rush for a terminal truth' (Peters 2009: 163). Connecting research to music making and rethinking the role of methods beyond the systematic application of pre-established rules paves a way to say a bit more about the relation between arts, theory, and methods.

The experiment as method

The previous paragraphs have made clear that a rigorous application of (dogmatic) methods has been under attack in (Western) philosophy, science, and the arts. Is it possible to rethink the role, position, and implementation of a specific research method in order to formulate an alternative? And how can the arts and research done by sound artists contribute to such a reorientation? In order to start answering these questions it is good to see that what connects Peters's erring, the artistic practice of Dick Raaijmakers, the theoretical insights of Hans-Jörg Rheinberger, and several contributions in this Handbook is the practice of experimentation as the key method to gain access to something new, be it art or scientific knowledge. What was perhaps a very natural and self-evident way of searching and researching for Raaijmakers receives a more theoretical foundation in the work of Rheinberger. Experimental activities, Rheinberger states, generate not only awaited and predictable results but, crucially, also unexpected, unknown ones as well. Therefore, these activities are not methodological procedures made to confirm or reject certain hypotheses or to verify, refute, or modify theories; this would lead merely to knowledge that is already preformulated, and the experiment would play only a subsidiary role in rationalistic accounts of theoretical developments or changes. Instead, experiments should function as actual generators of knowledge that were previously obscure and unexplored (Rheinberger 1997: 138; Assis 2018: 108). Rheinberger warns the reader not to interpret this as either a paradigm shift, the replacement of one theory by another, or an irregularity within an established conceptual frame. Experimental processes open up unforeseen directions; they are generators of surprises. After Royston Roberts he calls the results of such processes *pseudo-serendipity*: they come as a surprise but are made to happen through the inner workings of systems (Rheinberger 1997: 133–134) and the quest of researchers.[5] Quoting Michel Serres, Rheinberger describes such experimental researchers as people who, paradoxically, are 'not yet quite sure what they are looking for, and yet blindly do know what they are after' (Serres in Rheinberger 1997: 14).

Researchers, according to Rheinberger's use of the term, tinker, linger, and are on an almost constant journey into the unknown; they are *bricoleurs*, unable even to formulate a clear research question at the beginning of their investigations. Experimentation as a method could then be described as creating conditions where the unexpected can happen. This description implies that a method – how research should be done – should at least partly be deprived of its functionality and teleology and be rethought as simply a way of doing. And it could be precisely art that is able to stretch the framings of a method as a doing beyond its axiomatization[6] – a doing that oscillates between what is defined and what is (as yet) undefinable, between limits and the unlimited, between the already known and the still unknown. In this context 'oscillating between' should be understood in the way Deleuze and Guattari have thought it: 'Between things doesn't designate a localizable relation going from one thing to the other and back again, but a perpendicular direction, a transversal movement that sweeps one and the other away, a stream without beginning or end that undermines its banks and picks up speed in the middle' (Deleuze and Guattari 1987: 25).

Experimental processes as methods of research perpetually lead to events that cannot be anticipated (in contrast to, for example, relations of cause and effect) but only appear in the making. As such, they are not only contingent but also irrevocably local and situated.[7]

Artistic research and method

Although rather prescriptive and still speculative, Rheinberger has shown that the traditional scientific methods, grounded on privileging neutrality and objectivity, are not the only valid knowledge-producing methods. Other modes of knowing – knowledge coming from experience, situated knowledge, embodied knowledge, or processual knowing – require other methods. Research done in and through art does not simply produce descriptive knowledge or logical schemas. It may ask questions that could not be anticipated within a more traditional epistemological horizon. And methodically, it is *performative*, a practice where theory and practice coincide – 'imbricated within each other' – based on an investigative strategy of *doing-thinking*. Artistic researchers *do* knowledge, meaning that 'thinking is not constrained to the abstract and propositional' (Nelson 2013: 62, 66).[8] In the artistic-experimental network of affective and interconnected forces, a network of objects, events, concepts, materials, practices, minds and bodies, spaces and places,[9] a transformation takes place from science-as-a-system to science-as-a-process: reflection becomes refl*action*. As Manning states, in artistic research 'making is a thinking in its own right' (Manning 2015: 53).

Taking these insights into the concrete realm of a contemporary musical performance practice, Lucia D'Errico states in this part of the Handbook that recitals of musician-researchers may become a locus of experimentation in which 'the known' is reconfigured: by considering musical works as dynamic reservoirs of forces, functions, traits, and materialities, artistic researcher-performers critically reflect on current modes of thinking and making music, not through theoretical approaches and detached observations but by actually performing music themselves. The reflection takes place in and through music making; it takes place before (e.g. while rehearsing or while analysing and reorganizing repertoire), after (e.g. while evaluating a concert), but – most importantly – also *during* a performance. Performing becomes an experimental strategy and embodied research method through concepts such as a musical work can be (re)considered.[10]

Paul Nataraj's contribution to this book testifies to a comparable methodical approach, albeit in a completely other musical domain. Interested in the relation between music – more specifically: vinyl records, considered as a site of convergence between many human and nonhuman agents – and memory, Nataraj not only interviewed several persons about their connection with a specific record, but carved a transcription of the respondent's story back onto the surface of the record. In and through his artistic practice he thus works on the record's materiality and resistance, deconstructing its prescribed model of sonic replication by adding the voice of his respondents as well as other sounds on top of the original music. Nataraj neither does away completely with old elements nor is he introducing something altogether new; rather, he reorients and rearranges given elements by exploring new relationships and thereby opening 'sonic possible worlds' (Voegelin 2014).

It permeates this whole Handbook: the privilege of neutrality and objectivity, both on an epistemological and methodological level, is questioned, and modest contours of other warrantable research methods are well within reach. Insights as well as methods are situated – the method is not predetermined by the researcher, nor pre-existent and ready to be used, but tailored to the situation; it is a singular pathway[11] hacked within the field of research[12] – but should nevertheless be cognitively apprehensible. As Nelson writes:

> If there is no secure, neutral basis for establishing objective knowledge in any discipline, and if there is no firm ground from which to make a 'truth language' claim of superiority for science, history or philosophy among competing micronarratives, it is incumbent upon all disciplines, including the sciences, to offer a reflexive account of their methodology and the rigour of its internal methods.
>
> (Nelson 2013: 55)[13]

Towards a sonic methodology *without method*

When artistic experimentation is indeed recognized as a valid research method in sound studies, this also implies that the total research process can be embedded in sound art, implying that this type of research can only be done by the artists themselves. Hence, experimentation as an important element in research that takes place in and through sound and sound art consists of a feedback loop between action and thinking, practice and theory, and, perhaps especially, mind and body or the intelligible and the sensible. More pervasive, more explicit, and more conscious than in most other scientific fields, the body belongs to the methodological toolkit of artistic researchers: their bodies play, rehearse, perform, walk, make, construct, repair, disassemble, etc. in order to contribute something new or extra to already existing discourses, knowledge, experiences, and intuition. A sonic 'methodology *without method*' rooted in sound art is experimental as well as performative, that is, embodied, embedded, and enacted. While interacting with the environment, the body is gathering information – information regarded here as having been (in)formed by corporeal knowledge, perception, affect, intuition, and sensibility – dispersing cognition partly away from the brain.[14] Next to and in addition to theoretical doing, artistic research takes place via an actual-corporeal participating. Bearing in mind Serres's description of experimental researchers who are 'not yet quite sure what they are looking for, and yet blindly do know what they are after', a methodology *without method* in sound art research could be called an unexpected but nevertheless expectant exploration of a network of nodes, agents, and their relations through making art. Methodology can then be understood in an extended manner, namely as making numerous and surprising connections, an exploratory means for the discovery of potentiality and contingency.[15] Methodology as a doing, a gathering, a connecting, or as a way to encounter *unknown unknowns* is then less concerned with matters of fact than with matters of concern.[16]

These matters of concern are singular and diverse, from *dis-covering* Vietnamese rural soundscapes by placing musical instruments in rice fields and on the top of mountains, thereby making the wind a co-player (see Östersjö and Thanh Thủy, in this volume), to gaining access to a recording studio by capturing the sounds of both the control room and the live room (see Thompson, in this volume) – both examples of a sonic ethnography; from Edwin van der Heide's trials and experiments with pneumatic valves in order to compositionally shape audible air pressure waves via the controlled release of compressed air, to Jana Winderen's explorations regarding optimal use of the most suitable hydrophones, depending on weather parameters, such as temperature and wind speed, and what one is seeking to record – both examples of research methods depending on technological knowledge; and from investigating possibilities to create sound art installations for public spaces that both interact with as well as improve the already existing sonic atmosphere (see Lacey and Højlund et al., in this volume), to turning sound art into a political instrument that makes us more aware of climate change and ecological pollution (see Gilmurray, in this volume) – both examples of how sound can be deployed as a method to raise awareness and increase well-being. To quote Gary Peters once more, research in and through sound art is 'methodical rather than methodological' as it is 'developed from work to work and from moment to moment' (Peters 2009: 162).

Notes

1. Historian and curator Sarah Cook and arts worker Beryl Graham explain in *Rethinking Curating: Art After New Media* (2010) that the term *laboratory* resonates with scientific models that emphasize process over product. Raaijmakers's practice certainly happened in spaces which deserve the name 'lab'.
2. Of course one could question whether the times of positivism are really over. More important, however, is Nelson's idea that the research methods belonging to positivistic or so-called objective theories can no longer be taken for granted, no longer be accepted as an incontestable standard.
3. Rheinberger's ideas echo those of Jean-François Lyotard, who writes in *The Postmodern Condition*: 'A Postmodern artist or writer is in the position of a philosopher: the text he writes or the work he creates is not in principle governed by pre-established rules and cannot be judged according to a determinant judgment, by the application of given categories to this text or work. Such rules and categories are what the work or text is investigating. The artist and the writer therefore work without rules, and in order to establish the rules for what will have been made' (Lyotard 1984: 81).
4. In *Ludic Dreaming* the Occulture (a collective consisting of David Cecchetto, Marc Couroux, Ted Hiebert, and Eldritch Priest) briefly touch upon the affordance and constraints of methods. They claim that methods by definition exclude and prioritize data and are thus 'constitutively incapable of representing within its framework that which is excluded or deemphasized in this method' (The Occulture 2017: 89).
5. For Rheinberger the term system does not refer to an enclosed, perfectly defined set of rules and axioms but to a loose network of technical and organic units, existing both in

time and space. A system consists of both experimental conditions, determined within given standards and operative within sufficiently stabilized procedures, as well as so-called 'epistemic objects', the latter being vague, as they embody what is not yet known.
6. In *The Artistic Turn: A Manifesto*, Kathleen Coessens, Darla Crispin, and Ann Douglas write that 'artistic research can be defined as knowledge of the process of creativity, not its outcomes. It offers an account of the search trajectories in artistic practices, not a real explanation and certainly not a "prediction" of where they will lead' (Coessens, Crispin, and Douglas 2009: 26). In fact, they present artistic research as a method here, yet strongly diverging from established scientific methodological models.
7. Writing about musical improvisation as a research methodology in this volume, Rebecca Caines claims that such a methodology is often eccentric, experimental, personal, and non-generalizable, which implies that the research conditions are most of the time unable to be repeated. Rheinberger would in this respect probably speak of 'nonidentical reproduction' as an inevitable consequence of a situated making of science. A situated method – *this* time, on *this* occasion, under *these* circumstances – is never settled in advance, but must be worked out, per-formed.
8. For pianist and artistic researcher Paulo de Assis, researchers thus appear as doers, not as enlightened academics delivering proof of given theory (Assis 2018: 112).
9. It should be clear that the researcher-as-doer is not external to this network but one force within it. In her contribution to this part of the Handbook, Elena Biserna makes this very concrete: sound artist-researchers always participate in the soundscape they listen to. Simultaneously with the analyses, evaluations, reflections, and – if applicable – interventions of these persons, they co-create the soundscape under study rather than purely functioning as non-involved observers and detached analysts.
10. In *The Reflective Practitioner*, the philosopher Donald Schön coins the terms 'knowing-in-action' and 'reflection-in-action', both referring to the idea that this type of thinking does not precede acting but coincides with it. It is a thinking about doing while doing it, of which the rules are difficult to describe; this thinking hinges on surprises and (thus) does not rely on categories or methods of established theories and techniques. According to philosopher Sher Doruff it is exactly this element of surprise (what we don't know we don't know) in *research-creation* that is the node of the indeterminate contingencies of artistic research practice (Doruff 2010: 7).
11. Elsewhere in this volume, Naomi Waltham-Smith writes that Derrida regards method as a *meta-hodos*, that is, following of a way or path. Therefore, method can only take shape in the process of its practice, hence in a singular event. At the same time, however, it is repeatable yet always already open to unanticipated modifications.
12. In her contribution to this part of the volume, Darla Crispin stresses the interactions of musical materials with the specific 'tone' of a researcher-performer, which creates a unique synthesis.
13. Nelson's words resemble those of Feyerabend when he states that 'one of my motives for writing *Against Method* was to free people from the tyranny of philosophical obfuscators and abstract concepts such as "truth", "reality", or "objectivity", which narrow people's vision and ways of being in the world' (Feyerabend 1993: 179).
14. In this second part, Højlund et al. call this the *attuning approach,* that is, the active engagement of the enactive user through practice-based experiments moving beyond the already familiar or expected.

15. The suffix '-logy' as in methodology could then be read in the Heideggerian sense as coming from the Greek λέγειν, letting-lie-before-us or gathering, assembling, and connecting (Heidegger 1968: 208).
16. In this sense, a sonic methodology against method might satisfy Douglas Barrett's concerns, expressed in this part of the Handbook, about the 'formalist tendencies' in sound studies. Although a sonic methodology of course hinges on sound – it can best be described as a method-without-methodology directed by sound – the overall idea presented and defended in this Introduction with its focus on research being done in and through art rather centres around a critical stance towards conventional ways of thinking about and applying research methods instead of emphasizing a medium specificity.

References

Assis, Paulo de (2018). 'Experimental Systems and Artistic Research'. In *Logic of Experimentation: Rethinking Music Performance through Artistic Research*, 103–118. Orpheus Institute Series. Leuven: Leuven University Press.

Coessens, Kathleen, Darla Crispin, and Anne Douglas (2009). *The Artistic Turn: A Manifesto*. Leuven: Leuven University Press.

Deleuze, Gilles and Félix Guattari (1987). *A Thousand Plateaus: Capitalism and Schizophrenia*. Trans. Brian Massumi. Minneapolis, MN: University of Minnesota Press.

Doruff, Sher (2010). 'Artistic Res/Arch: The Propositional Experience of Mattering'. *Acoustic. Space Journal* 9. Available online: https://sherdo.files.wordpress.com/2010/09/artistic-res_arch.pdf (accessed 2 July 2020).

Ferrara, Lawrence (1991). *Philosophy and the Analysis of Music: Bridges to Musical Sound, Form, and Reference*. Bryn Mawr, PA: Excelsior Music Publishing.

Feyerabend, Paul (1993). *Against Method*. London: Verso.

Gray, Carole and Julian Malins (1993). 'Research Procedures/Methodology for Artists & Designers'. Available online: https://www.researchgate.net/publication/237475054_Research_Procedures_Methodology_for_Artists_Designers (accessed 2 July 2020).

Heidegger, Martin (1968). *What Is Called Thinking?*. Trans. J. Glenn Gray. New York: Harper & Row.

Ingold, Tim (2015). 'Foreword'. In Phillip Vannini (ed.), *Non-Representational Methodologies: Re-Envisioning Research*, vii–x. New York: Routledge.

Lyotard, Jean-François (1984). *The Postmodern Condition: A Report on Knowledge*. Trans. Geoff Bennington and Brian Massumi. Minneapolis, MN: University of Minnesota Press.

Manning, Erin (2015). 'Against Method'. In Phillip Vannini (ed.), *Non-Representational Methodologies: Re-Envisioning Research*, 52–71. New York: Routledge.

Merriam-Webster.com Dictionary (n.d.). s.v. 'Method'. Available online: https://www.merriam-Webster.com/dictionary/method (accessed 16 July 2020).

Mulder, Arjen and Joke Brouwer (eds) (2007). *Dick Raaijmakers: Monografie*. Rotterdam: V2/NAI uitgevers.

Nelson, Robin (ed.) (2013). *Practice as Research in the Arts: Principles, Protocols, Pedagogies, Resistances*. Basingstoke: Palgrave Macmillan.

The Occulture (David Cecchetto, Marc Couroux, Ted Hiebert, and Eldritch Priest) (2017). *Ludic Dreaming: How to Listen Away From Contemporary Technoculture*. New York: Bloomsbury.

Peters, Gary (2009). *The Philosophy of Improvisation*. Chicago, IL: University of Chicago Press.

Rheinberger, Hans-Jörg (1997). *Toward a History of Epistemic Things: Synthesizing Proteins into the Test Tube*. Stanford, CA: Stanford University Press.

Schön, Donald (1983). *The Reflective Practitioner: How Professionals Think in Action*. New York: Basic Books.

Vannini, Phillip (2015). 'Non-Representational Research Methodologies: An Introduction'. In Phillip Vannini (ed.), *Non-Representational Methodologies: Re-Envisioning Research*, 1–18. New York: Routledge.

Voegelin, Salomé (2014). *Sonic Possible Worlds: Hearing the Continuum of Sound*. New York: Bloomsbury.

17

Ambulatory Sound-Making: Rewriting, Reappropriating, 'Presencing' Auditory Spaces

Elena Biserna

> If movement is itself a potentially transformative activity, then moving to sound is doubly so.
>
> —Michael Bull (2007: 47)

Walking as embodied situated spatial knowledge

Walking has featured in sound studies discourses first of all as a mobile, situated, and embodied methodology to explore and perceive auditory spaces. In other words, it has been interpreted primarily as a 'form of engagement integral to our perception of an environment' (Pink et al. 2010: 3). This interpretation, beyond aligning itself with some recent trends in urban anthropology, is linked to a whole tradition of thought that understands walking as a means to perceive, read, and comprehend the environment.

The origins of this tradition are far from recent and are rooted in the late nineteenth century. The flâneur recently back in vogue in a plurality of researches (Tester 1994; D'Souza and McDonough 2006; Elkin 2016, among others) – represents the archetypal figure of this possibility of exploration and observation of urban space from below by crossing it. As Mary Gluck states, the true prerogative of this literary figure, both a symbol and a symptom of the emergence of the modern city, is '[his] radical sensibility and innovative visual practices, which made him distinct from all other social types of his age. The flâneur's unique achievement was to pioneer a new way of seeing, experiencing, and representing urban modernity that privileged the everyday perspective of the man of the street over the bird's eye view of the rationalist or the moralist' (Gluck 2010: 272).

It is precisely this ability to observe the new metropolitan landscape from an internal perspective, by walking through it, that characterizes the *flânerie*.[1] The *flâneur*'s mobile gaze bonds the links between the heterogeneous elements that make up the urban fabric, tracing its fragmentary nature to continuity. This activity has a revealing potential: the possibility to perceive and experience the world outside of any banalizing frame, capturing the aesthetic dimension of everyday life.

If much of the literature on *flânerie* remains focused on visual perception, other contributions insist on the embodied and intersensory nature of walking (Sansot 2000; Thibaud 2008; Thomas 2010; among others). David Le Breton, for example, claims the centrality of the relationship between body and world unfolding through walking in opposition to the erosion of the sensory sphere in contemporary life. The city, for the French anthropologist, is an inexhaustible source of both physical and mental stimuli:

> The relationship existing between the walker and the city, with its streets and neighborhoods, whether one already knows them or discovers them on the way, is above all an emotional and bodily one. A sound and visual background accompanies his ambulation, his skin registers temperature variations and reacts to the contact of objects and space. He crosses layers of inviting or repelling odors. This sensory plot infuses the walk along the streets a pleasant or unpleasant shade depending on the circumstances. The experience of walking in the city solicits the body in its entirety.
>
> (Le Breton 2000: 14)

In these contributions, immersion and sensorial contact reinforce the flâneur's internal, 'bottom up' perspective and walking becomes a methodology for 'knowing the world through the body and the body through the world' (Solnit 2001: 29): a methodology producing a situated, affective, and embodied spatial knowledge.

Soundwalking as auditory spatial knowledge

This situated, internal, embodied, and affective methodology to experience space is fundamental to soundwalking. Anticipated in the mid-1960s by artists and musicians such as Max Neuhaus and Philip Corner, soundwalking has been defined and canonized within the World Soundscape Project (Westerkamp 1974; Schafer 1994) to spread, in the following decades, into a multiplicity of different research and aesthetic practices (see Drever, in this volume). What all these practices share is the will to develop often participatory experiences of exploration of space through a perceptual reorientation on hearing (Drever 2009; McCartney 2014). As stated by Hildegard Westerkamp, soundwalking can be done alone, in groups, with a map, recording, interacting with the environment, or simply listening: 'No matter what form a soundwalk takes, its focus is to rediscover and reactivate our sense of hearing' (Westerkamp 1974). In other words, soundwalking proposes to cross the environment to 'listen' rather than to 'hear', defining listening as a way to know and relate to the world.[2]

For Westerkamp, beyond orienting and establishing a dialogue between the walker and the environment, soundwalking has also an aesthetic potential and 'reveals the poetics of space' (Westerkamp 2010). However, in the World Soundscape Project, soundwalks are first of all used as a methodology to make a first acoustic cartography of studied sites (Paquette and McCartney 2012) and – in Schafer's pedagogical and ethical perspective – they are thought first of all as 'useful educational experiences for everyone' (Schafer 1977: 1), as exercises to refine our listening sensibility and to prepare the field for the development of sound design (82). In continuity with this pedagogical and research vocation, soundwalking is used today as a qualitative in situ research methodology in the fields of ethnography, sociology, geography, etc. (Pink 2009; Gallagher and Prior 2017; among others).[3]

Walking as rewriting and reappropriating space

Yet, walking is not only a practice that allows us to immerse ourselves in space, to know it from the inside, through contact and proximity. Walking – the first anthropic sign of demarcation, appropriation, and mapping of territories (Careri 2006) – is also a spatializing practice: a practice producing space (Lefebvre 1991).

This view is particularly important in French literature, where considerable emphasis is put on walking as a way to reappropriate and rewrite the urban. In *The Practice of Everyday Life*, Michel de Certeau starts the chapter devoted to walking in the city by comparing the view of Manhattan from the World Trade Center with the experience of the passer-by. He also underscores the role of the body by emphasizing its exclusion in the view 'from above':

> To be lifted to the summit of the World Trade Center is to be lifted out of the city's grasp. One's body is no longer clasped by the streets that turn and return it according to an anonymous law; nor is it possessed whether as player or played, by the rumble of so many differences and by the nervousness of New York traffic […]. His elevation transfigures him into a voyeur.
> (De Certeau 1984: 92)

However, for De Certeau, pedestrians' practices are not only a way to read and perceive, but also (and first of all) a form of *writing* of urban space. In his words, the city becomes 'an urban "text" [passers-by] write without being able to read it' (De Certeau 1984: 93). By establishing a clear dichotomy between planners and users and assigning to the latter the possibility of reshaping the spatial order imposed from above, De Certeau interprets walking as one of those resistance tactics through which users can reconfigure the dominant cultural economy. Accordingly, he explicitly refers to the linguistic system by comparing walking to a speech act:

> The act of walking is to the urban system what the speech act is to language or to the statements uttered. At the most elementary level, it has a triple 'enunciative' function: it is a process of appropriation of the topographical system on the part of the pedestrian […], it is a spatial acting-out of space […]; and it implies relations among differentiated positions, that is, among pragmatic 'contracts' in the form of movements.
> (De Certeau 1984: 97–98)

As a linguistic system, the functionalism of urban planning provides a system of use and control of spaces, setting up a number of possibilities, rules, and interdictions. Pedestrians actualize and put this system into use, but they can also redefine, reinvent, or deny its rules, thereby creatively rewriting urban space through their personal and social practices.

This linguistic metaphor returns in the writings of several authors and notably in Jean-François Augoyard's *Step by Step*. In this examination of the everyday walking patterns in a newly built borough in Grenoble, Augoyard proposes a rhetoric of walking and describes this practice as a form of interaction between the pedestrian's individuality and the organization of the built environment, as an act of articulation of the urban spatial structure, as a way to read and rewrite space:

> Walking resembles a reading-writing. Sometimes rather more following an existing path, sometimes rather more hewing a new one, one moves within a space that never tolerates the exclusion of one or another [...] the succession of steps effectively rewrites the space that opens before the walker, even when done in the slightest of action modes.
>
> (Augoyard 2007: 25)

From this perspective, urban spaces are not stable and inert formations, but are activated and actualized by the practices of those who cross them, by the 'legs' generative grammar' (Bailly 1992) put in place through walking. Therefore, walking becomes a methodology to read, but also to reappropriate and rewrite space.

Soundwalking as interacting with and rewriting auditory spaces

Brandon LaBelle translates this process in the auditory realm:

> The urban soundscape is itself a material contoured, disrupted, or appropriated through the meeting of individual bodies and larger administrative systems. From crosswalk signals, warning alarms, and electronic voices, the urban streets structure and audibly shape on a mass scale the trajectories of people on the move. In contrast, individuals supplement or reshape these structures through practices that, like de Certeau's walker, form a modulating break or interference.
>
> (LaBelle 2010: 92)

This interference, for the walking listener, can take the form of an interaction between her own sounds, those already travelling through the spaces she traverses and the acoustic features of the environment, transforming the city in a space-time multiplicity created and recreated in the contingency of her mobile experience.

By directing our attention to listening, soundwalking always also implies a relational way of experiencing or, better, of engaging with space. Listening provides information not only on the nature of the objects and subjects that inhabit the world but, first of all, on their mutual relationships, their constant becoming, their simultaneity. In other words, from the listening point of view, urban space is never static or inert, but is a field of dynamic, temporary, and processual relationships: 'a spatio-temporal geography, a dynamic geography of events rather than images, or activity rather than scene' (Rodaway 1994: 90). This geography is never independent of the one who crosses it. As Paul Rodaway writes in *Sensuous Geographies*: 'The soundscape moves with the sentients as they move through the environment and it continually changes with our behavioral interactions' (87). The mobile ear of the soundwalker experiences the city as a continuous series of events in perpetual movement, a dynamic multiplicity that is constantly generated in relationship with her changing positions and behaviours. In other words, she never has an external position but always participates in the soundscape she listens to, contributing to the ongoing constellation of sound events that is always already there in urban space and interacting with her environment through the reciprocal relationship that always exists between a sound event and the acoustic properties of the space where it propagates. Her own sounds spread through space and are reshaped according to its material and acoustic qualities creating different effects, such as resonance, reverberation, reflection, absorption. From an auditory point of view, then, we always both perceive and interact with auditory spaces as we traverse them. We always read and rewrite them.

Playing the city: Acoustic interactions

In current acoustic ecology's soundwalks, the most widespread and canonized 'format' is the listening walk: a silent exploration of a sonic environment guided by a leader who selects the itinerary and suggests the listening approach (McCartney 2010). However, as Barry Truax's *The Handbook for Acoustic Ecology* suggests, soundwalking can be reinforced through sound-making:

> In order to expand the listening experience, sound-making may also become an important part of a soundwalk. Its purpose is to explore sounds that are related to the environment, and, on the other hand, to become aware of one's own sounds (voice, footsteps, etc.) in the environmental context.
>
> (Truax 1999)

If we look at the World Soundscape Project's early definitions and related map-scores, sound-making appears to have an even wider scope. In *The Tuning of the World*, for example, after defining the 'sound-walk' and the 'listening walk', Schafer introduces several examples based on the idea of playing the environment and interacting with the context. He describes a walk where participants were asked to enter a store and tap on the top of tinned goods, thus 'turning the grocery shop into a Caribbean steel band' as well as another walk where participants had to 'sing tunes around the different harmonics of neon lights' (Schafer 1994: 212–213).

This interaction is further emphasized in several walks published in the *European Sound Diary* (1975): here, we find some map-scores conceived for different cities comprising an itinerary and instructions for listening to particular environments, but also for interacting with their soundscapes or their acoustic and architectural qualities. The *Vienna Soundwalk: Evening in the Old Town*, for example, starts in the St Stephan Cathedral asking us to listen to its quiet atmosphere and to imagine the sound of its bell, but continues by suggesting to enter a passageway and 'move through it playing with your fingers or a pencil on the shop shutters, grills on the floor, elevator grate, metal gate bars' (Schafer 1977: 83), to whistle while traversing an arch, to stomp on the wooden floor of a telephone booth. The instructions for Hildegard Westerkamp's *A Soundwalk in Queen Elizabeth Park in Vancouver* (1974) follow the same direction. Westerkamp invites us to play Henry Moore's metal sculpture, to listen to the sound of our own footsteps, or to clap our hands underneath a bridge to try to produce an echo and activate the acoustics of space.

In these early examples, soundwalking – an ambulatory auditory exploration of the environment – was equally understood as a practice of ambulatory sound-making to play the environment, to establish a process of reappropriation and interaction with sites through action and site-specific sound-making. However, in the World Soundscape Project practice, sound-making is often subordinated to listening, and 'playing the environment' while traversing it doesn't become a research or artistic methodology in itself. Conversely, this idea is often central in many ambulatory music and sound projects.

Michael Parsons's *Echo Piece (Canary Wharf)* (2009), for example, is a site-specific ambulatory composition for wind instruments in the homonymous district in London. This open piece is composed of short single notes played by a group of performers with trumpets, horns, and trombones 'moving around and exploring the acoustic properties of an open-air space' (Center for the Aesthetic Revolution 2009). Players activate the environment by evoking echoes from the sound-reflective surfaces of the glass buildings in this financial area, developing a performance that becomes also a one-hour walk through it. Thus, *Echo Piece* is based on an acoustic interaction transforming the city's architectural fabric into an expanded resonant chamber.

If *Echo Piece* makes use of traditional musical instruments, several projects use voice to enter into a dialogue with the environment. Voice as a vehicle of subjectivity and intersubjectivity, as the primary medium we use to communicate, to express ourselves, to look for reciprocity, to enter into relationships with the others and the world. In Viv Corringham's *Shadow Walks*, for example, the exploration of the environment becomes also a relational practice of vocal improvisation involving its inhabitants (Corringham 2010). For this series, which started in 2003, the artist asks local inhabitants to accompany her on their favourite walk to tell their personal histories and memories on the places they traverse. She records these conversations along the route and then comes back alone on the same path improvising with voice. Her improvisation is based on the acoustic qualities of the sites and their soundscapes, but is also the result of her meeting with the inhabitants and their stories, recognizing the polyphonic nature of places. Voice becomes here a powerful way to dialogue with the environment, thanks to its embodied materiality as well as to its 'paradoxical topology' (Dolar 2006: 73) between inside and outside, between self and the world. In other words, Corringham's site-specific ambulatory singing brings to the fore the negotiation of the interior and the exterior that is always already embedded in voice. At the same time, it makes use of voice in its 'post-linguistic' form: voice beyond language and *logos*, as pure sound, as vibration emanating from the body to diffuse and interact with space and, as such, as a 'sonorous self-revelation that overcomes the linguistic register of signification' (Cavarero 2005: 176).

This acoustic conversation between self and environment is also explored through extra-aesthetic or anti-aesthetic acts and gestures. The sound of footsteps features in many projects exactly with this function (see Biserna 2018). The step – the primary physical contact between the walker and the environment – establishes an embodied relationship with sites made audible through the sound of our footsteps. Generated through the interaction between our moving body and the materials and surfaces of the environment – between our feet and the ground and acoustic features of the surroundings – this sound projects our presence in space, activating it through sounding. As such, it continuously rewrites our auditory situation and actively interlaces with the many other rhythms and auditory dynamics taking place in urban space by means of our own personal rhythm, connected to our gait. katrinem's *SchuhzuGehör path of awareness*, for example, is a series of site-specific walks developed in different cities around the world using this sound to investigate their structures and their architectural and atmospheric qualities. Starting from an on-site study and observation of local walking patterns and habits as well as from repeated explorations of a chosen area, the artist plans a route – a 'path of awareness', as she calls it – designed to emphasize the self-consciousness of the walker with regards to his relationship with space through the interplay between sound events (footsteps) and the surrounding architecture. Through guided walks and site-specific scores inviting the public to wear their most resonant shoes, *SchuhzuGehör path of awareness* articulates a mode of subjective listening based on the acoustic interactions between our feet and the environment. The artist explicitly compares the shoes to musical instruments, asking us to refocus our attention on our footsteps *as if* we were listening or participating in a musical performance to recognize our dialogue with the city, to carve out an embodied ecology, to locate ourselves in space, and to infiltrate the urban polyphony with our own specific rhythm, along with those of other pedestrians.

Walking as sharing, reclaiming, and occupying public space

This 'amplification' of our presence in the environment and interference with urban space's polyrhythms finds another framework in a large body of works in which walking is discussed against the background of the dynamics of access to and sharing of public space.

In this perspective, walking is, first of all, a way of 'being present in the public space' (Gehl 2001: 135) and of reclaiming the 'right to the city' (Lefebvre 1996). Secondly, it becomes a way to encounter the 'Other', to expose oneself to the social, economic, and cultural complexity of urban life as well as 'all the difference of age, taste, background, and belief that are concentrated in a city – and aroused by the diversity around them' (Sennett 1992: 122). It is the 'walking "between"' as described by Franco La Cecla, 'the democratic walking of those who move in the city and meet both known and unknown people' (La Cecla 2011: 76); a way of being in the presence of the unfamiliar, of the unexpected, of the stranger, thus enriching our experience and experimenting with forms of living together as well as of friction and conflict.

The urban condition, here, is understood as a forum of interactions with strangers, a platform for facing, recognizing, and supporting difference and complexity in the public sphere, all interpreted as key elements of democratic life (Jacobs 1961; Sennett 1977; Young 1990; among others). The street becomes a platform for public life and walking a methodology to cultivate and reclaim it. As Rebecca Solnit states: 'Walking the streets is what connects [...] the personal microcosm with the public macrocosm. [...] Walking maintains the publicness and viability of public space' (Solnit 2001: 176).

This claim on the street as a primary site for sharing and participating in political life is also at the heart of the (often sonorous) campaigns of several groups fighting for the enlargement of the public sphere, such as Reclaim the Streets or Critical Mass. Their countercultural and sometimes carnivalesque street actions often explicitly refer to Situationism for which, through the *dérive*, walking became a dissensual practice aimed at analysing but also transforming the environment of everyday life through collective participation. For the situationists, walking and drifting were both a method for psychogeographic research and a revolutionary tool, a way to counteract the functionalism, rationalism, and alienation imposed on everyday life and urban planning (Sadler 1999).[4] Their programme aimed at the suppression of art into politics by radically transforming the city, its ordering, and its power dynamics to finally transform its life.

This permeability comes back very often in the arts, where walking emerges primarily as a privileged means to engage the urban in its many layers, crossing disciplinary boundaries, abandoning institutional venues, and infiltrating the everyday. From the first Dadaist excursions in Paris to the many contemporary walking artists, walking is above all an 'act of presence' (Ardenne 2004: 88) in public space.[5]

Playing the city: Occupying acoustic public spaces

Playing the city can also assume this role. In several music and sound projects ambulatory sound-making becomes a tool to subvert the urban's auditory order, to reclaim difference and dissonance in public space, to promote and make audible collective presence and action.

Urban soundscapes reflect wider principles of spatial organization that also correspond to dynamics of power, control, and privatization. The sociologist Rowland Atkinson uses the terms 'sonic ecology' to emphasize the power of sound and music to demarcate and connote space according to patterns related to use, to the social, functional, and cultural characteristics of the different parts of the city, as well as to their timing. In this way, the city is organized into 'acoustic territories', namely, 'spaces defined, owned or contested by those who, relatively speaking, control the soundscape of public and private spaces' (Atkinson 2007: 1910). In other words, the soundscape is not only organized but also 'socially organizing' (1907). Playing the city, making noise, in this regard, becomes a tactics to reappropriate, activate, or unsettle these territories, interfering with these patterns and disrupting the social organization of public space and life.

In the text/score *Suonare la città* (Playing the city), the Italian composer affiliated to Fluxus, Giuseppe Chiari proposes precisely this:

> playing the city is – can be – also playing through –
> in the city
> but playing the city can also be playing (direct object)
> the city. Where the city is the object that receives the action of playing
> where the city replaces the word violin
> in the expression to play the violin
> […]
> the city as an instrument
> as a musical instrument.
>
> (Chiari 1972, my translation)

For Chiari, the purpose of this intervention is very clear: he writes it six times in capital letters: 'to play out of tune'. The aim is to create interferences, to infiltrate the regulated rhythms of public space to disrupt the order imposed on everyday life and to open spaces for shared experimentation. 'To interrupt a concert, a concert of people playing conventionally the same score by heart' (Chiari 1972). A purpose that is perfectly aligned with the refusal of the autonomy of art for direct action in everyday life – or, better, the constant connection between art and life – proposed by Fluxus. In this text and in *Suonano la città* – an action presented at Campo Aperto in Como in 1969 –Chiari not only abandons the institutions of music to invest public space and collaborate with the inhabitants, he not only proposes extramusical gestures (such as tapping on shutters, metal gates, etc.), but he also directly questions the urban condition in its entirety: its limitations, its codifications, its regulations. Ultimately, playing the city is for Chiari a joyful subversion of the imposed (auditory) order, a contextual act rewriting the urban and its senses and the activation of a collective creativity diluted in social practices. The city, in its physical, social, and political dimensions, becomes a context and a material. It becomes an expanded instrument.

Between the end of the 1960s and the beginning of the 1970s, comparable actions were developed in England among members of the Scratch Orchestra. Inspired by John Cage and Fluxus, this collective, founded in 1969 by Cornelius Cardew, Michael Parsons, and Howard Skempton, experimented with improvisation, indeterminacy, and openness by further radicalizing the redefinition of the pre-established divisions between composer, performer, and public. The collective dimension and the participatory attitude of the Scratch Orchestra led the group not only to move away from institutional venues but also to adopt a much more explicit political agenda. In the Scratch Orchestra's practice, ambulatory sound-making appears in several verbal scores and itinerant projects, but the most striking event in this respect is the *Richmond Journey* (1970), planned by Stefan Szczelkun and presented in the general score as 'a day long concert as a journey throughout Richmond'. On Saturday, 16th May, the collective's members invaded this borough following a path

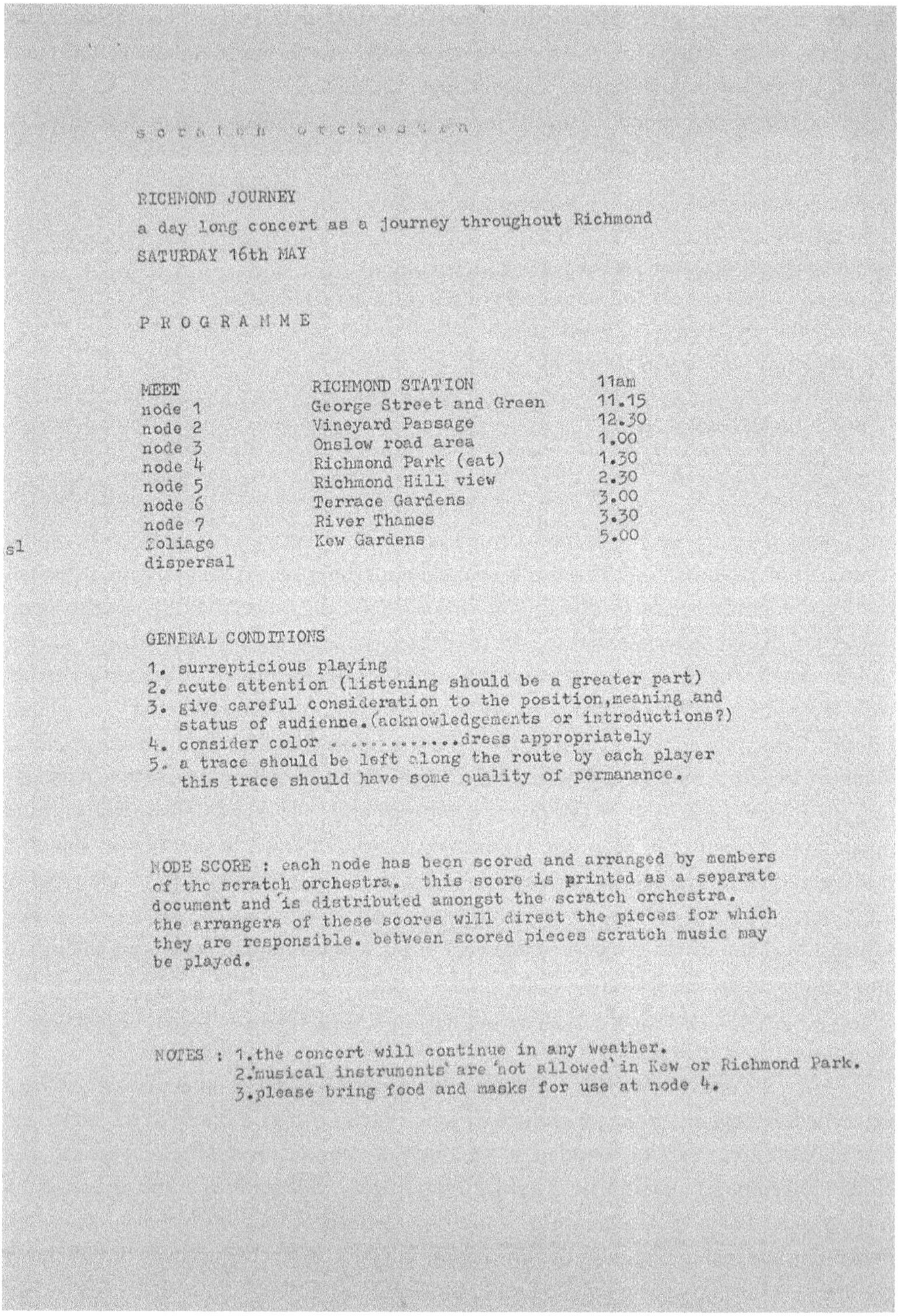

scratch orchestra

RICHMOND JOURNEY
a day long concert as a journey throughout Richmond
SATURDAY 16th MAY

PROGRAMME

MEET	RICHMOND STATION	11am
node 1	George Street and Green	11.15
node 2	Vineyard Passage	12.30
node 3	Onslow road area	1.00
node 4	Richmond Park (eat)	1.30
node 5	Richmond Hill view	2.30
node 6	Terrace Gardens	3.00
node 7	River Thames	3.30
foliage dispersal	Kew Gardens	5.00

GENERAL CONDITIONS

1. surrepticious playing
2. acute attention (listening should be a greater part)
3. give careful consideration to the position, meaning and status of audience. (acknowledgements or introductions?)
4. consider color ....dress appropriately
5. a trace should be left along the route by each player this trace should have some quality of permanance.

NODE SCORE : each node has been scored and arranged by members of the scratch orchestra. this score is printed as a separate document and is distributed amongst the scratch orchestra. the arrangers of these scores will direct the pieces for which they are responsible. between scored pieces scratch music may be played.

NOTES : 1. the concert will continue in any weather.
2. 'musical instruments' are 'not allowed' in Kew or Richmond Park.
3. please bring food and masks for use at node 4.

Figure 17.1 Scratch Orchestra, *Richmond Journey*, 1969, programme. MayDay Rooms Archives. Courtesy Stefan Szczelkun.

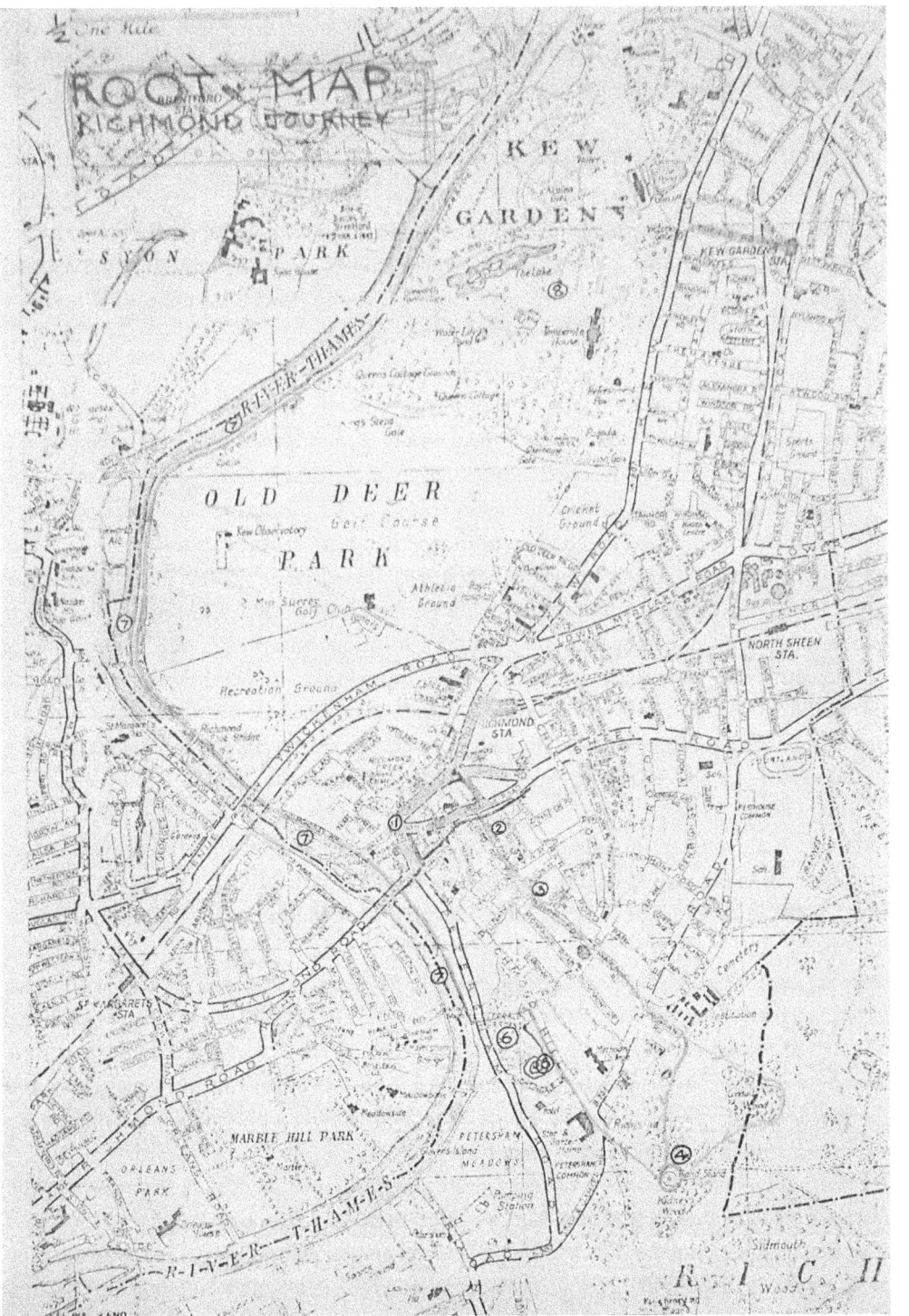

Figure 17.2 Scratch Orchestra, *Richmond Journey*, 1969, map with the itinerary. MayDay Rooms Archives. Courtesy Stefan Szczelkun.

imagined as an allegory of a revolt (Figures 17.1 and 17.2). The route was divided into eight 'nodes' corresponding to different areas and, on each section, a member proposed collective actions in the form of scores or instructions. Stefan Szczelkun recalls:

> We began by attempting to break the 'claustrophobic spell of capitalist normalcy': Richmond High Street was to be disrupted! We would then pay respects to our ancestors before climbing up through the residential district – recruiting deadened office workers. Our growing ranks would proceed to the top of the hill, to Richmond Park, to celebrate our connection to nature and reclaim the heights. After a break to eat we would descend through the steep Thames meadows and follow the great river on to our destination – that benign archive of the earth's flora, Kew Gardens.
>
> (Szczelkun 2019)

As Chiari's text, the event aimed to infiltrate the neighbourhood's life to subvert it through actions such as intervening in Richmond High Street's stores or awakening the residential area by ringing doorbells, knocking on doors, and so on.

At the end of the 1960s, then, ambulatory sound-making enters music experimentation to expand its contexts and practices. On the one hand, it becomes a paradigm of a wider interest in ordinary – 'infra-ordinary' (Perec 1989), I would say – actions that characterizes some of the tendencies of the period, from dance to visual arts. On the other hand, it becomes a way to radically abandon cultural institutions and plunge into the polyphony of the world, claiming public space as a field of intervention and embracing everyday life through actions that could be compared to the situationists' 'constructed situations': 'a moment of life concretely and deliberately constructed by the collective organization of a unitary ambiance and a game of events' (Situationist International [SI] 1958: 12).

This will to intervene in public space, to play the city while traversing it, continues today in a wide range of ambulatory performances, although sometimes with a less utopic tonality than in the projects mentioned before. Francis Alÿs's *Railings* (2004), for example, uses the city as an expanded instrument to be played while walking. For this performance in London resulting in a three-channel video projection and a series of maps, photographs, and sketches, the artist reveals a hidden urban rhythm and enters in resonance with space by using a pair of drum sticks to play the railings enclosing the city's squares and buildings. The action makes audible this previously purely visual rhythm and amplifies the subtle threshold between private and public space – between the space of the pedestrians and the space that remains inaccessible – thus investigating and emphasizing persisting power structures embedded in the city's urban fabric.

Marches is a performance and series of scores developed by the artist Lawrence Abu Hamdan in 2005 and presented in various cities, including London, Glasgow, Lisbon, and Santarcangelo. As katrinem's *SchuhzuGehör_path of awareness*, *Marches* uses the footstep as a means 'to exemplify the aural capacity to delineate space, treating architecture like dormant music, awakening it through the act of walking' (Abu Hamdan 2008: 3), but this project intervenes in the city by means of a collective performance. A group of performers meet and disperse in urban space according to the artist's instructions to interact with the architectures of the city and to create reflections, resonances, and other effects. The performers wear special shoes made by the artist in collaboration with artisans and designed to amplify this acoustic interaction. The performers' trajectories are also mapped out according to historical or sociological reasons: Abu Hamdan explored the histories of the cities to find stories of parades, marches, or protests that took place in the same area, thus allegorically echoing these past events. Therefore, the sound of the footsteps – emphasized by the customized shoes – allows the artist to rewrite and occupy the city, playing on its architectural and acoustic features as well as reverberating its cultural, social, and political history.

A project by Ligna – consisting of radio, media, and performance artists and theorists Ole Frahm, Michael Hüners, and Torsten Michaelsen – charges this collective unannounced presence in public space through itinerant sound-making with an explicit critique to its privatization and commercialization. *The Future of Radio Art* (2005) is an urban intervention that took place for the first time in one of Amsterdam's commercial areas, with several performers wandering around, carrying plastic bags and mixing with the crowd of shoppers. Yet their bags concealed ghetto blasters broadcasting a radiophonic monologue for one voice. This voice dispersed in public space became a denunciation of its increasing privatization, a commentary on urban space power dynamics as well as a materialization of the potential of radio to connote and rewrite the situation of reception.

If Ligna's project plays on the threshold between radio and public space and between the aesthetic and the everyday sphere, other projects directly transcend art in political organization and action. Take a Stand Marching Band, for example, is a street performance group initiated

Figure 17.3 Elana Mann, Take a Stand Marching Band, documentation of the Los Angeles May Day march, 1st May 2017. Photograph by Nateene Diu. Courtesy the artist.

Figure 17.4 Elana Mann, Take a Stand Marching Band, documentation of the Los Angeles May Day march, 1st May 2017. Photograph by Nick Popkey. Courtesy the artist.

by Elana Mann in 2017, on the occasion of the Los Angeles Women's March on 21st January. The group has joined marches and demonstrations amplifying their voice against Donald Trump's administration with the artist's sculptures 'hands-up-don't-shoot-horn' and the 'histophone': two prosthetic 'instruments for the human voice retooled from musical horns and megaphones in solidarity with social justice movements' (Mann n.d.). Cast from the human body, these sculptures cover the mouth of the speakers with one hand, while in fact amplifying their voice thanks to a trumpet, metaphorically and materially counteracting the silencing of people's voices (Figures 17.3 and 17.4). Playing the city, here, becomes a gesture of collective disobedience recalling a whole tradition of mobile sound-producing and vocal tactics enacted in demonstrations and parades to interrupt the power structures of public speech, to give voice to those who are usually unheard or to vocalize a radical disidentification with the status quo by being present and audible through public space.

'Presencing' urban space

An urban soundscape can be defined as a shared ecology where the city's material, cultural, social, and political dynamics are audible and always open to multiple individual and collective negotiations and interactions. In this context, the walking body always establishes a plurality of auditory relationships. It perceives and reads the urban becoming and its spatio-temporal multiplicity in a situated, affective, immersive way – such as in the flâneur's visual practice or in listening walks – but, at the same time, it always reappropriates and rewrites its structure by interacting with its polyrhythm, its acoustics, its difference, its ordering, its public and private territories and their permeability.

Ambulatory sound-making reinforces this interaction by amplifying a personal or collective presence in public space and becomes a platform for self-representation, agency, or transformation. This presence can be harmonic or disharmonic, producing consonances or dissonances. It can carve out an embodied auditory geography, inscribing the body in a conversation with the environment and establishing a dialogue with the city's architectural fabric and social life. It can project a radical dissent, challenging and subverting urban representations, functions, and practices, questioning and disrupting the sonic order or infiltrating and occupying the rhythms of public life. In any case, it is a matter not only of perceiving and knowing but of dwelling and participating: of 'presencing' urban space.

Notes

1. Visual practice is fundamental to *flânerie*. Although in *The Arcade Project* it is possible to find several observations on sound and the acoustic phenomena characterizing the modern metropolis, Walter Benjamin himself affirms, 'the category of illustrative seeing […] is basic for the *flâneur*' (Benjamin 1999: 419).
2. On the contrary, the role of listening and hearing is often marginal even for the thinkers who have most insisted on the bodily nature of walking. Le Breton, for example, devotes a paragraph of his *Eloge de la marche* to the auditory dimension of urban space but proposes a negative interpretation of it, while he recognizes the centrality of vision in the intersubjective relationships typical of urban experience: 'Urban sociability induces an excrescence of the gaze and a suspension or residual use of the other senses' (Le Breton 2008: 162). On the distinction between 'hearing' and 'listening', see Barthes (1982) and Kassabian (2013), among others.
3. Moreover, in the same years of the WSP, in France CRESSON began to develop peripatetic methodologies for research on sound effects in urban environments (Thibaud 2001).
4. The Situationist International (SI), formed in 1957, was an organization composed of writers, artists, intellectuals, and political theorists, and active in several countries in Europe up to its dissolution in 1972. The SI analysed and organized actions against the alienation and commodification of everyday life under the regime of the Spectacle and the capitalist mode of production. Accordingly, during the SI's first phase, several members worked on notions such as 'unitary urbanism', 'constructed situations', and 'psychogeography' conceived as methods to study and liberate everyday life. Psychogeography, in particular, was first proposed by Guy Debord (one of the founders of the SI) in his *Introduction to a Critique of Urban Geography* as 'the study of the precise laws and specific effects of the geographic environment, consciously organized or not, on the emotions and behavior of individuals' (Debord 1955). Debord theorized the behavioural and emotional impact of urban space on human beings and proposed to analyse its effects to lay the foundations of a new environment designed according to

the desires of its inhabitants. The main tool for this study was the *dérive*, described in the first number of the *Internationale Situationniste* as 'a mode of experimental behavior linked to the conditions of urban society: a technique of rapid passage through varied ambiances' (SI 1958). Therefore, drifting – walking without an aim or goal, following the attractions and desires arisen by the environment – became a method to experiment with, register, and understand urban atmospheres and their effects on human behaviour and affects, as well as the base of a new cartography and a subversive way to counteract the productivity and consumerism imposed on everyday life through a collective practice of disorientation, an experimental way of inhabiting places, and a different use of space and time.
5. The literature on walking in the visual arts is quite extensive: Hollevoet 1992; Davila 2002; Baqué 2006; Careri 2006; Evans 2012; O'Rourke 2013; Waxman 2017. Among exhibition catalogues: Arasse 2000; Horodner 2002.

References

Abu Hamdan, L. (2008). *Marches* [booklet]. Available online: https://www.thewire.co.uk/files/pdf/marching.pdf (accessed: 24 April 2018).

Arasse, D. (ed.) (2000). *Un siècle d'arpenteurs: Les figures de la marche*. Antibes: RMN-Musée Picasso.

Ardenne, P. (2004). *Un art contextuel: creation artistique en milieu urbain, en situation, d'intervention, de partecipation*. Paris: Flammarion.

Atkinson, R. (2007). 'Ecology of Sound: The Sonic Order of Urban Space'. *Urban Studies* 44 (10): 1905–1917.

Augoyard, J.-F. (2007). *Step by Step: Everyday Walks in a French Urban Housing Project*. Minneapolis, MN: University of Minnesota Press.

Bailly, J. C. (1992). *La ville à oeuvre*. Paris: Bertoin.

Baqué, D. (2006). *Histoires d'ailleurs: Artistes et penseurs de l'itinérance*. Paris: Regard.

Barthes, R. (1982). 'Écoute'. In *L'Obvie et l'Obtus: Essais critiques III*, 217–230. Paris: Seuil.

Benjamin, W. (1999). *The Arcades Project*. Cambridge, MA: Belknap Press of Harvard University Press.

Biserna, E. (2018). '"Step by Step": Reading and Re-writing Urban Space Through the Footstep'. In Caleb Kelly (ed.), 'Materials of Sound', special issue of *Journal of Sonic Studies* 16. Available online: https://www.researchcatalogue.net/view/456238/456239 (accessed 12 May 2018).

Bull, M. (2007). *Sound Moves: iPod Culture and Urban Experience*. London: Routledge.

Careri, F. (2006). *Walkscapes. Camminare come pratica estetica*. Turin: Einaudi.

Cavarero, A. (2005). *For More than One Voice. Toward a Philosophy of Vocal Expression*. Stanford, CA: Stanford University Press.

Center for the Aesthetic Revolution (2009). '*Echo Piece (Canary Wharf)* 2009 by Michael Parsons' [Blog] June. Available online: http://centrefortheaestheticrevolution.blogspot.fr/2009/06/echo-piece-by-michael-parsons-curated.html (accessed 24 April 2018).

Certeau, M. de (1984). *The Practice of Everyday Life*. Berkeley, CA: University of California Press.

Chiari G. (1972) 'Suonare la città'. *In | Argomenti e immagini di design* 6: 11–17.

Corringham, V. (2010). 'Shadow Walks'. Talk given at the Deep Listening Institute, Kingston, NY. Available online: http://vivcorringham.org/shadow-walks (accessed 24 April 2018).

Davila, T. (2002). *Marcher, Créer: Déplacements, flâneries, dérives dans l'art de la fin du XXe siècle*. Paris: Regard.

Debord, G.-E. (1955). *Introduction to a Critique of Urban Geography*. Available online: http://library.nothingness.org/articles/SI/en/display/2 (accessed 24 April 2018).

Dolar, M. (2006). *A Voice and Nothing More*. Boston, MA: MIT Press.

D'Souza A. and T. McDonough (eds) (2006). *The Invisible Flâneuse?: Gender, Public Space, and Visual Culture in Nineteenth-Century Paris*. Manchester: Manchester University Press.

Drever, J. L. (2009). 'Soundwalking: Aural excursions into the Everyday'. In James Saunders (ed.), *The Ashgate Research Companion to Experimental Music*, 163–192. Farnham: Ashgate.

Elkin, L. (2016). *Flâneuse. Women Walk the City in Paris, New York, Tokyo, Venice, and London*. London: Chatto & Windus.

Evans, D. (ed.) (2012). *The Art of Walking. A Field Guide*. London: Black Dog.

Gallagher, M. and J. Prior (2017). 'Listening Walks: A Method of Multiplicity'. In C. Bares and A. R. Taylor (eds), *Walking through Social Research*. New York: Routledge. Available online: https://orca.cf.ac.uk/93168/1/Gallagher%2C%20M.%20%26%20Prior%2C%20J.%20Listening%20walks%20-%20a%20method%20of%20multiplicity.pdf (accessed 16 July 2020).

Gehl, J. (2001). *Life Between Buildings: Using Public Space*. Copenhagen: Arkitektens Forlag.

Gluck, M. (2010). 'Flâneur'. In R. Hutchison, M. Aalbers, R. Beauregard, and M. Crang (eds), *Encyclopedia of Urban Studies*, 272–274. London: Sage.

Hollevoet, C. (1992). 'Déambulations dans la ville, de la flânerie et la dérive. L'appréhension de l'espace urbain dans Fluxus et l'art conceptuel'. *Parachute* 68: 21–25.

Horodner, S. (ed.) (2002). *Walk Ways*. New York: Independent Curators.

Jacobs, J. (1961), *The Death and Life of Great American Cities*, New York: Random House.

Kassabian, A. (2013). *Ubiquitous Listening Affect, Attention, and Distributed Subjectivity*. Berkeley: University of California Press.

Küppers, P. (2010). 'Moving in the Cityscape: Performance and the Embodied Experience of the flâneur'. In N. Whybrow (ed.), *Performance and the Contemporary City: An Interdisciplinary Reader*, 54–68. New York: Palgrave-Mcmillan.

LaBelle, B. (2010). *Acoustic Territories: Sound Culture and Everyday Life*. New York: Continuum.

La Cecla, F. (2011). 'Puzzle cartografici'. In L. Pignatti (ed.), *Mind the Map: Mappe, diagrammi e dispositivi cartografici*, 71–77. Milan: Postmedia.

Le Breton, D. (2000). *Eloge de la marche*. Paris: Métailié.

Le Breton, D. (2008). *Anthropologie du corps et modernité*. Paris: PUF.

Lefebvre, H. (1996). 'Right to the City'. In *Writings on Cities*, 61–181. Oxford: Blackwell.

Lefebvre, H. (1991). *The Production of Space*. Oxford: Blackwell.

Mann, E. (n.d.). 'The Assonant Armory'. Available online: https://www.elanamann.com/project/assonant-armory (accessed 24 April 2018).

McCartney, A. (2014). 'Soundwalking: Creating Moving Environmental Sound Narratives'. In S. Gopinath and J. Stanyek (eds), *The Oxford Handbook of Mobile Music Studies*, vol. 2, 212–237. Oxford: Oxford University Press.

McCartney, A. (2010). 'Soundwalking and Improvisation'. *ImprovCommunity*. Available online: http://www.improvcommunity.ca/sites/improvcommunity.ca/files/research_collection/458/soundwalking_and_improvisation.pdf (accessed 24 April 2018).

O'Rourke, K. (2013). *Walking and Mapping: Artists as Cartographers*. Cambridge, MA: MIT Press.
Paquette, D. and A. McCartney (2012). 'Soundwalking and the Bodily Exploration of Places'. *Canadian Journal of Communication* 37: 135–145.
Perec, G. (1989). *L'Infra-ordinaire*. Paris: Editions Seuil.
Pink, S. (2009). *Doing Sensory Ethnography*. London: Sage.
Pink, S., P. Hubbard, M. O'Neill, and A. Radley. (2010). 'Walking Across Disciplines: From Ethnography to Arts Practice'. *Visual Studies* 25 (1): 1–7.
Rodaway, P. (1994). *Sensuous Geographies: Body, Sense and Place*. London: Routledge.
Sadler, S. (1999). *The Situationist City*. Cambridge, MA: MIT Press.
Sansot, P. (2000). *Chemins aux vents*. Paris: Payot & Rivages.
Schafer, R. M. (1977). *European Sound Diary*. Vancouver: A. R. C. Publications.
Schafer, R. M. (1994). *The Soundscape: Our Sonic Environment and the Tuning of the World*. Rochester, VT: Destiny Books.
Sennett, R. (1977). *The Fall of Public Man: On the Social Psychology of Capitalism*. Cambridge: Cambridge University Press.
Sennett, R. (1992). *The Conscience of the Eye. The Design and Social Life of Cities*. New-York: W. W. Norton Company.
Situationist International (SI) (1958), 'Définitions'. *Internationale Situationniste* 1. Available online: http://debordiana.chez.com/francais/is1.htm#definitions (accessed 24 April 2018).
Solnit, R. (2001). *Wanderlust: A History of Walking*. London: Penguin.
Szczelkun, S. (2019). 'Some open Artists Collectives UK 1968 – 1997'. *Draft writings by Stefan Szczelkun*, 13 June. *The Scratch Orchestra*. Available online: http://stefan-szczelkun.blogspot.com/2019/06/artists-collectives-uk-1968-1997.html (accessed 18 July 2020).
Tester, K. (1994). *The Flâneur*. New York: Routledge.
Thibaud, J. P. (2008). 'Je, Tu, Il. La marche aux trois personnes'. *Urbanisme* 359: 63–65, http://doc.cresson.grenoble.archi.fr/doc_num.php?explnum_id=331 (accessed 22 April 2018).
Thibaud, J. P. (2001). 'La methodologie des parcours commentés'. In M. Grosjean and J. P. Thibaud (eds), *L'espace urbain en methodes*, 79–100. Marseille: Parenthèses.
Thomas, R. (ed.) (2010). *Marcher en ville. Faire corps, prendre corps, donner corps aux ambiances urbaines*. Paris: Éditions des archives contemporaines.
Truax, B. (1984). *Acoustic Communication*, Norwood, NJ: Ablex Publishing Corporation.
Truax, B. (ed.) (1999). *The Handbook for Acoustic Ecology*. Available online: https://www.sfu.ca/sonic-studio/handbook/index.html (accessed 22 April 2018).
Waxman, L. (2017). *Keep Walking Intently: The Ambulatory Art of the Surrealists, the Situationist International, and Fluxus*. Berlin: Sternberg Press.
Westerkamp, H. (1974). 'Soundwalking'. *Sound Heritage* 3 (4): 18–27.
Westerkamp, H. (2010). 'What's in a Soundwalk?'. *Talk at Sonic Acts XIII*, Amsterdam. Video available online: https://vimeo.com/12479152 (accessed 22 April 2018).
Wunderlich, F. M. (2008). 'Walking and Rhythmicity: Sensing Urban Space'. *Journal of Urban Design*, 13 (1): 125–139.
Young, I. M. (1990). *Justice and The Politics of Difference*. Princeton, NJ: Princeton University Press.

18

Sound Installations for the Production of Atmosphere as a Limited Field of Sounds

Jordan Lacey

Introduction

The term 'sound installation' was introduced by Max Neuhaus, who 'distinguishes the genre from music by indicating that, in sound installation, sounds are "placed in space rather than time"' (Ouzounian 2008: 6). By this, he infers that a sound installation requires no temporal structure, but instead facilitates the placement of sounds in space. Neuhaus was a percussionist who worked with leading experimental composers (including John Cage and Edgard Varèse), but he turned away from the concert hall becoming instead interested in how the introduction of sounds into everyday environments could impact listening (Neuhaus 2004). In doing so, Neuhaus took sound from the concert hall into the streets initiating a new form of listening-based public art. It is this positioning of sound installations as a spatial art form, and its concomitant recontextualizing of music, which is of interest to the methodology presented in this chapter.

It should be noted that sound installations can be understood more broadly than this. Ouzounian, who wrote a thesis on the theme of sound installation art, proposes:

> Sound installations may be site-specific or not [...]; they may include performance, recording, or broadcasting elements; they may be installed across multiple spaces and times [...] (or) installed in galleries, museums, electronic networks, and in myriad non-traditional spaces.
>
> (Ouzounian 2008: 33)

This is a rich but very broad definition of the term. To be useful for the methodology discussed here, a focusing of intent is required. Sound installations, as understood in this chapter, resonates most strongly with the third chapter of Ouzounian's thesis, 'Everyday Spaces + Social Spaces', which 'traces the beginnings of sound installation art in relation to early philosophies of everyday life and philosophies of social spatialization' (Ouzounian 2008: 39). This is also the tradition my book *Sonic Rupture: A Practice-led Approach to Urban*

Soundscape Design (2016b) is connected with. The 'sonic rupture' concept applies affect theory as a means to propose a method for creating networks of urban sound installations that rupture everyday spaces to create new experiences and encounters. However, the sonic rupture concept only briefly touched on the atmospheric and musical possibilities of sound installation art, which will be more rigorously pursued in this chapter.

It is important to note that sound installations are understood to be distinct from soundscape systems. Soundscape systems are multi-speaker electroacoustic arrays that play back compositions and/or sound art works in public spaces (Harvey 2013; Anderson 2016). Primarily concerned with the playback of pre-composed compositions, soundscape systems are not necessarily concerned with site specificity. They tend to be located in spaces used frequently by the public and are, as such, in danger of competing with existing spatial programmes (i.e. consumerism and recreation). Consequently, they can become sources of annoyance even though the compositions themselves may be thoughtful and well executed (Harvey 2013: 123). Compare this to well-known permanent sound installation art located in underutilized, and thus uncontested, spaces: *Times Square* located beneath a subway grill; *Harmonic Bridge* located underneath a traffic overpass; *Neville Street Refurbishment* located in a (once) highly reverberant traffic and pedestrian tunnel;[1] and, *Fluisterende Wind* (Whispering Wind) – by Edwin van der Heide and Marcel Cobussen – a real-time, generative installation that filters noise with human voice recordings to create 'moments when the wind seems to be whispering'.[2] These are sound installations that reference surrounding sounds to enhance typical listening experiences. They are intimately tied to site-specific sounds and the transformation of perception, making them distinct from soundscape systems that play back pre-composed sound works.

Sound installation as a soundscape design tool

The concept of soundscape design (*qua* acoustic design) was a central contribution of acoustic ecology, first proposed by Murray Schafer (1994) and the World Soundscape Project (WSP). As I have previously argued (Lacey 2016b: 75–76), acoustic ecology has been historically less successful in applying compositional techniques to the creation of publicly situated sound installation art, due to its perception of noise as a negative urban phenomenon. This is a consequence of acoustic ecology's anti-urbanist tendencies (Sterne 2013; Ouzounian 2017), which leaves little room for the consideration of experimental music and sound art techniques in the design of urban environments. However, in recent years, the important role that sound installations, and sound art interventions, can play in an urban design context has developed rapidly (Cusack 2012; Cobussen 2016; Lacey 2016a, b; Ouzounian and Lappin 2016).

It is interesting to note that Max Neuhaus's (and the world's) first permanent urban sound art installation, *Times Square*, was installed in 1977, which is the exact year that Schafer's book *Soundscape: The Tuning of the World* was published. Presumably, Schafer would have known of Neuhaus's sound installation works. (Neuhaus certainly knew of Schafer's book – see Neuhaus 2004.) Given his disparaging comments on Russolo's and Cage's experimentations with urban noise (Schafer 1994: 110–111), one suspects Schafer's thoughts about Neuhaus's work would have been equally dismissive. Indeed, we could ask where soundscape studies would be today if the founders of the WSP, particularly Schafer, had supported the possibility of site-specific sound installations such as *Times Square*, itself rooted in experimental music traditions, as an innovative means to combat the increasing domination of urban space by noise. Indeed, Neuhaus's *Times Square* deftly demonstrates how sound installations can be designed and located to create subtle, yet transformative, listening experiences. For a more accessible (and less encumbered) language for understanding the potential role of urban sound installation practices in the design of urban environments, I turn to another theorist.

Sound installation as a new spatial music

I turn to Gernot Böhme's atmosphere theory, and ways in which his concept helps rethink sound installations as a form of site-specific music.[3] Whereas Neuhaus sought to distinguish sound installations from music, Böhme opens the possibility of sound installations as a new way to think about and practice music. In so doing, his work locates the sound artist as an active participant in the production of city atmospheres to affect the emotions of city inhabitants (on this point, see Cobussen [2016] for further discussion[4]). In its simplest understanding atmosphere 'may be defined as tuned space, i.e. a space with a certain mood. From here two more traits of the theory of atmospheres can be advanced: atmospheres are always something spatial, and atmospheres are always something emotional' (Böhme 2017: 2). And more specifically, in relation to its role as music: 'Musique concrete and sound installations, in particular, forced a revision of music theory and, moreover, changes in the fundamental concepts of aesthetics in general (as concerned with) the notion of music as environment art' (168).

Taking these two quotes together, we can consider sound installations as a type of music that simultaneously acts upon the environment and perception, in between which emerges an 'atmosphere' or 'mood'. This remains consistent with Neuhaus's statement that sound installations are spatial; however, we now have the opportunity to consider a sound installation as a form of 'spatial music'. This is not music generated from notation or predetermined compositions, but music as an expression of the environment, or more accurately, intentional re-expressions of the existing environment, enacted to enhance our aural perception of the environment. This desire to rethink the meaning of music can be traced to the historical efforts of experimental composers and sound artists. For instance, Edgard Varèse argued for the liberation of sound (Varèse 1966); John Cage for the emancipation of sound (Cage 2011: 87); and, more recently, Marie Thompson applies affect theory to contextualize music as the 'organization of sound' (Thompson 2017). Similarly, sound installation art can be considered the (re)organization of site-specific sounds to evoke new perceptions.

Böhme foregrounds two sound installation artists (both from the philosopher's home country, Germany), Sam Auinger and Hans Peter Kuhn, who have discovered successful means for producing atmospheres. When speaking of Sam Auinger's work (including his collaborations with Bruce Odland), he writes that the use of resonant pipes 'reproduces in material form what might be regarded as the origin of music altogether: the transformation of noises into tones by tuning' (Böhme 2017: 187). My only concern with this account is that notions of a 'tuning of the world' risk reducing urban sound installation art to the expression of recognizable musical tones (to be clear, I am in no way ascribing this limitation to the rich repertoire of Auinger's work), which is too closely aligned with perceptions of what music should sound like (i.e. tonal and clean). However, in another passage Böhme speaks of Hans Peter Kuhn's method of sampling original sounds, which he then integrates into his spatial works. This he calls 'environmental art, which has moved music into the realm of aesthetics of atmosphere', which 'provides the simple answer that music as such is the *transformation of physically sensed space*' (Böhme 2017: 170–171, my emphasis). It would appear that Böhme has a broad understanding of what music, in the context of atmosphere and environmental art, can be.

Atmosphere as a limited field of sound

The concept of a 'limited field of sound' was introduced to me in a review of my book by Nikša Gligo: '[*Sonic Rupture* has] opened a new possible view on the development of music (as a limited field of sound)' (Gligo 2017). After some email correspondence with Gligo, and self-reflection in relation to my own practice, I have come to consider it thus: a 'field of sound' can be considered the audible locus within which a sound installation can be heard (its geographical reach), with the types of sounds radiated by the installation being 'limited' to a relationship with the environmental sounds originating in the space. This presents an approach to music that is relational, insofar as it is dependent upon a relationship between existent and introduced sounds – a (re)organization of sounds – to create augmented atmospheres that evoke new perceptions.

This reveals a working methodology for the sound installation artist, working within a soundscape design context, that can be understood as follows:

1. The purpose of a sound installation is to generate an atmosphere. Atmosphere is the space that emerges between environments that generate ephemera (sound) and human perceptions of that environment.
2. A sound installation applies a shaping or transformative technique, such as resonant pipes, sampling strategies, or perhaps some type of material intervention,[5] to generate a new atmosphere as a limited field of sounds.
3. The affected limited field of sounds becomes a site of difference or encounter. Within this transformed environment, new perceptions are evoked.

Böhme provides an interesting passage that contextualizes the ambition of this methodology:

> The specificity of atmospheres is best experienced when their characteristics stand out – not when they have lapsed into something which surrounds us uniformly and inconspicuously. They are experienced, therefore, through contrast, when one is in atmospheres which cut across one's own mood, or upon entering them, through the switch from one atmosphere to another. Atmospheres are then experienced as 'impressions' that is, as a tendency to induce a particular mood in us.
>
> (Böhme 2017: 184)

Thus, atmospheres are most acutely perceived as a transition experience. These encounters ignite new perceptions, by rupturing the typical uniformity of sounds experienced in the contemporary city. I want to propose that Bohme's 'transitions' suggest a new way to consider those sound art installations that rupture small spaces for the diversification of experience. It is the 'switching' of perception, upon transition from homogenous urban atmosphere into a ruptured (by cause of a sound installation) atmosphere, that affords alternative auditory experiences. The urban soundscape could consist of an interconnected network of sound installations that rupture highly localized city spaces, to enrich our sensory connections as we traverse the urban environment. I will now turn to two recently produced atmospheres – the first an intervention and the second a public art work – that act as case studies for the proposed methodology.

Two artworks generating atmosphere as a limited field of sounds

It should be stated from the outset that my own work as a practitioner, to date, has been focused on electroacoustic techniques; however, this is not the only means by which a rupture can be created. Sculptural processes that might take advantage of resonant properties and/or sounding materials are also powerful and should be considered as of equal value (Lacey 2016a). I won't be commenting on the details of the fabrication of the following discussed works, or their inception, which is adequately covered elsewhere (Lacey et al. 2017a, b). Rather, these descriptions are focused on the artistic intent to produce atmospheres as a limited field of sounds for the purposes of transforming perceptions of the urban environment.

Case study 1: Noise Transformation

Noise Transformation was a collaborative research project (for further details, see Lacey et al. 2017a) that developed a sound installation technique for transforming roadside traffic noise into a musical experience. It is an example of sound installations being applied for a practical purpose, in this case a creative intervention as a means of ameliorating noise annoyance. It is a working example of how soundscape design approaches can improve listening experiences in small, localized urban spaces.

The work demonstrates how an atmospheric approach enables the simultaneous focus on the spatial generation of sound and its perception, and encourages new thinking about noise management. Typically, background noise sources are quantifiably measured to determine if the sound pressure level (SPL) is above or below a predetermined amount. The project industry partner would measure the SPL level of any location where a member of the public reported traffic noise to be annoying.[6] If the SPL was measured to be above the minimum standards set by the state road network regulatory body (VicRoads, at the time of writing), then the industry partner would act by double glazing the windows of the occupant's house and installing air-conditioning. The unintended impact is that people become even more isolated from their outside environment by trapping themselves in what could be described as a small acoustic bubble. An atmosphere we might think of as defensive (though hopefully also homely) towards outdoor conditions.

Such a quantifiable approach demonstrates the limitations of treating sound at its source rather than at the level of perception. By treating sound at the level of perception our research team made an interesting discovery. By increasing the noise levels by 1 to 2 decibels more pleasing aesthetic effects were created, which led to positive associations with a sound environment typically considered to be unpleasant.[7] The method was to capture road traffic noise with microphones. The captured noise passed through a computer-based algorithm and/or sound design tool that reprocessed the audio signal. This reprocessed signal was then played back through an accompanying speaker array that increased the overall SPL by 1 to 2 deibels. At this point the acoustic engineers involved in the project were outside of their comfort zone, given that they only know one successful course of action: quantifiable reduction in SPL measurement. However, through conversations with local community members by sensory ethnographers[8] involved in the research (Lacey et al. 2017c) our research team found that the new sound fields reduced people's sense of anxiety. The same people stated that they would use outdoor spaces (parks and balconies) more often if the traffic sounded more like the transformed sound environments.

This discovery supports contemporary research by environmental managers and engineers who are exploring new approaches to noise management (Brown 2016; Kang 2016). For instance, Lex Brown, at a talk for a transport and noise control industry presentation I attended in Melbourne (2017), suggested (I paraphrase) that soundscape managers might better consider noise as a resource rather than a waste. In fact, the organization who invited him, VicRoads, were one of the many organizations who attended the *Noise Transformation* demonstrations, at which one of the representatives suggested the installations were 'recycling noise'. This is an interesting insight that in my view resonates with Brown's suggestion that noise might be better managed as a resource. It demonstrates a new creative way to treat noise: not just as an exterior phenomenon to be removed/attenuated, but as a recyclable material that can be transformed into aesthetically pleasing perceptions. It is an excellent example of how designing for mood/atmosphere – which emerges at the interface of environment and perception – can be an effective alternative for typical noise attenuation approaches, and an affirmation of Böhme's statement that 'music as such is the *transformation of physically sensed space*' (Böhme 2017: 170–171, my emphasis).

Case study 2: Touchstone: The Artwork Remembers

Touchstone is a publicly situated artwork completed by an interdisciplinary team of creative practice researchers including a landscape architect, industrial designer, interactive systems designer, and sound artist (for further details, see Lacey et al. 2017b). The project sought to work with a local council to discover how integrating an artwork into urban design approaches might encourage local populations to be more engaged with their environment. *Touchstone* is an interactive artwork that is integrated into the plaza of a community centre. It immerses listeners inside a sound field that is generated by two vibrating metal plates and two in-ground speakers playing local field recordings. Twice a day (dawn and dusk) the artwork creates a short performance for the community based on the amount of daily interactions with the artwork's central sensing stone.

To achieve this the entire suburb was considered to be the field of sounds with which the sound installation would engage. I communicated with community members via a social media page to determine what sounds they felt best represented their community. From the received feedback, I generated a list of community sounds broken down into six categories: environmental, rural, human, industrial, construction, and performative. A team of field recordists captured these sounds in the local suburb, which were then integrated into the artwork to play back during the dawn and dusk performance times. The phonographic playback combines with the sounds of the vibrating metal plates to immerse listeners in a limited field of sounds as determined by the geographical and audible reach of the artwork.

The dawn and dusk performances are determined by the amount of interactions that have occurred each day, during which the metal plates vibrate in various ways depending on the touching of a central sensing stone. The vibrating plates have the effect of physically passing sound into the body, while virtual sine-wave generators radiate sympathetic frequencies via the in-ground speakers into the ears. The effect has been described by the artists and visitors as akin to a spacious and comforting sensation; a haptic-auditory experience connecting ground with sky. A sense of expansiveness occurs as the atmosphere seems to stretch beyond the immediate vicinity. This is an effect achieved by compelling the sensing body to enter into a new relationship with its immediate environment: vibrating ground, windswept land, and vast cloudy skies combine as interconnecting motions, which weaves the body into its surrounding environment; and incoming suburban sounds mix with a concentration of audio samples (derived from field recordings) that radiate from the ground. Readers familiar with *Sonic Rupture* will hear in this description the notion of rupturing the urban crust to expand the affective potential of the earth (Lacey 2016: 50–1); indeed, poetically, it felt as if this had been achieved with *Touchstone*.

Touchstone is indicative of the possibilities of playing with the atmosphere approach. In this case the larger suburban environment is contracted and collapsed into a confined space, where an atmosphere as a limited field of sounds emerges to direct perception outwards, into an awareness of land and sky. The artwork points to the role public artists have in using sound to facilitate community engagement with atmospheric processes that simultaneously design for perception and environment, rather than simply installing the finalized outcome of a preconceived concept.

Conclusion

When applying sounds to urban spaces it is prudent for the practitioner to remember Neuhaus's statement that sound installation art is spatial rather than temporal. As such, sounds are designed and placed to respond to spatial conditions without necessarily being concerned about compositional narratives with beginnings and endings. This has been described in this chapter as spatial music, wherein introduced and existent sounds co-mingle and intertwine to reorganize environments to make new perceptual experiences possible. This is consistent with atmosphere theory that considers atmosphere to be a spatial mood that emerges at the entwined interface of environment and perception. In this respect the sound installation artist considers both the spatial qualities of the environment, particularly its sounds, and the possible perceptual response of the visitor. The benefits of this approach are provided by the two case studies. *Noise Transformation* transforms existing noise sources into new auditory experiences, which, as evidenced by the sensory ethnography assessment (see end note 8), improves aesthetic experience. *Touchstone* considers perception by assessing preferred sounds through community consultation which informs the environmental expressions of the artwork. Both examples are integrated approaches in which artist, environment, and community coalesce to discover ways in which sound installations can produce affecting atmospheres for those that encounter them. As such, the sound installation artist can be considered a new type of spatial musician who reorganizes the sounds of the city to simultaneously produce alternative atmospheres and affect new ways of experiencing.

Notes

1. For more information on these three works, see Lacey 2016a.
2. For more information on *Fluisterende Wind*, see Studio Edwin van der Heide n.d.
3. It should be noted that Böhme is featured in the first edition of the *Soundscape Journal*. His work has been long known to acoustic ecologists as well as public sound artists and atmosphere designers.
4. Cobussen states that 'sound artists and artistic researchers are very well equipped, indispensable actually, to the process of reimagining and co-designing public urban spaces as sites that simultaneously provide for daily needs as well as facilitate environmental comfort by affecting the moods and emotions of the ones traversing these spaces' (Cobussen 2016: 10).

5. See Lacey (2016a) for further discussion, in which I examine three approaches to sound installation art: resonant, electroacoustic, elemental.
6. VicRoads, the key agency responsible for road traffic noise management in the state of Victoria, Australia, seeks to limit noise next to new or improved roads at 63dB(A) L10 (18hr). For further information, see 'VicRoads – Traffic Noise Reduction Policy' n.d.
7. This is consistent with the public experiences of Bruce Odland and Sam Auinger's work, *Harmonic Bridge*. For further discussion, see Lacey 2016a.
8. Sensory ethnography uses techniques such as videography to collect qualitative data. The employed techniques capture immediate sensory experience rather than questionnaires or interviews that remove people from their immediate experience. For more information, see Pink 2015.

References

Anderson, S. (2016). 'The Incidental Person: Reviewing the Identity of the Urban Acoustic Planner'. *Journal of Sonic Studies* 11. Available online: https://www.researchcatalogue.net/view/243093/243094 (accessed 15 May 2018).

Böhme, G. (2017). *The Aesthetics of Atmospheres*, ed. Jean-Paul Thibaud. New York: Routledge.

Brown, L. and C. J. Grimwood (2016). 'Loci for Urban Soundscape Planning, Design and Management'. In *Proceedings of the 22nd International Congress on Acoustics*, Buenos Aires, 5–9 September, 2016. Available online: http://www.ica2016.org.ar/ica2016proceedings/ica2016/ICA2016-0233.pdf (accessed 15 May 2018).

Cage, J. ([1961] 2011). *Silence: 50th Anniversary Edition*. Middletown, CT: Wesleyan University Press.

Cobussen, M. (2016). 'Towards a "New" Sonic Ecology'. Leiden: Leiden University. Available online: https://cobussenma.files.wordpress.com/2011/10/cobussen-inaugural-text.pdf (accessed 15 May 2018).

Cusack, P. (2012). 'Berlin Sonic Places'. Available online: http://sonic-places.dock-berlin.de/ (accessed 15 May 2018).

Gligo, N. (2017). 'Review and Information of Publications: Sonic Rupture; A Practice-led Approach to Urban Soundscape Design'. *International Review of the Aesthetic and Sociology of Music* 48 (1): 153–156.

Harvey, L. (2013). 'Improving Models for Urban Soundscape Design'. *SoundEffects: An Interdisciplinary Journal of Sound and Sound Experience* 3: 113–137. Available online: http://www.soundeffects.dk/article/view/18444 (accessed 3 July 2020).

Kang, J. and B. Schulte-Fortkamp (eds) (2016). *Soundscape and the Built Environment*. Boca Raton, FL: CRC Press

Lacey, J. (2016a). 'Sonic Placemaking: Three Approaches and Ten Attributes for the Creation of Enduring Sound Art Installations'. *Organised Sound* 21 (2): 147–159.

Lacey, J. (2016b). *Sonic Rupture: A Practice-led Approach to Urban Soundscape Design*. New York: Bloomsbury

Lacey J., L. Harvey, S. Moore, X. Qiu, S. Pink, S. Sumartojo, S. Zhao, and S. Maisch (2017a). 'Soundscape Design of Motorway Parkland Environments: Transformation, Cancellation, and Ethnography'. In *Proceedings of Invisible Places – Sound, Urbanism and Sense of Place*,

Azores, Portugal, 7–9 April 2017, 52–63. Available online: http://invisibleplaces.org/IP2017.pdf (accessed 15 May 2018).

Lacey, J., R. McLeod, C. Anderson, and C. Khoo (2017b), 'Touchstone: A Discussion of a Digitally Integrated Artwork Designed to Facilitate Community Engagement'. *Active Public Space Conference Proceedings*, Barcelona, Spain, 13–14 November 2017. In *Responsive Cities Active Public Space Symposium Proceedings*, 33–40. Available online: https://drive.google.com/file/d/13pTlFEDEO5ElPIrEwhN64o2gqoty75mv/view (accessed 15 May 2018).

Lacey J., S. Pink, L. Harvey, X. Qiu, S. Sumartojo, S. Zhao, S. Moore, S. Maisch, and M. Duque (2017c). *RMIT Acoustic Design Innovations for Managing Motorway Traffic Noise by Cancellation and Transformation*. Melbourne: RMIT University, Melbourne. Available online: https://researchbank.rmit.edu.au/view/rmit:45636 (accessed 15 May 2018).

Neuhaus, M. (2004). 'Listen'. Available online: http://www.max-neuhaus.info/soundworks/vectors/walks/LISTEN/LISTEN.pdf (accessed15 May 2018).

Ouzounian, G. (2008). 'Sound Art and Spatial Practices: Situating Sound Installation Art Since 1958'. PhD Thesis, University of California, San Diego.

Ouzounian, G. and S. Lappin (2016). 'Editorial. Recomposing the City: New Directions in Urban Sound Art'. *Journal Sonic Studies* (11). Available online: https://www.researchcatalogue.net/view/236505/236507 (accessed 15 May 2018).

Ouzounian, G. (2017). 'Editorial: Rethinking Acoustic Ecology; Sound Art and Environment'. *Sound Art and Environment* 6 (2): 4–23.

Pink, S. (2015). *Doing Sensory Ethnography*. 2nd edition. London: Sage.

Schafer, M. ([1977] 1994). *The Soundscape: Our Sonic Environment and the Tuning of the World*. Rochester, VT: Destiny Books.

Sterne, J. (2013). 'Soundscape, Landscape, Escape'. In Karin Bijsterveld (ed.), *Soundscapes of the Urban Past: Staged Sound as Mediated Cultural Heritage*, 181–193. New York: Colombia University Press.

Studio Edwin van der Heide (n.d.). 'Fluisterende Wind'. Available online: http://www.evdh.net/fluisterende_wind/ (accessed 3 July 2020).

Thompson, M. (2017). *Beyond Unwanted Sound: Noise, Affect and Aesthetic Moralism*. New York: Bloomsbury.

Varèse, E. and C. Wen-chung (1966). 'The Liberation of Sound'. *Perspectives of New Music* 5 (1): 11–19.

'VicRoads – Traffic Noise Reduction Policy' (n.d.). Available online: https://www.vicroads.vic.gov.au/-/media/files/documents/planning-and-projects/environment/noise/trafficnoisereductionpolicy.ashx?la=en&hash=6C28650833D6FD178B03FC47E5C7B60F (accessed 3 July 2020).

19

Fragile Devices: Improvisation as an Interdisciplinary Research Methodology

Rebecca Caines

Basque-born noise improviser Mattin suggests that it is the place of the improviser to 'go fragile', exposing yourself to 'unwanted situations that could break the foundations of your own security […]. Once you are out, there is no way back; you cannot regret what you have done. You must engage in questioning your security, see it as a constriction […]. Keep going forward toward what you do not know, to what is questioning your knowledge and your use of it' (Mattin 2009). Being true to an improvisatory research methodology requires an ethos which I define as a perpetual state of fragility (Caines, Kenny, and Seibel 2014). It is a commitment to move through and, with mistakes, admit naivete, and to let go of control to create together with others. This commitment is something some researchers are loathe to put into action, especially in public. Yet improvisation is utilized in rigorous research projects across the globe. This chapter will discuss the dissonant and 'fragile' ways that improvisation can act as a research methodology, with examples from my own work using improvisation in socially-engaged sonic arts, creative technologies, and performance. I argue that the improvisatory qualities of risk, active listening, collaborative response, and the reconfiguration of mistake into creativity can form a strong basis for research; and can also trouble disciplinary techniques and expectations; as well as productively disrupting borders between art, research, and pedagogy.

Improvisatory beginnings

I come to improvisation as an artist, working initially in theatre, and then later in free improvisation/creative music and sound art contexts. More recently I have utilized improvisation in interactive installations, in creative technologies research, and in university- and community-based teaching. The philosophies of improvisation, and the integration of improvisation games, scores, and techniques from a range of disciplines help me to foment creativity, both in a generative compositional sense, and when I am forming the content/material of live improvised events. While my initial training was in theatre and performance studies, my interdisciplinary work is now increasingly identified and located within the emerging field of critical studies in improvisation (see Caines n.d.; and IICSI 2020).

I also frame my work as 'community-based' or 'socially-engaged', as all of my art and research is made in partnership, addressing simultaneous social and artistic goals (Cohen-Cruz 2000; Thompson 2012). I partner with other professional artists, researchers, and scientists; and with community organizations, schools, welfare organizations, social workers, activists, people trying art for the first time, those working outside art industries, and/or people whose art has been forgotten, ignored, or erased. This kind of dialogical, co-created work challenges ideas of authorship, and is predicated on a requirement that we 'respect the Difference of the other enough to question and make vulnerable [our] own *a priori* assumptions […]. Genuine dialogical engagement is at least a two-way thoroughfare' (Conquergood 1985: 9). This kind of practice requires negotiating the complex territories of social practice, public art, relational art, and activist and applied interventions (Kester 2004; Bishop 2006; Prentki and Preston 2009; Helguera 2011; Jackson 2011). This kind of work can explicitly or implicitly require elements of improvisation, especially adaptability, flexibility, and co-creation. Improvisation thus provides me with the approaches, tools, and techniques I need in all of the realms I work in.

Art/research methodologies: Tensions and contradictions

In my graduate classes, I teach my students who are just starting out in research a fairly standard approach: that 'research methodologies' *must* be rigorous, tested, accountable, peer-reviewed frameworks for doing research, that bring theory and methods together, and provide a strong, justified philosophical rationale for the choices being made. Methodologies are also, of course, discursive tools of knowledge testing: 'Methodology therefore legitimates and delegitimates, validates and invalidates, approves and disapproves, passes and fails, claims to knowledge and knowledge production. Methodology is the final court of appeal in judging what counts as bona fide knowledge of something' (Matsinhe 2007: 839). It is a key challenge for all of those working in contemporary research practices to find appropriate research methods that question, expose, and navigate ethical boundaries, and allow new knowledge to form.

The term 'methodologies' is often used to refer to a range of qualitative and quantitative systems for undertaking social study (Denzin and Lincoln 2000). Methodologies may incorporate statistical, experimental, ethnographic, analytical, and discursive practices; and they involve a wide range of practical and critical tools. Researchers utilize and combine tested methodologies, and/or join peer-reviewed debate to actively critique existing methods or propose new combinations and frameworks. Proposed research designs are usually evaluated against the strength and appropriateness of their methodology. Clear research questions or precise methodological steps or actions are proposed initially, and then tested. These questions and/or steps may be developed in advance, or can be created during the research process with research participants; and can be based on discovery, action, or intervention. New methodologies are rigorously debated within their respective fields before becoming established and accepted. Whatever the discipline, many would agree that 'research takes place when a person intends to carry out an original study to enhance knowledge and understanding'. It starts with 'questions or issues that are relevant in the research context, and it employs methods that are appropriate to the research and which ensure the validity and reliability of the research findings'. An additional prerequisite is that the research process and the research findings be 'documented and disseminated in appropriate ways', and in some cases, that studies are also able to be repeated by others to produce similar results (Borgdorff 2012: 54).

It can be problematic, however, to use these kinds of parameters to try to understand the research that takes place in artistic contexts. While arts-based data gathering methods in the social sciences, which are sometimes referred to as arts-based research (ABR), or a/r/tography, are established methodological approaches that use art, theatre, music, dance, film, and literature; the results are not the same as professional art (Finley 2005; Leavy 2008). A/r/tographic researchers do not need to engage in artistic training regimes; or satisfy professional art contexts, etc. in order to produce results that are recognized in the field of ABR. In contrast, in artistic research in fine arts contexts like the sonic arts, research questions sometimes arrive at the end, with the final artworks, or even afterwards; not in advance. Artists work in order to see what the work is about, and often aim to develop unique working processes that are not always transparent or transferable. Writing is not necessarily involved. The peer reviewers for art may be art juries made up of professional peers; funding bodies; reviewers, curators, and gallery directors; record labels/producers; festival committees; and professional colleagues attending the events. Artistic methods are often deliberately eccentric, experimental, personal, and non-generalizable, and research conditions are often unable to be repeated. When pressed many artists would acknowledge that concepts such as attention to process, rigour, interaction with their discipline, and peer review, are all still key to producing quality outcomes (Horowitz 2014). Frames such as 'research creation', 'practice-based research', or 'practice-led research' have been helpful for some artists to articulate their methods (Smith and Dean 2009). For others, terms such as 'research design' and 'methodology' will always remain alien to the work that artists do, even if this work is located within academic contexts, such as universities, or even if the work is funded by academic research bodies. For some, the utilization of research language is a symptom of giving in to the creeping centralization, standardization, and bureaucratization of universities; or can risk instrumentalizing art rather than allowing it to exist on its own terms (Elkins n.d.).

In recent years, there has been some momentum to find ways through these seeming contradictions, by acknowledging artistic research as a distinct methodological stream in academic contexts, to be considered in parallel to scientific or social science methodologies (Slager 2009b; *Journal of Academic Research* [JAR] 2010; Biggs and Karlsson 2012). Examining literature in this area, artistic research can be understood to share at least some of the following characteristics:

- Artistic objects as outcomes
- Non-linguistic outcomes
- Preverbal/tacit/pre-reflective/preconceptual work
- Intuition
- Emotion
- Embodiment
- Contradictory, unfinished, confusing, disturbing aims
- Attention on aesthetic considerations
- Epistemic/hidden knowledges and forms of expression
- Applied implementation
- Connection to artistic histories, professional contexts, and training.

Artistic research methods are sometimes divided into subcategories. While language may differ, it is common to differentiate research-for-creation (for example, developing or discovering techniques and materials, drafting and testing versions, etc.); from research-from-creation (producing art in order to allow for new conditions for analysis). Other categories include: creative presentations of research (such as alternative ways to show data); and creation-as-research (research undertaken through and by practical artistic work) (Chapman and Sawchuck 2012: 15–20). Borgdorff famously defines artistic research as 'the articulation of the unreflective, non-conceptual content enclosed in aesthetic experiences, enacted in creative practices and embodied in artistic products' (Borgdorff 2012: 47). I do not think Borgdorff's focus on 'non-conceptual content' adequately covers the breadth of modern artistic research, given the strong conceptual nature of many contemporary artist's works. Henk Slager provides a little more nuance, when he suggests that 'the most intrinsic characteristic of artistic research is based on the continuous transgression of boundaries in order to generate novel, reflexive zones' (Slager 2009a: 198).

Improvisation as a research methodology

Improvisation is a well-known artistic method, present across most cultures and practices (Bailey 1980; Nettl and Russell 1998; Albright and Gere 2003; Frost and Yarrow 2007; Landgraf 2011). Improvisation has been a core element in many significant movements in music and sonic arts, theatre, dance, visual art, literature, film, and interdisciplinary practices; ranging from those in the institutional canon, through to avant-garde outliers, and folk contexts (Dean and Smith 1997; Caines and Heble 2014). Improvising activity also takes place in everyday practices and non-arts contexts (Sawyer 2007; Peters 2009). In addition, growing literature in the discipline of critical studies in improvisation has emphasized the social impacts of improvisational models and approaches (Lipsitz, Fischlin, and Heble 2013; Lewis and Piekut 2016; Siddall and Waterman 2016).

Improvisation is only more recently being discussed explicitly as a distinct research methodology. Nisha Sajnani, for example, articulates her own research method as a combination of improvised theatre techniques drawn from Developmental Transformation, Playback Theatre, and Theatre of the Oppressed. She suggests: 'When situated as research, improvisation functions as a kind of "disciplined empathy", inviting researchers to engage in an iterative process of identifying emergent issues and to respond with a corresponding design that permits further exploration' (Sajnani 2012: 83). For Stephen Levine, it is improvisation's ability to resist certainty that attracts him to using it as his research methodology:

> Even so-called qualitative research now looks for results that are 'evidence-based,' i.e., conclusions that are clear and distinct and that can be proven beyond any doubt. The aesthetic attitude which is embodied in an essentially improvisational research method can never be validated in this way. This is both its limitation and its strength.
>
> (Levine 2013: 27)

The Canadian-based research network, Improvisation, Community and Social Practice (ICASP 2007–2013), and its subsequent formation, the International Institute for Critical Studies in Improvisation (IICSI 2013–present), have been providing a sustained laboratory context for building knowledge on improvisation as research, with research sites across the globe (ICASP 2009; IICSI 2016). One example of interdisciplinary IICSI research is the work of law scholar Sara Ramshaw and musician/sonic arts researcher Paul Stapleton, and their team, on the Translating Improvisation Project. Ramshaw and Stapleton used improvised music forms to help train professionals in child protection to be better at their jobs, while simultaneously working to understand the limits of both law policy and of improvised music scores (Ramshaw et al. 2018). As IICSI director Ajay Heble states: 'Improvisation has become a vital model for the analysis of political, cultural, and ethical action and dialogue' (Heble 2005). More and more academic and artistic research networks are emerging that actively ground their philosophies and practices in the improvisatory, highlighting the need for rigorous methodological inquiry in this area.[1]

Breaking down the steps

In order to aid researchers unfamiliar with improvisatory art practices, and artists reluctant or unsure about research terminology, I have sketched out some broad steps below for building a research design, and then I have mapped them on to what I see as the corresponding steps in improvisatory approaches. I hope this can aid in articulating a broad improvisatory methodology that applies to both written and artistic research settings. At each step, I have included examples from my own sonic and interdisciplinary research projects.

Developing research questions

Active listening/improvising with

To improvise is to *improvise with*, with others, with 'enabling constraints' (Stravinsky 1946: 49), with spaces, and with new systems or interfaces. The first step to this responsive work is to rethink what it means to listen. A distinction is drawn in scholarship between involuntary hearing (the reception of audible signals and comprehension of familiar aural contexts) and involuntary listening (the active, attentive, heightened state of awareness to new sonic information). Phenomenologist Jean-Luc Nancy suggests, 'if "to hear" is to understand the sense […] to listen is to be straining toward a possible meaning, and consequently one that is not immediately accessible' (Nancy 2014: 6). Listening is one of the key traits of the improviser, both for performers and for scholars who analyse improvisation. Improvisers cultivate active listening, in order to be ready to respond to unknown material and develop new forms. Improvising musicians lean in to listen for the moments that will form the basis of their collective sound-making (Fischlin 2009: 2), or listen to the acoustic environment for new kinds of sounds and sonic relationships (Oliveros 1995: 19); and improvising dancers and theatre artists listen for offers from space, body, and co-performers to build from (Foster 2003: 6).

Composer, improviser, and scholar Pauline Oliveros, made it her life's work to grow the practice of what she named 'Deep Listening', a practice that she believed applied to much more than just her artistic work. Deep listening is 'listening in every possible way to everything possible—this means one hears all sounds, no matter what one is doing' (Oliveros 1995: 19). Oliveros and her collaborators aimed to 'cultivate a heightened awareness of the […] environment, both external and internal, and promote experimentation, improvisation, collaboration, playfulness and other creative skills vital to personal and community growth' (see Deep Listening Institute 2014). Oliveros explained the practice of deep listening as simultaneously improvising with sound and 'expanding your listening to continually include more' (Oliveros 2015). Listening is thus seen across disciplines as a key element of any improvisatory practice, and is central to an improvisatory research methodology.

Broadly speaking, an improvisatory methodology requires that researchers first see who is there, find exactly what they offer and need, and then utilize a process of active, 'straining for more', through interdependent and expansive listening. Research questions, offers, themes, and prompts can then be built from this improvisatory listening activity, whether the outcomes are to be written, artistic, or multimodal. This listening activity continues throughout the research and dissemination process. This process may be aided by using listening tools, exercises, and practices drawn from improvisatory music and sonic arts, and it can also incorporate interdisciplinary approaches to listening that take in lessons learned from other research modalities (Back and Bull 2003). It is, then, the task of the researcher to improvise with their collaborators and the spaces they work in, responding to the precise needs and conditions, while making room for their input to completely change the research focus and direction. In theatrical terms, this openness is often articulated as continually 'accepting offers', or incorporating a stance of 'yes, and ... ' (Spolin 1969; Halpern and Close 1994), while in music and sonic arts it might be better understood as 'not reacting exactly, but being overwhelmed by what happens and thus breaking, releasing, splitting open [...] not reacting, inter-acting' (Ninh 2010: 66). In this way, improvisatory methodologies may have links with those social science methodologies that are also grounded, iterative, or participatory.

In my research projects, listening to develop collaborative research questions, and then improvising with these questions has taken many forms. In the sound art project *Community Sound[e]Scapes*, for example, I listened with communities in Australia, Northern Ireland, and Canada to develop new kinds of soundscapes responding to site/space/place. Researchers and participants used audio recorders, soundwalks, deep listening exercises, digital audio workstation (DAW) software, and a co-created online interface for improvising with acoustic recordings, to listen to each other. We also utilized other kinds of listening practices such as informal, collaboratively created community needs assessments, interdisciplinary sound and site-specific theatre workshops, indigenous storytelling, and even mixed-ability bike rides (Caines 2015). In all five community partnerships, and with the other collaborating artists, listening together and improvising with each other, the constraints of each community radically changed the research process. This produced at times unstable responses that productively challenged control, brought power relations into view,[2] and produced new creative and written research opportunities, guided by what Bull and Back call 'thinking with our ears' (Back and Bull 2003: 3).

In a more recent project entitled *ImprovEnabled*, I collaborated with my research partner, cultural anthropologist Michelle Stewart, and a range of community partners, in order to explore improvisation as a tool for recovering the lived experience of the complex disability, fetal alcohol spectrum disorder (FASD) (Caines and Stewart 2017). We initially aimed to use workshops in music and theatre with applied social science interventions, to push back at unjust, dated, and uneven systems that leave people with FASD and their parents and caregivers in untenable situations. Stewart and I, and our partners, however, had to learn how to listen across and between the differences that separate art, research, and applied approaches. This resulted in unexpected new methods and outcomes.

In our initial project, we worked with adult participants in a support group. Participants included those with lived experience of the disability, families, support workers, and staff. Stewart and I had to try to understand each other's ideas of what acceptable research outcomes could be, which differs widely between our research disciplines. In artistic research, for example, the production of new forms of improvised music could be an acceptable research outcome in itself, while in a social science context, musical workshops would be included in a range of methods for gathering data that would then need to be analysed and written up at a future stage. There was also significant scepticism from staff and agencies around whether improvisation and art would be useful in community settings which are so often focused around direct management of issues. We had to listen together to co-develop plans with our partners that incorporated all of these different perspectives. We held information sessions, and then co-designed the theme and approach with the agencies and frontline staff, and with members of the support group.

In one workshop we used improvised theatre games from Forum Theatre to listen to scenarios of lived experiences of stigma and isolation with the group. We then followed this with an exercise in free improvised music and drawing. The participants were given a verbal prompt, and some drew pictures, while others created music on tablets using a range of different apps as instruments. Those playing sounds were given assistance to try making long, short, sharp, or flowing sounds on their different instruments, and to listen to each other to build the piece together around the theme. As the exercise was repeated, those drawing would respond to what they heard played, and those creating music would respond to the pictures. All the prompts were drawn from thoughts that were expressed in the earlier theatre games. One example of a prompt was the question: 'What does it feel like to not be listened to?' All people in the room improvised, including staff members, support workers, and researchers. Stewart led a discussion at the end of the session to learn more about the experiences that people had been expressing in the artistic work. In interviews following the project, we were told that staff were particularly affected by the ways these exercises brought out stories and expressive capacities that had not been seen before.

The results of this project are now being directly applied, in order to create new supports and advocacy strategies. Recordings of the improvised music created in the session described above have been played back in other settings across Canada and internationally (including to government working groups) to share the lived experiences and inequities faced by those with FASD and advocate for change. A free downloadable community improv toolkit with games developed in these sessions was released in September 2018 with community organizations and is now being used globally.[3] One group is currently adapting the toolkit and findings from the first project to create peer-to-peer mentoring on advocacy and life skills for teens. Another group formed from this project has worked with researchers from the Sonic Arts Research Centre in Belfast (Northern Ireland), to explore the possibility for improvised music and theatre in immersive spaces both to aid in understanding sensory differences and to build on the strengths of those with different listening capacities. Listening and collaborating to each other, using improvisation, has radically changed the original research plans and enabled new kinds of research to happen.

Building a research design, testing, and implementation

Risk / real-time collaboration / integration of improvisation techniques

Improvisatory methodologies are based on risk and trust. Waterman suggests of improvised musicians:

> All their decisions are made in the moment [...]. The possibility of failure is always imminent, because the process demands such a high degree of self-exposure. Improvisation is most

satisfying when the conditions of trust exist that allow participants to risk everything in the moment of performance. This means that improvisation is an arena of social interaction and accountability.

(Waterman 2014: 59)

Of course, risky situations are often something researchers are encouraged to avoid. There are real risks to people to be negotiated in research, particularly in interdisciplinary or community engaged research where vulnerable people can be hurt. In universities, we work in risk-averse spaces and policies at all points, and severe consequences are held up for those who work recklessly. As improvisers, however, we simply cannot be risk-averse. The kind of risk we take when improvising is, however, immediately accountable to others, and is collaboratively created and supported. This risk is built on reciprocity that needs to be earned. The research design, testing, and implementation phases of an improvisatory project must make us sensitive, aware, and attune to risking together. Decisions must be made together, and both dissonance and responsivity are key.

Improvisation methodologies use new kinds of research and pedagogy tools drawn from art contexts and can combine them in new ways. In my recent research, for example, short-form improvised theatre games have been used to create new sound art performances (Caines, Kotowich, and Schenstead 2015); automatic writing exercises have shaped focus group interview structures (Caines and Stewart 2017); and artists have engaged live with audience's written notes at a conference to create paintings and music onstage about indigenous and settler truth and reconciliation issues in Canada (Brownridge 2018). In the Creative Technologies programme that I have been leading at the University of Regina, improvising artists working with computer scientist and engineers have demanded impossible technologies that push engineering into new directions, and technologists have demanded new kinds of 'enabling constraints' on artists through new invention, making different research outputs possible. Engineer Craig Gelowitz's work with improvising sound artist Kim Morgan on adapting sound projects to public spaces is an excellent example, as each of the researchers had to find innovation in their own discipline for the project ideas to be realized (see Gelowitz, Morgan, and Benedecenti 2008).

One pedagogical example that combines different kinds of improvisation approaches is the Creative Technologies class 'Introduction to Sound Art', which is co-taught by Gelowitz and me. In this class, a mix of students from software systems engineering and fine arts learn sonic improvisation techniques from artists such as Jon Stevens and Pauline Oliveros, alongside learning software and physics related to acoustic phenomena. Lessons include a mix of improvisatory artistic workshops exploring sound creation, manipulation, and performance, and lectures. Students often find it difficult at first to work together across disciplinary divides, but they produce sound art projects that have combined engineering and improvisation techniques in interesting ways. Student final projects have, for example, included new improvisation scores using Twitter, audiovisual interpretation software generating improvisations from user-inputted text, multichannel sound sculptures that record user vocal improvisations and reinterpret them, and a range of game-based art works using sound, based on trust, risk, and real-time decision making. Students report approaching their own disciplines differently after taking the class; many engineers acknowledge that exposure to, and experience in, creative, improvisatory activities has changed the way they approach collaboration and project design.

Another example would be The University of Regina iPad Orchestra research and teaching project. I began this project in 2012 with two colleagues, David Gerhard from the Department of Computer Science, and Pauline Minevich from the Department of Music.[4] The class is centred on teaching free improvisation techniques, graphic scores, conducted improvisation, and sound art. This learning is coupled with technological lessons about the iPad, lectures on the history of improvised music, and laptop and mobile device orchestras. Weekly jam sessions focus on teaching wider improvisation skills but contain material drawn from a range of sources. These include experimental music techniques and prompts, including Fluxus scores, and contemporary game and conducted improvisation structures; jazz and free improvisation; short- and long-form theatrical and performance art improvisation (see Spolin 1969; Johnstone 1987; Gomez-Pena and Sifuentes 2011); live drawing and painting (see Schlanger 2018); and hip-hop practices such as sampling, beatmaking, MCing, and beatboxing. Students, most of whom have never improvised before, create a concert of improvised music and sound installation at the end of the semester, where they play iPads and smartphones coupled with more traditional instruments, as well as utilizing laptops, visual art, theatre, and app and interface development. A connected research project has allowed the knowledge gained in the classroom to be taken into other contexts, including research projects in community settings where the iPad offers new capacities for those with limited mobility.[5]

These classes have required social improvisation to occur alongside artistic improvisation. At times in the classroom this has meant negotiating complex intercultural protocols to share traditional knowledges, learning from students in campus inclusion programmes, and sharing radically different political views. Finding institutional and pedagogical ways to run and sustain these kinds of programmes, and meeting such different curricula and research needs has been a rewarding challenge for the researchers/instructors, requiring constant adjustment and adaptation, and at times the deliberate disruption of everyone's expectations. Improvisatory research designs, testing, and implementation can be guided by improvisatory actions of give, take, lift, support, and disrupt.

Outcomes

Reconfiguration of mistake

The hardest part of an improvisatory methodology is the constant and deliberate awareness of mistakes. In many research methodologies, errors in process are to be announced and accounted for, bracketed, avoided where possible, or used as cautionary tales. In improvisation mistake is the fruit of the work. What is useful in an improvisatory research methodology is the idea of seeking mistakes as material. Rather than erasing them, we learn and highlight, interrogate, play with, poke at, create with, our errors, our failures, our disasters. Whilst most research is based in some way on experimentation, I would argue that not many methodologies actively seek mistakes, nor are keen to keep revisiting and building from them once 'lessons are learned'. A too narrow understanding of improvisation approaches, however, might imply that 'there are no mistakes ever, only material', but of course, errors can happen. People can be hurt, power abused, trust lost, and people's lives affected. People can also fail to improvise together, slip into easy, comfortable, and pedantic routines. They can step all over each other to dominate and belittle, ignore advice, or make truly horrible, context-blind work. A rigorous methodology based in improvisation requires an ethics of responsivity and accountability.

I have written elsewhere in more details about the ethics of mistake in community-based sound art projects, with reference to an extension of the Community Sound [e]Scapes project that was entitled Community Sound [e]Scapes: Northern Ontario. This project took place in 2012 with First Nations partners in Northern Ontario, Canada. I was complicit in many mistakes in that project. Yet communities have found ways to make the project sustainable and useful, including finding new ways to work with the sound techniques we developed together on future projects, such as mapping land and resource use in remote communities (Caines, Kenny, and Seibel 2014). One quote, by a young participant in the project who faces significant disadvantages, continues to remind me of the importance of being able to acknowledge and move with mistakes. She suggested: 'The more broken you are, the more you can fit in. You can find a different puzzle piece from a different puzzle and make it fit to another puzzle and it will still look cool, right? […] I like making mistakes, because mistakes […] they can't define you […]. That's survival' (Caines et al. 2013a).

Evaluation

Returning to active listening

One important area of methodological inquiry is being able to track when we have succeeded or failed in our goals. Improvisatory models suggest that to evaluate is actually to return to the start of the methodological cycle, once again making a commitment to active listening, and agreeing to continually 'strain for more' in order to understand what has happened and is happening. Evaluation in the arts, and especially in socially engaged arts, is difficult; perhaps improvisatory methods can offer new ways to think through how we understand 'success' and 'failure' through cycles of improvised listening, creating, and responding.

Some last thoughts …

A chapter like this can only briefly sketch out the ways that improvisation can form a methodological framework. In each step, however, this kind of approach can offer a methodological grounding to those working in research in both artistic and academic settings. It is clear that improvisatory methodologies cross back and forth between linguistic and non-linguistic forms, between genre and resisting genre. In the examples I have discussed, improvisation can also hold different needs, expectations, and desires together. Improvisatory methods in my work help me to expose uneven power relationships through collaboration based on mutual input and benefit. They undermine the idea that one artist, principal investigator, or team owns the knowledge or controls the parameters. Finally, improvisatory methodologies can help me to navigate through thornier, binary oppositions in thinking about research and art. I do not argue that all research should be applied or instrumentalized. As an improviser, I understand that leaving people alone to play, experiment, and be obscure, and do 'what they need to do' is an important way to ensure that 'innovation' can actually take place. This is as true in social or hard sciences as it is in art. Cultures of surveillance – in order to ensure that research has measurable, applied impacts – is increasingly understood to hinder rather than support quality research output (Spooner and McNinch 2018). However, like Norman K. Denzin, I *also* agree that the stakes for us working in universities and other research contexts should be very high. Research needs to matter.

> There is a need to unsettle traditional concepts of what counts as research, as evidence, as legitimate inquiry. How can such work become part of the public conversation? Who can speak for whom? […] Can we forge new models of performance, representation, intervention, and praxis. Can we rethink what we mean by ethical inquiry? Can we train a new generation of engaged scholars and community leaders? What counts as scholarship in the neoliberal public sphere? Can we imagine new models of accountability, how do we talk about impact, change, [and] change for whom?
>
> (Denzin 2017: 8)

Improvisation may be one methodological approach that can hold the desires for freedom and accountability in productive tension, and cross back and forth between play, rigour, and positive impact – perhaps a kind of temporary, fragile device for ensuring creative *and* critical response.

Notes

1. See the Transdisciplinary Improvisation Network (UK), and the Institute for Improvisation and Social Action (US/Mexico), as well as the ongoing work of the Centre for Musical Performance as Creative Practice (CMPCP) at Oxford University, as examples of emerging research networks based in improvisation research.
2. Regarding this topic, see Battesti, in this volume.
3. Over one hundred groups have downloaded the free kit 'Playing to Our Strengths', from locations in Canada, the UK, the USA, Japan, Singapore, Australia, and New Zealand. The researchers have also been working directly with groups in Canada, the USA, and Northern Ireland who are using the kit in long-term programming.
4. Since Minevich has retired from the university, improvising vocalist Helen Pridmore has now joined the project.
5. For a detailed description of improvisation in sound art pedagogy, with further examples from both these classes and community-engaged research projects, see Caines 2019.

References

Albright, A. C. and D. Gere (2003). *Taken by Surprise: A Dance Improvisation Reader*. Middletown, CT: Wesleyan University Press.

Back, L. and M. Bull (2003). *The Auditory Culture Reader*. Sensory Formations series. Oxford: Berg.

Bailey, D. (1980). *Improvisation: Its Nature and Practice in Music*. Boston, MA: DaCapo.

Biggs, M. and H. Karlsson (2012). *The Routledge Companion to Research in the Arts*. 1st edition. New York: Routledge.

Bishop, C. (ed.) (2006). *Participation*. Cambridge, MA: MIT Press.

Borgdorff, H. (2012). 'The Production of Knowledge in Artistic Research'. In M. Biggs and H. Karlsson (eds), *The Routledge Companion to Research in the Arts*, 44–63. New York: Routeldge.

Brownridge, M. (2018). 'Liquid Art: Exploring Live Painting and Mentorship'. [Video] Vimeo. Available online: https://vimeo.com/262308093 (accessed 1 April 2018).

Caines, Rebecca (n.d.). 'Recent Works'. Available online: www.rebeccacaines.org (accessed 5 July 2020).

Caines, R. (2015). 'Community Sound [e]Scapes: Improvising Bodies and Site/Space/Place in New Media Audio Art'. In E. Waterman and G. Siddall(eds), *Negotiated Moments: Improvisation, Sound, and Subjectivity*, 55–73. Durham, NC: Duke University Press.

Caines, R. (2019). 'Resonant Pedagogies: Exclusion/Inclusion in Teaching Improvisation and Sound Art in Communities and Classrooms'. *Contemporary Music Review* 38 (5). https://doi.org/10.1080/07494467.2019.1684062.

Caines, R. and A. Heble (eds) (2014). *The Improvisation Studies Reader: Spontaneous Acts*. London: Routledge.

Caines, R. and M. Stewart (2017). *ImprovEnabled*. Available online: http://improvenabled.ca (accessed 1 April 2018).

Caines, R., J. Campbell, N. Loess, R. Loess, and M. Waterman (2013). 'Community Sound [e]Scapes: Northern Ontario'. KO-K-Net; Ed Video: North Spirit Lake FN; Guelph. http:// media.knet.ca/cse (accessed 1 April 2018)

Caines, R., D. Gerhard, and P. Minevich (2013). 'The University of Regina iPad Orchestra: Engaging Mobile Audiovisual Technologies in Music Teaching and Learning'. *Teaching and Learning to the Power of Technology (Tlt) Conference*, University of Saskatchewan, Saskatoon.

Caines, R., C. Kenny, and F. Seibel (2014). '"Going Fragile": Exploring Place through Community-based Art Practices'. *Of Land and Living Skies* 1 (2): 13–21.

Caines, R., R. Kotowich, and A. Schenstead (2015). 'Improvising with iPads: A Partnered Inquiry into Technology-based Music Therapy, Improvisation and Cultural Expression in Health Settings'. Calgary, AB: Canadian Association of Music Therapists.

Chapman, O. and K. Sawchuck (2012). 'Research-Creation: Intervention, Analysis, and "Family Resemblances"'. *Canadian Journal of Communication* 37: 5–26.

Cohen-Cruz, J. (2000). 'A Hyphenated Field: Community-Based Theatre in the USA'. *New Theatre Quarterly* 16 (4): 364–378.

Conquergood, D. (1985). 'Performing a Mortal Act: Ethical Dimensions of the Ethnography of Performance'. *Text and Performance Quarterly* 5: 1–13.

Dean, R. and H. Smith (1997). *Improvisation Hypermedia and the Arts since 1945*. London: Routledge.

Deep Listening Institute (2014). 'Deep Listening Institute: About Us'. Available online: http://www.deeplistening.rpi.edu/ (accessed 1 April 2018).

Denzin, N. K. (2017). 'Critical Qualitative Inquiry'. *Qualitative Inquiry* 23 (1): 8–16.

Denzin, N. K. and Y. Lincoln (eds) (2000). *Handbook of Qualitative Research*. 2nd edition. Thousand Oaks, CA: Sage.

Elkins, J. (n.d.). 'Fourteen Reasons to Mistrust the PhD'. In *Artists with PhDs*. Available online: http://www.jameselkins.com/yy/ (accessed 1 April 2018).

Finley, S. (2005). 'Arts-Based Inquiry: Performing Revolutionary Pedagogy'. In N. K. Denzin and Y. S. Lincoln (eds), *The SAGE handbook of qualitative research*, 681–694. 3rd edition. Thousand Oaks, CA: Sage.

Fischlin, D. (2009). 'Improvisation and the Unnamable: On Being Instrumental'. *Critical Studies in Improvisation* 5 (1): 1–8.

Foster, S. L. (2003). 'Taken by Surprise: Improvisation in Dance and Mind'. In A. C. Albright and D. Gere (eds), *Taken by Surprise: A Dance Improvisation Reader*, 3–10. Middletown, CT: Wesleyan University Press.

Frost, A. and R. Yarrow (2007). *Improvisation in Drama*. 2nd edition. Basingstoke: Palgrave Macmillan.

Gelowitz, C., K. Morgan, and L. Benedecenti (2008). 'The Public Space as an Interface for Technology Research and Art: A Study and Implementation of Two Interdisciplinary Collaborations between Engineers and Artists'. *International Journal of Technology, Knowledge, and Society* 4 (3): 65–72.

Gomez-Pena, G. and R. Sifuentes (2011). *Exercises for Rebel Artists: Radical Performance Pedagogy*. New York: Routledge.

Halpern, C. and D. Close (1994). *Truth in Comedy*. Colorado Springs, CO: Meriwether.

Heble, A. (2005). 'Editorial'. *Critical Studies in Improvisation* 1 (2). https://www.criticalimprov.com/index.php/csieci/article/view/15/44 (accessed 1 April 2018)

Helguera, P. (2011). *Education for Socially Engaged Art: A Materials and Techniques Handbook*. New York: Jorge Pinto Books.

Horowitz, R. (2014). 'Introduction: As if from Nowhere … Artists' Thoughts About Research-creation'. *RACAR: Revue d'art canadienne* 39 (1): 25–27.

Improvisation, Community, and Social Practice (ICASP) (2009). 'About ICASP'. Available online: http://www.improvcommunity.ca/ (accessed 16 November 2011).

International Institute for Critical Studies in Improvisation (IICSI) (2016). 'The International Institute for Critical Studies in Improvisation'. Available online: http://improvisationinstitute.ca/ (accessed 5 January 2018).

International Institute for Critical Studies in Improvisation (IICSI) (2020). 'Rebecca Caines'. Available online: http://improvisationinstitute.ca/team-member/rebecca-caines/ (accessed 5 July 2020).

Jackson, S. (2011). *Social Works: Performing Art, Supporting Publics*. London: Routledge.

Johnstone, K. (1987). *Impro: Improvisation and the Theatre*. New York: Routledge.

Journal for Artistic Research (JAR) (2010). 'About JAR'. Available online: https://jar-online.net/journal-artistic-research (accessed 1 April 2018).

Kester, G. (2004). *Conversation Pieces: Community and Communication in Modern Art*. Berkley, CA: University of California Press.

Landgraf, E. (2011). *Improvisation as Art: Conceptual Challenges, Historical Perspectives*. New York: Continuum.

Leavy, P. (2008). *Method Meets Art: Arts-Based Research Practice*. New York: Guilford Press.

Levine, S. K. (2013). 'Expecting the Unexpected: Improvisation in Art-based Research'. *Journal of Applied Arts and Health* 4 (1): 21–28.

Lewis, G. E. and B. Piekut (eds) (2016). *The Oxford Handbook of Critical Improvisation Studies*, vol. 1. Oxford: Oxford University Press.

Lipsitz, G., D. Fischlin, and A. Heble (eds) (2013). *The Fierce Urgency of Now: Improvisation, Rights, and the Ethics of Cocreation*. Durham, NC: Duke University Press.

Matsinhe, D. M. (2007). 'Quest for Methodological Alternatives'. *Current Sociology* 55 (6): 836–856.

Mattin (2009). 'Going Fragile'. In Mattin and A. Iles (eds), *Noise and Capitalism*, 18–23. Donostia-San Sebastiá: Arteleku Audiolab.

Nancy, J.-L. (2014). 'On Listening'. In R. Caines and A. Heble (eds), *The Improvisation Studies Reader: Spontaneous Acts*, 17–26. London: Routledge.

Nettl, B. and M. Russell (1998). *In the Course of Performance: Studies in the World of Musical Improvisation*. Chicago studies in ethnomusicology. Chicago, IL: University of Chicago Press.

Ninh, L. Q. (2010). *Improvising Freely: The ABCs of an Experience*. English edition. Guelph: PS Guelph.

Oliveros, P. (1995). 'Acoustic and Virtual Space as a Dynamic Element of Music'. *Leonardo Music Journal* 5: 19–22.

Oliveros, P. (2005). *Deep Listening: A Composer's Sound Practice*. New York: iUniverse.

Oliveros, P. (2015). *The Difference Between Hearing and Listening*. Available online: http://www.tedxindianapolis.com/speakers/pauline-oliveros/ (accessed 1 April 2018).

Peters, G. (2009). *The Philosophy of Improvisation*. Chicago, IL: University of Chicago Press.

Prentki, T. and S. Preston (2009). *The Applied Theatre Reader*. London: Routledge.

Ramshaw, S., P. Stapleton, A. Marquez-Borbon, and S. Mulholland (2018). 'Hydra: A Creative Training Tool for Critical Legal Advocacy and Ethics'. *Critical Studies in Improvisation* 12 (1). https://www.criticalimprov.com/index.php/csieci/issue/view/220 (accessed 1 April 2018).

Sajnani, N. (2012). 'Improvisation and Arts-Based Research'. *Journal of Applied Arts and Health* 3 (3): 79–86.

Sawyer, R. K. (2007). *Group Genius: The Creative Power of Collaboration*. New York: Basic Books.

Schlanger, J. (2018). 'musicWitness'. Available online: http://www.musicwitness.com/musicWitness.htm (accessed 1 April 2018).

Siddall, G. and E. Waterman (2016). *Negotiated Moments: Improvisation, Sound, and Subjectivity*. Durham, NC: Duke University Press.

Slager, H. (2009b). 'Nameless Science'. *Art & Research* 2 (2): 1–4.

Slater, H. (2009a). 'Experimental Aesthetics'. In J. Elkins (ed.), *Artists with PhDs*, 197–210. Washington DC: New Academia.

Spooner M. and J. McNinch (eds) (2018). *Dissident Knowledge in Higher Education*. Regina, SK: University of Regina Press.

Smith, H. and R. Dean (2009). *Practice-led Research, Research-led Practice in the Creative Arts*. Edinburgh: Edinburgh University Press.

Spolin, V. (1969). *Improvisation for the Theater: A Handbook of Teaching and Directing Techniques*. [Evanston, IL]: Northwestern University Press.

Stevens, J. (2007). *Search and Reflect: A Music Workshop Handbook*. Teddington: Rockschool.

Stravinsky, I. (1946). *Poetics of Music in the Form of Six Lessons*. Cambridge, MA: Harvard University Press.

Thompson, N. (ed.) (2012). *Living as Form: Socially Engaged Art from 1991–2011*. Cambridge, MA; MIT Press; New York: Creative Time.

Wang, G., D. Trueman, S. Smallwood, and P. R. Cook (2008). 'The Laptop Orchestra as Classroom'. *Computer Music Journal* 32(1): 26–37.

Waterman, E. (2014). 'Improvised Trust: Opening Statements'. In R. Caines and A. Heble (eds), *The Improvisation Studies Reader: Spontaneous Acts*, 59–62. London: Routledge.

20

'The Music Comes from Me': Sound as Auto-Ethnography

Darla Crispin

The singer composes the end

> Elektra, having heard the figure long ago, here takes it as her own, transforms it into melody, into a musical phrase of great beauty; in transforming it, what is she but a composer, teasing substance out of a dormant and common seed?
>
> —Carolyn Abbate (1989: 126)

'Ob ich die Musik nicht höre? Sie kommt doch aus mir' (Do I not hear the music? It comes from me). So sings Elektra in her final moments; the vengeance she sought having been achieved, this utterance stands as the ultimate linguistically articulated gesture of her existence. Yet, this music that she claims comes from her does not stop upon her falling silent but continues as a manifestation of her state of being. She dances to her death in a stylized state of transport, with the music 'coming from' her in so reified a form that the final cadence of the opera, a resolution from E flat minor to C major, is replete with evil: the 'whiteness' of C major awash with the blood of Orestes' avenging purge.

Much has been written about *Elektra* as a work that epitomizes the 'language crisis' of the twentieth century. Strauss's opera is itself a trope of modernity, a modernity which – ironically – has generated a wealth of extraordinary works, from Hugo von Hofmannsthal's *Lord Chandos Letter* to Arnold Schoenberg's incomplete *Moses und Aron*. Its significance is not confined to the way it handles structures related to language; it also has implications in terms of the institutionalization of art, which was generative of increasingly reflexive art works, alongside politicized materiality articulated through the genre of the manifesto.

The concern of this chapter is not to recount the particular set of modernist landmarks referred to above, nor to reread *Elektra* itself, but to respond to some of its cues in order to explore aspects of reflexivity, reflection, and auto-ethnography as they are currently being transformed within the field of artistic research in music. In light of current geopolitical events, it would seem that we are lapsing into a new kind of 'language crisis' for our own century, manifested in the problematizing of words and their truth content. Through the advent of 'fake news', and coincident with a rise in nationalist sentiment and party-political ruptures, it is becoming increasingly difficult to maintain a meaningful discourse other than within the echo chambers of one's own belief network. In thinking about how such global phenomena might relate to the seemingly narrow and protected world of artistic research, it is informative to compare the travails of artists in the last years of the 'long nineteenth century' with those of their counterparts operating in a new millennium, both to detect commonalities and to remark upon their differences.

Elektra's final claim, her assertion of performative authorship from within a work that her own utterance and action is about to bring to an end, is far from the only example in Western classical music of its 'fourth wall' being pushed at from within. Music possesses a special capacity to comment upon itself, and this has made it a site of exposure for irony and incongruity, with musical language critiquing from within the forms which contain it. But the twentieth-century language crisis also brought the particular spectre of silence in its wake. If Theodor W. Adorno could assert that Elektra articulated an affirmative formula for a negative world, manifested in a divided musical spirit and ruined by fixity and externality, then we might conclude that the failure of communication in *Elektra*, in the non-connection of its characters' rhetorics, is the opera's ultimate success (see Adorno 1965: 14–32, 1966: 113–129). As Carolyn Abbate states: 'Hearing, not hearing, hearing lies, silence: the opera *Elektra* is a play upon sounds, voices, and music itself, a play upon the collected utterances of its characters' (Abbate 1989: 107).

Hearing, listening, and reflexivity

In Abbate's essay on *Elektra*, in which the aspects listed above are explored alongside a concise musical analysis, she points to the 'obsession with hearing' as integral both to the creation of the opera and to the viability of a critique that departs from the textual and moves towards the performative. As she suggests in her study, hearing involves a cognizant presence that operates in the light of language, the possession of a species of understanding that unfolds via both a textual literalness and a trans-textual irony. Elektra's power, her right to the claim of 'composer', stands alongside our willingness to understand her as such, something that relates to our sense that she transcends the opera itself:

> What sets *Elektra* apart is, perhaps, our sense of another power, the 'polyphony beyond counterpoint' of the voice that shadows the text, the voice of the music, the narrator's semiotic voice, the vocal personae. Above all, *Elektra* is a clamor of tongues, and Elektra's voice among many, sings in animated, sonorous congress.
>
> (Abbate 1989: 127)

The idea of 'sonorous congress' highlights a paradox: that the resistant, refractive language of Elektra's voice remains part of a larger entity – one that functions as it should; her destruction 'works' in the context of a piece that must be about that destruction as something emergent from her own being. The imperative of 'hearing' constantly brings forward its opposing state, that of the silence which is Elektra's ultimate destination. More disturbingly, it conjures up shadows around deafness as both the involuntary and the willed inability to hear. As such, this inability can be, at certain times, an accompaniment of resistance, at others, a denial of freedom of choice.

Reflexivity – our need to 'hear' things on our own terms – is stylized in *Elektra* and, in a wider sense, has come to represent aspects of our being in the world. The interactions of Elektra and her sister, Chrysothemis, their lack of connection during their dialogues, shed light upon our contemporary predicament of being unable to hear any voice that is unlike our own; their non-identical languages indicate a high degree of self-reflexivity, although Elektra's treading beyond the compositional boundary could also be read as a positive gesture, a movement beyond her own, merely reflexive, language. In a wider sense, such a move potentially remakes the relationship of character and performer in the light of sound itself, something with implications beyond the opera – and perhaps even beyond 'composed' music altogether.

It is surely not unreasonable to locate the examination of a remaking in the light of sound itself within a field of study that gives prominence to sound. Sound studies has emerged as an important field in contemplating sounded phenomena; it enables us to conceive of music in a more wide-ranging way than has been possible within the more traditional disciplines of theory, analysis, and historical musicology that have dominated musical scholarship in the West.

The field of sound studies was, in part, developed to address the effects of a broadening conception of what constitutes music, and to mitigate some of musicology's perceived shortcomings when applied to music outside the Western classical tradition. It has encouraged new views and approaches to come to light when addressing work beyond the confines of the Western musical canon, many of which operate outside the forms and conventions of standard music science. But it has also provided insights into contemporary practices *within* Western art music, such as those of free improvisation, that are moving away from score-reliance towards non-text-based realizations:

> Music studies are numerous, of course, but audio culture is a diminutive interloper, a newcomer adding conceptual breadth and perspective through its concentration on aspects of auditory experimentation (in all fields of music), so-called sound art, the voice, experimental sound work, the phenomenology and philosophy of listening and so on.
>
> (Toop 2016: 70)

A complementary and superficially contrary trend to the growth of sound studies is the rise of historically informed performance, which seeks to recover 'authentic' past practices, drawing upon whatever evidence takes us closer to the ideal of experiencing earlier repertoire in the manner it was received by contemporary audiences. But sound studies and historically informed performance are more accurately seen as different manifestations of a common emphasis upon sound, its generation and reception. In wanting to hear earlier music as closely as possible to the ways in which it resonated in the buildings where it was first performed, in the ears of its first listeners and in the mind of the composer who conceived it, we are asserting the primacy of sound among the elements that make up its very identity.

Of course, authenticity has become a notorious territory of controversy, both in terms of the reliability of historical sources and, perhaps more fundamentally, because even if we can faithfully recreate the external parameters within which the music originally sounded, there is nothing we can do to recalibrate our own consciousness to that of the earlier listener. In this case, it is not so much a matter of 'the music comes from me' and more that 'the music must resonate through and within me'. The inescapable intervention of the self gives rise to an auto-ethnographical dimension to sound, and this, in turn, accounts at least in part for the ethically charged discourses that surround authenticity.

As a result, the discipline of sound studies has a contribution to make to an emerging debate around ethics in performative musical scholarship, questioning not just what kind of music is included or excluded and, in academia, what kind of work is deemed to merit valorization, but also which sounds and, inextricably interwoven with this, whose voices are heard in this new milieu and which remain stifled. Sound studies has not offered fluent solutions; instead, it has brought a new and necessary layer of complexity to the music sciences. The conceptual application of its orientations – beyond script and form, towards sound itself and the resulting ethical implications – has the potential to elicit quite different readings than those analytical practices that are oriented towards score-based practices.

It is important to note that, within music theory, aural analysis has become normalized, particularly as a mode of working within performance studies. As a rule, however, this still returns the readings to the score, and still relegates the performer to the third-party role. A focus upon sound as a more abstract, in the sense of de-textualized, phenomenon allows us to take greater account of reflective practices; auto-ethnography, in turn, gives disciplinary tools for this kind of work, although reflection itself has merit even prior to the application of an ethnographic apparatus.

Reflection has also emerged as an important tool within the sphere of artistic research and there, too, auto-ethnography is beginning to gain ground as a concept which artist researchers find useful in their work. But, as will be discussed, this has only exacerbated the tensions between traditional research and its artistically rooted counterpart. There is, perhaps, the potential for a revitalized retrospective musicology informed by sound studies to generate a rapprochement, retaining a form of disciplinary distance, but having this inflected – and even opposed – by practices adopted from artistic research, in which the artist must give an account for her- or himself. An emphasis upon sound, in contrast to opposing approaches that rely either upon text or upon human agency, with the one excluding the other, may have the capacity to merge reflection, auto-ethnography, and more familiar methods of analysis, thereby generating an inflected hermeneutics that can both embrace and go beyond the sound worlds of Western art music. Within sound studies itself, calls have been made for this fusion of approaches:

> Anthropologist Tim Ingold has welcomed the new flourishing of sound studies, but with certain reservations. He argues that the term 'soundscape' coined by R. Murray Schaefer for educational purposes in the 1960s and now in common usage, is flawed, because it places the listener at an objectified distance from what is in fact immersive and so reinforces the artificial divide commonly erected between mind and matter.
>
> (Toop 2016: 70)

Why does the music of the past, arguably saturated with hermeneutics, with interpretative readings, need such a multifaceted inflection of approach? One of the reasons is an ethical one. The application of methods from the more recent past may allow us to recover or renew certain kinds of access to more distant pasts. This becomes one way to address more recent problems in music academia around both 'deskilling' and the denigration of 'expertise'. If the discipline of sound studies prompts us to consider the auditory landscape as we encounter it here-and-now, its focus upon experienced sound opens doors to new historical teleologies, rather than merely recounting their anachronistic aspects. This is an important opportunity for music scholars, whether their disciplinary focus be upon sound, score, or both.

Hearing, listening, and artistic research

If Elektra's claim of self-authorship could be seen as enmeshed in the twentieth-century language crisis as exemplified in music, then its potential foreshadowing of the performative turn could correspondingly be seen as having a counterpart in the rise of artistic research in music. After all, artistic research has become concomitant with innovations around musical language and notions concerning its 'truth content' and has also been a driving force behind various innovations in art-making. Western art music is sustained but also challenged by the standard locations of its developmental, pedagogical, and professional practices: music conservatoires and music departments affiliated with colleges and universities, orchestral concert halls, and opera houses. These institutions, and many others, are woven into entrenched cultural and social spaces and, as such and in their different ways, all have influences and impacts upon the sociopolitical structures of which they are a part.

At the time of writing, the institutions of Western art music share in common, and across national boundaries, a sense of their work being scrutinized by those eyes that seek utility, profitability, political advantage, and ideological control. Yet, what appears to be a vulnerable position actually affords researchers affiliated to such institutions and their practices opportunities to develop exemplary work in a microcosm, addressing precisely the most contentious areas with a view to challenging essentialism. Indeed, in doing so, they can enrich a wider interdisciplinary evidence base in which it is argued that, far from being 'non-essential', the arts and humanities are needed more desperately than ever to safeguard the evolution of just societies. Just as *Elektra* is an opera that critiques the genre of opera, artistic research in music institutions can critique those institutions, going deeper, and potentially interrogating the very aspects of musical materiality which form the basis for core structures in musical culture.

Listening to the voices: Institutional ethics

The paradox of this critique from within, especially in the conservatoire setting, is that the fundamental material for this kind of substantial claim emerges from very specific, often self-reflexive, work that finds echoes in more universal arguments: the 'personal becoming political' (Hanisch [1970] 2006),[1] as first articulated within the second-wave feminist discourses of Carol Hanisch, which became a touchstone for the feminist movement in the 1970s.[2] It is significant that music conservatoires have, in some instances, become a focal point for the exposure of the networks of interpersonal power and its abuse that can arise in cultures where the individual, with their own unique qualities – one might say their 'voice' – is foregrounded. The debates, both internal and more public, that such exposures have prompted are not merely about the righting of past injustices; they concern the very nature of institutionalized knowledge and activity, which is why the close temporal proximity of the discourses on artistic research in music and the #MeToo debates has high educational, social, and ethical significance.

Within arts training institutions, the development of artistic research brings with it a set of concrete educational questions that necessarily challenge the paradigms of master-apprentice training, whereby the learner receives and accepts the established wisdom of their teacher rather than progressing by the formulation and resolution of their own artistic hypotheses. If the performer is, in part, the author of their own performances, the same applies to the music student's authorship of their own learning. Questions concerning the extent to which students might co-create and lead their own learning are not susceptible to satisfactorily resolution but, on the contrary, must remain open as an evolving approach to how curriculum development is to take place within a 'no longer so young', rather volatile disciplinary background. If artistic research is to substantiate its propositions for re-conceptualizing claims to authorship, for example, then this ethical aspect must come to the fore.

Similar dilemmas await performers upon entry into the profession. Consider the work and the roles of contemporary music performers today. Many belong to ensembles presenting highly complex music in which the *Werktreue* model is maintained to an almost exaggerated degree through the often-lively interventions of the composers with whom they work – sometimes to such a degree that the performers become more like composerly extensions. While this is far from the case in many such collaborations, it happens often enough that resistance has emerged, especially as new music opens up to improvisational and composed interventions by the performers themselves in, for example, the development and execution of specific extended techniques. This has resonance in the compositional manifesto of Jennifer Walshe, whose text 'The New Discipline' claims, controversially, that music now belongs in a wider, more theatrically-based conceptual space:

> While Kagel and others are clear ancestors, too much has happened since the 1970s for that term to work here. MTV, the Internet, Beyoncé ripping off Anne Teresa De Keersmaeker, Stewart Lee, *Girls*, style blogs and yoga classes at Darmstadt, Mykki Blanco, the availability of cheap cameras and projectors, the supremacy of YouTube documentations over performances. Maybe what is at stake for the New Discipline is the fact that these pieces, these modes of thinking about the world, these compositional techniques – they are not 'music theatre,' they *are* music. Or from a different perspective, maybe what is at stake is the idea that all music is music theatre. Perhaps we are finally willing to accept that the bodies playing the music are part of the music, that they're present, they're valid and they inform our listening whether subconsciously or consciously. That it's not too late for us to have bodies.
>
> (Walshe 2016)

One of the tools needed for the kind of enquiry that can generalize out from the specifics of *Elektra* and begin to address the agenda set by Walshe and others is a more considered view of the nature of 'hearing', one that goes beyond the 'obsession with hearing', in Abbate's phrase, which pervades *Elektra* and which finds articulation as an adjunct to a manifesto for progress. This needs to be set alongside deeper discussion of views on the differences between 'hearing' and 'listening', discussions that address these terms not merely from an etymological standpoint but, in particular, in the light of human experience, since 'ways of hearing are notoriously subjective' (Toop 2016: 67). Of interest here is Anahid Kassabian's study, *Ubiquitous Listening – Affect, Attention and Distributed Subjectivity*, which develops the topic of '*ubiquitous musics*, these musics that fill our days, [and] are listened to without the kind of primary *attention* assumed by most scholarship to date' (Kassabian 2013: xi, emphasis in the original). Interestingly, for the purposes of the current study, the launching point of Kassabian's argument is the redefinition of *listening* as:

> A range of engagements between and across human bodies and music technologies, whether those technologies be voices, instruments, sound systems, or iPods and other listening devices. This wipes out, immediately, the routine distinction between listening and hearing that one often finds, in which the presumption is that hearing is physiological and listening is conscious and attentive. I insist, instead, that all listening is importantly physiological.
> (Kassabian 2013: xxi)

This core claim is not without its problems. 'Listening' may, indeed, imply the active, cognizant engagement of the mind in contrast with the more passive phenomenon of 'hearing'; in another sense, however, this apparently passive receptivity could also be read more favourably as generating the vital potential for a deeply embedded and genuinely empathic understanding that brings sound into embodiment prior to the crystallization of judgement. In music, this prelinguistic realm is one site of the tacit knowledge that has the capacity to both carry and communicate. From this perspective, we might wish to dispute the claim that 'we could probably all agree that *hearing* is somehow more passive than *listening*' (Kassabian 2013: 8, emphasis in the original). Furthermore, the sharp divide delineated and deconstructed by Kassabian is already under scrutiny and deconstruction within Western art music itself; one of the points about *Elektra* is that it problematizes precisely this malleable space as part of its musical language via the potential composerly claim of its central character; cracks and crevices are already appearing in the delineation of roles and this has consequences for what 'hearing' and 'listening' might become.

Kassabian's study, however, does open up the field not only to a reconsideration of the 'ubiquitous musics' that are its central concern but also to the matter of the identities of their makers. I would argue that it also prompts a contemporary consideration of how the work of artistic researchers engenders a call for a better, ever-evolving understanding of the reflexive work that often accompanies their art-making but equally often embodies it. Some kind of rapprochement between 'hearing', 'listening', and 'aural reflection' would seem to be apposite, not least because of the embeddedness of each in identity formation. If, indeed, 'identity is the trace of affect' (Kassabian 2013: xxvii) then an enquiry into these three activities in relation to both the artistic research project and its associated practices of reflection (and, even, of auto-ethnography) may assist us in understanding developments in the field. Perhaps more importantly, such an enquiry may be an important adjuvant to a call for a more trenchant criticality as the field of artistic research matures.

If we ask how reflection, as a tool to be wielded by the artistic researcher, might articulate excellence without disregarding the marginal, we see that such a question is not merely about research: it is really about how we become more fully realized human beings. The twist is that, in artistic research, we are primarily looking at ways in which personal reflection, auto-ethnography, and self-reflexivity can continue to be developed as viable approaches to the conducting of musical research. Ends and means are reversed. By giving an expository voice to those artists who have previously felt divested of viable languages and by more readily admitting to academia those who experience themselves as being on the periphery – and by doing so *without* precipitately regarding these actions as passing harsh judgement upon the knowledge, practices, and practitioners of the past – artistic research may enrich the personal lived experience of such individuals; but it does so in the name of advancing and deepening our collective understanding of the art form at which it is directed.

Taking 'sound' and 'tone' personally

But, again, the collective and the personal are interdependent in a whole range of subtle ways. The idea that it is the *sound* of Elektra's voice that composes a new reality is contingent upon the vocal specificities of the singer giving voice to her character in the drama; this foregrounds the idea of the importance of a personal sound, a specific 'tone' through which a performer is able to make a unique contribution to the world-creation of a musical composition. In this context, listeners are also involved in their apprehension of sound as a specific kind of personal exposition. This aspect of performance, namely the nature of 'tone', has been the subject of a great deal of debate. The notion of the development of 'tone' – that is to say, of an attractive, desirable sound, the kind which Glenn Gould described with irony in an interview as 'burnished, singing, tone' (Gould quoted in Payzant 1997: 109) – is still fundamental to many of those involved in the training of musicians in the institutions of Western art music. In the nineteenth century, the development of good tone with respect to pianism was enshrined in numerous 'method' books in which 'tone' was explored as a specific topic, and the achievement of its desirable attributes was made possible by various means, from finger exercises to using arm weight effectively. The development of this kind of thinking with respect to how the pianist must make the instrument 'sound well' may be said to have reached an apogee with the publication of the chapter 'On Tone' within Heinrich Neuhaus's highly regarded and still influential tome, *The Art of Piano Playing*. The continued relevance of this book is significant for the current argument in its assertion of the centrality of 'tone' both for musical creativity and as an element that is revelatory of a performer's personality:

> I repeat: both the average pianist and the great pianist, if they only know how to work, will acquire their own individual tone quality which corresponds to their psychological, technical and physical make-up, and will never be a warehouse of 'universal' tone or any kind of technical perfection. There are, luckily, no such phenomena.
>
> (Neuhaus 1993: 80–91)

To highlight that this view is far from unanimous, one only needs to consult Payzant's account of Glenn Gould's more architecturally conceived pianism:

> It may come as a surprise that Gould is not seeking beautiful tone. This, after all, is what most musicians look for when they go shopping for an instrument. They play a note and listen to it and pronounce it beautiful or not. But Gould says that he is not interested in beautiful tone in this sense: 'As long as the piano has a good action, the sound isn't too important'.[3]
>
> (Payzant 1997: 109)

Gould's rationale has resonances with contemporary scientific investigations that debunk the idea that tone can be radically altered through remaking gestures and touch; this argumentation is linked more generally to a critique of the performer as authoritative contributor in Western classical music. But to ignore the aspect of tone as an adjunct of musical expression seems problematic; for classical performers, bypassing it denies us insight into aspects of the subjective nature of music-making that can actually be integral to our better understanding of how a musical performance is conceived. If we believe that musicians can not only be banally identified but actually take on their very identity through the sounds they create, it is a logical progression from this to conclude that the tone possessed by a performer is not merely something overlaid upon the musical material which they perform but becomes fused with that material to create a unique synthesis. Such an idea foregrounds the notion of the performer as co-creative with the composer, with tone standing as a prime agency in terms of the contribution that the performer brings to the creative 'table': the performer through her sounds co-creates the compositional world; 'the music comes from me [the performer]'.

It is, perhaps, in an example such as that of Gould's pianism that we can argue for a sounded quality so specific that its implications go far beyond the reproductive, the mimetic, to something more like original creation itself. The evidence of this is in the implication of 'copying'. Those pianists who attempt to emulate Gould will be 'found out'; his sounded world is specific and hyper-personal, far beyond the 'family resemblances' that emerge when generations of pianists are part of a single 'school'. For example, while many fine pianists emerged from the studio of Theodor Leschetizky, the Polish pianist and teacher, they were as remarkable for the diversity of their personalities as for the similarities in their physical approach to the piano, based around economy of means, the elimination of inessential movement, and the common heritage identifiable in the resulting sound (Brée 1997).[4] Leschetizky's methods, and his sound, could engender and sustain a rich genealogy of family likenesses. Gould's sound, however, has a creative urgency and an authorial absolutism that renders it resistant to emulation in an art form that requires originality of sound and tone. A fascinating question, that cannot be explored in depth here, is the extent to which the detected 'copying' of his sound approach on recordings forms a kind of 'plagiarism' or whether it is merely a product of poor artistic originality, and thus a qualitative fault.

The discipline of sound studies has played an important and necessary role in detaching musical creation from various performative frameworks of the past that have seemed relentlessly confining, limiting legitimate authorship to specific groups while impeding the creative expression of others. This contribution is evidenced in important collected texts that sketch out the widened terrain for inquiry.[5] However, something has been bypassed in this move, such that there remains an impediment to the ongoing discussion of the performer as author within the context of the burgeoning field of artistic research. Still, the very fact that scholars are engaging in this kind of work demonstrates an attitudinal shift that has taken place over a number of years and for a variety of reasons. For example, transformations in musicology itself have catalyzed moves in this direction. On the one hand, they have done so by re-problematizing the identities of those whose stories have hitherto been exposed as exemplary, suggesting that the musical past ought to be regarded as permeable, open to infinite rereading. On the other hand, these transformations have stimulated a similar problematizing of our contemporary identities, proposing that those doing the 'reading' ought to more overtly acknowledge their presence and power as readers.[6]

One of the early signs of this shift was the emancipation of 'biography' as a vehicle for remaking musical identities and for exposing the identities of those who had previously been invisible and on the margin. This has refreshed the view of the musical landscape and enriched our sense of the interdependent cultural 'eco-systems' from which canonical artists generally emerge while, at the same time, reinforcing a sneaky feeling that some of these artists are canonical for good reasons.[7]

The 'tone' of discipline-formation

There is, however, a danger. When one is part of the process of discipline-formation, one sees how initially idealistic hopes are all too easily traduced for reasons of practicality and, alas, of competition for often-scarce financial resources. It was hoped that those directly engaged in the creation of art would find ways of articulating their questions and sharing their knowledge, rather than remaining the largely mute subjects of scholarly study by others. The idea was not that artistic research should replace music analysis or musicology, but that its questions – and therefore its methods of answering them – would simply be different. There was, we thought, room for all and scope for each kind of research to enrich the others. This is the idealistic frame with which most of us are by now familiar.

But to appropriate such a loaded word as 'research' is to raise a whole series of assumptions about the kind of activity that is being envisaged; moreover, insofar as these assumptions may then be confounded by the new paradigms being proposed, it is a challenge to decades, indeed generations, of privileged status, carefully guarded standards, and vested interests. The proponents of artistic research were challenged to provide explanations for what they thought they were doing and this had the effect of moving the terrain of engagement inexorably from practice to discourse. Thus, there arose a fresh battle in the eternal war around words and how they are used.

Some practitioners struggled with the conventions of scholarly discourse; they had a vocabulary for their practice but it was rougher and more intuitive than that of the scholars. Crucially, it was rooted in the reality of their direct practical experience; for them, there was no need to be squeamish about articulating their arguments in the first person and with reference to their subjective experience. Meanwhile, music scientists looked upon these efforts with increasing disdain, uttering the term 'me-search' with thinly disguised scorn. For the reflective practitioners, personal experience is capable of illuminating the inner recesses of their creative actions; for sceptical scholars, its beams are simply reflected back off the image of the subjective 'I', dazzling the beholder and therefore serving not to illuminate but to distract and obscure.

This polarizing of the terrain of discourse led to an increasing defensiveness and created the conditions for the next wave of enquiry, which, I believe, is happening now, on the European continent at least. This began with artistic researchers venturing into the territories of social science and philosophy, and finding verbally resonant and therefore reassuring validation for their ideas in respected writers from these fields; now, it is taking the reverse form: that is, the taking over of the artistic research space by social scientists and Continental philosophers. Once again, practitioners are fearful of having others speak on their behalf.

'Tone' beyond boundaries

That being said, many of these kinds of writing are illuminating and have an artful poesy about them – even aspects of performativity that blur further the boundaries. Consider this reflection on the pianism of Thelonious Monk:

> Let us say for a moment that Monk evokes or *conjures*, at the very same time he perverts, the reflection of a powerful form of reasoning hailing from the dawn of ages. He is playing with its shadow, its ghost. I can only offer, as is, the gripping hypothesis of one who has listened to Monk very closely.
>
> (Szendy 2016: 43, emphasis in the original)

Peter Szendy's musings on Monk, within his book *Phantom Limbs: On Musical Bodies*, demonstrates both the *pas de deux* involved in a philosophical investigation of 'hearing' and 'listening' in light of a rereading of history that allows the body (human and otherwise) to be taken into account, and an idiosyncratic style of writing that is reflexive and autobiographical. His well-researched critique of organology is 'embodied' within the unorthodox prose of self-reflection and analysis – and this lack of orthodoxy, for me at least, becomes part of its authority.

So, we see how highly charged the issues of self-reflexivity in music research work can be. It is clear that, while self-reflexivity opens up research to new kinds of conceptualizations and researchers to new challenges, it is also risky; even its better manifestations will be dismissed generically by some, while poor examples will only serve to reinforce prejudices about its inability to measure up to a dominant perception of what research should be. Furthermore, if one is to contextualize this kind of work in light of the specifics of auto-ethnography, the particular rigours of that research process need to be kept in mind:

> First, like ethnographers, autoethnographers follow a similar ethnographic research process by systematically collecting data [...] analyzing and interpreting them, and producing scholarly reports, also called autoethnography. In this sense, the term 'autoethnography' refers to the process and the product, just as 'ethnography' does. Second, like ethnographers, autoethnographers attempt to achieve cultural understanding through analysis and interpretation. In other words, autoethnography is not about focusing on self alone, but about searching for understanding of others (culture/society) through self [...]. The last aspect of autoethnography sets it apart from other ethnographic enquiries. Autoethnographers use their personal experiences as primary data.
>
> (Chang 2008: 48–9)

The developing corpus of work in artistic research has yet to fully subject itself to the disciplinary structure that is delineated above; in general, work of an auto-ethnographical leaning remains situated in a more amenable frame that 'invites the reader into the lived experience of a presumed "Other" and to experience it viscerally' (Boylorn and Orbe 2016: 15). This foregrounds the qualitative over the quantitative, with the further complication that, within artistic research work, reflection is generally enmeshed in the material nature of the art itself.

A brief case study on reflective practice

Although problematic, this aspect is of importance to the field in offering scope for disciplinary innovation. Indeed, there is, at least, one country in which personal reflection is not only encouraged as part of the research process; it is a mandatory element of the f – the artistic research PhD in Norway. Among the Nordic countries, Norway was the most emphatic in enshrining such artistic development – without any justificatory apparatus of additional research connotations – as not just a right but an obligation of artistic practitioners working in higher education. As part of this thinking, an 'artistic fellowship programme', distinct from the ramifications of the PhD, was evolved. Recently, this programme has been adapted to enable it to be granted full PhD status. Ratified in January 2018, the PhD in *Kunstnerisk utviklingsarbeid* nevertheless continues to reflect Norway's emphasis upon developmental processes embedded in socially conscious educational philosophies and practices. This PhD does not require a thesis; it consists of the making of art, coupled with personal reflection upon the processes that lead to that art. And this reflection does not have to be in the form of writing, although few in the music field have yet ventured far beyond text in their submitted reflections.[8]

Reflexivity, subjectivity, autobiography – these have all coloured artistic research work in Norway, but the results have been only variably successful. Thus, the matter of reflection itself has become a research question for the National Programme – a programme that pays its fellows a salary and offers generous opportunities for post-doctoral group projects. A report commissioned by the Norwegian Artistic Research Programme (NARP) authored by Eirik Vassenden in 2013 revealed that the reflective work of the research fellows generally emerges in the form of practical consideration of three areas, with the relative emphasis upon these areas being different according to the work of those writing them:

1 The relationship between their own artistic practice and the surrounding field;
2 The relationship between their own artistic practice and the problem of articulation;
3 The relationship between their own artistic practice and their personal experience of theoretical work and reflective work. (Vassenden 2013: 31)

Vassenden articulates the challenges; many involved with artistic research would find this kind of discussion familiar:

> How [do we] put into words the experience of developing an artistic project or doing artistic work? All such attempts at articulation involve the writer […] finding a good and expedient language with which to describe his or her experience, a language that will also make it possible to share this experience theoretically and cognitively. A language that enables not only the sharing of experience, but also the discussion and problematization of the experience, so that the creative practice, filtered through a different medium, also becomes visible to the creative subject. In this perspective, the attempts at articulation are based on an underlying literal interpretation of 'reflection' which can function as a mirror, but also as a contrasting element.
> (Vassenden 2013: 4–5)

So, we can see that the NARP has moved to develop a critique of what reflection might be, recognizing that this, in itself, is important research work. But this does not mean that its research candidates find negotiating their studies plain sailing.

A future for listening, reflection, and an evolving artistic research attitude

When we request self-reflexivity, what we are asking is difficult. It entails nothing less than a 'search for voice', in addition to the other challenges of method and rigour which it raises. It is Vassenden's view that, within the NARP at least, while the self-reflexive, auto-ethnographical texts of the artistic research fellows can 'display new and unorthodox ways of reflecting', he also finds that 'few candidates have really succeeded in situating their projects within a larger reflection space' (Vassenden 2013: 32). His specific reference to 'a reflection space' is important here; it is, perhaps, a signpost towards what must happen next in the development of discourses around artistic research work: to understand the importance of the reflective elements in disclosing aspects of the work that are essential to a deepening of perception concerning that work, while – paradoxically – removing the focus upon the 'self' of that work, and even in reflective work, in order to consider how it 'speaks' in the wider world and as itself, rather than as its maker:

> To know the limits of acknowledgement is a self-limiting act and, as a result, to experience the limits of knowing itself. This can, by the way, constitute a disposition of humility, and of generosity, since I will need to be forgiven for what I cannot fully know, what I could not have fully known, and I will be under a similar obligation to offer forgiveness to others who are also constituted in partial opacity to themselves.
>
> (Butler 2001: 28)

This 'listening attitude' within reflection would appear to be integral to the evolution of artistic research work towards a new stage in which the matter of the self is simultaneously central, yet non-essential. It also reminds us that the nature of such 'knowing' has elements of 'humility' and 'generosity' that are sorely needed within academic discourses and beyond.

Notes

1. The essay is accessible online with a new introduction by the author written in 2006. It forms a locus for reflexive research questions, something that has ramifications for certain kinds of auto-ethnographical research work.
2. The current relevance of this motto is placed into sharp relief by the emergence of the #Metoo movement, which has developed first as an articulating platform for survivors of sexual harassment and abuse and, more recently, as a significant site for a new kind of resistance to discrimination more broadly.
3. Payzant also cites Jock Carroll: "'I don't think I'm at all eccentric,' says Glenn Gould" (*Weekend Magazine* 6 [27]: 11).
4. Brée, an assistant of Leschetizky, wrote down the principal aspects of his piano instruction, but the book is perhaps most suitably approached as a memorial to the master than a definitive guide to his approach to pianism, which is much less specific than the notion of a 'method' implies.
5. Such volumes include Michael Bull and Les Back's *The Auditory Culture Reader*, and Georgina Born's *Music, Sound and Space: Transformations of Public and Private Experience*.
6. This set of transformations has included the emergence of the 'New Musicology' in which musical study is placed within a more culturally contingent context, as articulated by such exponents as Joseph Kerman (who is viewed as one of its originators), the feminist writers Susan McClary and Marcia Citron, and the ethnomusicologist Bruno Nettl, amongst an increasingly large and influential array of scholars dealing with numerous vanguard topics, from mainstreaming queer theory to developing pop music studies as a serious sub-discipline. Artistic research, with its links to the 'practice turn', parallels these developments.
7. See, for example, Tatjana Marković and Vesna Mikić's edited volume *(Auto)Biography as a Musicological Discourse*, in which the diverse range of contributions is inflected by the particular situatedness of the editors and thus sheds light on the re-emergence of the musicology of their region following military conflict and ethnic divisions, meaning that the editorial work has an auto-ethnographical underpinning.

8. An index with links to artistic research reflections of research fellows on the KUST programme may be found online: https://www.researchcatalogue.net/view/314097/314098 (registration required).

References

Abbate, Carolyn (1989). 'Elektra's Voice: Music and Language in Strauss' Opera'. In Derrick Puffett (ed.), *Cambridge Opera Handbooks: Elektra*, 107–120. Cambridge: Cambridge University Press.
Adorno, Theodor W. (1965–66). 'Richard Strauss: Born June 11, 1864'. *Perspectives of New Music* (Fall–Winter): 14–32; (Spring–Summer): 113–129.
Born, Georgina (ed.) (2013). *Music, Sound and Space: Transformations of Public and Private Experience*. Cambridge: Cambridge University Press.
Boylorn, Robin M. and Mark P. Orbe (eds) (2016). *Critical Autoethnography: Intersecting Cultural Identities in Everyday Life*. London: Routledge.
Brée, Malwine (1997). *The Leschetizky Method: A Guide to Fine and Correct Piano Playing*. New York: Dover Publications.
Bull, Michael and Les Back (eds) (2016). *The Auditory Culture Reader*. 2nd edition. London: Bloomsbury.
Butler, Judith (2001). 'Giving an Account of Oneself'. *Diacritics* 31 (4): 22–40.
Chang, Heewon (2008). *Autoethnography as Method*. Walnut Creek, CA: Left Coast Press.
Hanisch, Carol ([1970] 2006). 'The Personal Is Political'. In Shulamith Firestone and Anne Koedt (eds), *Notes from the Second Year: Women's Liberation*, with new introduction by Carol Hanisch, 1–5. Available online: http://www.carolhanisch.org/CHwritings/PIP.html (accessed 16 November 2018).
Hofmannsthal, Hugo von (1995). *The Lord Chandos Letter*. Trans. Michael Hofmann. London: Penguin.
Kassabian, Anahid (2013). *Ubiquitous Listening: Affect, Attention and Distributed Subjectivity*. Berkeley, CA: University of California Press.
Marković, Tatjana and Vesna Mikić (eds) (2010). *(Auto)Biography as a Musicological Discourse*. Beograd [Belgrade]: Fakultet Muzicke Umetnosti.
Neuhaus, Heinrich (1993). *The Art of Piano Playing*. London: Kahn & Averill.
Payzant, Geoffrey (1997). *Glenn Gould: Music and Mind*. Toronto: Key Porter Books.
Szendy, Peter (2016). *Phantom Limbs: On Musical Bodies*. Trans. Will Bishop. New York: Fordham University Press.
Toop, David (2016). 'Each Echoing Opening; Each Muffled Closure'. In Michael Bull and Les Back (eds), *The Auditory Culture Reader*, 63–71. 2nd edition. London: Bloomsbury.
Vassenden, Eirik (2013). 'What Is Critical Reflection? A Question Concerning Artistic Research, Genre and the Exercise of Making Narratives About One's Own Work'. For the Norwegian Artistic Research Programme. Available online: http://artistic-research.no/wp-content/uploads/2012/09/What-is-critical-reflection.pdf (accessed 20 August 2018).
Walshe, Jennifer (2016). 'The New Discipline'. Notes for Borealis Festival, January 2016. Available online: http://www.borealisfestival.no/2016/the-new-discipline-4/ (accessed 10 November 2018).

21

Sound beyond Representation: Experimental Performance Practices in Music

Lucia D'Errico

There is a specific moment in the everyday practice of music performers which contains, as if compressed and ready to be propelled, an enormous creative potential: that in which they are just about to perform the first note of a score. In this moment, something is going to be *produced*, a materialization of sonic waves is about to be generated by the friction between physical bodies, and in turn rebounded, amplified, reverberated by a concatenation of other bodies. The instrumentalists' bodies are in contact with other physical bodies assembled with different materials: wood, metal, varnish, plastic, anodes, and cathodes. Several bodies are attending and influencing this moment – the body of the room, the bodies of possible listeners. In this suspended hiatus, in the tangency with the sonic event, music performance appears in its fully productive function, an aggregate of material components and quasi-material forces directed towards the generation of the unexpected. At the same time though, in spite of the cruciality of this hiatus, the instrumentalists are also far from facing the unexpected. The performance starts, the first note is produced: imagine that from this moment on all of these bodies are not anymore hovering on the brink of the possible, now rather seeming to move across the performative space as if anchored to a fixed track, where any incident or accident modifying the prescribed path is an undesirable disruption. What has happened between the two moments, between the density of *possibilities* swarming in the moment before the first note is produced and the secure set of *probabilities* characterizing the moment just after that?

Thinking of the space that a performer is facing before the production of a sonic event as an empty one is to a certain degree a misconception. On the one hand, the performer

---

The reflections contained in this chapter would not have been possible without the research done as part of the project 'Experimentation versus Interpretation: Exploring New Paths in Music Performance in the Twenty-First Century' (MusicExperiment21 n.d.). The reader who wishes to further explore some of the notions contained in this chapter will find more material in D'Errico (2018) and Assis (2018).

is indeed entering a space of silence that promises to unfold as complete openness to creativity. On the other hand, this space is not empty – it is actually overfull. Even before the first sound is produced, an array of conventions, relations, traditions, habits, norms, rules, vetoes (most of which are unspoken) coagulates in the space, tracing a path to be followed ahead of time. If, on the one hand, this balance (or tension) between predictability and emergence is common not only to any kind of music-making but also to the constitution of art in general, on the other hand, in the performance of Western notated art music such process takes on specific and explicit contours. This specificity is linked to the polarity between the two sides that are at the core of notated music – what Igor Stravinsky in the last lecture of his *Poetics of Music* calls 'two states of music: potential music and actual music' (Stravinsky 1947: 121), or more simply said, composition and performance. Crucially, each of these sides or states responds to operational modalities and forms of inscription that show no conformity with one another. What Stravinsky calls 'potential music' is mainly epitomized and expressed by the score – then, it takes place through mental and algorithmic categories (notation) and on material inscriptions that are visual, linear, perceived as bi-dimensional, and most importantly, *not sounding*. By contrast, music that is 'actualized' in the performative act comes about in and through a form of materiality that is incommensurable to any kind of mental category: the vibratory, haptic, n-dimensional spatial and corporeal dimensions of sound and gesture.

In traditional approaches to Western notated art music, the role of sound and of performance is regarded as the execution, recitation, transmission, or interpretation of an already sedimented musical text. At the intersection between notation and the above cited unspoken array of stabilized conventions, the performer detects precise spots to channel sound, punctual parameterizations to reduce the infinite set of possibilities allowed by the multifariousness of the sonic event. From this perspective, performance gets deprived of much of its creative potential, its function appearing as the concrete sonic *representation* of an already known structure. This chapter proposes a different vision of music performance of past musical works, where sound and gestures are not anymore regarded in their reiterative, reconstructive, or representational function, but they rather become a locus of experimentation, where 'what we know' about a given musical work is problematized and reconfigured.

In what follows, I will investigate how the current modes of music performance are governed by a specific way of enacting a relationship between the incommensurable dimensions of (mental) notation and (material) sound. In the first two paragraphs, a critical stance on Stravinsky's considerations about performance will be useful to take two fundamental steps. First, it will be necessary to unmask the delusion according to which it is possible to establish a stable and bi-univocal relation between notation and sound. We will see how sound always exceeds and saturates the categories of notation, and how such a stable relation, far from being a given, is culturally, historically, and aesthetically fabricated. As a consequence, there cannot be such a thing as a 'neutral' or 'right' way of relating sound to the algorithmic system of notation, but instead potentially infinite creative choices. In this respect, I will propose the existence of a multitude of 'regimes'[1] of music performance, each with its own specificities, operating ways, and modes of relating to the existent. The second step will be the observation of the two main paradigms of music performance: that of 'execution' (which Stravinsky advocates for) and that of 'interpretation' (which he denounces as almost fraudulent). I will argue that both paradigms are indissolubly linked to the practice of interpretation (in a broader, non-strictly musical sense), even if they enact different relationships to it. Finally, after examining the characteristics and implications of each paradigm, in particular as regards their relation to the production of sound, I will propose a third, more productive, step: a move beyond both paradigms, in the constitution of a performance practice that takes sonic materialization not as a form of *representation* (however complex and refined) of already-known structures but as a means for *experimentation* and *reconfiguration* of the known.

Several regimes of music performance

According to Igor Stravinsky, the division of music into 'potential' and 'actual' states is articulated in such a way that actualization requires from the performer 'the strict putting into effect of an explicit will that contains nothing beyond what it specifically commands', given that 'the composer's will [be] explicit and easily discernable from a correctly established text' (Stravinsky 1947: 122–123). In other words, this vision implies a direct correspondence between musical text and sound, between mental and material, between visible/readable and audible. But how is it possible that two dimensions that are incommensurable with each other have a relationship of identity? What dictates direct *resemblance* between a score and its performance, given the fact that their inscription occurs through materials and modalities that show no conformity with each other?

The situation portrayed by Stravinsky seems to suggest that sound is already present in the written score, nakedly available for a faithful portrayal by a neutral and transparent 'executant'. He actually execrates the arbitrariness of the kind of performer that he calls the 'interpreter', who takes the liberty to deviate from the explicit command of the musical text and to wander off in self-indulgent 'sin[s] against [the] letter' (Stravinsky 1947: 122–124). Even the acknowledgement that music notation fails to unambiguously express all aspects of its sonic enactment ('no matter how scrupulously a piece of music may be notated […] it always contains hidden elements that defy definition, because verbal dialectic is powerless to define musical dialectic in its totality') does not put in question his belief in the possibility for the 'executant' to provide a verbatim 'translation into sound of his musical part' (Stravinsky 1947: 123–124). Notwithstanding this problem, in advocating for the neutral executant and in condemning the interpreter, Stravinsky is accepting for granted the delusion of representativity: what is commanded on paper can have a bi-univocal relation to materiality, what is 'represented' (in the score) and what this 'represents' (sound) can be identical to one another. Though, the neutrality that Stravinsky wishes for is far from being a given.

Sonic reality is incommensurable to the fixed parameterization implied by notation – incommensurable in an almost mathematical sense: no matter how refined, detailed, enriched the notational system, there will always be a sonic remainder that does not lend itself to segmentation and compartmentation. Sound and notation miss a *common submultiple*. Thus, there is always the need for a system of references external to notation to reduce this irrational (again, in a mathematical sense) remainder, engendering a conventional way of shaping sounds according to given parameters. The illusion of a neutral transmission between the two states is therefore made possible only by the already cited array of former practices, rules, vetoes, traditions, all of them stratifying into a convention so strong as to substitute itself for the only thinkable possibility, for 'reality as it is'. Stravinsky's discourse on transmission and neutrality seems to ignore (perhaps deliberately so) two fundamental passages: first of all, that such a convention needs to be fabricated, and thus it implies a creative and productive act non-dependent on that of the composer, and rather encapsulating it; second, that this convention is by no means the only one afforded by the musical text, and that instead there are potentially infinite ways of relating the two 'states' of music to each other. Stravinsky himself is already proposing two possibilities, that of the executant and that of the interpreter – even if the latter is for him unacceptable.

Instead of proposing one single, supposedly 'neutral' possibility of relating notation and sound, I put forth the existence of a large and potentially infinite number of 'regimes' of music performance. Each regime is not absolute and abstract, but rather linked to specific cultural, historical, and geographical conditions, characterized by specific rules, but also by the possibility to change, become mixed with, and transform into other regimes with changing circumstances. This said, it is important to note that at a given time one regime can acquire such a pervasive power as to come to coincide with its own epistemic landscape, and therefore, within such a landscape, to substitute itself illusorily for an absolute reality. This is the case with the regimes of execution and interpretation, which have detained (and still today detain) such a power over the field of the performance of Western notated art music. This is why, before actively constructing and proposing an alternative to them, it is important to analyse and understand their model.

Execution and interpretation

Despotism and neutrality: The regime of execution

With its supposed faithfulness to the score and to the will of the composer, the regime of execution is the one promoted by Igor Stravinsky. This regime is characterized first of all by a vision of the musical text as an enclosed, self-sufficient system. The composer that does not tolerate unfaithfulness from a performer will write a kind of music that 'seeks to express nothing outside of itself', whereas 'the musically extraneous elements […] invite betrayal' (Stravinsky 1947: 125). What is explicitly stated through notation must be transmitted to the audience by the performer, who is regarded as a neutral mouthpiece of the intentions of the composer. This enclosed musical system must not lend itself to being anchored to any knowable portion of what lies outside of music. However, since what is commanded on paper by the composer has no sound, an executant wishing to pursue faithfulness will try to shape the sonic matter by intersecting other layers of signs to the notational level: the signs of the instrument (its musical-anatomical interface), the signs of anatomy (fingering, breathing, etc.), the signs of the metronome, the signs of the tuning system, the signs of the tone system, the signs of the performance treatises, those provided by music analysis, by music history, by organology, etc. The problem, though, is that those circles of signs, never reaching the unclassifiable complexity of sound, potentially expand in an infinite entropy, referring only to one another and unhinged to any form of sounding materiality. At the same time, this abstract network of signs, however detailed and exhaustive, will never measure up to the infinite variability of the materiality of sounds and gestures that constitute performance. That is probably why Stravinsky admits that behind notation there are 'hidden elements that defy definition' (Stravinsky 1947: 123). But these hidden elements are far from being acknowledged by him as the immanence and incommensurability of the sensuous occurrence of sound – what would amount to granting the performer infinite liberty and infinite perversion of the musical text. This hidden core is rather postulated as a kernel of truth concealed behind the text, contemporaneously (and paradoxically) both clearly inferable from it and infinitely disguised behind it. What orientates the executant is then a sort of oath of loyalty towards the composer, or 'a point of conscience' (Stravinsky 1947: 123). The executant has to satisfy the composer's will, which ultimately coincides with *the absolute sense* of the musical work, hidden beneath the infinite layers of signs and unreachable because always defying definition. The composer, then, is like an absent god, whose face is hidden and unrepresentable through signs, and as such remains obscure, unknowable. That is why a secondary system is needed by the executant: the system of interpretation,[2] where a sign or a group of signs is connected to some sort of materiality, thus interrupting the continuous proliferation of circles of signs which menaces to lead away from the centre of the musical system (the composer's will). A network of secondary authorities invoked by the executant assures the 'correct' interpretation, to even out arbitrariness and provide orientation in indecision. In his article 'Beyond the Interpretation of Music', Laurence Dreyfus provides a possible list of these authorities. Following '(1) the composer who creates the work' and '(2) the musical text which is commonly a stand-in for the composer himself', he identifies *authorities* such as:

> (3) the teachers and music directors who transmit the authority of the composer or the text; and (4) superior, usually older musicians whom one emulates […] (5) performers' traditions, as in the assertion that this is the way we have always done it; (6) musicological rectitude (if one is so inclined to defer to it); (7) musical structure (as defined by music theorists and analysts); and something called (8) musical common sense. All these authorities conspire to validate interpretations, to assure us that we are doing the right thing, and to help pass on interpretative practices to the next generation.
>
> (Dreyfus 2007: 254)

Here we have some concrete examples of the network of relations overcrowding the space of performance and securing the executant to the absolute predictability of what is commanded (already-said) by the score. Thanks to all these authorities, the perilous gliding from one sign to another is blocked by a group of signs disclosed by the interpretative guarantors.

The escape into subjectivity: The regime of interpretation

Beyond execution, there are performers who take liberties from what is commanded by the musical text, escaping its enclosed system of signs. Those are the ones anathematized by Stravinsky as 'interpreters': the regime of interpretation in music performance is characterized by this move away from the centre/composer's will, a move regarded as unacceptable in the regime of execution. Interpretation manages to give a positive value to arbitrium and to the infinite entropy of the circles of signs. The interpreter's relationship to the musical work becomes passional and subjective, and subjectivity gains precedence over the centre of the supposed composer's intentions, substituting it. Instead of faithfully repeating what is explicitly commanded through notation, the interpreter becomes a 'subject'. As such, he or she allows extramusical elements to enter the performative scene: emotions, personal experiences, relationality to spaces and to social practices, all elements that affect and shape the sonic and gestural utterance. A new reality is fabricated in the escape from the closed system of notated signs, and a subject-interpreter constitutes itself who is capable of identifying with such reality. In this subjectification, though, the interpreter fails to acknowledge the artificiality and arbitrariness of such reality, exchanging it for 'reality as is'. The interpreter shapes a musical world that is made to coincide with the score; in turn, the musical text is internalized, no longer regarded as an infinite layer of signs concealing the kernel of truth of the composer's intentions. The score gets inscribed in the interiority, becoming a terrain of passion: the interpreter assimilates it, makes it 'breathe', gives it a pounding 'heart'. From this internalization of reality follows a completely different function of interpretation (in the non-musical sense), which not only dispenses with the composer's will but also with secondary authorities, to become direct interpretation. Inside and outside are made to coincide, the subject infinitely reflects itself onto – and is in turn a reflection of – its own internalized outside.

In giving pre-eminence to subjectivity, the interpreter accepts to 'betray' the will of the composer. This betrayal could even happen at a not-so-conscious level, as in the Wagnerian performer as *alter ego* or proxy (Dreyfus 2007: 264), where the interpreter substitutes themselves for the composer claiming an ability to capture the composer's 'spirit' by means of a superior identification. According to this delusional model, 'Wagner recommends intuition and empathy to decipher intentions lying behind the musical notation, a method which – though unacknowledged – is still by far the most widely practised in the classical musical world at large' (264). It is as if the performer-subject was doubled by its own reflection in its inner reality: in obeying one's interiority, the interpreter ascribes it to the musical work, and vice-versa, the reality of the musical work (for example in the form of musical common sense) is internalized and substituted for one's own interiority. This kind of performer may believe they are more faithful to the composer, in that they are invested of the composer's 'spirit'. If the secondary authorities of the regime of execution resembled pedantic and bureaucratic priests, the interpreter is rather like a prophet who has heard the composer's voice in the desert of their interiority.[3]

By liberating the sonic utterance from the self-referential system of the centralized composer's will and secondary interpretational authorities, interpretation seems to provide a creative and productive alternative to the despotic system of the regime of execution. However, the liberation of creativity that it pursues is not fully fledged. This regime keeps reinstating a dialectic relationship between the 'objective' side of notation (even if interiorized) and the 'subjective' side of the performer's interiority: a sort of continuous self-reflection, by which the performer recognizes themselves in a musical text that they are mirroring. The regime of interpretation is in turn a closed one, where the infinite potentiality of the materiality of sound and performance is again reduced and domesticated. The only difference is that this time the interpreter, rather than obeying an absent external authority, is obeying oneself.

A locus of experimentation

So far, I have limited myself to observing the current modes of performance in Western notated art music. But there is a fundamental difference between tracing an existent status quo and operating a change in it. The reflection on regimes of music performance, on the conditions in which they are generated, and on their specificities and operative ways is a first fundamental step. Understanding that regimes are not normative but contingent, and that they can transform into one another, is a second, more profound step, which however is still not enough for producing a change in the modes of conceiving and making performance. One more step is necessary: the constitution of a different mode of music performance, one that, instead of representing the existent, constructs an existent that is not yet there. This new mode establishes a different relationship with the musical givens expressed by the score: but not because it institutes a new system, or a new 'regime' made of different but equally stabilized and bi-univocal relations, but because it faces the materiality of performance, and its incommensurability with systems of signs, as a terrain of experimentation.

The word 'experimentation' refers here neither to the historico-geographical acceptation of 'Experimental Music' nor to scientific or parascientific approaches to performance, based on data collection and observation of quantifiable phenomena. Rather, it designates a general but specific attitude, that of the artist who is not content of inhabiting pre-given systems, and of carrying out partial shifts in them, but rather aims at reconfiguring 'what we know' about the field of knowledge and of practice in which he or she is operating. Importantly, such an experimental attitude is particularly hard to achieve inside the mainstream modes of music production. For a musician inhabiting given paradigms regulated by established social, political, and historical structures, it is difficult to exit, to reconfigure, or even to acknowledge them as 'regimes'. In its metastable definition, in its fluidity between artistic practice and conceptual thought, in fostering unfinished thinking rather than definite knowledge, the field of *artistic research* is a particularly fertile ground for experimentation, contributing to a fundamental transformation of the role of the musician, and allowing them to exit the postures and disciplinary territories inherited from tradition.

In traditional regimes of music performance, both 'states' of music – notation and sonic enactment – are regarded as already stabilized entities, and by virtue of this an executant or an interpreter can make of sound a 'representation' of notation (or a composer make of notation a 'representation' of sound). This stability, this low degree of motility in the system of performance, describes the situation I portrayed at the beginning, where the performers are channelling into an already formalized system the different bodies and forces that are at play in the moment before performance. A view of the musical text as fixed, and of the musical work as an ontologically defined entity, cannot but envision a performer-as-executant, approaching it as a closed, self-sufficient system (the ideal work described by Stravinsky) in which the relation between score and sound/gesture is bi-univocal and stabilized. Approaches to music performance that mobilize the fixity of the musical work by setting it in dialectic dialogue with the subjectivity of the performer will generate only a performer-as-interpreter, segmenting the process of performance by engendering the infinite recoiling of the musical text into interiority, and vice-versa. In the experimental approach proposed here, musical works as codified by notation cease to be considered as instructions to be obeyed, or as fixed structures to be interiorized and mirrored. Rather, they become *dynamic reservoirs of forces, functions, traits, and materialities*. As a dynamic system, the musical work is indeed 'embedded' in its physical and semiotic formalizations and codifications, but at the same time not coincident with them. From such a system, it is possible to extract elements (material or functional) that in the moment of performance get reconfigured, reassembled, and rethought. The performer is no longer concerned with what the work (or the composer for it) 'thinks itself to be', and stops relying on primary or secondary authorities – in a few words, he or she *stops interpreting*.

A concrete example of my practice in this sense is the research project *Powers of Divergence* (D'Errico 2018),[4] where I amplified the unbridgeable divergence between codification and materiality, rather than minimizing it according to an existent convention, or even by fabricating a new convention. In this artistic research project, I addressed the performance of past musical works (among which, pieces by Giulio Caccini, Claudio Monteverdi, Ludwig van Beethoven, and Robert Schumann) through sounds and gestures unrecognizable as belonging to the original works as an interpreter or executant would approach them. Instead of relying on the culturally constructed regimes through which symbolic categories are bi-univocally connected to material events, this practice exposes the arbitrariness of such regimes, together with the boundaries of its epistemic implications. Thus, performance becomes a sonic 'image' that relates to what is different from it (the score) *by means of difference*, and not by attempting to construct a (supposed) identity. In this process, internal resemblance is negated, together with the idea of composition as *origin* (and creation) and performance as its *telos* (and reiteration). Instead of inhabiting the subject–object relationship of interpretation, where sounds mirror the score and the score already encompasses its sonic result, the musical work and the performer affect each other by means of externality. From this mutual affection, the sonic result emerges as a third state, neither subjective nor objective, in which the performer mobilizes themselves towards the work, but also the work gets transformed in turn by the encounter with the performer.

This renunciation of the representational mode of music performance implies a rethinking of the role of notation in relation to sound, or in other words, of the whole relationship between symbolic and material dimensions. In traditional interpretation, the performative act – including its sonic manifestation – has already been 'thought' by notation: pre-shaped by the symbolic, performance makes a 'text' of itself.[5] Renouncing the textual power of notation immediately involves a set of aesthetic and mediatic implications. The performances realized in this experimental framework are designed to elicit an unheard dimension, to bring about forces and traits embedded in, but not expounded by, the musical work regarded as a dynamic system. This unheard dimension does imply a creative musical act that has some similarities to a compositional act – and, as I trace this similarity, the limit of the current mode of thinking about creative hierarchies in music becomes evident: the creation of unheard sonic combinations is currently taken to be the prerogative of the composer, whereas performers have to limit themselves to reproduction and representation. In order not to fall back into either 'execution' or 'interpretation', however, such an act of musical creativity cannot take place through notational means, but it has to be carried out *as performance*, through the materials and operative ways that are part and parcel of the performative act. In *Powers of Divergence*, I faced the necessity to make use of modes of inscription that would bypass – at least partly – notational codification. The experimental performances have been carried out as a complex mixture of practices including electronic sound generation and processing, sampling, and semi-improvisatory actions, and of strategies such as playback and the design of sonic scores. Moreover, the sonic quality of these performances is also fruit of a specific aesthetico-epistemic choice. The sounds produced have to defy to some extent the intelligibility that is associated with the 'sonic texts' of interpretation, and therefore also the highly codified sonic parameters characteristic of what can be named, in David Davies's (2011) term, the 'classical paradigm': intonation, loudness, stability of sound emission, minimization of parasite noises, etc. By contrast, the sounds that are part of these new, divergent performances swarm with micro-variations and instabilities. The performer willing to embrace this notion of experimentation is therefore ready to carry out performance as a truly constitutive act, but preventing the reproducibility and transmissibility of a traditional notated 'composition' – and with it also the authorial and authoritative role of the 'composer'. The performer is an 'active' agent, by dint of their creative choices and material actions, but also 'passive', as they let themselves be affected by the relationship with the musical work, being *traversed* by it, as it were. In this recursive state, which is neither action nor passion (and both), predefined agencies are blurred and suspended, interpretation and subjectification jeopardized.

It is actually crucial to underline, as a final remark, that in this experimental mode the segmentation between composer as creator and performer as reproducer ceases to be effective. This traditional division inevitably leads to a fixed musical regime, be it despotic (execution) or dialectic (interpretation). The performer who surpasses these categories becomes instead an *operator*, the agent of a critical act that becomes a constitutive (and therefore purely creative) act. Critical because it observes and problematizes current modes of thinking and making music; constitutive because it operates not through theoretical approaches and detached observations, but through the concrete design and material making of performances. The operator is thus a figure where multiple roles converge. Not only are they encapsulating both the creative and performative figures of music-making; they are also capable of exiting the traditional terrain they come from and at the same time still inhabiting it in order to understand and criticize it. The operator *exits music through music*, in a trans- and metadisciplinary approach able to become *anti-disciplinary*, by disrupting and renegotiating pre-existing disciplinary boundaries. Beyond representation, execution, and interpretation, beyond closed and self-sufficient musical regimes, the operator reconfigures materials and forces into unexpected sonic and performative events. They leave behind the vision of performer as observer, reproducer, recitator, reconstructor, or enunciator of the existent, to instead reconfigure musical works into new configurations, producing *a new existent*.

Notes

1. The term 'regime' is adopted from the fifth plateau of *A Thousand Plateaus* by Gilles Deleuze and Félix Guattari, titled '587 B.C.–A.D. 70: On Several Regimes of Signs' (1987: 111–148). In particular, my reflections on the regimes of 'execution' and 'interpretation' have been deeply influenced by the considerations on what Deleuze and Guattari name the 'signifying regime' and the 'post-signifying regime'.
2. In the general sense of the word, interpretation is a fundamental component of the musical regime named here 'execution'. In the strictly musical sense, the word 'interpretation' acquires a more specific meaning. For a thorough discussion of the terms 'execution' and 'interpretation' in music, and of their history, see Danuser 2015.
3. For the distinction between the function of the priest and of the prophet and their relation to interpretation, cf. the fifth plateau in Deleuze and Guattari (1987). The priest is the secondary authority providing interpretation in the 'signifying regime', whereas the prophet covers a parallel but different function in the 'post-signifying regime' (Deleuze and Guattari 1987: 114–118 and 123–125).
4. For an online multimedia documentation of the project, see Orpheus Institute n.d.
5. Herman Danuser proposes that 'interpretation is based on texts and leads to texts', where the case of recording would even lead to the production of a 'sonic text of a work' (Danuser 2015: 187). I would take Danuser's suggestion even further: not only when a musical work is being recorded does interpretation lead to sonic texts. The idea that sound can entail a textual dimension is precisely what makes interpretation possible in performance (D'Errico 2018: 15).

References

Assis, Paulo de (2018). *Logic of Experimentation: Rethinking Music Performance Through Artistic Research*. Leuven: Leuven University Press.

Danuser, Herman (2015). 'Execution – Interpretation – Performance: The History of a Terminological Conflict'. In Paulo de Assis (ed.), *Experimental Affinities in Music*, 177–196. Orpheus Institute Series. Leuven: Leuven University Press.

Davies, David (2011). *Philosophy of the Performing Arts*. Foundations of the Philosophy of the Arts. Malden, MA: Wiley-Blackwell.

Deleuze, Gilles and Félix Guattari (1987). *A Thousand Plateaus: Capitalism and Schizophrenia*. Trans. Brian Massumi. Minneapolis, MN: University of Minnesota Press.

D'Errico, Lucia (2018). *Powers of Divergence: An Experimental Approach to Music Performance*. Leuven: Leuven University Press.

Dreyfus, Laurence (2007). 'Beyond the Interpretation of Music'. *Dutch Journal of Music Theory* 12 (3): 253–272.

MusicExperiment21 (n.d.). 'About'. Available online: https://musicexperiment21.eu/ (accessed 6 July 2020).

Orpheus Institute (n.d.). 'Powers of Divergence'. Available online: https://orpheusinstituut.be/en/powersofdivergence (accessed 6 July 2020).

Stravinsky, Igor (1947). *Poetics of Music in the Form of Six Lessons*. Trans. Arthur Knodel and Ingolf Dahl. Cambridge, MA: Harvard University Press.

22

Performing Centrifugal Sound

G Douglas Barrett

Introduction: Front, back, side to side

Sound encircles and is encircled by movement, language, desire, bodies, and history. Between 1982 and 1984, the artist Adrian Piper staged a series of events in which she taught small and large groups of participants how to dance to funk music while pedagogically intervening into and counteracting commonly held attitudes towards the art form especially as they related to race and class. Taking place initially around small dinner party tables with one to seven guests and later expanding to larger university groups of up to sixty participants, these *Funk Lessons* used the performance-lecture format while combining music, dance, and performance through an incipient form of what we might today call 'social practice' art (see Bishop 2012). Presented alongside interviews, news footage, and other musical performances, one such event staged in 1983 at the University of California, Berkeley, was the subject of Piper's 15-minute video work *Funk Lessons with Adrian Piper* (1984). Roughly ten years before Piper's first *Funk Lessons*, John Baldessari stared into a video camera as he sang Sol LeWitt's *Sentences on Conceptual Art* (1969), a work that consists of thirty-five written declarations intended as a kind of meta-commentary on the then-burgeoning movement of conceptual art. Addressing the viewer, he explains the premise of the work and refers to his singing of LeWitt's *Sentences* as a 'tribute' to the artist. He then proceeds to sing LeWitt's *Sentences* each set to one of several existing pieces of music including the American national anthem, a series of folk tunes, and various popular melodies. When he gets to sentence nine, he proceeds to the tune of *Camptown Races*, 'The concept and idea are different, DOO DA, DOO DA ... ,' but messes up and restarts it, twice. Repetition, circularity: as a song *cycle*, *Baldessari Sings LeWitt* is *of the refrain*.[1] Finally, roughly a year after Baldessari sang LeWitt, Vito Acconci followed suit with his video performance *Theme Song*, which comprises a close-up of the artist as he lies on a living room floor smoking a cigarette singing along to a series of songs by The Doors, Bob Dylan, Van Morrison, and Kris Kristofferson. As the music plays in the background, Acconci speaks with and over each song's lyrics as he deploys a series of sexual advances directed at the viewer while he curls his body around the frame. Not unlike some of his other works of the era, Acconci reveals a confrontationally scopic dimension of the song, already a historically significant expression (in both vernacular and high art song guises) of heterosocial desire. Enveloping the viewer, Acconci wields such a libidinal-gravitational pull through the coercive violence of a musicalized gaze.

Sound (art) theory appears perhaps as a fitting context in which to analyse these three artistic works, as indeed several other works by these artists have been considered in this context. Throughout these accounts, the sonic is circumscribed as an alternative to contemporary visual art's supposed privileging of visuality which these works, by various means and to varying extents, are said to undermine. Fred Moten, in his study of Piper's work, for example, reaches the somewhat counterintuitive and synaesthetic thesis that, 'sound gives us back the visuality that ocularcentrism had repressed' (Moten 2003: 235, emphasis removed). Indeed, Piper's *Funk Lessons* draws our attention to one of the most important yet disproportionately denigrated achievements of African-American culture, a musicality whose proper sensory register, despite for instance Piper's focus on movement and dance, is often understood as belonging to sound. In his pathbreaking *Phonographies*, along these lines, Alexander G. Weheliye argues not for African-American musical practice per se, but for a broadly construed sonic blackness encompassing artistic traditions ranging from slave spirituals to literature to film to R & B and contemporary hip-hop, as central to (yet also paradoxically outside) of modernity (Weheliye 2005: 5). Meanwhile, Seth Kim-Cohen recently announced a call for submissions to his 2016 College Art Association panel 'Singing LeWitt: Sound and Conceptualism', taken on the whole as an homage to Baldessari. In the accompanying text, although sound is 'constitutive of its cultural, historical, and political lifeworlds', it is also often embraced 'because it evades visual art's deferrals to representation and signification, sidestepping the appropriations and reifications of commodity' (Kim-Cohen 2016: 11). Despite Baldessari's ongoing engagements with music and LeWitt's less frequently acknowledged references to composition,[2] Kim-Cohen appeals not to the work's musicality but rather, alongside the work's conceptual status, suggests a tension between sound and its purported distance from the visual.

Sound (art) theory, through its generalization of artistic works and cultural practices and objects into sensorial strata, furthermore, can be read as an attempt to avoid the thorny problem of the above-described works' fluidity between and across the cultural-artistic registers of high and low. After all, sound stripped of its significance in artistic practice seems not to carry such baggage. In his seminal sound art theory text *Background Noise*, Brandon LaBelle discusses Acconci's work from this period in terms of the acousmatic voice but also in relation to pop music. Referring to Acconci's notorious *Seedbed* (1972) and *Claim* (1971), works that explore related forms of violently confrontational masculine sexuality, LaBelle contends, 'it is not about a pure jouissance of speech but a libidinal sociality that aims to blare out, like pop music from a car stereo, echoing Acconci's own statement that "the new model for public art is pop music"' (LaBelle 2006: 119). Yet rather than pursuing popular music itself as a site for such public art, or indeed for an analysis of Acconci's work, it is the art form's spatial resonances, emblematic of a broader trend in Anglo-American sound art theory (Engström and Stjerna 2009: 11–18) that LaBelle foregrounds. Above and beyond the spatial or musical, though, it is the sonic that these analyses privilege. Distributed along its strictly sensorial axis, sound becomes the gravitational centre around which a plurality of art works and cultural practices variously rotate. Sound, in such accounts, requisitely oscillates centripetally.

The problem of how to counteract such a centripetal pull, indeed how to overcome the formalist/formalizing tendencies sound art (theory) inherits from the modernist discourse of medium-specific formalism has been an important effort especially in recent sound art theory texts. For instance, Kim-Cohen has proposed an 'expanded sonic field' as an adaptation of Rosalind Krauss's schema from her seminal 1979 essay 'Sculpture in the Expanded Field' (Kim-Cohen 2009: 39–40), itself adapted from the semiotic square invented by French-Lithuanian linguist and semiotician Algirdas J. Greimas in 1966. Although it suggests a migratory expansion of medium opposed to the inward pull of medium specificity, such a model nevertheless seeks to *conserve* medium – sculpture in Krauss's case and sound in Kim-Cohen's – as its primary site of artistic organization. Meanwhile, in his oft-cited argument for a materialist/realist sound art, 'Beyond Representation', Christoph Cox insists that the sonic arts, due to a quality of concreteness over abstraction – and indeed contra the kinds of textuality and conceptuality for which Kim-Cohen advocates – necessitate 'not a formalist analysis but a materialist one'. Cox further calls for an account, extended from and in line with post-minimalism and conceptual art, not of the sonic in isolation but of 'sound *and* the other arts' (Cox 2011: 148–149, emphasis in the original). Ostensibly contradicting one another in terms of privileging either sound's conceptual or material status, Kim-Cohen and Cox nevertheless align in their differently articulated rejections of a contracted sonic field. Yet despite these thinkers' general self-identification *within* sound art theory as anti-formalist, when seen within the larger context of contemporary art, these frameworks, I want to argue, ultimately adhere to a formalism at a historical remove from yet strictly homologous to medium-specific modernism. In this sense, to draw on Krauss's later elaborated category, sound (art) theory remains by definition pre-'postmedium'.

What kind of sonic methodology, then, does not depend, however tacitly, upon the concept of medium? Whether materialist, idealist, or based on affect, sound (art) theory seems axiomatically subject to a kind of centripetal pull towards a medium-specific ontology of art based on the formal division between mediums. The notion of sound as an independent artistic medium first emerged during the 1950s and 1960s,[3] just as contemporary art began its radical critique of medium in works of canonical conceptual art, which further contributed to what Krauss (2000) labelled the 'postmedium condition'.[4] Against medium-specific formalism, contemporary art proposed, and in many ways achieved, a radically *generic* art whose governing concept was no longer based on the division of materials bound to discretized sense modalities (de Duve 1993: 145–198; Osborne 2013: 76; Osborne 2017: 279). In addition to conceptual art, performance participated in this shift away from centripetal medium and, along with music, paved the way for intermedia, Happenings, Fluxus, and other heterogeneous forms such as social practice. This chapter proposes the category of 'centrifugal sound' as a way to understand art practices that include sound but reduce its centrality through extra-sonic materials deployed in performance. Through readings from art history, contemporary art theory, musicology, and sound art theory, alongside a consideration of Piper, Baldessari, and Acconci, the chapter listens from the periphery and finds not an absence but a movement towards conceptuality and the social.

Around in circles

Sound, as oscillatory, comes in waves; fundamentally of the refrain, it returns us to the past. The historiography of medium-specific Modernism, and its relation to sound, then, is our first point of departure. As early as 1939, critic Clement Greenberg would argue in his notorious 1939 essay, 'Avant-Garde and Kitsch' that the artist must turn 'his attention away from common experience', which was merely representational, and 'in upon the medium of his own craft' (Greenberg 2003: 532). Through a generalized reinvention of the concept of 'medium', which Greenberg extended a year later in 'Towards a Newer Laocoon' from the work of eighteenth-century German writer Gotthold Ephraim Lessing, Greenberg sought not only to avoid the discursive, or 'literature', but to map the specificity of each medium to human sensory experience. 'There is a common effort in each of the arts to expand the expressive resources of the medium,' Greenberg contends, 'not in order to express ideas and notions, but to express with greater immediacy sensations, the irreducible elements of experience' (Greenberg 2000: 65). Each medium was to be matched up to a corresponding sensation, for example painting to sight, sculpture to space, music to sound, film to movement, etc. Such a positivistic schema seeking to catalogue the 'totality of the aesthetic' (Osborne 2013: 80) appears less than surprising given the trajectory staked out in subsequent structuralist-inflected reconceptions of medium seen decades later, for instance, in Krauss's 'sculpture in the expanded field'. Greenberg was, according to Osborne, 'a structuralist of medium' (80). The historical triumph of this schematization would ultimately set the stage for its defeat in conceptual art's radically generic concept of art, while also leading to medium specificity's afterlives in various revivals of formalism and in the more recent discourses of sound (art theory).

Sound, conceived as a category of artistic practice (or even as a non-artistic cultural activity), appears then as a kind of belated species of medium specificity's taxonomic construction of art as a genus divided into discrete mediums organized according to the separation of the senses. One reason for this belatedness was that medium specificity was *already*, in Greenberg's account, an extension of the 'sonic' in music, more specifically in the ideology of absolute music. Beginning in the early nineteenth-century writings of the group of German Romantic thinkers (Wackenroder, Tieck, Novalis, Jean Paul, Schlegel, and E. T. A. Hoffmann), instrumental music, which was later termed 'absolute music', had been elevated 'from the lowest to the highest of all musical forms, and indeed of all the arts in general' (Bonds 1997: 387). In the process, the movement effectively constricted the pre-modern 'intermedial' form of music as *harmonia*, *rhythmos*, and *logos* (harmony, rhythm, and language) to a monosensorial art form based on pure instrumental sound (Dahlhaus 1989: 8).[5] Discussing absolute music's role in the development of the avant-garde and medium-specific Modernism – and echoing Walter Pater's notorious 1888 statement concerning art's supposed striving towards the condition of music – Greenberg contends

> music as an art in itself began at this time to occupy a very important position in relation to the other arts. Because of its 'absolute' nature, its remoteness from imitation, its almost

complete absorption in the very physical quality of its medium, as well as because of its resources of suggestion, music had come to replace poetry as the paragon art.

(Greenberg 2000: 65)

Already by the 1810s, absolute music discourse had envisioned a sound art *avant la lettre*. Nevertheless, these roots of medium specificity in early nineteenth-century music discourse didn't prevent artists, curators, and theorists roughly a century and a half later from 'discovering' sound, again.[6]

Another reason for sound art's belatedness is that it first emerged in the 1960s just as conceptual art had begun its radical dismantling of medium in the work of artists such as Joseph Kosuth, Sol LeWitt, John Baldessari, and Adrian Piper. Conceptual art, along these lines, sought to move beyond the constraints of particular mediums, not only in the sense of their 'dematerialization' (Lippard 1973: vii) but also in an attempt to escape the gravitational pull towards medium emblematic of Greenberg's medium-specific Modernism. 'If one is questioning the nature of painting', Kosuth contends, 'one cannot be questioning the nature of Art' (Kosuth 1991: 18). A tension inheres in Kosuth's invocation of a privileged and deracinated 'Art' to support his critical negation of painting, a move that metonymized his attempt to transcend the material fixity of artistic mediums.[7] The postmedium condition was Krauss's term for the impasse generated by medium-specific Modernism on the one hand, and the complete obliteration of medium heralded by a certain strand of conceptualism on the other. As a response to this impasse, Krauss proposes that artists must reinvent or rearticulate mediums through a form of 'differential specificity' (Krauss 2000: 56). In the process, she takes issue with Kosuth's overarching category of 'Art', while also substantially breaking with Greenberg's medium specificity. Ultimately, the movement of contemporary art has been not in the direction of a re-entrenchment of medium but, rather, in its overcoming in the form of a radically generic art at once both post-conceptual and postmedium.

Against medium-specific Modernism, contemporary art has pursued a radically *generic* art as a decisive departure from the structural mapping of artistic materials to discretized sense modalities. Drawing from his reinterpretation of Adorno's *Aesthetic Theory*, Osborne defines contemporary art's rearticulated genericity as

> determined (1) historically and negative-dialectically, as the ongoing retrospective totalization, from the standpoint of the present ('the latest art'), of the multiplicity of individual works making up the system of negations of previous works which is the history of art: negations of the ontological significance of (especially, conventionally received) mediums standing to the fore within this multiplicity, since the mid-to-late 1950s. It is determined (2) by the speculative 'idea of art as such' – in its opposition to empirical reality – as the regulative conceptual unity of the aforementioned permanently ongoing process of retrospective unification; a process that is governed by 'laws of movement', rather than a simple selective aggregation, the laws of movement of a multiple collective singularity.
>
> What stops this generic 'art' from functioning as a bad abstraction is an emphasis on its collective, internally differentiated, relational and concretely historical characteristics. This unity is 'not abstract', because as Adorno put it, it 'presupposes concrete analyses, not as proofs and examples but as its own condition.' In this respect, the idea of art is given through each work,

but no individual work is adequate to this idea, however 'preponderant' that idea may have become over individual works themselves. The ongoing retrospective and reflective totalization of distributive unity is necessarily open, fractured, incomplete and therefore inherently speculative. It must include the afterlife of mediums within a post-medium condition.

(Osborne 2017: 279)

A radically generic contemporary art, then, emerges first as a negation of the ontological significance of medium specificity, that is, a negation of art defined as a closed taxonomy of mediums, in favour of an ongoing process that reconstructs an open totality from this closed multiplicity. It avoids the problem of becoming a 'bad abstraction' through a reflexive and internally conflicting process of concrete self-analysis which upholds its requisitely fractured and incomplete status. As suggested in the tension between the concepts of the 'generic' and 'specific' themselves, contemporary art's genericity is explicitly at odds with medium specificity. As Osborne explains, 'medium-specific modernism *ontologizes* the plurality of the arts as mediums in such a way, seemingly, as to block the very possibility of attributing significant *critical* meaning to the concept of art in general' (Osborne 2013: 80, emphasis in the original). Finally, a process of becoming-generic must therefore include a retrospective analysis of medium as the postmedium condition since contemporary art's genericity is one result of its ongoing critical negation of modernist medium specificity.

One may nevertheless ask whether such a conception of a radically generic contemporary art is an accurate description. Indeed, the example of sound (art theory) on its own is perhaps enough to contradict Osborne's model. If contemporary art has, in fact, attained such a radically generic condition, then why do we see such a proliferation of sound – as one of art's supposed species – across its discourses and institutions? Add to that the problem that sound itself appears as an internally fractured discourse wherein, although often variously claiming and claimed by contemporary art, many of its theorists and practitioners see their work as a species not of art but of music. It's interesting to note, then, that the chapter of Osborne's *The Postconceptual Condition* from which the above-quoted excerpt is taken, 'The Terminology in Crisis: Postconceptual Art and New Music', appeared first in response to an invitation to speak at a conference on new music (*Wirklichkeiten*, Stuttgart, Germany, 19–21 May 2016) to address the problem of music and post-conceptual art. Against an enclosed, quasi-medium-specific 'music *qua* music', he proposed that a 'music *qua* art' would indeed resemble a post-conceptual music (Osborne 2017: 271). This was around the same time that I proposed my own solution to the problem of post-conceptual music as *critical music after sound* (Barrett 2016). To the objection that a post-conceptual music would be subject to the same criticism as contemporary sound, namely as an anachronistically conceived species of a radically generic contemporary art, the more fundamental and important paradox should first be acknowledged wherein, again, music itself provided a model for the medium specificity out of which contemporary art's current post-conceptual genericity had initially emerged (Barrett, forthcoming).

Here it is important to note that Osborne's schema is not descriptive nor merely prescriptive, but rather exists as a *critical* determination that turns on art's contemporaneity. This is not simply a topological account of what one might expect to find across today's heterogeneous field of art – for instance, who's included in this year's biennials, etc. – but an actual judgement about what, among all the noise out there, should be included in the category of actually existing contemporary art, what makes art genuinely contemporary. What does make contemporary art contemporary is its *criticality*, when art is 'addressed in some way or another to the contemporaneities of the present'; here these are, in terms of a dual periodization, 'artistically post-1960s and geopolitically post-1989' (Osborne 2017: 269, 337):[8] on the one hand, the victory of conceptual art's radical genericity over medium-specific Modernism, on the other, the end of the Cold War and the victory of global neoliberal capitalism. Today, sound, inclusive of its dual life in new music and contemporary art, has become omnipresent in the proliferation of sound art Master of Fine Arts programmes, sound art awards, sound art exhibitions, and sound art theories, not to mention the continued zombie-like march of academic new music institutions' similar dedication to sound as a medium. Setting aside new music, the insistence of sound art on becoming a species of a contemporary art that has otherwise mutated radically beyond such a medium-specific taxonomy appears not only as a symptom of sound's belatedness, but sorely evinces its lack of contemporaneity. Sound leads us in circles.

Back around again

Centrifugal sound seeks to counteract this circularity as an incipient effort to account for contemporary art's audibility while simultaneously guarding against the centripetal pull towards medium as an anchoring categorical determinant, especially but not exclusively of the kind found in medium-specific Modernism. It allows for a discussion of sonic *features* (e.g. the soundtrack of a film or the incidental sounds of a performance) without foregrounding such phenomena as a kind of evaluative rubric. Centrifugal sound therefore oscillates in a direction opposite sound art theory's movement towards establishing sound as an independent medium of contemporary art. Spinning outward from this lure of categorical stability, the centrifugal enacts not amplification of an input signal but destructive interference up to full phase cancellation. Virtually equivalent to an imaginary number, centrifugal sound is a kind of conceptual-rhetorical placeholder, an oxymoronic category that by definition pursues an escape from its own categoriality. Not unlike Deleuze and Guattari's (1987) notion of the rhizome, it seeks, through a form of self-cancellation, to uproot the categorical hierarchies implied in its own coining: a line of flight from a kind of semantic whirlpool whose centripetality is set on seducing its inhabitants (recall Acconci's *Theme Song*) towards specificity. Rather than grouping together a set of practices, entities, objects, events, senses, histories, concepts, discourses, persons, etc., it suggests the opposite: the objects and phenomena to which it is applied do *not* necessarily belong together. Centrifugal sound only hints at fleeting intersections of shared characteristics. The centrifugal is an archetypically *weak* theory that demands supplementation across a broader artistic, sensory, and discursive field. A trojan horse on/in the field of sound, centrifugality not only skips out from the groove-locked repetitions of modernist medium specificity but, at an extreme, dislodges itself from the sounding apparatus entirely.

The flipside of the centrifugal, its counterspin, is the contention that sound, and by extension medium, remains viable as a *strong* organizing category for both artistic practice and as a site for the cultural analysis of non-artistic/cultural practices. Its most tenacious artistic proponents accord with what Seth Kim-Cohen criticizes as 'sound art fundamentalism' (Kim-Cohen 2013: 1242) which describes limit cases that argue (through practice or theory) for sound art as a rigidly defined medium ultimately indistinguishable from the epitome of Greenbergian Modernism. Along these lines, Kim-Cohen is right to criticize such fanatical strains of centripetal sound in artists such as Francisco López. Meanwhile, those studying non-artistic sound culture in sound studies often argue for sound as a kind of counter-hegemony, especially with respect to the visual. These arguments tend to go something like this: the visual has dominated Western (or even global) culture for so long that we must, in response, become attuned to the sonic culture that such visual dominance occludes. In its place, we must establish the primacy of sound because, for any number of reasons, it allows us to experience the world differently and thus produce more profound, or at the very least different, theorizations about it. Yet rather than representative of a breakthrough epistemological movement, such forms of anti-visuality, as Martin Jay (1993) has outlined, have existed for at least a century. Fred Moten's otherwise fascinating study of Piper evinces an exemplary form of such anti-ocularcentrism ('Sound gives us back the visuality that ocularcentrism had repressed') (Moten 2003: 235). The ways that racism works sonically as well as visually, racism's 'visual pathology', have doubtless been central to Piper's work (Piper 1999b: 177, quoted in Moten 2003: 234).[9]

Yet Piper has also been formative to conceptual art's radical negation not only of the visual (and indeed the aesthetic as such) but also to the overcoming of a medium-specific Modernism inclusive of both visuality and sound.[10] These accomplishments themselves, far from merely producing formal distinctions, have helped to establish the conditions of possibility for her critical work on race: specifically, a transdisciplinary, (proto-)postmedium artistic practice foregrounding, yet not limited to, the conceptual. In her 1988 essay, 'On Conceptual Art', in addition to discussing her work's resonances with that of LeWitt, Piper describes her practice not in terms of medium but as a turn to language. Within that category she includes 'typescript, maps, audio tapes, etc.' in order to explore 'objects that can refer both to concepts and ideas beyond themselves and their standard functions', while also 'draw[ing] attention to the spatiotemporal matrices in which they're embedded'. Such an approach, although articulated years after she initially presented *Funk Lessons*, has ultimately underpinned her work since the 1960s, which has variously confronted the multiple material and conceptual strata along which race and racism have operated, especially in the United States. In the same essay, indeed, she concludes that, 'Racism is not an abstract, distanced issue out there that only affects all those unfortunate people. Racism begins with you and me, here and now, and consists in our tendency to eradicate each other's singularity through stereotyped conceptualization' (Piper 1999a: 240–241). One of the various ways Piper has combated such stereotyped conceptualizations is through her extensive body of work based on the politics of racial passing and performativity.

Along with conceptualism, performance has also assisted the broader shift away from medium's centripetal pull. Not unlike music, performance can be said to lack an essentializable sensory register or even material substrate, except for perhaps the body (as we've encountered especially in body art). We see performance often as much as we hear it, smell it, or on occasion touch it and taste it. But a categorization of performance based on the separation of discrete sense modalities, as was the case with medium specificity, seems to miss the mark. One needn't explain the awkwardness of calling Rirkrit Tiravanija's events in which the artist cooked and served Thai food 'taste art'. More pointedly, such a determination can be said to appear at the wrong level of mediation, since such work was ultimately read as intervening into art's very relationality (Bourriaud 2002). In the video-mediated performances of Piper, Baldessari, and Acconci, sound appears as one element among many and lacks the centripetal pull of medium, especially as related to the senses, as an organizing concept. Performance not only yokes together various sense modalities but is often construed across divisions between the fine and performing arts.

Yet, like sound, a residual logic of medium specificity can be found in recent theorizations of the result of performance's transformation into post-relational aesthetics and social practice art. In her widely influential *Artificial Hells*, Claire Bishop risks a somewhat catachrestic formulation of social practice art as using 'people as a medium' (Bishop 2012: 2, 39, 284). (Compare this to Henry Flynt's 'concept art', a relatively under-discussed prefiguration of conceptual art, which Osborne describes as a 'medium-based conception of conceptual art' [Osborne 2013: 103; see also Joseph 2008: 153–212]). Despite a certain accuracy in describing social practice works in which people are indeed *used* in a manner not unlike paint, the concept of medium here seems somewhat out of place. Such a phrase may nonetheless point to the kinds of configurations social practice inherits from the performing arts and pedagogy, both sites of social distillation at a certain remove from medium.

One of the primary establishing contexts of social practice, as Bishop defines it, is the critical pedagogy movement, an initiative that has worked since the 1960s to transform the structure and policies of education through open and participatory strategies and alternative pedagogical structures (see also Freire [1968] 2014; and Rancière [1987] 1991). Note that Bishop included Piper's essay 'Notes on Funk' (1983–1985) in her 2006 Whitechapel volume *Participation*, an important precursor to *Artificial Hells*. Along these lines, Piper (1999a: 196) uses her 'pseudoacademic' performance-lecture format to counter-narrate common misperceptions about funk, and by extension black culture, by working against stereotypes and biases distributed across racial and socio-economic divisions. Piper thus foregrounds music's sociality in addition to its polysensory and transmedial dimensions. Another of Bishop's points of departure for social practice art is music and the performing arts. She locates the roots of social practice, not unlike Shannon Jackson (2011), in theatre practices – Bishop goes as far as to situate her project as 'rethinking the history of twentieth-century art through the lens of theatre rather than painting (as in the Greenbergian narrative) or the ready-made' (Bishop 2012: 3) – focusing on avant-garde groups such as the Italian futurists. But of equal interest are the ways in which music figures in this art-historical narrative: Russian music theorist Arseny Avraamov's *Hooter Symphonies*, in which factories were 'conducted' from rooftops; and the Russian Persimfans, conductorless orchestras that began in the 1920s, which sought to rethink the hierarchical relationships between ensemble performers established by the existing orchestral music canon (Bishop 2012:

63–66). Piper's *Funk Lessons* unites the sites of music – inclusive of its ties to movement and dance – and critical pedagogy in a conceptualist (proto-)social practice.

In addition to performance, music can also figure as a site for centrifugal sound if music is rearticulated along the lines of contemporary art's radical genericity. Such a reconception of music departs from both new music and sound art through a rejection of sound as an anchoring medium and through a privileging of criticality. Post-conceptual music, what Osborne calls 'music *qua* art' and what I have defined elsewhere as critical music after sound, is one way of describing such a rearticulation of music that accounts for the centrifugal status of medium in light of contemporary art's postmedium condition.[11] At the same time, such a notion of post-conceptual music would subject its objects of description to the test of contemporaneity as art's historically specific conditions of criticality. The three works respectively from Piper, Baldessari, and Acconci, while they contain centrifugal sound, can also rightfully be described as post-conceptual music. Piper's proto-social practice work mobilizes musicality's social/pedagogical form as a vehicle for combating racism using lecture-performance, video documentation, and collective movement. Meanwhile, *Baldessari Sings LeWitt* unearths the latent musicality of early conceptual art in his song cycle through an homage that oscillates between the solipsistic and the social. Finally, Acconci's *Theme Song* further probes the song form by exposing a coercive sexual violence in the pop music hits of the 1960s and 1970s. Again, sound appears as an element of each of these works, but of equal importance are the ways language, video, and performance give way to a critical contemporaneity. Sound is present yet only centrifugally.

We have seen then how centrifugal sound, despite its tentative and temporary status as a strategic placeholder notion, can figure as a viable concept for contextualizing sound's appearance in contemporary art without returning us, however wittingly or not, to the centripetal pull of medium. First acknowledging Greenberg's role in establishing medium specificity as an ontology of art based on the separation of sense modalities, we then considered sound's prefiguration as music within this discourse: the influence of absolute music on the very construction of medium specificity, and hence sound art's belatedness when it first appeared in contemporary art in the 1950s and 1960s. A further source of belatedness, it turns out, was sound's insistence on becoming a species of art just as conceptualism began a process of overcoming medium specificity in favour of contemporary art's radical genericity. It is in this context, as I've argued, that sound must sustain a form of peripherality by consistently spinning outside of itself, destabilizing its axiomatic tendential pull towards the stability of medium as a governing concept. Performance has paralleled this unseating of medium through its own decentring of the sensory in favour of social practice's critical engagements. Finally, the centrifuge cannot operate within a vacuum and must appeal to a contemporaneity, with all of its implicit risks and precariousness, that is the manifest outcome of our historical present: that is, on the one hand, a contemporary art still working through the radical transformations of the post-war avant-garde and conceptualism into contemporary art's post-conceptual postmedium condition; and, on the other, the triumph of the neoliberal world order (with its burgeoning neo-fascist spawns), which itself demands a discursive and radically generic art up to the task of undoing the present's most vexing aporias. As a sonic methodology, centrifugality begins this paradoxical process marked by destructive interference and tactical phase cancellation. Centrifugal sound comes full circle by spiralling outside of itself and into the now.

Notes

1. For my full argument on *Baldessari Sings LeWitt*'s status as an art song cycle, see Barrett 2016: 77–86. The reference here (which isn't an explicit a part of that argument) is to Deleuze and Guattari (1987: 310–350).
2. In 1969, the same year he authored *Sentences on Conceptual Art*, LeWitt writes of his practice, 'I think of it more like a composer who writes notes' (LeWitt n.d.).
3. Seth Cluett (2013) lists roughly ten sound-themed exhibitions between 1966 and 1972.
4. Krauss's *A Voyage on the North Sea: Art in the Age of the Post-Medium Condition* is based on her 1999 Walter Neurath Memorial Lecture at Birkbeck College, London. The 'post-medium condition', Krauss's critical target in the book, refers to the 'international spread of mixed-media installation [that had] become ubiquitous' (Krauss 2000: 20). Her corrective is 'differential specificity', which calls for the reinvention or rearticulation of mediums (56).
5. 'Oration, harmony, and rhythm', or *oratione, harmonia, rhythmus*, appears in Marsilio Ficino's 1491 Latin translation of *The Republic* (see Bonds 2014: 11; see also Chua 1999).
6. Georgina Born criticizes this paradigm of 'discovery' in discussing sound (art) theorists as 'portray[ing] sound as awaiting discovery, as a radical new paradigm, as incipient in existing work (etc.): as a year zero' (Born 2015).
7. Despite authoring numerous textual works, Kosuth has never completely abandoned material forms. His *Art as Idea as Idea* series, which began in 1966, consists of single words and their respective definitions rendered as photostat prints. Nevertheless, for Kosuth the material form remains secondary, as he claims, 'I never wanted anyone to think that I was presenting a photostat as a work of art […]' (Kosuth 1991: 30).
8. Contemporary art's contemporaneity, however, is a much more complex problem that Osborne addresses elsewhere and elaborates more fully (Osborne 2013: 15–36, 2017: 14–89; see also Barrett, forthcoming).
9. For a recent historical study of race and racism's aural dimension, see Stoever 2016.
10. On conceptual art's negation of the aesthetic, see Osborne (2013: 37–70). Moten does discuss Piper's conceptuality but almost exclusively in relation to the problem of objecthood and Michael Fried's position on minimalism, for example, 'in the end, Piper's conceptualism allows her rich historical animation of the minimalist object' (Moten 2003: 242).
11. Elsewhere I propose the term 'musical contemporary art' (see Barrett, forthcoming).

References

Barrett, G. Douglas (2016). *After Sound: Toward a Critical Music*. New York: Bloomsbury.
Barrett, G. Douglas (forthcoming). 'Contemporary Art and the Problem of Music: How Contemporary is Contemporary Music? Or, Toward a Musical Contemporary Art'. In

Patrick Valiquet (ed.), 'Contemporary Music and its Futures', special issue of *Contemporary Music Review*.
Bishop, Claire (2012). *Artificial Hells: Participatory Art and the Politics of Spectatorship*. New York: Verso.
Bishop, Claire (ed.) (2006). *Participation*. Whitechapel: Documents of Contemporary Art series. Cambridge, MA: MIT Press.
Bonds, Mark Evan (1997). 'Idealism and the Aesthetics of Instrumental Music at the Turn of the Nineteenth Century'. *Journal of the American Musicological Society* 50 (2/3): 387–420.
Bonds, Mark Evan (2014). *Absolute Music: The History of An Idea*. New York: Oxford University Press.
Born, Georgina (2015). 'Sound Studies / Music / Affect: Year Zero, Encompassment, Difference?'. *Current Musicology – 50th Anniversary Conference*, New York, 28 March.
Bourriaud, Nicolas (2002). *Relational Aesthetics*. Dijon: Les Presses du réel.
Chua, Daniel K. L. (1999). *Absolute Music and the Construction of Meaning*. Cambridge: Cambridge University Press.
Cluett, Seth Allen (2013). 'Loud Speaker: Towards a Component Theory of Media Sound'. PhD dissertation, Princeton University.
Cox, Christoph (2011). 'Beyond Representation and Signification: Toward a Sonic Materialism'. *Journal of Visual Culture* 10 (2): 145–161.
Dahlhaus, Carl (1989). *The Idea of Absolute Music*. Trans. Roger Lustig. Chicago: University of Chicago Press.
de Duve, Thierry (1993). 'Part II: The Specific and the Generic'. In *Kant After Duchamp*. Cambridge, MA: MIT Press.
Deleuze, Gilles and Félix Guattari (1987). *A Thousand Plateaus*. Trans. Brian Massumi. Minneapolis, MN: University of Minnesota Press.
Engström, Andreas and Åsa Stjerna (2009). 'Sound Art or *Klangkunst*? A reading of the German and English Literature on Sound Art'. *Organised Sound* 14: 11–18. https://doi.org/10.1017/S135577180900003X.
Freire Paulo ([1968] 2014). *Pedagogy of the Oppressed*. New York: Bloomsbury.
Greenberg, Clement (2003). 'Avant-Garde and Kitsch'. In Charles Harrison and Paul Wood (eds), *Art in Theory, 1900–2000: An Anthology of Changing Ideas*, 529–541. Malden, MA: Blackwell.
Greenberg, Clement (2000). 'Towards a Newer Laocoon.' In Francis Frascina (ed.), *Pollock and After: The Critical Debate*, 60–70. New York: Routledge.
Greimas, Algirdas Julien (1983). *Structural Semantics: An Attempt at a Method*. Trans. Daniele McDowell, Ronald Schleifer, and Alan Velie. Lincoln, NE: University of Nebraska Press.
Jackson, Shannon (2011). *Social Works: Performing Art, Supporting Publics*. New York: Routledge.
Jay, Martin (1993). *Downcast Eyes: The Denigration of Vision in Twentieth-century French Thought*. Berkeley, CA: University of California.
Joseph, Branden W. (2008). 'Concept Art'. In *Beyond the Dream Syndicate: Tony Conrad and the Arts after Cage*, 153–212. New York: Zone Books.
Kim-Cohen, Seth (2009). *In the Blink of an Ear: Toward a Non-Cochlear Sonic Art*. New York: Bloomsbury.
Kim-Cohen, Seth (2013). *Against Ambience*. New York: Bloomsbury.
Kim-Cohen, Seth (2016). '2016 Call for Participation CAA 104th Annual Conference'. Washington, DC, 3–6 February.

Kosuth, Joseph (1991). *Art After Philosophy and After: Collected Writing, 1966–1990*, ed. Gabriele Guercio. Cambridge, MA: MIT Press.

Krauss, Rosalind (1979). 'Sculpture in the Expanded Field'. *October* 8 (Spring): 30–44.

Krauss, Rosalind (2000). *A Voyage on the North Sea: Art in the Age of the Post-Medium Condition*. New York: Thames & Hudson.

LaBelle, Brandon (2006). *Background Noise: Perspectives on Sound Art*. New York: Bloomsbury.

LeWitt, Sol (n.d.). 'Art by Telephone (Chicago: Museum of Contemporary Art, 1969)'. Specific Object, David Platzker. Available online: http://www.specificobject.com/projects/art_by_telephone/#.VaV2axaNuRk (accessed 22 January 2015).

Lippard, Lucy R. (1973). *Six Years: The Dematerialization of the Art Object from 1966 to 1972; A Cross-Reference Book of Information On Some Esthetic Boundaries*. Berkeley, CA: University of California Press.

Moten, Fred (2003). *In the Break: The Aesthetics of the Black Radical Tradition*. Minneapolis, MN: University of Minnesota Press.

Osborne, Peter (2013). *Anywhere or Not at All*. New York: Verso.

Osborne, Peter (2017). *The Postconceptual Condition: Critical Essays*. New York: Verso.

Piper, Adrian (1999a). *Out of Order, Out of Sight*, vol. 1, *Selected Writings in Meta-Art, 1968–1992*. Cambridge, MA: MIT Press.

Piper, Adrian (1999b). *Out of Order, Out of Sight*, vol. 2, *Selected Writings in Art Criticism, 1967–1992*. Cambridge, MA: MIT Press.

Rancière Jacques ([1987] 1991). *The Ignorant Schoolmaster: Five Lessons in Intellectual Emancipation*. Stanford, CA: Stanford University Press.

Stoever, Jennifer Lynn (2016). *The Sonic Color Line: Race and the Politics of Listening*. New York: New York University Press.

Weheliye, Alexander G. (2005). *Phonographies: Grooves in Sonic Afro-Modernity*. Durham, NC: Duke University Press.

23
How to Cut Up a Record?
Paul Nataraj

Sound records, records sound

The tension between sound as transient matter and its commodification through different physical and now digital formats has always been my particular fascination. In trying to apprehend meaning from a sonic experience, we have to be aware of listening in the present moment, whilst simultaneously reaching for sound's becoming, and equally trying to remember what we have just heard. An active listening then requires us to experience the now-ness of sound. There has been an argument that the recording process is reductive to this experience, because much is lost in an attempt to capture the natural transience and space of the sonic. John Cage famously said that 'records are like postcards to the extent that both "ruin the landscape" – they destroy the experience of one's surroundings' (Grubbs 2014: 46–47). For Cage and others the beauty of sound is inherent in its immediate presence, in its now-ness. Repetition in the form of a record for example, could be conceived as being the death of the musical experience. Theodor Adorno, for example, wrote, 'the phonograph record is an object of that "daily need" which is the very antithesis of the humane and the artistic' (Adorno [1934] 2002: 58). For Cage the postcard is never as good as the holiday, but what if it was possible to vividly experience a different beauty of that holiday through imagination and memory. Indeed, what if replaying the memory of the holiday makes it extra special over time? What if repeatedly looking at the beauty of the postcard across the years of someone's life can render the holiday ever new, exciting, and different, through simply engaging with its representation? Seen through this metaphor, the record becomes so much more than a simple sound carrier. As David Grubbs writes, 'it isn't about the tune so much as how it gets across, and what previously unimagined sounds wind up in the grooves of a record' (Grubbs 2014: 96–97). For the majority of people recordings are their way into the world of music, and music provides a foundation of how they listen to and through the world, and so should be studied with commensurate respect.

Besides these 'previously unimagined sounds' the record's initial purpose is to transcribe the mysterious *Ur*-language of sound itself sealed within its surface. It is a groove drawn from the physicality of sound, in a form that is semiotically other; outside of the literary perception of human communication, yet able to transmit the beautiful detail of a sonic gesture across time and space. As different formats have come and gone, it seems fitting that the record has endured as a system for capturing the materiality of sound. Yet the form itself is a manifest paradox, the durable nature of the material is seemingly at odds with the ephemeral object that it holds. Yet both, sound as floating phenomenon and its aesthetic trace as the groove, have a mythic, indefinable, somehow magical quality, meaning that on a record, content and form are also perfectly matched. These interesting attributes, peculiarities, and qualities that this marriage has produced mean that the record is filled with contradictions, in its social, political, economic, and aesthetic contexts. This coupled with its multifold functionality as print artwork, book, sound storage, and sound reproducer, give the record a unique 'thing' power that still contains many secrets about the ways we interact with music's materialities. Because of these varied uses and, through the social significance of its creation, rule, demise, and resurgence, the vinyl record has proved itself to be a popular cultural sign of great complexity. I would agree with Dominik Bartmanski and Ian Woodward who argue that the record be elevated to the level of icon, things they describe as being 'potent objects whose surfaces do not simply represent hidden data and communicate information but constitute and transmit sensuous experience without which no culture can exist' (Bartmanski and Woodward 2015: 175). Due to its longevity the record has been an important sensuous sidekick in the cultural and sonic journeys of a great many people, building their cultural competencies and capital over their lifetimes. As Brandon LaBelle states, 'sound studies takes such ontological conditions of the sonic self and elaborates upon the particular cultures, histories, and media that expose and mobilize its making' (LaBelle 2010: xx). My own practice led research project You Sound Like a Broken Record (YSLABR) was a practice led interrogation of the ontological resonances of vinyl record culture, and draws from LaBelle's assertion to focus on the 'sonic self'.

Personal records

Jean Baudrillard makes the point that 'human beings and objects are indeed bound together in a collusion in which the objects take on a certain density, an emotional value – what might be called a "presence"' (Baudrillard 1968: 14). So behind the sign, or the object, which, for Baudrillard is always transient, fleeting, momentary, and often insubstantial, there exists the meaningful energy of relationships. As Jane Bennett has shown in her work, one can see materiality as being 'as much force as entity, as much energy as matter, as much intensity as extension' (Bennett 2010: 20). We could say then, that objects are produced as entities in communion with our daily lives, our practices and us. In other words, we are knitted into feedback loops of multifaceted significations alongside the things we use. The object is our marker of time; it is the depository of self, and remains as partner and griot, helper and confidante through the chaos and constantly changing intrigues of our lives. As we collect the objects around us, so significantly they collect us.

The record is an evocative object that allows us to see the exertion of Bennett's 'force' or vibrancy. It is a cultural hobo that holds a dialectical position as both a symbol of cultural subversion and a product of the mainstream music industry, remaining equally totemic in both paradigms. The record has an energetic multiplicity for its users and takes on new relevancies over the years of ownership. It presents each listener with an individuated space through which to engage with their personal version of a musical experience. The record produces new shapes of listening for us, as Evan Eisenberg writes, 'when a record is fitted over the platter, a transparency or slide is fitted over a segment of space and time. The effect is a double exposure' (Eisenberg [1987] 2005: 206). The record does this, not only through providing a private point of listening but also by acting as a transportation device taking us back to the time of writing, playing, and production, through the hands of all those who have held, owned, and listened before, it is a private yet collective dialogue.

In 1986 Kittler warned that 'data flows once confined to books and later to records and films are increasingly disappearing into black holes and boxes that, as artificial intelligences, are bidding us farewell on their way to nameless high commands. In this situation we are only left with reminiscences, that is to say stories' (Kittler 1986: xxxix). In YSLABR I went out to find these stories in order to expose a flip side of the record. Much of the literature on vinyl has thought about the record in relation to connoisseurship, expertise, musical marketing strategies, generic subversion, or the more recent 'vinyl is back' zeitgeist (Harvey 2017: 586). All these approaches are valid and essential to better theorize the continuing multistrand narrative of vinyl's ongoing story, but they all commonly look to the voices of experts to give insight into the way in which vinyl operates, or has previously operated. My interest, however, was in the way in which personal memory can be inscribed by the object, and to explore the ways in which the individual is able to consolidate or even perhaps gain agency during this process. The Popular Memory Group working in the early 1980s at the Centre for Contemporary Cultural Studies in Birmingham, UK, wrote that

> There is a common sense of the past which though it may lack consistency and explanatory force, none the less contains elements of good sense. Such knowledge may circulate, usually without amplification, in everyday talk and in personal comparisons and narratives. It may even be recorded in certain intimate cultural forms: letters, diaries, photograph albums and collections of things with past associations.
>
> (Popular Memory Group [1982] 2006: 45)

It is exactly this unamplified knowledge that the YSLABR project was trying to plug into. I was interested in those people whose few records languished in a bedding box at the bottom of the bed, and no matter if they were ever played, they were never getting thrown out. It was these stories and the correspondent records that I wanted to find. I hoped that by unearthing the stories locked into these grooves I might hear some trace of the interweaving voices that make up the complex ontology of the vinyl record.

Michel de Certeau has posited, very much in the lineage of Roland Barthes' (1977) thinking, that 'the text has a meaning only through its readers; it changes along with them; it is ordered in accord with codes of perception' (Certeau 1984: 170). With a record, it is not just the musical text that is assembled at the moment of listening, because, as

Osbourne (2012) points out, the song pressed onto a vinyl record becomes thought of as that vinyl record. Indeed *that* vinyl record is the conduit for the moment of listening to take place, and we have to embrace it in order to get to the sound. Consequently, through this amalgamation of song and material, the processes of meaning making for music has historically been bound into the space of ownership and usage. Yet, in its materiality the record cannot be disentangled from the industry that produces it, and the mechanizations that allow for its production and distribution. Those indelible traces of industrial power relations remain as it moves into the private space of listening, but are reinvigorated and transformed by the presence of the self, reformulating the object through its everyday use. The record becomes a tool of identification, a personal space for reinvention, a platform for a renegotiation of our relationship with the world around us, a mode of resistance. Just buying a record can be an act of rebellion; it is commensurate with a certain worldview. Where or who you buy it from, if it is new or second hand, swapped, saved up for, gifted by someone, whether you have a turntable to actually play it or not, if it is your first or your 2,500th, all these factors add to the narrative and give an insight into our relationship with the world and the sounds that we choose to fill it with. It opens up questions as to how the sounds that we purchase impact on the way in which we live our lives, either in resistance or acceptance, in protest or compliance? Also, how do these particular sounds find us, and what kind of practices do we use to try and understand them?

The vinyl record is a meeting place, the point of convergence between a constellation of actors. Each one of these voices, including our own and that of the artist, is in dialogue with those of our family and friends, our rooms, that particular sound system, an argument, a conversation, our hopes, dreams, successes and failures, not to mention the sound of that individual piece of plastic. Alongside this we have conversations with radio presenters, journalists, graphic designers, TV shows, documentaries, authors and a host of others whose expertise we use to better understand the sounds that emanate from this mysterious piece of disc. All these dialogic elements make up the sound of that piece of vinyl. The songs that come spiraling off the record's surface are intertwined with the multivalent soundings of these memories, actions, people, spaces, and places. As our attitude and relationship to each of them changes over time, each voice becomes repositioned, the changing the sound of the mix of the record. The record might flatten musical hierarchies but the dynamics of our personal topographies can be exhumed from a dig beneath the surface of those uniform cardboard jackets. As Bartmanski and Woodward point out, 'physical records record more than just sounds. As their obdurate condition allows them to last and outlast their owners, they can record history, personal and collective' (Bartmanski and Woodward 2015: 176). Using the process of ethnographic interviews I wanted to try and get to the collective through the very personal. I felt that the best way to approach this and previous questions would be to speak to individuals about a record that they felt held a special place in their heart. I wanted to find the story of that one record that had been saved from or might have saved its owner from the wreckage.

I placed a series of advertisements, in various real and virtual spaces, to ask for volunteers who would be willing to be interviewed and importantly give me a prized record. My request garnered records and interviews from fourteen respondents. All interviews were recorded then transcribed, with all transcripts being checked by the respondent, to ensure the accuracy of each account, before I continued with the practice element of the work.

Record breakers

After completing the interviews and transcripts I engaged in a sculptural practice, using an awl to carve a transcription of the respondent's story back onto the surface of the record to create a unique object, a vinyl palimpsest. In so doing I hoped to better expose the multifarious strata of significations that make up this complex and vibrant object. As Sarah Dillon states, often times there can be a 'lack of clarity in unearthing the ontology of someone's relationship to the object so instead the palimpsest provides us with a sense of "merging"' (Dillon 2007b: 4). I would argue that this 'sense of merging' deepens our understanding of the social, political, historical, and poetic context of the object and allows us to hear 'several people writing together', in the words of the authorial murderer, Barthes (1977: 144). The beauty of using the palimpsest as metaphor is a gestalt complexity where each individual participant – I think here of participants also as sounds, writings, videos, and the explanatory paratexts connected to the record – are nominally sealed beneath the next user. Yet throughout, each participant's voice is faintly audible through the morass of enmeshed experiential fibres that make up its ever-changing surface.[1]

This practice can be seen in a direct lineage of compositional approaches that Caleb Kelly refers to as 'cracked media' works. For Kelly, a cracked media artist is 'the experimentalist who is prepared to extend his or her instrument to the point at which it breaks, perhaps never to be used again in the manner in which it was intended'. They embrace, 'this risk of sometimes great loss', and turn it 'into great gain as traditional and commonplace sound practices are themselves transformed, extended and expanded' (Kelly 2009: 6). My work is also based on the Burroughsian premise of the cut-up: to physically explore and unlock the materiality of the record so that it might reveal something about the secrets of ownership trapped inside its grooves. To bypass the sounds that already exist there, or at least to bring the object and its story into relief, the inscriptions form a new sonic companionship, the scratch along with the present sounding material. The records play out the combined voices vying for position, the tension between music and the industry, sound and materiality, and themes of resistance through personal ownership, freedom, and constraint. It shows artistic practice as a way of unseeing and re-seeing the object through a radical intervention. As Kelly writes, 'for new meaning to be created a crisis or catastrophe must occur, or perhaps an accident, that will focus the elements of chaos into a singular focused emergent menacing. Noise is then filled with all future possibilities' (81). In the case of YSLABR, my respondents have stored up this noise, it has been waiting in the wings, in readiness for its starring role.

Cutting across the record's groove is like damming a valley, stopping the natural flow of information down this channel, interfering with the signal, pinching the recorded voice in mid flow, upsetting the rhythm, preventing the player, or the instrument, or the machine from reaching its rightful place down the line, rudely interrupting. The click of the scratch breaks the music's linearity; it becomes an anathema to the once inviolable communion between the listener and the ghostly representation of the disembodied voice. The scratch flips the subject back into reality, exposing the media, upsetting the illusion of transparency. This break in the system speaks to the agency that is shared by both object and its intervener, and represents a transgression against a prescribed model of playback.

I see all of my respondents as having and exercising such agency. It is evident in the stories they tell about the gifted records. Through the YSLABR project this agency became clearer, and the inscriptions instantiated its essence through the action of the artist. As Claude Levi-Strauss explains of the bricoleur, 'he speaks not only with "things" […] but also through the medium of "things" giving an account of his personality and life by the choices he makes between limited possibilities' (Levi-Strauss 1966: 21). It could be argued that by working with the materiality of the record as bricoleur, I was privy to the 'moment when the object becomes the Other, when the sardine can looks back, when the mute idol speaks, when the subject experiences the object as uncanny' (Bennett 2010: 2). These were private, invisible, and transparent moments, shared with me through the conversation I had about the records, and it was my job to expose them through the transformative transcription.

A shelf of records is filled with stories, and indeed when we look towards them we can imagine, or possibly even hear the voices of, these 'mute idols'. My respondents told me stories of family histories, tales of how and how not to live, messages of morality carved into the psyche of each listener and communicated through the grooves of each record in their collections. The stories very clearly illustrate how the 'pure' groove is compromised by the concrete external noise of context and aggregating tales. Each record is a bricolage, a composite, a fluidly increasing rhizome, a construct of a series of struggles occurring in different times, places, and spaces overlaid onto the surface of the object, all of whose noises are sucked in, pressed, and trapped by the groovy hieroglyph. Hidden in there are specific human value systems, political leanings, compassion, empathy, fear, anger, and all the experiences of growing up and trying to find and maintain your place in the chaos of the world. My palimpsestual intervention, my scratch, my introduction of noise, may just, as Eisenberg writes,

> derail the music's progress; but surface noise could turn any piece of music into such a struggle of order against chaos, of the human spirit against the flailing of the blind, but far from mute universe. There are works whose plot line remains […] hopelessly intertwined with the subplot of a long, insistent scratch.
>
> (Eisenberg [1987] 2005: 212)

I would argue that every work that has been owned and used by someone somewhere is 'hopelessly intertwined' with our noisy interventions. For the music can never escape our social and emotional specificities, it is constantly in dialogue with them. With every listen we replay the environment, invoke our memories, and are experientially moved, and it is the record that collects all this metadata, and in some way helps us to make sense of it all.

'Popcorn' tactician

Kelly, Certeau, and Levi Strauss, in 'cracked media', 'tactics', and 'bricolage', respectively, are thinking about the small, non-engineered, non-institutional, creative acts of subverting the status quo. For Certeau this could simply be a rhetorical or discursive gesture as 'the thin film of writing becomes a movement of strata, a play of spaces' (Certeau 1984: xxii). By focusing on the fleeting and apparently inconsequential, he provides the means of resistance for everyone. In so doing he outlines a model with which to negotiate one's own material environments in order to temporarily free oneself from the oppressive world of capital. According to Certeau, one does not enter into this negotiation chanting overtly political slogans but, rather, through 'making do'. One is able to adopt the dominant language and subtly change it to fit one's own needs, and in so doing attacking the hegemonic platform upon which the language one is using originally stood. This is a notion that chimes with the actions of a bricoleur who 'addresses himself to a collection of oddments left over from human endeavors', reinventing 'the material means at his disposal' (Levi-Strauss 1966: 19). Through this 'making', 'assembling', and 'cracking' we assert individuality and at least some semblance of freedom, for 'the weak must constantly turn to their own ends forces alien to them' (Certeau 1984: xix). We take that which oppresses us and we make it our own. In doing so we are able to remodel our lives and ameliorate the process of alienation.

For example, Andreas, a 53-year-old sound engineer originally from Italy, shows how the record itself can act as a tactical weapon in the practice of everyday life, against the orthodoxies of tradition. Below, Andreas tells the story of how he acquired the record 'Popcorn' by La Strana Società that he donated to the YSLABR project (Figure 23.1):

> In my home it was strictly forbidden to listen to pop music. It was absolutely a sin. In my house you could only listen to Mozart, Beethoven, Bach and so on. This was a good thing because actually at seven, eight, I could sing all four symphonies of Brahms, and so on. Anyway, it was really a sin, you couldn't even know, not to say the name of Beatles, because 'what are you saying?' So I got a small AM radio at ten and then I began to discover a real world outside with good music, not Handel, not Bach, but it was good music. Suddenly I discovered 'Popcorn'. Now 'Popcorn' was more a way to demonstrate the new Moog instrument, the new synthesizer instrument, rather than the song itself, but it had success. The first group that recorded 'Popcorn' was the group Hot Butter and then many others made covers, in Italy it was made by the group La Strana Società, The Strange Society, and I fell in love with the song. It has some qualities, some contrapuntal qualities, it's not so bad, it's very well written, it's very well played. So at twelve I decided I wanted to have this disc, and so I went to my father and I told him, 'Pappy, I would like to buy this record,' and it was a disaster. Everybody in the house was suddenly sad, and what are we going to do with this guy? It is incredible they phoned to family friends and they came to our house, and I could

listen and they said, 'Well be patient it's just a disc, it will not destroy his life.' A disc, a single disc, it was that. And my mother said, 'but you have a recorder why don't you record it?'

(Andreas, interview with author, 2013 [hereafter Andreas 2013])

Eventually his parents did allow him to buy the treasured disc although, they 'did not speak to me for a week' (Andreas 2013). This was a price worth paying however, because the record was 'the key to access the outside world, because otherwise this would not have been possible' (Andreas 2013). This purchase represents a tactical act of subversion in the everyday, from the everyday. A way of separating Andreas from the draconian rules of his family environment, for 'my family was a very sad family, everything was heavy' (Andreas 2013). For Certeau a tactic 'must vigilantly make use of the cracks that particular conjunctions open in the surveillance of the proprietary powers. It poaches in them. It creates surprises in them. It can be where it's least expected. It is a guileful ruse' (Certeau 1984: 37). The

Figure 23.1 'Popcorn', La Strana Società – inscribed record by Paul Nataraj.

crack that appeared for Andreas was the gift of an AM radio. His parents could not have imagined that he would use it to access this new 'popular' world outside of the confines of the family home and traditional ideology. Yet Andreas had found a way to expose this crack by physically manifesting the world that he had heard in the ether, purchasing the record and bringing the offending object into the home. It is extremely important when thinking about the resonances of the vinyl disc and its symbolic power to note that Andreas's mother was happy for him to have a recording of this music, but not actually own the disc. To have the record was to actually buy into a stratum of society with which his family did not want to be associated, La Strana Società indeed. As Andreas explains, pop music 'was something which could ruin the good society. The good society listens to Mozart, bad boys listen to the Rolling Stones, this was the grounding' (Andreas 2013). Certeau explains, however, that the gains made by employing these tactics are short lived, 'producing a sudden flash' (Certeau 1984: 87–88). For Andreas also, this would seem to have been the case. For now he is working in the field of classical music, as a producer and engineer, and although he describes it as 'boring', it would seem that the sugary 'Popcorn' did not compromise his taste so completely as his parents might have thought.

Crushed grooves

In the first instance then YSLABR collects records and their connected stories. It then takes these stories and reinterprets them through sculptural practice to produce a piece of visual art. These palimpsests are also 'cracked' instruments, and provide the sounds for compositions that make up the final YSLABR works. I will explain below some of the thinking behind this final part of the practice.

I recorded the playback of each record and then sifted through these sound files for fragments of sound that I felt spoke to the stories of my respondents. As the compositions are sound poems attempting to express the feeling of the stories that were shared with the record, its owner, and now myself, all these materials are brought together in the creation of the final sonic pieces. These are intertwined with my own personal relationship to music and a very clear acknowledgment of my own limitations as a composer, as I have no formal training in music. Below I outline the compositional process in just one of the tracks, 'Badgewearer'.[2]

I have engaged with a bit of musical borrowing to realize the musical aesthetic for the 'Badgewearer' sound piece. This is a technique well used by many composers down the years. Charles Ives is a famous example, as is the inimitable Erik Satie. For example, in his *Pièces froides* written in 1897, he 'takes as his source material the well-known Northumbrian folk tune "The Kneel Row". In lieu of simply quoting the jaunty melody, Satie adopted its easily recognized rhythms, then recomposed and reharmonised the melody as if to conceal the source' (Davies 2007: 67–68).

In 'Badgewearer', the rhythmic backbone of the whole work is based on the drum pattern from the 1987 track 'Top Billin' by Audio Two. This song placed at number one in 'Hip-Hop's Greatest Singles by Year' for 1987 in *Ego Trip's Book of Rap Lists* (Jenkins et al. 1999: 321). I absolve my guilt for this particular theft by placing the blame with my respondents because their stories contribute to my compositional motives and the conceptual intentions of each piece. So in this instance I chose 'Top Billin' as my source template because my respondent Tony was a huge punk fan during the early 1980s, a musical choice that had caused his Dad, an opera fan, a few sleepless nights. He recounts that his Dad 'was pretty Draconian you know, he went through the records and there was a couple of things that had dodgy covers and he just wasn't happy with me listening to it', but he goes on to note that 'it wasn't banned, it was kind of vetted' (Tony, interview with author, 2015 [hereafter Tony 2015]). He then tells me that his own son is listening to the French hip-hop artist Black M, and 'it was just horrific stuff, it was like I cannot, I just don't want it in my house for a start' (Tony 2015). Tony clarifies that the misogynistic lyrical content was the thing that he was averse to, adding that his attitude to it is 'more of a parenting question, than a music thing'. He goes on to say that because of his own involvement in punk, 'you know "pretty vacant" and all this kind of nihilism, I can understand the concern [from my own parents at the time]' (Tony 2015). So, in my role as participant observer, and to maintain the dialogic relationship of the material in this work, I started to think about my own experiences of the music I listened to that annoyed my parents. 'Top Billin' by Audio Two was right up there, alongside another family favourite 'Straight Out the Jungle' by, as my Dad used to call them, 'The Jungle Buggers', or for purposes of accuracy 'The Jungle Brothers'.

One could say that the form of 'Top Billin' makes it a difficult listening experience for those not versed in the sonic language of hip-hop. The song is quite simply a repeated beat with a rapped vocal line and echoed vocal samples chanting the phrase 'Go Brooklyn!' playing throughout: no melody allowed. The factor that made 'Top Billin' a standout work in the 1987 hip-hop landscape was the structure of the beat. Instead of having an accented snare drum on the second and fourth beat, which was the convention with most hip-hop at the time, 'Top Billin' plays a kick drum on one, two, and three, then lets the snare hit on four. At the opening of the track there is no hi-hat, so the work is made of a very sparse, heavy hitting, funky, loquacious, and memorable beat pattern that is the complete antithesis to the popular music's melodic tradition. The track is 'raw' hip-hop: a beat and an MC. This is further emphasized by the drum track filling the sonic field. My parents hated it so I felt it to be perfect as template for the 'Badgewearer' piece.

After I had decided on a rhythmic template derived from 'Top Billin', I started to structure the track according to a story that Tony had told me from his own musical past. The record that Tony had given me, and which was now providing the sonic material for this work, was the first album from his own band, Badgewearer. Based in Glasgow and a product of the post-punk movement, they had some critical acclaim during the early 1990s; 'it was our press that made us stand out from the other bands. Cause we'd kind of been recognized by others, by the music press' (Tony 2015). In describing Badgewearer's music, Tony says, 'we wanted to be, it's an old cliché you know, we wanted to be original' (Tony 2015). Structure was one of the ways they attempted to do this. 'We wanted changes on three and five rather than four, no middle eight, we didn't want choruses, basically we wanted [...] to write songs in a different way so that the audience didn't know what to expect' (Tony 2015). But as Tony points out, 'the thing with that is, you can only go so far before it becomes totally contrived I think' (Tony 2015). I began to structure 'Badgewearer' with this in mind and I attempted to achieve a balance between changes that somehow made sense, but were unexpected and challenged the pounding repetition of the Audio Two inspired beat. In some ways the fragmentary playback of the record itself made this slightly easier because it broke the linearity of the songs anyway. In the drums themselves I made micro changes: in every bar notes have been pushed just out of time by milliseconds, either before or after the beat, to replicate the human inaccuracies that would be heard in someone playing their instrument live. So the piece moves through a series of ever changing sections, with the intention of never allowing the audience to become comfortable in their listening, but to stop the track from feeling 'contrived'.

In 'Badgewearer' I am also invoking a tension between this present use of the record, the fallibility of his memory of the record, and the materiality of the record itself. I use Tony's recorded voice to narrate this relationship after three and half minutes of the 'Badgewearer' track, which has by this point gone through a number of different changes itself. He recounts the journey from his first Leo Sayer record to actually having a vinyl record of his own music saying, 'from Leo Sayer to that moment was fucking significant. You know to have it in my hand, you know to think I'm now contributing to the library of stuff that's out there [...] so when it arrived, well that was a great feeling' (Tony 2015). Just before Tony starts to speak, we hear a loop that occurred whilst the record had been playing as it got trapped between engraved words. At the start of this loop the singer screams 'Why?' so this word occurs at the start of every repetition. The loop then fades away behind what sounds like a synth pad, but is in fact just a very short fragment of a guitar note that has been passed through a reverb unit that pulls out the inherent frequencies to create washed out tones. I used this same sound earlier in the work to bring about a sense of calm and balance to a group of quite hard and unforgiving sounds. I wanted this 'synth' sound to lift us away from the screaming invocation of the previous phrase and introduce listeners to our protagonist. Tony's voice is itself abstracted as I recorded it directly from the phone during our conversation. The high-pass filter of the telephone receiver gives the voice an otherworldly quality and is a signifier of distance in and of itself, also laden with resonances of the connection between recording and telephony in the nascent days of its development. Due to the distortion in the recording, it is at times difficult to hear exactly what Tony says. However, this can add to, rather than diminish, the overall communicative potential of the track. As musicologist Maarten Beirens notes, cutting and manipulating speech brings about 'more direct access to the expressive content, transmitted by the voice, but hidden by the semantic system of the words' (Beirens 2014: 221). In order to segue back to the music and maintain the idea of the rhizomatic development of these pieces, I used an echo chamber on Tony's voice saying 'feeling', which is then matched by an echoed and delayed drum fill. These echoes morph into one another as both music and voice have equal opportunity and worth in meaning creation, as is the leitmotif for all the stages of the YSLABR work.

As the track ends we hear Tony's voice again, saying 'but the feeling evaporated as soon as I put it on because the sound was terrible'. The word 'terrible' echoes off into the distance, it disappears into a space where we can't follow. We are left to contemplate the statement without ever being either literally or emotionally close to it. As he and his frustration drift away from us on the wave of the echo, the discontent in Tony's voice is palpable. I wanted to include this not only to bring a loose resolution to the track but also to provide a sense of ambiguity as to which record he was talking about. For the listener could assume that he is referring to the track of which he is part, and in some ways they would be right, because in this case he has been a part of the sound of this record before it was even a record. This injection of ambiguity plays with the combinatory authorship that is at the heart of the YSLABR work.

I also wanted to use Tony's quote to give an extra layer of emotional synergy between the elements of this piece. Luc Ferrari states that 'after being processed in the studio, a conversation with someone is not recognizable as discourse, yet it retains its discursive value […] the feeling that can transpire from a word trembling faintly in the voice: to me all that carries meaning' (Ferrari quoted in Caux 2002: 48). Here we can hear the disappointment and frustration that Tony felt when first playing his record. This emotional content is imbued with new significance as he is surrounded by the sounds of that very record, expressed in significantly new textures and shapes.

The play-out groove

Maybe Cage was right about records, perhaps they are as useless as postcards. It could well be that they are reductive for the artist who conceives of their art as only existing in the now. But YSLABR has found the record to be rich and dynamic in both production and reproduction. It is full of fluid reassessments of its own position, and its lapidary potentials have provided many with the opportunity to fulfil a desire to remake the apparatus. Not only the artist but, as I have shown, the owner too, has been able to take the record off the shelf, out of the store, and in that very moment of connection, reappraise its status as commodity, opening it up to the possibility of becoming resistant material. In this way the record has provided a mirror to reflect our inner anxieties across its surface tension. Even as one of the most static of the technologies that facilitate sonic replication and copying, many small but significant victories have been won inside the sleeves and grooves of our record collections. The format provides a different set of affordances for resistance, ones that are once again becoming prevalent in new contexts. If the music industry is the wicked witch, trapping and enslaving music, with inbuilt obsolescence and unabating newness, then YSLABR contributes an original approach to its ontology, teased apart from the layers of the collaborative palimpsest of the record. This new approach reveals a fluid and ever-evolving rhizome that is apotropaic to such industrialized and canonical constraints. My new approach to the vinyl record could pull us out of a streaming wormhole, to re-evaluate the richness of our listening world, and personal listening histories. We are resistant together.

Notes

1. For a more detailed discussion of YSLABR and the notion of the palimpsest please see Nataraj, P. (2018) 'Surface Tensions: Memory, Sound and Vinyl', in M. Bull (ed.), *The Routledge Companion to Sound Studies*, pp. 257–268. London: Routledge.
2. To hear this and all the other YSLABR tracks, see Nataraj 2019.

References

Adorno, T. ([1927] 2002). 'The Curves of the Needle'. In *Essays on Music*, selected by R. Leppert, 271–277. Berkeley, CA: University of California Press.

Adorno, T. ([1934] 2002). 'The Form of the Phonograph Record'. In *Essays on Music*, selected by R. Leppert, 277–283. Berkeley, CA: University of California Press.

Barthes, R. (1977). *Image, Music, Text*. London: Fontana Press.

Barthes, R. (1999). *The Pleasure of the Text*. 25th edition. New York: Hill and Wang.

Bartmanski, D. and I. Woodward (2015). *Vinyl: The Analogue Record in the Digital Age*. London: Bloomsbury.

Baudrillard, J. (2005). *The System of Objects*. 2nd edition. London: Verso.

Beirens, M. (2014). 'Voices, Violence and Meaning: Transformations of Speech Samples in Works by David Byrne, Brian Eno and Steve Reich'. *Contemporary Music Review* 33 (2): 210–222.

Benjamin, W. ([1936] 2008). *The Work of Art in the Age of Mechanical Reproduction*. London: Penguin.

Bennett, J. (2010). *Vibrant Matter: A Political Ecology of Things*. London: Duke University Press.

Caux, J. (2002). *Almost Nothing with Luc Ferrari*. Berlin: Errant Bodies Press.

Certeau, M de. (1984). *The Practice of Everyday Life*. Berkeley, CA: University of California Press.

Davies, M. (2007). *Erik Satie*. London: Reaktion Books.

Dillon, S. (2007a). 'Palimpsesting: Reading and Writing Lives in H. D.'s "Murex: War and Postwar London (circa A.D. 1916–1926)"'. *Critical Survey* 19 (1): 29–39.

Dillon, S. (2007b). *The Palimpsest: Literature, Criticism, Theory*. Continuum literary studies series. London: Continuum.

Eisenberg, E. ([1987] 2005). *The Recording Angel: Music, Records and Culture from Aristotle to Zappa*. 2nd edition. New Haven, CT: Yale University Press.

Grubbs, D. (2014). *Records Ruin the Landscape: John Cage, the Sixties and Sound Recordings*. Durham, NC: Duke University Press.

Harvey, E. (2017). 'Siding with Vinyl: Record Store Day and the Branding of Independent Music'. *International Journal of Cultural Studies* 20 (6): 585–602.

Jenkins, S., E. Wilson, C. Mao, G.Alvarez, and B. Rollins (1999). *Ego Trip's Book of Rap Lists*. New York: St Martin's Press.

Kelly, C. (2009). *Cracked Media: The Sound of Malfunction*. Cambridge, MA: MIT Press.

Kittler, F. (1986). *Gramophone, Film, Typewriter*. Stamford, CA: Stamford University Press.

LaBelle, B. (2010). *Acoustic Territories: Sound Culture and Everyday Life*. London: Continuum.
Levi-Strauss, C. (1966). *The Savage Mind*. Chicago: University of Chicago Press.
Nataraj, PG (2019). 'Roman Tams Umbrella'. Available online: https://paulnataraj.bandcamp.com/releases (accessed 16 July 2020).
Osborne, R. (2012). *Vinyl: A History of the Analogue Record*. Farnham: Ashgate.
Popular Memory Group ([1982] 2006). 'Popular Memory: Theory, Politics, Method'. In R. Perks and A. Thomson (eds), *The Oral History Reader*. 2nd edition. London: Routledge.

24

Directing Listening: Sound Design Methods from Film to Site-Responsive Sonic Art

Ben Byrne

Introduction

Sound design is ubiquitous in contemporary urban life. It is paired with soundtracks in audiovisual media of all kinds, from theatre, movies, television, and radio, to podcasts, games, online media, and public screens. It is also used in all manner of devices and apps to draw our attention and communicate information, from mobile phones and social media that buzz, ding, and whistle to get our attention, to fire alarms, tram bells, and so much more. These uses of sound design may at first seem disparate but they share a focus on drawing attention and directing response, be it to a particular character on screen, a message online or a vehicle headed towards you. Sound is expected to ground you in what is going on and tell you how to feel about it – that the character just entered from the left to the surprise of the others on screen, that a welcome missive from a friend has landed in your pocket or that a car is coming and you best move, now. However, sound design often involves much more than this and there exists a body of methodological writing and established methods of practice. Although most often developed and employed in and around film, such as in the work of theorist Michel Chion and sound designer Walter Murch, these approaches are intimately linked to sonic art practices, both in their histories and because of how they are employed by contemporary artists.

Despite its increasing ubiquity, sound design as a term suffers from a lack of definition, describing a vast array of sonic practices and roles. There exists a number of key texts around sound design and film, including in particular Elizabeth Weis and John Belton's *Film Sound: Theory and Practice* (1985), Rick Altman's *Sound Theory, Sound Practice* (1992), Michel Chion's *Audio-Vision: Sound on Screen* (1994), and David Sonnenschein's *Sound Design: The Expressive Power of Music, Voice and Sound Effects in Cinema* (2001). Leo Murray's *Sound Design Theory and Practice: Working With Sound* (2019) surveys the field, attempting to connect theoretical discourse with practical experience and knowledge. There exists also an ever-increasing number of texts that explore different approaches to and applications for sound design. Andy Farnell's *Designing Sound* (2010) is one example, which takes a focus on using procedural audio – programmatically generated, that is – for the production of *sound objects* rather than specific recordings, sounds that can be procedurally generated as desired rather than existing as linear, temporally determined tracks. Jean-François Augoyard and Henry Torgue's *Sonic Experience: A Guide to Everyday Sounds* (2006), meanwhile, provides a catalogue of *sonic effects*, or auditory experiences, and details the use of key examples of these across a variety of disciplines, linking sound design to architecture and urban design, psychology, music, and more. However, there is little work beyond that of Holger Schulze's *Sound Works: A Cultural Theory of Sound Design* (2019) that seeks to examine the diversity of contemporary sound design practices and their associated labours. In the face of this, Murray (2019: 11) suggests addressing sound design's lack of definition by taking it to mean simply, and broadly, the 'deliberate use of sound'.

In this chapter I will set out some key principles of sound design for audiovisuals, using the film *Apocalypse Now* (1979) as an example. I will then link audiovisual sound design and sonic art practices, using my own works to demonstrate ways in which sound design effects and methods can be employed in sonic art. In my installation works *Murmur* (2016) and *The Flood* (2017), as well as with my release *Malfeasance* (2017), I employ a variety of techniques to create specific sonic effects, in particular *extension* as defined by Chion, as well as *delocalization* and *ubiquity* as defined by Augoyard and Torgue. Here I will articulate how the staging of these effects serve as methods that can be employed not only in narrative audiovisual contexts but also in site-responsive art, in which it serves to direct the listening of audiences, facilitating encounters with existing sonic ecologies.

Sound design in film

Michel Chion's theories are especially influential where the role of sound design in audiovisual work is concerned, for film in particular. In the foreword to the English edition of *Audio-Vision,* celebrated sound designer Walter Murch declares of sound design for cinema: 'Whatever virtues sound brings to the film are largely perceived and appreciated by the audience in *visual* terms' (Chion 1994: viii, emphasis in the original). Chion, however, sets out *audio-vision* as he understands it, addressing it as a specific mode of perceptual reception that is necessarily multisensory (xxv). He then goes about constructing a language with which to understand the role of sound, of sound design, in audio-vision.

Chion proposes a number of key principles and sonic effects for thinking and practicing audio-vision, including *added value, synchresis,* and *extension* (Chion 1994: 5, 87, and 129). Added value, for Chion, is a concept that refers to the 'expressive and informative' detail that sound confers upon an image, producing an effect seemingly natural to what is seen (5). Synchresis denotes the perception of an immediate relationship between what an audience hears and sees, an audiovisual effect in which added value is particularly significant, and which is very commonly employed as a deliberate sound design method (5). Extension, meanwhile, refers to the effect created when sound is employed to broaden the concrete space of the film beyond the borders of the frame (87). Chion also points out that sound in cinema is predominantly *vococentric:* it privileges the voice over other sounds (5).

Murch's work in the opening sequence of director Francis Ford Coppola's film *Apocalypse Now*, and its subsequent versions *Apocalypse Now Redux* (2001) and *Apocalypse Now Final Cut* (2019) is an excellent example of Chion's sound design principles. Murch contributed, and is credited for, the 'sound montage and design' as well as serving as one of the editors, purportedly the first instance of a credit for sound design in cinema. Interviewed by Michael Ondaatje in *The Conversations*, Murch explains of his role in the opening sequence: 'That became my job: to create an abstract, dynamic, visually arresting scene' (Ondaatje 2002: 61). The film opens with a slow-motion shot of a jungle in flames, helicopters circling, fading into a close up of Martin Sheen's Captain Willard in a hotel room in Saigon, lying on a bed looking up at a ceiling fan. A synthesized version of the helicopters' chopping rhythms serves as their sound, and in turn becomes the sound of the fan overhead before merging into a recording of a helicopter seemingly flying past outside of the room and out of shot as Willard gets out of bed and walks to the window. This is *delocalization*, as Augoyard and Torgue term it, in which the ubiquity effect they define, and I will discuss later, through which the sound seems to come from everywhere and nowhere at the same time, leads to moments in which 'the listener knows exactly where the sound seems to come from, while at the same time being conscious that it is an illusion' (Augoyard and Torgue 2006: 38). Demonstrating Chion's principles, effects, and methods clearly, the synchresis, or conjoining, of the synthesizer sounds with first the helicopters and then the fan, give added value in that they connect what Willard can hear and see in the room – the fan – with his experiences in the Vietnam War, during which the film is set, before explicitly extending the sonic environment when they morph into the sound of a helicopter outside. Chion claims that the point of audition in film, like the point of view, can have both a spatial and a subjective designation, that is where the audience is hearing from, which can be difficult to pinpoint given sound's atmospheric characteristics, and from which character's perspective they are hearing (Chion 1994: 89–90). In *Apocalypse Now* the audience hears what Willard hears, almost exclusively, which thus directs their listening.

Willard's narration, which anchors the film's story line, then begins, supporting Chion's claim of vococentrism despite the greater role of sound design to the film. 'Saigon, shit, I'm still only in Saigon.' The narration is a kind of internal monologue, as the phrasing of even its opening line indicates, a monologue that the audience listens in to, hearing from Willard's perspective as we do in the rest of the sound design of the film. The sound of Sheen's voice in the narration indicates this too, with its intimate quality achieved via deliberate very close microphone placement in the studio and mixing of the vocal track to all three speakers behind the screen, where other dialogue is placed only in the centre speaker (Ondaatje 2002: 64–65). Interestingly, in 'Apocalypse then and now', based on sections from the then forthcoming *The Conversations*, when asked about whether the original script for the film called for narration Murch is attributed as commenting 'Willard had an internal voice' (44), while in the book Murch comments 'Willard spoke to us' (63).

Both are accurate where the film is concerned, and this is significant because we, as an audience, listen *with* Willard rather than just *to* him.

The scene also features the use of the sound of crickets to depict the psychological presence of the jungle for Willard, leading to a chaotic climax in which The Doors' song 'The End', that had scored the jungle-shot opening, re-emerges and pushes the action to a traumatic peak. The sound of whistles being blown authoritatively outside among the bustle of the city – presumably directing traffic – slips away and is replaced by the initially similar crickets. Sound is used here and throughout the film to guide the listener, directing their gaze but also their ears, one sound leading into another. The sound of crickets is gradually filled out, creating what Chion calls an extension of the sonic environment of the film, extending the concrete space of the film well beyond the borders of the screen and Willard's visual field, but also inward, into the character's experience and memory (Chion 1994: 87). Chion credits Murch with speaking to his use of a version of this extension effect in the film as well, though not using the same term (87). Indeed, Murch provides a thorough description of the sequence in an interview with Frank Paine, 'Sound Mixing and *Apocalypse Now*: An Interview with Walter Murch' (Weis and Belton 1985: 356–360). At the same time, as Murch has himself pointed out, if we question the location of the crickets in the logic of the film 'They're nowhere. They're in Willard's head' (Ondaatje 2002: 245). The crickets we hear, and the jungle they precipitate, are, therefore, sonic effects, as Augoyard and Torgue define them, effects created by way of specific sound design methods.

Murch endorses the importance of manoeuvres such as those articulated in Chion's and his own work, arguing that it is necessary to '*stretch* the relationship of sound to image wherever possible' because 'the danger of present-day cinema is that it can crush its subjects by its very ability to represent them' (Chion 1994: xix, emphasis in the original). Film, Murch claims, lacks the 'sensory incompleteness' of other forms, such as music, radio, and literature (xix–xx).

I find that the issue Murch identifies persists in sonic art, regardless of its perceived relative incompleteness. Sound reproduction – from recording to sound design – is used to *represent* sonic experiences, to allow us to hear them in ways that we couldn't otherwise, but that can also supersede them, creating reproductions that usurp their sources in our lives. Sound design – the deliberate use of sound – can, however, be used to craft audio-visual experiences for listeners that go beyond representation, as we can hear, and see, in the work of Murch and his collaborators. Similarly, sound design can be used in sonic art to stretch the relationship between sounds. In this way, sound design methods can be deployed in site-responsive sonic art to create environmental encounters.

Sound design in site-responsive sonic art

Ubiquitous as sound design is in contemporary life, sonic art – be it experimental or avant-garde music, radio art, or sound art – is not generally something that comes to mind when sound design is raised. A prevailing sense of sound design as functional and art, especially sonic art, as anything but, divides the two conceptually. It stands to reason, however, that design is ideally artful, and much successful art is well designed, and further that a broad approach to sound design, such as Murray's (2019: 11) 'deliberate use of sound', leaves room for its use in the arts more widely. Moreover, while sound design and art seem separate fields there exists a long and rich history of contact between them.

Italian futurist Luigi Russolo, for instance, known for his *intonarumori* (noise-generating instruments) and for his accompanying manifesto written in the early twentieth century, 'The Art of Noises' ([1913] 2009), is suggested as the first noise artist. His instruments, though, were built to replicate modern sounds, the noises of industrial Europe, in keeping with the practice we now know as sound design. Also, Suzanne Ciani, a very early adopter of the Buchla synthesizer and a celebrated electronic musician, is at the same time the creator of the Coca-Cola 'pop 'n pour' sound logo, for which she used her instrument to produce a now iconic bubbling effect, a story told in Michelle Macklem's *Lost Notes* documentary podcast episode *Sonic Sculptor: Suzanne Ciani* (Macklem 2019). 'We didn't know the words "sound designer" back then,' Ciani reflected in an interview with Hannah Nemer (2017) for the Coca-Cola website. What is more, Murch details hearing the *musique concrète* of Pierre Schaeffer and Pierre Henry around age ten and being profoundly influenced by it, as well as attending some of experimental composer John Cage's concerts with his father (Ondaatje 2002: 7–9).

I have myself produced a series of works, across a number of media, which employ sound design methods to produce site-responsive works that direct listening towards surrounding sonic ecologies – *Murmur* (2016), *The Flood* (2017), and *Malfeasance* (2017). This approach has become intrinsic to my practice following earlier work, sound designing a short sound work for theatre, *Too* (2012), written by Cynthia Troup and performed by Carolyn Connors, for which I used a ten-channel speaker system to gradually shift the sound of a detuned radio onstage into a torrent of noise from all corners of the theatre, and giving site-responsive performances, such as *Merri Creek Drain* (2015), in which I performed from within a storm-water drain to an audience across the creek. Where Chion's theories are key to Murch's cinema, Augoyard and Torgue's are to analysing my work, though the interplay between all is significant. Eschewing the vococentrism Chion identifies, the works all employ pink noise to produce subtly shifting bodies of sound that merge with the sound of the environments in which they are presented. These environments, ranging from an urban art gallery to a remote riverside and the variable listening environments chosen by record listeners, are thus brought forward in the mix, so to speak, drawn into the foreground for audiences.

Pink noise is the name given to noise containing every frequency theoretically audible to humans, from 20 hertz to 20 kilohertz, with an inverse relationship between frequency and volume. It is also known as $1/f$ noise, where f is a given frequency, as this fraction expresses the power of each frequency in the noise. Bass frequencies are loudest, with volume decreasing with increasing frequency. It thus differs from white noise, the sound of, for example, an old analogue television that hasn't been tuned, or is set to a station that is no longer broadcasting, snow covering the screen. White noise is a sharp sound, notable for its high-frequency content. It contains every frequency from 20 hertz to 20 kilohertz at equal power but we hear the high frequencies as louder, and thus the noise as sharper, because humans don't perceive all sounds equally. Pink noise, with its inverse relationship between frequency and volume, sounds, at least to us humans, more balanced in its mix of high and low frequencies. While white noise is often called static and sounds harsh and electronic, pink noise sounds like the ocean, rain, or the roar of a city. It sounds like everything at once and nothing in particular.

Murmur was a stereo procedural audio installation presented at the RMIT Spare Room gallery in inner city Melbourne, Australia, from 15 July to 18 August 2016. Custom-coded in SuperCollider, which was then run on a Raspberry Pi computer, the piece produced a delicate but continuous sound through subtractive synthesis. It generated a pink noise signal and used band-pass filters and envelopes to create layers of sound that moved only very slowly and were at times barely perceptible, both due to the low volume and because I designed the overall morphology of the sound to mimic that of the noise spilling into the gallery space from the urban environment within which it is situated, from the crowd at the opening to traffic and rain outside.

The Flood, in contrast, was a durational procedural audiovisual installation developed and presented at Bogong Centre for Sound Culture (BCSC), in Bogong Village in the high country of Victoria, Australia. Included in the BCSC's PHANTASMAGORIA festival and exhibition for a month around Easter 2017, it was staged at an outbuilding next to the East Kiewa River that runs below the village. Bogong Village was created to support the Kiewa Hydroelectric Scheme. It is surrounded by a national park and has no permanent residents, but is home to a hydro-electric power plant. The plant is built into the ground beneath the village, which overlooks Lake Guy, created by the damming of the river as part of the scheme and into which the river runs. Developed during a residency at BCSC in late 2016, the work stages a flooding of the outbuilding with sound and light. I attached transducers all over the corrugated metal building, so it would effectively function as a speaker, but also filter and modulate the frequencies of the generated noise, and I used projectors to rear-project into the windows, three in all. This audiovisual system was driven by a Raspberry Pi running a program custom-coded in the software Processing, which generated synchretically connected abstracted images of an unstable water surface and pink noise, which would rise and fall up and down the windows and in volume unpredictably each night from sunset until dawn.

Malfeasance, meanwhile, is a seven-inch record released on my Avantwhatever label in 2017, available only by gift or direct request, which is composed of the fluctuating pink noise of a found data file. The sound fades in gradually and is of generally low volume such that it blends with its surrounds when played. The release was produced using a hand-cut lathe method, meaning there are subtle differences between each copy. The records will wear faster than pressed records, adding further layers of manipulation to the noise.

All these works employ what Augoyard and Torgue term *ubiquity,* a term distinct from Chion's use of ubiquity. In their book *Sonic Experience: A Guide to Everyday Sounds*, they describe the effect as one that 'expresses the difficulty or impossibility of locating a sound source' (Augoyard and Torgue 2006: 130). In the most common version of the effect, they detail, the sound seems to come from both everywhere and nowhere simultaneously.

Less common, they continue, is a variant in which sound seems to come from both a particular source and many sources (130). It is the more frequent type of the effect that can be observed in *Murmur*, *The Flood*, and *Malfeasance*, though arguably they can produce moments of the latter too. In each it is possible to sense the work itself as a source of sound, but at the same time the sound is heard as coming from the surrounding environment; it seems to come from everywhere and nowhere in particular, directing listening outward.

The use of broad band noise to create a sense of ubiquity is a key method in my work. Interested not to compose in a traditional sense so much as to create spaces for listening in which the sonic events of the world may be encountered, sculpting pink noise offers me a way to mimic and blur the boundaries between existing sounds. This involves using filters and envelopes to shape the sound, a method of composition more focused on removing material than adding it. This is a process that I deliberately seek to control only lightly too, be it via software-based stochastic processes such as I employed in the coding of envelopes for *Murmur*, the use of the materials of a site to filter sound as occurred with my use of transducers for *The Flood*, or the indeterminacy of working with found sound and especially degradable media as I did for *Malfeasance*.

Augoyard and Torgue note that ubiquity is well served by sound, given it cannot be seen, but specify 'diffused, unstable, omnidirectional sound' as presenting an intrinsic tendency towards the effect (Augoyard and Torgue 2006: 130–131). Spatialized and filtered noise, such as that in my works, fits the bill. Although Augoyard and Torgue point out that while many background sounds are ubiquitous in a particular situation, including the noise of a city, listeners tune them out and

> for the ubiquity effect to occur, we must consciously look for the source location of the sound, and fail, at least for a moment, to identify it [...]. The listener is in search of information. The ubiquity effect is based on the paradoxical perception of a sound that we cannot locate, but which we know is actually localized.
>
> (Augoyard and Torgue 2006: 131)

Murmur, *The Flood*, and *Malfeasance* all produce sound that is known to be localized – emerging from speakers, an outbuilding, and records respectively – but which at first cannot be located, and which even when located blurs with the surrounding sounds to delocalize listening once more. The sound they produce is diffuse, unstable, and omnidirectional due to the use of filtered noise to produce a constantly shifting mix of frequencies, as well as multiple speakers or transducers, especially in the case of *The Flood* and its use of the materials of the building itself to propagate the sound.

Farnell details how listeners localize sound in his book *Designing Sound* (2010: 79). He cites three 'general rules': high-frequency sounds with short attacks are easier to localize than low sounds with longer attacks; it is easier to localize sounds outside than in small rooms where there are many reflections; and listeners locate sound more successfully if they are able to move their heads around to get different 'takes' on the sound, principles echoed in the work of Augoyard and Torgue. All three of the works I discuss here work with continuously shifting pink noise, with layers of filtering fading in and out with relatively long attack times, making localizing the sound difficult in keeping with Farnell's first rule.

Murmur also used a very small room, making the most of the delocalizing effect of the reflections produced by the close walls, as outlined in Farnell's second rule. In all three cases the listener is free to move, indeed not only their head but by moving around the space, which according to Farnell's third rule will help listeners to locate the sound but at the same time exposes them to the environmental sounds with which the works blur. The works are thus designed to disrupt listeners in their attempts to localize sound.

This, in turn, contributes to the works achieving delocalization. In the case of *Murmur*, the visual cue of the wall-mounted speakers suggests a source that is then contradicted by a sense that the sound is coming from outside, through the entryway to the space. With *The Flood*, the window projections suggest the outbuilding as a source but the multiple transducers disperse the sound so that it is hard to pinpoint if it indeed is the source, diffusing the sound into that of the nearby river. *Malfeasance* stages a source with the record, player, and sound system, but the merging of the sounds produced with environmental noise may lead listeners to question that assumed source.

Augoyard and Torgue identify urban and architectural environments as obvious locations for the experience of ubiquity effects due to their multiple surfaces and spaces contributing to the delocalization of sound sources (Augoyard and Torgue 2006: 131–132). This was evident in the urban environment surrounding the RMIT Spare Room gallery in Melbourne, where *Murmur* was presented and also, to an extent, in that of Bogong Village where *The Flood* was installed, as, though it is a remote location and could not be described as urban, it is nonetheless a heavily built environment due to the power station. *Malfeasance*, however, clearly does not control its environment, though its release on a record certainly guides it. In each case, the works extend the listening space of audiences through a deliberate delocalization of sound and accompanying sense of ubiquity, which is supported by the material realities of their locations.

The delocalization of sound sources and accompanying sense of ubiquity I achieve in these works further contributes to an extension of listeners' sonic fields. Engaging peripheral listening that might otherwise be unconscious, the works then draw links between sounds in a given location, constructing a sense of interconnection and ecology. That simultaneously brings to the foreground sounds listeners might otherwise tune out, and develops an awareness of the sounds around and beyond those. This is the same method of extension as that theorized by Chion and employed by Murch. However, I use it not to extend narrative space but to shift the listeners' attention, engaging them with the expanded auditory space they inhabit.

Augoyard and Torgue distinguish between two listener positions in which the ubiquity effect might be experienced: from within a given sonic environment and from outside of it. They argue that in the first situation the experience of ubiquity is linked to a 'multiplicity', real or imagined, of surrounding sound sources, while in the second situation it is the distance from sources and propagation effects that cause sounds to become delocalized (Augoyard and Torgue 2006: 132–133). I propose here that my works demonstrate a third possibility, that of the effect positioning a listener in an extended sonic environment, as Chion (1994: xix) proposes is possible in film, connecting their immediate surrounds, be they a small room or a riverside, with a sense of distance beyond the sensible field of vision and audition. These works stretch the relationship between the sound produced and its assumed sources, in line with Murch's call for creative uses of sound design methods. Rather than communicating what is going on by giving emotional cues, sound is used to complicate and highlight the listening experience. The works do not represent sonic environments but create sonic effects by way of sound design methods that refer listeners to the surrounding environment. They engage listeners directly with the sounds around them and their interrelationships.

Sound design is increasingly common, in contemporary life and art, and sound design methods and effects such as synchresis, delocalization, extension, and ubiquity can be employed in site-responsive sonic art. This is demonstrated in my own work with the use of pink noise, subtractive synthesis, and indeterminate processes. As Augoyard and Torgue note, 'ubiquity, by its very definition, supposes active listening' and so 'if there is a "sound object," it cannot be immutably perceived by a passive receptor organ; it is constructed and "realized" by an active ear that creates it as such' (Augoyard and Torgue 2006: 135). Addressing audiences as active listeners is key to my work and use of sound design, as it is to that of sound designers such as Murch in film. Delocalization, extension, and ubiquity, understood as both sonic effects and methods, can be used to induce an awareness of this active listening, encouraging listeners in their sonic explorations. As Murch claims 'listening to interestingly arranged sounds makes you hear differently' (Jarrett 2000: 4). Beyond representation, in site-responsive sonic art this supports a directing of listening towards listeners' sonic environments, engaging them with the multiplicity of complex sonic ecologies in which we live.

References

Altman, R. (1992). *Sound Theory, Sound Practice*. New York: Routledge.
Apocalypse Now (1979).[Film] Dir. F. F. Coppola. USIUA: United Artists.
Augoyard, J.-F. and H. Torgue (eds) (2006). *Sonic Experience: A Guide to Everyday Sound*. Trans. A. McCartney and D. Paquette. Montreal: McGill-Queen's University Press.
Chion, M. (1994). *Audio-vision: Sound on Screen*. Trans. C. Gorbman. New York: Columbia University Press.
Farnell, A. (2010). *Designing Sound*. Cambridge, MA: MIT Press.
Jarrett, M. (2000). 'Sound Doctrine: An Interview with Walter Murch'. *Film Quarterly* 53 (3): 2–11.
Macklem, M. (2019). 'Sonic Sculptor: Suzanne Ciani'. *Lost Notes* [Podcast]. Available online: https://www.kcrw.com/culture/shows/lost-notes/sonic-sculptor-suzanne-ciani (accessed 22 August 2019).
Murray, L. (2019). *Sound Design Theory and Practice: Working With Sound*. New York: Routledge.
Nemer, H. (2017). '"Pop 'n Pour": This Electronic Music Pioneer Created the Sound of Coke's Beloved Bubbles'. Coca-Cola Company. Available online: https://www.coca-colacompany.com/stories/meet-suzanne-ciani-the-legendary-creator-of-cokes-pop-n-pour (accessed 22 August 2019).

Ondaatje, M. (2001). '*Apocalypse Then and Now* (Interview)'. *Film Comment* 37 (3): 43–47.
Ondaatje, M. (2002). *The Conversations: Walter Murch and the Art of Editing Film*. London: Bloomsbury.
Russolo, L. ([1913] 2009). 'The Art of Noises'. In L. Rainey, C. Poggi, and L. Wittman (eds), *Futurism: An Anthology*, 133–139. New Haven, CT: Yale University Press.
Schulze, H. (2019). *Sound Works: A Cultural Theory of Sound Design*. New York: Bloomsbury.
Sonnenschein, D. (2001). *Sound Design: The Expressive Power of Music, Voice and Sound Effects in Cinema*. Seattle, WA: Michael Wiese Productions.
Weis, E. and J. Belton (1985). *Film Sound: Theory and Practice*. New York: Columbia University Press.

25

Sound, Space, and Pneumatic Valves: Using Pneumatic Valves as Sound Sources to Create Spatial Environments

Edwin van der Heide

One day a very simple idea came up in my mind: to use pneumatic valves as sources for sound production. I imagined that it would be possible to use the valves as a kind of loudspeaker. A loudspeaker moves its membrane forwards and backwards in relation to its rest position. While the movement of the membrane follows the waveform of the sound signal to be produced, local air pressure changes are being created that propagate as sound waves through the air. I envisioned that it would be possible to create a comparable effect by releasing pressurized air through electrically controlled pneumatic valves. By controlling the valve, the amount of air that passes through is being varied. When an air compressor is used in combination with a pressure tank a reservoir of compressed air is created with a pressure higher than the environmental pressure. This compressed air is then supplied to the inlet of a pneumatic valve of which the outlet is open, meaning that the compressed air is released into the environmental air. The idea is that this would result in a pressure increase similar to the membrane of the speaker moving forward.

An idea and its consequences or realization are two different things. My experiments and tests to use pneumatic valves for sound production have resulted in two different sound installations that are both presented outdoor. The installation *Pneumatic Sound Field* uses a large horizontal grid of pneumatic valves above an audience; the installation *Schwingungen – Schwebungen* uses pneumatic valves to drive acoustic horns that are placed in a field that the audience enters.

The valve versus a loudspeaker

In order to create a situation that is somehow comparable to a speaker, the valve needs to be able to open and close very fast, just as the membrane of the speaker can move very fast. The opening and closing of the valve results in a changing air pressure at the outlet of the valve that corresponds to the waveform of the to-be-produced sound. With this set-up there are nevertheless a number of differences to how a loudspeaker behaves. First of all, we can only create pressure increases and no pressure decreases. This is not necessarily a problem but it means that there is a continuous 'offset' resulting in a part consisting of continuous wind instead of sound pressure waves propagating through the air. Secondly, there is no good coupling between the pressure released by the valve and the pressure in the air surrounding the outlet of the valve. This means that most of the pressure coming out of the valve results (again) in wind and doesn't translate to, and result in, sound pressure waves in the air. It is possible to create a much better transfer of the energy by placing a horn in front of the valve. The shape of the horn makes it possible to connect a big surface of air to the small outlet of the valve. This way most of the present energy is transferred into sound pressure waves.

The human voice is also pneumatic

Originally, I didn't think of this but the use of a pressurized tank in combination with a pneumatic valve has similarities to the human voice. Our lungs provide air pressure to our vocal cords that open and close with the frequency of the pitch that we produce. This is similar to a valve opening and closing at that same frequency. After the vocal cords follow the throat, nose, and mouth to shape and filter the sound and transfer the energy to the surrounding air. The wind 'offset' that is the result of the valves only being able to increase the surrounding pressure is equally present in the principle of the human voice. While we sing or speak we also release air.

The ideal of a loudspeaker is to be a universal and thereby generic sound source. In *Between Air and Electricity*, Cathy van Eck, a composer, sound artist, and researcher in the arts, describes that loudspeakers (and microphones), when used for reproduction, 'should act like transparent devices, adding no sound of their own' (Van Eck 2017: 38).[1] The human voice is the opposite of this: a voice is unique to a certain person. This can be attributed to the personal properties and qualities of the involved organs and also to the particular way they are controlled. We could say that the properties and qualities of the involved organs correspond to the qualities of an acoustic instrument and that how they are controlled is comparable to the personal (style and) skills of a musician.

Two types of valves

While exploring the possibilities of pneumatic sound production I came across two different types of valves: discrete ones that have just two states – open and closed – and valves whereby the opening can be proportionally controlled. The binary nature of the first would make it comparable to a speaker where the membrane only has two possible positions. We could call this a one-bit speaker. The valve is only able to produce pulses with a variable length including square wave-like tones with a variable pulse width. On the other hand, the proportional valve is able to continuously change its opening and thereby vary the amount of pressure that is being released. This behaviour is more similar to a speaker that can continuously change the position of the membrane in time without being limited to specific positions. Where the membrane of a speaker has a rest position and can move forwards and backwards the proportional valve cannot create pressure decreases and will therefore have to be half-open as its rest position. From there on the opening increases and decreases following the amplitude of the to-be-produced signal. As mentioned before this results in a permanent flow of wind.

The valve as musical instrument?

We can make a distinction between musical instruments that have unique and specific features and loudspeakers that are supposed to have universal and generic properties. If we would have to characterize the discrete and proportional valves in this context it would be logical to say that the discrete valve has a very specific behaviour and the proportional valve can be used in a (more) generic and universal way. A difference between musical instruments and loudspeakers, that we have not touched upon yet, is that musical instruments not only (re)produce sound but also generate sound. Speakers, however, are supposed to produce sound that is not generated by themselves. The electrical signal separates the sound generation and the sound production from each other. By doing this, there is no inherent interaction between the two. (However, this doesn't mean that no interaction can be created; think, for example, of feedback.) The generated sound is 'fed' to the speaker(s) in a one-directional way. This observation allowed me to play a bit with how I could interpret the role of the discrete valve. Yes, discrete valves have a very specific (and limited) behaviour that is thereby similar to musical instruments but I could still argue that they act as speakers because we use them to (re)produce sound and not to generate sound themselves. Playing a rich sound signal with a complex waveform on the discrete valves would result in a heavily distorted production of it. However, since there is no interaction between the generation of the signal and the sound production, I would argue that the valve is not fully playing the role of a musical instrument. I could even take another step in my argumentation and say that the role of the two-state valve is comparable to the role of a loudspeaker when I intentionally send a signal that already has a discrete two-state nature itself. In this case the valve is fully capable of producing the generated sound and acts like a 'perfect' loudspeaker. Perfect, really? Well, in the installation *Pneumatic Sound Field* it is not so perfect because the air travelling through the valves meets a lot of small and sharp corners that create friction, which results in a form of hissing. It is possible to reduce this hissing with special pneumatic dampers, but for this installation I decided to keep it. I would say that since the hissing is generated by the valves themselves I cannot just claim that the valves only produce sound. They play both roles, and following this reasoning they act both like speakers and musical instruments.

Acoustic and visual localization

One aspect that I really like about the discrete two-state valves that I have chosen to use for *Pneumatic Sound Field* is that they are very small (37 millimetres × 29 millimetres × 12 millimetres) (see Figure 25.1). They produce a sound that has a certain strength and directness to it, resulting in an impact that one wouldn't expect from such a tiny object. We have a natural tendency to trace where a sound comes from, independently whether it comes from an acoustic source (such as the human voice, an acoustic instrument, or a car driving by) or from speakers. Speakers are usually relatively large and thereby visually really present. Although there are many moments in the composition for *Pneumatic Sound Field* where it is easy to localize the sound as coming from the positions of the valves, the audience often asks where the sound comes from since they cannot visually understand that the sound is produced by such a small object. I believe that not knowing where the sound comes from, or having an auditory experience that has a strong (spatial) presence that doesn't relate to what we see, emphasizes our perceptual focus on listening, the spatial perception of sound, and the auditory perception of space. In other words: when we know the sound comes from speakers with fixed positions, we will focus less on the spatial information of the sound.

Figure 25.1 The pneumatic valve used in *Pneumatic Sound Field*. Photograph courtesy of Studio Edwin van der Heide.

The speaker as spatial source versus an acoustic window

Speakers can be used in various ways, for example to position the sound (and represent the produced sound) in space. In this case the speaker is not only producing sound but the sound is also intended to be perceived as coming from the position of the speaker. The speaker forms the spatial source of the produced sound. A well-known example of such speaker use is the guitar amplifier. Speakers however, can also be used to mediate a spatial experience. Think, for example, about listening to an orchestra recording on a stereo speaker set-up. In this case the speakers do not act as the spatial positions where the sound is intended to be positioned and spatially originates from; the speakers are used to mediate the spatial experience of the orchestra playing, including the acoustics of the hall it is playing in. In other words, the acoustic behaviour of the recording (or simulated) space is part of the mediation. In this case the sounds are not intended to be perceived as coming from the speaker but the speakers function as a window to another space. This other space can be a completely different one than the listening space (i.e. listening to an outdoor space or a concert hall in a living room, or listening to an indoor space while being outdoors), but it can also be similar to, or even be, the listening space itself. We could say that there are two extremes regarding the function of a speaker: situations in which the speaker plays a sound in the space and functions as a spatial sound source similar to an acoustic sound source, and situations in which the speaker plays exactly what the listener is intended to hear. The last situation takes place in its extreme form when listening to headphones.

Pneumatic Sound Field

The installation *Pneumatic Sound Field* consists of a rectangular grid of seven lines with six valves each, at a height of 4.5 metres above the audience. The spacing between the valves on a line is 1.70 metres while the distance between the individual lines is 3.33 metres. Together this results in structure with a width of 10 metres (the lines are a little longer than the space needed for the six valves) and a depth of 20 metres (see Figure 25.2). Each of the valves is controlled individually and forms its own channel. Together the valves form a 42-channel installation. Every sound or impulse can originate from any position inside or outside of the grid and propagate through the plane of valves by using individual delay times for each of the valves. The idea behind this set-up and approach is the possibility to play, in a structural way, with the arrival times of the played sounds in relation to the audience members' (changing) positions, and to make this part of the compositional approach. These differences in arrival time result from a software-based system in which each of the valves has its own delay time. However, the perceived differences for the audience also depend upon their position in relation to the valves.

Figure 25.2 *Pneumatic Sound Field* during DEAF07, Museum Boijmans van Beuningen, Rotterdam, 2007. Photograph courtesy of Studio Edwin van der Heide.

Sound localization and interaural time difference

One of the reasons for our ability to hear and localize sound in the space around us is that small differences in time occur between a sound arriving at our left ear and that same sound arriving at our right ear, depending on whether the sound source's position is to the left, middle, or right of us. These differences are called *interaural time differences* (Blauert 1997: 204–206). A sound originating from a position to the right of our head will arrive at our right ear before it arrives at our left ear because it has to travel a little further to reach the left ear (the speed of sound in air is approximately 340 metres per second). In a stereo loudspeaker set-up, small time differences between playing a sound from the left and from the right speaker can be used to spatialize that sound and create a phantom source location in between the two speakers (De Boer 1940: 50–51; Franssen 1964: 32). This principle is called *time delay stereo*. Using time differences between the left and the right speaker up to 3 milliseconds can make one experience the phantom source originating from the left speaker, anywhere in between the two speakers or originating from the right speaker (Franssen 1964: 32).

The precedence effect

The *precedence effect* (also called 'Haas effect' or 'law of the first wavefront') is the effect that, although a sound produced in a reverberant environment reaches us via many different paths, each with its own travel time, we will localize the sound at the origin of the first wavefront that reaches us (Ruth et al. 1999: 1633). Most often this is the direct sound (the sound that travels from the source directly to the listener) since the reflections of the sound take a longer time to reach the listener. Not only do we localize the sound at the origin of the first wavefront reaching us, the reflections seem to fuse with the original sound. In other words, we usually don't hear the reflections as separate sounds but as part of the sound. However, when the delay time between two instances of a sound exceeds 30–50 milliseconds, we will start to perceive them as separate from another in the form of an echo (Blauert 1997: 224–225).

Wave field synthesis

Wave field synthesis (WFS) is another approach to sound spatialization using loudspeakers. It aims to 'synthesize' and (re)create the spatial shape of the wavefront in the air. By using lines of loudspeakers with small distances between them it is possible to synthesize (and thereby recreate) the spatial shape of the waves of sound sources both behind and in front of the line of speakers (Berkhout 1988: 979–981). Where time delay stereo aims to create differences in arrival time between the two ears, WFS uses slight time differences between the many speakers, not to directly address our ears but to synthesize and thereby physically (re)create the spatial shape of the waves in the space. This principle only works when the spacing between the speakers is smaller than one quarter of the wavelength of the frequencies present in the to-be-produced wavefront. A larger spacing results in what is called 'spatial aliasing', meaning that the synthesized wavefront has a different shape than intended.

Back to *Pneumatic Sound Field*

The grid of *Pneumatic Sound Field* is not intended to function as a generic spatialization system but as a field in which spatial phenomena take place that are based on differences in arrival time of impulses and tones. The perception of these phenomena depends on the position of the listeners since they are closer to certain valves and further away from others and thereby influencing the differences in arrival time.

In *Pneumatic Sound Field* impulses are generated that originate from a certain position inside or outside of the grid of valves and travel with a certain (variable) propagation speed through the surface of the grid. A way to imagine the propagation is looking at a stone falling in the water and watching the circular waves expand around the origin of the stone hitting the water surface. The propagation speed of the impulses can correspond with the speed of sound in air, but it can also be faster or slower. When the speed is infinitely fast, all the valves receive exactly the same signal and open and close simultaneously. The reason I just referred to wave field synthesis is that it uses the same metaphor of the stone falling into the water to simulate sound sources at positions behind and in front of the line of speakers. The set-up of the valves in *Pneumatic Sound Field* is not suitable for WFS because the distance between the individual valves is too large and spatial aliasing will occur. Besides that WFS requires the use of the speed of sound in the air for the exact calculation of the delay times for the speakers and *Pneumatic Sound Field* applies a variable propagation speed. I could say that *Pneumatic Sound Field* is intentionally misusing the physical foundation of WFS and exploiting the obtained artefacts as sonic and spatial material.

Let's take a look at some different scenarios:

1. An infinitely high propagation speed (all valves play the same signal and at the same moment in time)

When the audience stands in the middle of one of the long ends of the grid and the impulses originate from the middle of the other side of the grid I can play with their perception when I change the propagation speed (the simulated speed of sound). As mentioned before, all valves will receive exactly the same signal when the propagation speed is infinitely high. This means that the sound that comes from the valve that is the closest to an audience member will arrive first, and the further away a valve is, the later it will arrive. What became clear from the above-mentioned precedence effect is that we will localize the sound source at the origin of the first wavefront that arrives at us, meaning that we will localize the sound at the valve that is the closest to us. From the time delay stereo effect one could learn that when the sound of one valve arrives to one of our ears a tiny bit earlier (or later) than the sound of another valve to the other ear that we can perceive a phantom sound source that has a position in between these two valves. When all the valves open and close synchronously, the largest time difference between the closest and the furthest valve is the travel time between the two corners of the grid that are diagonally opposed to each other. The distance between these valves is (8.5 metres × 20 metres = 21.73 metres), and with a speed of sound of 340 metres per second this takes 0.064 seconds. As mentioned, as part of the precedence effect we perceive two identical sounds as separate from each other when they are more than 50 milliseconds apart. In this example this would never happen because 64 milliseconds is the maximum possible time between the first and the last impulse; the impulses from all the valves arrive in between this interval, meaning that perceptually they all merge together. Wherever the audience is or moves, they will localize the sound at the valve that is the closest to them. This means that one could say that the position of the sound source keeps on following them while they move under the grid. Since all the valves play the same sound I could argue that the sounds from the valves that arrive after the sound of the valve that arrives to the audience first can be seen, and therefore perceived, as reflections of the sound that arrives at us first. Following this reasoning, combinations of the impulses arriving to us at different times could be interpreted as different acoustic effects.

2. A propagation speed of the impulses in the grid that is identical to the speed of sound in the air

If the audience stays in the same position as in the previous scenario and the impulses are played again from the side of the grid opposite to where they stand but now with a software-generated propagation speed that is identical to the speed of sound in the air, what will happen is that the impulses from the valves that are the closest will arrive at one's ears at the same time as the impulses from the valves on the other side of the grid. This means that a listener is exactly at the point where the perception of the position of the source turns over from the valve that is the closest to the valve where the impulse originates from. When the software-generated propagation speed is decreased a little bit more the sound of the valve from where the impulse originates will always arrive before the impulse is played by the valves close to the listener (including all the valves in between the origin and our position) and one will localize the sound at its origin (as opposed to the valve that is the closest). When listeners now change their position in the grid without changing the origin of the impulses that propagate through the grid the arrival times of the various impulses will change accordingly, but the impulse from the origin will nevertheless always arrive first. When one listens at the origin itself the impulses are travelling away from instead of towards oneself. This means that the arrival times at one's location is the sum of the actual speed of sound in the air and the software-generated propagation speed.

The two examples above show that one can play with how we spatially perceive impulses propagating through the grid and where the origin of these impulses is localized. In both examples, we will perceive an impulse propagating through the grid as a single event meaning that the individual impulses are associated with each other and not heard as individual events in time.

3   Slowing down the propagation speed

When the software-generated propagation speed is slowed down more, a point is reached where the impulses propagating through the grid are heard as individual events instead of a single event with a certain spatial quality. This happens when the time differences between the individual impulses arriving at one's ears are larger than 50 milliseconds. As a result, the propagation will be perceived as a rhythmical pattern in space since the individual impulses of the individual valves are perceived and localized as separate events. A spacing between the valves of 1.7 metres translates to a generated propagation speed of (1.7/0.05 =) 34 metres per second (or less) which corresponds to 1/10 of the speed of sound in the air.

The three examples above show that varying the software-generated propagation speed leads to distinctly different perceptions of the impulses propagating the grid. We are used to interpret what we hear in relation to the actual speed of sound in the air. When the propagation speed is changed in the software, a continuum is being created between aspects involving both the spatial and rhythmical perception of sound. Furthermore, what we hear partially depends on the listening position in relation to the generated pattern.

So far, I have been speaking about individual impulses that have an origin in- or outside the grid and that propagate with a certain speed through the grid of valves. The transition between perceiving two identical impulses that have a short delay time (less than 50 milliseconds) between each other as one sound event and with a longer delay time (more than 50 milliseconds) as separate events is not only valid in the context of the spatial perception of sound and sound localization, but also in the context of the perception of pitch versus the perception of rhythm. When an impulse is repeated at a frequency lower than 20 hertz (20 times per second), we perceive it as a rhythmically repeating pulse. When it is sped up to frequencies faster than 20 times per second, we will hear it as a tone with a certain pitch. In our perception the individual impulses merge into a single continuous tone. This transition happens at the same time interval as the hearing of individual impulses in space. A frequency of 20 hertz corresponds to a repetition cycle of 50 milliseconds.[2]

The installation *Pneumatic Sound Field* uses the above phenomena and transitions in perception not just to create compositional material but as a perceptual grammar that is explored in a compositional context. Impulses propagating the grid can perceptually transform into rhythmical spatial patterns and thereby create a continuum between the spatial perception and localization of quickly propagating impulses and the rhythmical perception of impulses that propagate slower. Furthermore, the composition exploits the perceptual difference between repeating impulses in time and perceived continuous tones.

Important questions were: Can this approach result in aesthetically meaningful and interesting material? How do we compose for such a system? What are the different sonic possibilities and how should they be classified? How do we create developments, oppositions, and surprises? Over the years multiple pieces have been composed for the system discovering different aspects of the installation without the need of changing or expanding the system itself.

Schwingungen – Schwebungen

The installations *Schwingungen – Schwebungen* and *Pneumatic Sound Field* share the use of pneumatic valves for their sound production. At the same time, the installations are opposites in what sounds the valves produce and how the spatial behaviour and experience of the sound is approached.

Horns in front of the valves

The valves used in *Pneumatic Sound Field* are very small and have a direct and relatively strong sonic impact that one would not directly associate with such a small object. The energy source for the produced sound is a 1,000-litre tank filled by a compressor that keeps the pressure in the tank at about 7 bar (100 psi). With a pressure regulator at the output of the tank the pressure supplied to the valves can be set. Both the tank and the compressor are placed away from the installation, invisible to the audience. Although the sonic impact of the valves is relatively high, the efficiency is low, and a lot of the air pressure that gets released does not result in sound pressure waves, it becomes movement of air (wind) instead of a temporary displacement (sound pressure waves). This is because all the energy released by the valve only touches a very small surface of air since the opening of the valve is very small. The mismatch between the two is called an impedance mismatch. The same problem would occur if we had a subwoofer with a tiny diameter but with a large amplitude. One way of getting a better coupling between the released pressure and the air surrounding the valve is to place a horn in front of the valve outlet. The use of the horn not only improves the efficiency of the translation into

Figure 25.3 *Schwingungen – Schwebungen*, bonn hoeren, Bonn, 2015. Photograph courtesy of Studio Edwin van der Heide.

sound waves: the bigger the output surface of the horn, the lower the frequency at which this efficiency is being achieved. This means that the bigger the horn, the better it is at producing low frequencies. When the valves in *Pneumatic Sound Field* produce a low frequency, the higher harmonics of the tone have more energy in them than the lower harmonics, and there is little energy present in the fundamental itself. For *Schwingungen – Schwebungen* I designed two types of horns: the first type is an exponential wooden horn with a length of 3 metres and an output surface of 3 metres wide by 1.2 metres high; the second type is a circular aluminium horn with a diameter of 1.25 metres and a height of 30 centimetres at the output (see Figure 25.3). The circular horn is driven by a valve placed at the centre of the horn and radiates the sound in a full horizontal circle. Where the large wooden horn can produce frequencies down to 50 hertz, the circular aluminium horn produces frequencies down to about 120 hertz.

The proportional valve

For *Schwingungen – Schwebungen* I chose to use (the aforementioned) proportional valves, which can have any position between being fully closed and fully open. This makes it possible to produce signals with a continuously changing amplitude, similar to what a loudspeaker can produce by moving its membrane. Also, where the hissing resulting from the friction of the air passing the discrete two-state valve in *Pneumatic Sound Field* forms an important part of the installation, I decided to use pneumatic dampers at the connection between the valve and the horn for *Schwingungen – Schwebungen*. These dampers don't fully dampen the hissing but reduce it substantially.

Pure tones

Where I was using sharp impulses and square waves containing a lot of overtones as sound material in *Pneumatic Sound Field*, I was particularly interested in using pure tones, also called sine waves (sine waves are called pure tones because they do not contain any overtones) for the installation *Schwingungen – Schwebungen*.[3] The choice to use the proportional valves in combination with the horns did not originate so much from the ambition to be able to produce 'any' sound by means of a generic pneumatic loudspeaker as much as to be able to produce pure tones by means of pneumatic sound generation.

Sound localization of pure tones

Our ability to localize sound is dependent on a combination of factors. We are good at localizing attacks and impulses especially when they have energy in the middle and higher range of the frequency spectrum. This counts for single impulses but also for tones that have many overtones like square and sawtooth waves. Pure tones on the other hand are (more) difficult to localize especially when they don't have an attack (sharp onset). The Dutch physicist Nico Franssen wrote in his book *Stereofonie* (1962) about a perceptual illusion in which a pure tone that starts in the left speaker and gets immediately panned to the right speaker was localized as continuously coming from the left speaker. This perceptual illusion was afterwards called the Franssen effect. The physicist William Hartmann and communication scientist Brad Rakerd further researched the experiment and showed, among other findings, that the Franssen effect only works in reverberant environments and fails in an anechoic environment (Hartmann and Rakerd 1989: 1366–1373). This is relevant because the acoustic properties of an outdoor environment are in between a reverberant and space and an anechoic situation.

Standing waves

One of the causes of the Franssen effect are the standing waves that build up because of the signal from the speaker(s) reflecting in the room. The direct sound and its reflections interfere with each other in the space, resulting in standing wave patterns with nodes and antinodes at specific positions throughout that space. These standing waves don't build up in an anechoic room nor in an outdoor space that has no nearby walls or objects, since there are no reflections and therefore no interferences. When the same tone is played from two speakers, each with its own position in space, standing waves occur because the signals from the two speakers each arrive at different times at each position in that space. When two identical tones coming from two individual speakers have a slightly different pitch the result is a beating between the tones. Furthermore, positions in space where the tones add up and positions where they cancel each other out are shifting. Both positions form (moving) lines in space that originate from a position in between the loudspeakers. The composer Alvin Lucier has used standing waves and beating between pure tone oscillators and/or instruments extensively. In an interview with Douglas Simon about his piece *Still and Moving Lines of Silence in Families of Hyperbolas* he describes the following situation:

> Each person in the audience perceives the waves moving by at a different time. One other fact I have to tell you is that the direction of the movement is toward the low speaker. For example, if you have one oscillator at 1,000 Hz and the other at 1,001 Hz, the hyperbolas [lines of cancellation] will move toward the 1,000 Hz speaker.
>
> (Lucier and Simon 1980: 133)

Schwingungen – Schwebungen is using the same standing wave principle to create spatial and moving interference patterns in an outdoor space. It does so by generating a pair of tones each corresponding to its own horn with its own position in space. The difference is that the sounds used in *Schwingungen – Schwebungen* are not only pure tones but do sometimes contain some overtones. If two identical tones with overtones would be slightly detuned in relation to each other, each of the overtones would beat with its own speed and there would not be a clear standing wave pattern. In order to overcome this, the two signals are not detuned but one signal is shifted in frequency linearly in relation to the other. This means that all the overtones have the same frequency shift and thereby the same beating speed.

Sound localization, standing waves, and wavelength

From the Franssen effect I learned that, in the case of pure continuous tones, the standing waves are in the way of localizing the sound at its source. This results in a perceptual effect that could be described as if the sound is floating in the air. The distance in space between the lines where the two sounds add up (antinode) and the lines where the same two sounds cancel each other out (node) depends on the wavelength of the produced frequency. Higher frequencies have a shorter wavelength resulting in smaller distances between the alternating lines of nodes and antinodes. At a frequency of about 1,000 hertz the distance between the lines corresponds approximately to the distance between our ears. That means that it can occur that we hear the tone at its loudest at one ear while it is at its softest at the other ear. When this pattern is moving in space because of a small frequency shift between the two signals (or when we move) it seems as if the sound is moving through or floating around our head. The standing waves occur in regions where the loudness of the two sources is relatively similar. At locations where one of the sources is much louder than the other, for example when we stand in front of one of the two horns playing the signals, we will localize the sound as coming from that horn instead of floating in space.

The composition

Schwingungen – Schwebungen has a generative nature and is created in real time according to certain rules. It is build-up of phrases with a variable duration that overlap each other. Each phrase consists of a tone with a certain pitch developing over time and a frequency shifted version of this tone where the amount of this shift also develops over time. The tone and its frequency shifted version are played on two separate horns thereby creating the moving interference patterns in space.

The installation *Schwingungen – Schwebungen* is set-up in a rectangular field. Two large wooden horns are being used, one on the far-left side of the field and another on the far-right side. In the field between these two horns the five circular aluminium horns are placed. As mentioned before the horns are always used in pairs. A phrase can either be played on the two wooden horns or on a combination of two cylindrical horns. A horn that is in use by one pair cannot simultaneously be used by another pair. This means that a maximum of three phrases can overlap: one phrase on the wooden horns and two phrases on the circular horns. Hence there is always at least one circular horn that is not playing. When a phrase is using the wooden horns, the resulting interferences occur in the field between the two horns, that is, where the circular horns are placed. The interferences of a phrase played on a pair of circular horns fills the space around these horns.

Discussion

Both *Pneumatic Sound Field* and *Schwingungen – Schwebungen* are installations that were not envisioned before. They came into being as a result of many experiments, thoughts, and reflections originating from my fascination for the idea that pneumatic valves could be used to produce sound. What intrigued me was that using air pressure to create air pressure changes seemed to be a more direct and fundamental form of creating sound than moving a membrane in a loudspeaker.

While experimenting I have followed two opposite approaches. I have been studying the specific qualities and limitations of different valves and I have been questioning their generic qualities: does a valve exhibit specific behaviour that is interesting to explore and use in a musical context or can it be seen as seen as something universal, similar to what we expect from a loudspeaker? Here we touch upon the question whether a valve can be regarded as a musical instrument or not, and whether it is meant to be audible (present) or inaudible. The opposite approaches were used as a method to study their specific qualities without the intention to result in two different and complementary artworks as it eventually did. The first version of *Pneumatic Sound Field* originates from 2006 and *Schwingungen – Schwebungen* from 2015. For *Pneumatic Sound Field* I have chosen to use the specific qualities and limitations of the discrete valves as point of departure for the contents of the work. In *Schwingungen – Schwebungen* the use of the proportional valves in combination with the dampers and horns are means to create a more generic form of sound production; it is closer to the ideal of being able to produce any sound and mediate the to-be-produced signal in a more transparent way.

The discrete and binary nature of the valves in *Pneumatic Sound Field* implies that I could only use time as its domain for the creation of content. Time describes the duration of a discrete pulse or the frequency and pulse width of a to-be-produced square wave. Fast repeating impulses result in the perception of pitched sounds, while slow repeating impulses are perceived rhythmically. The extreme limitations of the valves resulted in the idea to use a grid of multiple valves and to develop a model in which patterns spatially propagate through the grid at different speeds. In this way different relations between time and space could be created. The propagation speed is variable and defines how fast (in time) an impulse travels through the grid of valves. Fast propagation speeds result in the perception of single events with specific spatial qualities; slower ones result in rhythmical patterns in space. Small time differences between the valves determine where and how the audience will locate the impulses produced by the valves. Changing position in relation to the valves changes the arrival times at which the sounds from the valves arrive at one's ears and thereby how the event will be perceived.

The compositional content of *Pneumatic Sound Field* has developed from the qualities of the discrete valves and the behaviour and possibilities of the propagation model. It is the process of experimenting and reflecting that has led to the development of the installation and its content.

Despite the more generic possibilities the combination of the proportional valves, dampers, and horns offers in *Schwingungen – Schwebungen*, this installation uses the set-up in a very specific and limiting way. Just as the limitations in *Pneumatic Sound Field* functioned as a driving force for the development of its content, the same counts for *Schwingungen – Schwebungen*. The sound material is limited to the use of pure tones and triangle waves. Furthermore, the principle of spatial interferences (standing wave patterns moving in space) between combinations of horns is being used to place the audience in the middle of the spatial interferences that take place in between and around the horns. The possibilities and limitations of this approach are used to develop the content of this work.

It is interesting to observe how different the two works have become. *Pneumatic Sound Field* creates a very strong, and I would say, concrete spatial presence – a sensation of presence that is so strong that it is almost physically present and gives the idea that one can almost touch the sound. The spatiality in *Schwingungen – Schwebungen* is much more ephemeral. The sound seems to float in the space around the listener but does not get a physical quality. This is an interesting paradox: whereas the valves in *Pneumatic Sound Field* are visually hardly present, the sound has a strong sense of presence. The horns in

Schwingungen – Schwebungen, on the other hand, are visually very present, but the sound, resulting from the spatial interferences between them, is more ephemeral and not directly recognized as coming from the horns.

Let me end with a methodical remark. The artworks and principles addressed in this chapter are not just the results of prior knowledge but are mostly developed while working with the materials used in the installations and while reflecting upon my experiences.

Notes

1. It is important to mention here that Van Eck aims for the opposite in her book. She investigates the use of microphones and speakers as musical instruments instead of reproduction.
2. The same transition happens when looking at a sequence of images in a movie. When the frame rate is slower than 18 images per second we will see them as individual still images but when we play them faster we will perceive a moving image. The individual images merge into something that we perceive as continuous.
3. There are two ways of describing a sound containing overtones. We can speak about the fundamental frequency in combination with overtones or about harmonics. In the last case the first harmonic corresponds to the fundamental and the second harmonic corresponds to the first overtone. When a sound contains overtones all these overtones can be regarded as individual pure tones.

References

Berkhout A. J. (1988). 'A Holographic Approach to Acoustic Control'. *Journal of the Audio Engineering Society* 36 (12): 977–995.

Blauert, J. (1997). *Spatial Hearing: The Psychophysics of Human Sound Localization*. Revised edition. Cambridge, MA: MIT Press.

de Boer, K. (1940). 'Stereofonische Geluidsweergave'. PhD thesis, Delft.

Franssen, N. V. (1964). *Stereophony*. Eindhoven: Philips Technical Library.

Hartmann, W. and B. Rakerd (1989). 'Localization of Sound in Rooms IV: The Franssen effect'. *Journal of the Acoustical Society of America* 86: 1366–1373.

Litovsky, R. Y., H. S. Colburn, W. A. Yost, and S. J. Guzman (1999). 'The Precedence Effect'. *Journal of the Acoustical Society of America* 106: 1633–1654.

Lucier, A. and D. Simon (1980). *Chambers: Scores by Alvin Lucier*. Middletown, CT: Wesleyan University Press.

Van Eck, C.(2017). *Between Air and Electricity*. New York: Bloomsbury.

26

The Overheard: An Attuning Approach to Sound Art and Design in Public Spaces

Marie Højlund, Jonas R. Kirkegaard, Michael Sonne Kristensen, and Morten Riis

Introduction

Recent research in sonic interaction design and sound art has called for an ecological and enactive methodology that entails investigations of actors and the interrelations with their respective environments as attuning ecosystems (Franinović 2012). We take up this challenge by proposing an *attuning approach* as a methodological framework capable of accommodating both the multisensory atmosphere and the active engagement of the enactive user through practice-based experiments. Through the project *The Overheard*, we explore and develop the attuning approach in practice and present selected perspectives unfolding strategies of 'ecological overhearing'.

Atmospheres and attunement

There is a growing interest in atmospheres as a field of research, related to the broader field of nonrepresentational research ranging from philosophies of atmospheres (Bollnow 1943; Böhme 1995) over analyses of urban sensory environments (Thibaud 2011; Stenslund 2012; Edensor 2016) to the applied orchestrations of architectonic settings (Wieczorek 2013; Kinch 2014; Stidsen 2014). The existence of atmospheres in everyday experiences is often taken for granted and thus remains unnoticed. Yet, in recent years scholars have argued that atmospheres are vital to our everyday experiences of places and situations (Pink and Sumartojo 2018). The attuning approach to sound art and design takes its starting point in an understanding of atmospheres as affective attunements.

One of the central figures in the research area of atmospheres, the French sociologist Jean-Paul Thibaud, suggests that we do not perceive atmospheres as such. Rather, we experience atmospheres as 'a sensory background that specifies the conditions under which phenomena emerge and appear' (Thibaud 2011: 212). In this sense, atmospheres are comparable to the weather or light. As such, we grasp the atmosphere of a place 'before' or 'beneath' our rational understanding and intellectual assessment of a place (Pallasmaa 2012). Atmospheres constitute our immediate all-encompassing experience of 'finding oneself in environing worlds' (Böhme 1989: 9), involving our entire range of senses but often unnoticed by direct attention, although still experienced in an immediate and yet complex manner. In other words, an atmosphere should be understood not as an abstract concept, a metaphor, or as the property of things, but as a relational 'thing' or 'quasi-thing'.

As atmospheres are mostly something one is not aware of, something affective and prior to will and cognition, how can designers and artists work with them actively? As one cannot simply decide to delete atmospheres, Heidegger suggests that agency towards existing 'Stimmungen' must be exerted tactically in a mediated fashion through 'Gegenstimmung' or 'counter-attunements' (Heidegger 2002: 136). Thus, a consideration of attunement and counter-attunement in the listening domain is instructive for working with atmospheres as a design and artistic practice. According to Heidegger, a counter-attunement can only emerge if it resonates with an existing attunement. If one is to be affected by a counter-attunement, this counter-attunement needs to connect with the existing attunement. In this way, facilitating counter-attunements demands a disjunction between one 'there' (an existing attunement) and another one 'here' (a counter-attunement). A counter-attunement can thus only be achieved by creating an object or event of affective attachment in an existing attunement that challenges it from the 'inside'.

Literature scientist Jonathan Flatley suggests how such objects of affective attachment can be brought into being by drawing on Daniel Stern's identification of affective attunement between parent and infant (Flatley 2008: 504). In the interplay between parent and infant, mirror expressions are essential building blocks in the infant's experience of intersubjective relatedness and communion with the world and other humans (Stern 1998: 138). Characteristic of affective attunement is that this mirroring cannot be based on identical mimicking, but it must give the impression of imitation, though translated into similar gestures in other modes or senses, for example from sound to movement. This cross-modal aspect is important because it shows the infant that the parent not only understands what the infant does, but also how it feels, and therefore the cross-modal attunement is focused not so much on the outside behaviour of the infant but on the quality of the inside feelings. Due to its cross-modal character, affective attunement happens by way of amodal equivalences such as intensity, shape, and rhythm (Flatley 2012: 516). Getting close to patterns of intensities and rhythms between the different objects and actors in a situation sets the ground for a sense of being-together. Flatley concludes that promoting affective attunements through sensible exposures and rhythms can therefore facilitate counter-attunements.

However, following Heidegger, counter-attunements would require a disjunction in the existing attunement. Thus, we argue that Flatley's suggestion that affective attunement can set the ground for counter-attunements needs to be revisited. For that purpose, we propose to turn our attention to Stern and his second category of attunements termed 'purposeful misattunement'. When a parent engages in purposeful misattunement, the aim is to be with the child through affective attunements, but then to change the level of activity or affect through intentionally over- or under-matching the intensities and timings of the infant. Such attunement demands that the parent 'slips inside' the infant's state of feeling 'far enough to capture it', and then misexpress it enough to alter the infant's behaviour, 'but not enough to break the sense of an attunement in process' (Stern 1998: 148–149). This two-way process corresponds to Heidegger's claim that in order to set the ground for counter-attunements we first need to capture the existing attunement. Second, Heidegger's idea of disjunction by offering a new object or event of affective attachment in the existing attunement corresponds to Stern's concept of purposeful misattunement as a way to intentionally change the existing attunement by slightly over- or undermatching the amodal equivalences.

Later in this chapter we will present examples of how this is employed in practice. Before doing so, we unfold how to stage counter-attunements in the listening domain through the concept of 'ecological overhearing'.

Ecological overhearing

'Ecological overhearing' has its basis in perceptual theory and the notion of peripheral attention. Distracted listening based on peripheral attention[1] has the potential to integrate someone in space and the events taking place in that space. This is also described in Gernot Böhme's (2000) article 'Acoustic Atmospheres: A Contribution to the Study of Ecological Aesthetics', which suggests a peripheral mode of listening characterized by a feeling of 'Ausser-sich-sein' (being outside yourself).

Artist, writer, and theorist Brandon LaBelle highlights the sonic background as participating in overlaying ambiguity on built space, acknowledging how the senses are always navigating through different layers that require different types of attention, which sometimes can be productive, while at other times not. This overlapping ambiguity is denoted as 'productive mishearing' (LaBelle 2010: 180) or 'overhearing' (LaBelle 2015), constituting a generative and constructive ground on which new relations and surprising encounters can take place. Overhearing introduces a shared and messy space and adds multiple perspectives, as there is always sound outside the frame of attentive listening. Echoing Michel Serres's elaborations on background noise, LaBelle argues that overhearing forms an essential part of our experience of everyday environments. Overhearing is not to be considered a passive hearing as opposed to an active listening to a non-conscious background, but a necessary ground on which a signal is heard, and therefore part of the relation and a productive component in any information transmission.

Common to Böhme's and LaBelle's accounts is that peripheral sonic experiences resituate our relation to figure and ground, to foreground and background, by operating on the peripheries of perception. This view promotes an understanding of overhearing that is more complex than just 'not-hearing'; it suggests an ecological understanding of overhearing as our peripheral and attuning mode of listening through which we vaguely sense our surroundings, thus constituting a way to feel part of those surroundings. Both Böhme and LaBelle emphasize that we need to relearn this skill for ecological overhearing, as most people today tend to focus on a simple signal/noise dichotomy. Thus, we propose to facilitate such learning process by opening the affective zone in the listening domain to counter-attunements through a two-step process. First the listener catches the existing attunement, and then a purposeful misattunement is introduced by way of over- or undermatching the sonic intensities in order to reconfigure the habitual background/foreground relations through ecological overhearing.

The philosophical and conceptual framework for the attuning approach combines theories on atmospheres, attunement, and ecological overhearing based on a relational ontology. This integration couples the entanglements of the complex atmospheric quality of a space with an understanding of human beings as embodied, enactive, and situated. The human and the world are co-constituted through exploratory attunements based on prior experiences and resulting in refined attunements such as new appraisals and habits. From an ecological perspective, attunement becomes a key concept capable of accommodating the combination of the enactive and the focus on atmospheres.

The Overheard

The purpose of the project *The Overheard* is to explore and develop the attuning approach in practice and present selected perspectives as unfolding strategies of ecological overhearing. Put briefly, the project comprised six publicy available, site-specific sound installations across a central region of Denmark during *Aarhus 2017 – European Capital of Culture*. The sound installations could be visited both physically on site and virtually through a website streaming high-quality, real-time audio from the different locations. The website allowed listeners to create their own soundscape by blending the sound installations (Figure 26.1).

All of the sound installations were very different, as were the locations and the artists commissioned to make them. In the following, we will describe the project's framework, its objective, and how the concepts of ecological overhearing and affective attunement can be experienced in the resulting interventions and assessed as design strategies in future projects dealing with public sound installations as a means of transforming an existing sonic environment.

The Overheard was initiated not as a research project but as an artistic, cultural project with a broad public impact. The involvement of Aarhus University brought the research dimension into the project in the early stages, but the project's identity was kept in the artistic, experience-oriented domain. This frames an initial methodological point in relation to balancing research and art. Many research projects that apply different sorts of

Figure 26.1 Screenshot from *The Overheard* website. By The Overheard.

artistic methods and interventions have a structure where the research method delimits and dictates the design of the intervention to first and foremost fulfil the goal of data harvesting, and only secondly to solve the problem that is actually being investigated. Starting the other way around – trying to 'solve the problem' instead of trying to do research – has been an important driving force in this project and may actually have pushed the research outcomes closer to applicable results than a traditionally controlled experiment would have done. Therefore, collecting massive amounts of empirical data was not the key focus of the project. Rather, we wanted to investigate whether combining artistic and design-research in a process called *curation* would at all yield solutions that evoke attunement and ecological overhearing, and if so, how these attuning mechanisms were revealed.

The first part of the curation process was a dialogue with the involved municipalities where they suggested possible locations for the interventions. The main criteria was that each environment should include complex sonic qualities, with traces of both technology/civilization and nature. The curation was carried out through a selection process in which the works of sound artists were critically assessed against the qualities of each of the chosen places. The process was a reversal in comparison to curation in the traditional sense, as the purpose was not to display a piece of sound art in a somewhat 'neutral' exhibition site, but to use sound art to create and condition an atmosphere with the potential to establish experiences of ecological overhearing and to supply agency to engage in affective attunement. A key element in the process was to ensure complete artistic freedom, thus trusting the skills, experience, and intuition of the artist; if artistic processes are to be integrated in a research project, they should not be governed by non-artistic agendas.

In the following we describe and analyse two of the installations, *Forest Megaphones* and *4140 Voices*, in terms of ecological overhearing and affective attunement. Furthermore, we describe the web-based audio stream, as that part of the project adds further to the staging of the attuning mechanisms.

Forest Megaphones

Forest Megaphones, by Estonian artist Birgit Öigus, comprises three wooden megaphones, 2.5 metres tall and 3 metres long. The place chosen is a glade on a small island, 80 metres across, in a beautiful forest area with lakes and streams. The place is flanked by a relatively busy road, adding a sonic component of tyre and engine noise to the sonic environment of birds, wind, water, boats, and people (Figure 26.2). What separates the megaphones from most other sound art is that they don't produce any acoustic signal themselves but nevertheless transform the sound of the place.

While it is not discussed in detail as to how the skill of overhearing is to be acquired anew, *Forest Megaphones* proposes a way of enacting this type of overhearing by design. The symbolic shape of the cone signifies sonic amplification, that is, the iconic old hearing aids, gramophone funnels, volume control icons, etc., hinting for people to listen to the environment. The busy road is visually hidden behind trees and thus would typically fall into the category of overheard sounds that do not so much bring to mind their representational value but appear more like noise. As the megaphones invite listening more carefully and directing the attention to all sounds, both the ones within and the ones outside normal attentional focus, the reduced signal/noise dichotomy is revealed and questioned, and the noise from the cars is equalized with the birds and water sounds. Hence the installation challenges what is often termed the bifurcation of nature (Whitehead 1920: 20) and offers the listener an affective potential to engage in the perceptual (re-)acquisition of a relational appreciation of the atmosphere.

Forest Megaphones invites the listener to engage in a process of counter-attunement, and the first step of catching the existing tuning happens through ecological overhearing. Through an elevated attention to the atmosphere – by means of acoustic amplification and physical, tactile, and visual engagement – the foreground/background relation is shifted. The next step of counter-attunement is created by the possibility to invert the engagement and use of the megaphones for amplifying one's own sound and echoing them back into the existing soundscape as over- or undermatched responses that shape the soundscape anew. This process may reveal the atmosphere as a dynamic and ever-changing assemblage of sounds in which the listener does not only listen, but also has agency to contribute to and transform the sonic ecology. The process of attunement and counter-attunement facilitates new actions and behaviours and the possibility to relearn a refined attunement and experience the different layers of listening through an

Figure 26.2 *Forest Megaphones* by Birgit Öigus. Large wooden megaphones placed in the city of Silkeborg, Denmark. Photograph by Malte Riis.

engagement with the megaphones. This can happen in many ways, and from watching people interacting with the megaphones it became clear that the approaches were as diverse as the number of participating visitors. Individuals engaged differently than groups, kids engaged differently than adults, and first-time visitors engaged differently than returning visitors. The refined attunements also revealed themselves in different forms, either through articulated dialogues and discussions, tacit knowledge, positive memories, confusion, or bodily experience.

4140 Voices

The sound piece *4140 Voices* by Marie Højlund and Morten Riis is a composition written especially for a memorial monument at Mindeparken, Aarhus. The material is based on the recordings of citizens of Aarhus reciting the names of the 4,140 Danish soldiers who fell during the First World War. These names are engraved in the stone walls of the memorial monument. Through the composition of voices, a new sensory connection is created between the present-day inhabitants of Aarhus and their history. Reciting creates a vibratory sensibility that enables the audience to relate in a sensorial manner to the names engraved in the limestone. The names become more than mere letters when they are activated through sound (Figure 26.3).

Merely from looking at the picture of the monument, one can imagine a rather strange acoustic phenomena arising from the echoes and reverberations occurring between the circular stone walls in the monument. Twenty-four equally spaced speakers – mounted on the top rim of the monument – pour the sound of the endlessly reciting voices into the acoustic space. As one enters into this massive, translucent sonic ambience, it is impossible to separate out singular voice components. It is only possible to identify vague contours of names or syllables, with the individual, recited names concealed and withdrawn. Since

Figure 26.3 The memorial monument at Mindeparken, Aarhus. Photograph by Malte Riis.

it is impossible to register the actual meaning of the words, the sonic impression must be sensed by means of an overhearing strategy. The exact names flee the centre of attention and thus become part of the present atmosphere, just like traffic noise in the street or a humming radio next door. The sound is present, perceived, and registered as a part of an ecological sonic reality, peripherally overheard and yet constituent of our sense of being here now. The lack of direction or centre leaves one without an obvious focus of attention and brings about a mode of overhearing in which one is 'forced' to give in to not-hearing the sonic present.

When entering the monument, the promoted mode of listening, the overhearing, is also a state in which one attunes to the place. The sonic atmosphere is vibrant, shimmering, and omnipresent, and has an out-of-this-world feeling due to the unusual acoustics. But what does it add to the monument? How is it transformed?

The monument itself (without the sound of the voices) has been there for more than eighty years. It has a devout character as it stands in silence through time carrying the names of the fallen soldiers. It is a very well-designed monument for contemplation and reflection, but as generations grow up in peacetime without any first-hand experience of war, the monument also fades away as something historical, pointing back in time at how things were. In that light, the addition of *4140 Voices* challenges the historicity of the atmospheric mode of the monument by translating the names carved in stone to words spoken by people living from one modality to another.

This process can be understood as a counter-attunement that takes place between the monument and the voices, and that is what visitors can tap into when entering. The sound of the voices has something to tell the audience. There is a message or an urgency, something that may be experienced as an intentionality from the monument directed towards the listener. Now the names are not just to be read from the walls at will, but they call out and bring about a relational potential that visitors can act upon. This is where the affective attunement comes into play. The visitors' agency is locked to their bodies and their ability to move around, as they realize that the sonic texture of the voices changes with every slight movement. The sonic phenomena invites visitors to move around along the walls of the monument where the names of the soldiers can be read. Getting closer to the walls also means getting closer to individual speakers, and suddenly the full pronunciation of names can be traced in the stream of voices. By changing physical position, one can shift to the foreground what was initially ascribed to the overhead background. This effect, that a sound source stands out when it is approached, is obvious, but it is a clear example of the aforementioned effect where sensory stimuli promote bodily movement in a process of affective attuning. The overhearing mode of listening thus initiates a cycle: one catches the tuning of the atmosphere and moves towards the sound; the movement changes the sonic experience and creates a reconfigured point of overhearing. The implementation of the voices seems to have made visitors more prone to touching the engraved names and sometimes even, when sinking into their own thoughts, moving their lips and whispering the names of the fallen, as a way of joining the choir of the atmospheric voices.

Reflections on the paths ahead

The attuning approach seeks to facilitate an ecological overhearing that increases possibilities for engaging in place making by affecting the collected sensory stimuli that define a certain place. It thereby provides a powerful framework that does not build on a polarity between the environment and the human perspective. The attuning approach offers an expanded capacity for coexistence through ecological overhearing as a way to support potential solutions to a variety of problems relating to well-being and health. One example of a pertinent problem, to which we address our attention in a final note, is sound control from a quality of life perspective.

Over the last century, most research on noise in public and shared spaces has focused on explaining how people become stressed and fall ill owing to noise pollution. The most common strategies focus on noise abatement through regulations based on quantitative assessment. However, in recent years the growing field of sound studies and the soundscape approach have prompted a shift in terminology from *noise* to *sound related annoyance*, and from *noise reduction* to a *quality of life* perspective. This shift should be seen as a response to the limitations of earlier regulative approaches, for example corresponding to the noise reduction approach in hospitals that has not been successful in solving general noise problems (Andringa et al. 2013). A significant example of this shift towards a quality of life perspective is the *EU COST* project 'TD0804 Soundscape of European Cities and Landscapes', which introduces a soundscape approach as a new paradigm (see Fiebig and Schulte-Fortkamp, in this volume). This change in focus connects to the growing interest in the broader field of sound studies, where sound design is not only considered to be a technical sound-internal skill but also must be thought of in its broader entanglements with cultural, historical, philosophical, and technological contexts.

The growing focus on sound environments in research seeks to account for the increasing sound-producing artefacts in everyday life, and a lack of education and knowledge on how sound plays a significant role in the constitution of our current state of mind, and not only as externally existing stimuli supplying more or less useful or enjoyable content. The quality of life perspective emphasizes the crucial roles that empowerment versus confinement plays in forming our overall appraisal of sound environments by either enhancing or preventing us from acting upon intrusive sounds. In this way, the emerging soundscape approach seeks to account for a substantial part of noise annoyance explained by so-called non-acoustic factors or higher-level cognitive factors (Davies et al. 2012: 15). Studies have shown that an important non-acoustic factor is the feeling of being in control of – or being exposed to – noise as an explanatory factor in people's coping mechanisms (Bijsterveld 2008: 254; Andringa et al. 2013). The paramount concern has thus moved from noise reduction to providing a wide diversity of acoustic opportunities, for example sound art installations in public spaces, focusing on design parameters such as control, motivation, and empowerment (Andringa et al. 2013: 17).

Many recent and ongoing research projects seek to develop ways of realizing these effects in practice. However, although the theoretical perspectives are shifting and developing rapidly under vivid debates, there is still a massive lack of concrete operationalization and evaluation of real-life experiments (Kang et al. 2013: 9). While our demonstration of the attunement approach has been unfolded in an aesthetic context, the fundamental principles and arguments apply readily to health-related topics. For instance, sound-related annoyance in hospitals is deeply intertwined with a negative affective attunement and a lack of ecological overhearing possibilities. Counteracting the experience of noise should therefore also take a starting point in counteracting the negative affective attunements by facilitating counter-attunements and providing possibilities of ecological overhearing.

The attuning approach also points to the potentials of enactive technology, art, and design as tools for accommodating zones of overhearing by creating direct encounters with existing atmospheres in which our engagement with them can set the ground for affective attunement. Technology should therefore not be hidden or considered neutral as it offers a direct way to make the environment responding and malleable to our dwelling practice. Art and design can work together with technology, not as decoration but as tangible places for such encounters to unfold. In this way, the division between technology as a supposedly neutral and transparent tool and art as providing space for reflection in the imagination disengaged from the current lived experience becomes brittle.

When art, technology, and design become places for encounters and not only act as representations or media for some other use, they can operate as ruptures in our habitual modes of being and in our habitual subjectivities. The rupturing encounter contains a moment of affirmation, the affirmation of a new world, which leads to new ways of experiencing this world (O'Sullivan 2006). The zones of overhearing present places for such ruptures to occur, as tangible and malleable interfaces that operationalize the co-creations of atmospheres for those involved. In this regard it is important to notice that the rupture is not experienced as violent or transgressive, as the term might imply. On the contrary, the transformation that it introduces might not even be noticed as it operates on the affectual level as something intuitive and poetic (Lacey 2016: 16). Take as an example the aforementioned *Forest Megaphones*. Engaging in the attuning/counter-attuning process is the rupture that the megaphone offers, and the transformation that it leads to embeds itself in new habitual ways of being.

The attuning approach can thus be employed as a methodology to demonstrate how the creation of counter-attunements involves breaking down habitual experiences through affective attunements in order to help people reconfigure auditory background and foreground relations. Exposing people's own agency to shift perceptual perspectives, to discover the peripheral mode of attention, to attune to an existing atmosphere, and to even activate counter-attunements as a way of transforming and interacting, supplies a level of control and acts as a strategy for coping with the atmospheres of everyday life.

Note

1. We refer here to the Finish architect Juhani Pallasmaa's idea of peripheral vision.

References

Andringa, T. C., M. Weber, S. R. Payne, J. D. Krijnders, M. N. Dixon, R. V. D. Linden, and J. L. Lanser(2013). 'Positioning Soundscape Research and Management'. *Journal of the Acoustical Society of America* 134(4): 2739–2747.

Bijsterveld, K. (2008). *Mechanical Sound: Technology, Culture, and Public Problems of Noise in the Twentieth Century*. Cambridge, MA: MIT Press.

Böhme, G. (1989). *Für eine ökologische Naturästhetik*. Frankfurt: Suhrkamp.

Böhme, G. (1995). *Atmosphäre: Essays zur neuen Ästhetik*. Frankfurt: Suhrkamp.

Böhme, G. (2000). 'Acoustic Atmospheres: A Contribution to the Study of Ecological Aesthetics'. *Soundscape: The Journal of Acoustic Ecology* 1(1): 14–18.

Bollnow, O. F. (1943). *Das Wesen der Stimmungen*. 2nd and extended edition. Würzburg: Königsgausen & Neumann.

Davies, W. J., M. D. Adams, N. S. Bruce, R. Cain, A. Carlyle, P. Cusack, D. A. Hall, K. I. Hume, A. Irwin, P. Jennings, M. R. Marselle, C. J. Plack, and J. Poxon (2013). 'Perception of Soundscapes: An Interdisciplinary Approach'. *Applied Acoustics* 74 (2): 224–231. https://doi.org/10.1016/j.apacoust.2012.05.010.

Edensor, T. (2016). *Geographies of Rhythm: Nature, Place, Mobilities and Bodies*. New York: Routledge.

Flatley, J. (2008). *Affective Mapping: Melancholia and the Politics of Modernism*. Cambridge, MA: Harvard University Press.

Franinović, K. (2012). *Amplifying Actions: Towards Enactive Sound Design*. Plymouth: University of Plymouth. Available online: https://pearl.plymouth.ac.uk/bitstream/handle/10026.1/1496/2013franinovic10048253phd.pdf?isAllowed=y&sequence=1 (accessed 6 July 2020).

Franinović, K. and C. Salter (2013). 'The Experience of Sonic Interaction'. In K. Franinović and S. Serafin (eds), *Sonic Interaction Design*, 39–75. Cambridge, MA: The MIT Press.

Heidegger, M. (2002). *Being and Time*. New York: Harper.

Kang, J., K. Chourmouziadou, K. Sakantamis, B. Wang, and Y. Hao (eds) (2013). 'Soundscape of European Cities and Landscapes'. Available online: http://soundscape-cost.org/documents/COST_TD0804_E-book_2013.pdf (accessed 5 July 2020).

Kinch, S. (2014). 'Designing for Atmospheric Experiences: Taking an Architectural Approach to Interaction Design'. PhD dissertation, Aarhus School of Architecture, Aarhus.

LaBelle, B. (2010). *Acoustic Territories: Sound Culture and Everyday Life*. New York: Continuum.

LaBelle, B. (2015). 'Lecture on Overhearing, Shared Space, and the Ethics of Interference'. *Nordic Sound Art LP-katalog fra afgangsudstilling 2014* [LP]. Copenhagen: Det Kongelige Danske Kunstakademi.

Lacey, J. (2016). *Sonic Rupture: A Practice-led Approach to Urban Soundscape Design*. New York: Bloomsbury.

Pallasmaa, J. (2012). *The Eyes of the Skin: Architecture and the Senses*. 3rd edition. Chichester: John Wiley & Sons.

Pink, S. and S. Sumartojo (2018). *Atmospheres and the Experiential World: Theory and Methods*. New York: Routledge.

O'Sullivan, S. (2006). *Art Encounters Deleuze and Guattari: Thought Beyond Representation*. New York: Palgrave Macmillan.

Stenslund, A. (2012). 'The Whiteout of Smell: Experiencing and Exhibiting Aesthetic Epiphanies'. Paper presented at the *Ambiances in action/Ambiances en acte(s) - International Congress on Ambiances*, Montreal, Canada, 19 September 2012.

Stern, D. N. (1998). *The Interpersonal World of the Infant: A View from Psychoanalysis and Developmental Psychology*. London: Karnac Books.

Stidsen, L. (2014). 'Light Atmosphere in Hospital Wards'. PhD dissertation. Aalborg: River Publishers. Available online: http://www.riverpublishers.com/book_details.php?book_id=246 (accessed 6 July 2020).

Thibaud, J.-P. (2011). 'The Sensory Fabric of Urban Ambiances'. *Senses & Society* 6(2): 203–215.

Whitehead, A. N. (1920). *The Concept of Nature: The Tarner Lectures Delivered in Trinity College November 1919*. Cambridge: Cambridge University Press.

Wieczorek, I. (2013). 'From Atmospheric Awareness to Active Materiality'. Paper presented at the *19th International Congress of Aesthetics*. Aesthetics in Action.

27

Sound on Sound: Considerations for the Use of Sonic Methods in Ethnographic Fieldwork inside the Recording Studio

Paul Thompson

Introduction

The recording studio has been overlooked as a potential site of study within the field of sound studies and across other academic disciplines too. Consequently, there is a limited amount of published research involving ethnographic fieldwork inside the studio (some notable exceptions include Hennion 1990; Fitzgerald 1996; Meintjes 2003, 2004; Porcello 2004; Gibson 2005; Williams 2007, 2009; Bates 2008; Thompson 2016, 2019). A critical reason for this dearth of study inside recording facilities is because the recording studio is designed to be isolated both acoustically and socially (Thompson and Lashua 2014). In her rich ethnographic study of Downtown Studios in Johannesburg, South Africa, Louise Meintjes noted that: 'the studio is remote and exclusive. It is closed to outsiders except for haphazard, enticing ingressions like mine and those of friends of the music-makers who might drop in for a session or a moment' (2012: 270) and so researchers' first challenge is in gaining access to a recording studio session.

Once inside, researchers are challenged to consider how the medium of sound can be used to represent the multimodality of recording studio fieldwork. In so doing, researchers need to 'rethink the ocularcentrism through which anthropology has generally constructed knowledge about culture' (Kheshti 2009: 15) and find new ways to explore the cultural field of the recording studio through sound. The following chapter draws upon the author's experiences of conducting ethnographic fieldwork in recording studios in the UK, Canada, and the USA and offers some useful insights into the recording studio as a space for sound studies and suggests a number of pragmatic approaches in capturing the aural ecology of the recording studio.

Sound and the studio architecture

The majority of recording studios have two principal areas of operation: the control room and the live room. The control room typically houses the vast array of recording equipment needed to capture the performances of the musicians, which includes the mixing console, speakers (referred to as monitors), computers, tape machines, and sound processing equipment that the engineer can access during a recording session. The live room is often larger as it has to accommodate performing musicians (which in some cases may be an entire orchestra). The two rooms are acoustically separated from each other to avoid sound transference between them. Communication between the rooms is achieved visually, often through a large acoustically sealed window, and sonically through a talk-back microphone on the mixing console and the headphones of the musicians.

From an acoustic perspective, the control room and live room are normally designed differently reflecting their specific purpose. The control room is often significantly less reverberant than the live room as engineers and producers require a space that reduces the amount of sound reflections from the surfaces of the room to prioritize the direct sound from the studio monitors. In this way, engineers and producers can make critical judgements on microphone quality, microphone positioning, the sonic qualities of a musician's instrument or the accuracy of a musician's performance. The live room is typically more reverberant, often designed with more reflective materials such as wood, in order to help musicians deliver their performance, as acoustically dead environments can be very uncomfortable to perform in. Live rooms sometimes have moveable design features to alter the acoustics of the space with reversible or movable acoustic panels, carpets, or curtains to 'liven' or 'deaden' the acoustic depending upon the requirements of the recording session. It is through its characteristic architecture that the recording studio places an overtly sharp focus on the quality of sound:

> The acoustics mark the studio as a space out of the ordinary. But its distinction is not only derived from its focus around a sense other than the eye […]. The studio also draws enchantment from the very quality of the sense it privileges.
>
> (Meintjes 2012: 272)

This privileged medium operates within three distinct and interrelated sound worlds: (1) the control room, (2) the live room, and (3) headphones. For researchers interested in the aural ecology beyond these main sound worlds there are also often a series of 'backstage' areas such as the lounge, the kitchen, the hallway, the parking lot – these are the areas that don't appear to be directly related to studio work but where interpersonal dynamic interchanges occur and where a lot of rich sound material can come from.

Accessing the sounds of the studio

Although highlighted as one of the central issues within empirical research (Hammersley and Atkinson 1997), it is startling that the issue of accessibility isn't foregrounded in previous studies of the recording studio (with the exceptions of Meintjes 2003, 2012; and

Bates 2008), particularly as the issue of accessibility is ongoing throughout the entire process but is 'often at its most acute in initial negotiations to enter a setting during the "first days in the field"' (Hammersley and Atkinson 1997: 54). Issues of accessibility within my first recording studio study were evident from some of the initial exchanges of contact between the intended participants and myself. The social world of recording studios, and consequently their sound worlds, can be largely inaccessible for even the musicians who wish to record in them (both financially and socially), so gaining access to a recording studio both socially and physically can be a challenging task for an ethnographer. A fundamental reason for this is that recording studios are 'sealed' facilities in several ways: firstly, they are often constructed to be separated acoustically from their local environment so that sound doesn't disturb nearby buildings or residences and, most importantly, so that sound does not enter the recording studio and hinder the recording process. Secondly, recording studios are not public spaces (like a city square or a town library) and so physical access to them is limited and gained only by invitation from the engineer, producer, or from the studio manager or studio owner. Thirdly and fundamentally, the recording studio during a recording session is a place of work in which studio personnel and musicians require an environment that is private and free from distraction, which allows them to create an intimate setting and thereby maximize effective collaboration and communication.

Gaining access to a recording session can be the most challenging obstacle of all, particularly because a recording session is often limited to only those involved in the recording process. Any additional individuals in the recording studio may become a distraction, affect the flow of the session, or disrupt communication between the studio personnel. In her ethnographic study of female popular musicians Mavis Bayton (1990) identified the importance of privacy in the practice room in order to enable effective collaboration and to resolve any issues. In a similar way, recording studios are intentionally secluded with access limited to only those involved in the recording process. For an ethnographer who is not directly involved in the recording process there is little opportunity to gain access to the sounds of the recording session.

My own initial attempts to gain access to a recording studio session began with an exploration of my personal network of recording engineers and record producers who were either previous colleagues of mine or friends of these colleagues. I assumed that having a background as a practitioner would prove to be useful when seeking permission to conduct ethnography in the recording studio and my first contact was a commercially successful record producer working at a recording studio in Liverpool. He showed some interest when discussing the intended research, however, when I asked him if he would mind me observing an upcoming recording session the response was tentative and he expressed a preference for me to only observe bands that were not signed to a record label. He was concerned that if signed bands were involved there might not only be an infringement of copyright but additional people from the recording studio might need clearance from the record label or management company concerned. In response I suggested that I could perform menial tasks in the studio, such as setting up microphones or coiling cables, which might help to remove the explicit role of the 'observer' or 'listener' in the room. He was adamant that he would already have enough personnel for the session but agreed to a follow-up phone call to arrange a meeting and discuss the project further. However, after numerous failed attempts to get in touch, I decided to explore other possible contacts.

In an attempt to learn from my initial attempt to gain access to a studio session, I emphasized to other potential record producers that observation wouldn't get in the way of the record-making process. These responses were also justifiably hesitant because 'unfortunately, my studio is too small to have an extra person' and 'the bands I work with are signed so the label won't want anyone else involved' (personal communication, 2011). It was evident through my efforts to gain access to a recording studio session through my network of engineers and record producers, that as the ethnographer I had been positioned as an outsider. The role of outsider in the recording studio has an identifiable tradition in popular music where at best you are surplus to requirements and at worst you are considered to be negatively impacting the flow of a session. Engineer Dan Turner explains that:

> The studio is often such a private, intimate place that any outsider inevitably changes the way you operate, often directly influenced by the circumstances of the session [...] it can completely ruin your day [...] any outsider can change the whole atmosphere and often get in the way.
>
> (personal interview, 2012)

Record producer Phil Harding adds that:

> Having an outsider in the room when I'm working with an artist in the studio would be too compromising. You want to give your client and your artist the best performance from your side and you're going to feel compromised if there's an outsider in the room. I have had situations where I've had to ask the artist to either get their friends to leave or not come in next time. On the other hand, I can certainly remember for instance Toyah Wilcox, when I was engineering with a producer and a group of her session musicians, her boyfriend would often come in and constantly make comments [...] that's so difficult, who's going to say to Toyah 'don't bring your boyfriend?'
>
> (personal interview, 2012)

Both the responses from engineers and producers, and the initial failure to gain access to a recording studio session, highlighted that although engineers and record producers facilitate the needs of the musician and act as intermediaries between the artist and the industry, they are not the gatekeepers to a recording session. As identified by both Dan Turner and Phil Harding, the gatekeepers to a recording studio session are the musicians who are recording in the studio. Sonic ethnographers should therefore begin their search for a recording studio by approaching the musicians on the session first as this will ease negotiations with engineers or record producers at a later date. If the musicians are the main client and are paying for the studio time then it's even more likely that access will be granted once you've gained permission from the recording musicians. Creating a rapport with musicians before the studio session may also help in creating a more cohesive atmosphere in which the researcher isn't adversely affecting the flow of the session. The challenges of gaining access to a recording studio session described above not only show some of the mechanisms of recording studio practice but also highlight some of the power relations and social hierarchies that can operate covertly within a recording studio context.

Sound and social relations

The social and physical issues that surround gaining access to a recording studio session serve to illustrate the unique social imperatives that govern recording studio practice. Because ethnography demands immersion into the social context of interest, the ethnographer's position within the recording studio session, both physically and socially, in other words 'the ethnographic self' (Coffey 1999), must also be addressed. The primary intention of any research is often to avoid influencing the natural processes that occur in the setting, which in the instance of sonic ethnography, means attempting to maintain a primarily 'listener' position. This however can prove difficult as it isn't always possible (or desirable) to be a continual 'fly-on-the-wall'. The close proximity of the studio participants means that avoiding conversation or social interaction could adversely affect the atmosphere and the natural social exchanges that occur during a studio session. This is no more acute than when the researcher may be asked, 'What do you think?' after a particular take of a performance. For this reason, discussing the expectations of the research and explaining the researcher's position to the participants is necessary before the fieldwork begins.

In my own research (Thompson 2016), I was able to explain my researcher position during a pre-production meeting between the band and the record producer. Pre-production is typically the stage before the musicians enter the studio and allows the band and record producer to sketch out what they plan to do over the course of making the record. Pre-production 'serves as a vital preparatory stage during which an image of the record's shape and tone is developed, even if only in a rough form' (Zak 2001: 137). During the pre-production meeting, I invited the band to ask questions about the research, which allowed their role and the researcher's role to become less ambiguous and dispelled the band's initial assumption that they would have to behave or perform in a particular way to avoid any contact with me or my sound recording equipment. Without discussing this during pre-production, the participants may have found the presence of a researcher in the recording studio unsettling, which in turn could have undesirably altered the flow of the recording session. This is commonly referred to as 'observer effects' and these interactions with the field and its participants have historically been viewed as a negative attribute of ethnographic research because:

> They indicate a 'contamination' of the supposedly pure social environment being studied (Hunt 1985). Some methodologists advise qualitative researchers to hone an awareness of possible observer effects, document them, and incorporate them as caveats into reports on fieldwork (Patton 2002). Others encourage ethnographers to seek out explicitly evidence of observer effects to better understand – and then mitigate – 'researcher-induced distortions' (e.g., LeCompte and Goetz 1982; Spano 2006) […]. The possibility that the ethnographer can both have an effect and by doing so tap into valuable and accurate data is seldom explored in contemporary literature on methods.
>
> (Monahan and Fisher 2010: 358)

Building relationships through social interaction during a recording session has proved to be an important aspect of my research in the recording studio. Rather than ignoring the participants and minimizing observer or 'listener' effects, developing a rapport with those involved can allow greater access to their thoughts and ideas that would not be possible through listening to a sound recording alone. In addition, discussing other artists' work, technologies, and practices, can also help to frame the participants' musical references, musical influences and importantly their musical performances. A non-participatory perspective on field relations may restrict access to 'rich data in the field' (Monahan and Fisher 2010: 370), may lead to 'failing to understand the orientations of the participants' (Hammersley and Atkinson 2007: 87). The underlying role of the researcher is therefore 'not to determine "the truth" but to reveal the multiple truths apparent in others' lives' (Emerson et al. 1995: 3–4).

Whilst participation and interaction can prove fruitful in gaining greater insight into the sound world of the recording studio it should also be considerate to the social situation and the established conduct of the recording studio. This is commonly referred to as 'studio etiquette'. Etiquette is described as a 'collective social knowledge – "no one taught us these rules" – the rules are learned through long years of socialization' (Sawyer 2000: 18) and studio etiquette is a general expectation of all recording studio personnel who support the recording process. Signature Sound Studios offers the following on studio etiquette:

> Knowing when it is appropriate to communicate in the studio is perhaps one of the most important concepts to grasp […]. On the other hand, knowing when to be silent is also very important. For example, when an engineer is in the middle of a recording or mixing session – even if he or she is just listening back and not hands-on doing something – do not interrupt by asking questions, making comments, or any other unnecessary noise. Any of these actions might break the engineer's concentration and he or she will probably not be very pleased with you. Your best bet when you find yourself in a recording session is to be silent, observant, and readily available if your help is needed.
>
> (Signature Sound 2011)

The expectations and recommendations described above are not only relevant to studio apprentices; they are suitably applicable for listeners conducting sound research in the recording studio. Observing studio etiquette is necessary to allow all the participants to communicate effectively between each other, for the engineer and record producer to make critical judgements on the musicians' performances and to maintain a degree of naturalness in the field setting. Observing studio etiquette is not only an essential part of effective social integration during a recording session, it also governs the timing and opportunity for informal interviews and exploratory conversations. Knowing when to ask a question becomes a useful skill that develops as the researcher becomes more familiar with the working practices of the participants throughout the process.

Capturing the sound of the studio

Conducting sonic research in the recording studio presents some unique social and logistical challenges that are fundamentally related to the distinct architecture of the recording studio and the social setting of a recording session. The construction of a typical recording studio creates a division between the control room and the performance space 'with a glass window that isolates the sound of one world from the other' (Williams 2011). This presents a challenge to the researcher who is only using one microphone to record the sound of the recording studio. If listening is taking place in the control room as the musicians are recording then it is only the sound of the control room that is captured and not the actions and interactions in the live room. Therefore to fully appreciate the sound worlds of the recording studio capturing the sound of both the control room and the live room can help to gain a perspective on what the performing musicians experience, and similarly in the control room in order to record the sonic experiences of the engineer and the producer.

In visual anthropology, the point of view offered by a single camera invites questions of 'where shots are to be taken, whether the camera should be fixed or mobile, whether a single focus is to be adopted or whether the focus should vary; and if so when and how' (Hammersley and Atkinson 2007: 148). It also questions the representation of a single view based on the researcher's relationship to the field and their fieldwork: 'the ethnographic self' (Coffey 1999). Capturing the sonic ecology of the recording studio too presents the same logistic as well as political and social aspects of representation, which Roshanak Kheshti labels 'aural positionality' and 'although sonic representation could be said to be less reductive and more ambiguous than visual representation, sonic representations of culture nonetheless include an imposed layer of meaning mediated by the body and ears of the ethnographer, recordist, editor and producer' (Kheshti 2009: 15). In choosing what to focus a microphone on, researchers knowingly or otherwise are therefore engaged in aural positionality, which can often be influenced by the type of microphone or recording techniques used.

Binaural recording is a method that uses two microphones arranged to capture sound in a similar fashion to the human ear. There are expensive and inexpensive versions of binaural recording, from using a dummy head with two, omnidirectional precision microphones placed inside a moulded set of pinna that models a human head, or using a stereo pair of microphones that can be positioned either side of the researcher's head. In Kheshti's case, using a microphone attached to the researcher situates her own aural positionality as 'the vantage point from which my body and the attached microphone hear the sounds that are recorded and re-presented in the context of my ethnography impacts what listeners hear when they listen' (Kheshti 2009: 15). In a recording studio situation binaural microphones placed near to the researcher's ears may provide a lifelike representation of the acoustic space of the studio but may limit where the researcher can capture sound based on the size of the studio or the particular situation during the studio session.

Using a single microphone that can be extended away from a recorder may offer more flexibility to the researcher to capture parts of the studio's acoustic ecology beyond where the researcher can reach. Although limited in its single perspective, it allows a greater exploration of sound in the studio space and, in addressing aural positionality, the position of the researcher's microphone may be determined in consultation with the engineer, producer, or musicians. This may help to both remove some of the researcher's representational-bias and directly involve those whose sonic world you are attempting to capture. Consulting the engineer may also help the researcher gain some insight into their particular process for positioning microphones during a session and some of the things they consider when doing so.

Finally, although it may present a technical challenge to the researcher, another useful way of addressing aural positionality is through the use of multiple microphones at the same time. Using a computer, audio software, a series of microphones positioned around the studio space, and an audio interface (that converts microphone signals into digital signals), the researcher can effectively capture different perspectives of the acoustic environment in a single recording session without the need for the researcher to move between the studio's multiple spaces. One distinct advantage of using multiple microphones to capture the acoustic ecology of the recording studio is that they are naturalized within the space; that is, studio participants expect to see microphones throughout the space and therefore a researcher's microphone wouldn't be considered out-of-place or particularly conspicuous.

Sound on sound in the studio

The permanence of recorded sound is a distinct affordance as the entirety of each recording session can be repeatedly played and replayed, allowing our focus or attention to be changed each time the audio is played. This can also serve to identify sonic events that may not have been evident in situ. Multi-perspective microphones provide an opportunity to capture different aspects of the acoustic ecology of the recording studio and, once reassembled for listening, may offer the researcher an alternative to solely writing about the culture of studio recording. Kheshti labels the practice of focusing on cultural acoustics 'acoustic ethnography' or 'acoustigraphy', which 'like ethnography, is a form of writing culture, with an emphasis on sound over other media, or sound alongside other media with a particular sensitivity to sonic culture' (Kheshti 2009: 15).

Using the medium of sound to capture the sonic interactions of a space that privileges sound over any other sense has some distinct advantages, not least that it is a space designed for recording, controlling, and processing sound and therefore allows the researcher to capture high-resolution sound recordings with reduced extraneous noise or sound reflections that can mask speech intelligibility. Rick Altman reminds us though that 'according to the choice of recording location, microphone type, recording system, postproduction manipulation, storage medium, playback arrangement, and playback locations, each recording proposes an interpretation of the original sound' (Altman 2012: 229). Analysing the recorded sound of the studio therefore requires consideration for the context, the situation, the positionality of the microphone and the researcher and the ethical implications of recording audio in the studio. Firstly, contextualization of the recordings is needed to highlight particular details because:

With a camera it is possible to catch the salient features of a visual panorama to create an impression that is immediately evident. The microphone does not operate this way. It samples details. It gives the close-up but nothing corresponding to aerial photography.

(Schafer 2012: 99)

Because of the lack of visual information from a sound recording it may not be possible to know who is present during the recording and including a map of the studio space, the location of the microphone (or microphones) and a general layout of where participants were can help to provide important contextual information for both the researcher and the listener. The position of studio participants can change over the course of a recording session, which can then in turn alter what is captured, and so updating maps and diagrams as a recording session progresses can help to provide both a visual record and some useful context to the sound recordings.

Analysis of the situation is also key to contextualizing the recorded sound captured in the studio. Sound can tell us a lot about a recording studio situation; there may be times of intense sonic activity or periods of almost total silence and this can be dependent upon the time of day, the purpose or type of recording session, and whether or not the audio was captured towards the beginning of a session or towards the end. Long periods of silence, for example, where no one is listening to playback, discussing another take, or generally interacting may underline a particularly tense atmosphere. Laughing and general joviality may indicate that things are going well – having an understanding of the studio participants and their personalities can help significantly in these assessments and developing a social rapport will go some way to help these analyses.

The type of sonic interactions can also tell us a lot about the positionality of the researcher or the recording device. For example, collecting sound in the control room is likely to relate to sound engineers and record producers; musicians do enter the control room throughout a studio session but a lot of the time the control room is the domain of the sound engineer, the record producer and associates of the process such as record company representatives, band management, partners of the band. Conversely, whilst engineers and producers enter the live room to adjust microphone positions or discuss alterations to performances, arrangements, or lyrics, etc. with performing musicians, sound in the live room will typically relate to musicians and their sonic experiences of the recording process.

Importantly, there are ethical implications for capturing the entirety of a recording session both prior to gaining ethical approval form participants and after the data has been gathered. Audio recordings capture conversations and the overall sonic environment of the recording studio but, because of the naturalization of the microphone in the studio, participants often forget that recording is taking place and can sometimes reveal intimate details, offer private information, or make remarks about other participants that aren't intended to be heard. It is therefore imperative that any of the recorded audio is scrutinized before it is replayed to any of the other participants to avoid any unnecessary harm or distress. This is most important where instrumental or vocal performances are being discussed and care must be taken to introduce the background of the discussion in order to contextualize the comments of the participants.

Conclusion

The recording studio is an exciting and varied acoustic space in which to conduct sonic ethnography and capture the sound of the rooms, the equipment, and the interactions of its inhabitants. There are three main sound worlds in the recording studio as well as a series of 'backstage' areas where a lot of rich sound material can come from. The recording studio however is fortress-like both acoustically and socially and therefore gaining access to a recording session is a challenge for researchers. Although engineers and record producers facilitate the needs of the musician and act as intermediaries between the artist and the industry, it is musicians that are the gatekeepers to a recording session and permission should be sought from them first. If the recording musicians are paying for the studio time then it is even more likely that access will be granted from other participants, such as engineers or producers, once permission has been granted from the recording musicians.

Once inside the recording studio, conducting sonic research presents some unique social and logistical challenges because of the recording studio's architecture and the social setting of a recording session. A single audio recorder, a single microphone, or a binaural recorder attached to the researcher can adequately capture the sound of a single sound world of the studio but using multiple microphones allows the researcher to capture the acoustic ecology of the recording studio from various perspectives and, because microphones are naturalized within the studio space, researcher's microphones wouldn't be considered particularly conspicuous.

Finally, analysing the recorded sound of the studio requires consideration for the context, the situation, the positionality of the microphone/researcher, and the ethical implications of recording audio in the studio. In reassembling the captured sounds for playback, sound ethnographers should consider each of these aspects in turn to creatively and responsibly present the 'acoustigraphy' of a recording studio session.

References

Altman, R. (2012). 'Four and a Half Film Fallacies'. In J. Sterne (ed.), *The Sound Studies Reader*, 225–233. Abingdon: Routledge.

Bates, E. (2008). 'Social Interactions, Musical Arrangement, and the Production of Digital Audio in İstanbul Recording Studios'. PhD dissertation, University of California, Berkeley (unpublished).

Bayton, M. (1990). 'How Women become Musicians'. In S. Frith and A. Goodwin (eds), *On Record: Rock, Pop and the Written Word*, 201–219. London: Routledge.

Bull, M. and L. Back (eds) (2004). *The Auditory Culture Reader*. Oxford: Berg.

Coffey, A. (1999), *The Ethnographic Self, Fieldwork and the Representation of Identity*. London: Sage.

Emerson, R. M., R. I. Fretz, and L. L. Shaw (1995). *Writing Ethnographic Fieldnotes*. Chicago, IL: University of Chicago Press.

Fitzgerald, J. (1996). 'Down Into the Fire: A Case Study of a Popular Music Recording Session'. *Perfect Beat: The Pacific Journal of Research into Contemporary Music and Popular Culture* 5 (3): 63–77.

Gibson, C. and J. O'Connell (2005). *Music and Tourism: On the Road Again*. Clevedon: Channel View Publications.

Hammersley, M. and P. Atkinson (1997). *Ethnography: Principles in Practice*. 2nd edition. London: Routledge.

Hammersley, M. and P. Atkinson (2007). *Ethnography: Principles in Practice*. 3rd edition. London: Routledge.

Hennion. A. (1990). 'The Production of Success: An Anti-Musicology of the Pop Song'. In S. Frith and A. Goodwin (eds), *On Record: Rock, Pop and the Written Word*, 185–206. London: Routledge.

Hunt, M. (1985). *Profiles of Social Research: The Scientific Study of Human Interactions*. New York: Russell Sage Foundation.

Kheshti, R. (2009). 'Acoustigraphy: Soundscape as Ethnographic Field'. *Anthropology News* 50 (April): 15–19.

Lashua, B. and P. Thompson (2016). 'Producing Music, Producing Myth? Creativity in Recording Studios'. *IASPM@Journal* 6 (2). https://doi.org/10.5429/2079-3871(2016)v6i2.5en.

LeCompte, M. D. and J. P. Goetz (1982). 'Problems of Reliability and Validity in Ethnographic Research'. *Review of Educational Research* 52 (1): 31–60.

Meintjes, L. (2003). *Sound of Africa!: Making Music Zulu in a South African Studio*. Durham, NC: Duke University Press.

Meintjes, L. (2004). 'Reaching Overseas: South African Sound Engineers, Technology and Tradition'. In P. Greene and T. Porcello (eds), *Wired for Sound: Engineering and Technologies in Sonic Cultures*, 23–48. Middletown, CT: Wesleyan University Press.

Meintjes, L. (2012). 'The Recording Studio as Fetish'. In J. Sterne (ed.), *The Sound Studies Reader*, 265–282. Abingdon: Routledge.

Monahan, T. and J. A. Fisher (2010). 'Benefits of "Observer Effects": Lessons from the Field'. *Qualitative Research* 10(3): 357–376.

Patton, M. Q. (2002). *Qualitative Research and Evaluation Methods*. Thousand Oaks, CA: Sage.

Porcello, T. (2004). 'Speaking of Sound: Language and the Professionalization of Sound Recording Engineers'. *Social Studies of Science* 34: 733–758.

Sawyer, K. (2000). 'Improvisational Cultures: Collaborative Emergence and Creativity in Improvisation'. *Mind, Culture and Activity* 7 (3): 180–185.

Schafer, R. M. (2012). 'The Soundscape'. In J. Sterne (ed.), *The Sound Studies Reader*, 95–103. Abingdon: Routledge.

Signature Sound (2011). 'Proper Studio Etiquette'. Available online: https://www.signaturesound.com/proper-studio-etiquette/ (accessed 25 October 2018).

Spano, R. (2006). 'Observer Behavior as a Potential Source of Reactivity: Describing and Quantifying Observer Effects in a Large-Scale Observational Study of Police'. *Sociological Methods & Research* 34(4): 521–553.

Thompson, P. (2016). 'The Scalability of the Systems Model in the Recording Studio'. In P. McIntyre, J. Fulton, and E. Paton (eds), *The Creative System in Action*, 74–86. London: Palgrave.

Thompson, P. (2019). *Creativity in the Recording Studio: Alternative Takes*. Basingstoke: Palgrave Macmillan.

Thompson, P. and B. Lashua (2014). 'Getting It on Record: Issues and Strategies for Ethnographic Practice in Recording Studios'. *Journal of Contemporary Ethnography* 43(3): 46–769.

Williams, A. (2007). 'Divide and Conquer: Power Role Formation and Conflict in Recording Studio Architecture'. *Journal of the Art of Record Production* 1 (1). Available online: https://www.arpjournal.com/asarpwp/divide-and-conquer-power-role-formation-and-conflict-in-recording-studio-architecture/ (accessed 25 October 2018).

Williams, A. (2009). 'Charged Encounters: The Mercurial Nature of Role Formation in the Recording Studio'. *Proceedings of the 2009 Art of Record Production Conference*, University of Glamorgan, UK, 13-15 November 2009. Available online: https://www.arpjournal.com/asarpwp/putting-it-on-display-the-impact-of-visual-information-on-control-room-dynamics/ (accessed 25 October 2018).

Williams, A. (2010). 'Celluloid Heroes: Fictional Truths of Recording Studio Practice on Film'. *Proceedings of the 2010 Art of Record Production Conference*, Leeds Metropolitan University, UK, 3–5 December 2010. Available online: http://arpjournal.com/1412/celluloid-heroes-fictional-truths-of-recording-studio-practice-on-film/ (accessed 25 October 2018).

Williams, A. (2011). 'Putting it on Display: The Impact of Visual Information on Control Room Dynamics'. *Proceedings of the 2011 Art of Record Production Conference*, San Francisco State University, USA, 2–4 December 2011. Available online: http://arpjournal.com/1845/putting-it-on-display-the-impact-of-visual-information-on-control-room-dynamics/ (accessed 25 October 2018).

Williams, A. (2012). 'I'm Not Hearing What You're Hearing: The Conflict and Connection of Headphone Mixes and Multiple Audioscapes'. In S. Frith and S. Zagorski Thomas (eds), *The Art of Record Production: An Introductory Reader for a New Academic Field*, 113–128. Farnham: Ashgate.

Zak, A. (2001). *The Poetics of Rock: Cutting Tracks, Making Records*. London: University of California Press.

28

Ecological Sound Art

Jonathan Gilmurray

Throughout the course of human history, the arts have always reflected the prevailing concerns of their time; and in the twenty-first century, humanity's growing awareness of ecological issues such as biodiversity loss, pollution, environmental justice, and climate change has been expressed in an explosion of ecologically concerned works across every area of arts and culture. In literature and the visual arts, this trend has been recognized in the establishment of new fields of practice, such as eco-art (Weintraub 2012; Brown 2014; Cheetham 2018), ecofiction (Dwyer 2010; Levin 2011; Raglon 2012), and ecocinema (Gustafsson and Kääpä 2013; Rust, Monani, and Cubitt 2013; Murray and Heumann 2017). As in other art forms, sound artists have also been creating works that address contemporary ecological issues; however, an equivalent ecological movement has yet to be recognized in sound art, resulting in its exclusion from the wider discourse surrounding the cultural response to ecological issues. The current chapter sets out to address this issue with an examination of this growing contemporary movement, which it will term *ecological sound art*.

In other areas of the arts, the defining characteristic of 'ecological' or 'eco-' works is their central focus upon ecological concerns; thus, in keeping with this, the following may be given as a basic definition of the field under discussion:

> Ecological sound art describes a modern movement within sound arts practice comprising works whose form, content or subject matter demonstrates an active engagement with contemporary ecological issues.

In order to expand upon this basic definition, this chapter will proceed to analyse a representative sample of ecological sound works, with the aim of identifying some of the common characteristics and core methodologies of this urgent and flourishing new field.

Perhaps the most pressing ecological issue being faced today is that of climate change, one consequence of which is the melting of the ice at the North and South poles, leading to the rise of global sea levels. This issue is addressed by a number of sound artists whose work utilizes the sounds made by melting glaciers, such as Douglas Quin's *FATHOM* (2010), which features underwater recordings of glacial ice at the Arctic and the Antarctic; a number of Jana Winderen's works, including +4°C (2007), *Evaporation* (2009), *Energy Field*

(2010), and *Spring Bloom in the Marginal Ice Zone* (2018), in which her recordings of Arctic glaciers are underpinned by drones, which give the feeling of being submerged beneath the icy waters; or Daniel Blinkhorn's *frostbYte* cycle (2012–2015), a suite of compositions whose microsound aesthetic invites comparisons between the cracking and popping of the polar ice and the sonic palette of glitch music. One of the key aspects of all of these works is the way in which sound communicates the dynamics of the glacial melting process. Approached from a purely visual perspective, a glacier appears to be literally 'frozen', its melting too gradual to be perceived in real time. In contrast, the sonic dimension of glacial melting facilitates our understanding of it as a dynamic process, happening ceaselessly and progressively, encouraging the sense of urgency with which it is so vital that we learn to regard and respond to climate change.

Other sound artists have created installations or performative works which enable audiences to bear witness to the process of glacial melting happening right before their ears, in the here and now. Katie Paterson's *Vatnajökull (the sound of)* (2007) provides audiences with a telephone number connecting them to a hydrophone submerged beneath the glacier of the title, enabling them to listen to it melting in real time, with the very personal medium of a 'one-to-one' phone call collapsing the geographical distance between the listener and the glacier. Paterson's related work *Langjökull, Snæfellsjökull, Solheimajökull* (2007), meanwhile, involves recordings of three melting glaciers being pressed onto records made from their own refrozen meltwater, which are then played on turntables until they completely melt, re-enacting the glacial melting and reminding us that, while recordings can be played back to give the illusion that their subject is in the room with us, when the glacial ice itself is gone we will be unable to simply return the needle to the start.

Another work to physically re-enact the melting of an Arctic glacier is Max Eastley's *Glacial Soundscape* (2005), just one of a number of sound works resulting from his involvement with the Cape Farewell climate arts project. Eastley's work consists of two large blocks of ice with stones embedded in them suspended over an amplified aluminium sheet, creating a constant dripping sound as the ice melts, with the sporadic loud bangs created by the falling stones communicating that the melting of a glacier is not always a gradual, steady process but can also be sudden and violent in nature – something which also creates for the audience a constant feeling of nervous anticipation, enacting the anxiety which we perhaps should be feeling when we contemplate the melting of the Arctic ice. Finally, Cheryl E. Leonard's performative work *Meltwater* (2013) from her suite *Antarctica: Music from the Ice* (2009–2014) combines field recordings of the disappearing Marr Ice Piedmont glacier with the sounds made by a number of icicles suspended over amplified Pyrex beakers and Petri dishes, creating steadily increasing patterns of rhythmic dripping as they melt in the warm, man-made climate of the concert hall, mirroring the real glacier melting as a consequence of the global warming caused by climate change. Further, Leonard and other performers play instruments made from materials gathered from the Antarctic, including Adélie penguin feathers and bones, referencing the decline in penguin colonies being caused by the changing climatic conditions and progressive disappearance of the ice.

As well as enabling us to bear witness to the dynamic process of ecological disaster unfolding before our ears, some works of ecological sound art also feature the sounds being made by those who bear at least partial responsibility for the damage being done. Max Eastley's *ARCTIC* (2007), another outcome of the Cape Farewell project, comprises twelve works composed from sounds captured by Eastley on the expeditions, including not just the natural sounds of the Arctic but also the industrial noise of the Barentsberg coal mine, something which both challenges the common notion of the Arctic as an empty, silent wilderness and draws attention to the tragic irony of its being a location for the mining of fossil fuels, our use of which is one of the major causes of the climate change which is causing it to melt. Moving away from the polar regions, meanwhile, *Dark Sound* (2016) by Mikel R. Nieto documents the impact of the oil industry upon the Amazon rainforest in Ecuador, with the sounds of insects, frogs, and birds becoming gradually replaced by those of generators, pumps, and drills, mirroring the way in which the oil industry is progressively obliterating the rainforest, and raising the spectre of a future in which the rainforest, along with all of the animals and people that live there, may one day no longer exist.

Another source of the destruction of the earth's forests is revealed in David Dunn's *The Sound of Light in Trees* (2006), which provides a window into the complex soundworld of the pine bark beetle, whose rapidly increasing numbers due to climate change are causing the decimation of the population of piñon pines in New Mexico. However, the CD release of their squeaking and clicking sounds constitutes only one aspect of Dunn's work: he discovered that playing the recordings back to the beetles and combining them with nonrepeating synthetic sounds had a profound effect on their neural system, causing them to cease their tunnelling and feeding behaviours and even shutting down their reproductive cycle, enabling him to use them as an environmentally friendly form of pest control, and resulting in a work of ecological sound art which also constitutes a remedy for the problem it highlights.

While the above works showcase the sounds of ecological destruction, others highlight the sounds of the living creatures that are threatened, or have already disappeared, as a consequence. The sounds of various extinct and endangered species form the basis of works such as *Suspended Sounds* (2006), created for the inaugural Ear to the Earth festival in New York, which features eight-channel compositions by artists Joan La Barbara, Joel Chadabe, Alvin Curran, David Monacchi, Aleksei Stevens, and Rama Gottfried; and Maya Lin's sound sculptures *The Listening Cone* (2009) and *Sound Ring* (2014), two of the works in her multifaceted 'last memorial' project *What is Missing?* (2009–present). The many aquatic species which live in the earth's coral reefs, meanwhile, are the focus of both Jana Winderen's *Silencing the Reefs* (2011–2014) and Leah Barclay's *Sonic Reef* (2017–present), two ecological sound art projects which each explore the changing soundscapes of the reefs and their ecosystems in a variety of ways including field recordings, compositions, concerts, installations, and workshops, promoting greater understanding of these environments and how they are being negatively impacted by human actions.

Other sound artists have turned their attention to the plight of the earth's birds. Krista Caballero and Frank Ekeberg's installation *Birding the Future* (2013–present) involves the creation of an immersive soundscape from the calls of extinct and endangered bird species specific to each region in which the work is presented. In the work, the calls of the extinct species feature as straight field recordings, presented as a simple memory of what has been lost, while those of the endangered species are modified to form Morse code messages warning of their impending fate, their density steadily decreasing over the course of the work in line with projected rates of extinction to the end of the century. Sally Ann McIntyre, meanwhile, has realized several sound works focused upon the extinction of birds native to New Zealand: *Collected Silences for Lord Rothschild* (2012) consists of recordings of the silences of stuffed specimens of five species of birds rendered extinct as a result of the European colonization of New Zealand; *Huia Transcriptions* (2012) is a recording of a music box placed in a forest playing a reproduction of the song of the Huia bird, based on a written transcription made shortly before it became extinct in 1907; and *Huia Notations (like shells on the shore when the sea of living memory has receded)* (2015) features a recording of a piano playing the bird's song cut to wax cylinder (the only sound recording medium available while the bird was alive) and played back on an Edison phonograph, a fragile medium which will eventually destroy the sound through its own playback.

Another creature whose numbers are in steep decline due to human actions is the honeybee – a fact which has serious consequences for a multitude of other species, including our own. *Amhrán na mBeach* (Song of the Bees) (2014) by the duo Softday is a performative sound work focused upon the ecological threat posed by colony collapse disorder, combining eight-channel recordings from inside a hive with a musical score based on scales which are 'in tune' with the bees; while *Resonating Bodies* (2008–present), a multifaceted arts project conceived by Sarah Peebles and realized in collaboration with a number of other artists and ecologists, combines multimedia installations, community outreach projects, and educational initiatives focused upon pollination ecology, with works including *Bumble Domicile* (2008), an installation combining a live audio and video feed from inside a hive, a soundscape made from audio transformations of the bees and live data visualizations based on the DNA sequences of the pollen; and *Audio Bee Booths* (2009–present), a series of nesting cabinets for wild solitary bees incorporating headphones through which visitors can listen in to the sounds they are making.

While the work of these sound artists encourages an attitude of concern for the future of other species, others have focused upon promoting a greater awareness of the ways in which our own species also suffers the consequences of our abuses of the earth's ecosystems. Peter Cusack's *Sounds from Dangerous Places* (2006–2012), a 'sonic journalism' project which investigates the soundscapes of sites that have undergone major ecological damage, features at its heart a number of recordings from the Chernobyl exclusion zone, revealing it to have developed a somewhat complex ecological identity. On the one hand, we hear the soundscape of what appears, in the absence of humans, to have become an ecologically thriving wilderness, home to a wide variety of animals; however, the other side of the story is also represented by the poems and songs performed by the town's evacuees, who are unable to return home. Another of Cusack's sonic journalism projects, *Soundscapes of Water Use and Abuse* (2012–present), explores the implications of various ways in which humans make use of this precious natural resource, including the ecological damage caused by the dredging of the Thames Estuary, and the consequences of the damming of the Tigres and Euphrates rivers in Turkey by hydroelectric companies.

The practice of damming major rivers to produce electricity – something which is being done in a number of places around the world – brings with it serious ecological justice issues, something which has also been explored by works of ecological sound art. *The Dam(n) Project* (2011–present) is a collaborative, multidisciplinary project chronicling the damming of the Narmada River in North India by hydroelectric companies, and the impact this has had upon the many communities – comprising over thirty million people – who depend upon the river and who have consequently been displaced. Underpinning the project are Leah Barclay's sound works, which combine recordings of the sounds of the river with testimonies and songs from the local communities gathered in non-violent protest, and which have been presented both as a site-specific sound installation, and as the soundtrack to a contemporary dance work and a documentary film, both entitled *Zameen*. In Graciela Garcia Muñoz's *El Sonido Recobrado* (2014), meanwhile, the sounds of Chile's largest and most powerful river, the Baker, which faces exploitation from a proposed hydroelectric plant, are played back over twenty-eight speakers set in the bed of the Petorca, a river which has been completely dry since the 1990s as a consequence of being illegally dammed and drained by mining and agriculture companies, causing ongoing problems for local communities.

Returning to the issue of the melting Arctic ice, Holly Owen and Kristina Pulejkova's audiovisual work *Switching Heads: Sound Mapping the Arctic* (2015) investigates the impact of this phenomenon upon the human community of the Norwegian city of Tromsø, located within the Arctic Circle. Over the course of the work, the city is explored by an ice sculpture of a human head with binaural microphones implanted in its ears, listening in both to the soundscapes of the place, and to interviews with local people talking about the ecological, social, and economic impacts that the melting of the Arctic ice will have on their lives. The use of the binaural format, meanwhile, means that the ears of the ice head become our ears, creating an embodied connection and personal identification with it; and as it appears in various states throughout the film, from its fully frozen and perfectly rendered form to an unrecognizable block of slush, we find ourselves personally invested in its well-being, willing it not to melt, creating a powerful conceptual resonance with the fact that the well-being of humankind – most immediately in places such as Tromsø – is reliant upon the Arctic ice remaining frozen.

Another sound work to combine polar soundscapes with interviews on the theme of climate change is *Sonic Antarctica* (2009), in which Andrea Polli juxtaposes her Antarctic field recordings with the voices of climate scientists discussing their research, as well as their own personal feelings about its implications. The data being studied and discussed by the scientists, meanwhile, is also integrated into the work through the technique of sonification, with Polli translating it into a sonic palette of electronic bleeps and rhythms, something which functions on a conceptual level to enable a sensorial encounter with the data which, as the scientists explain, holds dire implications for the human race. Polli has also realized a number of other sonification-based works of sound art utilizing different types of ecological data, such as *Heat and the Heartbeat of the City* (2004), an interactive work which allows the listener to explore a sonification of projected temperature increases in Central Park caused by global warming; and her *Airlight* trilogy (2006–2007), which involves the sonification of data from air quality monitoring stations in Taipei, Socal, and Boulder to create real-time aural representations of the levels of various atmospheric pollutants. Taking a similar approach to this last issue, meanwhile, is Wesley Goatley and Tobias Revell's installation *Breathing Mephitic Air* (2017), which sonifies air pollution data gathered from the area surrounding the site of its exhibition at Somerset House in London, transforming it into a surround-sound soundscape whose different sonic elements convey the levels of different pollutants in the air which visitors themselves have just been breathing.

Various approaches to the sonification of climate change data also form the basis for James Wyness's *If We Do Nothing* (2017–present): the first of the work's 'models' combines a slowly rising electronic tone mapped to CO_2 emissions with a falling tone mapped to glacier ablation spanning the years 1880 to 2050, with the predicted tipping points coinciding with the limits of human hearing; while the second model maps recordings of stories and myths in the native languages of the Arctic region and scientific texts on climate change onto scientific and ethnographic data. Two other models, in the planning stage at the time of writing, will also involve the sonification of data measuring the disappearance of Arctic sea ice, and comparisons of rural and urban pollution levels in Scotland. Softday, meanwhile, use sonification in a pair of works exploring issues of water pollution in Ireland: *Nobody Leaves till the Daphnia Sing* (2009) focuses upon the issue of contaminated drinking water supplies, combining live sonifications of the activity of daphnia magna water fleas, insects commonly used for the analysis of water and soil toxicity, with the performance of a score generated from eighteen years of data from water samples around Ireland; while *Marbh Chrios* (Dead Zone) (2010) takes a similar approach to the investigation of the dramatic increase in oceanic 'dead zones' (areas of seafloor with too little oxygen for most marine life), featuring the live sonification of ecological data from two such zones combined with the performance of a score based on eight years of related marine and environmental data.

Another ecological sound work which combines sonifications of ecological data with scores performed by live musicians is Matthew Burtner and Scott Deal's *Auksalaq* (2012), a multimedia opera featuring a number of Matthew Burtner's 'ecoacoustic' works, in which he employs a variety of strategies to translate ecological materials and processes into sound and music. The work's subject matter centres upon the effects of climate change upon Alaska and the Arctic; however, another key component is its wider philosophical theme of ecological interconnectedness. Different parts of the work are performed simultaneously in a number of different locations and brought together over high-speed internet networks, functioning as a metaphor both for being impacted by geographically distant events and for people in different places working together to achieve a common goal. The audience is also invited to actively participate in this cooperative staging, as a specially designed app called NOMADS (Network-Operational Mobile Applied Digital System) enables them to contribute to aspects of the performance via their own laptops and mobile devices, such as during the work's opening movement, when each member is given control of the speed and pitch of the sound of a single droplet of glacial meltwater; thus together, the whole audience forms a melting glacier. As well as making the performance an interactive experience, this forms a powerful metaphor for the way in which we are all responsible, in a small way, for a tiny element of our changing climate; and how, when all of our seemingly insignificant individual actions are combined, it can result in the melting of an entire glacier – or even, eventually, the entirety of the Arctic sea ice. *Auksalaq* thus becomes a work which is not just about highlighting ecological problems but also about emphasizing a sense of personal connection, both with the earth's threatened ecosystems and with each other, thereby fostering a sense of empowerment, and encouraging us to take collective action in order to help preserve them.

This encouragement to consider our personal connection to the earth's threatened ecosystems also lies at the heart of David Monacchi's *Fragments of Extinction* (2002–present), an ongoing project which utilizes high-definition ambisonic recordings of the soundscapes of the earth's three remaining areas of primary equatorial rainforest, both for their preservation as a form of 'sonic heritage', and for use within Monacchi's own 'eco-acoustic' sound works. The project has been the subject of a documentary film, *Dusk Chorus* (2017); and its latest development utilizes Monacchi's patented 'eco-acoustic theater', a domed space designed for immersive listening which exists both as a portable touring structure and in a permanent version installed at Denmark's Naturama Natural History Museum. Within the theatre, Monacchi's three-dimensional soundscape recordings and compositions are paired with live spectrogram visualizations that demonstrate the principle of Bernie Krause's 'Acoustic Niche Hypothesis' (1987), which states that in a healthy ecosystem the calls of each species sound within their own specific frequency or temporal niche so that all can be heard, much like the different sections in an orchestra – a delicate balance which is increasingly being destroyed by the encroachment of humans upon the rainforest ecosystems and their soundscapes. However, Monacchi's work also strives towards the positive and hopeful principle of healthy coexistence in the form of his own contributions to the soundscape, performing electronic improvisations by way of infrared sensors reading the movements of his fingers. Crucially, the sonic content of these improvisations is carefully confined to the available acoustic niches in the rainforest soundscape, thus creating a powerful metaphor for how humans might learn to approach the earth's natural ecosystems guided by the principles of listening, of understanding, and of learning how to humbly and harmoniously integrate into an environment, being guided by the healthy operation of the ecosystem as a whole.

Another sound artist who utilizes technology as a tool to help us learn to live more responsibly and conscientiously in our relationship with the earth's ecosystems is Leah Barclay, whose work comprises a number of ongoing, multifaceted ecological sound art projects. *Sonic Explorers* (2012–present) engages young people in ecological sound art through field recording and composition workshops, live performances, and sound mapping; while *River Listening* (2014–present) takes the form of an interdisciplinary art-science project, combining the scientific study of the soundscapes of various rivers as a means to measure their health and biodiversity with sound art which enables audiences and communities to learn about and connect with the rivers where they live through talks, workshops, and interactive soundwalks delivered via their mobile devices, with recordings, compositions, and live hydrophone streams geotagged to specific locations along the length of the river. *Rainforest Listening* (2015–present) takes a similar approach, with geotagged soundwalks made up from over one hundred individual recordings of the

Amazon rainforest embedded in prominent city locations such as Times Square and the Eiffel Tower, encouraging listeners to consider the ways in which our actions in the city might impact the rainforest in another part of the world, in a place other than 'here', and in a time other than 'now'. Of course, this principle applies not just to humanity's ecologically irresponsible and destructive actions, but also to the positive action we choose to take, something which Barclay's work crucially facilitates by enabling listeners to use their mobile device to donate to the Rainforest Partnership – an non-governmental organization (NGO) which works with rainforest communities to help conserve the forest – whilst still in the moment of engagement with the work. Finally, on an even larger scale is *Biosphere Soundscapes* (2012–present), a multifaceted project exploring the soundscapes, ecology, and environmental health of UNESCO Biosphere Reserves, pivoting on three core systems: BioScapes Residencies, which bring together selected artists, researchers, and scientists to explore and map the soundscape of a biosphere reserve, share knowledge, work on creative outputs, and engage with local communities; BioScapes Labs, a series of workshops exploring specific research questions around the soundscape ecology of biosphere reserves; and the BioScapes Community, an online resource which acts as a platform for the dissemination of the project's outputs, including an interactive sound map and educational resources. In their scope and ambition; multifaceted, interdisciplinary, and collaborative nature; combination of a local and global focus; holistic ecological outlook combining artistic, scientific, and political concerns; community engagement; educational benefit; multiplatform delivery; global reach; and overall artistic and ecological integrity, Leah Barclay's works represent perhaps the most powerful and effective model for future ecologically concerned sound artists to aspire to.

Common characteristics of ecological sound art

Having explored the topics, tools, and techniques of a representative selection of works of ecological sound art, a number of common characteristics may now be identified, indicating some of the core methodological principles of the field.

1 The use of listening as a pathway towards greater ecological understanding.

Listening to the sounds of an environment or ecological phenomenon is frequently used within works of ecological sound art as an intuitive, experiential means of learning more about the functioning of the earth's natural ecosystems and the problems they are facing.

2 The promotion of an ecological mode of listening.

The soundworld found in works of ecological sound art directs us towards an ecological mode of listening, in which our focus moves away from isolated objects or things in favour of the dynamics of ecological processes, interactions, and interrelationships, and which thus facilitates a sensorial awareness of the principles of the interconnected ecosystem.

3 A prioritization of listening over sounding.

Ecological sound art is commonly centred upon the artist's own investigative listening, with the work functioning as a channel through which to share their listening and learning with others. This principle involves a degree of humility on the part of the artist that equates to a tangible ecocentrism, in which the focus of the work is shifted away from the artist, and onto the environment, ecosystem, or ecological phenomenon being explored.

4 The use of sounding as a metaphor for ecological coexistence.

When ecological sound art does incorporate the sounding of humans, or of man-made instruments and technologies, this tends to be used as a metaphor for learning how to positively coexist with other elements of the environment within a space of ecological commons, guided by the principle of the healthy and harmonious operation of the ecosystem as a whole.

5 A form that functions as an ecosystem.

The ecological principles explored in works of ecological sound art are often exemplified by a work which itself functions as an ecosystem, comprising a number of interconnected parts whose interaction determines the final form of the overall work. This principle also involves an element of ecocentric humility on the part of the artist, as it necessitates relinquishing a certain degree of control over the precise form and content of the work.

6 A blend of art with science, of ecology with environmentalism.

Works of ecological sound art tend to be inherently interdisciplinary in nature, blending the investigation and experimentation of science with the creativity and expression of art, and combining the principles and dynamics of ecology with the causes and concerns of environmentalism.

7 A combination of the educational with the philosophical.

In their engagement with contemporary ecological issues, works of ecological sound art frequently combine the informative or educational with the emotive or philosophical, thus leaving audiences with both an enhanced understanding of contemporary ecological issues and the emotional impetus to do something about them.

Conclusion

While it would be absurdly optimistic to claim that sound art is the answer to solving climate change, halting deforestation, or preventing pollution, it is also imperative that we do not merely relegate these urgent issues to the fields of science and politics, but that we learn to engage with them in every aspect of our lives, in which regard our arts and culture have an immensely significant role to play. Ecological sound art, in its wide variety of approaches towards this end, represents a vital and thriving contemporary field of practice, with a growing number of artists creating sound works which aim to help us to open our ears to ecological issues, to listen to and understand the warning signals, and to explore ways in which we might learn to live more harmoniously within the ecosystems of which we are a part, and upon which we all depend.

References

Brown, A. (2014). *Art and Ecology Now*. London: Thames and Hudson.
Cheetham, M. (2018). *Landscape into Eco Art: Articulations of Nature Since the '60s*. University Park, PA: Pennsylvania State University Press.
Dwyer, J. (2010). *Where the Wild Books Are: A Field Guide to Ecofiction*. Reno, NV: University of Nevada Press.
Gustafsson, T. and P. Kääpä (eds) (2013). *Transnational Ecocinema: Film Culture in an Era of Ecological Transformation*. Bristol: Intellect.
Krause, B. (1987). 'Bioacoustics: Habitat Ambience and Ecological Balance'. *Whole Earth Review*, 57: 14–18.
Levin, J. (2011). 'Contemporary Ecofiction'. In L. Cassuto, C.V. Eby, and B. Reiss (eds), *The Cambridge History of the American Novel*, 1122–1136. Cambridge: Cambridge University Press.
Murray, R. L. and J. K. Heumann (2017). *Ecocinema and the City*. Abingdon: Routledge.
Raglon, R. (2012). 'A Green Turn: Western Canadian Writers and Ecofiction'. *Canadian Literature* 214: 132–134.
Rust, S., S. Monani, and S. Cubitt (eds) (2013). *Ecocinema Theory and Practice*. Abingdon: Routledge.
Weintraub, L. (2012). *To Life! Eco Art in Pursuit of a Sustainable Planet*. Berkeley, CA: University of California Press.

29

Hydrophonic Fields

Jana Winderen interviewed by Stefan Helmreich

In this interview, musician and sound artist Jana Winderen talks with anthropologist Stefan Helmreich on how she thinks about – and technically apprehends – Earth's contemporary undersea realm, a realm that, in the age of climate change, is transforming in its temperature, its creaturely ecologies, its acidity – and its sounds. Winderen is well known for her recordings and compositions in and about underwater realms, celebrated for pieces of sound art and music that transport listeners into unfamiliar submarine and subaquatic zones. Helmreich has written on sound and sensibility in submarine spaces and on the unusual history of underwater music (see Helmreich 2009, 2012).

Some of Jana Winderen's notable compositions and releases include:

- *The Noisiest Guys on the Planet* (Ash International # Ash 8.1). Cassette, 2009/2010.
- *Energy Field* (Touch # TO:73). CD, 2010.
- *Debris* (Touch # Tone 45.4). Vinyl, 2012.
- *The Wanderer* (Ash International # Ash 11.8). USB, 2015.
- *The Listener* (Ash International # Ash 12.5). USB, 2016.
- *Spring Bloom in the Marginal Ice Zone* (Touch # Tone 65). CD, 2018.

Her multichannel installation works include: 'UltraField' (2013), 'Dive' (2014), 'bára' (2017), and 'Spring Bloom in the Marginal Ice Zone' (2017 and 2018).

Many of Winderen's pieces/recordings are available through Touch (n.d.) or through Winderen's bandcamp site (Winderen n.d.). It should be said that attending a performance or installation of these works will offer a much more dimensional experience than these digital documents.

Stefan Helmreich (SH): How did you arrive at underwater recording/composition — both as a mode or genre you wanted to explore and as a practice to which you wanted to dedicate yourself?

Jana Winderen (JW): I have always been occupied with the ocean, and also with fresh water. I was a child in the 1970s and grew up in the 1980s, and I lived by a lake, the largest lake in Norway, Lake Mjøsa, which I watched become almost strangled by algae overgrowth over the years.

I originally thought I wanted to become a marine biologist and so I studied mathematics, chemistry, biochemistry, and fish ecology at the University in Oslo. But I then studied Fine Arts at Goldsmiths in London. In my second year, I decided to stop making physical, bounded art objects and to start to make art pieces using more 'immaterial' materials, like sound, light, and air – which of course had a long history in art practice. When I started doing this, though, I was unaware that such work might also fall into a genre of *music*. That came later.

By the early 2000s, I had begun to work with interactive sound installations, driven by different kinds of sensors and triggers. Through the use in my installations of piezoelectric sensors (which can detect sound pressure differences), I became aware of the piezoelectric technology inside of hydrophones, which are a kind of underwater microphone. As I began to use hydrophones, I became interested in testing the limits of these devices. I would freeze them in ice and record the melting of the ice around them. I went on an early recording trip with musician and sound recordist Chris Watson and the director of Touch, Mike Harding, during which we explored an ants' nest with hydrophones. I recorded in all sorts of places – under the sand as waves washed over a beach, in mud, inside trees (where I could hear woodpeckers and insects) – but mostly underwater, at different depths, under and inside glaciers down to 90 metres under the sea ice in Greenland.

I remember once, quite a long time ago, during one winter, in more than a metre of snow, we brought a boat to the edge of the ice-covered shore of Lake Mjøsa and rowed out into deep water. I had two hydrophones at that time, and I placed them at different depths. At 30 metres down, a singing sound showed up, surprising me. It was the first time I had experienced the extreme difference of sound at different depths. Exploring sound in water is almost always more interesting in the vertical direction than it is in the horizontal direction. This is probably something all creatures living underwater know. It was around then that I started to think about invisible but audible landscapes under water, sound spaces that I believe fishes, crustaceans, and mammals use to orientate themselves. We know that whales have a sound channel underwater, where the sound waves they produce and receive bounce up and down, without losing energy. That channel exists at a particular depth, depending on things like temperature, salinity and pressure – and it means that whales' sound and hearing can carry over enormous distances. Humans seem to be ignorant of the damage and destruction we inflict with our engines, our sonar, or with our air guns for seismic testing – to say nothing of the detonation of bombs. Many of these sounds can be extremely destructive to the creatures living in the sea. I am trying with my work to point towards ecosystems that are often overlooked because of human ignorance or because we are missing a language for talking about them. Thankfully more and more people are becoming aware of the ocean world and its sound. People now know that fish also communicate and experience sound.

So, my mode is exploration. I don't think of hydrophone recording as a 'genre'. It has simply been through investigation that I have come upon the hydrophone as a tool for finding different, less obvious sounds, the sounds of creatures we do not think of in terms of sound, or that are so small we do not hear them, in frequency ranges we do not perceive […]. There are endless areas to explore: listening to insects underwater, exploring fish perception, thinking about how all these things are affected by rising carbon dioxide levels, what will happen to undersea populations with sea ice melt […]

SH: Right. And your work ranges really fascinatingly along these different domains and in these different registers. I guess my question about genre was animated by my own listening to things like David Dunn's 'Chaos and the Emergent Mind of the Pond', a collage of recordings of aquatic insects in ponds in North America and Africa, Erik De Luca's *[In]*, a collection of compositions of underwater recordings around the Florida Keys, and Annea Lockwood's *Sound Map of the Danube River*. And because I've gotten kind of obsessed with tracking these things down, I also think, with your mention of ice melt, of Peter Cusack's *Baikal Ice*, Andrea Polli's *Sonic Antarctica*, and Wendy Jacob's *Ice Floe*. My gathering of all these things together as 'genre' is probably much more about how I encounter these pieces – which is as a *listener* – than it is about what you and I are talking about right this moment, which is about how you *create* your works. But staying in 'genre' for a spell, are there key composers/recordists/sound artists whose work you find inspiring or appealing as you think compositionally?

JW: In Vienna in the 1980s there was a great underground scene, people building their own instruments, making concerts. In 2003, I got to know Carl Michael von Hausswolff through a workshop called 'Sound as Space Creator'. Later, this became the project 'freq_out', which I have been part of ever since and through which I met fantastic artists and composers. I also met Mike Harding, one of the directors of Touch, with whom I later worked and through whom I got to know Chris Watson, BJ Nilsen, Philip Jeck, and other artists on the Touch roster. I'm of course aware of other artists working with underwater materials, and I appreciate and listen to their work, but my first inspirations are usually from scientific papers, which I can say more about in a moment.

SH: OK, we'll get back to that for certain! I do like this trajectory you've just mapped out, from *underground* to *underwater* music and from art to science […]. Let me leap over to *technology* for a moment then, to the question of how you create your works. What field recording methods have you found essential for capturing underwater sound? What do you listen for in the recordings? And what is the most surprising thing you have found in the marine settings you explore?

JW: You develop as you go along, through the challenges you meet. You learn how to pack your equipment in minus 40 degrees Celsius, or when it is 99 per cent humidity, for example. You also learn work habits that permit you to acquire the sounds for which you're searching – not being interrupted by human-made sounds, for example. Sometimes, I need to go out at 4.00 a.m., when most people are sleeping.

My technique will also be very different depending on whether it is boiling hot or very cold (and, in Arctic regions, whether I am in danger of being attacked by polar bears!). In the Caribbean, the most sensitive hydrophones are too much for the singing humpbacks and will overload recording levels completely, and one can hardly hear details of toadfish and crustaceans. The same hydrophones will be perfect in still water environments, though. The best will be a time when there is no wind, when I can concentrate completely on small sounds underwater, and start to listen to the richness and complexity in fish environments.

What do I listen for? I listen for things I haven't heard before. Let me tell you about a sound that freaked me out, though – the sound of the 'seal scarer'. I was travelling with music students from Glasgow to record on a flameshell reef off the Isle of Skye in Scotland. I was expecting to hear the crackling and popping sounds of crustaceans, the sounds of water filtrating through shells, cod grunts, haddock knocking, but when I put my hydrophones under the surface, I immediately heard a sharp, metal-against-metal kind of sound screaming in my headphones. I had to turn the recording level down quickly. I asked the guide who was in the boat what on Earth this sound was, and he told me that it was the 'seal scaring' audio device. These devices are situated there, he told me, to scare seals away from fish farming pens. It turns out that in Scotland, there exists a law that once the fish farmers have tried everything to scatter seals from around fish farms, they can shoot them. To my mind, it would make more sense to have an anti-predator net installed, though I have heard that these are expensive to maintain. Still, it is quite incredible that the horrendous sound of the 'scarer' is permitted to play 24 hours a day, seven days a week, in the seal habitat, where the seals have been living for far longer than the fish farms have been in place. I later contacted a maker of seal-scaring audio devices and he was devastated to learn how the devices were being used, since he had hoped that they would be used to save seals from being shot – not used as an excuse or warrant for *when* to shoot seals.

SH: That's really distressing. But it does open up to the next question I wanted to ask you, which is about what you do with the sounds you collect. What messages do you want them to relay? How do you think about reanimating, composing, and reproducing underwater sound? And how do you think about doing that for the generally (airy) settings of human listening?

JW: I think in terms of stories. I try to give the compositions a sense of travel, movement – of moving from above to under water, for example, or from warmer to colder areas. Of course it's best to take people there, and even to go underwater with them. At the moment I am working on a project where I am emphasizing bone conduction through the technique of putting an oar to the head to listen to underwater sound environments.

A while ago, I read a story in the journal *Nature* about how amphibian hearing, developed to be able to listen underwater, evolved, in some lineages, towards hearing in air. When ancient polypterid fishes (which some people have called 'Darwin fishes') began to move onto land, they started to breathe through the tops of their heads, with their breathing apparatus eventually becoming an ear (see Graham et al. 2014). I became interested in the possibility that our inner ear is similar to the inner ear of fish.

I have just started to investigate this again. I grew up close to my grandfather, who was an ear, nose, and throat specialist doctor, and I remember him putting a tuning fork on my head to check the hearing through bone conduction. I still have those tuning forks, and I find it fascinating to test your inner ear like this, through the bone structure. In fact, my project for the Thailand Biannual this year is based on this knowledge. I'm calling it 'Through the Bones', and it is based on listening (through bone conduction) through an oar to the sea around Greenland as well as through an oar to the sea around Thailand.

SH: Mention of evolution and of the physiology of hearing prompts me to return to your mention earlier that some of your first inspirations for work are often drawn from scientific papers. What role does reading into various scientific literatures on ocean ecology play for you?

JW: This is essential for me to understand and learn. I used to study science, so I recognize stuff from biochemistry, mathematics, fish ecology. I read papers online, or ask scientists questions, through mail, phone, or in in-person meetings. Also important are the collaborations and contacts I have in the industry that makes the equipment, the developers of hydrophones for example.

SH: That's really interesting, that meeting of science with technology and of both with sound creation. One thing that is exciting about your work is the way you seek out and gather hydrophonic sounds and then employ them in composition. I take it to be very different from what someone like Francisco Lopez did on his 1993 CD *Azoic Zone*, which, while it promises 'a soundscape journey to the life and environment of abyssal organisms', is, from what I can infer from the liner notes, largely made of synthesized and treated sounds, not hydrophonic recordings (though Lopez is an ecologist, so may have a sense of what things could sound like so deep). What sort of practice – documentary, aesthetic, scientific? – is composing with hydrophonic recordings for you?

JW: The compositions are not documentary. Though I tell stories, I do not want or expect listeners to follow the same story as I do. Most people have not heard underwater sounds before, at least not this implied or literal.

But much depends on the format in which I present the sounds or compositions. If I do a talk, I will identify the sounds: 'This is cod,' 'This is a Sperm whale,' or 'This is a bulldog bat echolocating.' I approach all that differently in releases – like *Energy Field* – or with an eighty-channel installation like 'Dive'. It's different again if I do a commission for radio.

Though I am inspired by science, my method, really, is not a scientific method at all. I am far too disorganized for that! I ask questions but I am not expecting to find answers. I propose hypotheses but am not systematically trying to find evidence in the classic sense, but am rather trying to encourage more questions and curiosity.

One hypothesis around which I have done several works proposes that one can listen to the health of a river through listening to underwater insects. Freshwater biologists are now counting underwater insects in order to say something about the condition of the water, discovering that some will survive certain pollutants, while others will not. Same with coral reefs; in my project with TBA21 Academy, 'Silencing of the Reefs' (2011–2014), I suggested that it is possible to listen to the sound of fish, crustaceans, and mammals on and around a reef to determine whether it is a healthy or dying reef.

SH: It's useful, as you've started to do, to think through specific works. Maybe we can turn to specific pieces that you've been working on and presenting lately. I know that you had an installation up in Vienna in 2017 – which has also been shown in 2018, in LeFresnoy – an installation called 'bára,' where 'bára' is Norse for 'wave'. Tell me about that.

JW: For the 'bára' installation, I was commissioned by TBA21 to use the whole of two large spaces at Augarten, a Vienna venue, so I thought of letting the piece work as a sonic portrait of an ocean wave that washes through two rooms. The content of the piece is based on recordings I made on various field trips from the North Pole to the Caribbean.

Since this was a group show with many other audiovisual works and objects in the space, I decided that the wave of sound would come in once a day, timed according to the wave of tidal change nearest to Vienna. I hoped to remind listeners of the relationship between humans and the tide, and of how the moon influences our bodies, which consist of around 70 per cent water. It was a practical solution to have the sound piece coexisting with all the other sounds in there, but the timing worked well, with other pieces turned off once a day when 'bára' washed through the space.

I mixed it specifically for those two rooms in Vienna, with an Ambisonic Decoder by my collaborator Tony Myatt, made particularly for that speaker setup. We had twelve speakers in each room plus four SUBs (subwoofers). We used these technologies to create a 'sphere of sound' around the audience. Interacting with the acoustics of the space, the idea was that one becomes less aware of the speakers – the set up lifts the sounds up and off the speakers, letting sound move more freely in the room. I wanted to create a sense for the listener of being immersed in the sound – and of not necessarily needing to be in a 'sweet spot', a concept with which I have problems, since it introduces a hierarchical sense into the space. I rather try to make pieces work in different spots of a room, to encourage people to explore their surroundings. I can make the mix fit into different speaker set-ups, too, whether I'm using four speakers or one hundred. I do make the compositions beforehand in my studio, though I know from experience it will change very much when I am on site.

Thinking back to 'bára', the two rooms in the Vienna space were very similar physically, but the feel of each space was quite different. One was very edgy and unsettling. The other was calmer. I think this had to do with the installed art objects. I had to change the mixes for each space drastically from what I had prepared in my studio, with one room needing to have fewer sonic details, and the other having more. In the end, the rooms worked well together. For the opening, I made a live version of the piece, but that was again a very different mix, since it was such a different situation, with lots of people moving around.

Just as working in the field often means that I have to seek out times and places where humans are not making a lot of noise, so too did I think about that here in the performance space. There was always so much going on – it is a continuous problem, to get enough time by oneself to work on the composition for any particular space. Even if you tell people you need it to be quiet, they start to whisper. I often end up working at nights in gallery spaces.

In the setup of 'bára' in Le Fresnoy, for the exhibition OCEANS, I had a fantastic time. It was a wonderful space to work in, and I had two whole days to work alone in this fantastic huge space with the twenty-four speakers plus four SUBs set-up. I was able to fine-tune the piece, so it was a total joy!

SH: One of the things that strikes me about 'bára', which I've been listening to in the admittedly suboptimal setting of my own home, is that you mix sounds from both below and above the water (the waves crashing, for instance). How do you think about 'where' you'd like listeners to 'be' as they listen to a composition that has their (airy) ears tuned to a range of above-and-below water sounds?

JW: I move between under and above water, and I try to keep some feeling of this movement, and some pulse in the pieces. Since I am not an experienced diver, I need to go up for air, and so I build breathing breaks into my pieces. Still, many underwater sections in my pieces can last longer than I can hold my breath, in reality.

In 'Dive', in the Park avenue tunnel in New York, I knew people would enter the tunnel from one direction, so I wanted to have above water sounds crashing to the beach in the beginning and in the end of the tunnel, so people would have the feeling they were leaving the familiar sound of waves crashing on the beach behind them as they moved through eight different underwater environments going to deeper and deeper waters in the middle, then they would rise to the surface again, reaching the end, and getting out of the water leaving the tunnel through the crashing waves.

In Energy Field you start on land, then dive in through the ice, swimming underwater, then coming up for air diving again and so on – this is how I am thinking, but it does not mean that the listeners need to think the same story as I do. It so much depends on the format of how I make the composition, if it should be more like a collage, a more static situation you enter into, or if it will move and have a beginning and an end, as a more concert or release kind of situation.

And it depends on how much 'human' sound I bring in. When a ship passes in one of my pieces, or the listener can hear the seal-scarer audio device I mentioned before, these sounds are present to contrast the small sounds of the fish and crustaceans – giving you an idea of the difference in intensity – and also to help me talk about the issues they raise or the particular problems they point to.

SH: Yes, I was wanting to get back to that, after hearing your harrowing story of the seal scarer and about the health of rivers. It does seem that part of the exploration and storytelling you're seeking to do is about alerting people to the increasingly damaged and at-risk aquatic world. So, in that connection, it's interesting that 'bára' is about the cyclical, daily repeating rhythm of the tide, whereas I know that other pieces you've done direct attention to ecological dynamics that are changing and may be irreversible. I'm thinking of course of climate change. Can you tell me about a piece that treats that?

JW: A little while ago, I was working on a piece called 'Spring Bloom in the Marginal Ice Zone', dealing with the spring bloom of phytoplankton and zooplankton near what Norwegian politicians like to call the 'Ice Edge'. There is nowadays a debate about oil drilling in the Barents Sea, near this so-called Ice Edge. Since the edge is moving north, many politicians want to permit oil companies to search and drill for oil further north. But to call it an 'edge' is quite a mistake – it is, rather, a seasonal zone of ice, and it constitutes a very fragile ecosystem, one that may be at great risk. If some species disappear in this ecology there will not necessarily be another to take over their job. Such areas need to be protected against human activities. Last year, I joined a research ship called the R/V *Helmer Hansen* to learn more about the plankton in the area, under, near to, and in the sea ice (see Winderen 2016). Twenty researchers from different parts of the world joined, looking into the whole water column, from the sea ice to the benthos. It was very interesting. I interviewed everyone and we were able to get off the boat two times (but, again, it was hard to get it quiet enough! There is a lot of stuff happening on the boat and many people need to get their work done. I dream of a research ship primarily concerned with sound under water! One day, I hope …).

Sometime earlier, I went to the North Pole with the Mamont Foundation to familiarize myself with what the sea ice felt like and sounds like (see Winderen 2015). Again it was impossible to get away from human created sounds. I tried to walk away from the camp as far as I felt was safe, but the sound from the generators reached me there through the ice and the water. Also, on the geographical North Pole point that day, happy English skiers were celebrating that they had made it to the North Pole and were screaming and shouting. I would have needed to be skiing for days on my own or with another quiet person to record the ambience without humans. I realize that we have colonized the whole planet with our sounds, and I'm not sure if it is possible anymore to find somewhere without human created sounds.

SH: It's interesting and maybe even ironic that these quotidian sounds of people are so disruptive, since part of what your work seeks to do is call attention to some of the larger human-caused damage to the sea! I almost wonder whether doing a recording or piece one day that emphasizes constant human interruption might be an interesting thing to do! It could make a kind of complement to the work you've done to make audible some of the creatures that live in the sea. Maybe you can tell me about one of those pieces?

JW: In work with the commission Classified, for the Borealis festival in Bergen, I wanted to put focus on how human created sounds, like military sonar, seismic testing with air guns, shipping traffic, detonation of bombs, are interrupting and are destructive to the creatures living in the ocean. In the final piece, which was performed on a multichannel setup in the marine institute storage warehouse, in addition to the recorded human-made sounds I used in the piece I created a low-frequency sound on the SUBs, which shook the building. This felt very physical to the audience, especially since I dropped it in quickly, so one could feel a sudden release of low-frequency sound pressure. When I did the rehearsal for this performance, the warehouse's neighbours complained. It was interesting, thinking that the sound pressure *I* generated was far less than what we put into underwater environments, seriously hurting the creatures living there. If I had used the same sound level above water, there would be uproar, and it would have been dangerous for our bodies [...]. It can be good to make comparison to what we do above water, then you start to understand the brutal way the oceans are treated by humans.

Another installation I worked on parallel was the piece 'Transmission', for the V-A-C foundation in Moscow. Here, I was invited by the curator for the show 'Geometry of Now', Mark Fell, to use a large industrial space that I wanted to just 'activate' with the sound of crustaceans, fish, and whales. It was a former plant for making gas into electricity, bought by a private company, now making it into a huge art gallery and project place.

SH: That's really interesting, again, this movement of underwater sound into these air-y places of human audition. What do you hope people do when they listen to such work of yours?

JW: When people ask me about listening, I try to explain a more active perception with all of us connected to the environment. In other words, taking the focus away from the ears towards listening. It is becoming clearer and clearer to me that it is about concentration and paying attention; I also find myself not really able to concentrate without closing my eyes these days. The whole of the body takes part in the listening process. For example, it is impossible to listen if you are too cold, too warm, restless, stressed, hungry, and so on. The core of my interest is the connectedness to our environment, no matter where we are. I try to avoid the distinction between *us* and *them*, between other species and humans.

References

Graham, J., N. C. Wegner, L. A. Miller, C. J. Jew, N. C. Lai, R. M. Berquist, L. R. Frank, and J. A. Long (2014). 'Spiracular Air Breathing in Polypterid Fishes and its Implications for Aerial Respiration in Stem Tetrapods'. *Nature Communications* 5: 3022.

Helmreich, S. (2009). 'Submarine Sound'. *The Wire* 302: 30–31.

Helmreich, S. (2012). 'Underwater Music: Tuning Composition to the Sounds of Science'. In Karin Bijsterveld and Trevor Pinch (eds), *The Oxford Handbook of Sound Studies*, 151–175. Oxford: Oxford University Press.

Touch (n.d.). 'Touch.' Available online: https://touch33.net/ (accessed 5 July 2020).

Winderen, Jana (n.d.). 'Bandcamp'. Available online: https://janawinderen.bandcamp.com (accessed 5 July 2020).

Winderen, Jana (2015). 'Field Trip to the North Pole, Mamont Foundation'. Available online: https://www.janawinderen.com/field-trips/field-trip-to-the-north-pole-mamont-foundation-18-23-april-2015 (accessed 5 July 2020).

Winderen, Jana (2016). 'Iskanten, Field Trip to the Marginal Ice Zone in the Barents Sea for the Spring Bloom of Plankton'. Available online: https://www.janawinderen.com/field-trips/iskanten-field-trip-to-the-marginal-ice-zone-in-the-barents-sea-for-the-spring-bloom-of-plankton-17-29-may-2016 (accessed 5 July 2020).

30

Melt Me into the Ocean: Sounds from Submarine Spaces

Yolande Harris

The Möbius strip of expanding awareness moves out from one's own body to immediate place, to other phenomena, on to remote environments and back to the self in Harris's rich body of thought and art. These are powerful works, in concept and realization. The sense of interdependence which they evoke and encourage is vital to our transformation into good stewards of our environmental neighborhoods.

—Annea Lockwood (2015)

Shifts in time and space, and undulations in daily perception, are active elements within the work of artist Yolande Harris, and brought forward through a deep curiosity for the world. A gap or fissure seems to appear, to break in – between seeing and believing, between material fact and poetical imagining, and between the near and the far, along with the ineffable and animate threads that may also connect and therefore disrupt such dichotomies. The gap, and the threads that traverse and link, and which invite us to enter their subsequent web of associations and slippages, disorienting layers and close-ups, and from which new perspectives are generated.

— Brandon LaBelle (2015)

Sonic consciousness

How can our conscious listening affect the world around us? How can learning to listen to underwater sounds transform us, and transform our relationship to the environment? My process of making art celebrates the coexistence of multiple ways of knowing, and feeds off moments of heightened awareness that arise in everyday experience. I explore my own personal experiences listening to underwater sounds, and emphasize not only the sounds themselves but also the context in which I find them. I am curious about the many ways we listen – technologically, intellectually, culturally, physiologically, psychologically, spiritually, and socially – and the different motivations listeners have. While science is revealing an image of the submarine environment that is endlessly complex, I am also interested in many ways of knowing the ocean, particularly how it exists in human imagination and culture, the arts and psychology. Part of my research is learning to embrace and incorporate forms of knowing that happen beyond my rational intellect, like those generated and stored in my body, and my intuition, allowing my mind to recognize and integrate these forms of knowledge. My creative process combines many hours of close listening, the rigorous analysis of those sounds, and freely following intuitive threads of ideas in the moment of making. For someone experiencing my work, I aim to provoke a sense of wonder in my audience, asking their imagination to actively make sense of the experience. My work activates the senses, drawing us back to the body as the central vehicle with which we interface with the world around us. There is a moment when the individual opens to receive these sounds into their system, in a way that is beyond information. When imagination, investigation, and the senses are combined, an expanded form of presence can arise. When something is given presence, it can no longer be ignored. I am particularly interested in how sound and listening can help facilitate this presence in the context of the ocean.

Over the past ten years my research has focused on sound in the submarine environment, with numerous works focusing on the dislocation between visually seeing surface and sonically experiencing depth.[1] Sound activates relationships between humans, animals, and our shared environments; and it is these relationships my work strives to reactivate and renew. To this end, I situate sound within a broader sensory context (walk, video, braids, bubbles) recognizing the interdependence of the body's own systems with the environment it is in. My performances, installations, workshops, walks, lectures, and writings combine deep listening, field recordings, live performance, and sound technology with images, particularly video, to explore remote places of the environment and the mind. My goal is to develop what I call 'sonic consciousness', a deepening of awareness through an attentiveness to sound, and 'techno-intuition', combining technological and intuitive ways of knowing and being in the world. My influences include: bioacoustic marine science; psychology of memory and trauma; sound healing practices; Pauline Oliveros's Deep Listening; human-animal relationships; acoustic ecology, deep ecology; and ecofeminism.

A range of underwater listening experiences

I lower a hydrophone, an underwater microphone, through the surface of the water. I pull it up and let it down again, listening to the change in sound as my extended ear crosses between air and water.

I tune in to live streams from hydrophones from deep ocean observatories around the world.[2] Most of the time there is the ocean hiss of white noise, faint or loud hums from distant boat engines, and the occasional voice of a whale or dolphin.

I am in the lab of a university oceanographer, listening to the first sound recordings of a sea-glider collecting data in the Pacific Ocean.[3] We hear constant machine sound from the glider, occasional bubbles and faint whale call. Later I ask her which species were brought into harbour by the whaling fleets of Dundee in Scotland. My ears open up to a soundscape full of bowhead whales, bearded seals, beluga whales, and narwhals, sounding in the creaking and squeaking of the Arctic ice caps.[4]

In a local natural history museum on the coast of the Monterey Bay in California, I set up a hydrophone for visitors to listen in the touch pool.[5] I watch as they move the hydrophone around, carefully exploring the acoustic spaces where the anemones live, mostly sonic shadows created by rocks that hide the ongoing reverberation of the pump.

I go whale watching and we come close to humpback whales feeding in the Monterey Bay. The engine is shut off, people on the boat gradually grow calm, the chattering stops, and as a whale dives deep showing its fluke, I hear a collective gasp and sigh from the crowd.

I listen to an interview with a 'dolphin ambassador' and participate in her guided meditation of dolphin sounds. I sense a change in the energy of the room.

I am watching the spectrogram of recorded sounds that I have been given by a scientist working with a hydrophone deep in the Monterey Bay and I see the shapes of sounds well outside my hearing range.[6] I listen in detail over and over, zoom into specific features, sometimes I recognize sounds. I shift sample rate, change speed, filter, remove ocean hiss, replace ocean hiss, equalize so I only hear certain bandwidths, check how sounds correspond to each other. I am getting to know the material and sonic space from many angles. Gradually the deeper ocean, while still out of my reach in so many fundamental ways, comes closer, or rather, I come closer to it. I begin to melt into it.

Listening to marine mammal voices in Dundee Harbour, Scotland

In *Whale* [...] visitors are invited to walk, letting the sounds of whales envelop them in their watery, deep murmuring. These sounds are at once distant from our earthly territory, our terrestrial senses, while they in turn immerse us within their sudden proximity: the immensity of the sounds – the great depth and dimension of their sonority – are brought right up against our skin, delivering all this depth and resonance into our listening.
—Brandon LaBelle (2015)

In 2017 I was commissioned to make a project in Scotland focusing on Dundee's history as a whaling port.[7] The resulting piece, *Whale (Dundee)*, is a 30-minute soundwalk using custom-made headphones with an audio player, situated along the historic harbour walls of the city and on the historic ship, RRS *Discovery*. Dundee has an interwoven history of whaling and jute production, whale oil was used in the production of jute, and jute ropes rigged the whaling ships. Although the whaling industry is over, jute production continues today in Dundee. Using a traditional Celtic braiding technique – a sixteen-strand hollow braid that enables a solid core to be wrapped in rope – I wove Dundee jute around the headphones for the soundwalk. The rough feel of jute in one's hands, along with the braided pattern in two colours, provided a material-physical reference to Dundee's whaling history that participants experience when they touch the headphones. The ongoing entwining of the cultural lives of these species – human and cetacean – fascinates me.[8] The braided headphones are a way of acknowledging the complexity of these cultural references in both the cetacean and human communities.

In response to my inquiry about the species of whales hunted by Scottish whaling ships, University of Washington oceanographer Kate Stafford sent me sound recordings that she had made under the Arctic ice in the vicinity of Fram Strait and Davis Strait, whaling areas close to Greenland. In addition to hunting bowhead whales, bearded seals, and beluga whales, Stafford suggested that Scottish whalers also hunted narwhals. Her recordings of the animals in their Arctic habitat are extraordinary to listen to. Bearded seals make long downward wavering whistles of pure tones that fall endlessly at differing speeds, some fast, some slow, some starting at a very high pitch, others at a low pitch, with no apparent 'breath' to break the sound. The sounds are rich with reverberation and appear to go on into an indefinite background. The bowhead whale has a squeezing multiphonic vocal sound, in repeating phrases alternating high and low, that have a compressing and decompressing effect on my body when I listen to them. The Arctic ice is irregularly creating a background of squeaks, creaks, and loud bass cracks that fill the sound field. These recordings induce in me an uncanny sense of connection with the direct physicality of another body making sound. I feel as though I could reproduce these sounds and as I try, I realize that they require me to have internal pressure inside my body, more like bodily resonations than our open-mouthed projections.

When I work with these sounds I go through a period of deep immersion in them, through repeated listening often for hours at a stretch. The high quality of the recordings enables me to work with them to reveal qualities I might not otherwise hear, by changing the speed and shifting the frequency into my hearing range. I also work with the visualization of the sounds in spectrograms both to help me listen, and to edit, at times zooming in on certain frequencies to get more clarity in my mind as to what this sonic environment consists of. After this immersion, my editing process consists largely of selecting portions of the recordings. I listen for clarity of the voices, as well as a richness of voices and background contextual sounds. I am not listening for one perfect voice, but for passages that clearly reveal the array of sounds happening simultaneously, and the sonic context within which they occur. I do not remix the sounds into a composition. Rather, I try to present the sound environment in a way that is minimally edited while maintaining the listener's interest. When listening through the headphones on the historic walls of Dundee Harbour, to these oceanic voices – bowhead and beluga whales, bearded seals and narwhals – we can imagine them speaking directly into our ears from Arctic waters. They become the agents, the guides, the voices of knowledge beyond our immediate experience. *Whale (Dundee)* creates a resonance with the past and our changing relationships to the ocean and the animals that inhabit it, while preserving a sense of wonder and inspiring a greater sense of empathy.

People who participated in this walk were invited to share their reactions in a guest book. Many of the comments point to the sound experience in combination with the site as revelatory or transformative.

A moment to reflect on this history in the landscape.
Beautiful and meditative, left a lot of space for thoughts to wander and [be] still.
Really thought provoking.
Was aware of smells and feet walking. Was very emotional, quite upsetting toward the middle and then calmed down at the end. Otherworldly and floaty.
Who would think all those sounds came from the sea!
Haunting sounds, feels like heaven, Thank you.
Sublime, altered perception of the waterfront.
Haunting. Made you reflect about our relationship with animals and think about the future and our role in preserving their environments.
Beautiful sounds of the ocean! Sometimes I felt like I was on another planet! Closed off from my surroundings and transported into the deep and wild oceans.
I've joined the whales. Send a note to my family…
Want to buy the headphones.

Sound, body, and oceanic states of mind

These are not unusual reactions to my work. Words such as *meditative, aware, haunting, heaven, sublime, altered perception, another planet, transported* are commonly used to describe a state of mind that is activated through the listening experiences I create. In my desire to build on the knowledge I gain through such exchanges with participants of my work, I look for answers in complementary forms of experience, bodily systems of knowledge that are a step removed from theory, a step closer to directly being in the world. Again, the questions arise: How can our conscious listening affect the world around us? How can learning to listen to underwater sounds transform us, and transform our relationship to the environment?

Composer Pauline Oliveros challenges us with her mantra 'listen to everything, all the time and remind yourself when you are not listening' (Oliveros 2000: 38). Listening to other beings, other phenomena, other machines, strengthens relationships between us and our environment. It demands our attentiveness to other beings, and calls into question distinctions between human and nonhuman, sentient and insentient. Sound asks for and requires us to pay attention, to enter into a relationship with it. Oliveros notes that sonic attention is focused and diffuse at the same time, and so it has the effect of situating us within a field of relationships, a field of varying qualities. This practice helps shift our state of awareness from one consumed with *doing*, to one attuned to *being*. I use the term 'sonic consciousness' to describe such a heightened awareness of sounds around us and of the materiality of sound (Harris 2011).

With Oliveros as inspiration, I explore different sonic experiences and ways of listening, particularly in relation to immersion and oceanic states of mind. In her Deep Listening practice, Oliveros draws inspiration and techniques from her lifelong study of meditation and her work can be considered as a form of sound healing. During one of Oliveros's deep listening retreats, I was first introduced to the Taoist practice of Tai Chi and the healing art of Qigong. I found these practices to be deeply complementary to my sound art practice and I have continued to study them. 'Chi', which is roughly translated as 'life energy', moves through the body and extends beyond the surface of our skin, which, although a physical threshold, is not the limit of the body's energy. I notice parallels between such a practice of energy flow through the body and the way sound behaves in space, moving through boundaries such as walls. This is a much more sonic than visual way of experiencing one's body, offering an awareness of oneself that is not bounded by the surface of the skin.

This more expanded 'sonic' notion of the body informs my understanding of echolocation by animals underwater. Cetaceans use echolocation to make sense of their environment. There are anecdotes of sperm whales and dolphins 'scanning' the bodies of human divers where the sound conveys information, not only of their outside shapes, as would a visual scanning, but of the interior of the body as the sound reflects differently depending on density (Hoare 2016). This ability gives them a conscious awareness of the interior as well as the exterior of other bodies. We visualize and see their physical body shape, but this is most likely not their experience of themselves or others. Such a sonic awareness of bodies de-emphasizes the apparent impermeability of surface, and reduces a lot of the power of human visually centred discourse limited to external surface appearances.[9] And in addition, much as sound can travel through walls behind which we cannot see, the visual surface of the skin or ocean need not limit us in our relationship and understanding. As a result of visual dominance and suppressed sonic perception, we tend to 'see' only the surface of the body and skin. A sonic sense would 'hear' or create mental maps or images that are not limited by surface.

So how does sound affect the body and one's sense of relating to the world around us? Can listening be considered a form of healing as Oliveros's work suggests? Sound healing therapy takes many different forms, but all relate to a notion of how sonic vibrations affect the energetic patterns of the body. In sound immersions or sound baths, the participant is a passive receiver (unlike Oliveros's active listening) of dense layers of continuous sound, created live, usually with bowls, gongs, bells, chimes, voices, and sometimes electronic sounds. These sound baths aim to immerse one in an 'ocean of sound'; the vibrations of the sound field directly affecting the vibrations within ones fluid body, and thus shifting blockages and patterns in ways that allow the energy to flow more healthfully. This experience is not so much a question of listening, as of allowing the physicality of the sound to work on one's body. Such sound-healing events have become enormously popular. I have been in a large space with nearly a hundred people calmly lying on the ground while the sound healers create waves and washes of sounds and vibrations. Afterwards my body and mind became acutely sensitized and aware of my environment. I noticed the clear light and shadow of moonlight far more intensely than usual, the pace of footsteps behind me, my own impact on the ground, moving in a heightened mental state induced by sound.

Moving beyond the surface, whether the skin of the body, the surface of the ocean, or the conscious mind, depends on the experience of energy exchange. In Jungian psychology, the vastness of the subconscious is comparable to a submerged iceberg, where the iceberg tip visible above water level is comparable to the conscious mind. Clarissa Pinkola Estés, a Jungian psychoanalyst and traditional storyteller, links the imaginative journey into the underwater world, with a dive into the subconscious. She interprets the Scottish mythological beings of the Selkies, seals that can become human and return to being seals, as an example of this ancient psychological pull. She refers to the notion of a 'medial woman' outlined by Antonia Wolff (1956), a psychoanalyst who worked closely with Jung, as one 'who stands between the worlds of consensual reality and the mystical unconscious and mediates between them' (Estés 1992: 288). In her interpretation, 'the seal-woman, soul-self, passes thoughts, ideas, feelings, and impulses up from the water to the medial self, which in turn lifts those things out onto land and consciousness in the outer world. The structure also works conversely' (289). In my own work I draw parallels between making conscious the unconscious through dream work (inspired in part by Oliveros's deep listening collaborator Ione, the Dream-Keeper), and making audible the inaudible using technological means, drawing the sounds up from underwater into our waking consciousness. Such a manifestation of the subconscious in a conscious or material form, is like listening to the remote deep ocean on land. Listening to this material when one's imagination is engaged and free flowing is an enriching experience. As Oliveros writes, 'Listening involves a reciprocity of energy flow, and exchange of energy, sympathetic vibration: tuning into the web of mutually supportive interconnected thoughts, feelings, dreams, and vital forces comprising our lives' (Oliveros 2000: 45).

Melt Me Into The Ocean: Santa Cruz and the Monterey Bay

What pulls us to the ocean? And when we arrive at its edge, what do we experience? *Melt Me Into The Ocean* (2018) imagines a deeper dive at the point where land and sea meet, bringing sounds up from under the surface, filling our air space with liquid motion from beneath the ocean. And while we look out at the surface of the ocean, feet on the land, ears underwater, how does our sense of place expand to integrate the submarine world into the presence of our imagination?

In Santa Cruz I live on the edge of a vast marine sanctuary, a submarine canyon deeper than the Grand Canyon, filled with life, resident and migrating through. When I look out I try to visualize in my mind the sheer depth of that canyon, the amount of space filled with ocean, and the proximity of vastness to this small city. I remember the first time I reached the edge of a desert canyon overlook in Utah, my stomach caved in as if punched hard and I burst into tears. I couldn't process that sudden vastness of scale dropping away beneath me. Now I imagine the darkness of the three-dimensional space underwater, and the sounds that flow through it, five times faster than sound moves in air and reaching greater distances, sounds above and below my hearing range. I imagine my body orienting itself in this cold, liquid, pressurized, sonic world. And then I feel my feet on the ground, in this place, looking out.

I listen through my body, I learn through my body, and sound helps to open up a sensitivity and vocabulary for integrating my bodily sensations with a sense of place and a sense of identity. Memory is stored in the body, sound acts on the body. Energy flows through and out of the body, and many times gets blocked in it. Healing is associated with allowing the energy to flow again.

Leaning over the edge of the wharf I am listening to a hydrophone I have lowered like the people fishing around me lower their lines. I can see it shining in a thick shoal of glistening anchovies hugging the wharf posts that are coated in muscles, barnacles, kelp and giant star fish, red, orange, and pink. I hear these creatures clicking and snapping and I can hear sea lions barking underwater. Somehow I am surprised that they can make these sounds underwater as I have seen them on the jetty throw their heads back and open their mouths showing big pink tongues and yellow teeth. I find myself thinking of their bodies in relation to mine, and whether I could make sounds underwater in the same way, without breathing or drowning. And then a sea lion comes darting through the anchovies chasing a larger fish, and all is flow and energy exchange.

What do I mean by the ocean as energy exchange? I link it to the flow of sound, the flow of energy in and through the body, the flow of energy within the liquid ocean, and the flow of energy from ocean to climate systems worldwide. If I think of the ocean not as filled with objects, but as a flow of energy exchanges, I come closer to it and enable it to move over me on land. Its influence in a coastal zone is always present, particularly in the air, in moisture, in smell, in sound. The coastal redwood forests thrive on the foggy moisture from the ocean system. Everything is softer, skin more supple, temperature steady. The coastal band absorbs the energy of the ocean by its proximity. The ocean has influence well beyond its surface, much as the energy of a human body has influence well beyond the surface of its skin. It makes me remember that our visual sense identifies surfaces as solid boundaries, as beginning and ends of objects, as identities and influence bound and contained by surface. In contrast our sonic sense hears beyond surfaces, it recognizes what is beneath and outside of the skin and ocean and beyond the walls of a room. Our sonic sense recognizes energy exchange and flow; it recognizes the ocean in ways that eludes our visual sense. Such a state of sonic consciousness brought about by listening to underwater sound, brings us closer to that awareness of *being* rather than *doing*.

Melt Me into the Ocean explores these ideas. It has included an evening installation in the Santa Cruz Museum of Natural History (SCMNH), and a day-long festival,[10] which included a soundwalk along the harbour jetty to Walton Lighthouse, and a sunset performance at Ocean View Park, which overlooks where the San Lorenzo river joins the Pacific. Siting the works in these public places, without an entrance fee, I invite a deeper awareness of the presence of the ocean environment through sound. These works use sound from the MBARI hydrophone, a high-resolution underwater microphone, placed on a ridge 300 metres down in the Monterey Canyon, and linked by high-speed cable to the shore.[11] Unlike most recorded whale or dolphin sounds that are publicly available, which are carefully edited to highlight just cetacean sounds, listening directly to the ocean environment is a more complex experience. The sounds emerge from the distance, blend with each other, echo off canyon walls, and are submerged in a loud ambient ocean hiss created by waves or rain on the surface. The distinct 'songs' of humpback whales repeat in long patterns, multiple overlapping echolocation clicks of perhaps dolphins, and even the lowest bass of the blue whale, which rumbles at the very edge of our hearing range, are all contextualized in the ambient sonic space of these recordings.

For the SCMNH installation I combined these sounds with video I recorded of humpback whales in the same location, projected in an atrium above our heads, inviting visitors to dive beneath the sea and hear the sound world of these and other animals via headphones. This technological extension of our ears asks us to learn to listen to a place we cannot directly experience. What are these sounds and how do they interact with each other? What do they teach us about this environment and how can this help transform our relationship to the ocean and its inhabitants? When I recorded the video of humpback whales in the Monterey Bay I was fascinated by the rhythmic punctuation of breathing, particularly the 'fluke dives' where the whale dives deep to feed for some minutes before returning to the surface. So the video remains on the surface, in intensified colours, leaving our imagination to dive with the whales and listen to the depths of their sonic environment.

The soundwalk to the lighthouse used the same headphones as *Whale (Dundee)* but with a soundscape that I created from my own hydrophone recordings combined with the MBARI recordings of the Monterey Bay. Walking out to the lighthouse along the harbour jetty the listener becomes increasingly submerged in the deep ocean sounds. The evening sunset performance at the park combined these sounds with my videos of a dreaming sea lion, ocean waves, and humpback whales, projected onto boulders in the park. At first the videos were very faint. As the light diminished the images came into focus and the sounds blended with the ambient sounds of the location. The audience that gathered was quiet and absorbed in being with this moment of changing transition from day to night, the emergence of the city lights, the distant roller-coaster, and the planets and stars overhead. The experience attuned us to being in the present moment, and in the presence of the nearby ocean through its sounds.

Listening to Whales: The Exploratorium, San Francisco Bay

The final presentation in this series took place at the Exploratorium, a science exploration museum located on a historic pier of the Embarcadero in San Francisco. The installation was situated in the upper-level Observatory space and exterior deck, surrounded by the water of the bay and with views of the city. I filled the spaces with sounds from the earlier works, including the Monterey Bay and the Arctic. During the evening, as the day turned to night, with city, boat, and bridge lights appearing through the fog, large-scale video projections of the ocean became visible across all the surfaces of the spaces. Within these activated spaces, visitors could put on the braided jute headphones and listen to another version of the underwater sounds in a more intimate way. The many visitors expressed surprise and delight at finding themselves immersed in underwater sounds of marine mammals, while simultaneously standing outside overlooking the environment of San

Francisco. Many stayed for long periods, others returned multiple times. This interweaving of direct bodily subjective experience with the remote submarine spaces through sound brings me back to Annea Lockwood's comment, 'The Möbius strip of expanding awareness moves out from one's own body to immediate place, to other phenomena, on to remote environments and back to the self' (Lockwood 2015).

Sound as the harbinger of a renewed relatedness

I began by considering the need for creating a space for the presence of the ocean within our land-bound lives. While many mechanisms are in place, from entertainment and tourism, to science, to spirituality, it is through sound – sonic consciousness – that a profound sense of presence can be experienced. I create the possibility for people to be in their bodies, in a way that enables them to extend their experience of the submarine spaces through their imagination, provoked by sound. An understanding of body-knowledge, imagination, and relatedness to environments needs to be embraced and practiced for the kind of deep empathy to emerge towards an uninhabitable environment such as the deep ocean. Sound is the harbinger of such a renewed relatedness. The technological advances that enable us to hear underwater with our feet on the earth, demand that we re-embody our sonic sense and our sonic intelligence towards the goal of a more conscious awareness of the remote, inaccessible environments of this planet. It is my hope, as an artist creating these opportunities for receiving ocean presence, that what leads out from the sonic imagination contributes to healing and re-balancing human relationships to our environments.

Notes

1. In addition to the works described here, to access *Sun Run Sun* (2007), *Pink Noise* (2009), *Fishing for Sound* (2010), *Swim* (2011), *Listening to the Distance* (2014), *Eagle* (2015), and *Bonneville Blue Whale* (2015), see Harris n.d. See also Harris 2012, 2018.
2. For examples, see Lido n.d.; and Orcasound n.d.
3. Lab visit with Kate Stafford, Principle Oceanographer in the Applied Physics Department, University of Washington, Seattle, 2014.
4. Email correspondence with Kate Stafford, 2017.
5. Installed during the special event 'Sensation: An Evening of Sensory Science Exploration' at the Santa Cruz Museum of Natural History, 10 February 2018.
6. Thanks to John Ryan, Senior Research Specialist at Monterey Bay Aquarium Research Institute (MBARI), see MBARI 2017.
7. Commissioned by Sarah Cook, North East of North Digital Arts (NEoN) for the Ignite Festival, Dundee.

8. That cetaceans are cultural animals is extensively reviewed in Hal Whitehead and Luke Rendell's *The Cultural Lives of Whales and Dolphins*. Cultural evolution in this context refers to knowledge and behaviours that are passed on and learnt within a group, allowing significantly more rapid adaptation than biological evolution. Humpback whales, for example, learn new songs each season, while orcas develop pod-specific hunting behaviours.
9. 'Seeing with sound would not be equivalent to seeing with light: the topology of inside and outside would be different […]. Bodies without opacity: an oxymoron for us, but perhaps mundane for dolphins' (Peters 2015: 68–69).
10. Commissioned and curated by Santa Cruz experimental music platform Indexical (2018).
11. Made available to me for this project by lead researcher John Ryan, MBARI.

References

Estés, Clarissa Pinkola (1992). *Women Who Run With The Wolves: Myths And Stories of yhe Wild Woman Archetype*. New York: Ballantine Books.

Harris, Yolande (n.d.). Available online: http://www.yolandeharris.net (accessed 5 July 2020).

Harris, Yolande (2011). 'Scorescapes: On Sound, Environment and Sonic Consciousness'. PhD dissertation, Leiden University.

Harris, Yolande (2012). 'Understanding Underwater: the Art and Science of Interpreting Whale Sounds'. *Interference: A Journal of Audio Culture* 2. Available online: http://www.interferencejournal.org/understanding-underwater/ (accessed 16 July 2020).

Harris, Yolande (2018). 'Listening to the Ocean in the Desert'. *Leonardo Journal* 51 (2): 193–194.

Hoare, Philip (2016). 'Of the Less Erroneous Picture of Whales: The Whale in Myth, in Industry, and in Futurity'. Lecture at Rhode Island School of Design, 14 January.

Indexical (2018). 'Melt Me Into the Ocean'. Available online: https://www.indexical.org/events/2018-08-11-melt-me-into-the-ocean (accessed 5 July 2020).

LaBelle, Brandon (2015). 'Yolande Harris: Aesthetics of Intensity'. In *Yolande Harris: Listening to the Distance*, 47–51. Woodbury Art Museum.

Lido (n.d.). 'Listening to the Deep Ocean Environment'. Available online: www.listentothedeep.com (accessed 5 July 2020).

Lockwood, Annea (2015). 'Integrated Circuits'. In *Yolande Harris: Listening to the Distance*, 63–65. Woodbury Art Museum.

Monterey Bay Aquarium Research Institute (MBARI) (2017). 'Ocean Soundscape'. Available online: https://www.mbari.org/technology/solving-challenges/persistent-presence/mars-hydrophone/ (accessed 5 July 2020).

Oliveros, Pauline (2000). 'Quantum Listening: From Practice to Theory (to Practise Practice)'. *Music Works* (Spring): 38.

Orcasound (n.d.). 'Listen for Whales'. Available online: www.orcasound.net (accessed 5 July 2020).

Peters, John Durham (2015). *The Marvelous Clouds: Toward a Philosophy of Elemental Media*. Chicago, IL: University of Chicago Press.

Whitehead, Hal and Luke Rendell (2014). *The Cultural Lives of Whales and Dolphins*. Chicago, IL: University of Chicago Press.

Wolff, Antonia (1956). *Structural Forms of the Feminine Psyche*. Zurich: C. J. Jung Institute.

31

Attentive Listening in Lo-Fi Soundscapes: Some Notes on the Development of Sound Art Methodologies in Vietnam

Stefan Östersjö and Nguyễn Thanh Thủy

This chapter is dedicated to the memory of the composer Vũ Nhật Tân (1970–2020).

Introduction

In this chapter we will outline the development of the sound art practices of leading composers and sound artists in Vietnam. The recent history of Vietnamese experimental music and the development of artistic methods for making sound art was largely set in Hà Nội. Our discussion builds on our own artistic experience from collaborative projects, involving all of the artists discussed here, but we also carried out interviews with each of these musicians and analysed some of their central works.

While the development of sound art practices around the world may largely be connected to the possibilities afforded by audio technology of capturing and monitoring environmental sound through microphones and recording devices, this seems less true for the early history of sound art in Vietnam. Rather, several leading practitioners refer to transformative experiences of listening to the soundscapes of the city of Hà Nội and of the countryside in ways reminiscent of the nineteenth-century vogue in England for what has been termed 'close listening' (Picker 2003). This practice predated the invention of audio technologies by several decades and created a novel 'recognition of ambient sound as ubiquitous and inescapable and its endowment with new material and figurative meanings' (Picker 2003: 6). For these Victorian listeners, this entailed a questioning of the listening attitude prompted by Western art music, and proposed an engagement with sounds in natural landscapes as well as with the noise of in the times of industrialization. The composer and sound artist Vũ Nhật Tân describes a similar listening experience when weaving together the sounds of his piano playing at home with the environmental soundscape, which, typical of a country like Vietnam, was also ever present indoors: 'I would hear the kids playing outside, and I would be inside playing my piano. I listened to them but I continued to play my piano. I just liked that combination' (Cunico 2015). Attentive listening became his method for turning an everyday experience into an artistic practice, which eventually would be further shaped into compositions and performances. Hence, this childhood experience from the 1980s became the impetus for a compositional practice which engaged with the noisy soundscapes of Hà Nội in various ways. In later years, an example could be a solo piano and electronics project premiered during the Hanoi New Music Festival in 2013. Here, Vũ recreated this childhood listening situation in a piano performance interacting with field recordings from Hà Nội's massive drilling noise as a recurring feature.

The composer and sound artist Lương Huệ Trinh describes similar experiences some twenty years later than Vũ. She refers to her youth, spent in the countryside in the Bắc Ninh region north of Hà Nội, as being 'very slow and close to nature' (Lương, personal communication, 2018). When she was thirteen she moved to the busy and noisy environment of Hà Nội to study jazz piano at the Hanoi Conservatory of Music. She describes how the experience of these contrasting soundscapes deeply affected her compositional practice:

> For me, each sound has its own content, message, context, and space. These soundscapes are extremely diverse. For instance, the sound of crowing roosters in early morning, cries of street vendors, sounds of the wind, of fires or rain on various materials, noises on streets, street music or insects in the fields, streams or rivers, and so on. These are enormous and rich sources for me to exploit.
>
> (Lương 2018)

Sound art methodologies in Vietnam have emerged out of such engagement with the soundscapes of the country through attentive listening, which has generated an awareness of their shifting qualities as a result of rapid processes of economic and social change since the 1980s. Here, the concept of a 'hi-fi soundscape', a central building block in acoustic ecology as introduced by R. Murray Schafer in the 1960s can be a useful figure of thought. Regarded from a Vietnamese perspective, it is perhaps best defined through its counterpart, the lo-fi soundscape of its modern cities. The quiet and the notion of clarity of signal – that is, with less overlap between foreground and background – are typical of pristine sites in a natural environment (Schafer 1977).[1]

In the 1980s, Hà Nội was a city of less than half a million inhabitants (Smith and Scarpaci 2000). With the shift towards a restricted market economy through the reforms referred to as Đổi Mới (renovation) in 1986, a process of rapid urbanization and industrial growth started. Today, Hà Nội is nearing a population of eight million, a development which also has resulted in an explosion of noise pollution, in particular from the increasing use of motorbikes (Phan et al. 2010). Besides the traffic noise, Vietnamese soundscapes, both on the countryside as well as in small towns and large cities, are characterized by the omnipresence of speakers in public places, transmitting governmental information and propaganda, interwoven with music broadcasted from local radio stations (Thế 2016). First created as a communication system during the war against the United States, these radio broadcasts have remained a constant in Vietnamese daily soundscape, contributing substantially its lo-fi quality (Östersjö and Nguyễn 2016). Arguably, these speakers also constituted the first audio technology to impact Vietnamese society, while record players remained rare. Hence, sound art methodologies in Vietnam developed less through interaction with technology and more from an engagement with the soundscapes through attentive listening. However, we will also discuss below how audio technologies eventually changed and diversified these methods, starting only in the early twenty-first century.

A short history of electronic music and sound art

Sound art has a short history in Vietnam. The country was largely isolated from external influence up to *Đổi Mới*. But even after 1986, the flow of information from the outside world increased only gradually. One of the first artists to engage with electronic music was the DJ, improviser, and composer Trí Minh. He describes his initial encounter with electronic music as follows:

> When the Soviet Union dissolved in 1991, many Vietnamese guest workers returned home [...] bringing along LPs released on the state Melodiya label. That is how I first got to hear electronic music and the disco group Zodiac's first album *Disco Alliance*. I was so amazed with how the songs had no lyrics, since I had been taught that only classical Western music is instrumental. I found the sounds so engaging I had to listen thousands of times [...]. This very first introduction to electronic music was just instrumental Euro disco.
> (Trí, personal communication, 2018)

During the 1990s a number of artists in diaspora visited the country for longer periods, bringing some of the first international impulses towards new directions in music and other arts. In 1994, the choreographer Ea Sola invited Nguyễn Xuân Sơn, aka SonX, a traditionally trained Vietnamese percussionist and composer, to perform in a piece titled *Drought and Rain*. With her extensive knowledge of experimental Western music and her roots in performance art, she not only opened up a window to American experimental music and European avantgarde but also to a choreographic and musical mode of expression which encapsulates the sense of loss and the activation of memories, drawn from an engagement with the sounds of traditional life and nature in Vietnam. As their collaboration continued through a series of internationally successful productions, SonX developed a compositional practice which gradually also incorporated electronic media. He describes how he likes to

> use everyday sounds from ordinary life, because they are natural and follow no rules or regulations. In society, and in particular in a country like Vietnam, there are a lot of constraints. The unintentional sonorities of ordinary life seem to refuse all of these constraints, denying the rules, existing by necessity, and therein I find their beauty.
> (Nguyễn, X., personal communication, 2018)

Here, SonX adds another layer to the discussion above: attentive listening to everyday sounds can become a method for articulating a politically informed critique through sound art. In the 1990s, the musical community in Hà Nội was conservative and had no interest in experimental work. Therefore, SonX turned to a group of visual artists whose interests also stretched to sound and performance.[2] However, the visual arts scene was also stale: the artist Nguyễn Mạnh Đức describes it as 'the choking experience of oneself being unable to transcend the stereotyped and one-dimensional direction of the field' (Nguyễn, M., personal communication, 2018). With the aim of 'creating a space where concepts and ways of doing art could be challenged', Nguyễn Mạnh Đức and his colleague Trần Lương founded the Nhà Sàn Studio which became the first non-profit experimental art space in Vietnam. The studio was housed in Nguyễn's family home, a Mường ethnic minority house on stilts which was moved from the mountains to a neighbourhood on the outskirts of Hà Nội.

If guest workers returning from the dissolving Soviet Union were the first carriers of influences from abroad, a more substantial impact was brought to Hà Nội from the composers Vũ Nhật Tân and Trần Thị Kim Ngọc, through their studies abroad, initially in Cologne, funded by DAAD (Deutscher Akademischer Austauschdienst) scholarships. Trần Thị Kim Ngọc studied composition with Johannes Frisch and improvisation with Paolo Alvarez, while developing a particular music theatre practice during her years in Europe. Vũ Nhật Tân went to Cologne in 2000 to study composition with Clarence Barlow and Johannes Frisch. This constituted his first introduction to experimental electronic music and computer music, and he describes how his encounter with the Atari 'changed me so much and opened to me a new vision of new music and sounds that I can create and compose and perform by the computer and electronic instruments' (Vũ, personal communication, 2018). He returned briefly to Hà Nội in 2002, but continued his education in the United States, where he studied with Chinary Ung. During his stay in the United States he developed a sound art practice that included the use of field recordings, thus transforming his original impetus towards the creation of sound art through attentive listening by introducing audio technology and computing.[3]

In 2009, Hà Nội saw the launch of two festivals that would host much of the further development in experimental music and sound art: the annual Hanoi Sound Stuff Festival, initiated and curated by Trí Minh, and the Hanoi New Music Meeting, which eventually lay the ground for the Hanoi New Music Festival, organized by Đom Đóm, an independent centre for experimental music and art, headed by Trần Thị Kim Ngọc, who is also the curator of the festival. Two years later, the Nhà Sàn Studio was forced to close down its public activities as a response to a performance which was interpreted as critical of governmental policies towards China. In 2013, the Nhà Sàn Collective was created, which since then has been moving between different spaces in Hà Nội.[4]

Soundwalks, field recording, documentary and ecological sound art

In 2003, the British Council funded a sound art project initiated by the British artist Robin Rimbaud, aka Scanner, involving Nguyễn Văn Cường, Trí Minh, Vũ Nhật Tân, and Nguyễn Mạnh Hùng. Titled *Street Cries Symphony*, this became the first substantial sound art work to be created from field recordings in the country. The following year, Kim Ngọc created the music for *Venus in Hanoi,* a large-scale dance performance featuring field recordings from the streets of Hà Nội. As mentioned above, field recording also became a central practice for Vũ Nhật Tân upon his return from the United States. Just as in the aforementioned pieces, he developed a particular focus on the multifaceted noises of Hà Nội. In an interview he describes how every time he goes out, 'I look around, and if I hear something interesting, I will follow it and I will go record it' (Cunico 2015).[5] The artistic results of this practice have been channelled into a series of projects called *Hanoise,* also involving the sound artist and composer Nguyễn Nhung, aka Sound Awakener. On the initiative of Trí Minh and with support from the Goethe Institute, the German sound artist Herbert Henke joined Vũ Nhật Tân and Trí Minh for a three-month project called *Hanoi Soundscapes,* presented during the very first edition of the Hanoi Sound Stuff Festival in 2009 (Henke et al. 2013).

In 2014, Lương Huệ Trinh was invited to participate in *Echoes,* a geolocative sound art project curated by Josh Kopecek, using software developed by Mathias Rossignol with which GPS coordinates can locate each user, enabling a soundwalk design in which the user can decide which route to take through the city. The project was divided into two parts, set in Hà Nội and Copenhagen, and was designed as a collaboration between artists from the two countries. It was proposed that Lương should collaborate with the Danish artist Hans Sydow. She describes the project as follows:

> My idea was to create an experience of a site, which brought the present and a distant or more recent past together through sound. This also entailed an interaction between sound distributed over headphones that participants had to wear during the audio tour and the sounds in the streets. The collaboration with Sydow developed through an exchange of field recordings and other audio, and through this exchange we created soundwalks for the Danish and Vietnamese sites combining our materials.
>
> (Lương 2018)

Lương made field recordings to capture the sound of contemporary Hà Nội, but she also collected recordings from the past such as 'the voice of a person who conveys policies of the state, propaganda songs that praise the Party and Hồ Chí Minh played back through speakers on every street corner, the sound of a tram, and cries of street vendors' (Lương 2018). In this way 'the participants could experience a dialogue, a connection between past and present, between stereo sound heard through headphones and the intrusion of the urban soundscape of Hà Nội' (Lương 2018).

While all of the above projects engage with urban soundscapes, also the sounds of life in the countryside have continued to fascinate composers. In 2013, the director of the Mường Cultural Space Museum commissioned a piece from Nguyễn Xuân Sơn. Nguyễn describes how, when he went to the museum to start working on the piece, he

> found that the soundscape where the museum is situated was amazing. It was there all the time but people were just unconscious about it. That was why I came up with the idea of using only environmental sound, recorded around the museum, and brought inside for it to be heard. My reward was not that people would remember my composition, but that they became aware of a beautiful soundscape that they had forgotten to listen to.
>
> (Nguyễn, X. 2018)

The piece, titled *Hòa Bình*, constitutes one of the few expressions of a classic soundscape composition practice in Vietnam in the spirit of Murray Schafer, and aims to highlight the increasingly rarefied hi-fi qualities of this particular sonic environment. SonX clearly situates his practice closer to the pole of documentation than to composition, a positioning which also actualizes the issue of representation in soundscape recording. Marinos Koutsomichalis observes that 'what a soundscape conveys cannot find its way into some electroacoustic representation of it, and even if, hypothetically, there were no technical constraints, the recording medium would fail to preserve the original semantics and subliminal significances of someone's encounter with an acoustic environment' (Koutsomichalis 2018). *Hòa Bình* constitutes an invitation to the listener to also return to the site of origin, outside the museum, with a novel awareness. However, despite this implicit invitation to a different mode of listening to the surrounding landscape, the aim of the field recording was to create a representation of place through an indoor installation.

A contrasting approach was developed within the ecological sound art project *Quê/Homelands*, which engaged the two authors of this chapter and the British composer and sound artist Matthew Sansom. It was premiered at the Manzi gallery in Hà Nội during the Hanoi New Music Festival in 2013. Rather than a representation through field recording – as is a characteristic of Schafer's World Soundscape Project, the Landscape Quartet, of which Sansom and Östersjö are members – it develops methods for a more participatory approach to sound art. This entails the creation of interactive systems, such as the aeolian guitar, developed by Östersjö. A guitar is stringed with fishing line around trees on a site. When the strings are brought to tension by the performer, the wind will excite harmonics on the extended strings. The assemblage of guitar-strings-tree-performer-wind becomes an interactive system, which allows for a different engagement with the affordances of the site.

Quê/Homelands was set part partly on one of the mountains surrounding the little village of Ngang Nội and partly in the rice paddies around the village. One of the central characteristics of both sites was the constant presence of noises from the surrounding villages. On the mountain, an aeolian lute[6] constituted the interactive element at the site, while in the rice field a *đàn tranh* (a Vietnamese zither) was hung upside down to be played by Nguyễn Thanh Thủy in interaction with rice and wind. While the interactive design involved only the participating artists, performative approaches can propose ways in which an audience can be further drawn into the sonic event.[7]

A project in which we explored methods for a more immediate interaction with an audience through the creation of performative ethnographies, drawn from documentary work, was carried out by our group The Six Tones.[8] This project, titled *Arrival Cities: Hanoi*, is a piece of experimental music theatre which weaves documentary and sound art as individual narratives in a collaborative and performative format. The project started out in 2014, when the Swedish director Jörgen Dahlqvist, The Six Tones, and composer Kent Olofsson started making field recordings and doing interviews in Hà Nội and on the countryside. By engaging with stories of migration from the countryside to Hà Nội, the project captures the experience of several generations of rural–urban migrant workers in Hà Nội – such as the emblematic street vendors, who will normally have their homes and families hours away from the city – and also of the radically shifting soundscape of a developing country.⁹ Ironically, while the cries of street vendors are advertised by the tourist industry, the same migrant workers are hunted by the police for carrying out their illegal trade. Interviews, traffic noise, propaganda through the speakers, and many other sources, contributed to the final staged production. Hence, sound art practice is here brought into a performative context, in between music theatre, documentary film, and electroacoustic music. Through the performative situation of the staged event the role of listening has taken on a political direction, as a form of engagement with the Other, and by engaging with the absolute Other (Derrida 2000) as represented by the migrant.¹⁰

The Vietnamese-American film-maker and theorist Trịnh T. Minh-hà's latest film *Forgetting Vietnam*, premiered during Cinema du Reel in Paris in Spring 2016, constitutes an example of a similar weaving together of sound art, field recording, and documentary. The making of the film, which started with Trịnh crossing the country by train and bike in 1995, chronologically encapsulates the entire history of sound art in the country and constitutes one of the first attempts at capturing the political and social change in Vietnam through its shifting soundscapes. The material for the film was recorded in several periods, with the final field recordings being carried out in the Spring of 2015. Given the historical outline above, the audio recordings made for the film in 1995 – for instance of the train ride from Hà Nội to Sài Gòn – may constitute the very first artistic approach to field recording in the country. The music in the film – developed in collaboration with The Six Tones – engages with the soundscapes of Vietnam in a number of different ways. The following section is a conversation between the two authors and Trịnh T. Minh-hà, with the aim of unpacking the methodical considerations which formed the basis of these early sound art practices.

Forgetting Vietnam: A conversation about sound art and memory

Stefan Östersjö [SÖ]: How would you describe your engagement with the sounds of cities in the process of making *Forgetting Vietnam*?

Trịnh T. Minh-hà [TTM]: Vietnam is very much an agricultural country with at least 80 per cent of its people living in rural areas. So although the three main cities, Hà Nội, Huế, and Sài Gòn (or Hồ Chí Minh City) serve as one of the referential threads in *Forgetting Vietnam,* marking the film's trajectories across the three main regions, rivers, and linguistic variations, its soundscapes alter drastically between urban and rural, and from one location to another.

For me, and perhaps for most musicians attentive to the sound environment, working with soundscapes, whether in the city or on the countryside, merely requires an active reception of what is perceived as 'noises', and an ear unconstrained by musical training. So for example, what loses pertinence are the very hierarchies set up between music and 'ambient sound', or between silence and sound. In the context of film and video, this would also mean a predisposition for a free play between synchronized, off-sync, and non-sync sound, as well as between musical composition, improvisation, and indeterminacy. Such a path of the 'musically unbound' actually requires a precise, demanding work of multiplicity.[11]

Cities have their own sonic imprints. One recognizes their soundscapes the way one remembers faces. Sometimes this ability comes with attentive listening, but at other times it catches one by surprise as it surreptitiously arises with memory. To give just one example, the frequency with which people resort to honking while driving, the way they beep, blow, blare, hoot, and whistle is quite distinct from city to city. It is a language of its own. Hà Nội traffic can be crazy and loud, especially in the old quarter. In their density and frequency as well as in their tone, dynamics, and speed, horns and bells of cars, scooters, mopeds, cyclos, and bikes play a unique role in defining the aural identities of cities like Hà Nội, Huế, and Sài Gòn.

In *Forgetting Vietnam*, you can also hear the differences of the clamour of the streets in terms of times and temporalities: the same city caught on camera in Hi8 in 1995 and then in HD in 2012 shows dramatic changes, not only visually, in the content of the images, and technologically, via the analogue and the digital, but also aurally, via differing sound fabrics that include fragments of popular songs dear to radio media during certain periods of Vietnam's history. Viewers are quick in identifying an era or a period through images, but they are much less attuned to how the sonic and the musical also designate time; and more specifically, how besides being a powerful sensorial experience, a film soundtrack is a multi-voice, multi-texture, and multi-text just as complex, significantly informative, and interpretive as the visual and optical.

In the city, whether one stays in a place or walks through a neighbourhood, if one remains all ears to the environmental soundscape, one would be struck by the overwhelming wealth of sounds in coexistence and co-formation. Short fragments of popular songs played on the radio mingle with animal, human, and mechanical noises of all sorts. Arising in the city's sonic fabric are the diversely textured street vendors' calls, children's laughter, dogs' yapping, and people's vocal interactions against a host of local labour sounds. These do not really clash with one another; they resonate as a multiplicity. In *Forgetting Vietnam*, there is, for example, a multilevel conversation unfolding both between the older and more recent soundscapes of the same city, and between the differing urban–rural, human–nonhuman, high-tech–low-tech, liquid–solid, and water–land soundscapes.

SÖ: It strikes me as one of the central characteristics of *Forgetting Vietnam* how it is as much a cinema for the ears as it is a documentary film. One component in the work of The Six Tones on collecting materials for the film was driven by the thematic threads related to water and land in Vietnamese history and mythology. We carried out fieldwork in the Bắc Ninh region to record Quan Họ songs relating to these themes. This is a tradition which is very much alive on the countryside north of Hà Nội, and when work is over for the day, people may gather in a master singer's house to join in. The songs heard in the film were all recorded in the house of Nguyễn Thị Bướm, a female master in her eighties, sometimes singing solo, and sometimes together with her daughter and granddaughter. In a small village like Ngang Nội,

Quan Họ singing still characterizes the soundscape in the evening, merging with the sound of cicadas, domestic animals, and occasional motorbikes.

TTM: Yes, I was happy to have Nguyễn Thị Bướm's Quan Họ in the film; it was a real gift of yours, and especially of Thanh Thủy and her aunt. Besides being a recognized cultural heritage, carrying with it the quality of the communal characteristic of a village's soundscape, the inclusion of Quan Họ is also personally dear to me. As per my mother's and grandmother's stories, this folk music and its historical 'antiphonal' nature used to be such a lively event – a call-and-response mode of singing exchanged between a pair of female and male singers during the rice planting season, which remains amazing, both in its improvisations and in its playful love-and-seduction undertones.

Each city, each neighbourhood has its own multi-sonic assemblage. By giving ear to it, I usually use music – mainly the music produced by your collective, The Six Tones – so as to prolong and highlight the sounds of everyday life and, further, to take them into the realm of an 'elsewhere within here'. For example, in the sequence of boats shown at the beginning of the film, it is the various motor sounds of the boats in motion, the rhythmic creaking of the boats anchored (or of metal moving with the lapping of water) and later, the cadenced rowing (wood against water) that motivates accordingly the choice of the musical sections put to use. What is heard as banal and familiar in an everyday soundscape thus subtly changes before the spectator realizes it and catches on in the process. Rendered fluid and mobile, the relation between composed music and the sounds of daily life is worked on, both so as to ground the viewers' relation to the subject on screen and to unsettle that relation, by drawing attention to hearing and to the dimension of the sonic and the musical in film experience.

In mainstream films, all is geared towards achieving the 'realism effect'. Music primarily performs a secondary role of underscoring and enhancing the images, and filling in what they lack, while sounds are reduced to Foleys or to the reproduction of everyday sound effects. Thus, to generate a change in conventional listening and to bring the spectator's awareness to the aural realm, what is often made to enact a structural role in my films is acousmatic sound and acousmatic listening (off-screen sounds or those sounds one hears without seeing the source cause). The ear is not made to serve the eye, and rather than providing the images with the realism they lack when shown in silence, the role of the soundtrack is to create a multisensorial, multifunctional space in which many relations between the seen (and unseen) and the heard (and unheard) are possible.

SÖ: There are some other emblematic sonic elements, one being the train, recorded in a journey from Hà Nội to Sài Gòn. How would you describe your work with the train recording? We also recall the signal processing we employed to transform these recordings when the scenario shifts in the film.

TTM: The train is a trans*[12] character in my work. It was already featured in two previous films of mine. In *The Fourth Dimension* (2001), it constitutes one of the two main characters, the other being the drum; these are the two rhythms that characterize the way Japanese culture negotiates between the ancient and the modern. In *Night Passage* (2004), a film inspired by Kenji Miyazawa's novel Milky Way Railroad, the train takes on a protagonist role in the life-and-death journey of two young women. In *Forgetting Vietnam*, the train again plays an important role. It is linked to the way the film unfolds as a conversation between land and water – the two elements underlying the formation of the term 'country' (*đất nước*) in Vietnamese. Here, the encounter happens between the ancient as related to the solid earth, and the new as related to the flow of liquid – the two forces that regulate life and can be traced back to every beginning and ending, more particularly to the geological formation of Vietnam's Mekong and Red River deltas. As first explored in my installation *Old Land New Waters*, this non-binary Two generates a third space: that of memory and forgetfulness in the realms of history, culture, and family.

The sound and rhythm of a train anchors one to the earth, while the sound and rhythm of the boat invites one to let go and float. In 1995, we mainly circulated by bicycle, moped, and pedicab, and took the train to travel from North to South via Huế. When I returned in 2012 we mostly travelled by car and bus, and flew from Hà Nội to Sài Gòn. But the boat was used extensively in both travels. A large part of Vietnamese culture is river born and Vietnam is, in its core, a water body. The necessary equilibrium between water and land therefore plays a fundamental role in the country's economics and politics. It was then important to include in the film the means of transportation that determine the ways one sees or hears the country. This was done by featuring the sound of the old train I took in 1995, which was clanging away along window views of the scorched landscape and of the extensive wet rice fields.

It is in this context of sound, rhythm, memory, and forgetfulness that we can further discuss the link between the recorded and the transformed that you raise in your question. One of the most exciting processes for me in this film was to work with The Six Tones' music. There is a happy encounter or something like a kin spirit between the way I work in sound-and-image montage and the way your collective combines the *arrière* and avant-garde – namely the experimental, electronic, and concrete in music with the ancient and traditional.

I remember when I first heard you, Stefan, and Henrik Frisk in a performance at the New Music Center in Berkeley. I was struck and very excited by the processes of 'metamorphoses' in your music, and this also came back to memory when I decided that the sound of the train should go beyond its recorded state. In working with the interconnected, non-binary between memory and forgetfulness, I first introduced the recorded sound of the old train so as to invite the audience to listen more intensely to the train sound in its amplified stabilizing cadences, and then in its transformation – how it 'liquefies' and 'etherealizes' with yours and Henrik's intervention – and further, to forget so as to remember more intensely how, ultimately, every sound image on screen is actually an image of memory. To remain creative and to keep the field of possibilities open, one listens to the becomings of reality's recorded sounds and soundscapes. And for this, The Six Tones' music plays a primary role.

SÖ: The director Jörgen Dahlqvist and composer Kent Olofsson have also collaborated extensively with The Six Tones. In a recent paper, they analyse the artistic strategies they have developed within experimental music theatre as a vertical dramaturgy – by reference to Ruth Finnegan's claim that writing is multimodal in ways similar to performance – and they advocate a multilayered conception of music theatre, in which layers of music, text, scenography, and acting are sometimes independent but also aligned, both horizontally and vertically, within the dramaturgy of the composition (Dahlqvist and Olofsson 2017). It seems to me that *Forgetting Vietnam* develops a similarly contrapuntal relation between its elements.

TTM: It is very relevant to introduce this dramaturgy of composition in relation to the multiplicities at work in film. Such a notion of vertical and horizontal dramaturgy can also apply in terms of substance and surface or what I often call the within, the between (inter) and the across (trans*) in arts and politics.

What is most questionable is the widespread, reductive tendency in mainstream productions to fold the space of the verbal over that of the visual or the musical, and hence to reduce these to the functions of illustrating, explaining, and duplicating. Rather than being one of subordination and domination, the relation between the verbal, the musical, and the visual could remain, at the core, one of multiplicity.

Viewers have asked me why in *Forgetting Vietnam* there is no voice-over as in my other films. But the 'voice' of the film is here already so richly manifested through a tapestry that weaves together Vietnamese classical music (the *đàn tranh* as performed by Nguyễn Thanh Thủy), popular songs, recited poetry, folk music, avant-garde composition, and everyday sound art, that I prefer to leave the verbal commentary to the eye, together with the selected translations of the singing and the local people's comments and interactions. Since *Forgetting Vietnam* takes the viewer through the three regions of Vietnam with a focus on the rich activities on the Red River in the North, the Perfume River in the centre, and the Mekong River in the South, the differently accented voices of the people on and off-screen – bus and taxi drivers and riders, street vendors, boat passengers, performers of traditional, popular and folk music, for example – are not only used for cultural and political reasons, but are also indicative for each of the regions through which the viewer is led. Film as music for the eye and editing as composing are bound to be at once visual, textual, and musical.

Discussion

The sound art practices of pioneering composers such as SonX and Vũ Nhật Tân have their origins in methods centring around the transformative nature of attentive listening. When sound art emerged in Vietnam, the access to audio technology was limited, as was the interaction with artists in other parts of the world. Sound art in Vietnam has a strong connection to the changing soundscapes of the country, and is reflected in changing modes of listening among composers and other artists.

The relation between the political and economic reforms and the development of sound art practices in Vietnam are linked in paradoxical ways. *Đổi Mới* unleashed a wave of urbanization and growth which has transformed the soundscapes in ways that have prompted many artists to respond through new ways of listening and making sound art. The governmental radio transmission in public places contributed in a similarly double manner, by constituting an early introduction to audio technologies through their mere presence and at the same time by becoming an iconic reference to noise pollution. But technologically driven methods for sound art emerged only in the beginning of the twenty-first century, introduced through composers who studied abroad.[13]

Most of the projects discussed in this chapter address the role of memory in one's engagement with sound and place through the use of field recording, often, as made clear by Trịnh in the interview above, using methods that explore the relation between the recorded and the transformed. Similarly, through the juxtaposition of past and present in Lương's soundwalk, or in Vũ's re-enactment of his childhood experience of listening to city sounds through the piano, a stronger awareness of sonic memories and of the political signification of sound in society can be achieved. An engagement with everyday soundscapes through field recording can thus hold a political dimension as is also expressed by SonX, who observes how unintentional sonorities of ordinary life seem to refuse the constraints that characterize daily life in a country like Vietnam (Nguyễn, X. 2018).

In our own practice, the development of participative methods through ecological sound art, and the creation of a more clearly articulated political dimension through the performative ethnographies which constitute the basis for *Arrival Cities: Hanoi*, suggest a development in sound art practices in Vietnam that can contribute to a further engagement with the sonic environment, and create new encounters between sound artists and audiences. Further, we see – in the conversation with Trịnh and with reference to the method development of Dahlqvist and Olofsson (2017) – the possibility of developing compositional methods for sound art that combine dramaturgy and musical composition, as a means for structuring what Trịnh refers to as the 'work of multiplicity'. Methodologies for sound art in Vietnam have developed in a particular sonic and political landscape, but the variety of methods and formats for output, as well as the focus on listening experiences, suggest that some of these findings may have a bearing on the development of the field more widely.

Notes

1. Paradoxically, such hi-fi soundscapes, today indeed rarefied in the country, may perhaps have been experienced even in Hà Nội when Vũ Nhật Tân was a child. He describes how, in 'the quiet Hà Nội, you could hear the birds chirping and there were bicycles on the streets'. His work builds on the experience of the shift from the sounds of Hà Nội from before the arrival of cars to its contemporary soundscape, which he characterizes as 'one of the noisiest cities in the world' (Cunico 2015).

2. Central figures were the painter Nguyễn Văn Cường; Nguyễn Mạnh Hùng, a visual artist who also developed a practice of experimental music performance with electric guitar and analogue pedals; Quách Đông Phương, working with voice, distortion pedals, and home-made drum kits; and Đào Anh Khánh, who developed a central practice as a performance artist. This group of artists was joined by Trí Minh and Vũ Nhật Tân in the early 2000s and created many sound art events at the Nhà Sàn Studio.
3. Vũ Nhật Tân hereby became the first composer to introduce technologically driven methods for sound art composition in Vietnam.
4. The history of sound art in Vietnam is short, but it has an important function within alternative cultures in the country. All artists discussed here are active in independent fields outside of institutions for music, art, and higher education. During the 1990s and up to 2013, the support from European organizations was substantial and created opportunities for experimental artistic work and for increasing exchanges with the outside world. However, this support has been in decline, which for instance can be seen in the radical cuts of funding for the independent centre Đom Đóm, which had substantial European funding during its first two years, which is now all gone. Unfortunately, the Vietnamese authorities do not provide any support, and music institutions do not teach or engage in experimental arts. Hence, while sound art in Vietnam is constantly developing, it exists in a fragile situation, in which initiatives of small independent organizations and individuals sustain the entire field.
5. This citation brings the early sound art methodology of Akio Suzuki to mind, who developed a practice in the 1960s which he called 'throwing (*nagake*) and following (*tadori*)', one of the first examples of a participatory form of sound art. Just as Akio Suzuki's practice could also be termed 'close listening', Vũ's field recording practice is a consistent development from his early listening experiments.
6. The choice of instrument was a three-stringed lute, the *đàn đáy*, which is exclusively used in a form of Vietnamese chamber music called *Ca Trù*. The affordances of a three-stringed instrument are different to that of the six-stringed aeolian guitar, since it is possible to obtain proper lute strings at any length. The lute produces less complex harmonic clusters, but is beautifully distinct when the low extended strings are played as well as with melodic figurations with the harmonics.
7. For a further discussion of this project, and some audio and video examples, see Östersjö and Nguyễn 2016.
8. The Six Tones are Nguyễn Thanh Thủy (who plays *đàn tranh*) and Ngô Trà My (who plays *đàn bầu*) from Vietnam, and the Swedish guitarist Stefan Östersjö. Since 2006, The Six Tones have been working on the amalgamation of art music from Vietnam and Europe, but the group also functions as a platform for interdisciplinary collaboration. Over the past ten years, The Six Tones have initiated many collaborations with playwrights, film-makers, and choreographers, and created a series of sound and video installations.
9. For a further discussion of this project see Östersjö and Nguyễn 2016; 2019, 2020.
10. Peggy Phelan refers to Levinas, and the notion of ethics as essentially expressed in the face-to-face relation to the other in her discussion of *The House with the Ocean View*, a piece of performance art by Marina Abramović. Through this piece, Phelan observes how 'live performance might illuminate the mutual and repeated attempt to grasp, if not fully apprehend, consciousness as simultaneously intensely personal and immensely vast and impersonal' (Phelan 2004: 574).

11. James Gifford's discussion of Lawrence Durrell's *Avignon Quintet* strikes a very similar note to Trịnh's reference to multiplicity here. Gifford points to how the first book, *Monsieur*, mirrors the structure of the entire cycle, by being 'divided into five independent sections where authorship of the book is brought into question. It is this constant remembering of the authorship – or constructed nature of fiction – which most strongly brings the reader's mind to the epistemological crisis, and refutes the concept of the work revealing a single absolute truth. *Monsieur* is a work of multiplicity' (Gifford 1999: 7). Gifford further suggests that Durrell's project aims to replace 'our sense of an absolute truth derived from our sensory experiences with a realization of the necessarily multiplicitous and unrealizable nature of absolute truths' (4), in a manner very similar to Trịnh's claim that there is no such thing as documentary: 'On the one hand, truth is produced, induced, and extended according to the regime in power. On the other, truth lies in between all regimes of truth. To question the image of a historicist account of documentary as a continuous unfolding does not necessarily mean championing discontinuity; and to resist meaning does not necessarily lead to its mere denial. Truth, even when "caught on the run," does not yield itself either in names or in filmic frames; and meaning should be prevented from coming to closure at either what is said or what is shown. Truth and meaning: the two are likely to be equated with one another. Yet, what is put forth as truth is often nothing more than a meaning. And what persists between the meaning of something and its truth is the interval, a break without which meaning would be fixed and truth congealed' (Trịnh 1990: 76).
12. Jack Halberstam proposes that the use of the asterisk 'modifies the meaning of transitivity by refusing to situate a transition in relation to a destination, a final form, a specific shape, or an established configuration of desire and identity. The asterisk holds of the certainty of diagnosis; it keeps at bay any knowing in advance what the meaning of this or that gender form may be, and perhaps most importantly, it makes trans* people the authors of their own categorizations' (Halberstam 2018: 4)
13. Such influence is still strong also in the younger generation. For instance, Lương Huệ Trinh has just finished her master's studies with Georg Hajdu in Hamburg, studies that have brought her sound art practice into an exploration of interactive media.

References

Cunico, K. (2015). 'Vũ Nhật Tân: The Noisiest City's Noisiest Musician'. *Contented*, 27 October. Available online: https://contented.cc/2015/10/vu-nhat-tan-the-noisiest-citys-noisiest-musician/ (accessed 15 November 2018).
Dahlqvist, J. and Olofsson, K. (2017). 'Shared Spaces: Artistic Methods for Collaborative Works'. *Studies in Musical Theater* 11 (2): 119–129.
Derrida, J. (2000). *Of Hospitality*. Trans. Rachel Bowlby. Stanford, CA: Stanford University Press.
Forgetting Vietnam (2016). [Film] Dir. Trịnh T. M.-H. USA: Moongift Films.
Fuchs, T. (2012). 'The Phenomenology of Body Memory'. In S. C. Koch, T. Fuchs, M. Summa, and C. Müller (eds), *Body Memory, Metaphor and Movement*, 9–22. Amsterdam: John Benjamins.

Gifford, J. (1999). 'Reading Orientalism and the Crisis of Epistemology in the Novels of Lawrence Durrell'. *CLCWeb: Comparative Literature & Culture: A WWWeb Journal* 1 (2): 1.

Halberstam, J. (2018). *Trans*: A Quick and Quirky Account of Gender Variability*. Oakland, CA: University of California.

Henke, R., Trí Minh, and Vũ Nhật Tân (2013). [CD] *Hanoi Soundscape*. Hanoi: Goethe Institute.

Koutsomichalis, M. (2018). 'On Soundscapes, Phonography, and Environmental Sound Art'. *Journal of Sonic Studies* 4. Available online: https://www.researchcatalogue.net/view/268080/268081/0/0 (accessed 9 May 2019).

Nguyễn, T. and S. Östersjö (2020). 'Performative ethnographies of Migration and Intercultural Collaboration in Arrival Cities: Hanoi'. *Journal of Embodied Research* 3 (1), 1 (25:10). http://doi.org/10.16995/jer.19.

Night Passage (2004). [Film]. Dir. J-P. Bourdier, and Trinh T. M.-H. USA: Moongift Films.

Östersjö, S. and T. Nguyễn (2016). 'The Sounds of Hanoi and the After-image of the Homeland'. *Journal of Sonic Studies* 12. Available online: https://www.researchcatalogue.net/view/246523/246546/23/0 (accessed 20 May 2019).

Östersjö, S. and T. Nguyễn (2019). 'Arrival Cities: Hanoi'. In C. Laws, W. Brooks, D. Gorton, T-T. Nguyễn, S. Östersjö, and J. Wells (eds), *Voices, Bodies, Practices*, 235–294. Orpheus Institute Series. Leuven: Leuven University Press.

Phan, H. Y. T., Y. Takashi, S. Tetsumi, and N. Tsuyoshi (2010). 'Characteristics of Road Traffic Noise In Hanoi and Ho Chi Minh City, Vietnam'. *Applied Acoustics* 71 (5): 479–485. https://doi.org/10.1016/j.apacoust.2009.11.008.

Phelan, P. (2004). 'Marina Abramovic: Witnessing Shadows'. *Theater Journal* 56 (4): 569–577.

Picker, J. M. (2003). *Victorian Soundscapes*. New York: Oxford University Press.

Schafer, R. M. (1977). *The Soundscape: Our Sonic Environment and the Tuning of the World*. Rochester: Destiny Books.

Smith, D. and J. L. Scarpaci (2000). 'Urbanization in Transitional Societies: An Overview of Vietnam and Hanoi'. *Urban Geography* 21 (8): 745–757. https://doi.org/10.2747/0272-3638.21.8.745.

Thế, K. (2016). 'Loa phường được phát thanh những nội dung nào?'. *Dân Trí*, 9 December. Available online: https://dantri.com.vn/xa-hoi/loa-phuong-duoc-phat-thanh-nhung-noi-dung-nao-20161209162829359.htm (accessed 15 October 2018).

Trịnh, M.-H. (1990). 'Documentary Is/Not a Name'. *October* 52 (Spring): 76–98.

Trịnh, M.-H. (2011). *Elsewhere, Within Here: Immigration, Refugeeism and the Boundary Event*. New York: Routledge.

Part III

Geographies, Politics, Histories

32
Introduction to Part III: Listening as Method
Marcel Cobussen

Introduction

Roughly speaking, investigating, collecting, and evaluating a diverse range of sonic methodologies that this Handbook contains takes two paths. On the one hand, the attention for sound, sound studies, and/or methods which are based on listening can form, inform, and transform certain (academic) disciplines and discourses. The invention of the *stethoscope* and *echography* have contributed to the development of medical diagnoses and treatments (see Eggermont, in this volume); *soundwalks* and *field recordings* serve, for example, anthropological, ethnographic, and biological research (see Biserna; Guillebaud; Battesti; and Halfwerk, among others, in this volume), influence the theoretical as well as practical orientation of architects (see Arteaga, in this volume), and contribute to the mapping of noise pollution in urban spaces (see Brown, and Fiebig and Schulte-Fortkamp, in this volume); *sonification* makes it possible to aurally access information stored in (big) data (see Vickers, in this volume); the (systematic) exploration and deployment of *sound archives* may shed a new light on specific historical periods and events (see Hoffmann, in this volume); etc. In short, in these examples sound has become a methodological tool through which new information and knowledge can be gathered, organized, disclosed, and/or presented – information and knowledge at the service of or benefitting these specific disciplines.

On the other hand, already established (academic) disciplines affect sound studies and discourses around sound, for example when the methods of the former are put to use in research strategies of the latter. The use of *questionnaires*, *participant observation*, *action research*, and working with *case studies* are tried-and-tested sociological and anthropological methods that can be applied in sound studies as well (see, for example, Waldock; and Bild, Huijsman, and Zentschnig, in this volume); *experimentation*, well known and often used in scientific research, has also become an established method in sound art projects (see also Van der Heide; and Teboul, in this volume) and sonic ethnography (see Thibaud, in this volume); *prototyping* has entered the domain and discipline of sound design (see Misdariis and Hug, in this volume); feminist *theories* and philosophical discourses affect the way sound studies develops (see Thompson; Di Bona; and Waltham-Smith, among others, in this volume); etc.

In both options – attention to sounds and sonic information influencing other (academic) disciplines as well as attention for how methods coming from various research fields enter sound studies – two basic forces seem to be operative: the first one is listening, the second one discourse. It will not come as a surprise that all methods and methodologies[1] described in this Handbook in one way or another deal with the issue of listening.[2] Rather than focusing on the intrinsic qualities of sounds themselves, sonic methodologies emphasize the role of perception.[3] Either directly (as in, for example, soundwalks and field recording) or indirectly (as in using questionnaires or by applying philosophical concepts), listening – in all its variety: its modes of attention, its intentionalities, its negotiations between activity and passivity, its balancing between determinacy and indeterminacy, its oscillations or tensions between the sonorous and the audible (Bonnet 2016) or between the perceiving senses and the perceived meaning (Nancy 2007: 2), etc. – as a method seems to be the glue that connects all chapters. However, the same holds for the second force, discourse. In this context, discourse should not be restricted to language and speech that create constructs of cohesive and interconnected concepts within which the world receives meaning and through which reality is formed. Besides and beyond this, discourse should also be regarded as material events and concrete practices connected to all manner of historically contingent procedures that exert power in and on a society and its citizens, thereby producing, organizing, and normalizing specific knowledges and meanings while disqualifying and excluding the ones that might destabilize or challenge the power of the dominant discourse. In short, I understand discourse here as an external force – although it remains to be seen how external it is – acting on sound, soundscapes, sound art, and sound studies.

What interests me here, and this is based on and informed by all of the chapters collected in this Handbook, is how listening and discourse can be related to one another: how the one is affected and inflected by the other. In other words, I do not consider 'regimes of listening'[4] as a priori subordinate to discourse: modes of listening act on and co-create discourse as much as discourse influences listening attitudes and the knowledge and experiences stemming from these listening attitudes. While discourse (the grammatical, the structural, the semantic, the signifying) confers legitimacy, authority, and direction on what we hear and how we listen, the *sens*(e) of sounds is never secured, as the way we affect and are affected by them also exceeds the cognitive structures of signification that discourse claims and aims to name and contain (Finn 2009). The one forms the other, but, simultaneously, both are un-formed, de-formed, trans-formed in and through their connectivity.

Below, two art works will be presented as starting points, as case studies, as more or less arbitrary markers to investigate and, simultaneously, to re- and de-territorialize the relationship between discourse and listening, the one being a movie, *Das Leben der Anderen* (The Lives of Others) (2006), the other a novel, *FOON* (Phon) (2018).

Das Leben der Anderen: Part I

As can be read throughout this Handbook, many professions require an acute ear – from the auscultation of patients by doctors to the designing of utensils and household appliances, and from conducting ethnographic research in megalopolises to tracing the various roles of sonic communication by animals. And spying! *Das Leben der Anderen*, a German movie from 2006 by the film director Florian Henckel von Donnersmarck, mostly takes place during the heyday of the Deutsche Demokratische Republik, the former East Germany, during the 1960s and 1970s. The protagonist, Gerd Wiesler, is a dedicated citizen of this socialist republic and a captain within the infamous security service *Stasi*. As such he is charged with controlling fellow citizens. The plot starts when Wiesler receives the order to eavesdrop on the internationally recognized and initially pro-communist playwright Georg Dreyman and his girlfriend Christa-Maria Sieland, as their loyalty to the communist principles is questioned by the Minister of Culture, a cynical man who intends to enter into a sexual affair with Sieland. Dreyman's apartment is stuffed with electronic listening devices, and Wiesler finds himself installed in the attic of the apartment block with headphones and a tape recorder; the relevant parts of what has been recorded must be typed up and forwarded to his superiors.

Ear-witness testimonies must become a crucial form of proof: sound leaks into the attic, affording an acoustic impression of what happens elsewhere. And this auditory information should thus provide evidence as to whether Dreyman and Sieland are guilty or innocent.[5] Recorded materials *should* not only tell the truth; they *will* tell the truth, if only the truth that pleases the East German authorities. In other words, what is made audible is not only constituted by the phenomenon that produces it but also by the ears to which it is addressed (cf. Bonnet 2016: 42). In the end, it is the act of listening that makes the sounds speak, that will give sounds their meaning. With the French philosopher François Bonnet, one could state that 'listening does not depend upon an audibilizing of sound – it *produces* this audibilization' (Bonnet 2016: 136). Hence, listening is not neutral and passive; it is a method that organizes, composes, modulates, and selects. Aural perception determines what can and cannot be heard, what should and should not be heard, what is desired and what is dreaded to be heard (for example manufactured as evidence that forces the defendant to plead guilty). In that sense listening is never neutral: it is never just connected to sounds, to the sonorous or the audible, but always already determined by a pretext, a context, a frame within which it takes place. Listening is teleological; it serves a goal. Bonnet summarizes this under the denominator 'discourse'. When listening is subjected to discourse, it is by definition subjected to 'mechanisms of control and regulation' (198). It is discourse that determines an aim for listening, a reason, thus making it instrumental, for example for heuristic verification; listening is disciplined, incorporated in (a belief in) the existence of an order. This is what spying does, this is what Wiesler does, this is what he is told to do by his superiors: to listen, to listen in to find the truth, better yet, *a* truth, a truth that corresponds to the dominant discourse of the East German leaders.

However, another aspect catches the eye – literally. Wiesler's chiefs only get visual information: they never listen themselves; they read! Their method to obtain evidence is not to lend an ear but to rely on reports; they read the notes Wiesler has made on the basis of what he has heard. The 'real truth' is written down. Their attitude seems to confirm what James Clifford already claimed back in 1986, namely, that Western culture is grounded in acts of inscription, reading, and interpretation, acts within the domain of vision and visibility (Clifford 1986: 25). Is cribbing in the end more reliable than eavesdropping, reading more trustworthy than listening, visual information a better proof than audible accounts?[6] Can listening only be regarded as a useful method when it is sustained and, ultimately, even replaced by the written word? And is listening always subordinated to and encapsulated in discourses which make it teleological? These questions are being raised by this movie but are also pertinent in the novel *FOON* to which I now turn.

FOON: Part I

In comparison to *Das Leben der Anderen*, *FOON*[7] (2018) starts from the opposite position. *FOON* is a book by the Dutch author Marente de Moor, and for over three hundred pages readers find themselves in the psyche of Nadja, an ex-student and the current spouse of Lev, an old professor who suffers from dementia. They live in primitive circumstances in a deserted and remote village in the forests of western Russia, near to their former, now disestablished, research centre and shelter for orphan bears. Years ago they fled from the (academic) world of technology and efficiency and the over-organized (Soviet) society with its obligations, regulations, and standardizations. However, what is real and what is a product of Nadja's imagination, the latter partly influenced by her alcohol consumption, remains vague and uncertain throughout the book. Nadja wallows in a world of shadowy stories; vague conversations; inner monologues and desires; talkative animals; ogres, ghosts, and witches; incomplete memories; and half hallucinations; only once in a while the outside, regular, and real world, seems to seep through, for example in the form of the delivery of unpaid bills and the visit of a few policemen as well as a Dutch woman who mixes ethnographic research with New Age sentiments.

Within this context, oscillating between reality and fantasy, between sensory impressions, memories, and imagination, it becomes clear that, for some time, both unstable Nadja and senile Lev suffer from hearing odd sounds – 'the Big Sounds' they call them.

> 'Have you heard anything tonight?' he asks.
> 'The usual alarming sounds.'
> 'No, I mean those from heaven.'
> 'Ah, those.'
> 'The Big Sounds.'
> 'Yes, yes.'

(De Moor 2018: 11)[8]

This is how the Big Sounds are first introduced, as coming from heaven. Nothing to worry about, so it seems, but it does leave the reader with an uneasy feeling. Invisibility and evanescence easily lead to insecurity, hesitancy, and uncanniness. And just as Nadja and Lev cannot predict when the sounds will (re)appear, where they come from exactly, and what they signify, the reader is also left without any further information or explanation. In fact, they have to wait until page seventy-three before a first, small corner of the veil is lifted and the sounds are described in more detail.

> It never sounded that loud. The first tone, sluggish and colossal, trails from east to west through the sky. As if God is moving furniture. Then a silence that, I already know, will not last long. Something builds to swell into a roar. One register lower this time, a rusty yowl runs above our heads. The ears don't hurt, because it is not directed to something in particular. It is simply everywhere. When it is over, we won't be able to copy it, not with our voices and not whistling; we will not even be able to describe it.
>
> (De Moor 2018: 73)

The description only magnifies this sonic mystery, even though listening to sounds becomes more secure and secured when the source can be detected, named, categorized, and localized. Here, the sounds at first seem to come from above, but De Moor immediately undermines this conclusion by claiming that they do not come from one particular direction: they move from 'east to west' and are 'simply everywhere'. The brief silence doesn't offer comfort either: at any moment its deceptive peace can be shattered. And even a mere description or sonic imitation of the sounds seems impossible: words and the human voice fall silent, unable to translate these uncommon sonic signs into common language. However, this uncontrollable network of imminent danger combined with the impossibility to put it into words didn't frighten Lev and Nadja initially:

> When we heard it for the first time, we leapt up, happily. We thought that someone was playing on an alphorn, or something alike. Finally a bit of life to the world! However, we rectified ourselves immediately, as this sound was too huge for human ears. It was not meant for us. It should be a side effect of something mysterious.
>
> (De Moor 2018: 75)

'There is often that sense of there being more to what I am hearing,' Brandon LaBelle writes in *Sonic Agency* (2018: 60), and this also seems to apply to Lev and Nadja. Behind or beyond what they can hear, something inhuman, something unthinkable must hide itself. But perhaps the reverse is also true: what they hear exceeds their imagination and transgresses the available discourses that name and frame the sonorous and put it into the domain of the audible; no discourse can get a grip on the sounds that haunt this couple.[9] However, at times, Nadja doesn't seem to care: "The eerie in the world was never enigmatic. For thousands of years, man lived in a wealth of mysteries, which you could disregard undissolved. Sometimes they took the shape of a wonder. Long live the enigma!' (De Moor 2018: 105)

It is here, in these phrases and in Nadja's simple and acquiescent meditation, that the fundamental difference between the role of listening in *Das Leben der Anderen* and *FOON* becomes most prominent. Both share the same point of departure, namely that listening gives access to a world otherwise hidden or veiled; in both cases listening is, one could say, the *method* through which world and man encounter each other, determine one another, create each other's identities, and give each other significance. But what separates the two is equally clear: In *Das Leben der Anderen* listening is a monitoring tool, an instrument to control and to be controlled, a means to subjugate and repress, a method to hold on to the dominant discourse, ideology, and politics. Conversely, in *FOON* listening can be considered as a tool to come into contact with the enigmatic, the unknown, the uncanny. Perceiving the Big Sounds makes one drift away from control and mastery: one can only absorb the sonic atmosphere as it unfolds. Listening becomes disengaged from the teleological as imposed on it by discourse. Nadja and Lev's are ears of refusal, hallucinating or ecstatic ears. Their experience is one of putting the ear out of place, of a coming to be of a *sens*(e) that never arrives as such, as if the audible truth is anyhow accessible to argumentation, exegesis, and analysis, instead of an irreducible and untranslatable excess or effect of the panphonic power and volatility of the sonorous itself (Finn 2009).

Das Leben der Anderen: Part II

While eavesdropping, while overhearing, the listener is in control. He decides when to listen, what to listen for, how to assess what he hears, and what to do with the information gathered. In *All Ears: The Aesthetics of Espionage* Peter Szendy speaks about this position as a providing *panacoustic* power, explicitly relating it to Jeremy Bentham's panopticon prison in which a guard can see into each prisoner's cell at any moment without being seen by the one who is being watched. However, Szendy also expresses a concrete reservation with regard to this sonic equivalent: compared to the prison guard, the position of the eavesdropper is not as sovereign (Szendy 2017: 48). First, he is destined to lose himself in what Szendy describes as 'the infinite finitude of the detail' (74). Second, he is confronted with an excess: sounds always carry with them more information than would lead to a single, correct interpretation. And third, he is restricted to

> a mere gathering whose horizon and general plan escapes [him]. [His ears] abandon their task of watchful surveillance at the threshold of a totality whose aim is not given to them, although it alone would be capable of transfiguring the collected details by retrospectively illuminating them with meaning.

(Szendy 2017: 109)

This undermining of the sovereignty of the eavesdropper is exactly what happens in *Das Leben der Anderen*. At a certain moment, Wiesler's assistant, who sometimes replaces him in the attic, picks up on a conversation between Dreyman and Sieland but completely misinterprets its meaning. Sieland's confession that she is being blackmailed by the Minister of Culture in exchange for sex is regarded by him as a first incentive for a new theatre play.

The auditory surveillance of those whom one wishes to master fails. Wiesler's superiors receive unsuitable and inferior information; that is, the eavesdropping does not lead to the desired and anticipated result, namely to have proof that Dreyman commits subversive actions. The perceptual will to extract something tangible and verified goes awry. Wiesler's role as a messenger, an intermediary, as the one who simply has to transmit something that someone before him has spoken appears to be quite complicated. At first, his own 'voice' seemed completely absent in the work he has to do. But now it becomes clear that his listening is not just doomed to be a mere instrument of a dominant and disciplining discourse. Through listening, sounds can be de- and re-territorialized, interpreted in various ways, and thus be projected into another discourse in which they take on other values, in which they tell other stories: 'a ruinous reassemblage of the system of signs', according to Bonnet (2016: 276). Listening (in) – as a method to gather information, to acquire knowledge, to find the truth – is not reliable but inevitably destined to fail. And perhaps this is not even an exception, but the very condition of listening (in). Failure is an essential risk of these kinds of operations. The possibility of a negativity is a structural possibility. Listening (in) opens the way for misunderstandings, misinterpretations, the possibility of hearing something with another intention, to hear something else or in a different way than how it was said or intended. Listening (in) as a tool to execute power and to establish and maintain a certain order can simultaneously undermine this power and subvert this order.[10]

FOON: Part II

I left my reflection on *FOON* by stating that an explanation of the occurrence of the Big Sounds failed to appear and that Nadja was celebrating the mysterious. However, the search for an explanation is already well under way, seeping through Lev and Nadja's ostensibly laborious or even laconic observations. Framing it as an alphorn or as God moving furniture are already (futile) utterances made in the attempt to get a grip on the ungraspable; after all, they are (still) scientists, trained to dissect the world or to extract the secrets of nature.[11] Hence, former colleagues are consulted. One claims that these are sounds coming from the sun; another speaks of radio waves propagating from the ocean; a third one says that NASA has been testing a new, secret device (De Moor 2018: 291). Even Lev, though suffering from dementia, immediately tries to find a clarification when confronted with the unfathomable. His conclusion: they are ultrasonic. However, once he realizes that this cannot be a real, solid, and decisive explanation, he falls prey to existential fear and tries to physically and mentally escape from this sonic horror haunting him.

Other clarifications are sought. Scientific theories – from increasing geodynamic activity and quickly moving magnetic fields to gliding tectonic plates – prevail, but more esoteric and religious hypotheses are also considered: the biblical End of Days, Time itself, which doesn't want to proceed in silence anymore, the background noise of Life (De Moor 2018: 290). Or, is it indeed 'only' a product of the imagination? Not everyone hears the Big Sounds; perceiving them seems only possible for those whose mind's eye is open in a specific way, for example the mentally confused or disabled. The desperate quest to solve this mystery, the uncomfortable feeling of not being in control, the inability to fully accept the incomprehensible – these reactions show how both the protagonists in *FOON* as well as its readers are longing for that which Wiesler and his assistant had to abandon: an overriding discourse that gives direction, sense, and meaning to their listening experiences.

Listening between discourse and dissonance

It is time to recapitulate, time to tie up the loose ends. Why have I chosen to, briefly and incompletely, describe and reflect on this German movie and Dutch novel in an introduction meant to present Part III of this Handbook on sonic methodologies? Why did I present these two stories that seem to take inverse journeys – the one progressing from the imperative to control to the reality of excess and precarity and the other proceeding from an encounter with the unknown to a desperate (but futile) search for clarity? The answer might be that I wished to make clear once again that the one (discourse and control) cannot exist without the other (excess and unknown), that the one is always already (latently) present within the other, that the one can never completely exclude and rule out the other, as that which needs to be excluded and ruled out, that which needs to remain on the outside, is *thereby* already incorporated in the inside.

Roland Barthes's seminal text 'Listening' from 1976 illustrates this very well. He distinguishes between three basic forms of listening. The first one, called indexical listening, is closely related to hearing as a physiological phenomenon, directed towards the appropriation of a (sonic) space, that is, to the act of paying attention to whatever might disturb this space and defending oneself against surprises. Perhaps this is the initial listening mode of Nadja and Lev when they are confronted with the Big Sounds. But it certainly also applies to Wiesler, as his listening attitude implies getting used to the 'normal' sounds in Dreyman's apartment and to be all ears when something extraordinary is detected. The second mode, called hermeneutical listening, could also be connected to Wiesler's eavesdropping activities, as it relates to an attitude of decoding that which is obscure, of enciphering and deciphering reality. However, as Barthes continues by writing that 'by her noises, Nature shudders with meaning', and 'to listen is […] to try to find out what is happening' (Barthes 1985: 250), this second mode of listening is also applicable to the story of the Russian couple. The third mode is called psychoanalytic listening and is an unbiased listening, attentive but not predetermined: 'The originality of psychoanalytic listening is to be found in that oscillating movement which links neutrality and commitment, suspension of orientation and theory […] an attention open to the interspace of body and discourse and which contracts neither at the impression of the voice nor at the expression of the discourse' (254–255). This third listening mode, Barthes continues, involves a risk: 'It cannot be constructed under the shelter of a theoretical apparatus' as it grants access 'to all forms of polysemy, of overdetermination, of super-imposition' leading to a 'disintegration of the Law' because 'no law is in a position to constrain our listening' (256–260).

Both Wiesler and the Russian couple as well as the audience of *Das Leben der Anderen* and the reader of *FOON* experience, first of all, that none of these listening modes supplants the others (as Barthes also makes clear in his essay): listening takes place between the indexical, the hermeneutical, and the psychoanalytical; it takes place between the sensible and the intelligible; it takes place between the acoustical and the grammatical – between, that is, going from one to the other and back again, with the one always already being affected, infected, and inflected by the other. Second, listening is always inadequate as well as excessive. Szendy stresses this inadequacy and calls it 'a dissonant listening', a phase 'where overhearing encounters its limit', where 'it ends up announcing the deconstruction or the disenchantment of […] a complete inspection'. What remains, then, is listening's 'unresolved duplicity, its fission that is not absorbed in a fusion' (Szendy 2017: 116). Whereas Szendy thus underlines listening's dearth, the philosopher Geraldine Finn puts more emphasis on the excess, on experiencing the effects of a non-discursive sonority, that is, on a surplus value generated by sounds that cannot be assimilated by discourse.[12] Next to systems of signification that can be mastered and explained by analysis, an in(de)terminable play with the materials of sound and sense is always already operative, to make them sound and make sense otherwise (Finn 2005).

This in-between state of listening, this listening in the service of discourse while simultaneously undermining it, this listening which disciplines, is disciplined, while at the same time subverting and undermining, of course influences the way in which it can be presented and defended as a method. If the precarity and indeterminacy inherent in any regime of listening can ever be considered as productive for a sonic methodology, this methodology should emerge from flexibility and adaptation rather than from mechanistic predispositions and fixed procedures. And precarity and indeterminacy will only be tolerated when methods can evolve in real time, in particular situations, in concrete contexts. What is interesting about this Part III is that all authors take as their points of departure concrete case studies to describe, analyse, and reflect on the methods being used and the role sounds play. So, although the sonic methodologies as such might be described in general terms (soundwalk, experimentation, action research, etc.), they receive very specific meanings and various implementations within the case studies described in each contribution. Thus, the methods and methodologies are emptied of the generic and given a particularity and singularity, a difference in each repetition.

What thus stands out are the various ways in which the relation between listening and discourse, the relation between sounds and information or knowledge, the relation between the audible and the sonorous are played out in this section. In other words – and this could also be concluded from the above reflections on *Das Leben der Anderen* and *FOON* – listening cannot be confined to a pure function of knowledge production, as it necessarily oscillates between perceiving sonorous vibrations without any signification and attributing semantic content to the audible. Moreover, listening implies a being-in-relation: there is a strong and reciprocal relation between that which sounds and the one who listens, if only because they share time and place. As such, sonic methods and methodologies can be simply described as strategies for engagement and exploration, at their best facilitating a rigorous and responsible perception in an in-between space.

Notes

1. For the difference between method and methodology, see my Introduction to Part II.
2. Here I will not assume a fundamental difference between listening and hearing. Although slightly criticized in Darla Crispin's essay (see Part II), I basically follow Anahid Kassabian's deconstruction of that opposition. As she writes: 'By listening, I mean a range of engagements between and across human bodies and music technologies, whether those technologies be voices, instruments, sound systems, or iPods and other listening devices. This wipes out, immediately, the routine distinction between listening and hearing that one often finds, in which the presumption is that hearing is physiological and listening is conscious and attentive. I insist, instead, that all listening is importantly physiological, and that many kinds of listening take place over a wide range of degrees or kinds of consciousness and attention' (Kassabian 2013: xxi-xxii). Listening is not always and only an intentional act, but also happens on a subliminal level: un- or subconsciously, sounds enter our body – that is, our body is listening, too. It can in fact be considered as a gigantic membrane.
3. See also the description of a soundscape in the international soundscape standard, the ISO 12913–1, as 'an acoustic environment *as perceived or experienced and/or understood* by a person or people in context' (my emphasis). Several authors in this Handbook refer to this standard and its definition of a soundscape.
4. I have introduced this term in another essay: see Cobussen 2020.
5. Whereas *Das Leben der Anderen* is 'just' fiction, albeit based on prevalent East German practices, a more recent story may remind us of the role ear-witnesses can play in reality. On 14 February 2013, a famous South African runner, Oscar Pistorius – also known as Blade Runner as his two legs have been amputated so that he has to run on two artificial legs – was arrested on suspicion of murdering his girlfriend Reeva Steenkamp. He was found guilty on the basis of the testimonies of ear witnesses who heard the two quarrelling, followed by several pistol shots. Steenkamp died of three bullets, and Pistorius has been sent to prison.
6. This reminds me of the incredulity of the disciple Thomas who refused to believe what the other apostles *told* him, namely that the resurrected Jesus had appeared to them: 'Except I shall see in his hands the print of the nails, and put my finger into the print of the nails, and thrust my hand into his side, I will not believe.' When Jesus appears once again, Thomas believes, but Jesus answers: 'Thomas, because thou hast seen me, thou hast believed: blessed [are] they that have not seen, and [yet] have believed' (Jn 20.24–29).
7. 'Foon' translated into English as phon, is a unit for the perceived loudness of pure tones. Human sensitivity to sound is variable across the frequency spectrum. For example, the ear is less sensitive to low and very low frequencies and most sensitive to middle and mid-high frequencies. The phon takes the perceived loudness of a 1 kilohertz tone as a reference. For a tone with another frequency the physical intensity is adjusted to result in the same perceived intensity and thereby the same phon value (see also Wikipedia 2020).
8. All translations of *FOON* are mine.
9. Bonnet makes a clear distinction between the sonorous and the audible. Whereas the sonorous refers to the sounds per se, to sound as an emergent and primordial force, the audible – that what 'gives itself to be heard' (Bonnet 2016: 8) – is always already

permeated by meaning and signification. Lev and Nadja seem to experience the Big Sounds in a space *between* the sonorous and the audible.
10. It would be possible to understand Darla Crispin's remark, in this volume, that sound studies has a contribution to make to the debate around ethics in this context.
11. The human yearning for explanation, logic, consistency, and control also became apparent during the solving of some mysterious sounds heard at the US Embassy in Cuba in 2018 and 2019. Employees suffered from headaches and became sick and lightheaded due to sounds, they believed, that had no clear source. Suspicions developed that the Cuban secret service was responsible, using a sonic weapon to cause these inconveniences. Many months later this suspicion turned out to be false: these strange sounds were actually produced by the mating of an indigenous bug.
12. The German philosopher Gernot Böhme also makes the connection between a sonic experience and excess. 'Hearing is being-outside-oneself', he writes, as one can be 'carried away by sweet melodies, knocked over by thunderclaps, threatened by droning noises, or wounded by a piercing tone'. And this being-outside-oneself while listening has more fundamental consequences. It is not simply so that the listening subject meets sounds while being outside; it becomes 'shaped, moved, modelled, nicked, cut, lifted, squeezed, widened, and constricted by those sounds' (Böhme 2017: 133). Hence, the subject does not precede its listening experience but only comes into being through and in the act of listening.

References

Barthes, Roland (1985). 'Listening'. In *The Responsibility of Forms: Critical Essays on Music, Art, and Representation*, 245–260. Trans. Richard Howard. New York: Hill and Wang.

Böhme, Gernot (2017). *Atmospheric Architectures. The Aesthetics of Felt Spaces*. Trans. Tina Engels-Schwarzpaul. London: Bloomsbury.

Bonnet, François J. (2016). *The Order of Sounds: A Sonorous Archipelago*. Trans. Robin Mackay. Falmouth: Urbanomic Media.

Clifford, James (1986). 'Introduction: Partial Truths'. In James Clifford and George Marcus (eds), *Writing Culture: The Poetics and Politics of Ethnography*, 1–26. Berkeley, CA: University of California Press.

Cobussen, Marcel (2020). 'Regimes of Listening … or … One Day in the Life of a Music Philosopher'. In Nanette Nielsen, Jerold Levinson, and Tim McAuley (eds), *The Oxford Handbook of Western Music and Philosophy*. Oxford: Oxford University Press.

Finn, Geraldine (2005). 'The Truth in Music: The Sound of *Différance*'. *Muzikološki Zbornik – Musicological Annual* 41 (2): 117–146. Available online: https://revije.ff.uni-lj.si/MuzikoloskiZbornik/article/download/5600/5339 (accessed 6 July 2020).

Finn, Geraldine (2009). 'Giving Place – Making Space – For Truth – In Music'. Available online: https://www.twu.ca/verge-conference/conference-archive/verge-conference-2009/conference-presenters/bios-abstracts-geraldine-finn (accessed 6 July 2020).

Henckel von Donnersmarck, Florian (2006). *Das Leben der Anderen*. Available online: https://www.bing.com/videos/search?q=film+das+leben+der+anderen&qpvt=film+das+leben+der+anderen&view=detail&mid=0421FEC96CC1B9740F840421FEC96CC1B9740F84&&FORM=VRDGAR (accessed 6 July 2020).

Kassabian, Anahid (2013). *Ubiquitous Listening. Affect, Attention, and Distributed Subjectivity*. Berkeley, CA: University of California Press.

LaBelle, Brandon (2018). *Sonic Agency: Sound and Emergent Forms of Resistance*. London: Goldsmiths Press.

Moor, Marente de (2018). *FOON*. Amsterdam: Em. Querido's Uitgeverij B.V.

Nancy, Jean-Luc (2007). *Listening.* Trans. Charlotte Mandell. New York: Fordham University Press.

Szendy, Peter (2017). *All Ears. The Aesthetics of Espionage.* Trans. Roland Végsö. New York: Fordham University Press.

Wikipedia (2020). 'Phon', last updated 30 June 2020. Available online: https://en.wikipedia.org/wiki/Phon (accessed 6 July 2020).

33

Auditory Diagramming: A Research/Design Practice

Alex Arteaga

This chapter presents a detailed and extensive account on an aural research and design[1] practice termed 'auditory diagramming'.[2] This practice has been conceived in the last few years and continues to be developed at the Auditory Architecture Research Unit and in the framework of different projects, seminars, and workshops.[3]

This chapter is split into three parts, realized respectively through three different practices of writing. In the first part, produced through a practice close to academic writing, I describe the conceptual context in which auditory diagramming is defined. On this basis, I outline the practice conceptually and specify its function. In the second part, generated by a kind of instructions-for-use writing practice, I explain how to perform auditory diagramming step by step. The third part is brought about through a writing style I have been developing in my latest projects, which I denominate 'exploratory essay writing'.[4] I consider this to be aesthetic research practices in the medium of written language without any a priori formal and style-related limitations. It is a practice of *slow observation* that mobilizes the inherent epistemic agency of the semantic, syntactic, and morphological aspects of written language in order to disclose the object of inquiry, in this case two different but intimately intertwined moments in the process of diagramming: *making* the diagram and *reading* the diagram. I explore these moments initially hypothesizing that the first is enabled by an *aesthetic* procedural manner – what I call *aesthetic action*[5] – and the second by a *poetic* one. In a closing postscript I briefly explain, performing a practice related to academic writing, how the practice of auditory diagramming establishes a continuum between researching and creating an agent for the transformation of an environment: an architectural design.

Conceptual framework and outline of the practice

Auditory diagramming is a practice conceived of and developed in the framework of *auditory architecture*, a new approach to the relationships between *aurality* and *environment*.

In this section I briefly outline the basic traits of this approach together with the conceptual framework in which it has been initiated and continues to be realized.

The concept and practices of auditory architecture mark a double and fundamental turn with regard to the lines of research initiated by Murray Schaefer and further developed by, among others, Barry Truax. The first discontinuity between the traditions of acoustic ecology and auditory architecture is the substitution of *sound as focus* of inquiry with *aurality as medium* of research. In these traditional frameworks sounds are the focus of awareness. In these contexts, processes of inquiry aim at figuring out how the sounds are present in a certain environment and how they relate to each other. In short, it is about sounds. In contrast, the research developed in the framework of auditory architecture is not focused on sounds but on practices conceived and performed in the aural medium.

Specific practices of hearing and listening actualize the potentialities provided by this medium within the limits and according to the constraints that it establishes. Aurality is understood here as a set of conditions of possibility for the performance of practices of hearing and listening.[6] As such the aural is not the focus of research,[7] but the medium in which the research takes place, that is, the field of potential agencies that enables and constrains the processes of research. The second aspect of the fundamental turn inaugurated by auditory architecture defines its object of research. Elaborating on the idea that the focus of attention in the context of acoustic ecology is sound, I posit that the object of research in these traditions is a part, component or feature of the environment: its acoustics. In simple terms: it is, mainly and firstly, about how an environment – or maybe a landscape – sounds. On the contrary, in the framework of auditory architecture, the object of research is the whole environment or, indeed, the environment as a whole – as a significant, enveloping, all-over presence, as the immediately surrounding world.[8] Auditory architecture, therefore, is environmental research performed in the aural medium, actualized through specific practices of hearing and listening.

The concept of environment that underpins the development of auditory architecture and that, in turn, auditory architectural research further develops, has been outlined in the phenomenology-based *enactive approach to cognition*.[9] The basic idea is that an environment emerges out of the interaction between living beings – biologically realized autonomous units – and non-living beings – heteronomous entities. The dynamic unfolding of the *structural coupling* between a specific spatio-temporal configuration of the members of these two classes sets the enabling conditions for an environment to come into being, namely as the environment for,[10] the living being. Furthermore, the environment and the living being to whom it appears are in a relation of *mutual conditioning*; or, expressed in enactivist terms, they *co-emerge*.

To summarize, the idea is that out of the inherent and inalienable interaction between living beings and their physicochemical surroundings, *selves* and *environments* co-emerge. Specifying this concept for the case of an auditory architectural researcher who investigates the environment through diagramming, it means, first, that both researcher and environment *as such* – that is, as the very specific researcher and environment they become through their interactions – are in a relational process of co-constitution. Secondly, that the practice of diagramming and the immediate result of it – the diagram – are *interventions* in and therefore constitute new constraints of their common process of coming into being in mutual conditioning, that is, of co-emergence. None of the participant entities in this process exist in isolation or are ultimately completed: they all – researcher, researched, and means of research – are in a shared process of emergence. Therefore, the research situation, as well as all its components are understood here as radically dynamic, relational, and transformative.

The fact that the environment is present for a living being does not necessarily mean that it *represents* its environment. Instead, the living being *enacts* it. According to the enactive approach, the environment, although it appears in this way, is not given to the living being. It is not simply out there to be caught. It does not exist in itself, independently of the actions of the self to whom it appears, waiting to be grasped. Accordingly, perception is not understood here as the process of taking in an 'outer reality' in order to represent it in the brain or in the mind. Instead, perception is conceived as a network of bodily acts, or to be more precise, of interactions that co-constitute what appears as real.

Adopting an enactivist position – instead of the realism-representationalism adopted by the traditions of acoustic ecology – has fundamental consequences for auditory architecture. These consequences become primarily manifest in the presentation of a new term as an alternative to soundscape – *Klangumwelt* – and in the conception and development of new research practices like auditory diagramming. The German term *Klangumwelt*[11] can be translated as 'sound environment' or, literally, as the surrounding (*um-*) aural (*Klang*) world (*Welt*). This word is a dense expression of the concepts succinctly presented so far: it refers to a surrounding, all-over and significant presence, a *life-world*, experienced primarily in the aural medium, that is, mainly conditioned in its process of emergence in and through the practices of hearing and listening. A Klangumwelt, therefore, is not a given configuration of sounds occurring in a given topology; it is not a soundscape. This clear definitional difference implies that, whereas a soundscape can be recognized or apprehended – the performance of the concept of perception as representation of a reality existing in itself – a Klangumwelt can be co-constituted through perception and other intentional acts – understood, thus, as an embodied and situated process of active participation in a system of co-emergence. The fundamental difference between both concepts also means that whereas a soundscape can be recorded and reproduced – a technological implementation of the realist-representationalist concept of reality and perception – a Klangumwelt cannot. Instead, the dynamic structure of a Klangumwelt can be *understood* through the performance of certain research practices not based on representation but on *mediation*, that is, on the generation and activation of a *medium* – like an auditory diagram – that allows the disclosure of the researched environment.

Although auditory architecture participates in the conceptual framework of the phenomenologically-based enactive approach to cognition, it is not the result of 'applying a theory in the practice'. Auditory architecture is not the implementation of a cognitive theory.[12] On the contrary, auditory architecture can be regarded as a further development of the enactivist way of thinking in the aural medium. Auditory architecture is, therefore, aural, phenomenological, and enactivist environmental research. As such, auditory architecture is architectural research. Auditory architecture participates in and further develops a specific concept of architecture. Auditory architecture is a contribution to the concept and the practices of architecture as *intervention* in the emergence of environments *through construction and related practices*.[13] Architecture is not conceived here as the design of big-scale objects, generically dissociated from any concrete surroundings; and, clearly, auditory architecture does not aim at designing how these objects should sound: auditory architecture is not acoustic design. Through the definition of new practices, auditory architectural research aims rather to conceive of constructive interventions[14] that expand the possibilities of mutual relationships between subjects and their environments. This should result in possibilities for those subjects exposed to 'the agency of these interventions' to *understand* themselves and their life-worlds differently, that is, of co-emerging for one another as significant and intimately intertwined phenomena in alternative, unforeseen ways. Hence, architecture is conceived here, as a medium that enables the emergence of radically alternative phenomenal selves and environments.

In order to serve this purpose, the practices developed in the framework of auditory architecture are, fundamentally, *aesthetic* practices. They organize and systematize actions performed in a specific variety of relationships between the one who carries them out and the environment with which one interacts. I call this mode of interacting *aesthetic conduct* and characterize it fundamentally as being based on a spontaneous – meaning not controlled by will – performance of basic relational skills: perception and emotionality.[15] This kind of conduct produces a redistribution of the agencies at work that enable the inception of a field of *shared agencies*. In this framework, the practitioner does not control, define, and lead the situation but *participates* in a non-hierarchical network of agencies in which all components that make the emergence of the environment possible take part actively. Auditory architecture, therefore, is *aesthetic research*[16] and as such aims at contributing to define and develop an *aesthetic architecture*, a network of aesthetic practices that facilitate the realization of constructive and/or construction-related interventions into the co-emergence of environments and selves, aiming at expanding their respective actualizations.[17]

As a practice – that is, as a systematized set of actions – auditory diagramming *tends to*[18] specific environments, understanding the environment as a topological restraint of a phenomenological notion of *world*: 'not [as] an object whose law of constitution I have in my possession [but as] the natural milieu and the field of all my thoughts and all my explicit perceptions' (Merleau-Ponty 2012: lxxiv). The subject matter of this practice, therefore, is not an object but one of the conditions of possibility for the emergence of objects. The non-objectual status of this practice's subject matter fundamentally constrains the definition of its function. As the topic of this practice is not an object, its function cannot be to represent it, to depict it, or to 'grasp' it in any way, since only objects can be represented, depicted, or grasped. An auditory diagram is not a representation of an intended environment. Instead, it is a medium outside of the regime of representation that allows the practitioner to *participate* in the emergence of an environment in a specific way or, rather, with a particular goal: to *understand* its dynamic and relational structure. Furthermore, the singular objects that configure the diagram are neither representations of things, facts, or states of affairs: they are formulations of phenomena constituted by diagramming. These singularities, as constitutive components of the diagram, are the means to an end: to identify the *dynamics* that relate them to one another. Whereas the performing auditory diagramming is an investigative intervention in the emergence of a specific environment, the diagram – the artefact – becomes a *medium* for the realization of the practice that produced it and, more fundamentally, a medium for the achievement of the aimed goal: to understand the environment *as* environment.

Diagramming aurally: The performance of the practice

This section delivers a detailed description of the realization of an auditory diagram preceded by some considerations about its subjective and spatio-temporal structure and the required technology required to realize it.

In order to diagram aurally, a device endowed with a touch screen, such as a tablet, is required. Although a large part of the process could be done using other support, the most fundamental part of the process, that is, the moment of the practice that defines it as diagramming, requires a technology that enables a fluent, unmediated, and precise modification, at least of the size and position of the written words. So far, there is no specific app for diagramming aurally. Therefore, any app that allows one to realize the operations described in the following lines can be used for this purpose.

Auditory diagramming is originally conceived to be performed by individual practitioners, obviously without excluding the possibility of sharing results in a group, of combining individual realizations of diagrams, and working in a team. Diagramming collectively is possible but in order to overcome the difficulties derived from the individual character of perception, imagination, and association, the identification of new strategies would be required. This is one of the necessary further developments of this practice.

Before the practice begins a preliminary exploration of the surroundings of inquiry is recommended in order to initially define a spatio-temporal structure for the performance. In relation to time, it is useful to identify intervals in which the environment presents itself in a 'normal' way. To use the criterium of 'normality' in a phase previous to the beginning of the research is, to a certain extent, contradictory: it is not possible to know what the 'standard state' is of an environment that we do not yet know. Nevertheless, since we are acquainted with different types of architectural environments, it is possible to intuitively differentiate between standard and exceptional, ordinary and extraordinary states. I recommend diagramming in those moments that we intuitively recognize to be standard, although an initial identification can be changed. However, a random selection of time frames is also an option.

These considerations manifest an inherent problematic aspect of this kind of research, derived from the temporal contingency of environments: they are constantly changing. My position in this regard is based on the idea that every phenomenon, although being a temporal entity, tends to *stabilization*.[19] Independently of the kind of intentional act performed – perception, imagination, association, etc. – the degree of variation of the resulting intentional objects decreases progressively until it acquires a steady presence. This can simply be proved by the possibility of being able to recognize them. On this basis, I recommend a repeated performance of the practice until the researched environment becomes stable. The same kind of reflection can be done in relation to space, since an environment is also contingent in this regard. In principle it would not be mistaken to assert that an environment presents itself differently in each spot or trajectory or, more radically expressed, that each spot or trajectory enables the emergence of a specific environment. Nevertheless, based on experience, it is reasonable to affirm that stabilization – maybe it would be better to talk about *continuity* – also takes place regarding space, that is, that we have a sense of being in the same environment across and along a certain extension of a field. On this basis, I recommend initially defining, again in an intuitive way, a structure of spots and/or trajectories in and through the surroundings to be inquired according to vague criteria of interest and difference which can be articulated through the following questions: what do I consider to be interesting spots and/or trajectories? In which spots/trajectories do I recognize differences inside the homogeneity that makes a 'unity' of the environment possible?

The practice of auditory diagramming is based on a fundamental phenomenological distinction between intentional acts and objects – in Edmund Husserl's terminology, *noema* and *noesis* (see Moran and Cohen 2012: 222–224). The act of perceiving differs from the perceptions that appear by virtue of this act. On the other hand, there is an obvious correlation or mutual conditioning between the act of perceiving and the phenomena that appear by and through perception. The practices of hearing and listening that underpin auditory diagramming are conceived of as attending to this distinction and correlation and, furthermore, to the differentiation between intentional acts. Although other practices can be identified and included in this framework, auditory diagramming is currently realized on the basis of four aural practices: *qualitative*, *imaginative*, and *associative hearing*, and *analytical listening*.[20] My distinction between hearing and listening can be adequately described on the basis of the differentiation made by Jean-Luc Nancy, that is, understanding hearing as the non-tense, 'natural' performance of our aural skills and listening as its tense, focused mode (see Nancy 2007). Accordingly, hearing is understood as having a wide focus or, rather, as having no focus: we don't hear *to* something, we *just* hear – a set of acts that appears as events, as 'something that simply happens' – implicitly reducing the meaning of action as to *deliberate* action, that is, as action lead by will towards a goal. By hearing we do not direct our attention to aural phenomena: they simply appear, without us tending to them or trying to avoid them.[21] Hearing, we could say, happens while we are doing something else or when we are focused on something else other than sound.

This double condition of hearing – the aural connection with the environment as well as the possibility of performing other actions – allows the definition of qualitative, imaginative, and associative hearing. During *qualitative hearing*, the practitioner establishes a non-tense aural relation with the environment while testing from time to time that this relation is still there, that is, that the aural presence of the environment continues to be the prime modality of appearance, and on top of that, trying to identify the *qualitative aspects* that emerge. On the background of hearing, the practitioner addresses the question '*how* is it here?' Qualifications such as 'boring', 'interesting', 'disturbing', or 'subtle' might appear. But also these terms might give rise to the commonly asked question: 'Is it about the environment or about myself?', 'Are these qualities of the environment or are they subjective emotions and feelings?' My position in this regard is to provisionally suspend the distinction underlying this question – the distinction between 'my environment' and 'myself' – in favour of the first, that is, to posit that all these qualitative presences are emerging qualities that belong to the environment's emergence. Although this position is questionable from different perspectives, it allows the practitioner to register a large number of significant phenomena. This registration takes place by writing down the words with which, or better *as* which, these qualities appear on the touch screen.

Imaginative hearing and *associative hearing* have the same structure as *qualitative hearing*. In all three cases, hearing is the background activity on which another, focused action is performed. In the case of imaginative hearing, the activity obviously is to imagine. While hearing, the practitioner asks questions like 'What could happen here and now?', 'Who or what could appear?', and 'Who or what couldn't?' I have not yet identified an adequate question to trigger or support the intentional act of associating. I understand association as the spontaneous, unexpected, and even surprising emergence of a phenomenon different from the one which is currently perceived but strongly connected to it. To associate is neither to remember nor to establish a logical connection between a perceptual and a non-perceptual phenomenon.

Although the order in which these three practices of hearing are performed is not necessarily fixed, a good way to begin is simply by identifying sounds without making an effort, without focusing – that is, hearing and not listening – and simply allowing the aural manifestation of the environment to become present. From there on, qualitative perception, imagination, and association can be activated interchangeably. In all three cases, the emerging phenomena should be written down on the touch screen, exactly in the way they linguistically appear, that is, as words or word clusters[22] without any kind of judgement or correction, and without attending to their distribution on the screen's surface. The result is a formless group of words that can be expanded by the results of *analytical listening*. I consider the three described practices of hearing to be nuclear for auditory diagramming and, therefore, analytical listening to be a complementary practice. This aural practice is based on establishing a tensed aural relationship with the environment, focusing the attention first on the identification of single aural phenomena – *single sounds* and their *own features* – second, on the discrimination of *general features of sounds* (respectively in close, mid, and far range), and third, on the *characteristics of all present sounds as a unit*. The phenomena that can be constituted by setting each of these different foci are listed in Annex 1. Whereas the results of the three practices of hearing can be decisive in understanding the inner dynamic structure of the inquired environment, and therefore in having a sense of 'how it is as a whole' and 'how it could or even should be',[23] the phenomena made explicit through analytical listening can contribute to identifying concrete interventions for transforming some of its aspects.

Nevertheless, none of this would be possible if the phenomena notated on the touch screen are not modified and organized through *diagramming*. The practice of diagramming begins, *sensu stricto*, after this first phase of notating and consists of a process of *in-formation* – of *form-giving* – of phenomena that allow the environment to emerge in their respective specificities through their constitution of that environment in this very specific environment. Although the whole group of phenomena is going to be informed, the attention should be oriented, firstly, towards each of them and, secondly, to the relationships between a small number of them. Facing each singular phenomenon, the question refers to their *degree of presence*. If the presence of a phenomenon is big, the practitioner should increase the size of the word that expresses it; if the degree of presence is small, the size should be decreased. Addressing now the relationship between phenomena, the question is one of the *degree of connection* – what I express in German with the word *Sinnzusammenhang*, which can be translated as 'context' but attending to its composition as 'a sense of hanging together'. If two phenomena are connected, they should be displaced close to one another; the weaker their interconnection, the further the distance between them. Two criteria are fundamental in order to realize these operations. First, the relationship between the practitioner and the environment should be *perceptual*: the environment has to be perceived – I would rather say *sensed* – and not inquired through logical operations. The degrees of presence of the phenomena and the connection between one another should be immediately apparent and need not be asserted through deduction or induction. These features of the phenomena and their interrelationships should *appear*, enabled by the intimate perceptual connection between practitioner and environment. Second, a*dequacy, coherence*, or *correspondence* are the criteria to be employed in order to realize these operations. The diagram does not represent the environment but constitutes a relational medium between practitioner and environment. Accordingly, the words on the screen do not represent phenomena, nor do their size or their relative position to one another represent their degrees of presence and connection. Instead, these graphic resources – size and distance on a plane – allow the practitioner to articulate in a coherent way states of affairs perceived as present in the environment. The size of each word should not be the 'correct' one but the *adequate* – or, as I formulate in German, the *stimmige*.[24] Accordingly, the size of a word and the distance to another word should be varied, until they reach the adequate magnitude which can only be fixed through perceptual interaction with the environment. A third relevant feature of environmental phenomena, their respective *degrees of significance*, has not yet been addressed in the development of the practice. Despite its degree of presence, a phenomenon can appear in different ways for the whole environment. The procedure for implementing this parameter graphically is in a phase of consideration and trial. The level of transparency of a word seems to be a potential solution.

These operations, the notation of new phenomena or the erasure of some of the notated ones, should be carried on until the whole diagram appears as *coherent* with the environment, that is, until the diagram and the environment *stimmen*, are 'tuned'. As I wrote regarding the spatio-temporal structure of the realization of the practice, the number of changes introduced in the diagram will progressively decrease until it achieves a rather stable state. This might be the end of the productive phase of the practice, the moment that is inquired as being *aesthetic* in the following section. However, it is not the end of the practice: the diagram must be 'read'[25]: the, possibly, *poetic* moment. Although producing and reading are intertwined moments along the process of diagramming, once the artefact is stabilized, reading becomes the pre-eminent action.

Auditory diagramming as an aesthetic/poetic practice: An exploration

Two moments. Two actions or, better, two ways of acting. Two ways of behaving – of facing the diagram, of realizing, and of relating to the process of diagramming and to the diagram – to the artefact. Two ways of understanding it, of performing it, of signifying it, of activating it – its potentialities, its agencies as a medium and as a practice.

Two practices? Definitely not. Two qualitatively differentiated performances of the same practice. The performance of *a* practice *as* aesthetic practice and *as* poetic practice. Two significations of the very same practice – of sense-making, of conditioning the emergence of sense, of trying to understand the emergence of sense in one of its specifications and in a particular case: its manifestation as an environment.

Two modes of practising – they are *modes* of the same practice, intimately related to one another. In different ways. Time-wise: following each other, alternating, excluding each other as simultaneous acts. Relating to each other in time as a sequence: first one, then the other, then, maybe, the first one again. Or, on a larger timescale, as a simple series: first one, the aesthetic mode – making the diagram – and then the poetic mode – 'reading' the diagram or, being more precise according to the specificity of a diagram which is to participate in the linguistic and iconic logic, reading/perceiving-as-an-image the diagram.

Or, in a more complex relation, while maintaining the sequence, performing one of the modes as pre-eminent, and shifting – shortly, from time to time – to the other as a subsidiary mode. Establishing, therefore, an occasional, provisory, and reversible hierarchy between them. A hierarchy that shows the functional relationship between them or, better, their mutually complementary functions in relation to the function of diagramming and the diagram. A hierarchy that shows the necessity of performing this practice, combining these two modes in order to fulfil the function of diagramming and the diagram: to understand an environment in its emergence.

On auditory diagramming as an aesthetic practice

To begin with: aesthetics here as *aesthesis* – as a specific way of understanding through a certain use of the senses or, better, through a certain way of mobilizing our sensorimotor skills – our sensorimotor self. More fundamentally: aesthetics as a way of relating to our surroundings, as a way of interacting with *others* – with other entities, with other units that we experience, however they appear, as *other*: as differentiated from ourselves.

Aesthetics as a mode of participating in the process of sense-making – as a mode of co-conditioning the emergence of sense – through a specific form of *con-duct*: of leading-(ourselves)-with-others.

Aesthetics as a mode of being, actively – unavoidably actively – in touch – in intimate, fluid, porous touch. As a way of conditioning the touch. As a way of enabling a certain kind of touch. As a way of disposing, activating, and bringing the enabling conditions into relation for the emergence of a certain dynamic quality of touch: relying on the performance of our sensorimotor skills and of our emotionality – of our capacities of being moved, of acting without the constrains of our will but rather conditioned by other agents, by the performance of other agencies, of the agencies of others.

Aesthetics, in this case, as acting in a field of agencies shared by the one who is diagramming and the other units that enable the emergence of the environment to be investigated, to be understood – as participating in the process of emergence of the environment with which the diagram will be produced, through acting with other actors, through performing one's own agency as a constitutive and therefore inseparable part of a network of intimately shared agencies. Aesthetic conduct instantiated here as a practice, an aesthetic practice, which systematizes specifically a mode of acting enabled by dynamically disposing one's own basic and not target-oriented skills of communication, of doing together, of allowing the environment to emerge together, and simultaneously, observing its emergence through diagramming: producing an artefact that mediates simultaneously the emergence of the environment and the observation of this process of emergence – the observation of the environment as emergent.

Diagramming as an aesthetic practice: as a way of intervening in an ongoing process without disturbing the process in which the practitioner intervenes but rather becoming part of it – participating through adapting to its dynamics, to the enabling dynamic conditions for a network of significant enveloping qualities to appear as an environment.

I am somewhere. I might know a name, the convention that designates the spot on which I am, but not *where* I am, not the world becoming present and significant – present as significance – to me being, now, here, where I am. And being here, now, I begin to vary the way I act. I vary the way I relate to my surroundings – the way I interact. I suspend the way of acting that allows me to arrive here, to reach the spot I am at now. I stop judging what I perceive as right or wrong in relation to the goal I am seeking: to arrive here. I stop moving in relation to these judgements. Or, if I knew the way, if I have been here already, I stop acting according to my habits, to the embodied processes of signifying my surroundings as 'the right way'. With my movements I stop reaffirming the value that each element of my surroundings acquires in relation to a target – this spot – and a goal – to arrive here. I suspend the significance of this spot as a target. I empty this spot of significance as the condition of possibility for this spot to become a network of agents. I – my will-based self – retreat.

In this sense, I also empty myself or, maybe better, I reconnect with the constitutive emptiness of myself, with its selflessness, in order to allow other-selves to act, in order to allow a distribution of agency, in order to conduct (myself) aesthetically. In order to allow the performance of a network of agencies to enable an environment to emerge here and now: a network of qualitative significances, of dis-tended qualities, of qualitative presences not arising out of the tension established by the definition of a target, by acting in order to achieve a pre-established goal but obeying a complex process of observation: of keeping-safe what appears before the one who observes.

I begin to diagram. I begin even before I grasp any tool, specifying my conduct again. Maintaining my aesthetic relation with my surroundings, acting without abandoning this mode of action – this disposition, this conduct – I set a focus or, more precisely, a double focus. First, I prioritize those actions that allow me to establish an aural relationship with my surroundings – those actions that enable the emergence of phenomena that appear as aural, as heard rather than listened. A disposition in a disposition – a specific set of actions, by acting aesthetically. I continue to not impose my agency over the agencies of my surroundings. I simply reorganize the way my organism – my body – acts in this field of shared agencies. 'Simply' because it is enough to activate the agency of the concept 'hearing' – to 'think' of hearing – in order to set a focus of my awareness on what appears as heard. It suffices to think 'and now I hear' in order to do it, in order to make it possible for this to 'happen'.

And in the frame of this specification of my aesthetic conduct, I set a second focus; or, better, once I have established hearing as my main activity, I displace the focus of my awareness to the qualities that emerge, enabled and constrained by this very specific set of conditions: the agencies of my surroundings, the agencies of my-self – maybe rather of my body, of the organic substrate of my-self – acting aesthetically and privileging the aural – the agency of my-self-hearing.

Now I take a device endowed with a touch screen, I activate the app that will allow me to diagram. As I activate this technology, I allow its agency to participate in this field in order to co-constitute a visual medium and begin to actualize its potentialities through a specific practice: auditory diagramming.

Acting in an aural specification of my aesthetic conduct, focusing my awareness on the identification of qualities emerging out of the field of shared agencies I am participating in, I begin to write down the names with which these qualities appear: new components of the environment, new presences, new agencies intervening, expanding the sphere of agency – new actors conditioning the emergence of the environment and allowing me, simultaneously, incipiently, to understand it – its emergence, it-as-emergence.

My aesthetic conduct allows the words appearing on the screen that I hold in my hands to unfold – immediately, from the moment they appear – their own agency. They do not appear to me as the result of me writing them. I do not deny that I am writing them, this might be clear but it is not significant now. They appear as presences modifying, potentially, what is happening here and now, conditioning my next actions, my next move in this common sphere, in the process that we, as actors, share.

My aesthetic conduct does not confer agency to these words but allows their potential agency to be actualized (agency, as any other feature, is not a constitutive trait of an entity: it comes to be by virtue of relationships, by virtue of the activating force of touch, of a specification of an in-betweenness).

I continue acting aesthetically – my sensorimotor and emotional I continues interacting spontaneously with its surroundings – and I continue hearing – privileging those actions that enable auditory phenomena to be constituted.

On this basis, I temporarily activate my will in order to introduce a variation in my connection with the environment – in order to modify the way I am constraining its emergence, in order to introduce a new dynamic condition. Without altering my basic aesthetic conduct, without ceasing to hear, I act supplementary in such a way that enables *imaginations* to appear, phenomena whose material correlates are not part of the surroundings I am interacting with and therefore are not possible to be constituted through perception.

Correspondingly, I shift my awareness in order to identify these new kinds of phenomena and I write them down. I generate, by writing, new phenomena with a strong and intimate relation of identity with the phenomena constituted through imagination: I write the names with which or, better, as which they appear.

And I repeat the same procedure but now substituting imagination with *association* – the way of acting that enables former experiences to be spontaneously re-enacted by the present experience.

A cloud of words in front of me now – between me and my environment. An incipient relational medium. A bunch of dots relating to a continuous whole, I could say.

No form yet, only an accumulation of potentially form-giving elements.

New surroundings in my surroundings – a new field of agency in a field of agencies. A set of potentially transformative potential forces.

A prospective correlate, a correspondent, a counterpart – a possible 'Entsprechung', a 'stimmige Entsprechung': a latent coherent analogon of the emerging environment.

An in-between to be inhabited. A bipolar, asymmetrical focus of awareness: one part relating to the other, intending the other – the diagram to the environment.

The mobilization of the agencies of the intended environment is now needed again in order to organize the cloud of words in a way that fulfils its function, that accomplishes its correspondence. The mobilization of the agencies of the environment is now needed again in order for the diagram to become a cognitive medium, the field of conditions of possibility for understanding the structure of the environment – in order to transform the cloud of names into a diagram, in order to begin to diagram with the cloud of words.

My aesthetic conduct now, again, provides the conditions for these agencies to operate – the empty but receptive space for these agencies to unfold.

Orienting now the unfolding of my sensorimotor and emotional skills – in intimate touch with the environment to be disclosed – towards identifying affinities between the qualities, imaginations, and associations, I displace some words on the touch screen. I bring those together which appear as 'belonging together' and I take those apart which, although participating to the same environment, establish different polarities.

And now, attending to the level of presence of each phenomenon present on the environment – and expressed on and through the touchscreen – I modify the size of the words.

I do that without knowing or, better, *not-knowing* how – without being able to generate an explicatory artefact, of formulating a reason or a set of criteria to differentiate or justify a differentiation – but instead sensing a common sense, a commonality on the level of sense – a literal 'Sinnzusammenhang' – rather than of meaning.

I do that – I can only do that – by acting aesthetically: moving by being moved, acting in intimate touch and, fundamentally, by virtue of other agencies.

Form appears, incipiently.

What I have in front of me now, is no longer an amorphous cloud of signs but, latently, a significant constellation of presences surrounding me, enveloping me, allowing my world-around – my immediate world, my environment here and now – not only to emerge but, principally, to be understood, to be not only present but intelligible.

The beginning of a poetic moment or, better, the beginning of the primacy of the poetic – since poetics has been at work already, 'on the other side', 'in the shadows' of each aesthetic move.

On auditory diagramming as poetic practice

To begin with: poetics here as *poiesis*, as open-ended procedures of radical cogeneration: of bringing about a new phenomenon, a thoroughly new state of affairs, out of its very roots (the body or bodies involved in its arising and to which the poetically enabled presence will appear; the practices, or at least the actions of these bodies; the agency of the media in or, better, with which these practices are performed; the other agents involved, the interacting agencies of others).

Poetics as a fulfilment of an always ongoing process of sense-making – of emergence of sense. In the case of the diagram, fundamentally, as a way of 'reading' it, of establishing two sets of relationships intimately intertwined with one another: between the phenomena collected on the touch screen, and between them and the environment with which they have been constituted.

Putting a finger on a word – an operatively effective although physically simulated touch – and displacing it – slightly, carefully, maybe even hesitantly – navigating the uncertainty of not-knowing – of not having the possibility of relying on the illusory certainty of clearly defined criteria or, at least, on an explanation but, instead, letting the move be guided by the coexistence of agencies – of my aesthetically disposed body, of the surroundings in touch with it – subsumed now in the agency of the emergent coherence between the co-emergent diagram and environment.

Creating coherent clusters: small, 'regional' wholes in the environment as a whole – as *the* whole. Actualizing the power of conditioning the emergence of sense with which each word is potentially endowed through the negotiation of its coherence – of the possibilities of holding together with other words.

'Seeing', I could say, expanding the reductive gesture of affirming that a diagram can be 'read', 'sensing' maybe better, how the potentiality of new agencies comes to be through the alteration of the in-betweenness that simultaneously joins and separates the words – the in-betweenness that regulates the poetic power of each written word, of each phenomenon, of their poetic agencies: their capacity of enabling the emergence of sense, specified in this case as environment, as an environment, to appear as ineligible in a disturbing way, in a way able to simultaneously transform the environment and the one to whom it appears – to transform, simultaneously, unexpectedly, in a flash, the environment and my/its-self.

Putting now two fingers on a word – a variety of virtual physical touch, preparing the gesture, the media-specific gesture of modifying the size of a digital object. Altering the actualization of one word's agency in relation to [by change] English? sizes and by virtue of the level of presence of the phenomenon it contributes to make present.

In both cases, sensing the emerging sense – the alterations of the senseful all-over presence I am trying to understand. Conducting the actions according to the presences of the modified words, according to the way these modifications intervene in the emergent whole – in the environment or, better, in my incipient understanding of the environment, in the environment as my understanding of it. Or, taking now an analytical way of explaining it, participating simultaneously in two processes of mutually conditioning emergence: the environment and my understanding (of it).

Sensing, noticing, letting what is happening configure my understanding, or more radically, letting it become my understanding: my *positioning towards* the thing or the event of understanding.

Poetic understanding is the act of letting myself be repositioned towards the object – of letting the object of understanding acquire a new presence, a new, fuller, broader, more dense significance for the one who is engaged in the act of understanding.

Two simultaneous, mutually conditioning moves coming to be out of poetic actions: navigations of a field of shared agencies enabling the emergence of new, transformative, open trajectories of sense.

Postscript: Notes on the relationship between research and design through auditory diagramming

An attentive reader of the book *Klangumwelt Ernst-Reuter-Platz* commented to me that he understood the description of the practice or auditory diagramming – described in the first chapter – as well as the design proposed by the Auditory Architecture Research Unit for the transformation of this Berlin square – exposed in the last chapter – but he did not understand or could not retrace how the practice of auditory diagramming leads to the design, that is, how the design is realized through this practice. This reader could not see the continuity between the research practice and the realization of an auditory architectural design. I do not think that he missed the point: by researching and designing through auditory diagramming there is no continuity between research and design, that is, no linear development linking what is usually considered to be two different phases of one process.

Auditory diagramming establishes another relationship between the operations habitually categorized as either pertaining to research or to design. In this case the relation between these two terms does not obey the standard sequential link: the process of research delivers results and they are used as a basis for the realization of a design – design follows research and, therefore, we can talk about 'research-based design'.

The performance of auditory diagramming, on the contrary, allows us not only to understand the environment in its current state but in other possible ones. Probably due to the intertwinement between perceptual, imaginative, associative, and analytic procedures, the inquired environment appears in its *actuality* and in its *potentiality*. A sense of how the inquired environment emerges under the present conditions and how it could appear in a varied field of contingency arise *simultaneously*. Research – the systematic process of understanding – and design – the process of defining conditions for transformation – coexist as interlaced procedures through the practice of auditory diagramming by virtue of the intimate connection between the agency of the practitioner and the agency of the environment to be transformed.

Notes

1. Instead of 'research and design' I prefer the formulation 'research/design' in order to express the particular relationship between these procedures enabled by auditory diagramming. For a brief explanation, see the last section of this paper.
2. For the first description of this practice, see Arteaga et al. 2016b: 12–55.
3. The two main research projects in which auditory diagramming has been developed are 'Auditory Long-term Observation Schlieren' and 'Klangumwelt Ernst-Reuter-Platz.' For a description of these projects, see Architecture of Embodiment n.d. Most of the seminar and workshops related to this practice took place in the MA Sound Studies and Sonic Arts at the Berlin University of the Arts (n.d.).
4. My first 'exploratory essay' was written in the framework of the research project 'transient senses' and has been included in the book with the same title (Arteaga 2016b). For more recent exploratory essays, see Arteaga and Langsdorf 2018; and Arteaga, forthcoming.
5. For a first description of this concept, see Arteaga 2017.
6. For the interpretation of aurality as medium I am referring to the so-called transcendental definition of media developed, among others, by Dieter Mersch. See, for example, Mersch 2004.
7. In this framework, aurality becomes the object of research when a necessary reflection on one's own methodology takes place.
8. The term 'world' is understood here as outlined in the phenomenological tradition especially by Maurice Merleau-Ponty. For a comparative description of this concept in phenomenological context, see Jacobs 2018.
9. For a first definition of the enactive approach to cognition, see Varela et al. 1991. For further developments, see, for example, Thompson 2007; and Gallagher 2017.
10. See Varela et al. 1991.
11. This term was coined in the first phase of research of the Auditory Architecture Research Unit realized in Berlin and in the German language. This research platform is hosted

by the Berlin University of the Arts and associated with the MA Sound Studies and Sonic Arts.
12. I think that the generally unreflectively accepted operation of 'applying a theory in the praxis' does not express adequately the relationship between these two concepts. First, because I consider 'theory' to be a field of practices performed mostly in the medium of written language and, accordingly, 'theories' to be artefacts generated through these practices; second, because 'application' is not a possible operation between practices and/or the artefacts they generate. Instead, the practices included in both fields – the so-called 'theory' and 'praxis' – and the artefacts they produced *influence* one another when they coexist in hybrid processes, for example, of research.
13. With the formulation 'related practices' I mean here all kinds of possible practices performed within the *horizon* of architectural construction, that is, connected to the final possibility – it does not matter how remote or imaginary – of a material realization of architectural-constructive processes.
14. 'Constructive interventions' are those realized through construction and 'construction-related practices'.
15. See end note 5.
16. With the term 'aesthetic research' I specify the concept of 'artistic research' on the one hand through the concept of the aesthetic I have briefly outlined here, and on the other hand situating this kind of research beyond the normativity of art as social system. For a different approach to the differentiation between both terms, see Vilà, forthcoming.
17. I coined the term 'aesthetic architecture' in the framework of my research project 'Architecture of Embodiment' (n.d.).
18. I use the formulation 'to tend to' referring to the phenomenological concept of 'intentionality'.
19. This term relates to Husserl's concept of 'sedimentation' (see Moran and Cohen 2012: 288–291).
20. In order to avoid misunderstandings on the use of the term 'practice' for both auditory diagramming and qualitative, imaginative, and associative hearing, as well as analytical listening, and thus affirming that auditory diagramming is based on practices, it would be possible to consider auditory diagramming as a *method*, understood as the systematic connection between practices, considering consequently the procedures of graphic articulation of the emerging phenomena and practices.
21. Speaking phenomenologically in a strict sense, the actions of hearing are also endowed with intentionality: they tend or relate to something, they are *about* something. But this is a specific form of intentionality that can be identified or at least related to the Husserlian concept of *operative intentionality*.
22. Sometimes the expression of a phenomenon does not manifest itself with the required precision or adequacy as only one word or a word cluster is necessary in order to formulate it.
23. For some comments on the relationship between these two manifestations of an environment, see the postscript.
24. The term *Stimmung* means, among others, 'tuning', that is, the adequate relation of frequencies between two sounds.
25. A diagram, even if it is configured by words, participates in two different epistemic modes: the 'own' of language and the 'own' of image. That is why the term 'reading' is only valid in order to designate the interaction with a diagram *in sensu lato*.

References

Architecture of Embodiment (n.d.). 'Framework'. Available online: www.architecture-embodiment.org (accessed 6 July 2020).

Arteaga, A. (2016a). 'Steps towards an Architecture of Embodiment: Thinking the Environment Aurally'. In A. Arteaga, G. Green, and B. Hassenstein (eds), *Klangumwelt Ernst-Reuter-Platz: A Project of the Auditory Architecture Research Unit*, 12–55. Berlin: Errant Bodies.

Arteaga, A. (2020). 'Aesthetic practices of very slow observation as phenomenological practices: steps to an ecology of cognitive practices'. RUUKKU, (14). https://www.researchcatalogue.net/view/740194/862241

Arteaga, A. (2016b). *Transient Senses*. Barcelona: RM.

Arteaga, A. (2017). 'Estètica corporitzada i situada; Un enfocament enactiu a una noció cognitiva d'estètica'. *Artnodes*, (20). http://doi.org/10.7238/a.v0i20.3155.

Arteaga, A. (ed.) (forthcoming). *Architectures of Embodiment: Disclosing Fields of Intelligibility*. Zurich: Diaphanes.

Arteaga, A. and H. Langsdorf (eds) (2018). *Thinking Conditioning through Practice*. Ghent: Art Paper Editions.

Berlin University of the Arts (n.d.). 'Sound Studies and Sonic Arts (Master of Arts)'. Available online: www.udk-berlin.de/en/courses/sound-studies-and-sonic-arts-master-of-arts (accessed 6 July 2020).

Gallagher, S. (2017). *Enactivist Interventions: Rethinking the Mind*. Oxford: Oxford University Press.

Jacobs, H. (2018). 'Husserl, Heidegger, and Merleau-Ponty on the World of Experience'. In D. Zahavi (ed.), *The Oxford Handbook of the History of Phenomenology*, 650–675. Oxford: Oxford University Press.

Merleau-Ponty, M. (2012). *Phenomenology of Perception*. Abingdon: Routledge.

Mersch, D. (2004). 'Medialität und Undarstellbarkeit. Einleitung in eine "negative" Medientheorie'. In S. Krämer (ed.), *Performativität und Medialität*, 75–96. Munich: Fink.

Moran, D. and J. Cohen (2012). *The Husserl Dictionary*. London: Continuum.

Nancy, J-L. (2007). *Listening*. New York: Fordham University Press.

Thompson, E. (2007). *Mind in Life: Biology, Phenomenology and the Sciences of Mind*. Cambridge, MA: Harvard University Press.

Varela, F. (1991). 'Organism: A Meshwork of Selfless Selves'. In A. I. Tauber (ed.), *Organism and the Origins of Self*, 79–107. Dordrecht: Springer.

Varela, F., E. Thompson, and E. Rosch (1991). *The Embodied Mind: Cognitive Science and Human Experience*. Cambridge, MA: MIT Press.

Vilà, G. (forthcoming). 'Aesthetics and Aesthetic Research'. In A. Arteaga (ed.), *Architectures of Embodiment: Disclosing Fields of Intelligibility*. Zurich: Diaphanes.

Annex 1

Analytical listening: Foci of attention and correlated phenomena
Focus 1: Single sounds and their own features
- Name
- Parametric properties:
- Loudness
- Time structure: steady or variable, if variable: periodic (patterns) or non-periodic
- Level: range and average
- Pitch
- Sound's pitch
- Time structure: steady or variable, if variable: periodic (patterns) or non-periodic
- Range and average
- Internal distribution of pitch (spectrum)
- Time structure: steady or variable, if variable: periodic (patterns) or non-periodic
- Range and average
- Time structure: steady or variable, if variable: periodic (patterns) or non-periodic
- Spatial structure: static (indicate position) or in movement (indicate lines or fields of presence)

Focus 2: General features of sounds (respectively in close, mid-, and far range)
- Differentiability between single sounds
- Identifiability of single sounds
- Identifiability of the acoustic source of single sounds
- Locatability of single sounds
- Diversity
- Diversity of time structures
- Diversity of spatial structures
- Time structure: steady or variable, if variable: periodic (patterns) or non-periodic
- Spatial structure: static (indicate position) or in movement (indicate lines or fields of presence)

Focus 3: Features of all sounds as a unit
- Parametric properties:
- Loudness
- Time structure: steady or variable, if variable: periodic (patterns) or non-periodic
- Level: range and average
- Pitch
- Time structure: steady or variable, if variable: periodic (patterns) or non-periodic
- Range and average
- Time structure: steady or variable, if variable: periodic (patterns) or non-periodic
- Time density
- Spatial structure: static (indicate position) or in movement (indicate lines or fields of presence)
- Wideness
- Spatial density
- Reverberation/Echo

34
Close Listening: Approaches to Research on Colonial Sound Archives
Anette Hoffmann

Colonial knowledge production has left an archival echo. Between the hissing and crackling of old shellac records and wax cylinders, voices can be heard that speak, announce, musick, whisper, chant, narrate, sing, and criticize. These recordings were produced by linguists, musicologists, anthropologists, folklorists, or laypeople, sometimes a century ago, often in asymmetric situations of knowledge production. Today, most of these recordings are trapped in archives where they have been received and configured as specimens. Speakers, who were seen as 'native informants', are often irreversibly absented from the archival catalogues. Textual contents of recorded speech and song rarely surface in the catalogues of the archives. The collaborative nature of knowledge production is perpetually concealed, and archives often keep quiet about the colonial, violent context of their making. This figuring of acoustic collections as archival objects is durable; it has both shaped the recordings and indelibly watermarked their form and content (Hamilton et al. 2002; Quijano 2007; Stoler 2009). Yet sound recordings are specific archival documents. They allow interested listeners to revisit acoustic traces that resonate with the very moment of their making. While the recorded voice is mediated, what may be heard in a recording often differs from what can be read in the transcription, or from what appears in the catalogue of the archive.

A striking example of this chasm between what is catalogued and what is audible appears in the collection of the Austrian anthropologist Rudolf Pöch, which was recorded in the Kalahari in 1908 as part of an attempt to document 'Bushman languages'. Pöch's recordings were published by the Phonogrammarchiv Wien in 2003 (Schüller 2003). The booklet that comes with the CDs describes recording no. 20 on CD 1 as 'Speech', spoken by Bushmen, recorded in 1908, in a place Pöch called Kxau, in the vicinity of Ghanzi, British Betchuana Land (now Botswana; Schüller 2003: 36). On this recording, almost drowned out by crackle and hiss, one can hear the faint voice of a man speaking German with a thick Austrian accent, first in English, then in Afrikaans:

We bring some wood for fire and then we go to the pit to drink some water, you know...?
|Kxara, waar is die |Kxara?...[|Kxara, where is |Kxara?]
|Kxara, môre vroeg jy moet mich wakker maak [|Kxara, tomorrow morning you must wake me up]
Als... uitkomt, ik moet wakker zijn [When ... comes, I must be awake]
Is julle bang? [Are you afraid?]
Denkt die |Kxara ik wil die |Kxara fressen? [Does|Kxara think I want to eat him?]¹
Filos! Filos, you come!
Filos, did you look after the horse?
...The bush...very far
You look at it that it *niet hardloop* [so it doesn't run]

The recording sounds staged, some spoken words are inaudible; it is hard to assess why Pöch even speaks on this recording. As is very often the case, the documentation is sketchy: Pöch's notebooks of the days in July and August 1908 when the recordings were produced are missing. The printed *Protokoll* that is supposed to provide information about the acoustic documentation of so-called 'Bushman languages' doesn't specify the content of this particular recording, nor did Pöch deliver a transcription. Here, as with other acoustic collections, it is the usual archival practice to retain the historical categorization, which still states: '*Bushmen speaking Bushman speech*.' How and when did the anthropologist become a Bushman? Did the label of '*Bushman speech*' induce spontaneous agnosia in listeners in a European archive?

This example may seem bizarre, yet colonial knowledge production frequently sounds different from how it reads. Research on the history of particular sound collections often reveals that these recordings were but one part of a larger collection that was amassed during a particular project of 'collecting' and study. I write 'collecting' in inverted commas here, because voice recordings or musical recordings are often referred to in this way but were, in fact, never collected. They were produced for archiving. Once the recordings are listened to, and re-connected to other results of projects of knowledge production, they appear as inscribed with information that speaks beyond their status as examples of languages, folklore, and music.²

The recording cited above becomes *unheimlich* only in connection with Pöch's aims for his journey. It is part of his enormous collection, which has been dispersed to at least five institutions in Vienna. The collection includes hundreds of photographs, ethnographic objects, life casts, cinematographic film, and a massive quantity of human remains that Pöch robbed from graves in southern Africa (Rassool and Legassick 2000). With this in mind, the faint echo of the question '*Is julle bang?*' (are you afraid?) posed to his assistant |Kxara, who may have witnessed some of those practices, resounds differently. These and other recordings allow researchers to revisit the politics of producing acoustic collections, methods of archiving, cataloguing, and practices of dissemination for an understanding of the sonic objects' archival biographies and the ways in which they speak or sound in the present.³

Phonography and the politics of archiving sound

Once one turns to listening, acoustic collections resound with technological achievements entangled with the idea that cultural expressions need to be salvaged. Researchers who (said they) worried about the disappearance of languages, types of music, and oral repertoires readily exploited opportunities they found in the subjugation of colonized populations which allowed them to produce specimens and acquire artefacts. This means that the very politics of colonization, which endangered the survival of groups of people and led to epistemicides, also made the production of acoustic recordings possible logistically.

The technical precondition for sound recording was the introduction of phonography in the late nineteenth century, which made acoustic conservation and the archiving of hitherto ephemeral sound possible for the first time (Brady 1999; Stangl 2000; Sterne 2003). In Europe and in the United States, phonography was picked up promptly as a means to record and archive specimens for linguistic studies, as material for comparative musicology, and in connection to projects of salvage ethnography. The introduction of phonography also allowed researchers to arrest the flow of music, thus enabling them to dissect and atomize acoustic musical elements (not merely the written scores) for analysis. Erich Moritz von Hornbostel, one of the founding figures of German musicology, described this process of analytic listening in anatomic terms, which speaks to the proximity of analytic listening to dissection, as Eric Ames (2003) has remarked (Von Hornbostel and Abraham 1904). Phonography thus played a crucial role in the development of a dissecting, analytic style of listening in the West, which came to be vital for the impulse to archive sound, and to establish regimes of listening in the production of knowledge on music and languages. This historical shift in audible regimes, as one landmark among others, speaks of the plasticity of practices of listening, which corresponds with research interests but also with a larger politics of salvaging cultural artefacts and expressions (Sterne 2003; Ochoa Gautier 2014; Stoever 2016).

The historical sound collections that result from these projects of recording speak of phonography's claims to technical objectivity, of epistemic practices, of the politics of collecting, and of the establishment of disciplines such as musicology and linguistics. Together with their written documentation (if available), these collections resound with the asymmetrical power relations at play in and beyond historical situations of recording.

The possibilities of conserving sound on collectible objects, together with the aim of establishing acoustic collections for research, induced the founding of several phonogram archives in Europe, the earliest if which were in Vienna (1899), Berlin (1900), and St Petersburg (1902) (see Lange 2018). Yet historical collections are also held in libraries, museums, and universities, mainly in Europe and the United States. The Berlin Phonogramm-Archiv is an example of the implications that imperial politics coupled with violent epistemic practices have for the constitution of an enormous collection of acoustic recordings (Simon 2000). Among many other sound carriers, this archive holds more than 15,000 wax cylinder recordings, which were produced between 1893 and 1954. About a third of these early recordings were created in African countries; another quarter originated in Asia (Ziegler 2006: 29).

In the early years of the Berlin Phonogramm-Archiv, its directors regularly visited the sites and performers of *Völkerschauen* (ethnic shows) in the region to produce musical recordings. Susanne Ziegler's catalogue of the wax cylinder recordings of the Phonogramm-Archiv in Berlin (2006) lists around thirty collections that were produced with performers in zoos and at fairs.[4] The majority of the archive's wax cylinder collection was recorded by researchers, missionaries, and travellers, who were equipped with portable phonographs and cylinders, and were commissioned to deliver the desired recordings to the archive upon return.

Especially in colonized countries, the favoured locations and institutions for recording music and languages included prisons, police stations, and pass offices – where researchers could easily coerce colonized people into being recorded, examined, photographed, and sometimes cast in plaster (Luschan 1906; Pöch 1910; Hoffmann 2009). Incarceration of thousands of foreign men in Germany during the First World War was also seen as an occasion for recording. The result was an enormous collection of recordings of languages and music, which was produced by the Königlisch Preußische Phonographische Kommission (Royal Prussian Phonographic Commission) with prisoners of war in German camps, between 1915 and 1918 (Lange 2013; Hoffmann 2018a). Phonographic recording, in contrast to the acquisition of ethnographic objects, did not take songs and stories away from speakers and singers. Nonetheless, projects of acoustic recording were informed by an overall coloniality[5] of attitude, which informed and justified violent epistemic practices and routines of racist anthropology (see Berner, Hoffmann, and Lange 2011; Hoffmann and Mnyaka 2015). These practices have watermarked the recordings and their documentation, affecting both their form and content indelibly, and cannot be unlinked from them.

The context of production continues to affect the accessibility of recordings as well as the availability of knowledge on their existence. The geographical distribution of recordings, which were often produced in colonized countries and then archived in European and US institutions, mirrors the intended routes of knowledge production. In the present, politics of access vary immensely: whereas several US institutions are granting free access to their collections, European phonogram archives, albeit engaging in processes of retribution, are reluctant to share their collections online.

Digitization and the persistence of colonial categories

Especially for fragile wax cylinders – which do not allow for frequent playing due to physical degradation and thus cannot be listened to directly, let alone taken out of the archive for research – digitization has a significant impact. Digitization has literally mobilized sound recordings: as digital files, recordings can leave the protected and often sequestered environment of the archive to be sent or taken abroad. This mobilization facilitates collaborative research, which is a precondition for any analysis of these transcultural archival objects in the present.

Yet digitization does not automatically remove the barriers that prevent access: online sound archive portals, when they exist, are often not easily searchable (SPK Digital 2020b). Many catalogues are only available in the languages of the collectors and the recordings themselves cannot, or can only partially, be accessed or listened to online (Humboldt University in Berlin 2020). Nor does digitization automatically undo racializing categories, correct the misspelling of the speakers' names, or replace derogatory names that were given to ethnic groups. In other words, digitization does not alter the archival order of things or undo epistemic violence. The persistence of historical categories that arrange recordings either according to ethnic groups or according to names of collectors, also complicate the search for semantic and performative contents, or for particular genres of speech and song.

An example is the collection of Hans Lichtenecker, produced mainly in a police station in German South West Africa (now Namibia) in 1931. The digital catalogue of the Stiftung Preußischer Kuturbesitz has retained much of the information that appeared in the printed catalogue of 2006. The collection still appears as 'Lichtenecker Südafrika', probably because at the time of the recording the country was under South African mandate. Lichtenecker's original specification 'Hottentot' as a language and ethnic group of the recordings in Khoekhoegowap does not appear in Ziegler's catalogue of 2006. This led to the omission of a third of the recorded languages of the collection in this publication. In the digital version some of these recordings are specified as 'Nama', yet the names of the speakers do not appear, and the spelling of places as well as the categorization of recordings is consistent with Lichtenecker's erroneous list (see SPK Digital 2020a).

Altogether, the possibility to access recordings as meaningful, performative utterances, or the attempt to relate such recordings to each other in and across collections – for instance in identifying people speaking of the situation of knowledge production or of specific historical events – continues to be complicated. In this way, archival practices add to the sequestration of sound recordings within the silos of their disciplines (Hamilton and Leibhammer 2016). Especially in the case of recordings of linguists, musicologists, and perhaps somewhat less so, anthropologists of the early twentieth century, a change in the 'terms of engagement' would need to take place so as to direct the attention of scholars from outside of these disciplines towards sound archives and their contents.[6] A recategorization could be the first step of a move towards understanding historical sound collections that stem from projects of colonial knowledge production as inscribed with the histories of colonial epistemologies, and as potentially carrying significant textual and performative content. The first methodological step I suggest for the study of these historical sound archives is thus a 'refiguring' (Hamilton et al. 2002), which subsumes these recordings to the colonial archive at large. This change in the terms of engagement may alter the way materials in these archives are listened to.

Sound recordings and the colonial archive

So far, the colonial archive is quiet. Subsuming sound collections within the colonial archive at large is a strategic move in at least two senses: it may facilitate the recuperation of pertinent, as yet barely known acoustic documents, which adds the specificity of recorded spoken or sung texts as well as music to the debate around what traces of colonial histories can be found in the colonial archive. Secondly, listening to historical voice recordings and music as semantically and performatively meaningful utterances may allow for a better understanding of the speaking positions of subaltern subjects located in a situation of colonial knowledge production. In some cases, acoustic collections may present alternatives to forms of representation in writing or visual depiction, in which the racialized Other surfaces in specific ways (Hoffmann 2014). The strategic move to explore acoustic archives may recuperate commentary and critique in recorded speech acts, songs, stories, and accounts that were recorded and archived, yet not noted as meaningful speech, text, nor as speaking beyond the circumscribed sphere of the disciplines for which they were recorded.

I tentatively define the colonial archive at large as an imbrication of the discursive sense of 'archive', which determines what can be said, what is of interest, and what becomes knowledge (Foucault 1981) together with 'archives', as specific collections and institutions. The colonial archive, in this sense, is based on the paradigms and epistemic constellations that were operative in exploring, describing, and inventorying subjugated territories, resources, and people. It is predicated on, shaped by, and thus intrinsically connected to imperial power relations and agendas (Stoler 2016). Both the archive as a discursive formation and archives as deposits of sources, collections of documents, and materialized knowledge, which have been created within and for the colonial project, actively direct the work of researchers who study colonial history. As the 'source' for crafting historiographies, the colonial archive has seen rigorous critique in recent decades. Who speaks, and who does not; what is included or excluded; violent forms of representation; the archive's productive power to create and organize semantic content; protocols of collecting, storing, or circulation: all these are topics of debate. Most importantly, the critique of the colonial archive has moved the debate from the archive as repository of evidence, to the archive as practice (Stoler 2009). Analysing the colonial archive's bias shows that the enunciative positions of colonized people are difficult to retrieve in written documents of the colonial administration or from historical ethnographies (Spivak 1999; Prakash 2000).

Turns towards photographic archives have shown that different media may have documented aspects of colonial history that differ from those found in the texts of administrators, missionaries, and explorers (Edwards 2000). Still, some histories rarely surface, they seem to be caught in the shadows of imperial imagination. Sound archives are yet to be included in the theoretical debate on archive.

In recent years, the interest in sound archives has been growing. Yet so far, projects of recording have been analysed mostly on the basis of their written documentation, and not much systematic research has been conducted that engages with the acoustic collections as specific sources in their own right (Ames 2003; Scheer 2010). A change in the terms of engagement requires more than a theoretical re-conceptualization. To sound out acoustic collections, methods of listening to recordings are needed, which analyse acoustic files beyond their archival status as specimens. One such method is related to what Rodney Harrison (2013) has called the 'reassembling' of museums as sites of epistemic practices and histories of amassing objects.

Regarding sound collections, reassembling entails a systematic reconnection of recordings with available documentation, which may, for instance in the case of notebooks and diaries, be kept elsewhere. Other objects – for instance travelogues, photographs, and film – that were produced during the same study or on the same journey can be instructive for the analysis of the acoustic files. In many cases these connected objects are distributed to various archives and institutions according to the logic of disciplines, or the politics of archiving.

In the case of the traces of |Kxara (no surname mentioned) who appears as a speaker on recordings from the Kalahari, this means that his photographs are found in three different collections in Vienna, his phonographic recordings are archived in the Phonogrammarchiv in Vienna, and appear in the publication of 2003, plus a cinematographic film that shows him is held at the Filmarchiv Austria. Close listening in combination with reassembling pertinent documents and archival objects can significantly reframe sound recordings, and may add to the understanding of textual content (Hoffmann 2009, 2011).

What I tentatively call 'close listening'[7] describes the attempt to know by ear, that is, to grasp as much as possible of the audible features of a recording. This includes attention to recorded features which do not appear on the label, for instance, the sound of the pitch pipe (which indicates the speed at which the recording should rotate); the noise of a rotating cylinder or scratched record (which can deliver clues on how often the record has been played); the recordist's announcement (the 'acoustic tag'); the languages, performative, and musical genres documented on the recording; the features of the voice of the speaker and singer, together with accent, pauses, and background noises. Apart from exposing that what can be heard on a recording may or may not be described or noted in the file's label or written documentation, this exercise in listening makes audible that archived sound files often speak beyond the object status attributed to them by the recordists and in the archival documentation. Sharp contrasts between the audible and what was registered in the written documentation often announce the logic of archiving, for instance in the frequent omission of what I call the 'acoustic tag' from the written description of sound files. By acoustic tag I mean the recordist's spoken announcement of what is to follow on the recording. From about 1900 this announcement was requested in manuals for ethnographic collecting, which included an instruction on 'collecting with the phonograph' (Luschan 1896; Ankermann 1914). The voiced tag was added to the recording to secure the identifiability – especially of recordings in non-European languages – in the case of the loss of written labels and documentation. In this recorded announcement one hears the voice of the recordist, his accent, and tone. Local and historical idioms are audible, but also attitudes, for instance in the case of a German recordist who yelled '*Achtung!*' and '*Achtung Aufnahme*' in a commanding tone into the funnel three times, before the person who was to be recorded started to speak (Hoffmann 2018b, 2019). A recording labelled, for instance, as 'narrative told by XY' may, in fact, comprise a composition of the sounds of a noisy technology, an announcement made by a recordist, the narrative of XY, and background noises of the audience or a train passing by. Listening to everything audible makes clear that most recordings are composite sound objects that can be heard in different ways.

Close listening also alerts one to the positionality of listening and the status of sound in a given environment or situation. Modes of hearing and listening are culturally and historically informed. They precipitate in languages and philosophies, and have histories and impacts in the present (Baumann 1997; Chernoff 1997). This also means that however 'closely' one listens, one cannot escape one's own position as a listener, which may lead to insensitivity, or heightened sensitivity, to certain aspects of the audible.

Sound archives keep what has been recorded for specific purposes and interests. Yet attentive listening may perceive that the rationale of the researchers did not entirely control what entered the archive. The excess of voice vis-à-vis the spoken or sung words – in pronunciation, tone, and the ability to communicate non-verbally, often slipped past censors, or evaded the registration of the archiving process. Thus the performativity, meaning, and messages of recorded sounds, words, songs, stories, or even example sentences are not always contained within the rules and practices of recording.

Marked by the archival politics that have produced them in the first instance, recordings may present a neuralgic nexus in archiving, because they constitute an interface of different cultural practices of conservation that become articulated – in the double sense of expressing and joining – by means of recording. What have been archived as examples of music and languages, often hold elements of repertoires, which can be fragments of record-keeping such as oral poetry or songs and narratives that are part of a body of historiology.[8] This means that the expressiveness of recorded voices and sounds must be listened to with cognizance of the politics of their production as recordings, yet the rules and practices of creations may not completely grasp nor direct the performativity, generic properties, and meanings of spoken, sung, or played recordings, both in respect of their utterance and reception. The complexity of oral genres, or the inability to understand languages, means that contents of various repertoires may have entered sound archives unidentified. What was said was often not understood by those who recorded, contents were outside of the field of interest in which the archivists operated, or the double meaning or performative sense of songs escaped the radar of regulation and censorship.

An example from a POW camp

Archived voice recordings often appear incomprehensible. Reassembling and close listening may help identify genres, and registers of speech or retrieve additional archival material that adds to the meaning of recordings. A change of the terms of engagement can create the space for the recorded voice to lay out a track in the network of archives, as the following example shows. The voice recordings produced with the imprisoned Senegalese soldier

Abdulaye Niang, who was recorded in November 1917 in the POW camp Wünsdorf by the Königlich Preußische Phonographische Kommission is an example for a possible analysis of an acoustic trace by means of close listening and re-assembling. Eight wax cylinder recordings were produced with Abdulaye Niang as specimens of the Wolof language. None of the recordings was translated until 2013. The first person who listened to the spoken and sung words and was able to understand and interpret them was Serigne Matar Niang (who happens to have the same family name). He translated Abdulaye Niang's recordings in Cape Town in 2013. On two of the recordings, Niang urgently requests not to be deported to Romania, where many African and Indian prisoners were sent in 1917. The official reason given for their deportation to Romanian camps was concern about the health of prisoners from the south, whose mortality rates were alarming. Niang's request seems to respond to this reasoning on the recording listed with the file number 1114/2 at the Berlin Lautarchiv:

> I am truly worried as where we are right now is freezing and very uncomfortable. We were on parade this morning and the lieutenant summoned the prisoners and searched our belongings with the intention of finding money. A guy called Alexandre had 16 francs in his suitcase *is that understood?*
>
> Right now, the cold is terrible, and as we are headed for [Romania], we are uncertain what we will find there. We prefer to remain where we are right now. Irrespective of our next destination, at least we are familiar with our surroundings and we are coping with the climate.
>
> May this war end, so that we can return to our parents, resume our duties and treasure them. This place is freezing; we don't know this kind of weather, we are also not used to it. (emphasis in the original)

Abdulaye Niang used the recording situation to petition against deportation. Yet this does not surface in the personal files that were filled in meticulously for each of the several hundred prisoners who were recorded by the Kommission. The file provides Niang's name, indicates his place of birth, his profession, religious denomination, age, and education. His repeated, urgent request is listed under the rubric *Art der Aufnahme* (kind of recording) as 'Erzählung' (narrative).

The content of this recorded speech act is not all that is missing. Abdulaye Niang expressed his request in a specific way, using a particular register of speech in Wolof, one that has been described as connected to the class of *griots* (Irvine 1990). This particular register of speech can also be employed to express a request among speakers of a different class. In this case it becomes a performative 'bow' that seeks to present a plea in a pleasant and socially acceptable form. In the acoustic recording, this resounds distinctively, because this register of speech entails particular prosodic features, a characteristic tempo, expressivity, and vocabulary.

With regard to the specificity of historical sound recordings, this example demonstrates what can be gained by close listening, that is, listening not only for the purposes of translation and understanding semantic content, but with an ear for rhetoric form and genres of speech. The drama of this historical moment in a prison camp unfolds in sound. In the recording situation the prisoner's urgent request fell on deaf ears, because the recordists were focused on recording examples of Wolof, a language which they did not understand but wanted to 'collect' as a specimen. Communication was suspended in the moment of recording, at least from the side of the recordists. The prisoner apparently was not aware of this.

Apart from its outstanding expressiveness and historical significance, Abdulaye Niang's recorded voice also became the connective tissue in a network of archives in Berlin, Vienna, and Frankfurt. Niang's archival echo, the acoustic trace of his presence in the POW camp Wünsdorf, prompted my search for traces of his presence in Germany and Romania, which I found in the anthropometric photographs in the publication of a racial study conducted by Rudolf Pöch and Joseph Weninger (1927), and in the archive of the Institute of Anthropology of the University of Vienna. With these photographs, Serigne Matar Niang and I were able to identify him on a group photograph that was taken in Romania, and is now kept at the Frobenius Institute in Frankfurt. Re-assembling, in this case, told us that he was indeed deported to the Romanian camp Turnu Magurele.

Not all recorded traces are as dramatic as this example. Others recordings transmit fragmented echoes of witty responses to the projects of knowledge production. 'Doesn't the book tell the Whities?' a speaker who was recorded in Pietermaritzburg, South Africa, in 1908, by the Austrian missionary Franz Mayr, asked.[9] The recording resounds with mocking critique. *Abelungwana* translates roughly as 'Whities': the etymology of the term relates to dirty white foam on the sea. Apart from clearly disrespecting the missionary, the question mocks hierarchies of credibility and refers to the limits of written knowledge. The reference to *incwadi* – the book – speaks of competing spheres of knowledge production within different genres, media, and societal groups. The question of 'what counts as knowledge?' that the speaker, whose name is archived as 'Pakati', so confidently brings up is crucial for the interpretation of the recordings and for their production (Schüller 2007).

The speaker's question also shows that what was of little importance at the time of recording may achieve retrospective significance today (Trouillot 1995: 58). As recorded echoes, commentary, requests, pleas and critique have survived as acoustically documented, performative texts in many archives. There is much to sound out.

Notes

1. This is an idiomatic phrase in German, which makes fun of somebody's fear.
2. For the concept of re-assembling museum collections, see Harris 2013.
3. Most of these recordings are spoken or sung in Naro, a language spoken by about 20,000 people in Namibia and Botswana. For more translations and interpretations around this collection, see Hoffmann 2020.
4. Examples are 'Archiv Siam' (1900); 'Archiv Samoa' (1910); 'Archiv Sudan' (1909); 'Archiv Somalia' (1910); 'Archiv Tunesia' (1904); 'Archiv India' (1902) (see Ziegler 2006).
5. On the concept of coloniality, see Quijano 2007; and Garbe 2013; on coloniality and sound archives, see Hoffmann 2018b.

6. The archaeologist Nick Shepherd speaks of 'terms of engagement' in his study of the practices of archaeology in South Africa, in which the contribution of black South Africans has systematically been elided from publications (Shepherd 2015).
7. The notion of 'close listening' as a method to engage with historical recordings first came up in conversations with Britta Lange, when we taught a seminar on sound archives at the Humboldt University together in 2012. It was developed further in workshops on 'knowing by ear', which I organized at the Archive and Public Culture Research Initiative at the University of Cape Town in 2013 and 2014.
8. With historiology I mean orally transmitted interpretations of history.
9. In the transcription in Zulu that Mayr provides this reads as: 'Incwadi ayibatsheli abelungwana?' *Abelungwana* is a diminutive form of the term *abelungu* for white people (singular: *umlungu*).

References

Ames, Eric (2003). 'The Sound of Evolution'. In: *Modernism/Modernity* 10 (2): 297–325.

Ankermann, Bernhard (1914). *Anleitung zum ehtnologischen Sammeln*. Berlin: Georg Reimer Verlag.

Baumann, Max Peter (1997). 'Preface: Hearing and Listening in Cultural Context'. 'Cultural Concepts of Hearing and Listening', special issue of *World of Music* 39 (2): 3–8.

Brady, Erika (1999). *A Spiral Way: How the Phonograph Changed Ethnography*. Jackson, MS: University Press of Mississippi.

Chernoff, John (1997). 'Hearing in West African Idioms'. 'Cultural Concepts of Hearing and Listening', special issue of *World of Music* 39 (2): 19–25.

Edwards, Elizabeth (2000). *Raw Histories: Photographs, Anthropology and Museums*. Oxford: Taylor & Francis.

Foucault, Michel (1981). *Archäologie des Wissens*. Frankfurt: Surhrkamp.

Garbe, Sebastian (2013). 'Das Projekt Modernität/Kolonialität – zum theoretischen/akademischen Umfeld des Konzepts der Kolonialität der Macht'. In Sebastian Garbe and Pablo Qintero (eds), *Kolonialität der Macht: De/Koloniale Konflikte zwischen Theorie und Praxis*, 21–46. Münster: Unrast Verlag.

Hamilton, Carolyn, Verne Harris, Jane Taylor, Michele Pickover, Graeme Reid, and Razia Saleh (2002). *Refiguring the Archive*. Dordrecht: Kluwer.

Hamilton, Carolyn and Nessa Leibhammer (2016). 'Introduction: Tribing and Untribing the Archive'. In Carolyn Hamilton and Nessa Leibhammer (eds), *Tribing and Untribing the Archive: Indentity and the Material Record of South Kwazulu-Natal in the Late Independent Colonial Period*. Johannesburg: University of KwaZulu Natal Press.

Harris, Rodney (2013). 'Reassembling Ethnographic Museum Collections'. In Rodney Harris, Sarah Byrne, and Anne Clarke (eds), *Reassembling the Collection: Ethnographic Museums and Indigenous Agency*. Santa Fe, NM: School for Advanced Research Press.

Hoffmann, Anette (2011). 'Glaubwürdige Inszenierungen: Die Produktion von Abformungen in der Polizeistation von Keetmanshoop im August 1931'. In Margit Berner, Anette Hoffmann, and Britta Lange (eds), *Sensible Sammlungen: Aus dem anthropologischen Depot*, 61–88. Hamburg: Philo & Philo Fine Arts.

Hoffmann, Anette (2014). 'Verbal Riposte: Wilfred Tjiueza's Performances of Omitandu as Responses to the Racial Model of Hans Lichtenecker'. In Hans D. Christ, Iris Dressler, and Christine Peters (eds), *Acts of Voicing*, 323–332. Leipzig: Spector Books.

Hoffmann, Anette (2015). 'Listening to Sound Archives: Introduction'. Anette Hoffmann (ed.), 'Listening to Sound Archives', special section in *Social Dynamics* 41 (1): 73–83.

Hoffmann, Anette (2018a). 'Echoes of the Great War: The Recordings of African Prisoners in the First World War'. In Leon Wainwright (ed.), *Disturbing Pasts: Memories, Controversies and Creativity*, 11–35. Manchester: Manchester University Press.

Hoffmann, Anette (2018b). 'Kolonialität'. In Daniel Morat and Hansjakob Ziemer (eds), *Handbuch Sound: Geschichte – Begriffe – Ansätze*, 387–390. Stuttgart: Metzler.

Hoffmann, Anette (2019). '"Achtung Aufnahme!" Akustische Spuren der kolonialen Wissensproduktion'. In Iris Edenheiser and Larissa Förster (eds), *Museumsethnologie – Eine Einführung. Theorien – Debatten – Praktiken*, 204–206. Berlin: Dietrich Reimer Verlag.

Hoffmann, Anette (2020). *Kolonialgeschichte hören. Das Echo gewaltsamer Wissensproduktion in historischen Tondokumenten aus dem südlichen Afrika*. Vienna: Mandelbaum.

Hoffmann, Anette (ed.) (2009). *What We See: Reconsidering an Anthropometrical Collection: Images, Voices and Versioning*. Basel: Basler Africa Bibliografien.

Hoffmann, Anette and Phindezwa Mnyaka (2015). 'Hearing Voices in the Archive'. Anette Hoffmann (ed.), *Listening to Sound Archives*, special section in *Social Dynamics* 41 (1): 144–165.

Humboldt University in Berlin (2020). 'Lautarchiv'. Available online: https://www.lautarchiv.hu-berlin.de/ (accessed 6 July 2020).

Irvine, Judith (1990). 'Registering Affect: Heteroglossia in the Linguistic Expression of Emotion'. In Catherina A. Lutz and Lila Abu-Lughod (eds), *Language and the Politics of Emotion*, 126–161. Cambridge: Cambridge University Press.

Lange, Britta (2011). 'South Asian Soldiers and German Academics: Anthropological, Linguistic and Musicological Field Studies in Prison Camps'. In Ravi Ahuja, Heike Liebau, and Franziska Roy (eds), *'When the War Began, We Heard of Several Kings': South Asian Prisoners in World War I Germany*, 149–186. Delhi: CRC Press.

Lange, Britta (2013). 'Die Wiener Forschungen an Kriegsgefangenen 1915–1918'. *Anthropologische und ethnografische Verfahren im Lager*. Vienna: Verlag der österreichischen Akademie der Wissenschaften.

Lange, Britta (2018). 'Archiv'. In Daniel Morat and Hansjakob Ziemer (eds), *Handbuch Sound. Geschichte – Begriffe – Ansätze*, 236–240. Stuttgart: Metzler.

Ochoa Gautier, Ana María (2014). *Aurality: Listening and Knowledge in Nineteenth-Century Colombia*. Durham, NC: Duke University Press.

Pöch, Rudolf (1910). 'Reisen im Innern Südafrikas zum Studium der Buschmänner in den Jahren 1907–1909'. *Zeitschrift für Ethnologie* 42 (2): 357–361.

Prakash, Gyan (2000). 'The Impossibility of Subaltern History'. *Nepantla: Views from the South* 1(2): 287–294.

Quijano, Aníbal (2007). 'Coloniality and Modernity/Rationality'. *Cultural Studies* 21(2–3): 168–178.

Rassool, Ciraj and Martin Legassick (2000). *Skeletons in the Cupboard: South African Museums and Human Remains, 1907–1917*. Cape Town/Kimberley: South African Museums.

Scheer, Monique (2010). 'Captive Voices: Phonographic Recordings in German and Austrian Prisoner-of-War-Camps of World War I'. In Reinhardt Johler, Christian Marchetti, and

Monique Scheer (eds), *Doing Anthropology in Wartime and Ear Zones. World War I and the Cultural Sciences in Europe*, 279–309. Bielefeld: Transcript Verlag.

Schüller, Dietrich (2003). *Rudolf Pöch's Kalahari Recordings (1908)*. Vienna: Verlag der österreichischen Akademie der Wissenschaften.

Schüller, Dietrich (2007). *The Collection of Franz Mayr Zulu recordings*. Vienna: Verlag der österreichischen Akademie der Wissenschaften.

Sheperd, Nick (2015). *The Mirror in the Ground: Archeology, Photography and the Making of a Disciplinary Archive*. Johannesburg: Jonathan Ball Publishers.

Simon, Artur (ed.) (2000). *Das Berliner Phonogramm-Archiv 1900–2000: Sammlungen der traditionellen Musik der Welt*. Berlin: Verlag für Wissenschaft und Bildung.

Spivak, Gayatri Chakravorty (1999). *A Critique of Postcolonial Reason: Toward a History of the Vanishing Present*. Cambridge, MA: Harvard University Press.

SPK Digital (2020a). 'Search'. Available online: http://www.spk-digital.de/index.php/showDetail.html?id=%22/bam/museum/smb/em/979788%22 (accessed 6 July 2020).

SPK Digital (2020b). 'Welcome to SPK Digital'. Available online: http://www.spk-digital.de/ (accessed 6 July 2020).

Stangl, Burkhard (2000). *Ethnologie im Ohr. Die Wirkungsgeschichte des Phonographen*. Vienna: WUV Universitätsverlag.

Sterne, Jonathan (2003). *The Audible Past: Cultural Origins of Sound Reproduction*. Durham, NC: Duke University Press.

Stoever, Jennifer Lynn (2016). *The Sonic Color Line. Race and the Cultural Politics of Listening*. New York: New York University Press.

Stoler, Ann Laura (2009). *Along the Archival Grain: Epistemic Anxieties and Colonial Common Sense*. Princeton, NJ: Princeton University Press.

Stoler, Ann Laura (2016). *Duress. Imperial Durabilities in Our Times*. Durham, NC: Duke University Press.

Trouillot, Michel-Rolph (1995). *Silencing the Past: Power and the Production of History*. Boston: Beacon Press.

von Hornbostel, E. M. and O. Abraham (1904). 'Über die Bedeutung des Phonographen für die vergleichende Musikwissenschaft' (On the Significance of the Phonograph for Musicology). *Zeitschrift für Ethnologie* 36 (2): 222–236.

Von Luschan, Felix (1896). *Anleitung zu wissenschaftlichen Beobachtungen auf dem Gebiete der Anthropologie, Ethnographie und Urgeschichte*. Berlin: Luschan Königliche Museeun zu Berlin.

Von Luschan, Felix (1906). 'Bericht über die Reise nach in Südafrika'. *Zeitschrift für Ethnologie* 38 (4): 873–895.

Weninger, Josef (1927). *Eine morphologisch-anthropologische Studie: Durchgeführt an 100 westafrikanischen Negern*. Vienna: Verlag der Anthropologischen Gesellschaft Wien.

Ziegler, Susanne (2006). *Die Wachszylinder des Berliner Phonogramm-Archivs*. Berlin: Staatliche Museen zu Berlin.

# 35
Sonic Feminisms: Doing Gender in Neoliberal Times

Marie Thompson

In this chapter I address the growing body of European and American activist projects seeking to encourage, advocate for, and celebrate women's participation in sonic and musical cultures. Through various methods, strategies, and approaches, these projects have sought to reveal and reconfigure the gendered make-up of industries, practices, and genres. While some explicitly identify themselves as feminist, others have intentionally or strategically avoided the term, choosing instead to signal their ambitions via feminized terminology and aesthetics. Nonetheless, many recent projects centring sound, music, and gender can be understood in relation to certain strands of feminist thought and action, and contextualized in relation to broader discussions of feminisms in the plural.

A central point of debate within recent feminist thought has been the relationship between the ideological and socio-economic formations of neoliberalism and the correlative transformations of gender relations. As a form of 'political rationality', neoliberalism 'casts the political and social spheres both as appropriately dominated by market concerns and as themselves organized by market rationality' (Brown 2006: 694). Individuals are approached as self-sustaining and competing economic units in a system that legitimizes and advocates for 'the equal right to inequality' (695). Meanwhile, a wide range of cultural phenomena – including self-help literature, mainstream television, popular fiction, food journalism, and music – share and thus work to reproduce the basic presuppositions of neoliberal thought via the celebration and dissemination of competition, individuality, and meritocracy as norms (Gilbert 2013). These social, economic, and political shifts associated with neoliberalism have also had significant implications for the formation and function of gender. Pertinent for feminists are the assertions of some neoliberal thinkers that the apparent alleviation of discrimination is conducive to the expansion of economic norms and values: for the benefit of the market, some manifestations of neoliberalism have adopted a socially progressive outlook in terms of gender, race, and sexuality. Consequently, the priorities of anti-discrimination and equal opportunities feminisms, and some of neoliberalism's advocates have converged (Watkins 2018). Neoliberalism's apparent pursuit of egalitarian goals, however, is limited by its grounding in fundamentally unequal economic relations. If neoliberal society values competition between individuals, then this requires that there are winners and losers; and women have frequently been amongst neoliberalism's 'losers' (Endnotes 2013). Apparent advances in gender equality have also gone 'hand-in-hand with soaring socio-economic *in*equality across most of the world' (Watkins 2018: 7, emphasis in the original). Thus, women's relative gains in terms of access to education, political representation, and changing social attitudes from the perspective of society and culture have been accompanied by growing economic pressures that shape and are shaped by gender relations.

Where certain formations of neoliberalism appear to encourage the overcoming of previous inequalities via an emphasis on individual empowerment, diversity, and inclusivity, neoliberal rhetoric and values have often resonated with particular manifestations of feminism – namely in its liberal, anti-discriminatory, or post-feminist configurations. Indeed, the explosion of interest in 'sonic feminisms' – used here to primarily refer to activist projects that seek to address the gendered make-up of sound and music production and consumption and their surrounding discourses – might be considered symptomatic of feminism's resurgent 'popularity'. Feminist scholars such as Clare Hemmings and Sarah Banet-Weiser have noted the emergence of a media-friendly 'popular feminism', which places emphasis on empowerment, visibility, and success. Popular feminism is not straightforwardly neoliberal inasmuch as it remains indebted to liberal feminism's critiques of gendered exclusions from public and corporate spheres. Like liberal feminism, popular feminism is often a call to bring more women to the table because they are women (Banet-Weiser 2018: 13). Nonetheless, popular feminism is also shaped by 'decades of neoliberal commodity activism, in which companies have taken up women's issues […] within neoliberal brand culture, specific feminist expressions and politics are brandable, commensurate with market logic' (13), and often reproduces neoliberalism's emphasis on entrepreneurialism, individualism, and meritocratic market values. Consequently, popular feminism remains consistent with neoliberal political rationality in pursuing parity within inequality (Watkins 2018).

It is against this social and economic backdrop that contemporary projects aimed at addressing the inequalities of sound and music cultures take place. As a result, there are numerous points from which the intersection of neoliberal gender relations, popular feminism, and sonic feminisms might be traced; for example, the pedagogic investments in 'the girl' as a means of securing auditory futures; the figuration of inclusion and diversity as 'good for the market'; the reiteration of entrepreneurialism, empowerment, and self-development as strategies for achieving 'success' within the sound and music industries; or the conflation of women's 'success' with gender equality or challenges to sexism within contemporary music cultures. For the purposes of this chapter, however, I focus on two interrelated methodological components of what might be described as a 'popular' sonic feminism: *quantification* and *amplification*. Although it is important to recognize that these components are not common to sonic feminisms in general – not all sonic feminisms are 'popular' – they nonetheless can be traced within and across a range of sonic feminist projects.

To discuss quantification and amplification in terms of sonic feminist methods requires an expanded definition of the latter and its constitutive terms. Quantification and amplification are 'sonic' inasmuch as they are used to reveal and reconfigure listening habits and formations of sound production. Yet, as will become evident, the seen and the heard are often folded in on one another; and numerical data and audibility are often conflated. Consequently, 'the sonic' in this context becomes difficult to hold apart from other processes of mediation. Quantification and amplification are 'feminist' inasmuch as they are used to intervene in and transform gendered hierarchies within the fields of sound and music production and consumption, and can be situated in relation to the methods, practices, and limitations of broader feminist formations – in this instance, popular feminism. Of course, quantification and amplification are by no means exclusively 'feminist', inasmuch as 'there is no research method that is consistently or specifically feminist' (Ramazanoğlu and Holland 2002: 15). Indeed, as will become evident, the extent to which quantification and amplification operate in alignment with feminist interests is questionable. Finally, quantification and amplification are discussed in proximity to 'method' inasmuch as they name particular, systemic approaches that produce knowledge about a field (quantification) and the consequent strategies for changing that field (amplification). They also carry with them methodological assumptions about how change is evidenced (i.e. through increased audibility/visibility and numerical balance). However, to understand sonic feminist deployments of quantification and amplification in terms of method requires the notion of methodology to be extended beyond its usual academic connotations as pertaining to scholarly research: many sonic feminist projects have existed on the margins of – if not outside – academia. Nonetheless, insofar as these projects have sought to generate and mediate knowledge about gender's relationship to sonic cultures, they can be understood as undertaking methodological work. Significant for this chapter, quantification and amplification also produce and reproduce particular conceptions of gender and gender inequality, inasmuch as they are necessarily approached as measurable. By situating quantification and amplification in relation to feminist critiques of popular feminism (Banet-Weiser 2018; Hemmings 2018), I interrogate the ways in which gender and gender inequality are configured in some contemporary sonic feminisms as oppositional (i.e. male/female), singular (i.e. non-intersectional), and generalizable (i.e. held in common across geopolitical, cultural, and economic differences). In doing so, I also aim to demonstrate the ways in which feminist critique can function *as* method, as well as being *about* method. This chapter utilizes critique *as* method insofar as it is a means of asking questions about relationships between gender, sonic cultures, social life, and knowledge production. It does so by offering a feminist critique *of* some 'popular' sonic feminist methods, arguing for the need to remain attentive to the conceptions of gender that these methods – and sonic methods more generally – reflect, reproduce, and naturalize.

Identifying sonic feminisms

In 2016, the Still Waiting Discussion Group – a student-led collective concerned with the lack of diversity of London College of Communication's BA Sound Arts and Design curriculum – launched the Tumblr 'S-A-D boyz'. The blog offers a playful critique of the course's reading list, which is described as 'full of pale, male and stales'. Featuring a series of labelled author images that expand as the user scrolls through, S-A-D boyz visualizes the curriculum's gendered and racialized construction: the user is visually bombarded with the faces of mainly white and mainly male writers and sound artists. S-A-D boyz is indebted to internet humour and its visual aesthetics: a number of the subjects appear in awkward or funny poses (economist Jacques Attali looking unimpressed, Soviet film-maker Sergei Eisenstein posing with a giant phallic cactus), while the Instagram artist Audrey Wollen's protest-meme flashes up towards the end of the list. Crossed out images of bare trees, the desert, a book of John Cage's music, and the corner of a room are accompanied by all-caps text: 'BEWARE MALE ARTISTS MAKING ARTWORK ABOUT EMPTINESS: NOTHING DOES NOT BELONG TO YOU: GIRLS OWN THE VOID: BACK OFF FUCKERS!!!!!'

Although addressing different contexts, the use of internet humour and aesthetics to highlight the gendered make-up of sonic cultures connects S-A-D boyz to another Tumblr site, 'Very Male Line-ups'. Active during 2015 – a year prior to S-A-D boyz – the blog targets male-dominated dance music events, 'highlighting all or mostly male club/gig/festival line-ups and helping bromoters do better'. The blog features male-dominated events line-ups, accompanied by a 'thumbs up' image of Rihanna and a congratulatory tagline.

S-A-D boyz and Very Male Line-ups are manifestations of the contemporary expansion of activity around the gendered formations of sound and music cultures. In recent years, there have been numerous discussion panels, festivals, exhibitions, symposia, publications, websites, music compilations, and workshops that have sought to draw attention to women's roles within the histories of music and sound art; to interrogate the barriers that have and continue to prevent their inclusion; and to encourage and celebrate women's creative activities. While there is a rich lineage of feminist scholarship and practice that has sought to address these issues (for example, Oliveros 1970; McClary 1991; Rodgers 2010), the surge of interest in the intersections of gender and sonic practice suggest that they are being considered with a reinvigorated sense of urgency.

Many of these recent projects have centred on women's participation in the music industry and related professional fields. In alignment with liberal feminist values, industry-oriented projects often focus upon 'empowering' individuals in their pursuit of equality and institutional reform. By facilitating their professionalization, these projects aim to enable women and girls to enter into the music and audio industries, thus resulting in a diversification of its workforce. PRS Foundation's Keychange initiative, for example, which 'empowers women to transform the future of music and encourages festivals to achieve a 50:50 gender balance by 2022' aims to 'accelerate change and create a better, more inclusive music industry for present and future generations' (PRS Foundation 2018).[1] The US non-profit organization SoundGirls, meanwhile, aims 'to inspire and empower the next generation of women in audio', providing 'supportive community for women in audio and music production' and 'the tools, knowledge and support to further their careers' (soundgirls.org 2018). Founded in 2013 by live sound engineers Karrie Keyes and Michelle Sabolchick Pettiano, the organization provides a range of resources for prospective and current women workers, as well as for those wanting to support women in the audio industries. These include guidelines for writing about women in sound, feature profiles on women audio workers, and advice for freelancers on tackling sexual harassment in the workplace.

In contrast to industry-oriented initiatives, some sonic feminisms centre on building a space of one's own – be it discussion-based, educational, or performance-based – that has different rules, structures, and values to dominant musical cultures. Such projects typically involve a 'do-it-ourselves' ethos, drawing inspiration from the feminist art-making and activism of Riot Grrrl. The Ladyz in Noyz project, for instance, has sought to emphasize non-hierarchy, cooperation, and resource sharing (by, for instance, lending equipment to one another). Similarly, the long-standing feminist media project Pink Noises is described as seeking to encourage critical consciousness through exploratory uses of sound and audio technologies. Like Ladyz in Noyz, it centres community-building and peer support, specifically through an online platform, and aims to make information on sound and music production more accessible.

Historical and archival projects have also been undertaken as a means of rethinking patrilineal histories of sonic practice (Rodgers 2010). The *Her Noise* archive, for example, is a physical and virtual resource of collected materials investigating music and sound histories in relation to gender. Initially forming part of the *Her Noise* exhibition in South London Gallery and curated by Lina Džuverović, Anne Hilde Neset, Irene Revell, and Emma Hedditch, the archive now exists within the University of Arts, London Archives, and Special Collections at the London College of Communication. However, although archival work has been a feminist strategy of building cultural and political alliances across different generations, curator Anne Hilde Neset notes that '*Her Noise* wasn't a result of focused feminist strategy, it "just happened" in the end, it sprang out of the hundreds of interviews and conversations we had conducted over four years and a lifetime's obsession' (Neset 2007). Lina Džuverović has also reflected on the project's complex relationship with feminism. Although she notes the feminist resonances of the project's curatorial method, Džuverović suggests that the project was originally thought of as 'postfeminist, believing that by curating an exhibition of sound-based work by women, yet not articulating it as a feminist project, we were going beyond feminism, going one step farther, thus avoiding the alienation from the visual arts establishment that an outwardly feminist project, at that moment, would have brought about' (Džuverović 2016: 90). For Džuverović, the simultaneous 'identifying yet not identifying' with feminism and the ambiguity of *Her Noise*'s politics was a product of the moment in which it emerged: 'At the height of the backlash against second wave feminism' (90). Consequently, the feminist voice of the project was silenced: 'We, the *Her Noise* curators, had been brought up not just on a diet of feminist music, literature, and art but also with a profound belief in equality and in finding ways to *live* feminism. Yet in a post–Spice Girls world of the early 2000s, somehow the term did not roll off our tongues easily' (91, emphasis in the original).

In her remarks, Džuverović highlights the relationship between sonic feminisms and their social milieu: feminism is both implicated in the *Her Noise* project and its methods but disavowed due to the affective and semantic resonances of the term at the time.

Angela McRobbie, from whom Džuverović draws, identifies how in the social and cultural landscape of the early 2000s

> elements of feminism have been taken into account, and have been absolutely incorporated into political and institutional life. Drawing on a vocabulary that includes words like 'empowerment' and 'choice', these elements are then converted into an individualistic discourse as they are deployed in this new guise, particularly in media and popular culture, but also by agencies of the state, as a kind of substitute for feminism.
>
> (McRobbie 2009: 1)

The emergence of 'popular feminism' since the early 2000s, however, means that feminism's resonances have changed since the initial formation of the *Her Noise* project. Clare Hemmings notes how the term feminism has shed its negative connotations within popular discourse, often becoming a 'universally desirable' signifier of the post-colonial modern. Hemmings considers the conditions under which this mainstreaming of feminism has taken place, suggesting that 'feminism can only be framed as desirable when it is untethered from its association with both masculinity and intersectionality, and sutured instead to femininity and singular understandings of women's oppression' (Hemmings 2018: 964). This 'popular' and 'universal' feminism maintains and reifies heteronormative oppositions between men and women, 'all the while appearing to challenge the limits such binarism represents' (973). The gender binary upon which popular feminism relies helps to naturalize the socio-economic relations that are 'enduring condition of gendered discourse. The desire to include women as full and equal participants in the public sphere (or to present them as already such participants) is *necessarily contradicted* by the creation of conditions that prevent them from being able to participate' (973, emphasis in the original). It is these enduring conditions that popular feminism – with its emphasis on empowerment, visibility, and 'success' – struggles to address. Banet-Weiser offers a similar thesis to Hemmings, arguing that

> popular feminism exists along a continuum, where spectacular, media-friendly expressions such as celebrity feminism and corporate feminism achieve more visibility; and expressions that critique patriarchal structures and systems of racism and violence are more obscured. As a result, 'seeing and hearing safely affirmative feminism […] often eclipses a feminist critique of structure, as well as obscures the labor involved in producing oneself according to the parameters of popular feminism.
>
> (Banet-Weiser 2018: 4)

In light of McRobbie's, Hemmings's and Banet-Weiser's analyses, it becomes pertinent to consider the ways in which sonic feminisms and their methods bear the traces of feminism's institutionalization, popularization, and universalization; and the extent to which they participate in the reproduction and reification of singular and oppositional notions of gender.

I now turn to consider the relationship between sonic feminisms, popular feminism, and neoliberal political rationality with reference to two complementary methodological aspects of what might be recognized as 'popular' sonic feminisms: quantification and amplification. Where quantification refers to a method of understanding, producing knowledge about, and representing the gendered make-up of sonic fields, amplification is understood as a means of creating change. However, the former is typically implicated in the latter and vice versa. Amplification as a strategy usually rests upon methods of quantification; and methods of quantification often provide the rationale for strategies of amplification. I contest that both quantification and amplification are shaped by popular feminism in particular and neoliberal rationality in general, and as a consequence, rest upon and reproduce understandings of gender as singular, oppositional, and generalizable.

Quantifying inequality: The metrification of gender

In March 2018 the music streaming platform Spotify, in collaboration with vodka brand Smirnoff, launched Smirnoff Equalizer – a marketing campaign and web-based user platform that sought to address the marginalization of women artists. Launched just before International Women's Day, the campaign was a response to streaming data that revealed that the top ten tracks on Spotify for 2017 were all by men. The platform was part of Smirnoff Equalizing Music, 'a 3-year global initiative to double female and female-identifying headliners and inspire the next generation of DJs' (Smirnoff Equalizing Music 2018). Through linking their Spotify account to the Equalizer, which reveals the ratio of men to women artists in their previous streaming data, users are able to discover how 'equal' or 'unequal' their listening habits are. The platform also offered the opportunity to create a more 'balanced' personalized playlist: by adjusting a slider that offered different percentages of music by men and women, users could 'discover more amazing music'. Marked 'M' on one side and 'W' on the other, the slider allowed users to select options between 50/50 women to male artists to 90/10 women to male artists. Promoted in partnership with Chicago DJ Honey Dijon, the platform was available through the summer of 2018 and could be accessed by those of legal drinking age in the United States, Great Britain, Ireland, Australia, Mexico, and Argentina.

Spotify's Smirnoff Equalizer platform produces gender as a metric: it is something that can be counted and represented as percentages and ratios. By extension, inequality itself is quantifiable: the pursuit of gender equality becomes an issue of fixing the numbers. In this regard, the Equalizer platform and its methods of understanding gendered listening habits are consistent with Spotify's data-driven approach to music-streaming and advertising, and, more broadly, a neoliberal political rationality that awards the market and its logics ontological and epistemological primacy. As David Beer notes, 'competition and markets require metrics, measurement is needed for the differentiations required by competition' (Beer 2016: 13). Thus, although 'metrics are themselves nothing new […] there is little doubt that these systems of measurement have escalated and intensified over recent years, especially with the rise of new data assemblages and their integration into the very fabric of our lives' (4). In other words, neoliberalism is an ideological context in which these methods of knowing become normative. These methods, likewise, play a key role in the reproduction of neoliberalism as a coherent political rationality. Spotify and Smirnoff's quantification and 'metrification' of gender inequality, then can be situated in relation to neoliberalism's marketization of social relations.

While the Equalizer platform is framed as a means of enabling users to discover the unacknowledged gender bias of their listening habits, the quantification of inequality and the metrification of gender obscures as much as it reveals. As Sally Engle Merry argues, 'counting things requires making them comparable, which means they are inevitably stripped of their context, history, and meaning. Numerical knowledge is essential, yet if it is not closely connected to more qualitative forms of knowledge, it leads to oversimplification, homogenization and the neglect of the surrounding social structure' (Merry 2016: 1). Although cited in the Equalizer's advertising materials, there is no real explanation offered for *why* the top ten tracks of Spotify in 2017 were produced by men. It is perhaps unsurprising that this question is not really addressed in the advertising campaign; however, it becomes particularly pertinent in light of Maria Eriksson and Anna Johansson's (2017) research into the gendering of Spotify's algorithmic structures, which suggests that Spotify's recommendation system over-represents male artists. Furthermore, in keeping with Hemmings's aforementioned critique of popular feminism, the Equalizer and its methods of quantification produce gender as singular and oppositional. With regard to the former, gender is isolated from other co-constitutive social divisions, relations, and identities, such as race, class, disability, and sexuality. With regard to the latter, the Equalizer posits gender as binary – artists are either male or female, and visually and numerically represented as such via the slider. In a smaller font next to the slider, the platform notifies users that it has 'included artists who identify as non-binary if they match your listening habits'. How these non-binary artists are categorized in relation to Spotify's binary gender metrics remains unclear.[2]

It might be tempting to dismiss the Spotify/Smirnoff campaign as a crude capitalization on sonic feminist efforts and their struggle for gender parity in musical cultures. Indeed, Spotify has been critiqued for its symbolic adoption of social justice causes, all the while sustaining the inequalities it aims to contest (Pelly 2018). Yet a range of sonic feminist projects share Spotify and Smirnoff's methodological treatment of gender and gender inequality as quantifiable, with the use of numerical infographics and data a common strategy for making visible the gendered make-up of sound and music cultures and industries. The international network female:pressure, for instance, has carried out a series of biennial FACTS surveys that have sought to calculate the gendered make-up of electronic music, resulting in a number of infographic representations of women's participation in label rosters and club and festival line-ups. In a 2013 press release the network states:

> The members of the female:pressure network operate within a seemingly progressive electronic music scene and its subcultures. However, we find that *women are notoriously under-represented in the realms of contemporary music production and performance.* The female:pressure group would therefore like to invite you to take a look at the facts and make the mechanisms of this specific market more transparent … *Female:pressure believes there*

> *is no justification for more male-dominated music events. We need – and paying audiences deserve – invigorating and entertaining diversity!*
>
> (female:pressure 2013, emphasis in the original)

The framing of the female:pressure FACTS surveys varies. Similar to the Equalizer platform, the 2013 survey makes use of a slider infographic, noting the percentage of male, female, and mixed artists on labels and festivals respectively, as well as providing an alternative visual representation of the gendered make-up of labels and festivals according to country. By comparison, the 2017 FACTS survey offers more detail in terms of methods, results, and discussion, as well as an acknowledgement of possible future directions and notable limitations. With the expansion of the festivals and countries surveyed in 2017, female:pressure suggest that 'we may see the extent to which inequity is a systemic issue. Structural sexism perpetuates inequality by creating barriers and disincentives for women, which limits women's success in the arts to genres and media aligned with the status quo' (female:pressure 2017).

There are significant differences between the quantification of gender inequality in the context of Spotify and female:pressure that should prevent a straightforward conflation of their methodological approaches. While the former is primarily associated with the marketing of Smirnoff (as well as the accrual of data), the latter is associated with and pursued by a grassroots network of women artists, DJs, and producers; and it is undertaken without any funding. And although market logics are evoked in the framing of female:pressure's work ('paying audiences deserve invigorating and entertaining diversity!'), there is also an attempt to recognize some of the underlying causal factors inasmuch as 'structural sexism' is referenced as both cause and context of numerical discrepancies between male and female artists. However, in approaching gender as a metric to be calculated, female:pressure also risks producing similar obfuscations through their attempts to uncover the gendering of electronic music. In female:pressure's surveys, gender is once again produced as singular and oppositional: artists are categorized as either male, female, or 'mixed', with the network acknowledging that non-binary artists are currently not easily represented within this framework. Likewise, despite some of the breakdowns offered in terms of geography, in counting artists in relation to binary and generalized notions of gender, other cultural, economic, social, and aesthetic differences are flattened. In doing so, the female:pressure report reproduces popular feminism's treatment of gender as a universal and central category of analysis. Gender and gender inequality, rendered calculable, countable, and generalizable, become unmoored from their specific and constitutive socio-economic contexts.

'Make Yourself Heard': Amplification and the economy of visibility

For both Spotify and female:pressure, the purpose of quantifying inequality is to improve the representation of women artists: by uncovering gendered disparities in playlists, line-ups, and rosters through methods of quantification, listeners, promoters, sponsors label owners, and producers are encouraged to 'discover more amazing music by women' or 'give more opportunities to women'. Likewise, 'boosting' women artists serves to address quantified inequality by 'fixing the numbers'. Amplification, then, can be understood as the intended outcome of quantification and a means of changing the gendered make-up of sound and music cultures.

Although the deployment of amplification as an auditory metaphor might suggest improving audibility, some sonic feminisms focus on improving *visibility* within sound and music cultures: in this context, equality, diversity, and inclusivity need to be *seen*. The US organization SoundGirls, for example, has partnered with Spotify to host the EQL Directory – a searchable database of women and gender non-conforming people in audio and music production. As with many sonic feminist initiatives, the terminology, ambitions, and framing of the project reproduces the neoliberal rhetoric of economic success, self-empowerment, and entrepreneurialism: the directory enables users to 'showcase their work, market their skills and reach out to each other for collaboration and networking', and acts as 'a testament to female audio professionals' resilience' (EQL 2019). Putting a celebratory 'spotlight' on the work of women and gender non-conforming audio practitioners is understood to challenge barriers to access and, as a result, serve to diversify the field:

> By amplifying the careers of these women and people, we'll soon see equal access to encouragement, equipment and opportunities within the industry, as well as equal recognition of these incredible professionals' work […] through celebrating audio professionals who are women, trans, and non-binary we believe the EQL Directory has the power to create a positive feedback loop – when more diverse people are seen running a recording session or commanding the mixing console, people of all genders will feel empowered to enter these fields.
>
> (EQL 2019)

In keeping with the corporate feminist dictum that 'you can't be what you can't see', the EQL Directory and its ambitions are predicated on the understanding that the heightened visibility of women, non-binary, and gender non-conforming practitioners will inspire future practitioners to strive for success, thus reconfiguring the make-up of the audio industries. By becoming more visible through strategies of amplification, these practitioners create a more diverse image of the audio industries. 'Seeing' diversity, meanwhile, is associated with more equitable working conditions.

As the EQL Directory exemplifies, sonic feminist strategies that are centred on making marginalized practitioners more visible typically cohere with the logics of a 'economy of visibility' (Banet-Weiner 2018). As Banet-Weiser argues, the economy of visibility is central to the form and functioning of popular feminism within a broader 'attention economy'. The logics of an economy of visibility are clearly manifest in the media landscapes that popular feminisms and sonic feminisms occupy, 'a technological and economic context devoted to the accumulation of views, clicks, "likes" etcetera' (Banet-Weiser 2018: 2). By contrast to a politics of visibility, which makes visible gender (or race, class, sexuality, or disability) as a marginalized *political* category for the purposes of social change (e.g. to demand rights, to change media representation practices, or to transform the ways in which identities matter and are valued), when economized, 'visibility becomes *the end* rather than a means to an end' (23, emphasis in the original). Such is the case with the EQL Directory: the aim is to make gendered practitioners more visible so as to 'empower' others. Furthermore, an economy of visibility transforms political categories such as gender and race 'so that the visibility of these categories is what matters, rather than the structural ground on and through which they are constructed' (23). Yet in a context where visibility is often predicated on circuits driven by market values of accumulation, competition, and consumption, increasing visibility does not mean that identity categories such as gender, race, and sexuality will be unfettered from sexism, racism, or homophobia. Thus, once again, then, an economy of visibility prioritizes surface over structure: where the politics of visibility places the spotlight on marginalized identities to highlight their disenfranchisement, within an economy of visibility 'the *spotlight* on their bodies, their visibility, the number of views, is in fact its politics' (29, emphasis in the original).

Amplification means that some signals are boosted while others are rendered background noise. Spotlights illuminate some things while leaving others in darkness. While the EQL Directory amplifies the (self-selecting) 'winners' (the EQL Directory invites professionals working in the audio industries to submit a profile), it leaves in the dark the structural issues and industry norms that serve to differentiate 'winners' from 'losers' – these include, for example, precarious working conditions, intern cultures, and irregular wages, the effects of which are exacerbated by neoliberal diminishments of welfare and social provisions. Thus, in keeping with an economy of visibility's attention to surface over structure, amplification directs attention to discrete signals over the complex infrastructures.

Not all strategies of amplification, however, straightforwardly conform to the normative logics of popular feminism's economy of visibility. In this regard, the S-A-D Boyz and Very Male Line-ups – mentioned at the beginning of this chapter – can be understood as offering an alternative method of visual amplification with rather different consequences. Like many manifestations of popular feminism, these projects deploy digital platforms and social media. However, instead of celebrating diversity or amplifying the work of gendered practitioners, S-A-D Boyz and Very Male Line-ups use these platforms as means of amplifying the gendered and racial *homogeneity* of educational curricula and dance music culture. In doing so, they draw attention to – and ridicule – the current gendered formations of their targeted fields. Consequently, they can be understood as operating as a negative critique in at least two senses. Firstly, these sonic feminist Tumblrs might be framed as negative with reference to their affective mobilizations, which are notably different from projects that centre on empowerment and celebration. In their use of sarcasm and humour, S-A-D Boyz and Very Male Line-ups might be understood as attempts to generate embarrassment and shame on the part of course convenors and bromoters. In doing so, they depart from the palatable positivity of popular feminism and its confident 'can do' attitude. Secondly, these projects are negative inasmuch as they draw attention to 'what is' – that is, the current formations of educational curricula and dance music culture respectively – and in doing so, they open a space to consider who and what is missing and why. In placing the spotlight on the centring of white, cis men, it is not assumed by these projects that what is therefore absent is 'women' in the singular.[3] Consequently, they can be understood to maintain a distance from popular feminism's oppositional conception of gender.

Conclusion: Popular sonic feminisms and the need for feminist critique

Although there are now many different sonic feminisms acting within and upon different fields, a significant proportion of these align with the principles and values of popular feminism. Symptomatic of neoliberal reconfigurations of social divisions, relations, and identities, popular feminism marks a combination of liberal feminism's emphasis on the inclusion of women in the public and corporate sphere with neoliberalism's emphasis on individualism, empowerment, and entrepreneurialism. In this chapter I have offered an assessment of quantification and amplification as methodological components of popular sonic feminisms. In using quantification and amplification to reveal and address gender inequality, these projects can be understood to reproduce gender in alignment with popular feminism: gender is produced as singular, general, and oppositional, while gender inequality is unmoored from its underlying socio-economic conditions. Consequently, they can serve to conceal as well as reveal the gendered dynamics of auditory cultures.

In her essay on gender inequality in electronic music, Annie Goh argues that the task of feminism is not only to struggle on the terrain of representation, 'but also to understand and critique the very categories and structures of power in which gender discourses operate' (Goh 2014: 57). Indeed, feminist *critique* might be understood to have an important role to play in discussions of sonic method and sonic culture more broadly. Critique, here, might be understood in at least two senses: both *as* method and *of* method. With regard to the former: the discussion of S-A-D Boyz and Very Male Line-ups demonstrates how critique works to produce knowledge about gender's production and circulation within different sonic cultures. This chapter, likewise, might be understood to employ critique-as-method, that is, as a means of asking questions about relationships between gender, sonic cultures, the structures of social life, and the production of knowledge. Yet it is also a feminist critique *of* particular methodological approaches and their limits as feminist tools. Sara Ahmed notes that critique has sometimes been admonished as a 'bad feminist' habit – a reactive, uncreative, and unproductive response; however 'what critique does depends on *where* – and *where not* – critique is directed [...]. In fact, much of what needs critiquing still seems to go unnoticed in our academic worlds' (Ahmed 2014, emphasis in the original). At a time where the political progressiveness of both feminism and the sonic is often taken for granted,[4] critique remains a necessary resource for feminist sound studies. In addition to developing more affirmative strategies, feminist interventions into sound and music cultures must remain attentive to the ways in which different methodological approaches reflect, reproduce, and naturalize categories and structures of gender.

Notes

1. PRS is one of the main funders of new music in the UK.
2. This coincides with Spotify's treatment of user's data: registration requires users to identify in relation to a binary category of gender. While Spotify has sought to make some changes to include options for non-binary users, it would appear that Spotify's algorithm still assigns them to a male/female category (Eriksson and Johansson 2017).
3. It is notable that when the Tumblr account was mentioned in a *Guardian* article 'Does Club Culture Have a Problem With Bigotry?' (Beaumont-Thomas 2015) the Very Male Line-ups twitter account responded saying that, while it was good to see the club culture's gendered dynamics being discussed in major outlets, they rejected the piece's ranking of bigotries 'placing sexism as worse/more prevalent than racism/homophobia. It's reductive and unhelpful because a) it's not a competition b) these things don't occur in isolation from one another and can't be solved in isolation from one another. You could add classism, transphobia, ableism and more to that list' (@malelineups, 9 June 2015).
4. For examples of how the sonic is positioned in relation to emancipatory, progressive or resistive politics see Brandon LaBelle's *Sonic Agency: Sound and Emergent Forms of Resistance* (2018).

References

Ahmed, Sara (2014). 'Feminist Critique'. *Feminist Killjoys*, 26 May. Available online: https://feministkilljoys.com/2014/05/26/feminist-critique/ (accessed 1 March 2019).

Banet-Weiser, Susan (2018). *Empowered: Popular Feminism and Popular Misogyny*. Durham, NC: Duke University Press.

Beaumont-Thomas, Ben (2015). 'Does Club Culture Have a Problem with Bigotry?'. *The Guardian*, 9 June. Available online: https://www.theguardian.com/music/2015/jun/09/does-club-culture-have-a-problem-with-bigotry (accessed 1 March 2019).

Beer, David (2016). *Metric Power*. Basingstoke: Palgrave Macmillan.

Born, Georgina and Kyle Devine (2015). 'Music, Technology, Gender and Class: Digitization, Educational and Social Change in Britain'. *Twentieth-Century Music* 12 (2): 135–172.

Brown, Wendy (2006). 'American Nightmare: Neoliberalism, Neoconservatism and De-democratization'. *Political Theory* 34 (6): 690–714.

Džuverović, Lina (2016). 'Twice Erased: The Silencing of Feminisms in *Her Noise*'. *Women and Music: A Journal of Gender and Culture* 20: 88–95.

Endnotes (2013). 'The Logic of Gender'. *Endnotes Issue 3: Gender, Race, Class and Other Misfortunes*. Available online: https://endnotes.org.uk/issues/3 (accessed 1 March 2019).

EQL (2019). 'About: A Woman's Place is in Music'. Available online: https://makeiteql.com/ (accessed 1 March 2019).

Eriksson, Maria and Anna Johansson (2017). 'Tracking Gendered Streams'. *Culture Unbound: Journal of Current Cultural Research* 9(2): 163–183.

Female:pressure (2013). 'Press Statement – 8 March 2013 – International Women's Day'. Available online: http://www.femalepressure.net/pressrelease.html (accessed 1 March 2019).

Female:pressure (2017). 'Facts 2017: Discussion'. Available online: https://femalepressure.wordpress.com/facts/facts-2017-discussion/ (accessed 1 March 2019).

Gilbert, Jeremy (2013). 'What Kind of Thing is "Neoliberalism"?'. *New Formations* 80/81: 7–22.

Goh, Annie (2014). 'Sonic Cyberfeminisms and its Discontents'. *CTM Magazine*: 56–58.

Hemmings, Clare (2018). 'Resisting Popular Feminisms: Gender, Sexuality and the Lure of the Modern'. *Gender, Place and Culture* 25 (7): 963–977.

LaBelle, Brandon (2018). *Sonic Agency: Sound and Emergent Forms of Resistance*. London: Goldsmiths Press.

McClary, Susan (1991). *Feminine Endings: Music, Gender and Sexuality*. Minneapolis, MN: University of Minnesota Press.

McRobbie, Angela (2009). *The Aftermath of Feminism: Gender, Culture and Social Change*. London: SAGE.

Merry, Sally Engle (2016). *The Seductions of Quantification: Measuring Human Rights, Gender Violence, and Sex Trafficking*. Chicago, IL: University of Chicago Press.

Neset, Anne Hilde (2007). 'Tangled Cartography'. *Her Noise*. Available online: http://hernoise.org/tangled-cartography/ (accessed 1 March 2019).

Oliveros, Pauline (1970). 'And Don't Call them "lady" Composers'. *New York Times*, 13 September. Available online: https://www.nytimes.com/1970/09/13/archives/and-dont-call-them-lady-composers-and-dont-call-them-lady-composers.html (accessed 1 March 2019).

Pelly, Liz (2018). 'Discover Weakly: Sexism on Spotify'. *The Baffler*. Available online: https://thebaffler.com/latest/discover-weakly-pelly (accessed 1 March 2019).

PRS Foundation (2018) 'Keychange'. Available online: https://prsfoundation.com/partnerships/international-partnerships/keychange/ (accessed 1 March 2019).

Ramazanoğlu, Caroline and Janet Holland (2002). *Feminist Methodology: Challenges and Choices*. London: SAGE.

Rodgers, Tara (2010). *Pink Noises: Women on Electronic Music and Sound*. Durham, NC: Duke University Press.

Smirnoff Equalising Music (2018). 'About'. Available online: http://smirnoffequalisingmusic.com/about (accessed 1 March 2019).

Soundgirls.org (2018). 'About Us'. Available online: https://soundgirls.org/about-us/ (accessed 1 March 2019).

Watkins, Susan (2018). 'Which Feminisms?'. *New Left Review* 109: 5–79.

36

Sound as City Maker: Developing a Participatory-Collaborative Process to Work with Sound as an Urban Resource; the Case of Mr. Visserplein (Amsterdam, the Netherlands)

Edda Bild, Michiel Huijsman, and Renate Zentschnig

Introduction

There is increasing awareness and understanding of how city sounds are connected to urban identity and dynamics. After decades of focusing on the negative effects of noise on human health, various calls for increasing sound awareness have been made to move beyond the sound-as-noise approach. The calls acknowledge and make use of the potential of soundscapes as urban 'resources' for city planning and design (Böhme 2000; Brown 2011; Maag 2013). The adoption of a new ISO standard demonstrates this apparent growing interest in approaching sound as more than noise,[1] as do various projects worldwide that engage in diverse forms of urban soundscape design (for example, Axelsson, Nilsson, and Berglund 2010; Lacy 2016). However, we contend that sound awareness remains low both among city users – namely, the people who use and engage with the spaces in a city – as well as among city makers – specifically, the ensemble of actors that intervene and make decisions about the city, ultimately influencing the way it sounds (Aletta and Jiao 2018; Steele 2018). This also holds true in a Dutch context where soundscape ideas have not yet been formally integrated in urban design or policy practices, and participatory and/ or collaborative, user-centred projects in relation to sound – in which the contextual knowledge of users about their soundscapes goes beyond noise complaints – have practically been non-existent (Derksen and Van den Bosch 2018). We propose that by increasing awareness on why the auditory dimension of the urban experience matters, we can encourage multisensory perspectives on urban living; the project presented in this chapter is a starting point in that direction, briefly introduced in Textbox 36.1.

> **Textbox 36.1.** 'Crowdsourcing Mr. Visserplein' project description.
>
> 'Crowdsourcing Mr. Visserplein' was a collaborative project between the University of Amsterdam and Soundtrackcity that began in late 2016 and developed out of the perceived need for concrete examples to demonstrate the benefits of designing cities with sound in mind to both Dutch city users and city makers, with the long-term goal of encouraging the integration of sound in city making. By focusing on the on-site auditory experience of users of a busy square in the center of Amsterdam, the project aimed to show how local sound artists, researchers, and practitioners could collaborate and act as moderators to help 'processing' site-specific experiential knowledge of local experts. It did so by engaging with a wide array of tools and methods to promote sound awareness and address auditory concerns in a local public space.

Here we will describe and reflect on our proposed participatory-collaborative approach (we distinguish between 'participation' and 'collaboration' to better illustrate the process of the project). We then outline the design of the project and its methods, synthesize its results, and reflect on the insights. We will first situate the project in the urban, scientific, professional, and artistic contexts that have inspired our approach. Next, we map our three-part iterative process of knowledge elicitation; interaction and exchange; and engagement and dissemination, and summarize the results of the model used to structure our findings. We will offer an overview of the participants' descriptions of their auditory experiences, their verbal exchanges, and their suggested proposals for the Mr. Visserplein square transformation, visualized and auralized with the help of an architect and a sound artist. We conclude with reflections on the collaboration and participation process (and its outputs) and our sound awareness efforts, with an eye on integrating them in current policies, community practices, as well as in planning and design initiatives.

Background

The approach for encouraging sound awareness discussed here includes the *participation* of city users as local experts, whose explicit and tacit knowledge on the auditory dimension of their everyday uses of spaces should be discussed clearly and integrated in various steps of the city-making process, and the *collaboration* between researchers, artists, and practitioners; these actors work on an equal footing and use their discipline-based tools and methods to engage in iterative processes of elicitation of users' auditory knowledge that can be further integrated in strategies to encourage sound awareness among all those involved in the city-making process. Artists are considered essential: trailblazers who have a history of engaging with the sonic aspects of urban spaces in unique and creative ways that enact new forms of urban listening.[2] This approach considers city users as an active source of experiential knowledge, rather than as 'naïve participants' who are passively subjected to noise, which tends to be the norm in current sound-related policies.

To contextualize our approach, we briefly show the increasingly common trend of both collaborative and participatory perspectives in decision-making processes that could be applied to integrate a sensory dimension to the aforementioned processes. Collaborative planning was envisioned as a socially interactive and inclusionary process bringing together a more diverse array of actors in order to contribute to policies and plans that would significantly affect their everyday lives (Healey 1992, 1997, 2003; Innes and Booher 1999; Margerum 2002). Within the broad array of interpretations of what was referred to as the 'collaborative turn' in planning initiatives, participatory planning emerged as a key paradigm emphasizing the importance of integrating the knowledge of the community in city-making processes (Forester 1999; Fischer 2000; Innes and Booher 2004).

The idea of participatory approaches might sound commonsensical in today's 'participation society' (Troonrede 2013; Tonkens 2014), and evidence on why and how integrating this knowledge matters has been extensively documented in academic literature. The focus lies specifically with so-called 'local' knowledge as the relational knowledge of everyday users of spaces, who have their own ways of engaging with, describing, and evaluating spaces and environments (Wilson 2008; Chilisa 2011; de Sousa Santos 2015). This approach can serve to add the experiential dimension to current ways of tapping into various forms of context-embedded knowledge, for example explicit and tacit knowledge (Nonaka et al. 1994). While explicit or codified knowledge is standardized and often articulated formally in, for example, text or maps, tacit knowledge is more personal, informal, embedded in experience and know-how, and situated in a specific context (Nonaka et al. 1994). This tacit knowledge is particularly relevant in relation to the ineffable sensory experience of (sonic) spaces, as members of many linguistic communities – including English and Dutch – lack a commonly agreed upon vocabulary to unambiguously articulate their (auditory) experiences (Corbin 1986; Dubois 2000; Dubois et al. 2014). To date, there have been efforts to valorize this city or space user-centred knowledge in initiatives attempting to capture some of the theorized benefits of the soundscape design strategy (Axelsson 2011; Brown 2012; Bild et al. 2016b). Various ways of using sound to improve the quality of urban public spaces for their users have been proposed, experimenting with integrating different experts and disciplines in the process, engaging with local decision makers, and involving the city users as participants to different extents (Axelsson 2011; Bild et al. 2016a; Maag and Bosshard 2016; Cerwén et al. 2017; Steele et al. 2017; Xiao et al. 2018; Lavia et al. 2018). However, these projects remain scattered and highly dependent on local dynamics and the willingness of local actors to support such initiatives. The challenge that remains is the transfer of collected knowledge (tacit or integrated tacit and explicit) from academia to practice, in a manner that can be 'translated' to specific action items, for example for design practices.

In the Netherlands, city makers focus overwhelmingly on noise and increasingly on what is supposed to be the antonym of noise, namely, quietness (Bijsterveld 2008; Booi and Van den Berg 2012; Licitra et al. 2014). While the political discourse and policies are centred on city users and their well-being, the methods used rely on acoustic measurements, embedded in noise abatement and management strategies; there is limited participation of city users, who are only consulted (usually 'after-the-fact') to provide input on noise complaints or the occasional completion of public health-related questionnaires.

With this in mind, capturing the resources of the auditory dimension of public spaces requires a more situated approach. The 'Crowdsourcing Mr. Visserplein' project aimed to show how a transdisciplinary approach could encourage both collaboration and participation, and generate knowledge that could increase sound awareness among city users and makers. The integration of different disciplines was tested to address the aforementioned difficulty in eliciting the various types of knowledge of city users. While researchers usually encourage verbal or written articulations, artists and other professionals (e.g. architects or designers) can prompt expressions that are also non-verbal, for example visual or aural, to stimulate exchanges rooted in multisensory on-site experiences.

In bringing together the individual auditory experiences of various local and professional experts, and encouraging exchanges and discussions, we provided a more holistic understanding of the urban auditory experiences and a pragmatic overview of users' needs and expectations in their spaces, thus also offering a starting point for public space transformations. To further demonstrate the relevance of this approach to urban designers, for example, part of the process was focused on suggestions for minimally invasive, cost-effective sound-based solutions that we contend could improve the quality of the square for its users.

Method

Peter Brook wrote that, in theatre, the empty stage evokes another world by the stories of the dramatic figures (Brook 1968), arguing that the stage changes with every dramatic performance and every world evoked by it. Through the stories and dialogues of characters that are part of a performance, the space itself changes and transforms. We have appropriated this notion in 'Crowdsourcing Mr. Visserplein': the square was but a busy, yet 'empty' space/stage that gained meaning through the stories and experiences of those using it in their everyday lives. We documented how people who are familiar with and use the square describe their on-site experience according to three topics: their perception of the space, its perceived patterns of use, and its sounds. We used this *space-use-sound model* to structure our data collection, processing, and analysis. We put particular emphasis on the extent to which the square responds to the users' auditory and behavioural needs and expectations, and how it could be redesigned to do so better.

The research process itself was not linear, but rather iterative, and had three key parts which partially overlapped: (1) knowledge elicitation, steps associated with the auditory and verbal data collection, processing, and analysis; (2) interaction and exchange, steps associated with encouraging social interaction and exchanges between participants on their on-site experience; and (3) engagement and dissemination, steps associated with strategies of encouraging sound awareness (see Figure 36.1). The process remained open, as additions could be brought to any core part, and it was also flexible, as there was overlap between the parts; for example, the discussions in workshops fall under 'interaction and exchange' but are at the same time activities that elicit knowledge. The methods employed to process and analyse the data collected, ranging from the transcription of recordings to thematic coding of the resulting written text, were also cross-cutting through the three elements. They allowed us to make the elicited knowledge explicit and 'visible', to facilitate social interaction, and to further integrate this knowledge in engagement and dissemination strategies.

In the next sections we will discuss each part in detail and provide details on the collaboration process as well as the site and participant selection processes. The first part of the data collection method as well as a detailed overview of the findings of that part have been detailed elsewhere, summarized here with the permission of the authors of that report (Van Kamp et al. 2018).

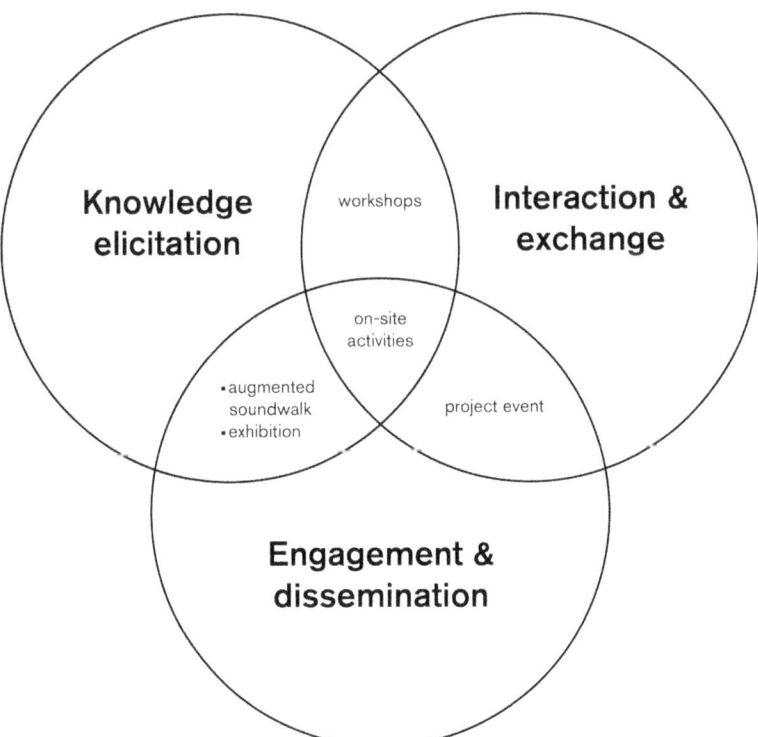

Figure 36.1 Visualization of the process of 'Crowdsourcing Mr. Visserplein' with its three core parts. In the overlapping areas, examples of specific activities.

Collaboration

The project was initiated through discussions and shared concerns at an event organized in September 2016 by one of the authors, at the University of Amsterdam (Bild et al. 2017).[3] The focus was on shared grievances on what, we, the authors of this chapter, considered to be insufficient collaborative and participatory approaches in sound-related practice in the Netherlands (particularly Amsterdam). Therefore, one essential goal of the Soundtrackcity–University of Amsterdam collaboration (see Textbox 36.1) was to integrate the work and expertise of sound artists and design professionals, and to support the participation of city users as local experts. We acknowledged the advantages and limitations of the methods and tools we used in our everyday professional activities, so we aimed to develop and test a process that brought together complementary, domain-specific methods to support participants in articulating their experiences (and thus their tacit auditory knowledge). The core team of 'Crowdsourcing Mr. Visserplein' included one visual artist and one theatre maker/director (members and initiators of Soundtrackcity[4]), one urban sociologist (from the University of Amsterdam), one sound artist/composer (Van den Broek 2020), and one architect (Blits n.d.). We had the support of other external collaborators (including one researcher/urban planner, a graphic designer, and a multimedia interaction designer).

Site selection

Mr. Visserplein is a square in the historic centre of Amsterdam, a place of contention and political debate (Figures 36.2 and 36.3), having undergone radical transformations in the last half century (Batjes 2008). It was the site for a number of unrealized plans made in recent history (Ten Dam 2009; Meershoek 2017). Being surrounded by traffic arteries from all sides (resembling a roundabout), the square faces environmental issues – including air pollution, noise complaints, and heat island effects – but also social concerns such as crowdedness in the city and tourism–resident tensions.

The specific urban form and seemingly hybrid function of Mr. Visserplein as an urban space, situated at the intersection of traffic arteries, of different urban communities, and of distinct urban neighbourhoods with different functions make it an appropriate case study to research the relationship between form, use of space, and perception of sound. Table 36.1 characterizes Mr. Visserplein in terms of space, use, and sound based on the observations of the project team.

Participant selection

We exclusively selected Dutch native speakers, to allow participants to communicate about and reflect on their auditory experience in a language they are comfortable in. Participant selection was completed according to the following criteria:

Figure 36.2 Historical photograph of Mr. Visserplein, 1983. View from the Moses and Aaron Church. Photograph by Jan van Boerkhoven. *Source:* The Amsterdam Municipal Archives.

Figure 36.3 Mr. Visserplein, aerial photograph, February 2016. © Thomas Schlijper.

- Relationship with the square: resident of the neighbourhood or worker in the area
- Age distribution: different ages, ranging from children to the elderly
- Personal background: different professional and educational backgrounds.

We also invited one participant with a severe visual impairment; one participant used to be homeless, having spent nights on the square itself. We did not aim to collect

Table 36.1 Characterization of Mr. Visserplein in Terms of Space, Use, and Sound.

Space	Use	Sound

• Historical importance
  • Buildings of cultural and historical significance
  • **Destination** space
• Large open space
  • Divided in half by a tram line
• **Transition** space
• Amenities encouraging lingering activities
• Surrounded by motorized and bicycle traffic on all sides
• Limited accessibility
• Underground level: TunFun – children's playground
	• Simultaneous mobility flows:
	  • Motorized traffic (including scooters)
	  • Bicycle traffic
	  • Pedestrians
	• Shared space
	• Diverse group of users
	  • Nearby residents
	  • Students
	  • Employees
	  • Visitors/tourists
	  • Passers-by
	  • TunFun users
	  • **Homeless people**
		• Reflections from buildings
		• Limited variety of sound sources
		• Dominant sound: mobility related
		  • Traffic: motorized vehicles and bicycles
		• Soundmarks:
		  • Static: audible crosswalk signals
		  • Dynamic: tram bells and tram crossing over metallic structure
		• Strong temporal patterns
		  • Strong temporal auditory 'rhythms'
		• High measured sound pressure levels both during the day and night (Government of the Netherlands n.d.)

a representative sample of square users but, rather, to have a maximally diverse group of participants, in order to find the distinct as well as shared aspects of their auditory experiences on site. We identified and developed connections with various local actors, both public and private, and accessed the existing local communities in the area to find a balanced distribution of participants who both live and work in the area. Twenty volunteers were willing to engage in the participatory process.

The project process

The process is the result of the intermingling of the three aforementioned core parts: knowledge elicitation; interaction and exchange; and engagement and dissemination.

Knowledge elicitation

The knowledge elicitation part consisted of the collection, processing, and analysis of on-site individual data (individual commented soundwalks with short-follow up interviews) and data from workshop discussions. We collected three different types of data: verbal, audio, and visual (i.e. photographic).

The on-site data were collected over a one-month period in the early summer of 2017, both on sunny and rainy days, during different times of the day and week. The average duration of a data-collection session (soundwalks and interviews) was 1.5 hours. The conversations took place exclusively in Dutch.

For the self-guided commented soundwalk participants were met either at their home or their workplace. They had chosen the route beforehand and acted as guides through their 'territory', where we, as knowledge elicitors, were mere guests (Figure 36.4). The participants were encouraged to describe what they heard and what they thought of the square, to share personal memories, and to reflect on their experiences, both at the time of the soundwalk as well as from memory. One project member walked alongside one participant, reminding them to focus on sound in various moments, as well as suggesting stopping at random points on the trajectory to focus on listening. While on the square, the participants were asked to stop, close their eyes and listen for a few minutes; afterwards they were asked to describe what they heard. The participants' commentaries were recorded, and a second project member, walking a few metres behind the participant, recorded the environmental sounds.

Immediately after the walk, participants were invited to conduct a short semi-structured interview, usually in a quieter space (a café or the participants' home). The interview was recorded and had four distinct parts: reflection on sound (and the auditory experience of the square), general evaluation of the square, own patterns of use of the square, and suggestions to improve the square (with a limited budget). Participants were asked to use one word or one phrase to describe the square, based on their current impression and previous experience.

Figure 36.4 Self-guided soundwalk: trajectory (left) and walk (right). *Source:* authors.

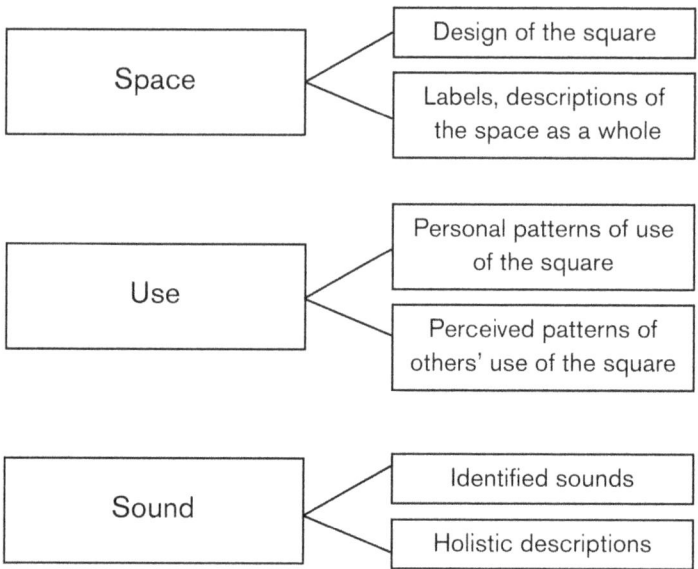

Figure 36.5 Space-use-sound model.

The recorded interviews and discussions were transcribed and the soundwalk trajectories were digitized. The audio data was processed, edited, and curated for further artistic, educational, and promotional use, with the participants' permission. We completed a thematic coding of the resulting written data, using the *space-use-sound model* visualized in Figure 36.5.

Interaction and exchange

To facilitate social interaction and exchange, we organized two workshops for the participants to come together and discuss their shared experience of the square. Workshop 1 focused on the participants characterizing the square as it was at the time, as well as what was needed in terms of space, use, and sound; Workshop 2 focused on the participants debating the proposals put forward by the architect and the sound artist to transform the square.

All participants were invited to take part in Workshop 1 in September 2017; seven of them joined the workshop, in which they discussed the characterization of Mr. Visserplein from three points of view: space/physical environment (the tangible elements of the space, infrastructure, buildings, materials, urban furniture), use, and sound, using keyword mapping. The discussion focused on what the participants thought the space was like *now* (i.e. the current state at the time of the soundwalks/workshop) and what they *needed* in Mr. Visserplein.

Based on a preliminary thematic coding of the transcribed data from individual soundwalks and interviews, we put together a list of one hundred keywords and phrases – direct quotations from the interviews. The participants were split into two groups; each group was given two sets of the same keywords, additional blank cards where they could write new keywords as well as modifiers ('too little', 'too much', etc.), and six sheets of paper where they could stick their keywords. The sheets indicated the topics that the participants were asked to characterize: space-now, space-needed; sound-now, sound-needed; use-now, use-needed. The participants were then invited to share their outcomes in an informal setting, where they focused on commonalities and differences in their keyword maps. The discussions were audio-recorded and photographed, with the participants' consent (Figure 36.6).

Based on the insights collected from the transcribed soundwalks and the discussions in this first workshop, the project architect and sound artist developed visualizations and auralizations of the participants' recommendations for redesigning the space (see below); the architect also developed his own proposal to redesign the space (building a suspended park on top of the square). In the end, four final proposals for physically redesigning the square were discussed (Figure 36.7): building a park on top, adding a green noise barrier with elevated cycling space, removing the two metallic roof-like structures in the middle of the square to expose the children's playground, and adding a fountain/water element. One additional proposal focused on adding a larger variety of activities on the square. The activity and water-centred proposals were auralized by the sound artist.

During the second workshop, four participants (three of which already participated in Workshop 1) discussed the two sound-centred and the four physical design-centred proposals in a moderated, round-table setting, with the entire project team present. These discussions were also audio-recorded and photographed (Figure 36.8).

Engagement and dissemination

The individual and shared knowledge of the participants was processed, disseminated, and used to encourage further engagement and sound awareness, using various strategies such as augmented soundwalks, exhibitions, and transdisciplinary events.

The augmented soundwalk allowed attendees to an event to listen, using headphones, to recordings of representative snippets of what the participants shared on their auditory experience while walking through Mr. Visserplein. The attendees were handed maps of Mr. Visserplein with various designated stops where they could listen to stories of the participants in the self-guided soundwalks and go-along interviews, recorded in that particular spot on the square (Figure 36.9).

The project exhibition offered a visual overview of the aim, process, inputs, and outputs of the 'Crowdsourcing Mr. Visserplein' project (Figure 36.10). Two iPods preloaded with

Figure 36.6 Workshop 1 on characterizing the space, use, and sound of Mr. Visserplein.
Source: authors.

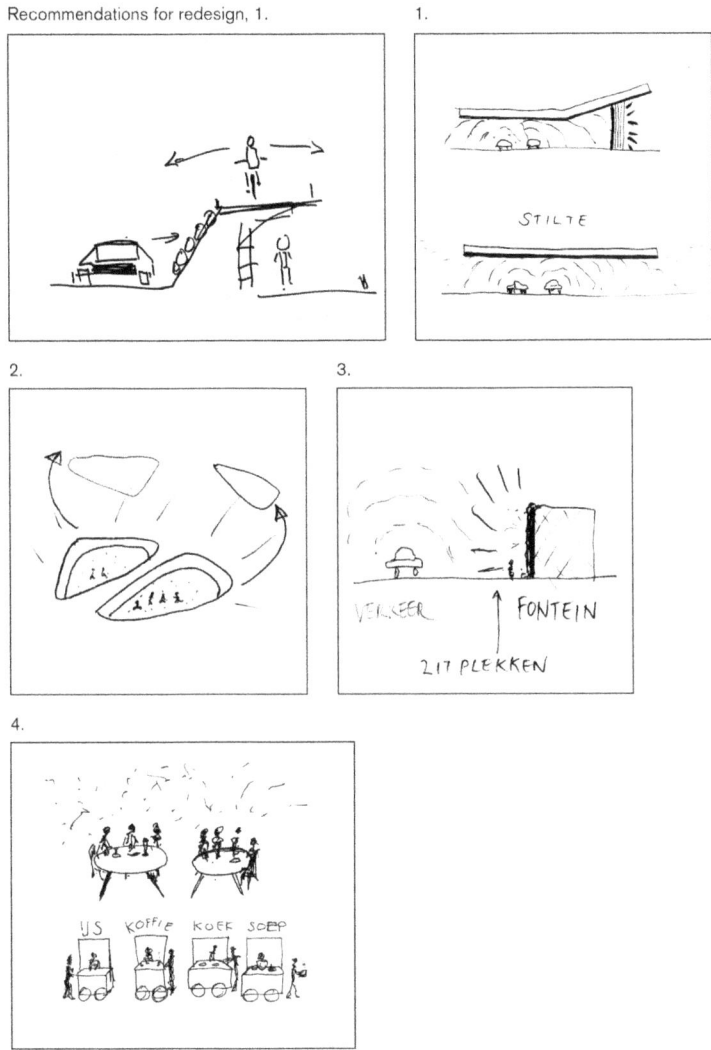

Figure 36.7 Square redesign proposals. Top left design (1) by Louwrens Duhen; other designs by Erik Blits.

the two auralizations of the water and user-centred proposals that can be listened to using headphones, were also part of the exhibition.[5]

The 'Geluid als stadmaker' ('Sound as a City Maker') event took place in November 2017 (Figure 36.11). It consisted of a guided augmented soundwalk on the square (see above), followed by presentations from various professional and local experts, and group discussions. It acted as a platform for engagement and dissemination on topics of urban sound and strategies for integrating auditory aspects in designing and managing urban spaces, with a focus on Dutch cities (particularly Amsterdam).

Figure 36.8 Workshop 2 on visualizations and auralizations of square redesign proposals.

Figure 36.9 'Geluid als stadmaker' event attendees on the augmented soundwalk.

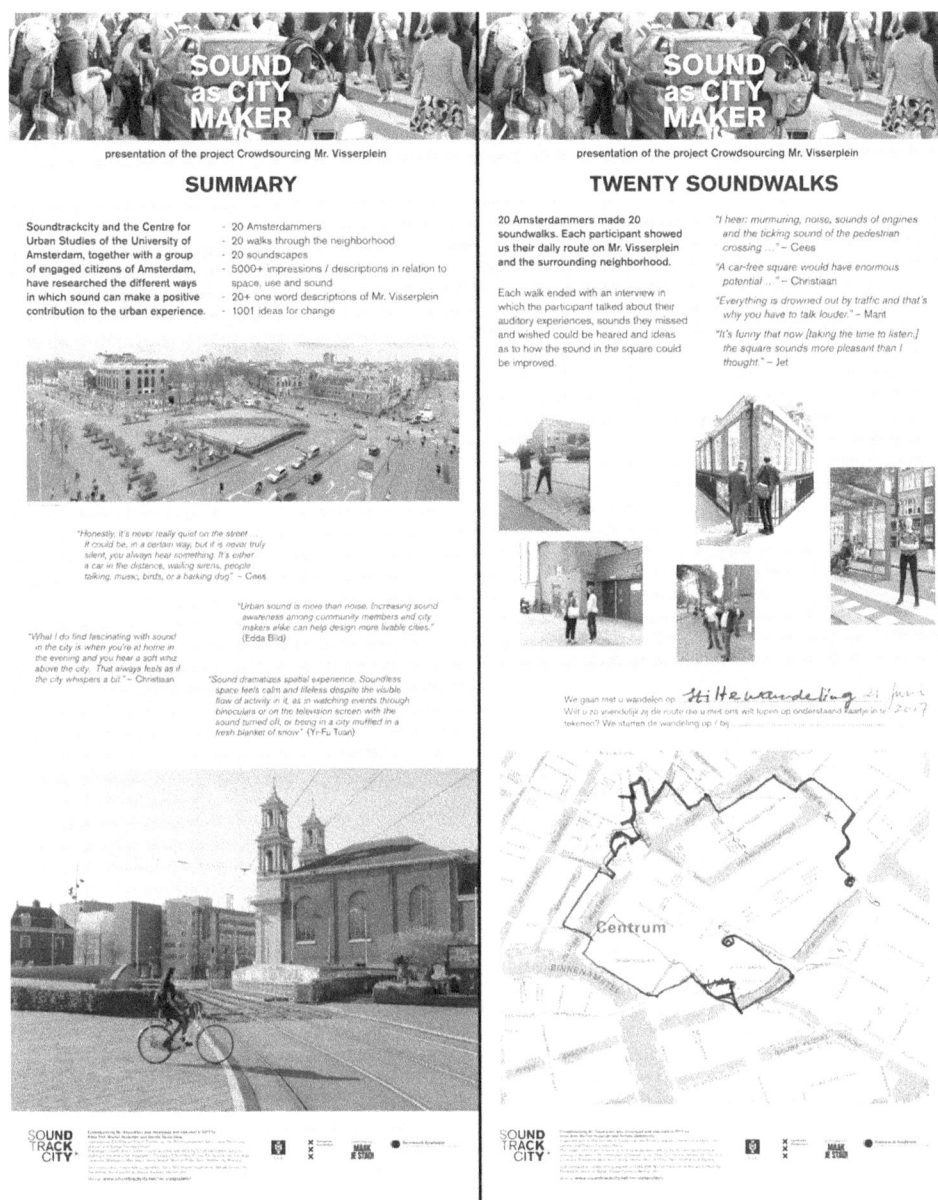

Figure 36.10 Project exhibition: banners.

Findings

In this section we summarize the output of the knowledge-building process, first reporting on the insights that were collected on how the participants evaluated the space itself, its perceived use as well as its sounds, based on our space-use-sound model (Mr. Visserplein – 'now'). Next, we report on the participants' recommendations on redesigning the square with sound and use in mind (Mr. Visserplein – 'needed').

Figure 36.11 'Geluid als stadmaker' – event brochure.

First, we analysed the participants' description and evaluation of the square and how it fit their environment and their expectations. For a number of participants, it was not a 'real square', one adding that it was not a 'user's square' (*gebruikersplein*) and it did not encourage 'square-appropriate' activities. The square was not described as a destination but, rather, as a point of transition, despite its amenities that were supposed to encourage lingering activities. However, for some participants, the square was a somewhat typical and expected part of the city, stating that the square was 'lively', particularly due to the many people passing it on bike or foot. The design of the square was mostly described as a 'collage' and 'incoherent', with one participant describing Mr. Visserplein as a leftover space (*restruimte*). For others, it was perceived as disappointing, considering the historical context in which the square is situated; some emphasized that the square lacked identity and 'a heart' in comparison with its surroundings, and that it was overall a 'missed chance'. Nonetheless, there were participants that enjoyed the square, 'as a space' (*als ruimte*), but felt distracted by the sound and sight of traffic.

Second, we focused on how and when the participants used the space, their satisfaction with it as well as how they thought others used it. The square was defined by many in relation to what it is not, which the participants connected to what types of activities it is not appropriate for. The square was perceived as being left unused, with the exception of the odd tourist out. Some participants entertained the idea of sitting and listening in the square (particularly on sunny days); however, the traffic was considered a distracting element that made the square unsuitable for sitting or recreation, but rather used as a transition/passage space, where people go through rather than stay. This idea of passage is related to the different types of mobility encountered in the space, emphasizing the nature of the square as a place dominated by continuous flows of people, both as part of motorized or bicycle traffic and as pedestrians ('pedestrian highway' – *voetgangerssnelweg*). For their own use of the square, participants stated that it was driven more by necessity than desire, as it is the only way to reach several other destinations they visit on a regular basis. The dominance of traffic was perceived to have repercussions on the participants' feelings of safety, as well as health-related effects, ranging from the air quality to the feeling of being overwhelmed/claustrophobic to downright getting a headache if one remained on the square for too long.

Third, we analysed the participants' impressions of the auditory environment and specific sounds and how that affected their experience. Unsurprisingly, the sounds that the participants referred to can be grouped according to types of mobility: motorized traffic-related, public transportation-related (tram), bike-traffic related, and pedestrian-related. However, participants did not heavily rely on words such as 'annoying' or 'nuisance' (*storend* or *overlast*), rather attempting to describe the 'layers' of sound. When asked to close their eyes and listen, respondents detailed their auditory experience through a larger diversity of sounds, compared to when participants only passed through the space. Some participants referred to the rhythm of the square (observable both in minute-long and daily cycles), with one participant likening it to a flow: 'Traffic light, tram, footsteps, heels, tram, scooter, child, bike, and now the traffic starts all over, engine […] a wave' ('stoplicht, tram, voetstappen, hakken, tram, scooter, kind, fiets, en nu begint het verkeer weer, motor […] een golf').

The recommendations for square redesigns were the result of both individual interviews and exchanges during the two workshops. We grouped the recommendations according to the physical and sound-related design they imply and labelled them according to what part of the space-use-sound model they addressed.

1. Large-scale redesigns of the square: a park on top of the square and a green noise barrier with elevated cycling space (space-, use-, and sound-related redesign)
2. Small/minimal physical changes to the square to enhance or minimize existing/audible sounds: removing the metallic structure in the square to expose the children's playground (space- and sound-related redesign)
3. Adding sounds by adding physical elements to the square: a waterfall or fountain (space- and sound-related redesign)
4. Adding sounds through added functions: introducing a food cart (use- and sound-related redesign)

The recommendations suggested by the participants go along the lines of the findings and work in literature on the principles of soundscape design, particularly in relation to the minimization of unwanted sounds (e.g. traffic through masking), enhancing wanted sounds (e.g. the sounds of other users), and introducing new sounds (e.g. the sounds of children and, to a smaller extent, the sounds of water) (see Axelsson 2011; Coelho et al. 2016). The sounds of others were considered to make the experience of the space friendlier, while still allowing users to 'feel like being in the city'. This justified the importance of adding 'reasons' for users transitioning the space to stay or regard the space as a destination as well. Some participants referred to these sound additions or enhancements as 'counter-sound' (*tegengeluid*), and the small redesigns were suggested to lead to a more liveable and dynamic auditory experience on the square: 'It becomes a place where more happens, more diversity, more life' ('Het wordt een plek waar meer gebeurt, meer diversiteit, meer leven'). One key finding was that participants did not seek to silence the space but rather to enhance its liveliness. As the participants stated that green spaces are not far from the square, and other quiet spaces are within reach, their idea was not to turn Mr. Visserplein into a quiet space. One participant argued that 'a city must remain a city' ('een stad moet een stad blijven') and square soundmarks such as the sound of the pedestrian crossing or the trams remind users that they are in a city, so they should be evaluated as pleasant and positive. The participants thus shared a pragmatic view to allow for a diversity of uses, with the changes suggested out of 'respect for the city'.

Discussion

The process behind the 'Crowdsourcing Mr. Visserplein' project was a participatory-collaborative one; it was developed to both encourage the participation of city users beyond consultation as well as to make full use of the approaches, tools, and methods of the different experts collaborating in the project. The project team thus operated with the shared goal of eliciting the participants' tacit knowledge on their on-site auditory experience, as well as supporting and moderating discussions and exchanges between them in order to make that knowledge explicit and shareable. While the methods used were not necessarily innovative (at least not in the context of the disciplines in which they were initially developed), the process of integrating them as part of this transdisciplinary project was insightful. It was particularly the overt interest expressed by city users in both the auditory dimension of their experiences as well as in being actively involved in city-making processes in general that encouraged us to develop a number of propositions or action items that city makers intervening in urban spaces can integrate in their practices and considerations:

1. *Being in the space*: Designers, planners, and other city makers should start by going into the space themselves rather than relying on plans and models of the space. The actual on-site experience can elicit their own knowledge – not just as professional experts but also as everyday users – to better gauge the potential consequences of policies, plans, or designs.
2. *Sharing the experience of the space*: Participation of city users is essential in city-making processes and city makers should visit the spaces with users. By visiting spaces together city makers and city users can reach a common (experiential) ground for discussion and talks. Having an equal and shared experience can allow also for mutual learning, and the knowledge elicited by being there can be more useful than off-site conversations about the space.
3. *Encouraging an active experience of the space*: Being in the space and experiencing it in an active manner, similar to everyday practices (e.g. walking) allows for embeddedness and the appropriation of a space through use and play (e.g. through soundwalking exercises).
4. *Listening together*: In the move towards embracing a multisensorial experience of spaces, it is essential to teach city users and city makers how to listen, and how to engage with spaces aurally. Being aware of sound can encourage a more meaningful relationship with a space and add or make explicit a previously tacit layer to an experience. This can, for example, represent the basis for previously ignored/not thought-of transformations or encourage a re-evaluation of certain dimensions of one's experience; if sound has just been 'negative' until that point – as noise or a source of annoyance – the experience of sound can become different by learning how to listen.

Particularly in relation to the auditory experience, insight collected in this project allowed us to articulate three lessons in relation to the added value of engaging in listening processes and increasing sound awareness:

- Listening together is a mutual learning process, as there are as many ways of listening as there are people. This implies not only listening with other city users but also collaborating with artists or other practitioners who may have different ways

of listening and whose methods and insights can elicit new forms of knowledge feeding into the city-making process.
- Listening together can encourage the creation of a local community.
- Listening and talking about sound can also elicit valuable linguistic insights on how everyday users of spaces talk about sound. This can help improve current common research instruments (such as surveys) or policy language.

Following these propositions and, more broadly, using methodologies such as the one described in this chapter, aiming to collect in-depth qualitative data, are labour-intensive but necessary to ensure that policies and interventions directed at improving the urban quality of life address the needs and expectations of users, mirroring their everyday discourses and experiences, and eliciting both their tacit and explicit knowledge. City makers must stay informed about the ways in which city users talk and experience sound in rapidly changing urban environments, for example their sources of complaint, annoyance, or enjoyment. This is particularly true in a political context that emphasizes the importance of public engagement and citizen participation with local- and country-level policies. Insight collected using such methods can also help update existing surveys in terms of content and language.

Both the outputs and the process described in this chapter can support a shift towards more multisensorial approaches to city-making. On the one hand, by using the dissemination and engagement tools described above, one can demonstrate to urban dwellers in different neighbourhoods what types of projects or knowledge could be collected or used in their own spaces to inspire more sound awareness and sound-driven change. On the other hand, by sharing a flexible process scheme with city makers, they can add, change, or use parts of it to collect necessary knowledge to transform cities with sound in mind. By doing so, they can avoid situations in which they need to 'fix' implemented plans rather than anticipate auditory issues or concerns.

To conclude, the participatory-collaborative process of 'Crowdsourcing Mr. Visserplein' is proposed to complement traditional methods and approaches used both in research and in professional practice or policy. This chapter advocates for engagement in processes of active consultation of local experts in both policy- and space-design initiatives as more than sources of complaints. The intention is to move towards more collaborative processes, in the sense that the insights of local experts should be integrated and used in multiple stages of the process, as part of an interactive process of co-production of knowledge.

Notes

1. See Fiebig; and Schulte-Fortkamp, in this volume.
2. See, for example, the work of Maryanne Amacher (1967), Max Neuhaus (1988), Akio Suzuki (2014), and Peter Cusack (2017).
3. The event focused on ensuring a platform for discussion and collaboration between scientists and practitioners on topics of public space research, design practice, and urban data gathering, with specific attention on sound.

4. Soundtrackcity is an artist initiative specialized in organizing soundwalks in order to make people more aware of their sonic ambiances (see Soundtrackcity n.d.).
5. For information, see Architectuur Centrum Amsterdam n.d.

References

Aletta, F. and J. Xiao (2018). 'What Are the Current Priorities and Challenges for (Urban) Soundscape Research?'. *Challenges* 9(1): 16.

Aletta, F., J. Kang, and Ö. Axelsson (2016). 'Soundscape Descriptors and a Conceptual Framework for Developing Predictive Soundscape Models'. *Landscape and Urban Planning* 149: 65–74.

Aletta, F., F. Lepore, E. Kostara-Konstantinou, J. Kang, and A. Astolfi (2016). 'An Experimental Study on the Influence of Soundscapes on People's Behaviour in an Open Public Space'. *Applied Sciences* 6(10): 276.

Amacher, M. (1967). *City Links: Buffalo, a 28-hour Piece Using 5 Microphones in Different Parts of the City, Broadcast Live by a Radio Station*. Available online: http://www.ludlow38.org/files/mabooklet.pdf (accessed 27 March 2018).

Architectuur Centrum Amsterdam (n.d.). 'Mr. Visserplein'. Available online: https://www.arcam.nl/en/mr-visserplein/ (accessed 7 July 2020).

Axelsson, Ö. (ed.) (2011). *Designing Soundscape for Sustainable Urban Development*. Stockholm: City of Stockholm. Available online: http://www.decorumcommunications.se/pdf/designing-soundscape-for-sustainable-urban-development.pdf (accessed 27 March 2018).

Axelsson, Ö., M. E. Nilsson, and B. Berglund (2012). 'The Swedish Soundscape-quality Protocol'. *Journal of the Acoustical Society of America* 131(4): 3476–3476.

Battjes, H. (2008). 'Het Mr. Visserplein, een gat in de stad: Vereniging Vrienden van de Amsterdamse Binnenstad'. Amsterdamse Binnenstad. Available online: http://www.amsterdamsebinnenstad.nl/binnenstad/230/mr-visserplein.html (accessed 26 March 2018).

van den Berg, F. and P. Huiszoon (2017). 'Nieuwe woningen en het geluid van Amsterdam'. *Plan Amsterdam* 2017 (4): 4–11 [in Dutch with English summary].

Bijsterveld, K. (2008). *Mechanical Sound: Technology, Culture, and Public Problems of Noise in the Twentieth Century*. Boston, MA: MIT Press.

Bild, E., D. Steele, and K. Pfeffer (2017). 'Creating Sound Awareness through Inter- and Trans-disciplinary Collaboration: A Workshop Report'. Presentation at *AESOP 2017*, Lisbon, Portugal, July 2017.

Bild, E., M. Coler, K. Pfeffer, and L. Bertolini (2016a). 'Considering Sound in Planning and Designing Public Spaces: A Review of Theory and Applications and a Proposed Framework for Integrating Research and Practice'. *Journal of Planning Literature* 31(4): 419–434.

Bild, E., K. Pfeffer, M. Coler, O. Rubin, and L. Bertolini (2018a). 'Public Space Users' Soundscape Evaluations in Relation to Their Activities: An Amsterdam-based Study'. *Frontiers in Psychology* 9: 1593.

Bild, E., D. Steele, K. Pfeffer, L. Bertolini, and C. Guastavino (2018b). 'Activity as a Mediator Between Users and Their Auditory Environment in an Urban Pocket Park: A Case Study of

Parc du Portugal (Montreal, Canada)'. In Aletta, F., and Xiao, J. (eds.) *Handbook of Research on Perception-Driven Approaches to Urban Assessment and Design*. Hershey, PA: IGI Global.

Bild, E., C. Tarlao, C. Guastavino, and M. Coler (2016b). 'Sharing Music in Public Spaces: Social Insights from the Musikiosk Project (Montreal, CA)'. In *INTER-NOISE and NOISE-CON Congress and Conference Proceedings* 253(5): 3657–3666.

Blits, Erik (n.d.). 'Home'. Available online: http://www.blits.nl/ (accessed 7 July 2020).

Böhme, G. (2000). 'Acoustic Atmospheres: A Contribution to the Study of Ecological Aesthetics'. *Soundscape: The Journal of Acoustic Ecology* 1(1): 14–18.

Booi, H. and F. van den Berg (2012). 'Quiet Areas and the Need for Quietness in Amsterdam'. *International Journal of Environmental Research and Public Health* 9(4): 1030–1050.

Brook, Peter (1968). *The Empty Space*. London: Mac Gibbon & Kee.

Brown, A. L. (2011). 'Advancing the Concepts of Soundscapes and Soundscape Planning'. In *Proceedings of the Conference of the Australian Acoustical Society (Acoustics 2011)*, November.

Brown, A. L. (2012). 'A Review of Progress in Soundscapes and an Approach to Soundscape Planning'. *International Journal of Acoustics and Vibration* 17(2): 73–81.

Cerwén, G., J. Kreutzfeldt, and C. Wingren (2017). 'Soundscape Actions: A Tool for Noise Treatment Based on Three Workshops in Landscape Architecture'. *Frontiers of Architectural Research* 6(4): 504–518.

Cerwén, G., E. Pedersen, and A. M. Pálsdóttir (2016). 'The Role of Soundscape in Nature-based Rehabilitation: A Patient Perspective'. *International Journal of Environmental Research and Public Health* 13(12): 1229.

Chilisa, B. (2011). *Indigenous Research Methodologies*. Thousand Oaks, CA: Sage.

Coelho, J. B. and J. Luis (2016). 'Approaches to Urban Soundscape Management, Planning, and Design'. In J. Kang and B. Schulte-Fortkamp (eds), *Soundscape and the Built Environment*, 197–214. Boca Raton, FL: CRC Press.

Corbin, A. (1986). *The Foul and the Fragrant: Odor and the French Social Imagination*. Cambridge, MA: Harvard University Press.

Cusack, Peter (ed.) (2017). *Berlin Sonic Places: A Brief Guide*. Hofheim: Wolke Verlag.

Davies, W. J., M. D. Adams, N. S. Bruce, R. Cain, A. Carlyle, P. Cusack, D. A. Hall, K. I. Hume, A. Irwin, P. Jennings, and M. Marselle (2013). 'Perception of Soundscapes: An Interdisciplinary Approach'. *Applied Acoustics* 74(2): 224–231.

de Sousa Santos, B. (2015). *Epistemologies of the South: Justice against Epistemicide*. New York: Routledge.

Derksen, T. and K. van den Bosch (2018). *Soundscape Projects in the Netherlands: A Survey of Soundscaping Interventions in the Physical Domain*. The Hague: I&W and RIVM.

Dubois, D. (2000). 'Categories as Acts of Meaning: The Case of Categories in Olfaction and Audition'. *Cognitive Science Quarterly* 1(1): 35–68.

Dubois, D., M. Coler, and H. Wörtche (2014). 'Knowledge, Sensory Experience, and Sensor Technology'. In C. Rangacharyulu, E. Haven, and B. J. S. Juurlink, *The World in Prismatic Views: Proceedings of the Second Interdisciplinary CHESS Interactions Conference*, 97–133.

Filipan, K., M. Boes, B. de Coensel, C. Lavandier, P. Delaitre, H. Domitrović, and D. Botteldooren (2017). 'The Personal Viewpoint on the Meaning of Tranquility Affects the Appraisal of the Urban Park Soundscape'. *Applied Sciences* 7(1): 91.

Fischer, G. (2000). 'Shared Understanding, Informed Participation and Social Creativity: Objectives for the Next Generation of Collaborative Systems'. In *COOP 2000: Proceedings of the 4th International Conference on Designing Cooperative Systems*, Sophia Antipolis, France, 23–26 May 2000, 3–15.

Foale, K. (2014). 'A Listener-Centered Approach to Soundscape Analysis'. PhD dissertation, University of Salford.

Forester, J. (1999). *The Deliberative Practitioner: Encouraging Participatory Planning Processes*. Cambridge, MA: MIT Press.

Government of the Netherlands. n.d. *Atlas Leefomgeving* (Environmental Health Atlas). Available online at https://www.atlasleefomgeving.nl/en (accessed 3 April 2018).

Healey, P. (1992). 'Planning through Debate: The Communicative Turn in Planning Theory'. *Town Planning Review* 63(2): 143.

Healey, P. (1997). *Collaborative Planning: Shaping Places in Fragmented Societies*. Vancouver, BC: University of British Columbia Press.

Healey, P. (2003). 'Collaborative Planning in Perspective'. *Planning Theory* 2(2): 101–123.

Innes, J. E. and D. E. Booher (1999). 'Consensus Building and Complex Adaptive Systems: A Framework for Evaluating Collaborative Planning'. *Journal of the American Planning Association* 65(4): 412–423.

Innes, J. E. and D. E. Booher (2004). 'Reframing Public Participation: Strategies for the 21st Century'. *Planning Theory & Practice* 5(4): 419–436.

ISO 12913–1 (2014). *Acoustics — Soundscape — Part 1: Definition and Conceptual Framework*. Geneva: International Organization for Standardization.

Lacey, J. (2016). *Sonic Rupture: A Practice-led Approach to Urban Soundscape Design*. London: Bloomsbury.

Lavia, L., H. J. Witchel, F. Aletta, A. Steffens, A. Fiebig, J. Kang, C. Howes, and P. G. Healey (2018). 'Non-Participant Observation Methods for Soundscape Design and Urban Panning'. In F. Aletta and J. Xiao (eds), *Handbook of Research on Perception-Driven Approaches to Urban Assessment and Design*, 73–99. Hershey, PA: IGI Global.

Lercher, P., I. van Kamp, E. von Lindern, and D. Botteldooren (2015). 'Perceived Soundscapes and Health-Related Quality of Life, Context, Restoration, and Personal Characteristics'. In J. Kang and B. Schulte-Fortkamp (eds), *Soundscape and the Built Environment*, 89–132. Boca Raton, FL: CRC Press.

Licitra, G., M. van den Berg, and P. de Vos (2014). 'Good Practice Guide on Quiet Areas'. *European Environmental Agency, EAA Technical Report* 4.

Maag, T. (2013). *Cultivating Urban Sound – Unknown Potentials for Urbanism*. Oslo: Oslo School of Architecture and Design.

Maag, T. and A. Bosshard (2016). 'The Sonic Public Realm: Chances for Improving the Sound Quality of the Everyday City'. In *INTER-NOISE and NOISE-CON Congress and Conference Proceedings* 253 (6): 2402–2411.

Margerum, R. D. (2002). 'Collaborative Planning: Building Consensus and Building a Distinct Model for Practice'. *Journal of Planning Education and Research* 21(3): 237–253.

Meershoek, P. (2017). 'Holocaust kan onder het Mr. Visserplein'. *Het Parool*, 1 December. Available online: https://www.parool.nl/amsterdam/-namenmonument-holocaust-kan-onder-het-mr-visserplein~a4542637/ (accessed 18 July 2020).

Neuhaus, M. (1988). 'LISTEN'. Available online: https://www.academia.edu/41490530/Listen_Max_Neuhaus (accessed 20 July 2020).

Nonaka, I., P. Byosiere, C. C. Borucki, and N. Konno (1994). 'Organizational Knowledge Creation Theory: A First Comprehensive Test'. *International Business Review* 3(4): 337–351.

Payne, S. R. and C. Guastavino (2018). 'Exploring the Validity of the Perceived Restorativeness Soundscape Scale: A Psycholinguistic Approach'. *Frontiers in Psychology* 9: 2224.

Radicchi, A. (2017). 'A Pocket Guide to Soundwalking: Some Introductory Notes on its Origin, Established Methods and Four Experimental Variations'. In A. Besecke, J. Meier, R. Pätzold, and S. Thomaier (eds), *Perspectives on Urban Economics*, 70–73. Berlin: Universitätsverlag der TU Berlin.

Schafer, R. M. (1977). *The Soundscape: Our Sonic Environment and the Tuning of the World*. Rochester, VT: Destiny Books.

Soundtrackcity (n.d.). 'Home'. Available online: http://www.soundtrackcity.net/ (accessed 7 July 2020).

Steele, D. (2018). 'Bridging the Gap from Soundscape Research to Urban Planning and Design Practice: How do Professionals Conceptualize, Work with, and Seek Information about Sound?'. PhD thesis, McGill University.

Steele, D., R. Dumoulin, C. Kerrigan, and C. Guastavino (2017). 'Sounds in the City Workshops: Integrating the Soundscape Approach in Urban Design and Planning Practices'. In *Proceedings of the AESOP 2017 Conference, Lisbon, Portugal*, 11–14 July.

Suzuki, A. (2014). 'oto-date: Bonn Hoëren leaflet/map'. Available online: http://www.bonnhoeren.de/wp-content/uploads/2014/06/bonn_festival_oto-date_web.pdf (accessed 2 April 2018).

Te Brómmelstroet, M. and P. M. Schrijnen (2010). 'From Planning Support Systems to Mediated Planning Support: A Structured Dialogue to Overcome the Implementation Gap'. *Environment and Planning B: Planning and Design* 37(1): 3–20.

Ten Dam, S. M. S. D. (2009). 'Het Meester Visserplein Amsterdam'. MA thesis, Eindhoven University of Technology.

Thibaud, J. P. (2013). 'Commented City Walks'. *Wi: Journal of Mobile Culture* 7(1): 1.

Tonkens, E. (2014). 'Herover de participatiesamenleving'. *S & D* 71(1): 85–95.

Van Algemene Zaken, M. (2013). *Troonrede 2013*. Available online: https://www.rijksoverheid.nl/documenten/toespraken/2013/09/17/troonrede-2013 (accessed 15 July 2020).

Van den Broek, Evelien (2020). 'About'. Available online: http://www.evelienvandenbroek.com/ (accessed 7 July 2020).

Van Kamp, K., E. van Kempen, E. Bild, K. van den Bosch, T. Derksen, R. Zentschnig, C. Baliatsas, and M. Huijsman (2018). *Perceptions of Local Environmental Quality: Comparing Methods of Measurement and Analysis - P_LOK*, RIVM internal report. Rijksinstituut voor Volksgezondheid en Milieu RIVM.

Wilson, S. (2008). *Research is Ceremony: Indigenous Research Methods*. Halifax: Fernwood Publishing.

Xiao, J., L. Lavia, and J. Kang (2018). 'Towards an Agile Participatory Urban Soundscape Planning Framework'. *Journal of Environmental Planning and Management* 61(4): 677–698.

37
Dropping Down Low: Online Soundmaps, Critique, Genealogies, Alternatives

Angus Carlyle

Key

Located between the compound noun 'soundmap' and its detached corollary 'sound map' is a stable definitional ground for the diagrammatic representation of acoustic place. It may be, however, that this territory is more narrowly circumscribed than necessary, is isolated from a wider genealogy, and contains some partially buried commitments which further compromise soundmapping's methodological potential. This chapter proceeds from a verification of my own bona fides in relation to this sonic practice, draws out that wider genealogy, then unearths those commitments – which are understood to involve the alignment with a specific technological apparatus and the adoption of a particular perspectival frame. I conclude by proposing three alternative approaches that are freighted with the means to deliver, in their diversity, formations of knowledge of the heard world that are transposed from the conventions of the cartographic to the experimental possibilities Samuel Thulin has recently designated the 'cartophonic':

> 'Cartophony' operates as a near-synonym of 'sound mapping' with the subtle difference that whereas 'sound mapping' suggests a qualified mapping and already carries associations with particular practices often involving a mimetic approach to [their] representation […] 'cartophony' is used as an attempt to speak to how practices of sound and mapping may feed into one another in a broad array of ways.
>
> (Thulin 2018: 193)

Orientation

My practice-orientated research seeks to establish sensory resonances between environments and their inhabitations. The sites in which my recording and listening strategies have found themselves include a suburban strip measuring 500 metres by 200 metres, the public galleries and backstage operations of a municipal museum, an organic small holding enmeshed in the concrete and steel infrastructures of a busy airport, a wind farm on a plateau, a series of footpaths in a national park, and a stretch of local woodland close to my home. I understand all these to be forms of soundmapping yet appreciate that they would to be excluded from definitions; nonetheless, at least two of my past projects more closely resemble the normative soundmap.

The first of these, *51° 32 ' 6.954" N / 0° 00 ' 47.0808" W*, was a contribution to the Sound Proof exhibition at E:vent Gallery, London, in 2008. This emerged from a three-month period of field work devoted to a 100 metre by 100 metre zone of the Lower Lea Valley, an area previously comprising light industrial units, transport networks, residential architecture from a variety of historical epochs, a canalized waterway, and liminal greenspaces but which was undergoing a process of demolition and construction for the London Olympics. *51° 32 ' 6.954" N / 0° 00 ' 47.0808" W* represented a 'mapping of intangible qualities, in the space where social and ecological, past and future intersect' (Biagioli 2018: 98) and comprised an hour-long audio composition of layered but otherwise unprocessed field recordings, a booklet of field notes and photographs, and an A0 scale map (that was reprised in A1 format for the exhibition catalogue). One of the curators of the exhibition recently addressed the work:

> Carlyle's contribution to a sense of social and spatial orientation is asserted through finely noted observation of activity and ephemera encountered during his visits to the site. These seemingly inconsequential events and objects are plotted diagrammatically on his map as key markers of the site, giving prominence to the vernacular components of this site in transition. His approach echoes the notion that whoever maps the space gives that landscape and location its territorial characteristics.
>
> (Biagioli 2018: 102)

The second project which bears closer structural analogy to how soundmaps are conventionally conceived, developed from field work of similar duration but differentiated by a more diffuse location in the rural environment of the Picentini mountains, in the hinterlands behind Naples and Salerno. Commissioned by the Fondazione Aurelio Petroni, an organization based in the small town of San Cipriano Picentino, it formed part of a multimodal account of the region entitled *Viso Come Territorio / The Face as Territory*. One of my contributions involved placing field recordings on an online map, deploying a platform devised by Peter Cusack for his *Favourite Sounds* project. The web-based mechanism enabled the *Viso Come Territorio* map to be gradually populated over the spring of 2012, to be accessed remotely (and retrospectively, since the soundmap remains online), and to function as the basis for an installation during the exhibition. Salomé Voegelin who engaged with this soundmap online and in her book *Sonic Possible Worlds* described the encounter:

> This geography is not that of San Cipriano Picentino and not of my living room either but that of their possibilities generated in my recentred listening, exploring the material that sounds there and bringing it back into the actuality of my present listening that is every thicker and pluralized for it. These sonic narratives do not share in the generality of the

visual map, nor in the image we might have of an area […] I am not following the map but mapping my own while listening.

(Voegelin 2014: 34)

This prior involvement in the creation of soundmaps has combined with experiences as a user that date from relatively early in the gestation of what we recognize as their typical form and my 2007 edited book *Autumn Leaves: Sound and Environment in Artistic Practice*, included documentation of several such soundmapping projects. Reflections on the creation and consumption of soundmaps have informed a previous conference paper on this theme, which sketched some initial problematizations of soundmapping as a sonic research methodology (Carlyle 2014), problematizations that are now deepened, are contrasted with alternative approaches, and are situated within a genealogical frame.

Scale

A conventional interpretation of a soundmap is announced in the very first line of the relevant Wikipedia entry as 'digital geographical maps that put emphasis on the sonic representation of a specific location'. By embarking on a speculative genealogy, I hope to demonstrate that what currently tends to pass for a soundmap – on Wikipedia and elsewhere – need not be as narrowly defined and that there are historical resources for mapping sound that prefigure something of the alternative approaches which this chapter finishes with, approaches which might be considered under the rubric of Thulin's 'cartophonic'.

Perhaps most recently presented in this lineage would be the noise map which, in its paradigmatic form, constitutes what cartographers call a choropleth, a thematic map where colour gradations reflect distributions of density, in this case measured (or simulated) acoustic intensity. In Europe, noise maps tend to be associated with the Environmental Noise Directive (END) (Council of the European Union 2002) and in a case study-based article offering the methodological innovation of triangulating noise maps with 'sound maps and soundscape maps', acousticians Francesco Aletta and Jiang King indicate that 'noise mapping is certainly one of the most relevant operational tools that the END relies on, providing visual representations of the yearly average noise levels in a selected area […] useful to assess easily the population's noise exposure and consequently to spot areas where noise action plans are required' (Aletta and Kang 2015: 1). Though we can enthusiastically acknowledge the imperative 'to listen beyond an exclusive focus on the quantitative regulation enshrined in noise pollution policies' (Di Croce 2016: n.p.), it might be, as we shall hear, that the choroplethic approach can be adapted to amplify other aspects of the sounded environment without succumbing to the conventions of the soundmap.[1]

Predating the choroplethic noise map and equally implicated in the substitution of visual for acoustic information was the audiospectrograph, the development of which is critically reconstructed by Joeri Bruyninckx as a complex process in which the impetus towards mechanical inscription (of bird song) involved a renegotiation of the competing legitimacies of sensorial and scientific knowledge and propelled 'recorded sound further into even flatter and more controllable, comparable units of sound […] deal[ing] with the dimension of time in a more felicitous way […] deal[ing] with the problem of noise, by erasing it visually' (Bruyninckx 2012: 145). Bruyninckx cautiously concludes that the audiospectrograph and such successor technologies as the spectrograph and the sonogram are taken to 'illustrate a scientific culture in which places of science are demarcated by sterility and silence and underscore the need to understand how this scheme is enforced by and reinforces wider cultural expressions of modern sound control' (147).

Valerio Signorelli has uncovered other such antecedents, dating, for example, 'the earliest documented examples of soundmap […] in the research conducted, in the 20s, by the Finnish geographer Johannes Gabriel Granö […] [who] developed a specific methodology to describe landscape features through direct multisensory observations of the proximate environment and a series of hatched maps' (Signorelli 2017: 155). It is possible to uncover two persuasive prior claimants to the title of earliest soundmap, each as imbricated within a scientific culture as the noise maps and the audiospectrographs, albeit a Victorian paradigm rather than Bruyninckx's Modernism. In 1888, the Royal Society of Great Britain reported on the eruption of the volcano Krakatoa, devoting extensive analysis to associated acoustic phenomena, of which two dimensions attract attention. First of these dimensions are the series of topographic maps plotting the global movement of the pressure waves that persisted for nearly 100 hours after the eruption, the data for which were recorded as traces detected by the network of meteorological barographs. Second is the report's tabulated accumulation of ear witness accounts from mariners, colonial administrators, weather scientists, and others which arguably constitutes another cartography, of which evidence from the Rodrigues Islands' Chief of Police – 4,777 kilometres from Krakatoa – is edifying: 'stormy […] heavy rain and squalls, […] wind […] blowing with a force of 7 to 10, Beaufort Scale […]. Reports like the distant roars of heavy guns'. The table of textual witnessing in the Royal Society report functions to map the perceived movements of sound across what was estimated to have been 'a 13th of the surface of the earth' (Furneaux 1964: 18), bringing detailed coloration to the more orthodox representations in the global plotting of pressure gradients on the projected continents and oceans.

The textual component intimates other pedigrees for a critical conceptualization of the soundmap as method. Ear witnessing reports, such as that derived from the Rodrigues Islands' Chief of Police, James Wallis, are perhaps less infrequent as mapping of acoustic phenomena within Victorian scientific culture than might have been anticipated. Archival analyses conducted by one of my doctoral students, Jennifer Allan, suggest the other conceivable candidate for earliest soundmap – some fifty years before Signorelli's Granö – in John Tyndall's *Report on Fog Signals* (Tyndall 1874). Alongside a bird's eye projection – a projection I will later call 'aerial' after Voegelin and R. Murray Schafer – that depicts the radial acoustic propagation from the fog signal and the 'sound shadow' in a bay neighbouring the coastal test site of South Foreland in Kent, there are circular graphics to demonstrate perceived loudness of a trumpet as it is directed to different points of the compass, where the intensity is given both in numbers and linguistically as 'weak, faint, good, very good'. In parallel to the Royal Society's report on Krakatoa, diagrams and maps are supplemented with a textual attention to the sonorous that can be comprehended as another mapping of the transmission vectors of sound, Tyndall's language rehearsing the words of Wallis and others in the tables of aural testimonies:

> The heavy rain at length reached us [...] the sound, instead of being deadened, rose perceptibly in power. Hail now added to the rain, and the shower reached a tropical violence. The deck was thickly covered with hailstones which here and there floated upon the rainwater [...]. In the midst of a furious squall both the horn and the syren were distinctly heard; and as the shower lightened, thus lessening the local noises, the sounds so rose in power that we heard them at a distance of 7 ½ miles.
>
> (Tyndall 1874: 22)

Although sensually attentive and allusive, the respective reports of Tyndall and Wallis, each bending their listening practices to the geophonic, remain within the parameters of gridded, choate writing. Other textual cartographies of sound have shown less obedience to the grid, and more proximity to that digital writing addressed by N. Katherine Hayles in which text 'becomes a process [...] "eventilized", made more an event and less a discrete, self-contained object with clear boundaries in space and time' (Hayles quoted in Carpenter 2017: 105). Two contrasting para-literary illustrations, separated by a neat century, that might also be accounted for as sonic mappings can be located in an example drawn from F. T. Marinetti's 'parole i libertà' (words-in-freedom) and in Christian Marclay's video work *Surround Sounds* (2015). Marinetti's poem 'Après la Marne, Joffre visita le front en auto' (1915) is described by curator JoAnne Paradise as 'unusual in that it borrows the basic form of a military map' to evoke what she dubs the 'surround sound' of the battle, depicting the physicality of landscape and troop dispositions and the sonority of combatants, their screams, machine gun reports, and artillery blasts rendered through an 'eventilized', dynamized combination of typography and paint strokes (Finkelaug 2006). Marclay's own Surround Sounds resembles an animated version of Marinetti's poem, deploying a graphic depiction of comic book sound effects to infer, in the gallery's own interpretation text, 'the acoustic properties of each word. "Boom", for example, is no longer static on the page, but bursts into life in a sequence of colorful explosions, while "Whooosh!" and "Zoooom!" travel at high speed around the walls. The work fuses the aural with the visual, and immerses the viewer in a silent musical composition' (White Cube 2018).

I am arguing that Marclay's and Marinetti's onomatopoeic 'eventilizations' constitute soundmaps and belong to a genealogy that can be obscured by the current conventional definitions. Room within the terminological scope of the soundmap can also be made for the more recognizable antecedents in choroplethic noise maps and audiospectrographs just as accommodation can be found for the diagrams of acoustic propagation encapsulated in Tyndall's fog horn tests or the Royal Society's plotting of the auditory aftermath of Krakatoa. The 'ear witness' textual testimonies that were also embraced by Tyndall and the Royal Society plant other roots from which new practices can grow. Indeed, in advance of the relatively recent access to digital and network technologies, the historical incarnations of soundmaps most closely approximated these two examples emerging from that (Victorian) scientific culture. Signorelli draws justifiable additional attention to the presence of a variety of soundmaps within the 1970s research of the World Soundscape Project (WSP) such as their 'isobel map of Stanley Park in Vancouver [...]; the Sound Profile Map of the Holy Rosary Bells of Vancouver; the yearly graph of natural soundscape in British Columbia coastline; a pictorial representation that shows the sound sources converted in textual notation; the acoustic horizon of Bissingen in Germany, and Lesconil in France' (Signorelli 2017: 156). If Signorelli's illustrative examples correspond to Thulin's *maps-of-sound* – a category in which he incorporates the Campaign to Protect Rural England's 'Tranquil Area' maps from the 1960s – other projects by researchers associated with the WSP would be classified in Thulin's taxonomy as *sound-as-map*. The *sound-as-map* is constituted as a 'sonic cartography [...] based on the richness of spatial and locational information that can be attained through listening, and approaches visual representation as secondary or, in some cases, unnecessary to the mapping of sound spaces' (Thulin 2018: 196).

Frame

With Thulin, I want to open up the genealogy of the soundmap to usher in instantiations of *sound-as-map* and I want to welcome both methods which can be configured in relation to the lineage of noise maps, audiospectrographs, and textual testimonies and methods which find forms external to those pedigrees. Yet the definition deployed by Wikipedia and others effectively bars entry to cartographies which do not bend to 'digital geographical maps that put emphasis on the sonic representation of a specific location'. In Jacqueline Waldock's early analysis of the form, she identifies 'a new interactive, publicly engaging medium [...] a social media application [...] the interactive soundmap' (Waldock 2011), which is echoed by Milena Droumeva in her later articulation of the soundmap as 'publicly engaging digital artefacts' (Droumeva 2017: 337). This commitment of the soundmap to a specific technological apparatus and the association between soundmaps and web-based distribution is, with perhaps inevitable circularity, one cemented by subsequent online discourse. In 2010, Merle Patchett posted 'Mapping Sound and Sounding Maps' to the blog Experimental Geography in Practice, introducing a field in which only two of the eight illustrative examples of 'sonic maps' are delivered offline (one of these being the work of the WSP); in 2015, Cities and Memory – 'which, though resembling traditional online soundmapping portals [...] defines itself as a global artwork, a participatory, yet curated portal for real and imagined soundscapes' (Droumeva 2017: 345) – published its 'Top 10 Sound Maps', all of which are internet platforms, as are the fourteen 'other sound maps well worth investigating'.

Cities and Memory cast their selection in an economy of personal taste, that same inspiration that has energized enthusiasm over more than two decades for the sound-triggering symbols that populate what must count as the hundreds of online soundmaps. Such preferences for what inspires are compelling and it is to be conceded that the potential benefits of online soundmaps include: inserting the tactics of crowdsourcing within sound arts practices; exploring auditory locality's relations to the live and the recorded (see Soundcamp 2020; and Locus Sonus 2020); investigating the ways in which the cartographic might be harnessed to diagrammatize what is perceived as negative and as positive in environmental sound (see Hush City n.d.; and Metcalfe 2013); coordinating acts of sensory conservation that approach the motivations of salvage ethnography (Samuels et al. 2010: 338); and discovering a mechanism in the internet-based map which makes the production and distribution of self-initiated, thematically coherent site-orientated sound projects as accessible as it does their consumption.[2]

These benefits aside, it is this mechanism of accessibility itself that is one of dimensions of online soundmaps that troubles their easy adoption as a sonic methodology. Whether it is Google Maps or any of the rival geographic information systems that are used as the base layer on which the online soundmap is created, each complex blend of cartographic material is derived from remote sensing systems and aerial photography from within an institutional environment favouring 'free trade, an open market and privately funded research and development […]. It requires a well-funded military–industrial complex that develops defense technology' (Lee 2010: 910). Whatever creative adaptations are made to these base-layers, 'it is important to recognize that they do so within a production environment where their emancipatory potential is always constrained by institutional forces that govern the production, storage, and provision of geo-spatial data' (Jetahni and Leorke 2013: 488).

In parallel to this question of 'the master's tools', attention needs also to be devoted to the mechanisms of access since the availability and cost of the high-speed internet necessary to upload material to an online sound map is far from equal. Jason Farman's 2010 article, for example, distinguished digital signal transmission costs in Japan as 6 cents per 100 kilobytes per second – a price that amounted to 0.002 of the average monthly salary – from the cost for the same data transfer rates in Kenya at twice the average monthly salary. More recent research substantiates the concern that 'internet exclusion coincides with other forms of marginalization […]. In the case of Africa, global digital inequalities have reinforced existing racial as well as economic chasms, shutting out a huge proportion of the continent from access to the internet. Although some 14% of the world's population resides in Africa, only 3% of the world's internet users live on the continent' (Robinson et al. 2015: 574).

The specific technological apparatus through which the conventional soundmap is delivered is the first of the two problematic investments I referred to earlier: the differential economics of access, the broader milieu of militarized and capitalized spatiality, and the imbrication within a logic of data collection where 'location-based services are being recognized as participative ways to normalize surveillance: a process through which leakages of personal information are seen as "normal" or "natural" in everyday life' (Diogo 2018).

This investment undercuts the 'seemingly neutral medium of the sound map' (Droumeva 2017: 339), as does the second problematic investment, which relates to a particular perspectival frame. The base maps are never faithful analogues of geomorphology nor indexical renditions of the built environment: they are the multitude of 'little white lies' that Mark Monmonier identifies as the basis of every map. These distortions are what 'suppresses truth to help the user see what needs to be seen' (Monmonier 1996: 25); without such deliberate down-scaling of information, a map would be rendered useless. This point is implied in Schafer's thoughts on what he calls 'soundscape notation', thoughts which inadvertently recall Jorge Luis Borges's anti-cartographic fable *On Exactitude in Science*:

> While everyone has had some experience reading maps [...] few can read the sophisticated charts used by phoneticians, acousticians or musicians. To give a totally convincing image of a soundscape would involve extraordinary skill and patience: thousands of recordings would have to be made; tens of thousands of measurements would have to be taken; and a new means of description would have to be devised.
>
> (Schafer 1994: 99)

In an extension of this argument, Schafer asserts that 'the microphone gives the close-up but nothing corresponding to aerial photography' (Schafer 1994: 99), a position that is echoed by Voegelin: 'Sound suggests a geography from within the depth of the place, rather than projecting an aerial view' (Voegelin 2010: 144). And yet, the 'aerial' of Schafer and Voegelin is precisely the perspectival frame which is projected by the conventional online soundmap. The aerial perspective which suspends the soundmap user at an abstracted height particularly chafes as a distortion because it is insensitive to the condition often ascribed to sound as that which engulfs the listener – like the 'surround' relationship attributed earlier to Marinetti and to Marclay. Moreover, the top-down aerial view, reiterates the militarized projection that is a function of the online technological apparatus, since it belongs to 'the cartographic imagination inherited from the military and political spatialities of the modern state' (Weizman 2002). Finally, although the aerial representation depends upon the vertical for its elevation above the visualized ground, it paradoxically occludes the vertical axis itself (Carlyle 2000, 2014). The nodes on a conventional soundmap are positioned on a flat plane where altitude can have no place, a diagrammatization that is particularly problematic methodologically at a historical juncture when political geographers are seeking to invest the vertical as a significant territorial dimension, when 'such a perspective neglects the three-dimensional politics of the worlds above, below and around borders' (Graham 2016: 3).

In addition to its symbolic evacuation of the vertical, reiteration of the militarized projection and dislocation from the 'surround' of sound, the aerial dimension of the soundmap risks disembodying its users through a process that parallels the sensory hierarchization evoked by Michel De Certeau in his famous meditation on the view from the 110th floor of the World Trade Center: 'one's body is no longer clasped by the streets [...]. Nor possessed, whether as player or played, by the rumble of so many differences. The city's agitation is momentarily arrested by vision' (De Certeau 1984: 92). The 'rumbles' are not entirely silenced, of course, since their amplification is the very functional imperative of the soundmap, yet there is an objectivized distance. Whether the map user moves the mouse and activates a virtual button from a rendered height or observes from afar a flattened, static visualization, both might be processes not entirely divorced from Bruyninckx's 'sterility', hailing the user at Donna Haraway's 'vantage point of the cyclopean, self-satiated eye of the master subject' culpable of 'seeing everything from nowhere' (Haraway 1988: 586 and 581).

The spatial demarcations involved in conventional soundmaps reach further still, the aerial perspective obscuring those complexities of urban soundscapes that relate to what Matthew Gandy has called 'the spatial porosity of atmospheres and the uncertain distinctions between what constitutes "inside" and "outside"' (Gandy 2017: 356). It is not simply that the construction of soundmaps' interfaces tends to isolate individual sources, and hence ignore the overlapping complexities, since some platforms, such as Cusack's *Favourite Sounds*, do allow multiple nodes to play simultaneously in a simulation of porosity. Rather, what is at issue is the prioritization of externalities, another consequence of relying on graphic base layers engaged from above and a problem that Waldock's research has been important in addressing:

> Within the short history of soundmaps, there has developed a cycle of otherness that obtains its clarity in the absence of the domestic. Within the sound maps there appears to be a trend to capture the public rather than the private moments of life [...] there are only a handful of recordings within the home.
>
> (Waldock 2011)

Just as Waldock's analyses have drawn our attention to soundmaps' capacities to silence auditory activities occurring under the roofs and behind the walls rendered on-screen and online – and how that suppression is a gendered one – Isobel Anderson has explored how only if the boundaries of the 'online gridded soundmap platform' are traversed can we access 'the peripheries of lived experience' and reveal 'the invisible "in-between-space" of personal relationships to sound, but also the unseen spaces of urban architectures' (Anderson 2016).

Projections

The final stage of this chapter is inspired by Anderson's projects that 'map sound in unconventional and creative ways' (Anderson 2016) and by Droumeva's suggestion of an 'alternative grammar' (Droumeva 2017: 346). It is informed by the wider genealogy of soundmapping and it has been alerted to the jeopardies of a specific technological apparatus and of the particular perspective frame that is the aerial.

Textual/graphic soundmaps[3]

In the parameters developed by Jacob Smith in *Eco-Sonic Media*, soundmaps which involve the inscription of written or graphic information have the potential to eschew complicity in a 'material culture that has caused so much environmental damage' (Smith 2015: 4); subject to the sustainability credentials of the paper and inks, these might instantiate Smith's 'no-wattage sound technologies' (6, 168). Such cartophonies can be threaded back to the genealogies of the textual components in Tyndall and in the Royal Society report and of the 'eventilized' words on Marinetti's page; equally, they are stitched into the histories of ornithological transcription that propelled the development of the audiospectrographs analysed by Bruyninckx and ultimately form part of the wider, complex, and antediluvian fabric of sound notation, as Schafer's reference to 'soundscape notation' underscores.

Although, in some of its manifestations, paradigmatic of Wikipedia's 'digital geographical maps that put emphasis on the sonic representation of a specific location', Cusack's long-running *Favourite Sounds* vehicle has also evolved in recent years to generate word clouds, in which solicited public preferences of acoustic place are rendered according to their statistical prominence in the sampled population. These word clouds – part of what Nicola Di Croce calls a 'sensitive attempt to represent personal feelings and build through them a collection of sensations which reflect everyday practices' (Di Croce 2016) – provide a contemporary example of the alternative approach to soundmapping which emphasizes the textual and the graphic. Introduced in the context of a community-engaging or pedagogic arts practice, this approach might avoid some of the intimidations of digital creativity and can construct a bridge for non-specialists to travel into the world of sound representation. A perspectival frame remains active, with elements of the aerial tending to persist in the more spontaneous initial efforts, but through guided iterations, the conventions of the top-down can be as challenged as any reflex elimination of porosity or 'in-between-spaces'.[4]

Cusack's word clouds correlate with the visual grammars we have come to associate with map-making; however, it may be that textual creativity can be entirely unhinged from the graphic, even from the 'eventilized', yet still retain a purchase on Thulin's cartophonic. Candidates can be identified within recognizable sound arts practices – Steve Peter's *Here-ings: A Sonic Geohistory* (2012) comes to mind, drawing as it does from a calendar year of listening devoted to a site in New Mexico and delivering sensed experience in spare, diaristic prose; so too, Voegelin's *Sound Words*, a mobile microsite that shuttles between personally authored and curated collaborations and has been hosted by various international festivals. Forms of locational writing that express an attentiveness to the sonorous yet fall outside genre delineations, such as those resonant passages of Nan Shepherd's *The Living Mountain*[5] that to my ear fashion for the Cairngorms a porous, ground-truthed soundmap that no screen with uploaded nodes could hope to match for nuance, might equally well qualify.

Desterilized soundmaps

For Bruyninckx, the audiospectrograph sought to flatten and control sound, it 'dealt with the problem of noise, by erasing it visually' (Bruyninckx 2012: 145). In common with the choroplethic noise map – the audiospectrograph is as attached to a specific, and loaded, technological apparatus as it is to a particular perspectival frame. There have been instances, however, of efforts to disentangle the spectrogram and the noise map from their silencing and sterile 'scientific culture', and to exploit their respective latent cartophonic possibilities (as opposed to their already materialized cartographic ones).

Bernie Krause's first encounter with spectrograms in the early 1980s is narrated in language dislocated from orthodox scientific culture – he evokes Turner's late seascapes and dares 'to think of the spectrograms as contemporary graphic musical scores' (Krause 2013: 87). Subsequent engagements with spectrograms enabled him to map the waxing and waning of biophonic and anthrophonic presences in particular places, both terrestrial and marine, that have been returned to over successive decades, and to represent these dynamics in the characteristic diagrammatic form. The spectrograms compliment the recordings from which they derive, sometimes coexisting spatially, as in the 2016 installation in the basement of the Cartier Foundation, Paris, which the exhibition designers explained within a representational schema of mapping, repeating the rhetoric of the 'surround' we have heard before: 'A cohesive, immersive experience that three-dimensionalises Krause's recordings and suggests scenes from the natural world […]. The spectrograms form an abstract landscape, an interpretation of the various global locations and times of day that Krause made the original recordings in a way that envelops the audience and encourages them to linger in the space' (United Visual Artists 2018).

The Spanish artist Edu Commelles decoupled the spectrogram further from its host scientific culture in *Spectre/A Secret Music* (2018) and, particularly, in *Espectrograma: Mislata* (2016) where a month of 'sound mapping the entire city' was concretized in two curved murals, each 2.6 metres high, one stretching out beneath low-rise tower blocks for 25 metres, the other spanning a neighbouring 20 metres. Although there is an audio dimension to the *Espectrograma* project, it is the soundless spectrogram structure, ironically given Bruyninckx's critique of silencing, which reverberates, since this, in the artist's own interpretation, 'aims to trigger imagination of the viewer to wonder which graphic correspond to each sound and to imagine those sounds […] sometimes, the imagined sound is the most powerful and compelling' (Commelles 2016).

Spectrograms or noise choropleths that are 'détourned' as singular soundmaps to address Droumeva's 'alternative grammar', are not entirely released from scientific culture (just as the grammar remains a grammar, however alternative). Rather, there are contiguities with what Eyal Weizman has to say of the forensic: '[We] use the term "forensics", but we seek, in fact, to reverse the forensic gaze and to investigate the same state agencies […] that usually monopolise it' (Weizman 2017: 9). A work by Lawrence Abu Hamdan, a colleague of Weizman, exemplifies the reversals that are possible here, reversals that can have the character of a counter-mapping, an expression that Nancy Lee Peluso introduced and which is highly instructive in the scope of the conventions of perspectival frame and specific apparatus: 'Counter-mapping can be used for alternative boundary-making […] for expressing social relationships in space rather than depicting abstract space itself' (Peluso 1995: 387).

Earshot (2016) is a multidimensional installation, its impetus derived from acoustic analysis previously commissioned from Hamdan by a charity who had sought to establish whether the Israel Defense Force had discharged rubber bullets or live rounds in an incident which left two unarmed teenagers shot dead in the occupied West Bank. An element of the research involved creating spectrograms of gunshots recorded on the day the youths died and these featured as evidence in newspapers and at a US Congress hearing; the same visualizations later deployed in gallery spaces in different configurations and draw, precisely, a soundmap of the lethal chaos of social relationships in space rather than a more abstracted, sterile cartography.

Compositional soundmaps

Few material constraints impede the construction of the low-wattage textual and graphic soundmaps, and this is one of their advantages in a workshop setting; the many resources which Steph Ceraso identifies in her inter-chapter in *Sounding Composition* facilitate the more environmentally impactful form of the online soundmap, and she has shown that they can be developed as part of curriculum projects. Although soundmaps, both online and offline, permit this relatively accessible assembly by use of relatively available software, some practitioners have distinguished themselves through the duration of their commitment: Cusack's twenty years of his *Favourite Sounds* project, Krause's many decades devoted to the spectrogram. Annea Lockwood's *Sound Map of the Hudson River* (1982) took a year and demanded recordings from fifteen locations along a 563-kilometre course; the fieldwork for her *A Sound Map of the Danube* (2008) generated some 80 hours of recordings.

Although Lockwood distinguishes her approach by insisting 'my intention is different from compositional work' (Lane 2013: 31), I see her work as emblematic of a kind of soundmapping that is usefully defined as compositional. Like Lockwood's work, projects such as Fernando Godoy's *Atacama: 22° 54 '24 'S, 68° 12' 25' W"* (2017) and Cathy Lane's *The Hebrides Suite* (2015) depend on sustained investments of time; deliver as multimodal combinations of images, texts, and sound; and foreground an adjudicatory, authorial listening that is active at the site, in the edit suite, and later governs the gallery installation or other form of dissemination. As Lockwood has it, the initial 'site has to be really satisfying to listen to and make my ears prick up' and subsequent choices of recordings amount to 'selecting sites that are really engaging and vivid to me – really alive' (Lane 2013: 33–34). This is not to say that Lockwood, Lane, or Godoy resist inscribing the testimonies of others within their sonic mapping. We hear vocalized witnessing in Lockwood and Lane. Witnessing contributes to Godoy's sound world too, not through audible speech itself but through the invocation of 'Atacama [as] a space of evocation, of memory, of overwhelming loss experienced by the mothers and women who have spent years searching for the remains of their loved ones [...] buried there during the extermination carried out by the Chilean dictatorship' (Pisano 2018). It functions as another of Peluso's counter-mappings, triggering 'a critical process that questions epistemological maps of knowledge by offering a possible renegotiation of the meanings of language itself' (Pisano 2018).

The compositional maps by Lockwood, Lane, and Godoy each offer their own answers to James Clifford's question 'but what of the ethnographic ear?' (Clifford 1986: 12). Rather than the aerial, their exploratory altitude drops down to ground level and sometimes lower still, as in the below-surface hydrophone recordings of Lockwood and the contact microphone recordings of Godoy and his Austrian collaborator Peter Kutin. These projects – that fall within the sound-as-map in Thulin's taxonomy – can have recourse to actual maps, such as those which appear in some of Lockwood's *Sound Map* installations and in Godoy's abstracted diagram that allows the audience to plug headphones into specific nodes on the gallery wall labelled with longitude and latitude in his contribution to the 2017 *Otros Sonidos, Otros Paisajes* exhibition at MACRO, Rome, curated by Pisano and Antonio Arévalo. However, in keeping with what Thulin says of this category, they approach 'visual representation as secondary or, in some cases, unnecessary to the mapping of sound spaces' (Thulin 2018: 196).

The compositional sound-as-map artists I have chosen to exemplify this third category of alternative approaches all use field recording, editing, and various dissemination formats and, taken together, these institute their own technological apparatus. This may attract different issues from those I associated earlier with the online soundmaps' own apparatus, yet, to measure their methodological robustness, a similar critical auditing is indispensable in parallel dimensions of economics of access (and Smith's 'wattage'), implied spatiality and questions of data collection (such as the privacy and property rights of those inhabitants of place who we hear vocalized).

Other projects, though still accountable as an individual artist's responsibility, demonstrate a relaxation of compositional control, and accommodate collaborative methods that endow participants with technical skills and equipment to enable a certain autonomy to map their own localities, perhaps engaging more directly with issues of access and data collection in the technological apparatus. Waldock's work in Liverpool's Welsh Streets repositions the researcher so that the domestic spaces are foregrounded and, for their inhabitants, 'instead of listening in on them, this methodology makes it much more possible to listen to and with them' (Waldock 2016: 67). A similar recalibration is discernible in Hong Kai Wang's *Music While We Work* (2011), where retired sugar factory labourers become recordists, soundtracking the multiscreen and multichannel installation, 'allowing them to work on their own, identifying and recording the sounds of their former work environment – their aural universe. Wang believes that whoever holds the microphone and what he or she records, delineates, from various viewpoints, the right to speak, the right to interpret and the power relationships between sound-maker and recipient' (Chang 2011).

Compass

One of the significant advantages of the online soundmap relates to its capacity to deliver a collective listening for its audience from the collective, recorded, and uploaded listenings of its curated contributors. Cities and Memory's recent project *Sounding Nature* incorporated the work of 250 artists who supplied some 500 recordings from 55 countries; Udo Noll's *Radio Aporee* has evolved into a platform for 1,665 contributors and to hear the totality of its collective cartography would demand 115 days and 23 hours. The peculiarities of how Cities and Memory or *Radio Aporee* or any of the other online collective soundmaps adapt the geographical information system's base layer to invite the rich, crowdsourced, material constitute a technological apparatus and a perspective frame. These stipulate an invoked territory, in terms of the militarized, capitalized, surveilled infrastructure on which they depend, an infrastructure that has not surmounted the issue of digital inequality nor escapes Smith's critique of environmental damage, and stipulate an impelled listening position, in terms of the aerial that disembodies as it plugs porosity and stresses externalities. Some online soundmaps have invested energies in engineering a distinctive interface, though there is often standardization within a project in terms of how acoustic content is presented, despite wide disparities in that content and between many separate projects there is a palpable presentational homogeneity, partly because their creators gravitate to similar software.

Less homogenous is the genealogy of the soundmap, a family history in which scientific culture's noise maps and audiospectrograms form one branch, textual ear witnessing another, diverse diagrammatic innovations a third, and the various alternative approaches a fourth. The three alternative approaches I provided could each have been deepened to draw in more exemplars, just as they could have been broadened to incorporate other cartophonic categories: the transmission works of Dawn Scarfe or Jiyeon Kim suggest the possibility of a live soundmap, the reverberation of interior or external spaces in projects by artists as different as Viv Corringham and Davide Tidoni imply a performative soundmap, and a potential classification of storied soundmaps arises out of the separate creative research endeavours of Isobel Anderson and Ultra-Red.

It is not that the alternative approaches have somehow evaded technological apparatuses and perspectival frames: they are still soundmaps after all, each with an invoked territory and an impelled listening position. Rather, the alternative approaches agitate the apparatuses and frames to critical motion, hazard counter-mappings, lower themselves from the aerial, admit the porous, and slip from the cartographic into the cartophonic.

Notes

1. Not fully cartophonic but nonetheless intriguing is the Chatty Map collaboration between Yahoo Labs, Bell Labs, and the universities of Turin and Sheffield where choroplethic maps are generated from tags on social media data to characterize acoustic perceptions organized across axes of chaos, calm, monotony, and vibrancy and to characterize sonic diversity (see Aiello et al. 2016).
2. Perhaps the emblematic soundmap project, in its breadth and depth, its balance of complexity and coherence, is Ian Rawes's London Sound Survey (2008–2020).
3. I am borrowing this forward slash from the work of Alison Barnes: 'The forward slash [in geo/graphic] is used to reconfigure the context of the word representation in discussion of creative outputs that endeavor to go beyond a one-to-one "mapping" of place. The use of re/presentation in relation to both the research and practice of this type emphasizes both "re" and "presentation" and again creates a productive interplay that enables one to move beyond the idea of the mimetic with regard to an image of place' (Barnes 2018: 4).

4. An insightful and inspiring account of the adaptation of soundmaps within classroom settings – and the compromises which emerge – can be found in Ceraso (2018).
5. The passages in chapter 4 of Shepherd's book are particularly resonant (Shepherd [1977] 2011: 22–29).

References

Aiello, Luca Maria, Rossano Schifanella, Daniele Quercia, and Francesco Aletta (2016). 'Chatty Maps: Constructing Sound Maps of Urban Areas from Social Media Data,' *Royal Society Open Science* 3. http://doi.org/10.1098/rsos.150690.

Aletta, Francesco and Jian Kang (2015). 'Soundscape Approach Integrating Noise Mapping Techniques: A Case Study in Brighton, UK'. *Noise Mapping* 2 (1): 1–12.

Anderson, Isobel (2016). 'Soundmapping Beyond The Grid: Alternative Cartographies of Sound'. *Journal of Sonic Studies* 11. Available online: https://www.researchcatalogue.net/view/234645/234646 (accessed 4 December 2018).

Barnes, Alison (2018). *Creative Representations of Place.* London: Routledge.

Biagioli, Monica (2018). 'Modeling the Organic: Cultural Value of Independent Artistic Production'. In Nancy Duxbury, William Garrett-Petts, and Alys Longley (eds), *Artistic Approaches to Cultural Mapping: Activating Imaginaries and Means*, 92–109. London: Routledge.

Bruyninckx, Joeri (2012). 'Sound Sterile: Making Scientific Field Recordings in Ornithology'. In Trevor Pinch and Karin Bijsterfveld (eds), *The Oxford Handbook of Sound Studies*, 127–150. Oxford: Oxford University Press.

Carlyle, Angus (2000). 'Beneath Ground'. In Nick Barley and Ally Ireson (eds), *City Levels*, 96–120. London: Birkhauser.

Carlyle, Angus (2014). 'The God's Eye and The Buffalo's Breath: Seeing and Hearing Web-Based Sound Maps'. In *Invisible Places, Sounding Cities, proceedings of Sound, Urbanism and Sense of Place*, Viseu, Portugal, 2014. Available online: http://invisibleplaces.org/IP2014.pdf (accessed 3 December 2018).

Carpenter, J. R. (2017). 'In the Event of a Variable Text'. *Convergence: The International Journal of Research into New Media Technologies* 23 (1): 89–114.

Ceraso, Steph (2018). *Sounding Composition: Multimodal Pedagogies for Embodied Listening.* Pittsburgh, PA: University of Pittsburgh Press.

Chang, Amy (2011). 'Music While We Work'. Hong-Kai Wang. Available online: http://www.w-h-k.net/mwww.html (accessed 4 December 2018).

Clifford, James (1986). 'Introduction: Partial Truths'. In J. Clifford (ed.), *Writing Culture: The Poetics and Politics of Ethnography*, 1–27. Berkeley, CA: University of California Press.

Commelles, Edu (2016). 'Artist's Statement'. Available online: http://www.educomelles.com/2016/11/espectrograma-mislata-2016.html (accessed 4 November 2018).

Council of the European Union (2002). Directive 2002/49/EC of the European Parliament and of the Council.

De Certeau, Michel (1984). 'Walking in the City'. In *The Practice of Everyday Life*, 91–110. Berkeley, CA: University of California Press.

Di Croce, Nicola (2016). 'Audible Everyday'. *Interference: A Journal of Audio Cultures* 5. Available online: http://www.interferencejournal.org/audible-everyday-practices-as-listening-education/ (accessed 30 November 2018).

Diogo, Gustavo Velho (2016). 'Google Earth, Surveillance, and the Power of Digital Cartography'. Institute of Network Cultures, 7 October. Available online: http://networkcultures.org/longform/2016/10/07/google-earth-surveillance-and-the-power-of-digital-cartography/ (accessed 4 December 2018).

Droumeva, Milena (2017). 'Soundmapping as Critical Cartography: Engaging Publics in Listening to the Environment'. *Communication and the Public* 2: 4.

Farman, Jason (2010). 'Mapping the Digital Empire: Google Earth and the process of Postmodern Cartography'. *New Media and Society* 12 (6): 869–888.

Finkelaug, Jori (2006). 'Marinetti: Oh, What a Futurist War'. *New York Times*, 27 August, Arts: 11.

Furneaux, Rupert (1964). *Krakatoa*. Englewood Cliffs, NJ: Prentice-Hall.

Gandy, Matthew (2017). 'Urban Atmospheres'. *Cultural Geographies* 24 (3): 356.

Graham, Stephen (2016). *Vertical: The Cities from Satellites to Bunkers*. London: Verso.

Haraway, Donna (1988). 'The Science Question in Feminism and the Privilege of Partial Perspective'. *Feminist Studies* 14 (3): 575–599.

Hush City (n.d.). 'Hush City App: Welcome to Hush City!'. Available online: http://www.opensourcesoundscapes.org/hush-city/ (accessed 4 December 2018).

Jethani, Suneel and Dale Leorke (2013). 'Ideology, Obsolescence and Preservation in Digital Mapping and Locative Art'. *International Communication Gazette* 75 (5–6): 484–501.

Krause, Bernie (2013). *The Great Animal Orchestra: Finding the Origins of Music in the World's Wild Places*. London: Profile Books.

Lane, Cathy (2013). 'Interview with Annea Lockwood'. In Cathy Lane and Angus Carlyle (eds), *In The Field: The Art of Field Recording*, 27–37. Axminster: Uniformbooks.

Lee, Micky (2010). 'A Political Economic Critique of Google Maps and Google Earth'. *Information, Communication and Society* 13 (6): 909–928.

Locus Sonus (2020). 'Locustream Soundmap | Live Worldwide Open Microphones | 2006–2020'. Available online: http://locusonus.org/soundmap/051/ (accessed 7 July 2020).

London Sound Survey (2008–2020). 'London Life in Sound'. Available online: https://www.soundsurvey.org.uk (accessed 7 July 2020).

Metcalfe, John (2013). 'Yo! I'm Trying to Sleep Here! New York's Wonderful Map of Noise'. Bloomberg City Lab, 15 April. Available online: https://www.citylab.com/life/2013/04/yo-im-trying-sleep-here-new-yorks-wonderful-map-noise/5279/ (accessed 4 December 2018).

Monmonier, Mark (1996). *How To Lie With Maps*. Chicago: University of Chicago Press.

Peluso, Nancy Lee (1995). 'Whose Woods Are These? Counter-Mapping Forest Territories in Kalimantan, Indonesia'. *Antipode* 27 (4): 383–406.

Pisano, Leandro (2018). 'Southscapes/Soundscapes: Other Spaces and Territories of Sound'. *Columbian Artistic Research Magazine*. Available online: https://carmajournal.org/artculo-8 (accessed 4 December 2018).

Robinson, Laura, Shelia R. Cotten, Hiroshi Ono, Anabel Quan-Haase, Gustavo Mesch, Wenhong Chen, Jeremy Schulz, Timothy M. Hale, and Michael J. Stern (2015). 'Digital Inequalities and Why They Matter'. *Information, Communication & Society* 18 (5): 569–582.

Royal Society of Great Britain (1888). *The Eruption of Krakatoa, and Subsequent Phenomena: Report of the Krakatoa Committee of the Royal Society*, ed. G. J. Symons. London: Trubner & Co.

Samuels, David W., Louise Meintjes, Ana Maria Ochoa, and Thomas Porcello (2010). 'Soundscapes: Towards A Sounded Anthropology'. *Annual Review of Anthropology* 39: 329–345.

Schafer, R. Murray (1994). *The Soundscape: Our Sonic Environment and the Tuning of the World*. Rochester, VT: Destiny Books.

Shepherd, Nan (2011). *The Living Mountain*. Edinburgh: Canongate.

Signorelli, Valerio (2017). 'Listen Through The Map'. In Barbara E. A. Piga and Rosella Salerno (eds), *Urban Design and Representation: A Multidisciplinary and Multisensory Approach*, 153–164. Cham: Springer.

Smith, Jacob (2015). *Eco-Sonic Media*. Oakland, CA: University of California Press.

Soundcamp (2020). '2 to 3 May 2020'. Available online: http://www.soundtent.org (accessed 4 December 2018).

Thulin, Samuel (2018). 'Sound Maps Matter: Expanding Cartophony'. *Social & Cultural Geography* 19 (2): 192–210.

Tyndall, John (1874). 'Report by Professor Tyndall to Trinity House, upon Experiments with Regard to Fog Signals; Letter to the Board of Trade, with Reference to Fog Signal at Cape Race'. UK Parliamentary Papers, 21 May. Available online: https://parlipapers.proquest.com/parlipapers/result/pqpdocumentview?accountid=14182&groupid=96842&pgId=87d24808-d8bc-4b8e-b093-e5384b5483a6&rsId=16828922E80 (accessed 1 January 2019).

United Visual Artists (n.d.). 'Great Animal Orchestra — Cartier Foundation'. Available online: https://uva.co.uk/works/great-animal-orchestra (accessed 4 November 2018).

Voegelin, Salomé (2010). *Listening to Noise and Silence*. London: Bloomsbury.

Voegelin, Salomé (2014). *Sonic Possible Worlds: Hearing The Continuum of Sound*. London: Bloomsbury.

Waldock, Jacqueline (2011). 'Soundmapping: Critiques and Reflections on this New Publicly Engaging Medium'. *Journal of Sonic Studies* 1. Available online: https://www.researchcatalogue.net/view/214583/214584 (accessed 3 July 2018).

Waldock, Jacqueline (2016). 'Crossing the Boundaries: Sonic Composition and the Anthropological Gaze'. *Senses and Society* 11 (1): 60–67.

Weizman, Eyal (2002). 'The Politics of Verticality'. openDemocracy, 23 April. Available online: https://www.opendemocracy.net/ecology-politicsverticality/article_801.jsp (accessed 4 December 2018).

Weizman, Eyal (2017). *Forensic Architecture: Violence at the Threshold of Detectability*. New York: Zone Books.

White Cube (2015). 'Christian Marclay'. Available online: http://whitecube.com/exhibitions/exhibition/christian_marclay_bermondsey_2015 (accessed 20 September 2018).

Wikipedia (2020). 'Sound Map'. Available online: https://en.wikipedia.org/wiki/Sound_map (accessed 18 July 2020).

38
Listening as Methodological Tool: Sounding Soundwalking Methods
John L. Drever

Soundwalking as an emergent practice

Amongst the interplay of competing commands and demands for our attention in daily life, multitasking attentive listening to the here-and-now with the bipedal locomotion mode of ambulation – along with an inordinate amount of other incessantly shuffling and intermingling of tasks – is considered by many as routine. Relentlessly endeavouring to attend to the sounds around you, whilst dwelling in and passing through everyday environments for an extended duration of time, by actively curtailing other customary cognitive tasks or behaviours, on the other hand, is an atypical activity. Prefiguring the developments of sensory ethnography (Pink 2015) and the 'sonic turn' (Drobnick 2004: 10), such a pursuit, under the overarching term, soundwalking, has been employed over the past forty years as a designated and dependable, even vital sonic method.

Approaching soundwalking as an emergent rather than a transplantable fixed practice with an ossified methodology, this chapter will feed off historical precedence and draw from the author's direct experience as a soundwalk facilitator in multiple situations, catering for participants with disciplinarily specialisms including acoustic engineering, architecture, ornithology, city planning, accessibility, social science, and arts practice, and extending out to school children and the general public at large – all stakeholders and individuals with diverse general and specific needs, concerns, and understandings. Attentive concentration on listening is an engrossing experience where one can become absorbed in the flow[1] of the enveloping soundscape. As it is beholden on the soundwalk leader to guide and to plan ahead to the safe and sound completion of the walk, whilst poised to attend to any pressing pragmatic issues that may transpire midst-walk, the actual emphasis on their listening tends not to be prioritized. But this in turn permits the participants to dedicate their entire attention to the task in hand. So, reversing roles, the author will also reflect on his various soundwalking experiences as participant – experience which encompasses dogmatic and more idiosyncratic approaches, in formal and performative, intimate and extrovert configurations. The chapter will critically reflect and evaluate on this multitudinous data set that endeavours to incorporate and verbalize sensuous experience and behaviour, whilst surfacing the practical, logistical, and ethical vagaries. It will unashamedly concentrate on soundwalks that do not incorporate audio playback via headphones or aspects of telepresent or augmented reality (beyond participants' regular use of audio prosthetics) such as audio walks by, for example, Janet Cardiff, Christina Kubisch, and Duncan Speakman; it is contended that soundwalking with the 'naked ear' is an already highly sophisticated and infinitely practicable and malleable methodology suitable for multiple research, training, and artistic needs.

Evolution

The dominant traits of soundwalking appear to coalesce in the 1960s around the Fluxus movement (where foregrounding, framing, and enacting forms of gait were a recurring theme) and the experimental music scene in part influenced by but departing from exemplars posed by John Cage, typified in *4'33"* (1952): in particular, the open air activities of Philip Corner, Max Neuhaus, and Ben Patterson, who in their own ways radically inverted concert hall conventions and aesthetics with the world outside. This attitude is most clearly exemplified by Neuhaus's rubber stamping the imperative 'LISTEN' (1966) on to the hands of a small group of participants, and leading them down West 14th Street, Manhattan, and in subsequent trips to out-of-the-way sites such as power stations.[2]

It was with R. Murray Schafer and the prodigious exploits of the handful of Vancouver-based researchers that constituted the aspiringly named World Soundscape Project (WSP) in the 1970s, that the soundwalk is pinned down and codified as a method: this is most clearly expressed and promulgated in a special issue of *Aural History* focused on 'Sound Heritage' (1974), in Schafer's instructive paper 'Listening' (1974) and from a more personal and motivational perspective, Hildegard Westerkamp's (an enduring practitioner and passionate advocate of soundwalking) paper 'Soundwalking' (2007). Echoing the pervasive uptake of walking in its many manifestations as core practice across-the-board (see Evans 2012; Smith 2014; Qualmann and Hind 2015), in the past decade soundwalking activities have mushroomed. In 2013 it was adopted in English primary schools as a recommended activity for Key Stage 1 (i.e. pupils aged five to seven) of the National Curriculum in England (Department of Education 2013), and in August 2018 it was enshrined as a scientific method for acoustic engineering in Part 2 of the ISO series on *Acoustics – Soundscapes* that is concerned with *Data collection and reporting requirements* (ISO/TS 12913–2: 2018).

Soundwalking

The conjoining of 'sound' and 'walk' to produce the compound noun, 'soundwalk', presents an immediately graspable and yet imaginative concept – I have tended to opt for the continuous tense form, 'soundwalking', indicating that it is an action that is in progress associated to time, space, and place, albeit on occasion vicarious or virtual. In the opening line of 'Soundwalking', Westerkamp articulates the soundwalk quite simply as 'any excursion whose main purpose is listening to the environment' (Westerkamp 2007: 49). For Westerkamp, and for the interdiscipline of acoustic ecology in general, this is no passive pursuit however, the practice demands practise, and in turn redoubles 'attentive listening' towards 'aural awareness on a wider scale' (52). From a preliminary survey of the rhetoric surrounding soundwalking you can find 'attentive' treated synonymously for other affirmative adjectives, each bringing its own inflection, on describing the kind of listening soundwalking may engender: 'critical', 'engaged', 'active', 'relational', 'meaningful', 'interactive', 'connective', 'deep', 'sensitive', 'purposeful'. What characterizes the soundwalk as a sonic method, however, is its alignment with the meta-concept of soundscape, again both a concept nurtured by the WSP and recently stamped by ISO, defined as an 'an acoustic environment as perceived or experienced and/or understood by a person or people, in context' (ISO 12913–1: 2014). Thus, the raison d'être of the soundwalk is in the interrelationship and intra-relationship (Barad 2007) between participant(s) and the prevailing acoustic environment that they encounter and experience. But, as we will examine, what actually constitutes a soundwalk and the motivation for soundwalking is a moot point.

Five Village Soundscape

The WSP made extensive use of soundwalking methodology in their *Five Village Soundscape* project that ran between February and June 1975, where they 'undertook to study the soundscape of northern Europe' (Schafer 1977a: 1). Fully aware of resource and time limits, they strategically decided to focus on a comparative soundscape study of five European villages, allowing a week to ten days of concentrated study in each location. On arriving in a new village, recuperating from their long journey in a rented Volkswagen bus, they would expeditiously get to work, the first activity being a walk: to provide them with 'an immediate initial sensory experience [...] which each village evoked' (11). This outsider's ear, even naïve listening is akin to Elias Canetti's resistance to prior knowledge espoused in his travelogue, *The Voices of Marrakesh*: 'I wanted sounds to affect me as much as lay in their power, unmitigated by deficient and artificial knowledge on my part' (Canetti 2003: 23). It could also be considered an enactment of an auditory take on the consumption of place parallel to John Urry's notion of the tourist's gaze: 'Places are chosen to be gazed upon because there is an anticipation, especially through day-dreaming and fantasy, of intense pleasure, either on a different scale or involving different senses from those customarily encountered' (Urry 1995: 132).

Moving on from their initial 'touristic' impressions, they analysed the 'acoustic rhythms and densities' (Schafer 1977a: 21) in a more systematic, quasi-statistical, and consistent fashion: along with traffic counts, 24-hour-long sound recordings and sounds preferences tests, they used their own hearing as a diagnostic tool. They were tasked with creating 'sound catalogues of all acoustic events heard by listeners in all areas of the village during half hour periods at five times between 7.00 am and 7.00 pm. To compile this the village was divided into sections and project field workers moved continuously through the streets listing to every sound heard' (21).

Deliberately focusing on man-made sounds, the team assigned what they heard into prearranged categories such as motor traffic, human traffic (e.g. footsteps, bikes), voices, indoor or outdoor human activity, domestic animals, and electro-acoustic (Schafer 1977a: 27–28). In the year 2000, the villages (with the addition of Nauvo in Finland) and the research methodologies were revisited in the Acoustic Environments in Change project, led by Helmi Järviluoma. Twenty-five years on the researchers found this specific task limiting, 'distracted from concentrating on the environment itself' (Vikman 2009: 63). Departing from a mechanistic process they found themselves inclined to acknowledge their auditory perception in situ: 'we distinguished between distances and directions of the sound sources, or the order in which a cluster of sounds were heard, so that chains of perceptions of each listener walker could be constructed later' (Järviluoma et al. 2009: 63).

Soundwalk/listening walk

Where Westerkamp regards soundwalking as an all-encompassing term that may include a wide variety of approaches which foreground listening, Schafer calls for a differentiation between a listening walk and a soundwalk, where 'a listening walk is simply a walk with a concentration on listening' (Schafer 1974: 17). The soundwalk, on the other hand, may be an elaborately devised affair, where specific modes of listening to the environment may be prompted by maps or scores and/or a greater level of performativity through sonic interventions or choreography by the participant or interlocutors, such as engineering 'a dialogue with a slat fence by dragging a stick across it' (17) – the kind of nascent sonic playfulness and openness displayed by children on entering a highly reverberant space.

I participated in such an active approach at the inaugural symposium for the International Ambiance Network hosted by CRESSON (Centre for Research on Sound Space and the Urban Environment) in Grenoble in 2009. Merging their expertise in dance, choreography, ethnology, and architecture, the Collectif Rendez-Vous led simultaneous soundwalks through the streets to prompt the delegates to identify, through in situ active listening and performative interventions, sonic effects. The 'sonic effect' is a pragmatic listening tool developed in CRESSON, presented as a repertoire of effects, geared towards apprehending the soundscape of the build environment 'that allows us to integrate the domains of perception and action, observation and conception, and analysis and creation' (Augoyard and Torgue 2005: 11). Echoing Situationist tropes, questioning the perception of the human scale in the design of the city and the way it influences our habitual deportment in an embodied manner, with a frisson of social disruption, the delegates were prompted to play spatial games and explore rhythmic variation of their steps, and unconventional deportment, including the creation of collective 'sculptures de corps' (Dugave and Regnault 2009).

The *European Sound Diary*

Not restricted to the villages, soundwalking in the Schaferian sense, was practised throughout the WSP's European tour. As they stopped off in cities to undertake preparatory research on the villages, they creatively adapted methods of soundwalking to the contexts they found themselves in. These activities are assiduously documented in the *European Sound Diary* (1977). As well as individual members' accounts of what they heard, the publication also includes detailed instructions and sound maps on carrying out place-specific soundwalks as 'useful educational experiences for everyone' (Schafer 1977b: 1).

The Paris Soundwalk acts as a stimulus to imagining the soundscapes represented or alluded to in selected paintings of the Louvre: 'Study the images, and let the genius of their execution speed your imagination to provide the appropriate soundtrack' (Schafer 1977b: 86). It also keeps the participant connected to the physical surroundings, drawing attention to the actual aural architecture of the gallery: 'Note marble stairway floorsounds on way up to 3rd floor – especially the clicking and ensuring reverberation' (91).

The Vienne Soundwalk: Evening in the Old Town, invites the participant to intervene in the soundscape; for example, on Backerstrasse and Dr. Innaz Seipel-Platz, the walker is asked to 'go to the telephone booth. Stomp on the wooden floor [...] whistle yourself through the arch' (Schafer 1977b: 84).

The London Soundwalk, which leads from Euston Square to Queen Mary's Gardens in Regents Park, introduces the notion of thresholds of comfort and discomfort:

- 'THRESHOLD OF COMFORT: find the transition point where the roadway sound gives way to the sounds of the park' (Schafer 1977b: 93).
- 'THRESHOLD OF DISOMFORT: the transition point where the sounds of the park are once more buried by the sound of city traffic' (94).

Whilst conscientious listening is encouraged throughout, soundwalking does not necessarily demand continuous ambulation. Once in the gardens the soundwalker is invited to: 'Sit on the bench nearby until someone crosses between you and the fountain. How do they affect the sound?' (93). The exercise goes on to highlight a highly subjective contextual factor for the soundwalker: 'Note the difference between the two threshold locations. Depending on how much the park has cleaned your ears, the second threshold will be farther from the outer streets' (94). An audiologist would refer to this kind of aural respite as recovering from auditory fatigue or temporary threshold shift (TTS), however the wording chimes with one of Schafer's central concepts, *Ear Cleaning* (Schafer [1967] 1976: 49–92), originally designed as a series of experimental workshops for music students to metaphorically open their ears: 'To induce students to notice sounds they have never listened to before, [...] the sounds of their own environment and the sounds they themselves inject into their environment' (49). Schafer, later expanding this concept from music education to the acoustic designer, regarded soundwalking as a principle exercise of ear cleaning, 'at the root of the acoustic design program' (Schafer 1994: 213). He also promoted ear cleaning for the whole society, starting with schools, which, as already noted, has now been picked up in England's National Curriculum.

Soundwalking methodology guide

The following is a fleshing out of the methodology of the much-trodden rudimental, orthopraxic soundwalk as prompted in Schafer's *No. 13 Listening Walk* of his *100 Exercises in Listening and Sound-Making* (Schafer 1992: 31) and *The Soundscape: Our Sonic Environment and the Tuning of the World* (Schafer 1994: 212–213). It provides a useful blueprint which one may elaborate from, deviate from, ignore, or work against. It is not quite, as suggested, 'simply a walk with a concentration on listening' (212) as to allow such 'concentration' requires the observance of series of strictures and structures.

1 Route

A route is prepared in advance, considering the specific needs and mobility of the participants. It is important not to be too prescriptive, allowing for some variation on the day; this requires research and ideally a recce of the potential routes. The scheduling of the walk is of course crucial, considering the rhythms of the day, week, season, tides, etc. You may aim to be in a specific location at a specific time to hear prominent soundmarks such as a church clock ringing out its Westminster Chimes on the hour. I often seek out aspects of urban soundscapes that have been consciously designed from a sonic perspective, such as water features and contrasting acoustic architectures. You may attempt to circumnavigate specific continuous or intermittent sounds radiating from fixed points, exploring the change in spectrum and directionality as heard from different sounding locations. Such activity should not be exclusively predicated on assumed auraltypical (Drever 2017) hearing of the participants – creative alternative methods are encouraged.

2 Leader

The walk will require a leader, which is a position of relative authority and trust. Taking inspiration from the *100 Soundscapes of Japan* by the Environment Agency of Japan (1997), and the *TESE* project on the Isles of Harris and Lewis, Scotland (1999–2002), when I directed a public soundscape study of Dartmoor, *Sounding Dartmoor* (Drever 2007), soundwalking was a key method of engagement, but unlike the WSP, with the help of Dartmoor-based arts organization, Aune Head Arts, the walks were all led by local inhabitants and stakeholders; they were regarded as the experts of the Dartmoor soundscape.

Emulating Max Neuhaus, the sound artist Christine Sun Kim has been leading soundwalks through the Lower East Side, a territory that she once inhabited. However, having been deaf since birth, hearing as a prerequisite for soundwalking is problematized; with the aid of graphic and text scores on an iPad, and imparting personal memories, listening 'is substituted, emphasizing layers of subjective, interpersonal, and technical mediation involved in non-verbal communication' (Kim 2016).

3 Appropriate footwear

Participants should come with appropriate clothing and footwear for walking in the specific environment the walk is set, and that does not generate excessive sound whilst moving. Perhaps after Isadora Duncan, barefoot soundwalking could be encouraged, providing a direct vibratory contiguity between ground and skin. Some innovative soundwalkers invite purposefully loud footwear or the acoustic embellishment of shoes: Davide Tidoni's *Exaggerated Footsteps* (2016), which consists of two metal plates, instructs: 'Fix the plates underneath your shoes and take a walk. When the plates touch the ground they activate the acoustics and magnify your own presence in space.'

As a leader my attention is often drawn to the sonic emanations of the participant immediately behind me – footsteps can provide an eloquent building acoustics reference tool akin to a geologist's rock hammer. During one walk, heralding his presence, the man immediately behind me unremittingly tossed and caught his large bunch of keys with impressive precision for the duration of the walk, the high-frequency content providing unparalleled acoustic illumination or echolocation of the space, expertly articulating the morphology of resonances and reverberations (the sound of which he apparently was blissfully unaware).

4 Proxemics

The guide leads at the front like a quasi-mute pied piper, and the group (which should be small in number, say twelve) follow on, one by one, leaving a wide enough gap between the participant in front so their footsteps are out of earshot of other participants; they should not crowd each other. I would also encourage the participants to spread out so as not to draw attention to the group, or to limit the group from becoming an invasive or an obtrusive presence.

5 Inter- and extra-communication

An idiosyncratic feature of soundwalking is the collective observance of silence; talking, whistling, humming, etc. during the actual walk is discouraged, saving up thoughts and insights for the debrief at the end. If participants want to catch those fleeting moments, they could jot them down. It is important to acknowledge that this facet shifts soundwalking into a ritualistic, performative mode, and can lead to some awkward moments as non-participants attempt to engage in conversation with soundwalkers mid-walk. In addition to the vow of silence, to help dedicate attention on the here and now, mobile devices are required to be set to airplane/flight mode or simply turned off. For practical and safety reasons the leader may talk (if necessary) and keep their mobile on. Schafer is also averse to sound recordings or videos being made by the participants, as I witnessed in a walk in Lisbon in 2005, as he regards it as a distraction for the focal task of listening.

6 Duration

A duration of 90 minutes including post-amble discussion time allows for a range of topography to be covered and, importantly, time for the participants to really tune into attentive listening of place. For the more elite soundwalker, longer durations are of course an option, such as Tony Whitehead's 12-hour overnight walk in Plymouth in 2010, to bear aural witness to a sequence of a day.

7 Pace

I am a habitually a fast walker, but soundwalking should not be rushed: it is not about journeying from A to B. The musical tempo designation, *andante*, referring to 'a walking pace' is a useful measure. It was commonly used by composers such as Johann Sebastian Bach and George Frideric Handel (Le Huray 1990: 36); there usage predates metronome markings, with *andante* today spanning from 76 to 108 beats per minute. This slowed down pace appears to help shift habitual listening practices and allows people to simply take their time. If you walk through a shopping mall and travel on an escalator, move at the speed that the escalator has set. You may of course be required to speed up on pedestrian crossings, likewise due to congestion you may be forced to go even slower. Go with the flow. Some artists have emphasized the slowness of the walk as a fundamental feature, for example Phil Morton's *Sonic Gaze*, which he refers to as 'a static soundwalk' (Morton 2019). The urban designer Jan Gehl reflects on his preferred gait of locomotion speed for walking and perceiving, albeit prioritizing sight:

> Our sensory apparatus and systems for interpreting sensory impressions are adapted to walking. When we walk at our usual speed of four to five km/h (2.5–3 mph), we have time to see what is happening in front of us and where to place our feet on the path ahead […]. At speeds greater than walking or running, our chances of seeing and understanding what we see are greatly diminished.
>
> (Gehl 2010: 43)

8 Caesura

When we walk, we move through the soundscape, but we can pause in opportune locations that give themselves to lingering (designed or otherwise), allowing the prevailing soundscape to move around us. This can also be helpful for refocusing listening attentiveness.

9 Meteorology

(Within reason) don't let inclement weather get in the way of appreciating the walk: a sudden gust of wind can sonically bring to life otherwise silent foliage; falling rain drops on surfaces, taking John Hull's heed, 'gives a sense of perspective and of actual relationships of one part of the world and another […] I am presented with a totality, a world which speaks to me' (Hull 1997: 27).

10 Safety

Soundwalking is potentially hazardous, as you are inviting people to slowdown and reorientate their senses in active everyday contexts. Therefore, prompt the participants to take extra care when crossing roads, etc.

11 Preamble

Once the group has assembled, the leader will need to prepare the participants and set the rules, along with imparting pragmatic information. What is said at this stage will prime predominant attitudes to listening, and this will of course depend on the agenda and motivation of the walk's impetus. The mantra-like instruction for soundwalking is: listen! – but this is vague, you may wish to explore concepts of listening, such as 'listening in readiness' and 'listening in search' (Truax 2001: 21–24). Introduce specific themes you may wish to draw attention to such as biophony or regeneration. Resist divulging the route, but reassure the participants that there is no need to worry: 'We will finish on time, at the designated location.'

These are the questions I primed participants with, as an activity associated with the 24th International Congress on Sound and Vibration in Westminster:

- We will be exploring the salient characteristics of the Westminster soundscape; is it congruent with your expectations?
- How are the sound sources modulated by this specific acoustic architecture?
- How much cognitive effort is required to listen attentively to the acoustic environment – is it pedestrian friendly?
- How does the actual prevailing acoustic environment shape the pedestrian experience of Westminster on a mid-week evening in July, and how does this experience impinge on your perception of the soundscape?

12 Post-amble

Allow ample time for open discussion in a safe and secluded location where the prevailing soundscape continues but voices are not masked. No contribution is invalid, insignificant, or incorrect. Allow time and space for the quieter voices to be heard.

13 Questionnaire and verbalization

When the aim of the soundwalk is to collect, compare, and evaluate specific data on the experience of the soundscape by the participant, different methods have been applied. The use of questionnaires in situ is a simple process and doesn't necessarily interrupt the flow of experiencing the soundscape completely. However, questionnaires may miss valuable nuance and contextual detail of that sensory experience. To capture more involved and meaningful data, researchers at CRESSON[3] developed a walking method, an elaboration of Jean-Paul Thibaud's 'commented city walks' (2013), where 'a researcher [equipped with directional microphone] accompanies the participant in order to guide them and to encourage them to speak if necessary' (Tixier 2002: 85). Building up a fuller picture of the location and the responses thereof, they repeat the route at different times of the day, weather, etc. The simple instruction is 'to say what one hears and to comment on it'. To add commentary to this information they are asked to 'qualify them and explain the relations they maintain with the city, the people or oneself' (86). Even for a soundscape studies expert it is hard to reflect on and verbalize one's experience of the soundscape as it unfolds around you, so the role of the researcher is key here in opening up a dialogue between participant and researcher. And the build-up of that relationship through sharing the walk is very much part of the process: 'The idea that walking with others – sharing their step, style and rhythm – creates an affinity, empathy or sense of belonging with them' (Pink 2015: 111).

The London Soundwalk: Re-enactment

Soundwalking promotes untrammelled listening in whatever location the participants may find themselves traversing. However, there are incumbent ethical issues, as such an attitude gives way to overhearing and verges on eavesdropping. On a Sunday morning in April 2009, I lead a re-enactment of the WSP's *The London Soundwalk*, thirty-four years on.[4] We adhered to the original route and instructions, with the addition of a circuit through the Euston Road train stations which were undergoing major redevelopment. There was one major alteration however on ethical grounds. The original walk also took place on a Sunday morning around Easter time with the inclusion of experiencing 'true calm' (Schafer 1977a: 92) by attending the morning meeting of the Society of Friends on Euston Road. Soundwalks have often taken in 'the inner ambience, reverberation and relative stillness' (92) afforded by religious spaces. On carrying out a recce of the route, I attended a regular Sunday worship which primarily takes the form of collective silence which is regarded by the Quakers as a mode of worship, a practice that parallels some attitudes to soundwalking (see below). I approached an elder of the group after the service and described what I had in mind. I quickly realized that bringing in our soundwalking group to listen to the Quakers' listening was obtrusive and unwelcome and verging on the unethical. Fundamentally, we would not be sharing the same orientation for silence and listening as the rest of the congregation – a kind of eavesdropping on the silence of others. As a compromise at the end of the walk we met in the Friends Meeting House for a debrief, allowing us to dwell in the original starting point.

The joy of soundwalking

Notwithstanding the health benefits of daily walking, and its accompanying boost of dopamine, serotonin, and endorphins, it can be a highly pleasurable activity. In his *A Philosophy of Walking*, Frédéric Gros, develops the *States of Well-Being* that the walking experience offers – 'to different degrees, on different occasions', as differentiated in antiquity – pleasure, joy, happiness, and serenity (Gros 2014: 139–146). For Hildegard Westerkamp, soundwalking affords the 'practical purpose of orientation in the environment' or can have a 'purely aesthetic purpose of creating a soundwalk' (Westerkamp 2007: 52), but much more than that, as shared or solitary daily practice, it is allied to the practice of meditation and mindfulness as it has the capacity for personal enrichment. On reflecting on many years of soundwalking practice – and resonating with Pauline Oliveros's practice of *Deep Listening* (2005) – she appraises 'soundwalking or any related ways of listening. Doing such a lifelong practice imbues a visceral, embodied knowledge of healing, calming, centering. It is in the doing that this knowledge emerges and the benefits are particularly relevant in this ever-increasing chaos and confusion of today's world' (acoustic-ecology@sfu.ca discussion list, 31 May 2018).

The potential for collective walking and listening to induce calm is astonishing; at the end of the walk there is often a reluctance across the group to break the silence back into the customary verbal mode of exchange. I have led a soundwalk around Goldsmiths' neighbourhoods in South London every year for the past decade, a route that takes in a wide range of social and topographic contrast. At the debrief one year a student announced that he had never felt so relaxed. Despite the frenzied and quantitatively loud and complex urban environment that we had traversed, the walk had imbued him with an inner silence, cocooning him from the physical acoustic environment. At the end of another walk on a cold and wet November evening in Leeds, a participant extolled on the most amazing 3D surround sound experience; the walk had rendered his listening experience of the physical acoustic environment into a highly mediatized hyperreal mode, detached from the everyday. Yet the urban soundscape is not rarefied or meticulously controlled like cinema sound design: ultimately it is haphazard, generative, unwieldly, and inherently complex, and most importantly, all sounds are indexical.

Walkability

I have observed some participants increasingly unable to block off the prevailing noise of the environment as walks have progressed. In feedback following a soundwalk of Plymouth city centre I led for the Geographies of Creativity and Knowledge research group from Exeter University in January 2015, which included the participants' intensification of their sense of smell (interestingly not an uncommon response), performance maker and director Paula Crutchlow explained:

> I was OK for a while and I was hearing things and following the source of the sound. Then it was the tuning in to listening to everything [that] made me feel anxious and overloaded. Like I could hear everything simultaneously. Not only hearing things coming from all directions, I felt like I needed to know where all the sounds were coming from and attach them to the source of the sound. I started to make up stories in my head for all the sounds and the snatches of conversation. In the end I felt like I was hearing everything all at once,

coming from all directions, and loudly – which was overwhelming. It was only when we sat down to talk about it that I realised how challenging the experience had been.

(email to the author, 6 September 2018)

Deliberating on this kind of dissonant reaction to soundwalking with Westerkamp, I learnt that she recommends participants take some time out following a walk, postponing the plunge back into everyday life. To help foster a potentially nourishing relationship with the acoustic environment, Westerkamp is careful in her choice of soundwalk locations and routes: 'It is best done in a place where we can hear ourselves and the more delicate sounds around us' (Westerkamp 2007: 52). Here there is pressing desire for a reorientation of urban soundscape design towards the human-auditory-scale in contrast to the preponderance of street design where the 'needs of drivers and motor traffic [are] put first' (Commission for Architecture and the Built Environment [CABE] 2008: 2). For Westerkamp, a judgement of human-scale can be simply the (in)ability to hear your voice or your footsteps due to masking: 'You cannot hear the sounds you yourself produce, you experience a soundscape out of balance. Human proportions have no meaning here' (Westerkamp 2007: 50).

Assessing the entirety of the human experience and behaviour within cities, with an eye to prioritizing the pedestrian (and cyclist) in the urban environment, Gehl and his team carried out comparative walking tours. Soundscapes tend not to feature too highly in their observations and concerns, however he makes a similar qualitative evaluation to Westerkamp's. On comparing 'pedestrian-friendly' Venice with London, Tokyo, or Bangkok, he pronounces: 'It is possible to speak quietly and pleasantly with others. At the same time you can hear footsteps, laughter, snatches of conversation, singing from open windows and many other sounds of life in the city. Both the possibility to hold a conversation and the sound of human activity are important qualities' (Gehl 2010: 152).

Lamentably, the uncrowded Vienna soundscape throws the soundscape of most urban agglomerations into sharp relief, which can be overwhelmingly hostile, alienating, and 'out of balance'. A briefing document by the Commission for Architecture and the Built Environment (CABE) on *Civilised Streets* acknowledges that 'most of our streets are not civilised, enjoyable places to be. They are mainly noisy, polluted, hazardous and unpleasant – with serious social and environmental problems the result' (CABE 2008: 2).

What much of the public realm is lacking can be best defined as 'walkability'. Articulated by pioneering soundscape researcher Michael Southworth, whose experimental research into accessibility and the senses included blindfolding participants – a method also carried out by Ben Patterson in *Tour* (New York, 1963) – and traversing them through urban environments on wheelchairs, walkability is 'the extent to which the built environment supports and encourages walking by providing for pedestrian comfort and safety, connecting people with varied destinations within a reasonable amount of time and effort, and offering visual [and aural] interest in journeys throughout the network' (Southworth 2005: 248).

Soundwalking in those auditory nourishing places is helpful for learning lessons about what constitutes good soundscape design, as well as for ameliorating walkability throughout the city, we also need to venture into the more challenging urban spaces, to understand what needs to be worked on, and to evaluate what extant features can be valorized and maintained. But here we have another ethical quandary: is it ethical to promote sensitive listening to a populous who unavoidably inhabit a potential stressful fight or flight inducing (corticotropin-releasing hormone and adrenocorticotropic hormone) noisy environment.

Conclusion

As I have shown, soundwalking approaches lie on a spectrum between soundwalking as a means to an ends and soundwalking for soundwalking's sake. Its methodology incorporates multiple practices of overlapping and divergent ideological, ontological, and epistemological underpinnings, the aims and objectives of which are inconsistent. Its form can be scrupulously prescribed and intentionally proscriptive, or aping the tradition of the *dérive* (drift); it can be open, generative, and improvisational. Today's versions of soundwalking can be found in multiple disciplinary contexts with a polyphony of converging and diverging, spoken and unspoken set of aims and motivations, and as such engender themes of participation, social context, aesthetic listening, environmental sensitization, interpretation, pedagogy, awareness raising, deep mapping, psychogeographic musings, and more recently the professional field of acoustics (ISO 12913–1: 2014). Whatever its orientation, soundwalking practices share the commonality of encouraging the prioritization of auditory perception(s) over the other senses outside of a lab setting, which might be understood as immersed in the everyday, the real world, in the field, or in situ. Hence it is inescapably and unashamedly context sensitive with all that may encompass. But it is not an activity that can be replaced. I would claim that if you have never participated in a soundwalk you will not be able to comprehend the profound experiential effect that an erstwhile prosaic activity can have.

Notes

1. 'The state in which people are so involved in an activity that nothing else seems to matter' (Csikszentmihalyi 2002: 4).
2. For a pre-history of soundwalking, see Drever 2009.
3. The salient research theme of everyday walking at CRESSON can be traced back to Jean-François Augoyard's formative study of the inhabitants of L'Arlequin, presented in *Step by Step* (Augoyard 2007, originally published in 1979 as *Pas à Pas*).
4. In collaboration with city planner Max Dixon, the UK and Ireland Soundscape Community, Noise Futures Network, and Sound Practice Research (Goldsmiths), and joined by Hildegard Westerkamp.

References

Augoyard, J.-F. (2007). *Step by Step: Everyday Walks in a French Urban Housing Project*. Trans. F. Choay. Minneapolis, MN: University of Minnesota Press.

Augoyard, J.-F. and H. Torgue (eds) (2005). *Sonic Experience: A Guide to Everyday Sound*. Trans. A. McCartney and D. Paquette. Montreal: McGill-Queen's University Press.

Barad, K. (2007). *Meeting the Universe Halfway: Quantum Physics and the Entanglement of Matter and Meaning*. Durham, NC: Duke University Press.

Canetti, E. (2003). *The Voices of Marrakesh*. Trans. J. A. Underwood. London: Marion Boyars.

Commission for Architecture and the Built Environment (CABE) (2008). *Civilised Streets*. London: CABE.

Csikszentmihalyi, M. (2002). *Flow: The Psychology of Happiness: The Classic Work on How to Achieve Happiness*. London: Rider.

Department of Education (2013). National Curriculum. Available online: https://www.gov.uk/government/collections/national-curriculum (accessed 18 July 2020).

Drever, J. L. (2007), 'Topophonophilia: A Study on the Relationship between the Sounds of Dartmoor and the People Who Live There'. In A. Carlyle (ed.), *Autumn Leaves: Sound and the Environment in Artistic Practice*, 98–100. Paris: Double Entendre.

Drever, J. L. (2009). 'Soundwalking: Creative Listening Beyond the Concert Hall'. In J. Saunders (ed.), *The Ashgate Research Companion to Experimental Music*, 163–192. Aldershot: Ashgate.

Drever, J. L. (2017). 'The Case for Aural Diversity in Acoustic Regulations and Practice: The Hand Dryer Noise Story'. In *Proceedings of ICSV24, The 24th International Congress on Sound and Vibration*, Ambiances International Network, Grenoble, 6. London: International Institute of Acoustics and Vibration and the Institute of Acoustics.

Drobnick, J. (2004). 'Listening Awry'. In J. Drobnick (ed.), *Aural Cultures*, 9–15. Toronto: YYZ Books; Baff, AB: Walter Phillips Gallery Editions.

Dugave, C. and C. Regnault (2009). 'Promenades d'Ambiances'. In *Ambiances* Newsletter 1. Grenoble: Réseau International Ambiances.

Evans, D. (ed.) (2012). *The Arts of Walking: A Field Guide*. London: Black Dog Publishing.

Gehl, J. (2010). *Cities for People*. Washington, DC: Island Press.

Gros, F. (2014). *A Philosophy of Walking*. Trans. J. Howe. London: Verso.

Hull, J. M. (1997). *On Sight and Insight: A Journey into the World of Blindness*. Oxford: Oneworld Publication.

ISO 12913–1 (2014). *Acoustics—Soundscape—Part 1: Definition and Conceptual Framework*. Geneva: International Standards Organization.

ISO/TS 12913–2 (2018). *Acoustics—Soundscape—Part 1: Data Collection and Reporting Requirements*. Geneva: International Standards Organization.

Jarviluoma, H., M. Kyto, B. Truax, H. Uimonen, and N. Vikman (eds) (2009). *Acoustic Environments in Change*. Trans. B. Johnson. Tampere: Tampere University of Applied Sciences (TAMK).

Kim, C. S. (2016). *(listen)*. Available online: http://christinesunkim.com/work/240/(accessed 3 April 2019).

Le Huray, P. (1990). *Authenticity in Performance*. Cambridge: Cambridge University Press.

Morton, P. (2019). 'The Sonic Gaze Static Soundwalk'. Available online at http://culture.org.uk/2017/08/the-sonic-gaze-static-soundwalk-27th-august/(accessed 3 April 2019).

Oliveros, P. (2005). *Deep Listening: A Composer's Sound Practice*. New York: iUniverse.

Pink, S. (2015). *Doing Sensory Ethnography*. 2nd edition. London: Sage.

Qualmann, C. and C. Hind (eds) (2015). *Ways to Wander*. Axminster: Triarchy Press.

Schafer, R. M. (1974). *Listening. Sound Heritage*, vol. 3, no. 4, Aural History, Provincial Archives of British Columbia, Victoria, 10–17.

Schafer, R. M. ([1967] 1976). *Creative Music Education: A Handbook for Modern Music Teacher*. New York: Schirmer Books.

Schafer, R. M. (1992). *A Sound Education: 100 Exercises in Listening and Sound-Making*. India River, ONT: Arcana Editions.

Schafer, R. M. (1994). *The Soundscape: Our Sonic Environment and the Tuning of the World*. Rochester, VT: Destiny Books.

Schafer, R. M. (ed.) (1977a). *European Sound Diary*, vol. 3. The Music of the Environment series. Vancouver, BC: ARC Publications.

Schafer, R. M. (ed.) (1977b). *Five Village Soundscapes*, vol. 4. The Music of the Environment series. Vancouver, BC: ARC Publications.

Smith, P. (2014). *On Walking: … and Stalking Sebald*. Charmouth: Triarchy Press.

Southworth, M. (1969). 'The Sonic Environment of Cities'. *Environment and Behavior* 1(1): 49–70.

Southworth, M. (2005). 'Designing the Walkable City'. *Journal of Urban Planning and Development* 131 (4): 246–257.

Thibaud, J.-P. (2013). 'Commented City Walks'. *Wi: Journal of Mobile Culture* 7(1). Available online: http://wi.mobilities.ca/commented-city-walks/ (accessed 3 April 2019).

Tixier, N. (2002). 'Street Listening: A Characterisation of the Sound Environment: The "Qualified Listening in Motion" Method'. In H. Järviloma and G. Wagstaff (eds), *Soundscape Studies and Methods*, 83–90. Turku: Finnish Society for Ethnomusicology.

Truax, B. (2001). *Acoustic Communication*. 2nd edition. Westport, CT: Ablex Publishing.

Urry, J. (1995). *Consuming Places*. London: Routledge.

Vikman, N. (2009). 'Changing Soundscape of Cembra Village'. In H. Jarviluoma, M. Kyto, B. Truax, H. Uimonen, and N. Vikman (eds), *Acoustic Environments in Change*, trans. B. Johnson, 56. Tampere: Tampere University of Applied Sciences (TAMK).

Westerkamp, H. (1998). 'Speaking from Inside the Soundscape'. In H. Karlsson (ed.), *Proceedings from "Stockholm, Hey Listen!" Conference on Acoustic Ecology*, 53–63. Stockholm: Royal Swedish Academy of Music.

Westerkamp, H. (2007). 'Soundwalking'. In A. Carlyle (ed.), *Autumn Leaves: Sound and the Environment in Artistic Practice*, 49–54. Paris: Double Entendre.

39
Sounding Wild Spaces: Inclusive Map-Making through Multispecies Listening across Scales

Alice Eldridge, Jonathan Carruthers-Jones, and Roger Norum

The conservation of wilderness is critical to the future of our biosphere, on both ecological and social levels. Scholars across disciplines have established the importance of wilderness as a key site for endangered species (Soulé 2014), human recreation and well-being (Milner-Gulland et al. 2014), as well as the wider network of ecological processes on which all life depends (Chan et al. 2006). Recognition of the value of wilderness across cultural, socio-economic, and ecological perspectives bolsters the conservation imperative, but the respective associated land uses rarely align with all of these perspectives. Imagine, for example, that you are the director of a national park. The park stretches from the edge of a small village, where unemployment is high, up wooded slopes, which provide habitat for the endangered wildcat, as well as a playground for local ramblers and ardent naturalists, before stretching up to the jagged peaks that are an international mecca for climbers and home to breeding pairs of golden eagles. The national government has just announced green incentives for the development of wind turbines; local government recognizes therein the potential to boost the economy and decrease unemployment in the local community, mandating development within the park. Your job is to decide whether wind farm construction is warranted and, if so, where to site the turbines. Current maps provide geophysical information such as access roads and landscape topology which indicate the optimum location in terms of power generation and distribution, but what about the impact of the turbines on flora, fauna, the local community, and the tourism industry? How would you visualize, analyse, and 'map' these other important perspectives, such as value and meaning for inhabitants, which are not readily quantifiable?

This scenario plays out in many national parks across the world: resolution of the conflicting needs of human stakeholders, and ecological and economic imperatives poses a significant challenge globally (Redpath et al. 2013; Vuceticha et al. 2018). Wilderness policy and planning, like all conservation decision making, must be evidence-based

(Sutherland et al. 2004; Adams and Sandbrook 2013; McIntosh et al. 2018). However, we currently lack the means to build evidence in a way that takes into account the needs of all abiotic processes and biotic beings living, working, playing, and otherwise becoming (Haraway 2008) in wild spaces. It is a complex, wicked problem (Rittel and Webber 1973; Elia and Margherita 2018), one which cannot be solved by researchers working within a single discipline.

One key obstacle is that current methods for mapping the landscape and the populations which inhabit them are incommensurable in scale and intrinsically prioritize one perspective over another. Land management decisions are predominantly based on maps created from satellite imagery which provide visual representations of broad vegetation cover and macrostructures of the built environment, yet these maps are blind and deaf to the details of the lives of the myriad critters (humans among them) which flourish in wild spaces. Site-based ecological surveys capture detail of which flora and fauna dwell at particular sites and times, but are intrinsically small scale and traditionally focus on nonhuman species. At the same time, participatory, ethnographic methods are increasingly being explored to access the knowledge, perception, and values of local human actors (Maginn 2007; Hollowell 2009). However, these are typically documented in such a way that any insights generated end up divorced from the geo-ecological contexts in which actors are situated (Pink 2010; Reed 2018). Each approach is de facto incomplete and spatially limited, and ipso facto fails to provide a comprehensive representation of wilderness across the local and landscape scales required. This limits their usefulness as a support tool in environmental decision making and planning which requires standardized, spatial data with homogenous coverage across their administrative remit.

A deeper epistemological issue is that we lack frameworks to synthesize the insights generated from these different methods. Each is born of distinct disciplines, between which there may be little interaction or communication, much less conceptual or methodological integration. In order to create maps that integrate empirical, ecological, and geophysical data at scale, with personal, particular existences, experiences, and knowledges of actors, we need a conceptual framework which resonates across attendant disciplines.

In an ongoing project we are exploring the potential for diverse forms of *listening* as a point of encounter between ethnographic and ecological perspectives, with the aim of integrating both within the standard geophysical, cartographic format, in order to create inclusive, multivocal wilderness maps. Our work builds upon contemporary research across ecological sciences, anthropology, and political geography which highlights the importance of the *soundscape* – understood broadly as all the sounds emanating from a given landscape – as a significant component of both ecosystem function and human experience, as well as a key factor in the politics of environmental justice and land management.

Walking and listening in the landscape

Existing approaches to wilderness mapping in Europe use imagery from satellite data to designate areas on a continuum from least wild (e.g. the centre of a large urban conurbation) to most wild (e.g. a remote corner of a mountainous region) (Carver et al. 2012; Müller, Bøcher, and Svenning 2015). Under this standard model the degree of wildness is designated by considering four key metrics, each inferred from satellite imagery: perceived naturalness; absence of modern artefacts; rugged or challenging terrain; and remoteness from roads and ferries. Similar multi-criterion approaches have been developed globally.[1] This remote, multi-criterion approach is attractive because it can be operationalized at scale using satellite data and geographical information systems (GIS) to create comprehensive maps that support decision making in landscape management in areas such as renewable energy development or protected area designation (McMorran and Carruthers-Jones 2015; Ma and Long 2019). However, in being constructed from information derived from reflected light (satellite imagery), these methods are inherently insensitive to necessary 'local' details: the subjective, multisensory subtleties of the human wilderness experience cannot be taken into account; similarly, whilst canopy cover may be documented through remote satellite surveys, any wildlife beneath the canopy is less detectable. Humans and other living beings are literally *not on the map* which informs the management of the lands in which they thrive.

Understanding the value and meaning that wild spaces have for particular human communities requires situated, ethnographic, and qualitative data. To this end, participatory research methods have gained popularity in recent years in the social and environmental sciences because they provide a means for incorporating the experiences, attitudes, and even ecological knowledges of local community members through research co-design (Calheiros, Seidl, and Ferreira 2000; Probst and Hagmann 2003). For example, participatory map-making mobilizes and produces knowledge, while maintaining relevance and legitimacy among actors in the field (Warner 2015). This not only complements high-level data with more finely grained local knowledge but also potentially can empower residents to envision improvements of spaces with which they have close, meaningful relations (Pain 2004).

Walking research offers an intuitive and compelling means of studying human relationships to landscape and place (Certeau 1984; Pink 2007; Edensor 2010). When walking methods integrate ethnographic interviews, the responses from participants have been found to generate deeper place-based narratives than sedentary research practices, particularly when considering narrative quality and spatial specificity of the study area (Evans and Jones 2011). However, structured approaches to walking interview methods have thus far focused primarily on urban spaces (Pierce and Lawhon 2015; Middleton 2018). Sound mapping in particular affords emplaced aural engagement (Westerkamp 1998; see also Carlyle, in this volume), and various forms of it are increasingly explored as a means of inquiry into spatial, geopolitical, and cultural issues (Droumeva 2017). However, most sound-mapping work has tended to focus on representation of and communication about the sound environments themselves rather than the personal responses of individual actors to the landscape (Droumeva 2017). A key challenge is thus to broaden the scope of these emerging, situated methods in a structured way that enables comparison between individuals and communities, and across different habitat types and landscape gradients. A related methodological question is how to design conceptual frameworks for combining rich qualitative data arising from these mobile methods and the quantitative data of remote sensing which forms the bedrock of current wilderness mapping.

Learning to listen to biodiversity

Despite the well-documented strategic importance of wild spaces for biodiversity conservation (World Wide Fund for Nature [WWF] 2008; Dudley 2009), biodiversity metrics have yet to be widely incorporated into wilderness mapping (Brown and Williams 2016). This is principally because traditional in situ point count methods – whereby experts walk transects so as to tally quantities of birds, bats, bees, barnacles, etc. – are time-consuming, costly, and require an expertise often simply unavailable in many situations. The result is that they are not operationalizable at the requisite landscape scale.

The need for cost-effective, scalable biodiversity monitoring tools is not unique to wilderness mapping, but is an urgent imperative for conservation efforts globally. The emerging field of ecoacoustics (Sueur and Farina 2015) proposes that *listening* to the environment may provide a solution. Just as doctors and physicians have listened to the health of our heart, lungs, and other bodily systems, ecologists are beginning to listen to ecosystems as an indicator of their ecological status. The rationale behind this approach can be understood in evolutionary terms. We are all familiar with the fact that sonic signals are critical to the survival of many species on land and in water. If your voice is masked – by other species, wind, rain, cars, planes, or motorboats – the chances of reproducing are reduced. If all your conspecifics live in a similar acoustic environment, eventually your species will die out. If, on the other hand, your voice carries well in the soundscape, you have a higher chance of mating, you might produce offspring, and their voices are likely to be rather like yours. For soniferous species, survival of the fittest subsumes singing the right song.

This has implications at higher levels of ecological organization. In a healthy, stable habitat (such as our idealized Wilderness Area), evolutionary theory predicts that competitive co-evolutionary forces will sculpt complex choruses in which each voice occupies a unique acoustic niche (Krause 1993). Much like the voices of a good orchestral arrangement, or well-mixed dance track, each voice has a specific place in a complex but coherent polyphony. In pristine areas, where there are hundreds or even thousands of vocalizing species, it follows that competition for acoustic niches will compose a more complex (but well-structured) soundscape, spreading the voices over a wider frequency range. To return to our musical analogy, consider the range of sounds in a string quartet (from cello to violin) compared to that of a symphony orchestra (from double bass to piccolo).

An evolutionary perspective highlights four productive ideas. Firstly, that sound is a core dimension in the evolutionary ecosphere, like food, water, and habitat. Secondly, beyond basic survival, sound is a significant component in the *Umwelt* (Uexküll 1926) of many species: within biosemiotics, a soundscape is investigated as a *cognitive medium* (Farina and Pieretti 2014). Thirdly, therefore, soundscapes *mediate*[2] *the interactions* between all soniferous and sonically sensitive species dwelling at a given place and time: like a global feedback delay buffer, soundscape is shaped by the past and shapes the future voices in a given biome. Finally, it follows that a soundscape is a *source of information* about the ecological status of an acoustic community. Just as the fossil record tells us something about hard-bodied things of the past, a soundscape tells us something about the lives of soniferous species in the present – so long as we can learn to listen in the right way.

The calls of soniferous species (biophonies) are not the only component of the soundscape, they interact with sounds that emanate from, and are shaped by, the landscape: wind, rain, or rivers (geophonies); and man-made sounds (anthrophonies), including those of industrial activity (technophonies). Understanding the causes and consequences of these soundscape components is the concern of soundscape ecology (Pijanowski 2011). Like its sibling landscape ecology, this emerging discipline addresses the dynamics of natural–human systems (Liu et al. 2007) and investigates the role of sound in mediating interactions between climate, landscape, human and 'natural' processes. We now recognize that sensitivity to sound extends to the plant kingdom (Gagliano, Mancuso, and Robert 2012), and recent studies reveal that sound-mediated relationships even traverse kingdoms, promoting consideration of the soundscape as an *interspecies media*.[3]

The theory and methods of ecoacoustics have inspired a sea change in ecological monitoring practices worldwide (Sueur and Farina 2015; WWF 2018). Empowered by the decreasing cost of robust, programmable sound recorders, researchers and land managers are establishing acoustic monitoring programmes at multiple spatio-temporal scales to assess biodiversity and other facets of the ecological status across a range of planetary ecosystems. Bioacousticians have long listened to the communication between the *individuals* of particular species, and machine-learning algorithms are being developed to automate identification of specific species calls (Stowell and Plumbley 2014; Turesson et al. 2016). In contrast, ecoacousticians deploy algorithms – known as acoustic indices – to assess the *global* composition of spatio-temporal patterns in a soundscape and, by inference, the structure of the acoustic community. Some indices are predicated on particular soundscape components being band-limited: the vocalizations of animals (birds, bats, insects, fish) tend to be high frequency (2–8 kilohertz) and intermittent, whereas the engines of transport and industrial machinery contain relatively low-frequency components and tend to be of constant amplitude. Under one approach, the 'health' of a soundscape is assessed in terms of the relative levels of biophonies and technophonies detected (Gage, Napoletano, and Cooper 2001). Where the composition of the acoustic community itself is of interest, the spectro-temporal arrangement of biophonic signals is assessed as a proxy for species richness or abundance. Over sixty computational acoustic indices have been proposed and evaluated to date (McKenna et al. 2017). These indices provide statistical summaries of the distribution of sound energy in short audio recordings (typically 1–5 minutes). Even relatively simple indices have been shown to reflect spatial heterogeneity of vegetation (Bormpoudakis, Sueur, and Pantis 2013), to correlate with observed changes in habitat status (Kasten et al. 2012) or biocondition (Eyre et al. 2015), and to strongly predict species richness across a wide range of terrestrial (Boelman et al. 2007; Eldridge et al. 2018) and aquatic (Bertucci et al. 2016; Harris, Shears, and Radford 2016) habitats.

Acoustic monitoring is attractive to the ecologist because it is non-invasive and sensitive to a wide range of species across media of the biosphere, including air, water, soil, and vegetation. It is cost-effective, removes human bias, and is unaffected by factors such as light level or vegetation density which hinder remote visual methods. Crucially, the technique is considered scalable at no loss of resolution: acoustic survey is embraced as a means to capture the intimate moment-to-moment dynamics of soniferous species interactions whilst being replicable at large spatio-temporal scales. For the development of multivocal, multidisciplinary wilderness mapping, the ecological conception of soundscape is attractive in providing a cost-effective method for biodiversity monitoring. It also provides a valuable *conceptual nexus* for understanding the interactions between anthropogenic, ecological, atmospheric, technological, and geophysical processes – as well as a transdisciplinary point of contact from which we might engender methodological interaction between ethnographic, ecological, and geographic mapping methodologies.

Case study: The WILDSENS project in Abisko National Park

In our ongoing project, WILDSENS: Sensing Wild Spaces: Integrated Participatory Mapping for Understanding Community Relationships to Dynamic Mountain Landscapes, we explore the potential for soundscape as a conceptual nexus for integrating methodologies across disciplines in order to make more inclusive wilderness maps. As a team we draw on a rich mix of disciplinary backgrounds relevant to thinking through the inevitable interdisciplinary challenges of integrating human, ecological, and geophysical perspectives on wild spaces: social anthropology, soundscape ecology, music, computer science, human and political geography, and the environmental humanities. In previous work, we have investigated the relationships between human perception and ecoacoustic metrics across wildness gradients in Scottish and French mountain regions (Carruthers-Jones, Carver, and McMorran, forthcoming; Carruthers-Jones et al. 2019), and have considered the importance of both mobile methodologies (Salazar, Elliot, and Norum 2017) and how mobility is mediated (Ramella and Norum, forthcoming) in interdisciplinary scholarship. The primary aim of WILDSENS was to further develop participatory mapping methods and explore ways to incorporate both qualitative ethnographic insights and quantitative ecological data within a cartographic frame. We designed a small pilot project to test a suite of ethnographic and ecological methods together in order to engage multivocally with the local landscape, wildlife, and key local actors in Abisko National Park in order to sketch a framework for inclusive wilderness mapping.

Abisko National Park lies at the edge of one of Europe's largest remaining wilderness, situated 250 kilometres inside the Arctic Circle in northern Sweden. Abisko provides an ideal study area as it encompasses a gradient of wilderness, from semi-urbanized areas to wild mountainous landscapes otherwise untouched by human activity. We were hosted by the Swedish Polar Research Secretariat as part of the EC-funded INTERACT network and were based at the Abisko Research Station, an established centre for ecological, geological, geomorphological, and meteorological research in Arctic and sub-Arctic environments.

Land use in this part of Sweden is extremely varied: in a single day's walk you may encounter indigenous agricultural and livestock farming, hunting and fishing, travel and tourism, commercial forestry, conservation, and research. This makes the area an ideal outdoor laboratory, offering representative examples of common environmental conflicts found globally. The location also provided access to a wide range of local actors, including the county government administrators, urban planners, tourist developers, tourists, and Sámi reindeer herders. The research station holds a wealth of historical landscape data as well as local ornithological expertise, which helped strengthen and supplement our own skill sets.

Research activities were choreographed around a transect walk along a wildness gradient. A transect provides a literal common path along which ethnographic and ecological surveys can be carried out in situ. The transect also provides a shared *methodological* path, being central to ecological, social science, and ethnographic methods, as well as arts practices and leisure activities (Carruthers-Jones, Carver, and McMorran, forthcoming; Carruthers-Jones et al. 2019). The Abisko transect was carefully designed through consultation with local community members and the study of extant maps to create a walk along a gradient of wildness. The path (see Figure 39.1) began at the Abisko tourist station on the road out of Sweden to Norway, diverging from the well-trodden Kungsleden trail up into the Kårsavagge valley and back along the lower slopes of Slåttatjåkka massif. This constituted a walk of 4–5 hours, which began in a car park, with the peri-urban sights and sounds of high-horsepower Nordic tourist vehicles, a bustling tourist shop, and a bar and restaurant, before crossing the roaring Abiskojaure River, heading up through silver birch woodlands and Salix scrubs, and into the windswept, bilberry-lined valleys at the heart of the national park. Five sample points were selected along this circular route, which characterized the range of habitats and landscapes traversed. At each point, acoustic and habitat surveys and ethnographic interviews were carried out in English, Swedish, or Finnish according to participant needs. A summary of data is presented in Table 39.1.

Participants were recruited using an iterative snowball method that began with local gatekeepers. Through social media, posting paper flyers, local press, and word of mouth, we identified and recruited a representative set of participants (see, for example, Reed et al. 2009; Colvin, Witt, and Lacey 2016). Over the course of our stay in June 2018, we took five separate groups of between three and six people out along the research transect. Groups included a team of officers from the regional land planning office, staff working at the area's tourist facilities, students on placement at the Research Station, and residents of Abisko (population eighty-five). We received introductions to multiple communities from several generous gatekeepers. Because the time of year of our field visit coincided with reindeer calving season, however, many members of the local Sámi population were unfortunately not available to participate in this methods development pilot study. We recognize both the critical nature of their voices to this conversation, and the time needed to build trust between researchers and indigenous community members prior to beginning such studies.

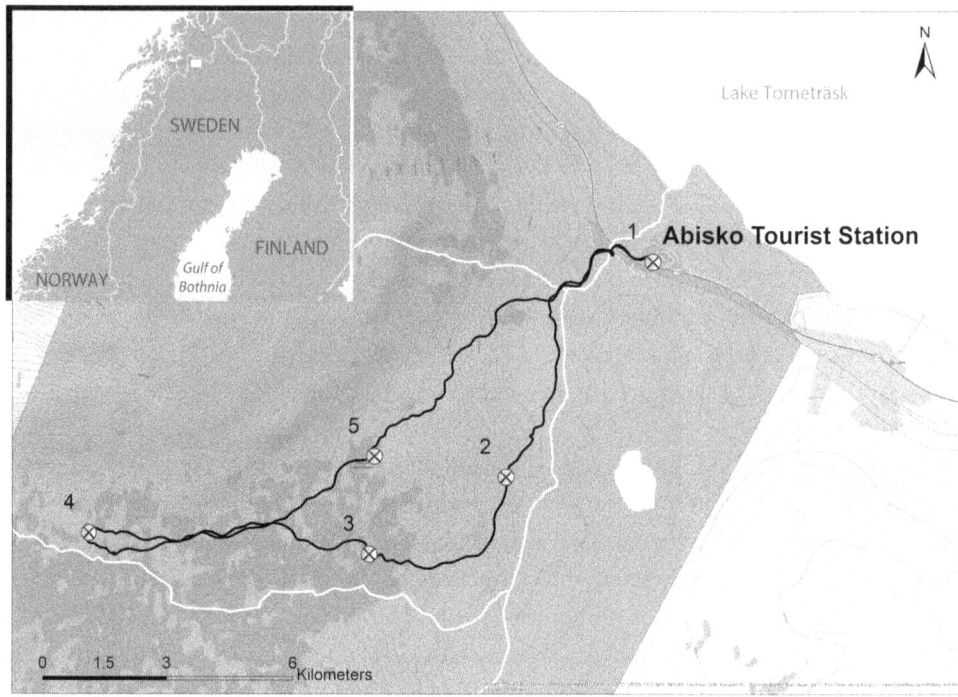

Figure 39.1 Research study site at Abisko National Park showing the walking transect (black) and waypoints (numbered crosses) and the river (white) running into lake Torneträsk. Esri, Garmin, USGS. © OpenStreetMap contributors.

Table 39.1 Data Types Associated With Each of the Surveys Carried Out in Abisko National Park

Geophysical	
• GIS data	• Roads, built environment, topology, macro-vegetation
Ecological	
• Habitat surveys	• Habitat data – range and cover of species (in situ quantitative)
• Acoustic surveys	• Acoustic recordings (audio) and acoustic indices (quantitative)
Ethnographic	
• Interviews	• Interviews and questionnaires (qualitative and quantitative)
	• (meta) Ethnographic observations (video and audio recordings)

Acoustic surveys were carried out at each of the designated waypoints. Programmable recording devices (Wildlife Acoustics SM2+) were attached to trees, and programmed to take short recordings at regular intervals throughout the day and night. These recordings were later analysed, giving numerical summaries of each file to provide an indication of the activity of vocalizing species at each site. Habitat assessments were made at the same points to complement the high-level GIS vegetation data with local detail: following a standard rapid habitat assessment procedure, estimates of vegetation structure were made and plants, mosses, and algae covering the ground identified within a 10-metre quadrant.

Structured interviews were carried out alongside these waypoints, also providing a welcome moment to rest and reflect at specific moments throughout the walk. Interviews were typically conducted in pairs in order to enable discussions to grow organically and without too much prompting from the interviewer. They lasted between 15 and 20 minutes, and focused on the proximate spaces around the participants. The questions we asked sought to solicit responses which were informed by sensory stimuli (e.g. sonic, olfactory, etc.) in the immediate environs, and how these related to participants' experience of 'wildness' at that given transect point. We also encouraged participants to discuss whether the immediate environment evoked any memories and/or affective responses regarding the past or anticipated future of the local environment. While the interviews were primarily qualitative, we also asked participants to offer an index (1–10) of how wild they perceived their immediate physical environment to be.

Even in this small-scale pilot project, the individual conversations brought forth a range of themes which both reflected (and contested) wider debates and assumptions around management of wilderness areas. For example, wildness quality maps (Carver et al. 2012) are constructed using 'perceived naturalness' as one component spatial layer, under the assumption that wilderness is associated with ecological intactness (and, by extension, higher biodiversity). Yet in listening exercises on the walk, ardent Arctic hikers explicitly identified bird song as a telltale sign that they had not yet reached the true wilds of the park, being still below the tree line. Similarly, the noise of helicopters above the park held different meanings according to participants' divergent interests and experiences. For tourists, the sound of a passing helicopter signalled a reminder of their proximity to the town and impinged on their 'wilderness experience'. For those working in the ski industry, however, the sound triggered various concern for a lost tourist (was it a rescue helicopter?), excitement at the promise of a Heli-ski experience, or a reminder that supplies were being transported during that month for the improvement of the forest walks.

We walked this transect under shifting conditions that spanned drizzling rain, bright sunshine, clouds of mosquitos, and the blustery winds of the Arctic summer. Walking the route each day, we were each struck by the extent to which variations in weather impacted our experience of the landscape. Walking away from the road into this great wilderness, even for a couple of hours, the magnetism and power of the wild was always strong, but of different character depending upon a multitude of other factors beyond weather and landscape – suitability of clothing, group dynamics, quality of snacks, sites, sounds, and smells specific to that particular visit. This reminded us how intrinsically multisensory and situated human experience is. As Feld has famously noted, 'place is sensed, senses are placed; as places make sense, senses make place' (Feld 1996: 90). Ethnographic methods that aim to garner data about being in a space must necessarily therefore also be situated and multisensory, which is to say that our methods were not exclusively sonic (Ingold 2007).

Toward a framework for co-design of comprehensive, multisensory maps

Our next step is to integrate these various data into a map which could be consulted for landscape policy and decision making. Taking inspiration from schema developed in urban community mapping (Warner 2015), we envision this developing as a composite map that integrates five distinct layers of information (Figure 39.2).

The *geophysical* base-layer is constructed from objective, remotely sensed, and publicly accessible satellite data that describes the macroscopic structures of the environment: the topology, broad vegetation cover, roads, and other structures in the built environment. Overlaid directly on this are local ecological details: the species data of the habitat assessments and biodiversity proxies derived from the ecological acoustic surveys. In our pilot project, single points were surveyed; larger spatial replicates would be needed to account for spatial variation and to provide links between local detail and macroscopic satellite data representations. The remaining three layers represent the *ethnographic* insights: the third layer represents the immediate multisensory experiences of people in the landscape, accessed through structured interviews carried out along the transect – what they see and hear (or smell) in situ. For example, when asked about her sensory perception at the first waypoint, one woman was particularly attentive to olfactory senses:

> And also smell. It's so […] well now it's been raining so it's a very rainy smell but it's still a very nature-y smell, you can't feel any pollution or fumes. It's very airy, it's very clean.

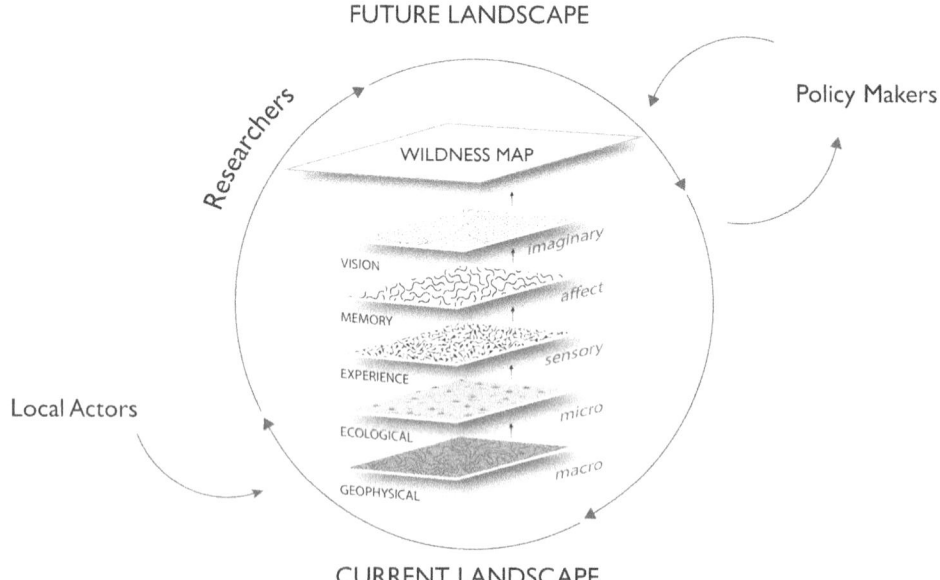

Figure 39.2 Schematic of proposed conceptual framework detailing co-design of mixed methods approach to inclusive wilderness mapping.

Another woman at Waypoint 2 commented:

> The river we can still hear quite loudly, which I thought was cars earlier. We've just passed that little trickle stream but I think that those sounds being so loud reminds you that it's really quiet here [...] there's no tractors, there's no strimming, [weeding], there's no drones, there's no people talking. We're getting these occasional helicopters and things [...] Maybe the roar of the river covers any other traffic residue that we might have from the road.

While these layers reflect fairly straightforward, perceptible phenomena, layers four and five are more complex, speaking not just to human perception but also to human feeling and imagination. The fourth layer reflects participants' *affective* responses to the landscape, their current thoughts, feelings, and emotions in relation to their memories of it. Finally, the fifth layer represents visions of the landscape for the future – hopes and fears for what may or may not come. Any of these ethnographic layers may directly reference the first two more perceptual layers – their sensory impressions of the landscape, absence or presence of sounds of traffic, flora or fauna. As noted by Warner (2015), these final two types of data are the most challenging to analyse and represent, as they are by far the most subjective and most variable – and, indeed, the most mutable.

The various survey methods outlined here are just a few of the active components in this methodology. The process of carrying out surveys itself provides an interface between landscape, policymakers, and local actors, each data collection method also being an opportunity for conversation and discovery. The act of taking part, of making space to listen to and reflect upon the landscape in itself modifies the ways in which stakeholders perceive and value wild spaces: human perspectives are as dynamic as the processes of ecological succession through which wilderness is restored. The next step in our broader project is to carry out a larger study in collaboration with land managers across Europe in order to complete the co-design cycle whilst revisiting and revising the constituent methodologies.

Soundscape as epistemological nexus

In this chapter, we have suggested that diverse forms of listening can offer a link between distinct perspectives and diverse scales, providing an interface between in situ ecological and ethnographic detail and landscape scale cartographies. In seeking to develop multivocal representations, we have described multi-aural methods which include: inviting participants to reflect on their relationship to landscape through listening; listening to stakeholder's responses to landscape and soundscape; and a technologically mediated *listening in* to the same soundscapes in order to assess biodiversity. These activities are consolidated by an evolving conception of soundscape which encompasses the sociocultural concerns of sound studies and the socioecological framework of soundscape ecology. Ecoacoustics serves to bridge scales in two important ways. Firstly theoretically, the foundational evolutionary perspective draws irrevocably reciprocal links between the voices of individuals and the overall global acoustic environment.[4] Secondly, in methodological terms, networks of acoustic sensors can be employed to listen in to the intimate inter- and intra-species communications of soniferous organisms whilst operating across vast spatio-temporal scales, arguably offering a technologically mediated response to Schafer's dream of listening up-close and at scale:

> To give a totally convincing image of a soundscape would involve extraordinary skill and patience: thousands of recordings would have to be made; tens of thousands of measurements would have to be taken; and a new means of description would have to be devised.
>
> (Schafer 1994: 99)

Furthermore, the development of machine listening methods and visualization tools which are necessary to integrate and make sense of this big audio data (Phillips, Towsey, and Roe 2018) themselves draw from cross-disciplinary efforts. Where sound mapping has been critiqued for drawing upon 'the cartographic imagination inherited from the military and political spatialities of the modern state' (Wilmott 2020: 346), the technological tools employed to make sense of big data from ecoacoustic research draw upon the sensitivities, insights, and skills of musicians, ecologists, indigenous land stewards, and computer scientists to create a fresh approach to eavesdropping that is motivated by care and concern, rather than surveillance and suspicion.

Through working across disciplines we are beginning to build a narrative that makes space for all voices, but methodological details require further development. Working in wild mountainous areas presents fresh challenges to this approach. Ecoacoustic methods are predicated on the co-evolution of individuals in dense acoustic communities and are known to operate best in terrestrial environments when the soundscape is dominated by birdsong (Eldridge et al. 2018). In the mountainous wilds of our studies, animal vocalizations are relatively sparse and vie with the howl of Arctic winds and roar of glacial rivers. New, low-level computational methods are needed to isolate and assess the interplay of sparse biophonies in the context of rich geophonies. Similarly, we need to develop new approaches that speak to other forms of data and analyses: conducting and representing insights from interviews, integrating participatory methods, and developing sound-mapping strategies to ensure that the concerns and knowledges of diverse human actors are brought onto the map and into consideration in the management of wilderness areas.

Through our interdisciplinary collaboration, we are developing a multispecies conception of soundscape which begins to ameliorate disciplinary divides. Understanding soundscape as that which mediates sonically responsive beings makes space for the rich and productive treatments of sound across the humanities and sciences disciplinary divide: sound as a distinct medium for knowing the world (Feld 1996); sound as 'the medium of our perception' (Ingold 2000: 265); sound as a productive and performative force (Augoyard and Torgue 2008; LaBelle 2010); geographies of sonic affects (Scrimshaw 2013); and sound as an ecological resource and significant semiotic component in the *Umwelt* of species both across and beyond the animal kingdom.

This broader techno-/geo-/socio-/ecological conception of soundscape attunes with our increasingly expanding understanding of listening – from being predominantly associated with human, conscious aurality to encompassing 'the responsiveness of bodies encountering sound' (Gallagher, Kanngieser, and Prior 2017: 620). In opening our ears and minds to other spaces and other species, we can more easily move beyond anthropocentricism, a mindset for which there is no place – intellectually, morally, or pragmatically – in the current climate. Our concern with inclusive wilderness mapping is emblematic of a larger imperative to take all species into account in imagining planetary futures, and managing the resources of the biosphere. It is our hope that listening across species and scales, and thinking across fields, might bring the critical perspectives of the humanities to the complex processes of evidence-based policymaking and conservation. Only by integrating ways of knowing across disciplines can we ensure that all species are represented on the map.

By figuring soundscapes as the locus of interaction between diverse actors, species, and disciplines we are investigating relationships with and responses to wildness through different forms of listening. Through this work we are developing a concept of soundscape as an *epistemological nexus* that affords disciplinary bridge-building for tackling some of the wicked problems we face, and for bringing both social and ecological matters of concern (Latour 2008) into earshot and onto the map for future generations of all living beings.

Acknowledgements

This project received funding from the European Union's Horizon 2020 project INTERACT, under grant agreement No. 730938, from the University of Sussex, Sussex Humanities Lab, and from the European Union's Horizon 2020 research and innovation program under the Marie Sklodowska-Curie grant agreement No. 642935.

Notes

1. Relevant examples can be found in the Australian national wilderness inventory (Lesslie and Maslen 1995), the wildness quality index for Europe (Carver et al. 2012), the human footprint index at the global scale (Sanderson et al. 2002), the map of Denmark (Müller, Bøcher, and Svenning 2015), and the Cairngorm National Park Wildness Quality map (Carver et al. 2008).
2. Note that this evolutionary perspective aligns closely with an anthropological conception of sound as 'the medium of perception' (Ingold 2000: 265) but expands this to an inclusive, interspecies media shared across species.
3. Numerous examples exist which illustrate the co-evolution of plant structure and function and animal and insect vocalization. Consider the carnivorous pitcher plant (*Nepenthes hemsleyana*) which has evolved what is essentially a parabolic reflector in order to attract a mutualistic bat species (*Kerivoula hardwickii*) to roost within it (Schöner, Simon, and Schöner 2016). A concave structure in the back wall of these

plants' pitchers strongly reflects the ultrasonic calls of the bats. This structure is missing in closely related Nepenthes species that do not interact with bats, suggesting that its principle function is to guide bats towards insect snacks and a comfy bed for the night, in return for high-quality nitrogen-rich manure. Recent studies further suggest that flowers are sensitive to the buzz of pollinators and use this information to increase the sugar content of their nectar as they fly past (Veits et al. 2019). The implication is that the flowers' shape is selected for 'hearing' ability and pollinators may evolve to make sounds that flowers can hear.

4. Note that this ecoacoustic conception of soundscape can be understood as a multispecies equivalent of the early usage of the term by Buckminster Fuller: 'When […] man invented words and music he altered the soundscape and the soundscape altered man. The epigenetic evolution interacting progressively between humanity and his soundscape has been profound' (Fuller 1966: 52).

References

Adams, W. and C. Sandbrook (2013). 'Conservation, Evidence and Policy'. *Oryx* 47(3): 329–335.

Augoyard J. F. and H. Torgue (2008). *Sonic Experience: A Guide to Everyday Sounds*. Montreal: McGill-Queen's University Press.

Bertucci, F., E. Parmentier, G. Lecellier, A. D. Hawkins, and D. Lecchini (2016). 'Acoustic Indices Provide Information on the Status of Coral Reefs: An Example from Moorea Island in the South Pacific'. *Scientific Reports* 6: 33326.

Boelman, N. T., G. P. Asner, P. J. Hart, and R. E. Martin (2007). 'Multitrophic Invasion Resistance in Hawaii: Bioacoustics, Field Surveys, and Airborne Remote Sensing'. *Ecological Applications* 17(8): 2137–2144.

Bormpoudakis, D., J. Sueur, and J. D. Pantis (2013). 'Spatial Heterogeneity of Ambient Sound at the Habitat Type Level: Ecological Implications and Applications'. *Landscape Ecology* 28(3): 495–506.

Brown, E. D. and B. K. Williams (2016). 'Ecological Integrity Assessment as a Metric of Biodiversity: Are We Measuring What We Say We Are?'. *Biodiversity and Conservation* 25(6): 1011–1035.

Calheiros, D. F., A. F. Seidl, and C. J. A. Ferreira (2000). 'Participatory Research Methods in Environmental Science: Local and Scientific Knowledge of a Limnological Phenomenon in the Pantanal Wetland of Brazil'. *Journal of Applied Ecology* 37(4): 684–696.

Carruthers-Jones, J., S. Carver, and R. McMorran (forthcoming). "Participatory Mapping of Wildness: Assessing the Potential of Mixed Methods Walking Research for Ground Truthing Wildness Mapping'.

Carruthers-Jones, J., Eldridge, A., Guyot, P., Hassall, C. and Holmes, G., 2019. The call of the wild: Investigating the potential for ecoacoustic methods in mapping wilderness areas. Science of The Total Environment, 695, p.133797.

Carver, S., J. Carruthers-Jones, and A. Guette (2019). 'À la recherche des derniers lieux sauvages de la planète'. *The Conversation*, 5 March. Available online: https://

theconversation.com/a-la-recherche-des-derniers-lieux-sauvages-de-la-planete-111446 (accessed 20 April 2019).

Carver, S., A. Comber, R. McMorran, and S. Nutter (2012). 'A GIS Model for Mapping Spatial Patterns and Distribution of Wild Land in Scotland'. *Landscape and Urban Planning*, 104(3–4): 395–409.

Carver, S., L. Comber, S. Fritz, R. McMorran, S. Taylor, and J. Washtell (2008). *Wildness Study in the Cairngorms National Park*. Leeds: University of Leeds.

Certeau, M. de (1984). *The Practice of Everyday Life*. Berkeley: University California Press.

Chan, K. M. A., M. R. Shaw, D. R. Cameron, E. C. Underwood, and G. C. Daily (2006). 'Conservation Planning for Ecosystem Services'. *PLoS Biology* 4(11): e379.

Colvin, R. M., G. B. Witt, and J. Lacey (2016). 'Approaches to Identifying Stakeholders in Environmental Management: Insights from Practitioners to Go Beyond the "Usual Suspects"'. *Land Use Policy* 52: 266–276.

Droumeva, M. (2017). 'Soundmapping as Critical Cartography: Engaging Publics in Listening to the Environment'. *Communication and the Public* 2(4): 335–351.

Dudley, N. (2009). 'Why is Biodiversity Conservation Important in Protected Landscapes?'. *George Wright Forum* 26(2): 31–38.

Edensor, T. (2010). 'Walking in Rhythms: Place, Regulation, Style and the Flow of Experience'. *Visual Studies* 25(1): 69–79.

Eldridge, A., P. Guyot, P. Moscoso, A. Johnston, Y. Eyre-Walker, and M. Peck (2018). 'Sounding Out Ecoacoustic Metrics: Avian Species Richness is Predicted by Acoustic Indices in Temperate but Not Tropical Habitats'. *Ecological Indicators* 95: 939–952.

Elia, G. and A. Margherita (2018). 'Can We Solve Wicked Problems? A Conceptual Framework and a Collective Intelligence System to Support Problem Analysis and Solution Design for Complex Social Issues'. *Technological Forecasting and Social Change* 133: 279–286.

Evans, J. and P. Jones (2011). 'The Walking Interview: Methodology, Mobility and Place'. *Applied Geography* 31(2): 849–858.

Eyre, T. J., A. L. Kelly, V. J. Neldner, B. A. Wilson, D. J. Ferguson, M. J. Laidlaw, and A. J. Franks (2015). 'BioCondition: A Condition Assessment Framework for Terrestrial Biodiversity in Queensland'. Assessment Manual. Version 2.2. Queensland Herbarium, Department of Science. Information Technology, Innovation and Arts, Brisbane.

Farina, A. (2013). *Soundscape Ecology: Principles, Patterns, Methods and Applications*. Dordrecht: Springer.

Farina, A. and N. Pieretti (2014). 'From Umwelt to Soundtope: An Epistemological Essay on Cognitive Ecology'. *Biosemiotics*: 7(1): 1–10.

Feld, S. (1996). 'Waterfalls of Song: An Acoustemology of Place Resounding in Bosavi, Papua New Guinea'. In S. Feld and K. Basso (eds), *Senses of Place*, 90–135. Santa Fe, NM: School of American Research Press.

Fuller, B. (1966). 'The Music of the New Life thoughts on Creativity, Sensorial Reality, and Comprehensiveness'. *Music Educators Journal* 52(5): 46–146.

Gage, S. H., B. Napoletano, and M. Cooper (2001). 'Assessment of Ecosystem Biodiversity by Acoustic Diversity Indices'. *Journal of the Acoustic Society of America* 109(5): 2430.

Gagliano, M., S. Mancuso, and D. Robert (2012). 'Toward Understanding Plant Bioacoustics'. *Trends in Plant Science* 17: 323–325.

Gallagher, M., A. Kanngieser, and J. Prior (2017). 'Listening Geographies: Landscape, Affect and Geotechnologies'. *Progress in Human Geography* 41(5): 618–637.

Haraway, D. (2008). *When Species Meet*, Minneapolis and London: University of Minnesota Press.

Harris, S. A., N. T. Shears, and C. A. Radford (2016). 'Ecoacoustic Indices as Proxies for Biodiversity on Temperate Reefs'. *Methods in Ecology and Evolution* 7(6): 713–724.

Hollowell, J. (2009). 'Using Ethnographic Methods to Articulate Community-Based Conceptions of Cultural Heritage Management'. *Public Archaeology: Archaeological Ethnographies* 8(2–3): 141–160.

Ingold, T. (2000). *The Perception of the Environment: Essays on Livelihood, Dwelling and Skill*. London: Routledge.

Ingold, T. (2007). 'Against Soundscape'. In A. Carlyle (ed.), *Autumn Leaves: Sound and the Environment in Artistic Practice*, 10–13. Paris: Double Entendre.

Kasten, E. P., S. H. Gage, J. Fox, and W. Joo (2012). 'The Remote Environmental Assessment Laboratory's Acoustic Library: An Archive for Studying Soundscape Ecology'. *Ecological Informatics* 12: 50–67.

Kesby, M., S. Kindon, and R. Pain (2004). 'Participatory Approaches and Diagramming Techniques'. In R. Flowerdew and M. Martin (eds), *Methods in Human Geography*, 144–166. London: Pearson.

Krause, B. (1993). 'The Niche Hypothesis'. *Soundscape* Newsletter 6: 6–10.

LaBelle, B. (2010). *Acoustic Territories: Sound Culture and Everyday Life*. London: Continuum.

Latour, B. (2004). 'Why Has Critique Run Out of Steam?: From Matters of Fact to Matters of Concern'. *Critical Inquiry* 30(2): 225–248.

Latour, B. (2008). 'A Cautious Prometheus? A Few Steps toward a Philosophy of Design (with Special Attention to Peter Sloterdijk)'. In *Proceedings of the 2008 Annual International Conference of the Design History Society*, 2–10.

Lesslie, R. and M. Maslen (1995). *National Wilderness Inventory Handbook of Procedures: Content and Usage*. 2nd edition. Canberra: Commonwealth Government Printer.

Liu, J., T. Dietz, S. R. Carpenter, M. Alberti, C. Folke, E. Moran, A. N. Pell, P. Deadman, T. Kratz, J. Lubchenco, and E. Ostrom (2007). 'Complexity of Coupled Human and Natural Systems'. *Science* 317(5844): 1513–1516.

Ma, S. and Y. Long (2019). 'Mapping Potential Wilderness in China with Location-based Services Data'. *Applied Spatial Analysis and Policy*: 1–21.

Maginn, P. J. (2007). 'Towards more effective community participation in urban regeneration: the potential of collaborative planning and applied ethnography'. *Qualitative Research* 7(1): 25–43.

McIntosh, E. J., S. Chapman, S. G. Kearney, B. Williams, G. Althor, J. P. R. Thorn, R. L. Pressey, M. C. McKinnon, and R. Grenyer (2018). 'Absence of Evidence for the Conservation Outcomes of Systematic Conservation Planning around the Globe: A Systematic Map'. *Environmental Evidence* 22(7): 1–23.

McKenna, M. F., R. Buxton, M. Clapp, and E. Meyer (2017). 'Rapid Extraction of Ecologically Meaningful Information from Large-Scale Acoustic Recordings'. *Journal of the Acoustical Society of America* 141(5): 3940–3940.

McMorran, R. and J. Carruthers-Jones (2015). 'Scotland's Wild Mountains: Addressing Key Challenges'. *Geographer* 4: 18.

Middleton, J. (2018). 'The Socialities of Everyday Urban Walking and the "Right to the City"'. *Urban Studies* 55(2): 296–315.

Milner-Gulland, E. J., J. A. McGregor, M. Agarwala, G. Atkinson, P. Bevan, T. Clements, T. Daw, K. Homewood, N. Kumpel, J. Lewis, and S. Mourato (2014). 'Accounting for the Impact of Conservation on Human Well-Being'. *Conservation Biology* 28(5): 1160–1166.

Müller, A., P. K. Bøcher, and J. C. Svenning (2015). 'Where Are the Wilder Parts of Anthropogenic Landscapes? A Mapping Case Study for Denmark'. *Landscape and Urban Planning* 144: 90–102.

Pain, R. (2004). 'Social Geography: Participatory Research'. *Progress in Human Geography* 28(5): 652–663.

Phillips, Y. F., M. Towsey, and P. Roe (2018). 'Revealing the Ecological Content of Long-Duration Audio-Recordings of the Environment through Clustering and Visualisation'. *PloS one* 13(3): e0193345.

Pierce, J. and M. Lawhon (2015). 'Walking as Method: Toward Methodological Forthrightness and Comparability in Urban Geographical Research'. *Professional Geographer* 67(4): 655–662.

Pijanowski, B., L. Villanueva-Rivera, S. L. Dumyahn, A. Farina, B. Krause, B. Napoletano, S. Gage, and N. Pieretti (2011). 'Soundscape Ecology: The Science of Sound in the Landscape'. *BioScience* 61. https://doi.org/10.1525/bio.2011.61.3.6.

Pink, S. (2007). 'Walking with Video'. *Visual Studies* 22(3): 240–252.

Pink, S. (2010). 'The Future of Sensory Anthropology/the Anthropology of the Senses'. *Social Anthropology* 18(3): 331–333.

Probst, K. and J. Hagmann, with contributions from M. Fernandez and J. A. Ashby (2003). *Understanding Participatory Research in the Context of Natural Resource Management – Paradigms, Approaches and Typologies*. ODI Agricultural Research and Extension Network Paper No. 130. London: Overseas Development Institute.

Ramella, A. L. and R. Norum (forthcoming). 'Mobile Geographie'. In T. Thielmann and M. Kanderske (eds), *Mediengeographie: Handbuch für Wissenschaft und Praxis*. Baden-Baden: Nomos Verlag.

Redpath, S. M., J. Young, A. Evely, W. M. Adams, W. J. Sutherland, A. Whitehouse, A. Amar, R. A. Lambert, J. D. Linnell, and A. Watt (2013). 'Understanding and Managing Conservation Conflicts'. *Trends in Ecology and Evolution* 28: 100–109.

Reed, M. S., A. Graves, N. Dandy, H. Posthumus, K. Hubacek, J. Morris, C. Prell, C. H. Quinn, and L. C. Stringer (2009). Who's in and Why? A Typology of Stakeholder Analysis methods for Natural Resource Management'. *Journal of Environmental Management* 90(5): 1933–1949.

Reed, M. S., S. Vella, E. Challies, J. de Vente, L. Frewer, D. Hohenwallner-Ries, T. Huber, R. K. Neumann, E. A. Oughton, J. Sidoli Del Ceno, and H. van Delden (2018). 'A Theory of Participation: What Makes Stakeholder and Public Engagement in Environmental Management Work?'. *Restoration Ecology* 26: S7–S17.

Rittel, H. W. J. and M. M. Webber (1973). 'Dilemmas in the General Theory of Planning'. *Policy Sciences* 4: 155–169.

Salazar, N. B., A. Elliot, and R. Norum (2017). 'Studying Mobilities: Theoretical Notes and Methodological Queries'. In A Elliot, R. Norum, and N. B. Salazar (eds), *Methodologies of Mobility: Ethnography and Experiment*, 1–21. Oxford: Berghahn.

Sanderson, E. W., M. Jaiteh, M. A. Levy, K. H. Redford, A. V. Wannebo, and G. Woolmer (2002). 'The Human Footprint and the Last of the Wild: The Human Footprint Is a Global Map of Human Influence on the Land Surface, Which Suggests that Human Beings Are Stewards of Nature, Whether We Like It or Not'. *BioScience* 52(10): 891–904.

Schafer, R. M. (1994). *The Soundscape: Our Sonic Environment and the Tuning of the World*. Rochester, VT: Destiny Books.

Schöner, M. G., R. Simon, and C. R. Schöner (2016). 'Acoustic Communication in Plant–Animal Interactions'. *Current Opinion in Plant Biology* 32: 88–95.

Scrimshaw, W. (2013). 'Non-cochlear Sound: On Affect and Exteriority'. In M. Thompson and I. Biddle (eds), *Sound, Music, Affect: Theorizing Sonic Experience*, 27–43. New York: Bloomsbury.

Soulé, M. (2014). 'The "New Conservation"'. In G. Wuerthner, E. Crist, and T. Butler (eds), *Keeping the Wild*, 66–80. Washington, DC: Island Press.

Stowell, D. and M. D. Plumbley (2014). 'Automatic Large-Scale Classification of Bird Sounds is Strongly Improved by Unsupervised Feature Learning'. *PeerJ* 2: e488. https://doi.org/10.7717/peerj.488.

Sueur, J. and A. Farina (2015). 'Ecoacoustics: The Ecological Investigation and Interpretation of Environmental Sound'. *Biosemiotics* 8(3): 493–502.

Sutherland, W. J., A. S. Pullin, P. M. Dolman, and T. M. Knight (2004). 'The Need for Evidence-based Conservation'. *Trends in Ecology and Evolution* 19: 305–308.

Turesson, H. K., S. Ribeiro, D. R. Pereira, J. P. Papa, V. H. C. de Albuquerque (2016). 'Machine Learning Algorithms for Automatic Classification of Marmoset Vocalizations'. *PLoS ONE* 11(9): e0163041.

Uexküll, J. V. (1926). *Theoretical Biology*. New York: Harcourt, Brace & Co.

Veits, M., I. Khait, U. Obolski, E. Zinger, A. Boonman, A. Goldshtein, K. Saban, R. Seltzer, U. Ben-Dor, P. Estlein, and A. Kabat (2019). 'Flowers Respond to Pollinator Sound within Minutes by Increasing Nectar Sugar Concentration'. *Ecology Letters* 22(9): 1483–1492.

Vuceticha, J. A., D. Burnham, E. A. Macdonald, J. T. Bruskotter, S. Marchini, A. Zimmermann, and D. W. Macdonald (2018). 'Just Conservation: What Is It and Should We Pursue It?'. *Biological Conservation* 221: 23–33.

Warner, C. (2015). 'Participatory Mapping: A Literature Review of Community-based Research and Participatory Planning'. Social Hub for Community Housing, Faculty of Architecture and Town Planning Technion. Available online: https://pdfs.semanticscholar.org/883f/768308cac12556b912750590e9720c42b146.pdf (accessed 8 July 2020).

Weizman, E. (2017). *Forensic Architecture: Violence at the Threshold of Detectability*. New York: Zone Books.

Westerkamp, H. (1998). 'Speaking from Inside the Soundscape'. In H. Karlsson (ed.), *Proceedings from 'Stockholm Hey Listen!' Conference on Acoustic Ecology*, 53–63. Stockholm: Royal Swedish Academy of Music.

Wilmott, C. (2020). *Mobile Mapping: Space, Cartography and the Digital*. Amsterdam: Amsterdam University Press.

World Wide Fund for Nature (WWF) (2008). *A Roadmap for a Living Planet*. Gland: WWF.

World Wide Fund for Nature (WWF) (2018). *Living Planet Report – 2018: Aiming Higher*, ed. M. Grooten and R. E. A. Almond. Gland: WWF.

40
The Emergence of Voices in an Indian Bus Stand: An Ethnographic and Acoustic Approach
Christine Guillebaud

The topic of sound is more relevant than ever in the contemporary world, and this is even more the case in countries of the Global South. Worldwide rankings of the most noise-polluted cities include the megalopolises of Mumbai, Cairo, and Tokyo. There as elsewhere, decibel counts are measured scrupulously according to standards set by national and international organizations.[1] Sound pollution has now become one of central governments' major concerns in the management of public spaces and infrastructure. If this conception of 'pollution' is legitimate from the perspectives of public health and the improvement of citizens' quality of daily life, it also touches on a number of anthropological issues. Ambient sound is produced and altered by a wide range of materials and surfaces, weather conditions, and media upon which its propagation depends. However, by nature it is also immaterial and part of everyday sensory experiences. This inherent complexity should be taken into account by treating sound as a composite material, its perception necessarily drawing from a vast spectrum of ways of paying attention, spanning simple inattention to ordinary sounds all the way to specific forms of listening, such as listening to acoustically prominent sounds that organize or prompt human activities. Indeed, at a local level, a simple physical decibel level count taken near an intersection, hospital, or school says nearly nothing of how residents and passers-by use and listen to the space. It says nothing of how they perceive ambient sounds or how they appraise and appreciate the sensory environment.

The notion of 'pollution' arbitrarily places thousands of daily commercial and ritual activities on the same level of acoustic reality, although their sounding characteristics are difficult to compare. Examples are not hard to find: festive uses of fireworks, various calls to prayer, loudspeaker systems, all manner of sound distortion, local sales methods (itinerant street vendors, bazaars, markets, etc.), dense transportation network signalling, commonplace ways of initiating interaction with others, or even multiple ways of conferring cultural identity on a place and its correlate, residents forming a community. The undertaking is all the while guided by social and cultural contexts that give local meaning, thus distinguishing it from collecting quantitative data to analyse 'nuisance' or 'pollution'. The ethnographic approach is primarily devoted to understanding the *sensory* modalities of the production of sound environments, decrypting the range of local knowledge and the imaginaries that they inspire in a given group or society.

My research[2] consists of observing (and recording) fragments of daily life in different public spaces (markets, temples, train/bus stations). By focusing on their sensory dimensions, one of the goals of this research is to restore in situ listening and observation. Different vocal and sound techniques are produced in these spaces to create effects that are perceived by inhabitants, clients, and passers-by, generally to attract their attention or even to shape 'ambiances'[3] that impose specific listening postures on these people. The present chapter focuses on an Indian bus station, considering it as a relatively autonomous *milieu*. By *milieu*, I mean a composite world made up of sounds produced, perceived, and listened to either intentionally or coincidentally. This simple definition, centred on the idea of experienced 'sound worlds' (Canzio 1992), is inscribed in a wider anthropological undertaking. Indeed, I consider the bus station as a site for everyday public interactions (Goffman [1959] 1973), which involve different procedures for sound perception as well as singular ways to manage the crowd.

I have previously described the sonic organization of the Saktan Tampuran Bus Stand located north of Thrissur, a city of the southern Indian state of Kerala (Guillebaud 2017). This sound environment seems very dense due to a number of concomitant sources and a rather high sound pressure level (decibels). Analysis of everyday interactions makes clear that instead of thinking of this sound space as a single, coordinated whole, it should be understood as consisting of multiple scales of listening mainly organized around the cries of ticket vendors announcing bus departures.[4] Criers' voices are the product of specifically acoustic work intended to draw the attention of passers-by by making their calls heard in a dense sound environment.

This chapter combines ethnographic observation methods with acoustic analysis (sonograms) to study sound in specific locations. Stations, which sociologist Antoine Hennion (2012) has called 'movements/places', offer high human and sonic density. The ethnography that follows will first present the triangle sound-perception-action, which regulates daily flows. Acoustic analysis will then demonstrate the acoustic signatures of prominent voices based on configuration types that have been observed in situ.

The bus station: A sound space and its scales of listening

At the Saktan Tampuran bus station, a space of mobility and circulation, travellers find themselves immediately thrust into a dense sound environment. The multiple events occurring simultaneously give the impression of a vast sonic chaos. The observer is struck by the number of (sound) events and the resulting sonic saturation, even more so because the spatial configuration of the station seems, on the contrary, to be particularly well defined. Following the stream of passengers, a central platform can be seen along which several dozen buses are parked, awaiting their departure. These vehicles are not parked randomly, as each has taken the space reserved for its destination. Among them is the town of Guruvayur, a holy place with its celebrated Krishna temple; Guruvayurappan, a highly sought-after destination, especially during the festival periods marking the Hindu calendar; Shoranur, a town where many transfers are made to northern districts; and Kuntakulam, an urban centre with a variety of traditional factories to which many workers commute daily. In the station, there are also many established small businesses, teashops, and grocery shops lined up in the middle of the platform. They mark the circulation space of travellers moving in opposite directions, up and down the platform which functions as an intermediary space reserved for passengers waiting to board.

The space concentrates sounds of traffic, motors, whistles, and the many cries of ticket vendors. The first impression of this sound space is one of extreme discord between what can be seen – the steady stream of passengers moving around the central platform – and the numerous sound actions taking place. There is no coordinated and overarching logic of this sound space, but rather different scales of listening that are mainly organized around the criers.

The crier: The man who captures more than he informs

Upon listening to the criers for the first time, one might think that their calls are informative – in other words, that they are simply making announcements. If this were so, their function would be similar to that of the vocal announcements – either pre-recorded or 'live' – broadcast via microphones and loudspeakers, as is the case in many bus and railway stations around the world. But in this case, no other informational system is associated with these voices, no visible signage or timetables, not even an information desk. There are as many criers as there are destinations, and just as many buses ready to depart. This is essential: the entirely acoustic character of the announcements and the numerous sound sources imply that there are also many ways of perceiving and locating the ticket vendors' voices.

From a semantic point of view, the criers' voices always indicate the destination – the name of a town. However, the striking feature of their calls is not what they explicitly state but their sonic form. Rather than choosing semantic clarity, articulation, and intelligibility, the criers deliberately shorten the names of towns and cities. The holy city of 'Guruvayur' becomes 'Guruyur', Kuntakulam is heard as 'Kulam', and the city of Thrissur becomes 'Shur'. The contraction of place names is combined with a prosodic principle of constant repetition and a melodic and tonal coloration that amplifies the phenomenon of personalization. Due to the simultaneous cries of the ticket vendors and the consequent extreme proliferation, the meaning of words is somewhat blurred in favour of specific sound effects.

With this type of sound production, one might wonder whether the term 'announcement' is truly relevant, as each crier develops his own singular way to capture passengers' attention. An announcement usually implies a certain element of schedule and time; it has a provisional nature. In many countries, a single voice broadcasting loudly and uniformly throughout the station is used to ensure the punctuality of departures. It relates information expressed in a future tense – the train or bus 'will leave' at such and such a time from this or that platform; it coordinates pre-scheduled actions to which travellers are invited to comply. In an Indian bus station, on the contrary, time is conjugated in the present; time is *immediate*. Vocalized information is transmitted when passengers are already on the move; it is not used to invite passengers to move. Furthermore, the acoustic propagation of these voices can only be heard in a limited area since they are not amplified, resulting in a dense sound space where they do not monopolize the full attention of the passengers. With his call, the crier simultaneously condenses three types of information. He indicates in situ the destination (semantically identifiable), the bus's location (by the fact that he is next to the vehicle), and the imminent departure (by capturing the attention of the potential client).

Passengers' modalities of perception

Now, from the regular passenger's (listener's) point of view, the general organization of the bus station is well known. Each space in the station is dedicated to a particular destination, and because this placement is rarely modified, travel habits become ingrained. The ticket vendor's action coincides with a very precise moment in the passenger's attention: the moment when passengers visually perceive the vendor (lateral vision) and acoustically distinguish his projected voice from the rest – that is, just before they get on the bus. Informal interviews conducted with vendors confirmed this observation. For the most part, the call is nothing other than a way to make a 'time announcement', an expression the vendors made in English to describe their work.[5] Behind the station criers' calls lie the principles of attraction and recognition. In this context, passengers are literally immersed in multiple sound spheres that impel them to listen in certain ways. The multiple projections of sound act upon their perceptive sphere and almost simultaneously drive them towards their bus. The very fact that the criers' voices do not compete according to a monopolistic principle makes it possible to understand the sound space as being organized in a fragmented system that is anything but random, but governed by the logic of *multiple attraction*.

Managing the flow of the crowd

Another important element must be taken into account: there is no space reserved for lining passengers up or putting them in order. The stops in the station are temporary and the ways of entering buses are subject to few rules. The passengers board when they present themselves at the bus and the departures happen when the vehicles are full. The throngs on platforms are commonly thought to indicate an inescapable wait or delay for the passengers, but in fact another logic is at work: the continuous flow of departures and the competition between buses favours a degree of fluidity in the crowds. This sonic and anti-monopolistic approach to managing crowds makes it less effective to study the station through the figures traced by an organized movement of people. For example, if waiting lines were imposed in such an acoustic economy, the sought-after fluidity would be hampered. In an article on the culture of the queue in India, Ajay Gandhi summarizes the line's defining feature as follows: 'It is a teleological and universal form; requires bodily self-containment; demands synchronicity with others; and inculcates a detached, disciplinary sense of place' (Gandhi 2013: 5). Gandhi justly underlines the quasi-emblematic nature of the line in contemporary Indian institutions, as a manner to 'normalize' crowds (and their bodies).

The absence of waiting lines in the bus station strongly contrasts with the example of the 'massified queues' imposed, for example, in the metro of the Indian capital New Delhi. It is clearly characteristic of the organizational methods applied by the private bus companies operating in the city, where business competition demands decentralized crowd management. The absence of synchronicity and apparent discipline (to use the terms in Gandhi's definition) do not equate with chaos and confusion. The logics that I have identified thus far emerge in all their singularity:

1) Acoustic salience of the voices
2) Triple semantics (destination, localization, departure)
3) Multiple attraction
4) Sound-action instants
5) Principle of flows

The principle of a multitude of prominent voices finds an echo in the visual logic of the buses. The colours, motifs, drawings of deities, and ornaments all distinguish each vehicle and contribute to a visual competition among them (Figure 40.1).

Acoustic analysis of prominent voices

The second phase of my study, executed in collaboration with Vincent Rioux, acoustician and computer scientist, and member of the research group MILSON, consisted in analysing the criers' voices by using the program Sonic Visualizer.[6] Here I will present three distinct examples, corresponding to a variety of individual ways of performing the cry: the content of the utterance, the vocal technique being used, and the rhythm of the utterance. The examples are also selected according to the number of audible individuals: one main vendor dominating the sound spectrum; one main vendor dominating two secondary vendors; seven vendors crying simultaneously.

Figure 40.1 Buses at Saktan Tampuran, 2015–2016. Photographs by Christine Guillebaud.

One main vendor

The first sonogram (see Figure 40.2a) represents a sequence where a ticket vendor is calling 'Palakkad', a destination in the north of the Thrissur district. It will first be read vertically, according to three frequency bands:

1. From ± 200 hertz to ± 700 hertz: this band marks the zone where spoken voices, motor noises, and all the sources of the ordinary *brouhaha* of the station can be found.
2. From ± 700 hertz to ± 2,800 hertz: this band marks the performance space of the vendors, who deliberately place their voices just above the first zone of the spectrum.
3. From 2,800 hertz to 3,800 hertz, and above: the third zone marks the frequency zone where whistle sounds are by far the most prominent. They are generally used when buses arrive or depart in order to guide the drivers' manoeuvres at a distance. In terms of perception, this zone is the most easily heard, and little energy is required to make these signals audible. This third band is inaccessible to the human voice.

In the vendor's zone (2), there is a clear strategy to produce a prominent sound. The fundamental of the human voice falls somewhere in the brouhaha zone (1), which is lower and less audible. The vendor's vocal techniques make it possible to strengthen the second harmonic (an octave above the fundamental). The energy is thus mostly concentrated around 800 hertz, which is just above the brouhaha and in the zone where the human voice generally performs the best.

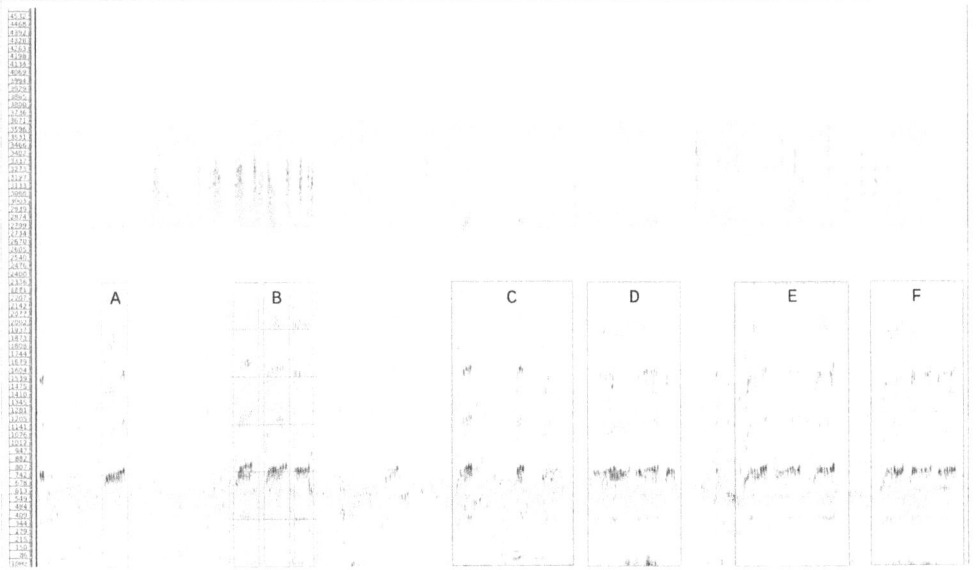

Figure 40.2a Sonogram of the main vendor making the utterance 'Palakkad' (duration: 2'43").

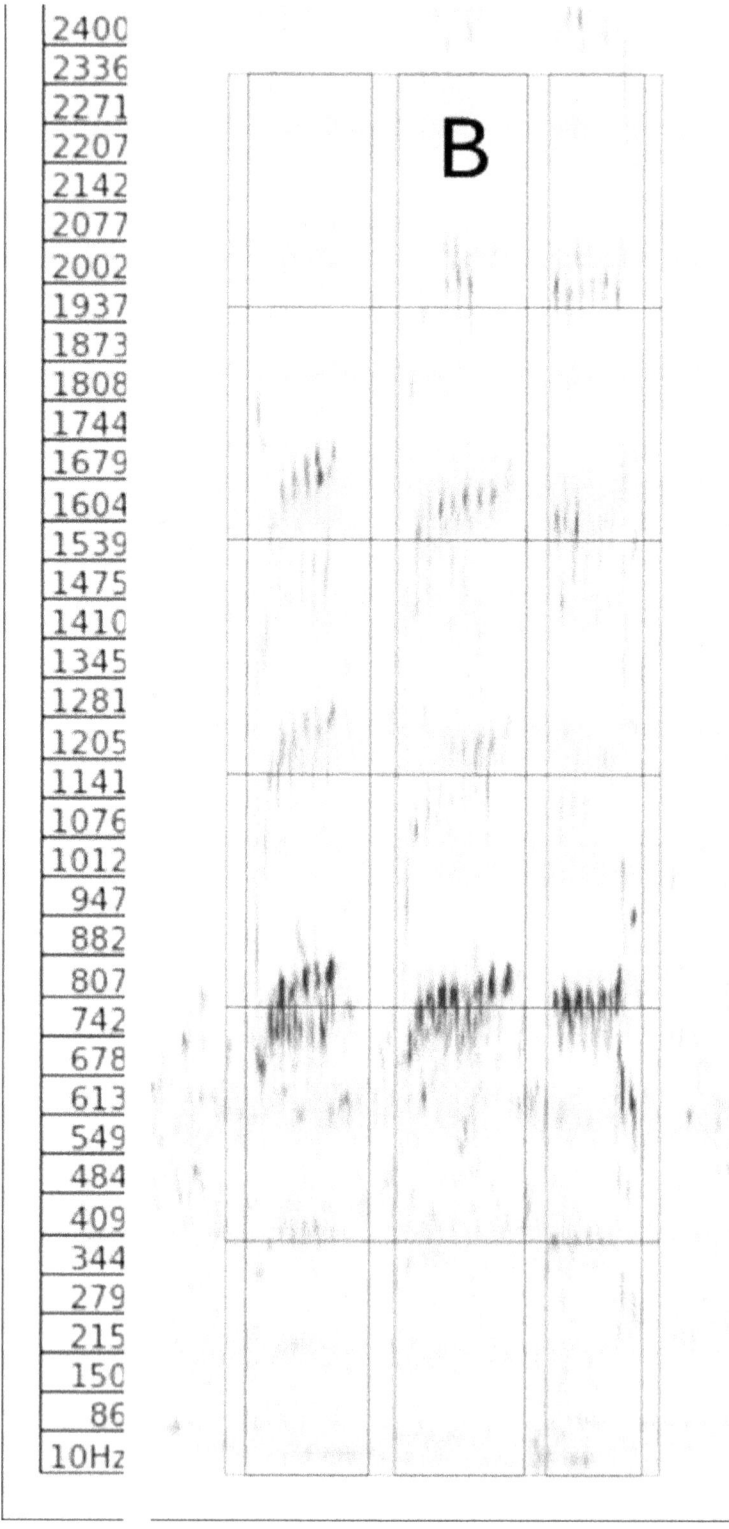

Figure 40.2b Sonogram focused on utterance B.

A horizontal reading of the sonogram (time) reveals six distinct blocks to which I have assigned letters (A, B, C, D, E, and F). They correspond to successive interventions by the same vendor. The timing is not entirely regular between the blocks: the vendor listens to what is happening in his environment and reacts accordingly.

A focus on block B (see Figure 40.2b) makes it possible to visualize the five harmonic components of the vendor's voice. On the rhythmic level, he utters the name of the destination city in a nearly cyclical manner, in three distinct phrases. Each of these phrases is composed of a continuous repetition of the name of the city (here, an average of eight times per phrase). However, the time intervals between the phrases in a given block are not of strictly equal duration. This irregularity is particularly visible in block C (Figure 40.2a), where the vendor adapts to the sounds around him to the point that the impression of cyclicality is lost for a moment, thereby clearly demonstrating his capacity for listening and adapting. In block D, another vendor starts to compete with the first (see the box located between 1'38" and 1'44"). The pitch of this voice is slightly lower than that of the main vendor, but he uses the same vocal technique. Over six seconds, the first vendor resumes his cry based on the second vendor, so that he can still be heard. The main vendor thus places his performance (and focuses his listening) in relation to zone (1), ensuring that it will be prominent in the spectrum. The concentration of harmonic energy (vocal technique), prosodic work (the continuous repetition of the name of the destination in three phrases), and temporal adjustments according to the performances of other vendors around him complete the desired attraction process.

One main vendor and two secondary vendors

The sonogram in Figure 40.3 shows the interaction between three vendors. The main one is announcing the town of Peecheedam, the second one Kuntakulam, and the third Palakkad. The first vendor largely dominates the spectrum in both the number of cries (four in all) and by the harmonic richness of his voice. A comparison with the previous sonogram makes it possible to mark off three frequency bands in exactly the same way. The fundamental of the voice of vendor 1 is in the brouhaha zone (1) while his vocal technique allows him to strengthen the second harmonic (cf. three boxes) practically in the same location as the vendor recorded in the previous example, which is around 700 hertz. In this example, the vendor's performance is shorter (the town name is repeated at an average of five times per block) and the vocalizations differ radically. In this instance, the main vendor systematically uses a rising frequency at the end of the phrase and works on the saturation of his vocal cords in the last 'Peecheedam […] Pee*che*', which make all the harmonics emerge quite clearly.

The other two vendors alternate their voices with the first. Their harmonic components are less visible in the sonogram, which is mainly due to microphone placement and their being at a greater distance from the microphone. However, the visual representation shows how these additional voices coordinate themselves over time, either by inserting themselves into pauses or by brief superimpositions.

Figure 40.3 Sonogram of the main vendor making the utterance 'Peecheedam', while the two secondary vendors call 'Kuntakulam' and 'Palakkad' (duration: 1'04").

Seven simultaneous vendors

In this final example, several vendors enter into a sonic competition. Seven of them can be made out, and each has been assigned a number. They are first of all distinguished by the content of their utterances (vendor 1: 'Kodungallur'; vendor 2: 'Vellur'; etc.) Unlike the preceding examples, the *brouhaha* zone (1) is nearly absent, which allows a much clearer visualization of the voices' 'F' fundamentals.[7] Because of this relative clarity, the vendors place their fundamentals a little lower than in the previous examples (here around 300 hertz), while using vocal technique to strengthen the second harmonic (around 600 hertz, 'H 2') and the fourth harmonic (around 900 hertz, 'H 4'), which are both quite visible.

From a temporal perspective, the sonogram in Figure 40.4 demonstrates the logic of aggregation of vocal occurrences. The sequence reveals first the successive entry and then the alternation between the first four vendors. One might hypothesize that two more vendors (5 and 6) then perceive this presence of the multiple voices and are drawn to the sonic aggregate, joining it with their respective voices in a concomitant way. After the sequence starting with four vendors, the opposite logic can be seen: most of them stop calling out to leave a less saturated space, and only vendor 3 continues his cry. This moment of relative pause makes the voice of a new vendor (7) burst forth, initially superimposed by the voice of vendor 3 and then alternating with him.

It is difficult to imagine these interventions – progressing through aggregation, disaggregation, and reaggregation in succession – as a conscious script planned by each vendor or as a collective predetermined composition, like a musical score. The logic of filling busses also influences when voices appear and disappear, just like the pauses for

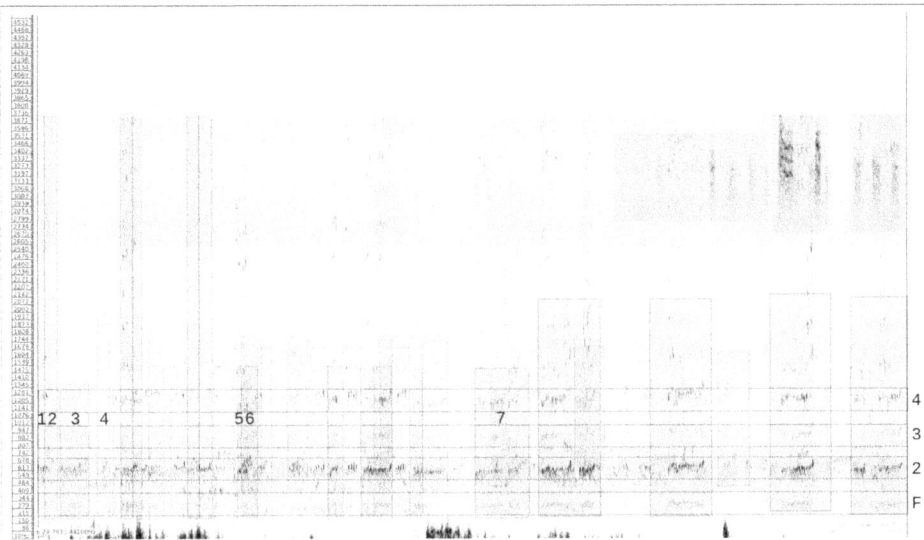

Figure 40.4 A multiple-configuration sonogram with seven vendors (duration: 2′00″).

breath imposed by the phrasing of each utterance and their inherent cyclicality. Here we can experience interactions that are built in situ and share a common frequency band. However, the performances are not static, and the individual performances are not simply juxtaposed. The space is organized according to a singular relationship between an individual and a collective performance based on coordinating attraction, competition, and timing.

Conclusion

This ethnographic and acoustic approach to an Indian bus station's sound space firstly illustrates that social interaction in this competitive setting is ultimately created by occupying different strata of the sound spectrum. The examples of the vendors' voices are to be considered as complex practices of sonic manipulation, seeking prominence through tone, prosody, and timing in addition to a merely modulating sound intensity. The acoustic events are organized at different scales in this sound environment, relying on auditory acuity to serve economic ends and efficiently manage the crowd on a daily basis.

By comparing such 'public voices', it becomes clear that, ontologically speaking, sound creates nothing in and of itself, but it does have the potential to create action and co-action, and this is what characterizes the users' experience. This approach is in line with social scientists' current efforts to address the perception–action dyad through affordance.[8] Speaking of the affordances of sound events not only means considering perceptions as situated achievements, as genuine actions, but furthermore assessing how certain events do much more than simply attract attention: they reorganize and affect one's activities. The example of the Indian bus station can easily be placed among 'complex' forms of action and reaction. It relies upon multiple forms of attention by travellers and weaves together focal points for perception that are produced by station sounds and reconfigured according to the effects of various events.[9] This reconfiguration also concerns the vendors. They are more than mere competitors meeting a challenge; within that challenge there is a dynamic of mutual change. This is not a face-off situation, as between card players, but rather a relationship that transforms each participating individual: the one is invested in the sound events he receives from the other, and the other invests in the response he gets in return.

There is, however, an additional dimension that seems to escape the category of affordance: the self-generating nature of these sound productions. Sound forms are regulated in situ, beyond the exclusive, individual intervention of each sound producer, but they are not the result of the simple sum of these contributions. This case study thus comes within the scope of collective coordination methods and what is commonly called 'theories of emergence'. Generally designating the appearance of new characteristics and behaviours beyond a certain degree of complexity, 'emergence' has acquired various meanings and has given rise to very different uses in the arts, humanities, and sciences. It has engendered research in each of these domains, leading to new considerations of interactions taking place between the individual and the collective, and highlighting the complexity of these interactions, that is, of *acting collectively*. This study of a bus station in India demonstrates how that collective production and creation is modified when the conditions for coordination among the vendors are changed. The self-organizing nature of these sound events raises questions about the methods and behaviours that arise as a result of sound density and a sonic heterogeneity that would not emerge if the individuals worked in more homogeneous conditions.

Notes

1. In particular the World Health Organization's (WHO) 'Guidelines for Community Noise', edited by Birgitta Berglund and Thomas Lindvall (1995), an updated version of the document published by the WHO in 1980 (see WHO 1999).
2. This research has been developed through the MILSON programme (Pour une anthropologie des MILieux SONores / For an Anthropology of Sound Environments) that I have had the honour of directing since 2011 with the support of the Fyssen Foundation. For further information, see MILSON 2020.
3. For an archeology of the notion of (sonic) ambiance, see notably Thibaud 2012.
4. A comparison was made with other public voices, such as the voice of the Kerala lottery (which is sped up and enhanced for acoustic prominence), or the announcements in French public transportation (Guillebaud 2017).
5. Malayalam is the language spoken in Kerala. However, in everyday life, a certain number of expressions are specifically in English, generally when the speakers wish to convey a certain element of modernity (as is the case here) or when the message should carry a certain emotional charge.

6. Freeware distributed under the GNU General Public License and developed at the Centre for Digital Music at Queen Mary, University of London (see Sonic Visualiser n.d.).
7. Below this band, in the lower range of the spectrum, the sonogram shows sounds that I made while implementing the recording. In the higher band, the sound of whistles generally remains prominent.
8. 'Affordance' is a term initially introduced in the psychology of visual perception by James J. Gibson (1979). The word plays on two meanings of the verb 'to afford': to offer and to supply. The affordance of an object refers to its capacity or ability to suggest a use or make a use possible as soon as the object is perceived. For a summary and analysis of the relevance of affordance to sound events, see Thibaud 2010, 2011; and Pecqueux 2012: 215–221.
9. For a characterization of the affordance of sound events according to two types, 'simple' and 'complex', see Guillebaud 2017: 95.

References

Canzio, R. (1992). 'Mode de fonctionnement rituel et production musicale chez les Bororo du Mato Grosso'. *Cahiers de Musiques Traditionnelles (Cahiers d'ethnomusicologie)* 5: 71–96.

Gandhi, A. (2013). 'Standing Still and Cutting in Line: The Culture of the Queue in India'. *South Asia Multidisciplinary Academic Journal*. https://doi.org/10.4000/samaj.3519.

Gibson, J. J. ([1979] 1986). *The Ecological Approach to Visual Perception*. Hillsdale, NJ: Lawrence Erlbaum.

Goffman, E. ([1959] 1973). *La mise en scène de la vie quotidienne*, vol. 2, *Les relations en public*. Paris: Éditions de Minuit.

Guillebaud C. (2017). 'Standing Out from the Crowd: Vocal and Sound Techniques for Catching Peoples' Attention in an Indian Bus Stand'. In C. Guillebaud (ed.), *Toward an Anthropology of Ambient Sound*, 77–97. Anthropology series. London: Routledge.

Hennion, A. (2012). 'La gare en action. Hautes turbulences et attentions basses'. *Communications* 90: 175–195.

MILSON (2020). 'For an Anthopology of Sound Environments/Pour une anthropologie des MILieux SONores'. Available online: http://milson.fr (accessed 8 July 2020).

Pecqueux, A. (2012). 'Les affordances des évènements; des sons aux évènements urbains'. *Communications* 90: 215–227.

Sonic Visualiser (n.d.). 'Visualisation, Analysis, and Annotation of Music Audio Recordings'. Available online: https://www.sonicvisualiser.org (accessed 8 July 2020).

Thibaud, J.-P. (2010). 'Towards a Praxiology of Sound Environment'. *Sensory Studies – Sensorial Investigations*: 1–7.

Thibaud, J.-P. (2011). 'The Sensory Fabric of Urban Ambiances'. *Senses and Society* 6 (2): 203–215.

Thibaud, J.-P. (2012). 'Petite archéologie de la notion d'ambiance'. *Communications* 90: 155–174.

World Health Organization (WHO) (1999). *Guidelines for Community Noise*. Edited by Birgitta Berglund, Thomas Lindvall, and Dietrich H Schwela. Available online: http://www.who.int/docstore/peh/noise/guidelines2.html (accessed 5 June 2017).

41
Historical Sounds: A Case Study
Aimée Boutin

If we could have recorded the soundscape of a street in Paris prior to the Industrial Revolution, what would it have sounded like? There would have been horse-drawn carriages rattling on cobblestone streets, rumbling of omnibuses, cries of peddlers resonating in narrow medieval streets, street musicians, muffled conversations and loud shouting, children playing, dogs barking, pigeons cooing, military parades, church bells, sounds of labour or machinery emanating from workshops, stalls and shops. By the Seine one might have heard sluicing and the burbling of the river. In the evening, in the theatre district, barkers and carriages would have filled the street with noise. There would have been a frightful din, but we have to rely on our imagination to tell the story of the city of noise.

In thinking practically about historical sounds in an era that predates sound recording technology, a number of methodologies can be successfully combined to catalogue and make sense of the sonic past. Historical sound studies, indeed, call for an interdisciplinary approach. Hearing the past is enhanced when the disciplines work in concert. This case study will consider the range of sources that can inform an understanding of the sonic past in the absence of 'direct' primary sources such as recordings (if such evidence ever is in fact unmediated). Bruce R. Smith (1999), Alain Corbin (1994b, 1998), Mark M. Smith (2001), John Picker (2003), and Jonathan Sterne (2003) – to name only a few scholars who significantly advanced the field of historical sound studies in the 1990s and 2000s – set forth methodologies that can serve as models for understanding the role of sound in past societies. As my research – notably the project that led to the 2015 publication of my book *City of Noise: Sound and Nineteenth-Century Paris* – has specifically been focused on urban noise and European modernity, I will draw many of my examples from my own work with the understanding that the methodology discussed can be applied to other contexts, both Western and non-Western.[1] Sources can yield a deeper appreciation of the rich sonic textures of past centuries, but how does the researcher go about selecting which sounds were most relevant? After discussing which everyday sounds best captured listeners' attention and resonated through the ages, this case study ends by considering whether the proper task of the historian of soundscapes involves the *reconstruction* or the *interpretation* of the sonic past, and sketches some examples of each approach.

There are many different kinds of sources that contain information about how sounds were perceived and how they shaped narratives of individual and collective identities in past societies. The first successful attempts to record sound date to the invention of the phonautograph by Edouard-Leon Scott de Martinville in 1857, and more famously, to the simultaneous invention of the phonograph by the Frenchman Charles Cros and the American Thomas Edison in 1877. Before the age of sound recording, however, sounds were defined by impermanence and evanescence. The question 'How did the past sound?' takes on a whole new dimension and requires a different methodology when we strive to hear the past before the invention of sound recording. In the 1970s, the sound ecologist R. Murray Schafer developed the notion of 'soundscapes' in an effort to sensitize people to the world's disappearing sounds and to raise awareness of the encroachment of sound pollution into formerly quiet zones. While some of this advocacy remains attached to the term soundscape, I use it here to describe the sonic environment without any direct reference to ecology, though, of course, it is worth stating that research into historical understanding of auditory cultures can usefully inform the present.

Sources: Musical compositions

Musical sources are the first, and perhaps the most obvious, source to draw from, to learn how the past was heard, specifically how urban soundscapes of earlier centuries sounded. Since the Middle Ages, Paris had been characterized by its chorus of street cries known as the *cris de Paris*. Inspired by what they heard in the streets, Renaissance composers such as Clément Janequin and Jean Servin, and nineteenth-century French composers such as Georges Kastner produced musical *cris de Paris*.[2] These musical compositions were not intended to represent the soundscape of Parisian streets with any accuracy, but their persistence as a musical motif over centuries suggests that street cries were culturally significant. Moreover, how the voices were arranged, opposed, and superimposed – in short their formal characteristics – reflects the period's desire to harmonize the streets and to orchestrate their sounds. Polyphony as a musical form can be heard as an attempt to balance what might have actually sounded like cacophony and to make it aesthetically pleasing.

Musical performance protocols, systems of musical patronage, and sonically marked spaces can also provide a record of sonic-spatial divisions, of listening practices, or of the relationship between music, sound, and power dynamics (Johnson 1995; Born 2013; Biddle and Gibson 2017).

Sources: Visual art

While visual art makes no noise in a literal sense, prints, photographs, and paintings prove to be highly useful sources of information on historical sounds. Interpreting the sonic through the visual is not only methodologically profitable but invites reflection on intersensoriality. In his fascinating book *Sinister Resonance*, David Toop uses the term 'clairaudience'[3] to define the hearing of inaudible sounds and examines paintings as 'silent recordings of auditory events, some more silent than others. Sound haunts their silence as a specter of history that can never be heard in full, yet its presence is buried within their creation' (Toop 2010: xiii). In my research on Parisian street cries, I examined medieval woodcuts as well as suites of street peddlers by eighteenth-century French artists such as Edmé Bouchardon. I found nineteenth-century French caricatures by Honoré Daumier, but especially his contemporary Bertall, to be excellent documentary sources on the social meanings of urban soundscapes precisely because of their humorous but incisive sociopolitical commentary. In fact, I discovered that Bertall included musical scores in his *Cris de Paris,* and realized that my methodology needed to be attentive to the ways that documents encoded sound events. The spatial layout of a document provides information about how bourgeois listeners attempted to harmonize street noises into visually orderly (and therefore pleasing or harmonious to the ear) sequences, suites, or grids. In examining works of visual arts for clues about listening practices and urban soundscapes, I sought out pictorial and musical analogies: the checkboard layout was the visual analogy of musical polyphony.

Other types of graphic works, especially caricatures, convey the social distinctions that divide the quiet-seeking from the noisemakers. Taking up a theme depicted by William Hogarth in *The Enraged Musician* (1741), J. J. Grandville illustrated a modern charivari as the ears of bourgeois men are assaulted in a public park by street musicians. Visual sketches such as Grandville's, which depict the interaction between performers and listeners, represent an act of listening that encodes class and gender bias, perhaps even xenophobia (street musicians were often foreigners). In *The Sight of Sound* (1993), Richard Leppert draws attention to the inclusion or absence of the body of the performer, the source of the music or noise, in paintings. Leppert argues, for example, that the visualization of acousmatic sound (e.g. sound whose source is not visible) in paintings such as *Listening to Schumann* by Fernand Khnopff represents a new, properly modern experience of intense listening and self-absorption, rather than the sociopolitical dynamics of music performance, patronage, or consumption more typical of early modern representations of musical experience in social contexts (salons, courts, country fairs). The latter is visible in an engraving by François Dequevauviller titled *The Gathering at the Concert* in which no one listens to the musicians tucked away in the corner despite one patron's attempts to quiet the room.[4] Visual art therefore can encode sonic experience either by representing the social context of listening practices (interaction between performers and listeners) or by focusing on the psychic interiority of the listener shown alone and separated from the sound source. I found that caricatures about street criers often depicted interactions between street performers and the passers-by who heard them; in contrast textual sources heightened the focus on the listening experience.

Sources: Historical and literary texts

I have primarily relied on textual sources to listen to the past. Sources abound on urban soundscapes given the satirical tradition of describing the unbearable din of the city that dates back at least to Juvenal's Satires of Roman noise written in the early second century CE and to the seventeenth-century French poet Nicolas Boileau's *Les embarras de Paris*. Indeed, I found that caricatures and satiric texts were very rich in descriptions of city noise – people of all ages are more likely to write about sounds that irritate them than those that do not. Likewise, outsiders are more attuned to their impressions of a new travel destination's soundscape than local residents, making travel writing and tourist guidebooks eloquent records, or what Schafer (1994) called 'earwitness' accounts, of nineteenth-century Parisian soundscapes. First-person accounts document what sounds were heard as well as provide a sense of how people felt about their sonic environment. Nineteenth-century French writers evoked in detail not just the sights but also the sounds of the city that shaped the feel of modernity. Acoustic impressions in Charles Baudelaire's poetry or Émile Zola's naturalist novels make sense of noise as an effect of the fragmentation of modern life, crowds, and disease but also as a creative inspiration for sound art. For a more inclusive perspective, one can combine fictional (literary, subjective) and factual sources. Novels, poems, diaries, memoirs, editorials in newspapers, as well as legislation, medical treatises, religious conduct manuals, civic archives, etc. can be mined for evidence of lost sounds and the sociocultural significance ascribed to auditory experience. City ordinances, sometimes even material culture (peddler's medallions which served as permits to sell in specific areas) provide evidence on which sounds were deemed sufficiently uncivil to prosecute, how the sonic environment was regulated, and what belonged in an ordered urban soundscape (Picker 2003; Hahn 2013). Medical literature, such as reports on noise-caused neurasthenia, can also substantiate the period-specific meanings and values ascribed to hearing (Trower 2012). Used together, fictional and factual sources provide a nuanced understanding of the social, political, economic, and aesthetic dimension of the city of noise (Smith 2004).

Sources: Architecture, maps, and urban planning

With respect to urban noise and specifically to the relationship between sound and space, the architect and sound studies expert Olivier Balaÿ (2003, 2017) has argued that material changes to the urban environment impact how the past was heard. The sonic past continues to resonate to some extent in architecture. Balaÿ draws on both historical analysis and on architectural and acoustic reconstruction to show that widening the narrow streets to make way for boulevards and removing overhangs and awnings impacted the perception of street noise. He has shown that urban renewal in nineteenth-century Lyon, as in Baron Haussmann's Paris, tended to lower frequencies, to produce more continuous noise, and to end high-pitched intermittent human sounds. Human sounds would be much less noticeable due to decreased reverberation in wider thoroughfares. If architectural acoustics helps determine how building spaces sounded, cartography usefully locates ambiances. Sound mapping can visually capture where sounds are located, how these locations evolve over time, and how the audible forms a key component of urban spatial experience, by charting the noise levels or sound qualities of specific spaces, or the affective responses attached to places (as in the psychogeographic maps of Guy Debord [1957] or the soundmaps of Norie Neumark [2015]).

There is therefore a variety of textual and non-textual sources on historical sounds, but at least two methodologies should be common to all: formalist and interdisciplinary approaches. Formalism focuses on the techniques and devices that construct meaning as much as on what meanings are produced. Attention to the way form and language determine meaning keep in check attempts to bring the sonic past fully back to life. Although peddling is an historically documented 'real' practice, peddlers' representation as cries are a discursive construction. Key formal and organizational features, for example, pervaded the musical, visual, and textual sources of the *Cris de Paris* that I used, always already shaping the historical reality to which I had tuned my ears. I could not fully hear nineteenth-century Paris, I could only listen to it orchestrated as cries. The merits of interdisciplinary methods cannot be overstated as the disciplines of musicology, art history, literary studies, history, architecture, and urban planning – as well as disciplines I did not address such as acoustic engineering, psychoacoustics, anthropology, and cultural geography – work best in concert to enhance clairaudience. The incompleteness of silent materials can be supplemented by combining sources and what Biddle and Gibson call 'methodological inclusivity' (Biddle and Gibson 2017: 2).

Selection: Which sounds are distinctive, tolerable, or noisy?

Given that unmediated, complete access to historical sounds is impossible, how does the researcher select which sounds resonated through the ages and captured listeners' sustained attention? What makes a sound distinctive enough not only to be heard but to be listened to, or conversely to be tagged noise? More than quantifying, we need some idea about thresholds of audibility. We owe the notion of thresholds of tolerance to the French historian Alain Corbin, who explained that 'the cultivation of sensory refinement was a means for the upper-classes to distinguish themselves from the lower-classes, thought to be cruder, louder, smellier and less capable of delicate sensory perception. The leisured needed the ear to be available for music appreciation' (Corbin 1994a: 19). In his book *Village Bells* (1998), Corbin shows the development of new thresholds of intolerance for noisy bell chiming in the early morning hours in the city when middle-class urbanites claim a new right to quiet. When Corbin identified a new horizon of expectation for silence developing in nineteenth-century Europe he made a methodologically important point: researchers do well to locate historical shifts in perception in time and in space so as to scrutinize social distinctions that might otherwise appear natural. Nevertheless, changes to auditory culture happen slowly, and old and new soundscapes often overlap.

It follows that sounds that cross the threshold of tolerance are historically determined in relation to changing listening practices (what and how attentively people listened in different eras, cultures, spaces). Corbin comments that 'the noise of traffic is today tending to disappear from the evocation or description of big cities, although it is not clear whether it is no longer noticed because of its omnipresence and the fact that no one heeds it, or whether its extreme banality leads insidiously to its being passed over' (Corbin 2005: 135). Once a catalogue of historical sounds has been established, the next step involves identifying which sounds went unnoticed and which provoked reactions among listeners. Traffic noise, which is of such concern today in big cities, may not have reached the threshold of disturbance in earnest before the end of the nineteenth century.[5]

Selection: Loudness and frequency, foreground and background

Is the threshold of tolerance a function of the physical characteristics of the sounds themselves? Unwanted sounds that are strident and loud are hard to ignore as they move to the foreground of the acoustic environment. This was one conclusion I drew from researching peddlers' cries: street cries rose above the city din to assault the ears of passers-by and that can explain why they left an indelible mark on how past ages imagined the city of noise. In textual sources, I frequently encountered earwitness accounts of intrusive noises and I found that high-frequency noises that were perceived as shrill or piercing were typically the kinds of sound that marked consciousness and left traces. Textual sources were rich in adjectives that connoted loud or intense, shrieking or strident sounds, and in the case of street cries, musical notation also coded the cry's high pitch. Notions of 'sound pollution' and 'quiet spaces' that have been part of the advocacy of sound ecologists also rely on the distinction among thresholds of tolerance, low-fi and high-fi environments, and the contexts of listening.

Undoubtedly, there are also sounds that barely registered as audible: rustlings of fabric, the tinkling of jewellery, the clicking of clocks, the soft treading of footsteps, the wind whistling in the trees, the hum of machinery, the mythical sound of the grass growing ... These sounds bring to mind the impermanence of sound, silence, and the sonic traces of things past. Novelists and poets are especially good at tuning their ears to the background sounds that barely crossed the threshold of audibility. Attention needs to be paid to ambient sounds in historical soundscapes and consideration must be given to the non-verbal, the animal, or the natural world – the power of the spoken word notwithstanding. A methodology that aims to hear the past will necessarily perform a certain amount of sound mixing and decide how much 'bandwidth' goes to ambient noise and how much goes to the signal, especially when it involves the human voice. In my own work I afforded a privileged place to the vocalizations of street musicians and peddlers, but other approaches target a less anthropocentric account of the historical past.[6]

Purpose: Reconstruction or interpretation?

Once sources have been mined for evidence, and selections have been made to determine which sounds were most meaningful in a specific context, then the historian of soundscapes must tackle the project's purpose: is the task to reconstruct or to interpret the sonic past? Today researchers can use multimedia to simulate historical soundscapes. A pioneering example of a reconstruction of the soundscape of eighteenth-century Paris called Projet Bretez was designed by musicologist Mylène Pardoen and a team of French historians, sociologists, and media experts at the Université Lumière Lyon 2 in France. Based on the Turgot map of Paris prepared by Louis Bretez in 1739, it draws on a variety of literary sources to recreate the sounds one might have heard at the heart of Old Paris, near the Grand Châtelet on the right bank of the Seine near the slaughterhouses. The purpose of such a 'restitution' is to facilitate researchers' analyses of historical soundscapes by objectively recontextualizing facts of sensory experience; for the general public, the benefits are in the immersive experience.[7] Although this example of 'soundscape archaeology' is without a doubt a fantastic project, it has its limitations which are important to consider if we are developing a methodology about historical soundscapes. The multimedia project cannot convey how people experienced city noise and how sounds were represented and discussed in the specific historical circumstances in which the sounds were produced. Did eighteenth-century Parisians actively listen to these sounds as we readers can (headphones are recommended), or did these sounds remain beneath a threshold of perception? What were the modes of attention? At what point, if ever, did eighteenth-century Parisians perceive these sounds as intolerable noise? Simulations of the past in the present do not directly address what street noise meant in its historical and cultural context, how sound reflected and produced distinctions of class, gender, religious, generational, regional, or national identities. These are the questions that propelled my research. Another aspect that Project Bretez sidesteps is vocal presence in cities: there are no human bodies in her soundscape so all sound is disconnected from its physical source, and our senses of hearing and sight are not always aligned. Nonetheless, historical soundscape archaeology projects such as this one – or other practice-based approaches that are part of museum displays[8] – are certainly worthwhile because they allow us to imagine or – better – to feel the past, even while we must remain mindful of the need for contextualization. Soundscape archaeology has its place alongside contextualization so that we can attend to the ways in which the perception of sounds and listening practices have changed over time.

The historian Alain Corbin has raised a number of these issues in his pioneering work on the history of the senses. Corbin refers to the work of Guy Thuillier (1977) who catalogued and quantified 'the noises that might reach the ear of a villager in the Nivernais in the middle of the nineteenth century' and offers a critical appraisal of this approach's 'immersion in the village of the past' at the expense of 'the historicity of that balance of the senses' and of 'the configuration of the tolerable and the intolerable' (Corbin 2005: 129–130). In commenting on Corbin's 'too charitable' assessment of Thuillier, Mark M. Smith is firmer about his stance on 'whether or not we can (or ought to) try to re-experience the sensate past' stating that 'without a dedicated and careful attempt to attach meaning to those noises, cataloging is not only of very modest heuristic worth, but is, in fact, quite dangerous in its ability to inspire unwitting faith that these are the "real" and unchanging sounds of the past' (Smith 2014: 20).

Purpose: The period ear

The relative merits of attempting to *reconstruct* or to *interpret* the sonic past can be further weighed when placed in the specific context of musicology. While debates about 'authenticity' in music and 'historically informed' performance are not our immediate concern,[9] consideration of the so-called 'period ear' could be constructive to the historian of soundscapes. The Israeli musicologist Shai Burstyn wrote a series of articles on the 'period ear' in the 1990s (when the debate was raging) in which she asked if listening was intuitively practiced and weighed 'the possibility that [music practitioners] might not hear music of the past as its contemporaries had done' (Burstyn 1997: 693). She foregrounded how cultural perceptions and attitudes towards time and space as well as other mental habits and interpretive skills are all factors that historically situate listening.[10] The history of the metronome as told by Alexander Bonus (2017), for example, shows how uncovering the historical constructed-ness of intuitive listening can involve challenging our assumptions about everyday sounds that we hold to be common sense. Bonus examines the 'metronomic turn' in the early twentieth century, a 'pivotal moment in the understanding of musical time' (Bonus 2017: 77) when precision-oriented, mechanically regulated rhythm supplanted the embodied or intuitive experience of rhythm as a positive ideal. Bonus argues that although the metronome was developed by Johann Maelzel for use as a musical instrument, the ideal of mechanical precision and accuracy spread from experimental psychologists to musicians, music educators, and composers, not the other way around. Psychologists institutionalized a laboratory standard that meant that 'divergence from such scientific sound-boundaries constituted human error, regardless of the cultural or historical circumstances' (90). Bonus cannot help but romanticize the past (justifiably so?) and he concludes that 'unbeknownst to those musicians today who subscribe to daily practice routines dictating that every printed note must synchronize to a metronomic sound (driving an unstoppable "motor unit" of musical time), their training methods mirror a nineteenth-century laboratory experiment first devised to measure attention span' (97).

The investigation of historical soundscapes is a rich interdisciplinary field. A range of sources from musicology, art history, literary and cultural studies, history, and architecture, can work in concert to help us hear the sonic past. While reconstructions of the historical sounds can usefully make us feel the past, their force is all the more cogent when accompanied by interpretation, to best contextualize the historically variable and ideologically determined nature of listening practices and thresholds of tolerance. The need to historicize sound is loud and clear even though we might at first think listening is innate or natural.

Notes

1. For the application to non-Western contexts, see Chapters 31, 34, 40, 45, 46, and 49, in this volume.
2. Clément Janequin, *Les Cris de Paris* (1547); Jean Servin, *La Fricassée des cris de Paris* (c. 1578). Jean Georges Kastner, *Les voix de Paris: essai d'une histoire littéraire et musicale des cris populaires de la capitale depuis le Moyen Âge jusqu'à nos jours; grande symphonie humoristique vocale et instrumentale* (1857). For a recording, see Ensemble Clément Janequin 2009.
3. Clairaudience is used with a different meaning by R. Murray Schafer, for whom it is a synonym of 'ear cleaning' (Schafer 1994: 4).
4. Fernand Khnopff, *En écoutant du Schumann* (Listening to Schumann) (1883), painting, Musées royaux des Beaux-Arts de Belgique, Brussels. Available online: https://www.fine-arts-museum.be/fr/la-collection/fernand-khnopff-en-ecoutant-du-schumann?artist=khnopff-fernand(accessed 15 April 2018). François-Nicolas-Barthélemy Dequevauviller, *L' Assemblée au concert* (A Gathering at a Concert) (late eighteenth century), photograph, Library of Congress. Available online: www.loc.gov/item/miller.0219a/ (accessed 23 April 2018).
5. Scholars debate the relative disturbance of traffic noise, see Baron 1982; and Bijsterveld 2008, for an overview.
6. For more on the power of the human voice, see Neumark 2017.
7. To consult Projet Bretez, see Bretez Site Officiel n.d.-a. For Pardoen's description of her goals ('restitution') and methodology, see Bretez Site Officiel n.d.-b. Selected other virtual reality reconstructions of interest, include: *Virtual St Paul's Cathedral Project: A Virtual Re-creation of Worship and Preaching at St. Paul's Cathedral in Early Modern London* (see Virtual Paul's Cross Project n.d.); and *Musical Passage, A Voyage to 1688 Jamaica* (see Musical Passage n.d.).
8. For example, Gétreau (2014), based on an exhibition held at l'Historial de la Grande Guerre, Péronne, France, 27 March 2014–26 April 2015.
9. For an overview, see Butt 2001.
10. See Pearse et al. (2017), who have moved beyond the entanglements of the authenticity debate to draw from early-music performance practice and reconstruct the sounds of the past in an artistic work that mediates between past and present.

References

Balaÿ, Olivier (2003). *L'Espace sonore de la ville au XIXe siècle*. Bernin Isère: À la Croisée.
Balaÿ, Olivier (2017). 'The Soundscape of a City in the Nineteenth Century'. In Ian Biddle and Kirsten Gibson (eds), *Cultural Histories of Noise, Sound and Listening in Europe 1300–1918*, 221–234. London: Routledge.
Baron, Lawrence (1982). 'Noise and Degeneration: Theodor Lessing's Crusade for Quiet'. *Journal of Contemporary History* 17 (1): 165–178.
Biddle, Ian and Kirsten Gibson (eds) (2017). *Cultural Histories of Noise, Sound and Listening in Europe 1300–1918*. London: Routledge.

Bijsterveld, Karin (2008). *Mechanical Sound: Technology, Culture, and Public Problems of Noise in the Twentieth Century*. Cambridge, MA: MIT Press.

Bonus, Alexander (2017). 'Refashioning Rhythm: Hearing, Acting and Reacting to Metronomic Sound in Experimental Psychology and Beyond, c.1875–1920'. In Ian Biddle and Kirsten Gibson (eds), *Cultural Histories of Noise, Sound and Listening in Europe, 1300–1918*, 76–105. London: Routledge.

Born, Georgina (ed.) (2013). *Music, Sound and Space: Transformations of Public & Private Experience*, Cambridge: Cambridge University Press.

Boutin, Aimée (2015). *City of Noise: Sound and Nineteenth-Century Paris*. Urbana, IL: University of Illinois Press.

Bretez Site Officiel (n.d.-a). 'Accueil'. Available online: https://sites.google.com/site/louisbretez/home (accessed 8 July 2020).

Bretez Site Officiel (n.d.-b). 'La conception'. Available online: https://sites.google.com/site/louisbretez/le-son/la-conception-1 (accessed 20 April 2018).

Burstyn, Shai (1997). 'In Quest of the Period Ear'. *Early Music* 25 (4): 692–701.

Butt, John (2001). 'Authenticity'. *Grove Music Online*. Oxford: Oxford University Press. Available online: http://www.oxfordmusiconline.com.proxy.lib.fsu.edu/grovemusic/view/10.1093/gmo/9781561592630.001.0001/omo-9781561592630-e-0000046587 (accessed 23 April 2018).

Corbin, Alain (1994a). 'Bruits, excès, sensations, discipline: tolérable et intolérable. Entretien avec Alain Corbin'. *Equinoxe Revue romande de sciences humaines* 11: 13–23.

Corbin, Alain (1994b). *Les cloches de la terre: Paysage sonore et culture sensible dans les campagnes au XIXe siècle*. Paris: Albin Michel.

Corbin, Alain (1998). *Village Bells: Sound and Meaning in the Nineteenth-Century French Countryside*. Trans. Martin Thom. New York: Columbia University Press.

Corbin, Alain (2005). 'Charting the Cultural History of the Senses'. In David Howes (ed.), *Empire of the Senses: The Sensual Culture Reader*, 128–139, Oxford: Berg.

Debord, Guy-Ernest (1957). *Guide psychogéographique de Paris, Discours sur les passions de l'amour, pentes psychogéographiques de la dérive et localisation d'unités d'ambiance*. Bauhaus Situationniste, Copenhagen: Chez Permild & Rosengreen.

Ensemble Clément Janequin, with Dominique Visse (2009). [CD] *L'Écrit du cri: Renaissance and 19th to 21st-Century Songs*. Harmonia Mundi, HMC 902028. Recorded in August 2008. Includes music by Janequin, Kastner, and Servin.

Gétreau, Florence (2014). *Entendre la guerre: sons, musiques et silence en 14-18*. Paris: Gallimard.

Hahn, Philip (2013). 'Sound Control: Policing Sound and Music in German Towns, ca 1450–1800'. In Robert Beck, Ulrike Krampl, and Emmanuelle Retaillaud-Bajac (eds), *Les cinq sens de la ville du Moyen Âge à nos jours*, 355–368. Tours: Presses Universitaires François Rabelais.

Johnson, James H. (1995). *Listening in Paris: A Cultural History*. Berkeley, CA: University of California Press.

Leppert, Richard D. (1993). *The Sight of Sound: Music, Representation, and the History of the Body*. Berkeley, CA: University of California Press.

Morat, David (ed.) (2014). *Sounds of Modern History: Auditory Cultures in 19th- and 20th-Century Europe*. Oxford: Berghahn Books.

Musical Passage (n.d.). 'Musical Passage, A Voyage to 1688 Jamaica'. Available online: http://www.musicalpassage.org/ (accessed 20 April 2018).

Neumark, Norie (2015). 'Mapping Soundfields: A User's Manual'. *Journal of Sonic Studies*, 10. Available online: https://www.researchcatalogue.net/view/219795/219796/2732/633 (accessed 20 April 2018).

Neumark, Norie (2017). *Voicetracks: Attuning to Voice in Media, and the Arts*, Cambridge, MA: MIT Press.

Pardoen, Mylène (2015–). *Projet Bretez*. Available online: https://sites.google.com/site/louisbretez/home (accessed 20 April 2018).

Pearse, Linda, Ann Waltner, and C. Nicholas Godsoe (2017). 'Historically Informed Soundscape: Mediating Past and Present'. *Journal of Sonic Studies* 15. Available online: https://www.researchcatalogue.net/view/408710/408711 (accessed 18 July 2020).

Picker, John (2003). *Victorian Soundscapes*. Oxford: Oxford University Press.

Schafer, Murray R. (1994). *The Soundscape: Our Sonic Environment and the Tuning of the World*. Rochester, VT: Destiny Books.

Smith, Bruce R. (1999). *The Acoustic World of Early-Modern England: Attending to the O Factor*. Chicago, IL: University of Chicago Press.

Smith, Bruce R. (2004). 'Listening to the Blue Wilde Yonder: The Challenges of Acoustic Ecology'. In Veit Erlmann (ed.), *Hearing Cultures: Essays on Sound, Listening, and Modernity*, 21–41. Oxford: Berg.

Smith, Mark M. (2001), *Listening to Nineteenth-Century America*. Chapel Hill: University of North Carolina Press.

Smith, Mark M. (2014). 'Futures of Hearings Past'. In Daniel Morat (ed.), *Sounds of Modern History; Auditory Cultures in 19th- and 20th-Century Europe*, 13–22. Oxford: Berghahn Books.

Sterne, Jonathan (2003). *The Audible Past: Cultural Origins of Sound Reproduction*. Durham, NC: Duke University Press.

Thuillier, G. (1977). *Pour une histoire du quotidien au XIXe siècle en Nivernais*. Paris: Mouton.

Toop, David (2010). *Sinister Resonance*. London: Bloomsbury.

Trower, Shelley (2012). *Senses of Vibration: A History of the Pleasure and Pain of Sound*. London: Bloomsbury.

Virtual Paul's Cross Project (n.d.). 'John Donne in 1622'. Available online: https://vpcp.chass.ncsu.edu/john-donne-preaching/ (accessed 8 July 2020).

42
Sonic Writing
Holger Schulze

An anthropological loop

A sound event is not a character. A sound you hear is not identical to a word or an ideogram or a newly invented or historically conventionalized symbol you write. As I write these words I sense the fresh air of an early Monday morning coming in through the open door of my home office, from the inner court of the new apartment house we moved into two months ago. I hear almost nothing; well, some birds singing. As I was in bed a few minutes ago, I seemed to almost kinesthetically sense their movements momentarily as movements in a calligraphic performance. In order to focus on my writing here, I found this playlist I am now selectively listening to, by the name of *Songs To Test Your Headphones With*. I am writing these sonic, material experiences, now. This writing continually embodies a switch from a sort of sound event to a character: a switch that happens apparently with great ease in the documented artistic practices and technical apparatuses of recording. It is one of the most common sound practices that us *humanoid aliens* – living entities like you or me (Schulze 2018) – love to perform. I will thus not speak for or about sound cultures of *extra-*humanoid aliens in various dimensions or alternate universes; they might actually be very different in their physical representations of sounds as well as their cultural artefacts having evolved from those material sonic experiences in their recent centuries or millennia. Yet I will try to speak and to write about those sound cultures documented in the archives of this tiny and tormented planetoid at the beginning of the twenty-first century. To start with, the humanoid aliens on this planet do apparently love to indulge in expanding a given sonic experience by writing down the sounds and the music they experience. They love to produce all sorts of scores, euphoric or sometimes disappointed, depressed ramblings, suggestive drawings, or psychedelic accounts of sensory experiences. In their various and globally dominating cultures, so it seems, an infinitely running feedback loop has been growing over centuries and millennia – that is still running, in you as well as in me. I am in turn quite sure that there were and are cultures where noting sounds plays no major role in everyday life; though apparently for the contemporary hegemonic lifestyle the various forms of writing sounds seem to be crucial from *ubiquitous listening* (Kassabian 2013) to *mobile music* (Gopinath and Stanyek 2014).

This anthropological loop I am speaking of connects one specific sonic experience in a given situated and environmental, material, physiological, and somatic constellation on the one side to an imminent urge on the other side to continue and to even expand and to permeate this experience by transposing, transcribing, or inscribing a mark or even a representation for this experience in some material, some lasting area and thing; and again to later perform after this very recipe for sound and music in order to generate even newer sounds, maybe never heard ones – that again might provide a different sonic material experience in a situated and environmental, material constellation immediately connected to an urge to expand and to permeate this experience – and so on, and so on, so forth, and so forth. The loop of experience and generativity cuts transversally through one's life's erratic experiences and one's idiosyncratic activities. A loop made out of material consistencies and tensions corporeally felt: 'A body is therefore a tension' (Nancy 2008: 134). This individual and corporeal tension generates and apprehends at the same time the material cohesion of an artefact. In such a truly *anthropological loop of practice and behaviour*, of tension and representation, of generating and connecting, of continuation and transposition, of expansion and compression, the one and only starting point to stop at is hard to detect. There might not even be one.

The track I have just listened to now went through a break that operated as a bridge to the next and last verse of that song. Nevertheless, one can indeed find quite different states in the ongoing feedback between experiencing and writing, sounding and performing, and again writing and experiencing: between listening, partaking, dancing and noting, scratching, scribbling, writing, steno-, calli-, or idiographically. Such numerous states represent the shifting stages and situations in which the process of sounding and the process of writing intersect or interfere with each other, through the activities of listening and reading. This situation of notation has recently – in the *longue durée* of cultural history at least – been expanded to, added with, and was further explicated by technology and its related practices of recording of various kinds between the phonautograph, the magnetic tape, and the codification in and conversion to digital audio files. The advent of modern acoustics since the nineteenth century and the research and development activities since then have led to a vast amount of new consumer products, a multitude of new implementations of sound recording and listening devices entering our everyday lives, every season a new assortment of devices or software suites can be purchased. Sounding and writing is, since then, at the same time a technologically enhanced and supported activity as it is corporeally anchored and individual practised.

The chapter has four parts: (1) 'An anthropological loop' discusses the recursive structure of writing and performing sound; (2) 'Leap into the nexus' explores the ontological distance between the perceptual substance of a sonic experience and the notational codes; (3) 'Sonic heuristics' proposes the various phenomenological options of writing as heuristics in sound studies; and (4) 'In a cohesive flux' discusses this proposed approach with the intricate stylistic and epistemological challenges that writers in sound studies face.

With these manifold multiplications of the options of writing it almost might seem that sound can – since the advent of media recording, storage, and reproduction – only be represented and can only have an effect in these new technologically performed forms of writing: be it in actually written signs, notes, words; in mechanically inscribed oscillations and movements of a materially connected needle as its stylus; or in a transduction starting from various forms of resonating membranes and surfaces transmitting their movements to electrical signals that then again get translated into digitized values representing the energy extruded from these sound events. I can hear the high-frequency sound of this text file being saved, as scheduled, to the hard disk of this computer. I did not hear though as it was being saved to a computer in the so-called *cloud*, a hard disk drive somewhere else on this planetoid.

There is now only writing and reading and nothing but writing and reading, now, right? This is what sound is: writing and reading, and reading and writing, nothing else. Any other operation connected to sounding and to listening, as executed by us humanoid aliens, might not even be imaginable, right? And yet, such an insisting claim represents more a compulsive obsession than any actual insight. This claim represents the compulsive obsessions implemented by and in contemporary media culture and its desire projected towards the apparatus and its performance. A crystallized artefact covering and shaping the material culture of humanoid aliens more and more every single day.

Leap into the nexus

How do I do this, actually? Only to take note and then to take notes of a sonic experience is a strange, an unsettling, and to some extent even a seemingly impossible and unreal operation to execute. How can one even start doing this as in a situation where a sonic experience is in itself a rather erratic and strange entity, hard to grasp. More often than not there are no clear and materially obvious limitations that could even define where a sonic experience starts and where it ends? How does this certain sensory experience on my or your skin, in your or my inner ear, in your or my intestines or legs that we might call a sonic experience even result in some more specific activity executed with your or my extremities: be it in writing or dancing, tapping or gesturing? What is one actually doing as the very activity of notating and writing sounds seems to incorporate at the same time an almost paradox combination of activities: it combines a receptive, malleable, and plastic *listening body*, an analytic and detailed attention directed at understanding sensory experiences on the one side – and on the other side an intensely generative and highly focused and active, corporeal performativity necessary to move one's body or certain bodily extremities in an orderly way to leave traces on some highly receptive surface on which one records these traces. This complex activity in itself is highly strange, seemingly impossible to undertake or to complete, let alone to reflect upon; nevertheless, it is incessantly done, probably at any given time of every day in at least some locations on this incessantly tormented solar planetoid.

Someone is, right now, for instance, excited to tell their friend about a song or a composition they recently heard. Right now, someone is writing down notes and rests, breaks and accidentals, time signatures and dynamics in order to provide one of the recipes this largely westernized global culture calls *a score* that musicians use to perform. And right here and in many places, studios, homes, and offices there are at the same time technological recording facilities that write down the codified representations of individual sound events and singular sound sources of a musical or sonic performance onto a given recording medium, transmitting a highly codified representation of its specific sequences and layering of frequencies. This apparently ubiquitous desire and general habit of engaging in a feedback loop of sensing and writing activities when hearing sounds and music, when thinking of music and sounds, these practices might be understood as yet just another form of hearing and listening, just as: 'Dancing is a way of hearing; singing is a way of dancing: Singing is a way of hearing' (Schulze 2018: 231). The nexus between kinesthetic continuation, mimesis, a sensation of being almost inextricably connected to the sensory events, the sonic events around one, and the following sensibilities, the reactions and activities, the *felt sense* (Gendlin 1992) that takes hold of one's *sensory corpus* as a whole – this nexus is not an accidental side-effect of sounding and listening. This nexus precisely marks the vibrant linkage between the corporeal and individually existential reality of humanoid aliens – and all the activities, events, processes, turmoil, and erratic instances that seemingly were not executed by one of us. As a humanoid alien one apparently lives and listens in this transient yet agile nexus.

However, this nexus requires an incredibly daring and dark effort to bridge this seemingly unbridgeable abyss: an abyss between two radically different and, as one might assume, almost completely disconnected materialities. On the one side humanoid aliens experience the percepts of oscillations and intensities, of pressure waves and loudness, of timbre, instrumental sounds, of impact and tenderness, desire and repetition, of erratic sound events and continuous ambient soundscapes, an intrusive power exerted onto you and me. This is the *sonic material* of an enveloped and situated experience humanoids can share. On the other side, though, one finds the material limitations and physical constraints, the cultural codifications and regulations, the transformative mappings, the legends and descriptions for how to write certain sounds in the framework of certain notational practices and in certain techno-cultural constellations and apparatuses of inscription and automated writing. This is the *sonic writing* that intends to represent or to crystallize the experience. But why repeat this sonic experience at all and why prolong it in the form of a lasting representation? What desire precisely drives this strange urge to solidify a malleable and ephemeral experience into a stable, archivable document?

Those two seemingly disconnected sides of *writing sounds* – the side of sonic material experience and of sonic material inscription – are not only representing different aspects or perspectives; both are also situated also in radically different approaches to materials, to the senses, to experience, to formulas, to patterns and schemata, to forms of organization, selection, and representation. Simply put: Both sides of sonic writing – *sonic material experience* and *sonic material inscriptions* – can appear almost incompatible on all levels one could think of. Between both one finds an abyss, a void, an erratic non-space of non-proximity. No connection exists apparently between sonic material experiences and sonic material inscriptions, so it might seem. Hence, it might be just precisely this disconnectedness that facilitates a contact. You and I, we are the actual contact zone: the sensory corpus of a humanoid alien. Its sonic sensibilities. This touching and almost impossible contact between a sonic material experience and sonic material inscriptions is what effectively incites the desire to perform this impossible act. The act of performing resonances, sensibilities, the act of listening: 'I'm going to prove the impossible really exists' (Schulze 2018: 132).

In starting to write sounds or in more general terms: in starting to write the sensory experiences one actually jumps into this abyss – expecting and awaiting a sudden connectedness, a surprising new nexus between these aforementioned separated materialities. '*But what if there is none? – What if I just fall? And if I crash?*' You simply jump. Being disconnected to any entity around you, is what apparently can carry you as soon as you jump. 'Sound writing is a gong resonating through bodies, sentient and non' (Kapchan 2017: 2). This gong of which Deborah Kapchan speaks resonates only as soon as one moves into the zone where it is able to activate resonances: one needs to jump into this zone of resonance. This daring and necessarily risky jump into the multiple emanations of sentient connectivities is exemplified by Kapchan with the flight, the rise, and the fall of the mythical figure of Icarus; according to a poem by William Carlos Williams, quoted by Kapchan, this fall is unheard, actually unwritten, maybe unsound – though probably not 'quite' as Williams writes:

> a splash quite unnoticed
> this was
> Icarus drowning.
>
> (Williams 1962)

But what if this jump is somewhat forced and strangely uncomfortable? What if it is only partly desired and partly coerced? What can such a jump then achieve and protrude? One jumps maybe nevertheless into this abyss – and I like to think of David Bowie's jump performed in the song and the music video by the same name, 'Jump They Say': Standing on top of a skyscraper the protagonist doubts if he should jump, yet various people seem to scream: 'JUMP!' – and they ask questions about this humanoid alien standing and doubting there:

> They say
> He has two gods
> They say
> He has no fear.
>
> (Bowie 1994)

Whereas Icarus indeed jumps and rises, flies and falls, Bowie's persona in this song does neither; he seems to have neither a pair of eyes, nor a mouth. He doubts and waits, he gets to be pushed and still doubts but in the end decides for himself to do this jump, this scary leap into the unknown. Still, in such an instance 'listening itself is a speculative method' (Kapchan 2017: 3). The jump into the abyss in between materialities remains speculative, insecure, partly impossible. This jump and flight and the minuscule horror in between sensing and codification, between receiving sounds, assimilating sonic percepts on the one side and the action of inscribing codes or representations on some surface on the other side, this jump into a sudden movement, into scripture, is a jump from the hard and unquestionable security of the architecture one is standing on to the soft and highly irritable qualities of gases and molecules, incessant movements and dynamics. One jumps hence *from the hard to the soft,* in the words of Michel Serres:

> The given I have called hard is sometimes, but not always, located on the entropic scale: it pulls your muscles, tears your skin, stings your eyes, bursts your eardrums, burns your mouth, whereas gifts of language are always soft. Softness belongs to smaller-scale energies, the energies of signs; hardness sometimes belongs to large-scale energies, the ones that knock you about, unbalance you, tear your body to pieces; our bodies live in the world of hardware, whereas the gift of language is composed of software.
>
> (Serres 2008: 113)

The bodies of humanoid aliens are enveloped in such invasive and intense impacts by sonic emissions on an *entropic scale: pulling your muscles, tearing your skin, stinging in your eyes, bursting your eardrums, burning your mouth*; whereas the ephemeral and transitory textures of language one weaves out of characters and strings, periods and samples, idioms and metaphors, signatures and symbols carry such *smaller-scale energies of signs*: they might just vanish, unknown, if no one cares for receiving, for taking up what might be woven into them, what meanings, what sense. The emissions around us probably do not leave us many *choices* than simply to be under their influence, maybe to give in, maybe to indulge in their vast radiance; yet the textures of meanings we weave and knit and lace, they need our ability to lace and knit and also to weave in order to understand, to receive, to detect, and to decode what is coded in it. However, as so as we started this knitting and lacing characters and strings, we might never even be able to stop doing so: we then will *have to* inhabit this realm of symbols and metaphors, the textures and references. It remains though a highly skilled, an incredibly refined task one exercises here. It is a bit like *chess boxing* – a highly hybrid activity that comes together nevertheless in one nexus that generates cohesion, movement, and drive: *impact and critique.*

Sonic heuristics

Here I am now, in a nexus between sounding and writing, connected by listening. So how do I do this? As troubling as this quite bland, starting question might sound: this is probably the best starting constellation for a leap into the nexus between sonic material, sonic experience, and sonic writing. It requires a jump, more often not from an actual skyscraper, but maybe from a high-rise, a pyramid, or the top of a rocket launchpad of affects, involvement, repulsion or desire. 'They say: Jump!' (Bowie 1994). With this jump one does not leap into codifications and automation of inscription that would be the mechanism, a *method* for an apparatus to record and write. Yet, one jumps into 'the realm of the irrational, absurd, the distracted, the melancholic, the obsessive, the insane' (Cave 2000). Whereas the author of this quote, the singer and novelist Nick Cave, is speaking about his strategy for writing love songs, exactly this jump into the very close, preferably into the too dark, the too intimate, into the erratic is what this leap into the nexus actually means: it is a heuristic, namely, a problem-solving strategy (cf. Schulze 2005) that claims to lead one applying it to a more specific focus and to a more specific way of production. Heuristics are *meta-methods*: methods to find the in each case specific and new method to explore and to scrutinize a given sonic situation. They are *sonic heuristics*, heuristics in sonic research. This research now begins, again, with a leap into the nexus: 'Not only is it the literary that's useless, all traditional theory is pointless' (Eshun 1998: 189). The too clever and too distant and too self-assured forms of knowledge keep you from actually registering and sensibly observing what happens in this dense and intense situation of a sonic experience. It is a nexus of the highest volume, it swallows you, this room volume – *right here! right now!* – full of sound and sensory percepts, full of movements and all the tactile and haptic interpenetrations of specific material consistencies:

> There is no distance with volume, you're swallowed up by sound. There's no room, you can't be ironic if you're being swallowed by volume, and volume is overwhelming you. It's impossible to stay ironic, so all the implications of postmodernism go out of the window.
> (Eshun 1998: 189)

As soon as the volume of sound swallows you, you can actually sense the sound and hence write sonically. This is the major prerequisite for sonic writing. '(And yet, we could all be wrong [...]. Wouldn't be the first time)' (Anderson 1984). Writing sound is a corporeal and a sensory performance. This performance though is only possible if connecting to *the irrational, the absurd, the distracted, the melancholic, the obsessive, the insane* as sources for its writing. The sensory corpus (Schulze 2018: 136–159) in its fullest and most intense presence needs to be activated. It gets activated by a reconnection to the *felt sense* (Gendlin 1992; Schulze 2018: 148–150) of the sonic persona who is writing. Opening up this felt sense in the sonic nexus refers actually to three main focuses in writing: on *idiosyncrasies*, *corporeality*, and *situatedness*. I begin with the latter.

Sonic writing does not begin with a historical or critical account of a given sonic artefact. It starts with the specific, narrow, personally experienced *sonic situation*, right here, right now. There are no details of such a situation that would be too banal or too irrelevant or too intimate or too dirty or too humiliating to narrate. The more insanely obsessive, distracted, and absurd the narrated details of an everyday listening situation are the better. 'Got to believe – somebody' (Bowie 1994). These details then obviously border and transgress into imagination, memories, tangible desires and fear, all of them always connected, of the sonic persona who does the writing. This complexly interwoven and interpenetrating *magma* (Castoriadis 1975), this *chaosmos* of desire in a specific listening situation is where sonic writing starts.

This writing then includes, apparently, individual corporeal sensibilities, personal preferences and experiences, inabilities and abilities, poignant formations and biographical formants that the writer has embodied in her or his sensory *corpus* (Nancy 2008). This corpus is not the mere physical body that might be abducted after its death, it is not an imaginary ideal body that one might find in school textbooks of biology; it is foremost a sensed, felt, and experienced body, structured by the felt sense of the sonic persona. This corpus resembles more the *corps sans organes* (Artaud 1947; Deleuze and Guattari 1972; Deleuze 1981), the *body without organs*, in that it is not limited to a notion of the civilized and docile, the domesticated body: 'Lorsque vous lui aurez fait un corps sans organes, alors vous l'aurez délivré de tous ses automatismes et rendu à sa véritable liberté' (Artaud 1947): 'When you have made him a body without organs, then you will have delivered him from all his automatisms and restored him to his true liberty' (Artaud 1992: 329). And in sensing and receiving the sounds around you, you start to dance, as dancing is indeed a way of hearing. 'Alors vous lui réapprendrez à danser à l'envers comme dans le délire des bals musette et cet envers sera son véritable endroit' (Artaud 1947). 'Then you will teach him again to dance inside out as in the delirium of our accordion dance halls and that inside out will be his true side out' (Artaud 1992: 329). And this dancing is a sequence of gestures that transgresses into writing the kinaesthetic, the sonic.

Finally, this process of writing the sonic incorporates the most daring vulnerabilities, idiosyncrasies and strange desires, unacceptable needs and fears of a sonic persona. A sonic persona as a sonic writer necessarily starts with these qualities of their own sensory experience. One does not start with the historically documented and critically acclaimed knowledge about sound and music. But, and that is the next step, all of this knowledge goes into the writing process. Sonic writing does neither stop in a self-harmed limitation on the immanence of one, highly subjective listening moment with all its momentary impulses; yet, it begins there in implicitly carrying a whole lot of knowledge and readings, epistemes and truisms, false assumptions and strange bits of collected factoids. In unfolding, in carrying forth the writing on sound, this knowledge again can and should be questioned, added to, and contradicted with new research findings and enhanced with contrary positions:

> All that works is the sonic plus the machine that you're building. So you can bring back any of these particular theoretical tools if you like, but they better work. And the way you can test them out is to actually play the records.
>
> (Eshun 1998: 189)

Sonic writing does not intend to limit itself without need to a stupid and ignorant fiction of oneself: all practices, experiences, insights, and doubts that the writer encounters before, during, and after the writing process can and should go into writing about a sonic situation – revising the written sonic stream of writing again and again; and again. And again. Until, at a certain point, the resulting text has an appropriate amount of texture of consistency, of sonic references, auditory epistemologies and ontologies in it. It might then be considered as finished. For the time being, at least.

In a cohesive flux

In the moment I am writing this, the interpenetration of moments, relations, things, and sensibilities, of thoughts and concepts, historical ideas, insights, and findings of a seemingly endless multitude of research fields, might seem to some readers to be too erratic, too arbitrary, too random to even constitute a piece of writing with a sufficient degree of consistency. Yet, this seemingly erratic flux in writing might be the only sufficiently complex and dynamic, the only appropriately responsive, relational, and transformative representation in written form that is indeed capable of giving an account of the experiential character of listening, sensing, and receiving the sonic material in a given sonic situation. This submerging into the immanence of a sensory experience is precisely what makes sonic writing possible at all:

> Writing *about* sound and writing sound are two different processes. The first maintains the positivist position of subject (writer) and object (sound). The second breaks out of duality to inhabit a multidimensional position as translator between worlds – the writer listening to and translating sound through embodied experience, the body translating the encounter between word and sound, sound translating and transforming both word and author.
>
> (Kapchan 2017: 12, emphasis in the original)

This daring jump into the nexus between experientiality and artefact, this dense area of immanence, is at the same time a risky jump out of the common limitations and definitions of acknowledged epistemologies of direct objectivity, of an anonymized and widely camouflaged researcher, and of the arbitrary as well as highly idiosyncratic motivations, discontinuations, transruptions, and detours that actually make research. When one jumps into this abyss between sonic materiality and codified artefacts one opens a whole realm that seems uncharted and void, if not scary and dangerously personal. And yet, precisely this sense of being lost, of being in a vacuum of academic practice, is immediately triggered and ad hoc structured by the very sensory corpus and the material resonances running through you or me, their repercussions and affecting effects, their surprising reconnections and new emerging structures, in this very instant. This sudden switch can take place when feeling lost and scared in a state and under pressure of articulation, one arrives at a moment, when the established and learned routines and clichés, the phrases and idioms of everyday articulation simply do not make sense anymore – and different, not so much verbal and semiotic references in history and in academic writing, but corporeal and sensory references in one's individual bodily situation provide the best starting ground for articulation: 'Bodily implying is a value-direction' (Gendlin 1992: 203).

This does not mean that what can be extracted from starting with this bodily felt sense would be infallible – yet, it is in some instances the only possible and viable starting point. In sonic writing this is precisely the situation: surprisingly for some, there are no pre-existing structures: they grow, they are there, they emerge as soon as one is performing these kinds of movements, inscriptions, writings. If one starts writing on a complexly layered situation of sonic experience, then the process of writing, this highly erratic and idiosyncratic flow of memories, sensations, fears, and desires generates out of itself a sequence, a flux, a kind of connection and consistency that is not a consistency in the sense of a logical argument: this consistency is a consistency of material – be they sonic or rhetoric. It is not necessarily a *coherent* text, that proposes a semantically coherent flow of propositions, relating to each other in a consistent sequence that provides an intelligible argument. It is more a *cohesive* texture, that offers a materially suggestive flow of sounds, signs, and words, emerging out of each other in a rhythmical and perceptual sequence that carries its readers and listeners, its audience through this time-based flux. In sonic writing one creates such a cohesive flux.

I hear the voice of Hildegard Westerkamp, during her narration that drives the famous soundwalking piece of hers, *Kits Beach*:

> In another dream, when I entered a stone cottage, I entered a soundscape made by four generations of a peasant family sitting around a large wooden table eating and talking: smacking and clicking and sucking and spitting and telling and biting and singing and laughing and weeping and kissing and gurgling and whispering.
>
> (Westerkamp 1996: 5:21–5:45)

I imagine this place, the sounds, the relation of the sonic personae in this situation by the sensory, situated description of Westerkamp. It is the material plasticity that represents all the sounds present in this situation. And suddenly, I hear the voice of Alvin Lucier in my head:

> I am sitting in a room different from the one you are in now. I am recording the sound of my speaking voice and I am going to play it back into the room again and again until the resonant frequencies of the room reinforce themselves so that any semblance of my speech, with perhaps the exception of rhythm, is destroyed.
>
> (Lucier 1990)

Here the narration itself effectively generates the piece of sound art; it activates the sonic formants of the space it was recorded in. But foremost, again, by the idiosyncrasies of Lucier's stuttering, his phrasing and vocalization: the material of his tender yet firm and distinct speaking. This memory triggers now the first six questions, still sticking in my head, from Pauline Oliveros's famous *Ear Piece* – a narrative opening for almost any sonic environment:

> Are you listening now? Are you listening to what you are now hearing? Are you hearing while you listen? Are you listening while you are hearing? Do you remember the last sound you heard before this question? What will you hear in the near future?
>
> (Oliveros 2005)

In the near future, listened to by Kodwo Eshun in the year 1998, I can hear that

> Futurhythmachines complexify the beat into what Kelly terms an 'alien power.' When polyrhythm phaseshifts into hyperrhythm, it becomes unaccountable, compounded, confounding. It scrambles the sensorium, adapts the human into a 'distributed being' strung out across the webbed spidernets and computational jungles of the digital diaspora.
>
> (Eshun 1998: 77)

I can hear these breaks and time stretches, aborted samples and accelerated beat particles. And yet I ask myself, with Oliveros:

> Can you hear now and also listen to your memory of an old sound? (Oliveros 2005)

References

Anderson, L. (1984). [CD] 'Kokoku'. Track 4 on *Mr. Heartbreak*. USA: Warner Bros.

Artaud, A. (1947). *Pour en finir avec le jugement de Dieu*: émission radiophonique enregistrée le 28 novembre 1947. Paris: ORTF.

Artaud, A. (1992). 'To Have Done with the Judgment of God'. Trans. Clayton Eshleman. In D. Khan and G. Whitehead (eds), *Wireless Imagination: Sound, Radio and the Avant Garde*, 309–330. Cambridge, MA: MIT Press.

Bowie, D. (1994). [CD] 'Jump They Say'. Track 5 on *Black Tie White Noise*. UK: Savage Records.

Castoriadis, C. (1975). *L'institution imaginaire de la société*. Paris: Le Seuil.

Cave, N. (2000). *The Secret Life of the Love Song*. London: King Mob.

Deleuze, G. (1981). *Francis Bacon: Logique de la sensation*. Paris: Editions de la différence.

Deleuze, G. and F. Guattari (1972). *L'Anti-OEdipe: Capitalisme et schizophrénie*. Paris: Éditions de Minuit.

Eshun, K. (1998). *More Brilliant than the Sun: Adventures in Sonic Fiction*. London: Quartet Books.

Gendlin, E. T. (1992). 'The Wider Role of Bodily Sense in Thought and Language'. In M. Sheets-Johnstone (ed.), *Giving the Body its Due*, 192–207. Albany, NY: State University of New York Press.

Gopinath, S. and J. Stanyek (2014). *The Oxford Handbook of Mobile Music Studies*. Oxford: Oxford University Press.

Grossan, A.-J. and M. Woodworth (eds) (2015). *How to Write About Music: Excerpts from the 33 1/3 Series, Magazines, Books and Blogs with Advice from Industry-leading Writers*. New York: Bloomsbury.

Kapchan, D. (ed.) (2017). *Theorizing Sound Writing*. Middletown, CT: Wesleyan University Press.

Kassabian, A. (2013). *Ubiquitous Listening: Affect, Attention, and Distributed Subjectivity*. Berkley, CA: University of California Press.

Lucier, A. (1990). *I Am Sitting In A Room*. New York: Lovely Music.

Nancy, J.-L. (2008). *Corpus*. Trans. R.A. Rand. New York: Fordham University Press.

Oliveros, P. (2005). 'Ear Piece (1998)'. In *Deep Listening: A Composer's Sound Practice*, 34. New York: iUniverse.

Schulze, H. (2005). *Heuristik: Theorie der intentionalen Werkgenese – Theorie der Werkgenese*, vol. 2. Bielefeld: transcript Verlag.

Schulze, H. (2018). *The Sonic Persona: An Anthropology of Sound*. New York: Bloomsbury.

Serres, M. (2008). *The Five Senses: A Philosophy of Mingled Bodies*. Trans. M. Sankey and P. Cowley. New York: Continuum.

Westerkamp, H. (1996). [CD] 'Kits Beach Soundwalk (1989)'. Track 3 on *Transformations*. Canada: empreintes DIGITALes.

Williams, W. C. (1962). 'Landscape with the Fall of Icarus'. In *Collected Poems, Pictures from Brueghel and Other Poems: Collected Poems 1950–1962*, 4. New York: New Directions.

43

Silence of Mauá: An Atmospheric Ethnography of Urban Sounds

Jean-Paul Thibaud

Mauá, according to ambiance

In 2015 a group of Brazilian and French researchers carried out a field investigation at the Condominío Barão de Mauá (Figure 43.1).[1] The condominium is located in the town of Mauá, some 30 kilometres from São Paulo, Brazil. The complex, built in 1996, consists of fifty-four buildings housing about seven thousand people and stands on land contaminated by industrial waste buried there. The residents became aware of the contamination in April 2000 when an explosion occurred during maintenance work on one of the pumps for the condominium's underground water reservoirs. It was probably caused by the presence of methane gas. A worker was killed in the explosion, another one badly burned. Legal proceedings have been underway ever since, leaving the residents exposed to the risk of contamination by carcinogenic substances and another explosion.

The purpose of the field work was to test the notion of *risk ambiance* and understand how one can go on living in a contaminated environment of this sort. We posit that a risk-prone environment is embodied in specific ambiances and that it lends itself to particular sensory experiences that can be documented, and that providing a suitable methodology for study can be found. Our aim is to focus on the emergence of a resident sensibility to risk, to investigate what might constitute a risk-sensitive culture, and to highlight the sociopolitical consequences of day-to-day vulnerability under conditions of this sort. In preparation for this methodological experimentation, we investigated the ongoing controversy over the situation, in order to understand what was at stake for the contaminated area, going back over its history and public life, to pinpoint the various players involved and to clarify the context of our field investigation.[2]

To approach Condominío Barão de Mauá in terms of ambiance entails studying the sensory contexts of the lived-in space and how it is experienced, practised, and perceived day to day. As a first approximation, an ambiance may be defined as a space-time experienced in sensory terms (Thibaud 2011: 203–215).[3] It is always situated (even if its

---

Chapter translated by Harry Foster and Sophie Provost.

Figure 43.1 A view of Condominío Barão de Mauá. Photograph by Sylvain Dubert.

outlines are fuzzy), it activates all the modalities of perception (sound, light, smell, heat, and so on) and proceeds by weaving together the material properties of an environment, social practices, residents' memories and narratives, and the affective tonalities that colour a situation. If ambiance gives us a grasp of the quality, singularity, and tonality of a situation, how does it enable us to experience a risk-prone territory? What can we learn through ambiental awareness? The present chapter explores the possible make-up of an inquiry in ambient mode, in short investigating risk by way of ambiance, *according to* ambiance and its sonorities.

An atmospheric ethnography

For this field investigation we carried out a five-day experiment in methodology that we might describe as *atmospheric ethnography*.[4] Before presenting our investigatory approach in greater detail, I should underline the three key features of ambiance that underpin this field action. Firstly, ambiance proceeds from a 'pervasive field', which places the emphasis on immersion rather than on face-to-face relations. With ambiance we are steeped in a sensory milieu which wholly envelopes us at all times. By attending to the order of feeling as much as perceiving, ambiance puts us in contact with the globality of a situation, by paying full attention to floating attentions, peripheral perceptions and the fringes of experience. Secondly, ambiance proceeds from a 'permanent background', a *basso continuo*, on which the world takes on a certain physiognomy, a certain affective tonality. In other words, ambiance sets the tone for situations and everyday territories. Rooted in the habits and practices of residents, it makes the forms of social life perceptible and embodies ways of being together in a particular setting. Thirdly, ambiance proceeds from an 'unobtrusive pregnancy'. It gains consistency through low-intensity phenomena, so slight as often to be imperceptible. Proceeding most of the time by light touches, small inflections, and micro-phenomena, it is of the order of the infra and of little perceptions, so tiny and mixed up with each other that they are hard to perceive distinctly.[5] These three characteristics – pervasive, habitual, and unobtrusive – make ambiance a particularly relevant notion for understanding how risk-prone situations impregnate residents' experience.

Our proposal for an atmospheric ethnography consists in coming as close as possible to lived situations in order to gain access to the discreet expressions of resident sensibility, to the pervasive sensations of the ambient environment, and to the everyday background to the perceptions and practices of residents. In so doing we aim to capture and describe what relates to sensory perceptions, material traces and social memory, everyday gestures, ordinary words, and to ways of being together. This exploratory approach seeks to study the embodied, situated, enacted, and shared nature of the sensory relation which residents entertain with their home. The main difficulty here is touching what usually passes unnoticed, barely voiced, the thousand and one little signs that play a part in everyday sensory life. It should be immediately apparent that resorting to the conventional methods of inquiry is of no use here.

Our investigatory approach is designed to make us as amenable as possible to all we encounter on the spot. Rather than starting from any prior assumptions or preconceived methodological guidelines, our purpose is to develop a posture of openness to our surroundings and adopt an attitude of 'disengagement' or 'letting go', coming as close as possible to a floating ear and unfocused attention. We refer to this as atmospheric ethnography because it is a matter of *experimenting with a device for paying attention to the pervasive*. To do this we carry out a form of inquiry that is immersed, plural, collective, and evolving.

- Our inquiry is *immersed* because it uses in situ investigation, right in the middle of the Barão de Mauá housing complex. In doing so we aim to take seriously our bodily presence and our own capacity to be affected by an environment of this sort. The investigators experience the ambiances of the Condomínio, sense their effect on themselves, and are transformed through contact with them. A sharable experience is thus possible for investigators and residents.
- Our inquiry is *plural*, drawing upon various lightweight yet complementary approaches. It involves techniques as varied as ethnographic observation, commented walks, floating listening by individuals or groups, audio recordings with sonic reactivation, photographic drifting, organized encounters with residents,

and a joint logbook.⁶ Such variety helps to broaden the field of perception while introducing variations to modes of attention. Resonances take place between open approaches that initiate the investigation (drifting, floating listening, informal observations and encounters) and more framed approaches that specify it (commented walks, sonic reactivations, in situ collective listening).

- Our inquiry is *collective*,⁷ bringing into play three dynamics, one inside the group of investigators, one among residents, and one between residents and investigators. In this context the joint logbook plays a key role. It is kept on a day-to-day basis, fed by regular encounters and discussions between all the investigators (all sessions are recorded). Each investigator's experiences are systematically shared and debated in such a way as to articulate and give an account of the atmospheric qualities of the place.
- Our inquiry *evolves*, lending itself to all sorts of modalization, reconfiguration, and branching, as encounters with residents and discussion among investigators progress. Regular debate on investigators' respective experiences and recurrent interplay between collective analysis and field investigation transform, with each day that passes, the sensibility to the risk-prone environment. A key feature of the investigatory toolkit, the joint logbook, enables us to register the ongoing process of exploring, making explicit and sedimenting knowledge.

This proposal for an atmospheric ethnography may be referred to as 'ambulatory knowledge' as formulated by William James ([1909] 2010).⁸ Its purpose is to explore how a group of investigators gradually allows themselves to be immersed and transformed by the ambiances of a place, in such a way as to reveal its tonality and agency.

Listening to an enigmatic silence

The ambiances of the Barão de Mauá condominium are largely impregnated by the risk of contamination. Sometimes this is manifest and very visible; witness, for example, the abandoned buildings standing empty after construction work suddenly stopped. They prompt a sense of desolation, conjuring up images of a territory laid waste by war or natural disaster. This feeling is reinforced by the fact that the residents do all they can to care for their living space, paying great attention to its upkeep. In other cases the impression is less obvious, more secret, as with the presence of odours which permeate the housing complex, a nagging reminder of the contaminated soil, the smell of the air and earth, more noticeable at some times than others. According to the residents the strength of this smell varies a great deal depending on the weather, more perceptible when it rains and the temperature rises.

But at an early stage we encountered an enigma: the strange sense of silence gripping the condominium. This prompted us to investigate the sounds of the site. The atmospheric ethnography then turned into a shared listening to local ambiances. From the outset, a collective impression emerged from our explorations and discussions, but it struggled to express itself clearly and was stated in a variety of ways. At a loss for words our exchanges hesitated on how to translate this shared yet largely elusive sensation. Our joint logbook helped us afterwards in reconstituting the progress of our listening, gradually bringing us into resonance with Barão de Mauá. As part of its regular meetings, the group undertook a phenomenological description. The polyphony of remarks, comments, translations, and reformulations that unfolded among the investigators sought to describe this experience as closely as possible, gradually attuning our collective ear to the neighbourhood and even shifting the direction of our inquiry. An interaction loop was set up connecting our lived, in situ experience and a collective return on experience. On the one hand, field work enabled us to flesh out our sonic experience, gathering the accounts of residents and additional ethnographic observations, test new forms of attention and in situ listening modes, and validate previously shared descriptions. On the other hand, group meetings enabled us better to share individual experiences, gradually composing a common way of listening to situations, opening up new ideas for description and interpretation; they also prepared for further site visits. In this way we tested in two ways the enigmatic silence that became a key motif in our inquiry: through a collective description, and through an ongoing in situ investigation. It took all the available time of our stay to gradually come close to this enigmatic silence, which proved remarkably revealing as to the singular condition of the neighbourhood.[9]

Textbox 43.1 is a polyglot composition of excerpts from the joint logbook. All the investigators contributed to this log, which also takes the form of a collective composition in the making.[10] The sequence of these extracts follows the timeline of the log and reflects the verbal dynamics between the investigators. Some topics recur during our exchanges and are reformulated in the course of meetings; various key ideas emerge gradually.

Textbox 43.1. Polyglot recomposition of excerpts from the joint logbook: 'The Silence of Absence'.

[…] a risk territory cannot be considered solely from the angle of risk […] there are inhabitants, there are people living and it is a living environment … therefore it also means care given to the place where we live […]

[…] yes, but we don't hear it […] I barely heard *any voices* […] I heard very very *few inhabitants* […]

[…] indeed […] I noticed the very *very calm* dimension of the neighbourhood […] at the beginning it was a peaceful quiet […] and then as I walked I noticed something '*between the peaceful quiet and the silence of desolation*' […]

[…] Anali says that it's the second time she comes here, and that she also has this sensation that […] uh, the people […] *it's lacking some people* […]

Birds and planes

[…] I was very attentive to the transition between interior and exterior, which happened for me with the soundscape and the birds […]

[…] I'd just add the … the *planes* in the landscape, they are really present […]

Noiseless interior
[…] it depends a lot on the trips that are made and there is […] it's true, there's much less noise inside the condominium than outside […]

[…] when we came down […] we walked along the stream and there is another neighbourhood […] there are small houses with fencing at the entrance, and at this point there is a lot of noise […] *it's much more lively*, we found noise, barking dogs, people speaking loudly, music […] we found a lot livelier, a lot louder […] just like we expected […] *the condominium is still much more silent than the average* …

Static and lifeless
[…] here inside, it's rather static […]. In São Paulo, in a Condomínio, you can hear the TV, smell food … here, you know that dwellers live here because you can see clothes, but it's static […] there are no children's voice […]

[…] when you enter the apartments, you start to see life […] but *outside, no* […]

[…] the first impression, that's the buildings […] abandoned […] after talking to the manager […] today, of course, there's the rain […] but it's lacking a welcoming place for people, you see […]

[…] outside, a lot of movement […] here […] *silence* […]

Absence of garden
[…] actually, I looked for clues from the absence of clues […] the things that wouldn't be here, actually […] in particular the gardens […] would there be vegetable gardens […] I didn't see any anywhere […] in many places, I thought to myself 'oh, here maybe there'll be one' because it looked like ideal conditions for a garden, vegetable garden, family garden […]

[…] yes, there was this story […] *they tried to create gardens* but actually, because of the contamination, the plants, well […]

Vacant apartments?
[…] there's still some […] there is a lot of vacancy still […] which means that the buildings that are inhabitable are not completely full […]

[…] some people are gone, you know […] *some are gone*, and others have leased again […]

[…] which could explain in part the silence […]

The 2000 explosion
[…] we were with a dweller from step four […] that's exactly where the accident happened […] his name is Elton […] and so he was there at the time of the explosion […] he was at his place when he heard a boom! […] the explosion […] and because there's the petrochemical hub (not far) there are always explosions […] and so he first thought that it was the petrochemical hub […] and then he saw the two people that they […] that were injured […] he saw a few people screaming […] and then he went down to see and found the one that was completely burned […]

Everything is stuck
[...] everything here is stuck [...] for example they wanted to build a place for parties, but they can't because it's stuck [...] they can't improve the buildings because everything is stuck [...]

[...] there is a problem because she wanted the approval from CETESB [Environmental Agency of the State of São Paulo] to build a garage [...] It's the CETESB that has to give its approval, but the thing is that we can't perforate [...] indeed there's always the threat of explosion, but, well, since *we don't dare perforating, drilling* [...]

A place without rhythm
[...] there were several people who were coming by car, were repairing their cars, two young girls who were playing in the play area and whose phones were used as speakers for music [...] so, all I'm talking about has a sound dimension ...

[...] here, it was [...] even with the sound, it was *super silent and quiet* and we couldn't hear the city in the background [...] I mean, in fact, it really feels like there's a *rhythm set by the city nearby* and like on site we have the sounds of punctual events, but there's no rhythm to this place [...]

Paralysis effect
[...] and so if I understand correctly, once more, since everything seems impossible [...] the paralysis effect we mentioned yesterday [...] this paralysis effect we mentioned yesterday about sensory experiences [...] here is translated similarly in the possibilities of dwelling actions, are we clear?

[...] yes, this isn't simply the impression given by being conscious of the risks, but objectively there's a *paralysis of any development* [...] and that paralysis, it can't not have an effect on people's state of mind, on mentalities, all of that [...] yes, so this statement is very interesting [...]

[...] for example, there are children, there are women who are doing stuff, there are young girls who are playing [...] that's what should punctuate the condominium [...] there a *life that should* [...] *that we should be able to hear* [...]

[...] there's also an *abandoned basketball court* because it was in the condominium that stopped [...]

The outside rhythm
[...] you said that the small sounds that you have [...] whereas there was sound [...] that the rhythm was coming from the outside, the horizon.

[...] but instead of that, if you want, it's more partial than that, less strong [...] it's more [...] we can feel it less [...] *what you can feel strongly, it's in the background*, the big boulevard, you know [...] that drives fast, the big petrochemical hub and all the little houses we saw, too, that are on the riverside, and *that's what creates animation* [...] whether it is about colors, living things [...]

The milieu is lacking
[…] in fact, the idea of […] in the sound field, there are three rather simple categories that are the […] the audible signal […] so that's an audible signal [knock knock on the table] […] the soundscape, which is when we're talking […] and then the background sound, maybe the wind tunnel, the outside noise […] here it's like there is the audible signal, which means the people when they're doing something we can hear […] the car, when it's driving around, we can hear it […] people when they slam the door or when they play football, we can hear […] and then, there is the background sound, which is the city in the background, which is another rhythm, very diffuse, visual, sound […] but the soundscape is lacking […]

[…] we're not […] *we're not into,* except in the silence, into […] into a soundscape

Pumping effect
[…] the nearby space is […] the nearby space is almost a desert […] there's a desertification of the nearby space of […] of the body […]

[…] *the activities are discreet,* in terms of space discretion […]

[…] it's pumped, it's completely pumped by a superior ambiance that is given by the […] I think there's a *pumping effect* […] like a blotting paper, the sound is blotted […]

Crushing feeling
[…] we already said it yesterday, the very strong presence of airplane, either during take-off or landing […] I didn't check that […] I can't tell, but they're flying low, and so they're creating a sound ceiling that is very very low […] It feels like being crushed by the ceiling […] but at the same time, am I completely trapped? Yes and no. Why? Because I can find something that Nicolas already mentioned […] that are the sound horizon in the background […]

[…] I heard the ring-road from below, which doesn't have a lot […] maybe a little punctuated, but not much, and I call it the belt […] *the fat belt* […]

The garden effect
[…] so, here I've got an opening […] but I have another one that is even more beautiful that I call the garden effect […] and freedom […] the sensation of freedom […] and what gave it to me? A bunch of dogs that seemed almost free […] country dogs that were doing their own thing […] they were barking a little in every direction, and the sound space was animated […] something like 'ahhh' [breathing] so here, almost a contradiction in the soundscape compared to the crushing feeling […] it was a nice change, to clear my mind […]

The echo from the construction site
[…] I was on the little wooden bench, on the promontory that is at the top […] so that's a little elevated […] and there, there was today, yesterday too, a small construction site [we will learn that it was in fact Geoklock's site, a society in charge of the neighbourhood's environment remediation] […] construction workers using

their tools, with metallic sounds, there were rods [...] metallic sounds [...] and where I was, there was an echo [...] but not reverberation, though, an echo [...] an echo effect, of very strong rebound that made me think of an acoustic feeling of confinement [...]

[...] so, in conclusion [...] yes, there are openings [...] the only real opening is surely given by a *soundscape coming from the outside* [...] from the confines [...] the horizon [...] and so this impression of being really [...] in a half-grave [...] I'm sorry, but that's a very impactful place, *very much imbued with death, still* [...]

From felt silence to thickened silence

The research consisted in gradually thickening the felt silence, loading it with social, historical, and political content, and associating it with a set of contextual elements that inform it. It was thus not simply a question of sharing an 'in situ' sound experience but also making explicit the frames around it and the conditions that made the experience possible. Therefore, the study became the building of bridges between the sounds of places and the form of social and material life that underlies them. The joint logbook represents the revelation of the way the sounds are embedded in the neighbourhood's social history, in the inhabitants' daily actions, and in the built space's material patterns. If the surrounding silence we were facing borrowed both from the action of silencing's *taceo* and the absence of movement's *silencio* (Barthes 2005), it can thus be tackled as a real analyser of the condomínio's problematic situation and the way to inhabit it. In some way, it is a question of giving silence a voice and researching what it unveils about this living environment.

This silence, described as voiceless, static, or inconsistent, appears like the expression of a way of living that is dominated by the inhabitants' risk and vulnerability. It demonstrates a neighbourhood that is left hanging, characterized by a depressive tone. This ambiance reveals the state of a slow-motion territory that is coming to a stop, of a living environment that is frozen and left out. With the incomplete state of the ghost-like buildings as the most noticeable trait, the surrounding silence also shows the absence of any internal dynamics. Everything is at a standstill as if time had stopped. We gradually learnt that it was difficult – even forbidden – to develop the condominium's physical structure. For fear of another gas explosion, it was now impossible to cover or rehabilitate the garages, to drill or perforate the ground, to build a collective space, or to start new construction works. Moreover, we noted the lack of vegetable gardens (contamination of plants and appearance of waste on the ground) and an abandoned basketball court (hardly operable, as it was located in a contaminated area). The development appeared to be stuck together with many of the activities that usually provides rhythm and animation to the local life. This is why the human voices' frugality and the activities' discretion within the condominium tended to produce an inconsistent ambiance, lacking a real internal dynamic, creating a strong influence from the outside's sounds: a very lively neighbourhood nearby, a petrochemical hub further away, a busy boulevard, and a ceiling of airplanes. We thus can talk about a 'pumping' and 'paralysis' effect to define this ambiance, which is mainly depressive and struggles to assert itself.

The dwellers' attention bears the mark of hypervigilance and of dispossession. The sound presence from the Geoklock company's works, responsible for the remediation within the condominium, keeps reminding the residents of the state of contamination, and it tends to disrupt the feeling of being at home. The most pregnant and audible sounds from the inside are precisely those produced by people who are not residents of the condominium. Moreover, many inhabitants remember the deadly explosion that took place in 2000. Nowadays, when they hear a sudden sound – a car accident, an unexpected noise by Geoklock or the petrochemical hub – they cannot help but think about the potentiality of another explosion. Everything unfolds as if the inhabitants are on high alert, expecting a forthcoming accident. The risk remains in the residents' collective memory and is a trace in this hypervigilant attention mode. The feeling of strangeness and the disruption of home characterize the ordinary sound experience. The condominium's silence sets itself as the trace of an everyday life that remains problematic and that struggles to be self-evident. Activities as trivial as playing, gardening, or tinkering lose their obviousness and keep reminding the inhabitants of their own vulnerability.

Last but not least, the research also taught us about how frustrating, harmful, and painful this situation is for the condominium's dwellers. A long-standing struggle has been fought by the inhabitants to uphold their rights, since this situation is hardly bearable day to day and causes severe illnesses (depressed residents, sick or ready to leave when they can afford it, and children suffering from cancer). This collective disarray struggles to be heard by public forces, and it reveals the lack of expression and vitality that characterizes the neighbourhood. Everything comes about as if the silence is the witness of the injunctions and prohibitions that the inhabitants are subjected to, like the various constraints on everyday activities. It is also the expression of the depressive state of a distressed group and the inhabitants' feeling of hopelessness. Furthermore, it is adding to the inhabitants' inability to be heard.[11] It thus becomes a real analyser of the controversy that is happening in this contaminated territory.

Conclusion

This outline of an *atmospheric ethnography* offers to explore what constitutes a form of social life from the standpoint of its ambiances. The hypothesis is that each way of living, each collective way of being, is embodied and deployed in particular affective tonalities.[12] In other words, a form of social life is always embedded in an atmospheric way of being, both intertwined. To ignore the social dimension of ambiance would turn the sensory experience into a purely subjective one, and would eventually lead to a solipsistic version of human existence. The sharing of an experience involves first the sharing of an ambiance. Moreover, to omit the atmospheric content of communal life would mean neglecting the ability to be affected by our surroundings and would lead to the building of a world devoid of any intrinsic vitality. Living together never goes without feeling the atmospheric existence of the world at the same time.

One of the major challenges offered by an atmospheric ethnography is to develop a sensitivity to the tones of situations, to their affective content and their existential value. In that sense, sound is particularly helpful in order to vibrate in accordance to the surrounding world, to get in touch and to resonate with an ambient situation. It could perhaps even present the ultimate medium of an atmospheric attunement (Stewart 2011). The silence that we came across in Condominío Barão de Mauá is a perfect example: similar to John Dewey's theory of inquiry, it works as a diffuse quality that initiates and guides the research from one end to the other. As the expression of an undetermined unrest, this silence made us attentive to the neighbourhood's surrounding sounds. It was less about looking for a way to describe a soundscape analytically and more about listening to the situations' tones and the atmospheric effects around everyday behaviour. Therefore, we tried to get closer to listening to 'significance', at least as much as listening to indexes and signs (Barthes 1991).[13] The question was not to identify and catalogue the sounds that we heard, but rather to grasp its gestures and rhythms, the intensities and paces from everyday life which embody the living experience. The in situ felt atmospheres were thus the object of a 'contextual rebound' aiming at making explicit their frames of existence and the conditions for their appearance.

Deeply exploratory, this sound approach consisted in grasping the interior's tones and in being in tune with the situations we came across. Between distracted and focused listening, articulating the *doing* phase and the *undergoing* one, it was a question of being available for the emerging sensory and motor phenomena, letting them come and following them in their specific dynamic. The researcher's body thus became a resonator of the site's atmospheric vibrations.[14] If this atmospheric ethnography indeed involved a work of genuine attention, it was also about implementing a *collective learning process on the sensitivity to field work*. The communing of everyone's experiences with the help of the joint logbook allowed to broaden our common knowledge of the field, to intensify and multiply the interactions with the inhabitants, and to refine our common ability to listen. This is how our listening never stopped evolving throughout the research, becoming increasingly sensitive to micro-variations and to the small threads that were weaving themselves day after day.[15] Contrary to the ethnographies that are often carried out in a solitary way, this approach relied from the beginning, and in a constitutive manner, on a common experience evolving in time. The relationship that was growing was not one between a researcher and a group of inhabitants, but rather between two groups that got together on the basis of diverse and varied modalities. In a provisory way, this atmospheric ethnography argues for an art of nuances, attentive to the various tonalities and attunements of social life.

Notes

1. This research is part of a dual framework: the CNRS programme PEPS FaiDoRA *Faibles Doses, Risques, Alertes* (CRESSON-UMR AAU, France) and the FAPESP programme *Da Cominicação de Riscos à Cultura de Risco* (CETESB, Brazil). The research team was composed of Silvia Regina Burzaca, Alvaro Florentino da Silva Jr, Anali Espindola de Campos, Sylvain Dubert, Carolina Poletti Maestri Ferreira, Jacques Lolive, Cintia Okamura, Norma Lucia Porto, Thiago Rigui, Patrick Romieu, Maria de Lourdes Pinheiros Simões, Jean-Paul Thibaud, and Nicolas Tixier.
2. This approach was led in particular by Jacques Lolive and Cintia Okamura (see Lolive and Okamura 2016: 152–172).
3. For a more extensive and thorough approach of ambiance, see Thibaud 2015.
4. This *atmospheric ethnography* can be related to two major trends of research: on the one hand 'atmospheric methods' (see Anderson and Ash 2015; and Schroer and Schmitt 2017), on the other hand 'sensory ethnography' (see Laplantine 2005; and Pink 2009).
5. A theory of 'little perceptions' has been developed by Gottfried Wilhelm Leibniz in his preface of *New Essays Concerning Human Understanding* from 1704. A very stimulating comment has been devoted to this notion by Gilles Deleuze in *Le pli. Leibniz et le baroque*.
6. For a description of some of those methods, see Grosjean and Thibaud 2001.
7. Collective ethnography is starting to develop in contemporary social sciences, see Buford May and Pattillo-McCoy 2000; and Laferté 2016.
8. James makes a distinction between 'ambulatory knowledge' and 'saltatory knowledge': 'Now the most general way of contrasting my view of knowledge with the popular view (which is also the view of most epistemologists) is to call my view ambulatory, and the other view saltatory; and the most general way of characterizing the two views is by saying that my view describes knowing as it exists concretely, while the other view only describes its results abstractly taken' (James [1909] 2010: 107). Ambulatory knowledge takes into consideration intersubjective experiences and intermediate experiential steps, and is played out step by step.
9. This task greatly benefitted from the auditory skills of Patrick Romieu and Nicolas Tixier.
10. Most discussions were in French, with some input in Portuguese and more rarely in English. Cintia Okamura, a Brazilian researcher on the team produced a translation. It would be worth devoting a whole article to the question of translation as part of investigatory methodology of this sort.
11. At the end of the investigation, Patrick Romieu, one of the French researchers of the team, organized a 'big scream', a collective moment in which the inhabitants were gathered to shout, together with the researchers – a big scream from the condominio's parking lot. This collective performance was both a way to thank the inhabitants for their hospitality and a way to acknowledge their rights to be heard. Silence was momentarily broken.
12. An abundant literature exists on the concept of affective tone, emanating essentially from phenomenology. For an overview of this concept, see Bollnow 1953. The notion

of affective tonality enables one to go beyond a purely subjective view of emotion and offers an alternative to the object-subject dichotomy. It relates to the overall quality of a situation.
13. The way of listening we developed may also be related to the psychoanalytical listening and listening with the 'Third Ear': 'It is not the words spoken by the voice that are of importance, but what it tells us of the speaker. Its tone comes to be more important than what it tells' (Reik 1956: 136).
14. Relying on William and Henry James, David Lapoujade develops the idea of a resonator as 'someone who makes audible the tones in the voices, but also in the places, the things or the atmospheres, the art of the *Stimmung*' (Lapoujade 2008: 52).
15. Regarding the approach about small links and attention to nuances in sensory experience in anthropology, see Laplantine 2003.

References

Anderson, Ben and James Ash (2015). 'Atmospheric Methods'. In Philipp Vannini (ed.), *Non-Representational Methodologies: Re-Envisioning Research*, 34–51. London: Routledge.

Barthes, Roland (2005). 'The Silence'. In *The Neutral: Lecture Course at the Collège de France, 1977–1978*, 21–29. New York: Columbia University Press.

Barthes, Roland (1991). 'Listening'. In *The Responsibility of Forms: Critical Essays on Music, Art, and Representation*, 245–260. Berkeley, CA: California University Press.

Bollnow, Otto Friedrich (1953). *Les tonalités affectives: Essai d'anthropologie philosophique*. Trans. L. R. Savioz. Neuchatel: La Baconnière.

Buford May, Reuben A. and Mary Pattillo-McCoy (2000). 'Do You See What I See? Examining a Collaborative Ethnography'. *Qualitative Inquiry* 6 (1): 65–87.

Deleuze, Gilles (1988). *Le pli: Leibniz et le baroque*. Paris: Minuit.

Grosjean, Michèle and Jean-Paul Thibaud (eds.) (2001). *L'espace urbain en méthodes*. Marseille: Parenthèses.

James, William ([1909] 2010). *The Meaning of Truth: A Sequel to 'Pragmatism'*. The Floating Press.

Laferté, Gilles (2016). 'Retours d'expériences: Plaidoyer pour l'ethnographie collective'. Ethnographiques.org 32. Available online: https://www.ethnographiques.org/2016/Laferte (accessed 9 July 2020).

Laplantine, François (2003). *De tous petits liens*. Paris: Editions Fayard, mille et une nuits.

Laplantine, François (2005). *Le social et le sensible, introduction à une anthropologie modale*. Paris: Téraèdre.

Lapoujade, David (2008). *Fictions du pragmatisme: William et Henry James*. Paris: Minuit.

Leibniz, Gottfried Wilhelm ([1704] 1996). *New Essays Concerning Human Understanding*. Cambridge: Cambridge University Press.

Lolive, Jacques and Cintia Okamura (2016). 'Quelle communication pour la société du risque? Des expérimentations méthodologiques pour développer une culture du risque'. *Cahiers de géographie du Québec* 60 (169): 152–172. https://doi.org/10.7202/1038668ar.

Pink, Sarah (2009). *Doing Sensory Ethnography*. London: Sage.

Reik, Theodor (1956). *Listening with the Third Ear*. New York: Groce Press.
Schroer, Sara Asu and Suzanne B. Schmitt (eds) (2017). *Exploring Atmospheres Ethnographically*. New York: Routledge.
Stewart, Kathleen (2011). 'Atmospheric Attunements'. *Environment and Planning D: Society and Space* 29: 445–453.
Thibaud, Jean-Paul (2011). 'The Sensory Fabric of Urban Ambiances'. *Senses and Society* 6 (2): 203–215.
Thibaud, Jean-Paul (2015). *En quête d'ambiances: Eprouver la ville en passant*. Geneva: Métis Presses.

44

Sound Design Methodologies: Between Artistic Inspiration and Academic Perspiration

Nicolas Misdariis and Daniel Hug

Introduction

This chapter addresses issues on sound design methods in the light of several case studies implemented in a range of artistic and scientifically grounded paradigms. The collection ('gallery') of case studies that are discussed parses a broad range of approaches and applications or industrial fields: scientific design of informational sound signals (watchmaking), innovative product sound design (automotive), and explorative sound design for interaction. The central aspects are discussed after an introductory presentation of the topic: what does sound design involve in terms of definition, concepts, and tools? And before a conclusive discussion: how do methodologies influence and address a *designerly* way of making sound design?

General framework

This section outlines the general framework of the topic; it will specify the nature of sound design and its practices, between art, sciences, and industry. A particular phenomenon, *fixation*, is discussed in order to highlight the relationships between creativity, tools, and methodologies, and to better comprehend the results of sound design, namely, the sonic artefacts or solutions.

Between artistic inspiration and academic perspiration

When discussing sound design methodologies, it is necessary to first discuss the notion of sound design. There is a need to clarify the understanding of design as such, as a particular activity for purposeful creation. Second, the rather novel domain and practice of sound design in a broad sense has only recently emerged as a discipline and profession, and is still in search of a clear identity and self-understanding in relation to other design disciplines.

Even for some senior professionals and leading experts in the domain, sound design as a discipline is quite difficult to demarcate. As a field of professional practice, it not only appears in the area of radio and film sound, where it originated, but also in the context of video games, digital information display, product design and engineering, interactive environments, scenography, communication and marketing, etc. Nevertheless, in the words of Louis Dandrel, we can postulate that 'making sound design is making design with sounds' (Rodriguez 2003). This perspective establishes sound design as a subcategory of design, and by its legacy, at the intersection of art, sciences, and industry. The convergence of practice and theory also makes clear that sound design is a 'project discipline' characterized by a dualism between conception (designing) and realization (making, producing). This requires a methodology that allows one to imagine the conceptual solutions before their practical implementation in order to control their complexity and to anticipate their potential implications for individuals and society.

Understanding sound design as a subdiscipline of design can also help to place sound design in the field of *science of design*, that is to say – following Andrzej Strzalecki's (2000) or Nigel Cross's (2001) use of the term – 'the study of design', its 'principles, practices, and procedures (53)'. This lays the foundations of, and motivation to develop, a conceptual framework for a *science of sound design* that inherently involves methodological issues. This convergence has multiple benefits. It gives access to thinking and formalizations in that domain (science of design) and makes it possible to consider sound design as a 'coherent discipline of study in its own right' (Archer 1979). It also allows one to transpose Cross's principles in design research, and to consider his 'three research loci' for sound design research: people, process, and products (Cross 2006). Still following Cross's ideas, it finally addresses the question of the existence of *designerly ways of knowing* and creating sounds in sound design, and to subsequently ask the question: how can sound design innovation be achieved and what is the relevance of the familiar, the conventional, and the expected versus the unprecedented, the surprising, and novelty in the experience of designed sounds?

The focus on the *process* further leads to the exploration of the notion of *fixation*, its relationships with creativity, and the role of external factors. In design practice, fixation is defined as 'a blind, and sometimes counterproductive adherence to a limited set of ideas in the design process' (Jansson et al. 1991). This phenomenon has inspired Nathan Crilly (2015) to conduct an experiment, based on field observations of professional designers, with the aim of identifying components that encourage or discourage fixation. *Prior art* (previously developed solutions), *initial ideas* (first inspirations), *constraints* (obstacles to an exhaustive exploration of the solution space), *blame* (fear of error and risk minimization), or *briefing* (predefined insights) are elicited to 'increase the risk of fixation or […] the severity of its effect' and seem to disrupt a common sense on innovation. On the other hand, *teamwork* (collective rather than individual efforts), *methods* (systematic methodology), *facilitation* (expertise of fixation), *making* (sketching and prototyping), *expectations* (needs for various solutions), or *experience of variety* (thematic culture integration) are in turn elicited as 'fixation-aware' or 'fixation-resistant' (Crilly 2015).

For instance, in auditory display design – the discipline concerned with the design of *functional* sounds for software and appliances – *design fixation* can be observed in the form of dominating design strategies, for example, *auditory icons* (Gaver 1986) or *earcons* (Blattner et al. 1989). At the same time, motivated by the use of sound specific ends – be it to communicate, to convey emotions, to convince and sell, to convey a brand image, or to support the interaction with a product – empirical approaches and scientific reasoning become important. This legitimizes a sound design process guided both by *scientific* and *designerly* elements or standards. However, the scientific method has a strong tendency to limit the design space to a small amount of well-defined and controllable parameters. Furthermore, it may lead to thinking in fixed, established categories (see the auditory display paradigms mentioned above). This results in a conflict in terms of the role and self-concept of the sound designer (Hug and Misdariis 2011).

Methodologies and tools in sound design

Consideration of the forces related to decision making in the sound design process make it clear that the key to understanding and dealing with the balance between creative openness and all sorts of fixation in sound design lies in the related *methods* and *strategies of knowing and making*. Crilly, for example, promotes the exploration of methods beyond the standard brainstorming or design review approaches and also proposes an approach to prototyping that allows one to 'detach from ideas that [are] not satisfactory and move to explore the alternatives' (Crilly 2015: 72).

Historically, design methodologies have existed in a field of tension between the 'artistic' inspirational aspects of creativity and the aim for rational, reliable, scientifically, and empirically grounded decision making (Mareis 2016). At one extreme, as represented by the design methods movement, the design methodology is aimed to deliver results as objectively and efficiently as possible. In parallel to this, the notion of participatory design emerged, which advocates the involvement of all stakeholders in the creative process (Schuler and Namioka 1993). More recently, the *design thinking* approach proposes design-based methodical progressions as means to enable creative, new, and even unexpected solutions to challenging problems of all kinds (Brown 2008). In general, the design process aims at providing a defined space for systematic creativity and innovation using methods to generate and synthesize new ideas, thus helping to deal with 'wicked' problems (Ritter and Webber 1973) or the need to identify and exploit new and unknown design potentials, always integrating sensory experience and abduction in the process (Kolko 2010). An initial design claim or hypothesis is followed by an explorative phase and an iterative process of prototyping and formative evaluations, gradually reducing the range of possibilities while increasing the degree of refinement. The result is not known in the beginning of the process and there is no conclusive initial requirement analysis and specification.[1]

Sound design for media

As opposed to other domains, highly standardized production processes of the respective industries have always characterized design methods for sound in media. For instance, in film sound design, points of reference for developing the sound generally exists beforehand, in the form of a script, storyboards, or edited scenes. Here, fixation may occur early in the process and is often the price paid for stylistic coherence, for example, to a filmic genre.

But film sound also provides an interesting example of how systematic, analytic, even scientific, and more spontaneous artistic creativity can be combined in sound design. The 'Movie Brats' of the late 1960s, such as George Lucas, Steven Spielberg, and Francis Ford Coppola started to treat sound in a new, non-conformist way, inspired by the musical avant-garde and technological possibilities alike. Ben Burtt gives an example of the attitude and practice of the 'New Hollywood' sound creators, which integrates rationality, scientific approaches, and artistic experimentation:

> Since I was trained scientifically, part of my attitude is first a literal one. I ask myself, 'If this sound-producing object really existed, what would it sound like?' […] I do always consider the literal aspect of it, because ultimately, you're trying to convince the audience of a certain truth.
>
> (LoBrutto 1994: 142)

It is worth noticing here that the scientific approach was not motivated by scientific goals but rather by the need to achieve plausibility in terms of an 'imaginary realism' and, ultimately, a suspension of disbelief in the audience. The means to achieve this plausibility, however, were explorative and unconventional, involving the use of broken equipment, unintended sound phenomena, and unorthodox recording techniques (cf. anecdotes given by Burtt in several interviews about creating the light saber of *Star Wars* or the numerous *Wall-E* sounds).

Sound design for products and interaction

When dealing with products, in particular interactive computational artefacts and the 'wicked' nature of real-life design problems mentioned earlier, sound design practice is confronted with less standardized and linear design processes, very diverse methods and tools, and a much greater range of influencing factors to deal with. Still, it is possible to identify specific procedures and methods.

Nykänen (2008) proposes that the product sound design process might be conceptualized along the lines of the industrial design process, running through stages of investigation of customer needs, conceptualization, primary refinement, further refinement, and finally concept selection, with the help of control drawings or models. Nykänen considers sonically addressing the conceptual and primary refinement stages as the biggest challenge in this process. Particularly for prototyping, no established tools exist yet. Following the notion of sketching design ideas, Ekman and Rinott (2010) and later Rocchesso and colleagues (2015) have proposed 'sonic sketching', using vocalizations as a form of early representation of a design. But there is no straightforward analogy between sketching in visual and sound design. In fact, Murphy et al. (2006) report that when presenting draft-quality sounds to a panel of participants they were immediately criticized for their poor quality. Similar problems were identified by sound design practitioners being confronted with clients in the early phases of a project (Hug and Misdariis 2011).

Moreover, the interactivity of the product itself poses several challenges to sound design, which traditionally is rooted in linear media. To begin with, the actual sequence in which sounds will be heard (and thus contextualized) emerges along the steps of an interaction process which may change with every user or situation. This also applies to the temporal structure of interaction which is highly relevant for the auditory experience. On the other hand, the sonic context, in which interaction sounds will be heard, is more or less beyond the control of the sound design – as opposed to sound in cinema or even in home television. On the production side, procedural sound generation and adaptive composition are still relatively young design techniques, far away from being industrial standards. Last but not least, as of today, there is still no coherent notion of a 'sonic interaction design' methodology. Rather, the methods are drawn from various disciplines according to the project at hand.

Following Krippendorff's proposition[2] for a science of design which draws from specific experiences related to actual design work, we will present a range of case studies, ranging from the rather conservative and conventional to the innovative and experimental: a data-driven product sound design work for the watchmaking industry, a pioneering sound design project for the automotive industry, and a study on design and interpretation of sounds for interactive commodities, conducted in an educational environment.

For each of these cases, we present the specific design issues raised, the method used to address them, the outcomes, and the resulting insights related to the topic of this chapter, in particular regarding the balance of the artistic-scientific dimension in sound design and the numerous connections between artistic skills and practice, scientific knowledge, and methodologies.

In general, the analysis will frame the design cases along three major phases: 'orientation and inspiration', 'creation and production', and 'evaluation and decision'. While this implies a strong simplification of the reality of the processes, it represents a common denominator among them that helps to comparatively address the influences and impulses provided by scientifically or artistically informed measures.

Case study 1: Functional sound design – Sound of time

Design issue

This project was a collaboration between the watchmaking industry and the composer Sebastien Gaxie.[3] Its aim was to design sounds of watch ringing, named a *minute repeater*, which allowed one to recall the time 'by ear'. Based on an hour/minute codification,[4] the ringing's character had to match with the brand's identity. It was produced by two mechanical devices (gongs hit by hammers), miniaturized, and inserted in the watch's body. Thus, this project required the consideration of micromechanical sound generation and a fresh approach to watch sounds that fulfilled the informational requirement.

Method used

The challenge was tackled with the following three-step methodology developed by the Ircam/Sound Perception and Design team (IRCAM 2020):

1. *Analysis* of the state-of-the-art. In this step, an initial sound corpus is formed. It allows an experimental approach to probe perceptual and cognitive mechanisms that underlie the perception of these sounds in terms of functionality, agreement, or aesthetics. This results in specifications and guidelines for sound design.
2. *Creation* starting from these specifications. It involves other constraints (e.g. technical) and produces sonic solutions. This step is usually supported by a composer or a sound designer.
3. *Evaluation* for identifying solutions. This step usually uses an experimental approach to compare specifications and solutions. If needed, this step is looped with the previous step to improve the solutions (Susini et al. 2014).

Within this framework, the minute-repeater project was then implemented as follows:

- Analysis and constitution of perceptual specifications. From initial data, which was derived from an analysis of sounds used by existing brands, a corpus of reference sounds was built. This material allowed the experimental construction of a perceptual space, further described by standard or ad hoc acoustic features (spectral centroid, spectral spread, etc.) (Peeters et al. 2011). This model was augmented by experimentally measured preference weights for each sound situated in the perceptual space. In addition, verbal attributes (previously collected by a verbalization experiments) were used to describe each corpus sound's properties with respect to timbre or source, and then drawing its semantic profile. Another semantic profile was created in order to map brand identity values (described by words provided by the industrial stakeholder) with the same corpus sounds. These two kinds of profiles (sounds ⇔ verbal attributes, and brand values ⇔ sounds) led to the building of a direct relation between *brand-words* (identity values) and *sounds-words* (sonic properties). This finally provided useful elements for a brand description by sound prototypes, which helped to guide the sound design process.
- Conception and creation of initial solutions. Composer Sebastien Gaxie integrated the whole set of previous specifications as well as the technical possibilities and limitations, and implemented strategies and tools to complete the creation work. He defined high-level composition principles: harmonicity (for the spectral contents) and duality between fusion and transposition (for the minute/hour changes). He also built a *workbench* by integrating several models for sound production: initial

 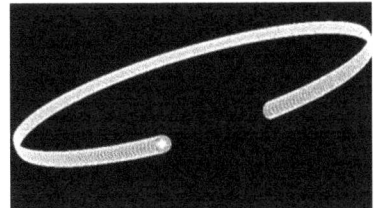 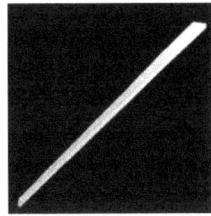

Figure 44.1 Image of the minute-repeater device (left) and its physical model: realistic (middle) and extrapolated (right). *Source:* left: https://www.apmostwatch.com/pink-gold-breguet-tradition-repetition-minutes-tourbillon-7087-replica-watch/breguet-tradition-re-pe-tition-minutes-tourbillon-7087-4/ (accessed 8 July 2020); middle and right: Modalys physical modelling sound synthesis outputs made by Nicolas Misdariis.

corpus sounds (additive analysis/synthesis), musical or environmental sounds (e.g. metallic percussions, jingling glasses, etc.), or physical models of the actual mechanical device (gong and hammer). These physical models could further be extrapolated with respect to geometries, materials, etc., for the sake of timbral exploration (Figure 44.1).

- Initial reduction of possible solutions. Subsequently, the approximately four hundred sounds produced by Gaxie as results from these models were again placed in the perceptual space as defined in the initial analysis by computing their respective acoustic features. This approach allowed us to predict each sound's position with regard to the reference corpus sounds, also taking into account their associated brand values and preference weight. This way, a subset of relevant and targeted solutions could be selected from the numerous sounds produced by Gaxie.
- Selection of optimal sounds. A final scientific experiment filtered the pre-selected designed solutions by building their semantic profile with regard to brand identity (*brand-words*). This protocol led to the selection of two (out of fifty-eight) best solutions, as the final project deliverable.

Outcomes and insights

The main outcome of the minute-repeater project were the two final ringing sounds. They were delivered as numerical specifications related to the spectral content of the sounds. The watchmaking partner was in charge of making the reverse process to reach the mechanical structures (gongs and hammers) responsible for the specified spectral properties. On that point, it is worth noting that the feasibility of recreating electro-acoustic sounds with mechanical means was implicitly integrated in the technical specifications given to the composer (frequency bandwidth, harmonic properties, etc.), but didn't form, per se, a selection criteria early on.

From a methodological point of view, we mainly learned – and demonstrated – that a deductive approach could efficiently address sound design issues. In other words, artistic, creative processes could actually be integrated within a scientific protocol, taking advantage of data analysis and computing (e.g. multidimensional scaling analysis or acoustic features processing) to inform or guide creative ideas.

We also learned that the three-step looped methodology that we intended to apply tended to become a linear rather than a retroactive process. In fact, the evaluation was turned into a simple selection, primarily due to time and cost constraints. The refinement process – expected from the loop – became a more funnel-shaped approach where propositions were successively filtered on the basis of objective data or subjective decisions. In summary, this evaluation appeared to be delivering advisory outputs rather than prescribing and guiding additional conception rounds.

Moreover, a striking property of this approach was the relatively clear delineation between 'scientific' and 'artistic' steps in the process. The artist/sound designer provided a kind of interpretational and creative hinge between the more scientifically grounded steps (requirement specification at the beginning, evaluation at the end of an iteration). Regarding the contribution of scientific and artistically informed impulses on the process, the orientation and inspiration phase was almost entirely informed by scientific procedures, resulting in a set of defined parameters for the artist to follow. The production, then, was to some extent an artistic 'black box', where the composer tweaked the provided parameters, finding new ways of combining and weighting them. This resulted in a surprisingly large amount of options, which then in the final phase were again reduced to a final solution by using scientifically inspired procedures.

Case study 2: Innovative sound design – Sound of silence

Design issue

This project dealt with sound design for silent electric vehicles (EV). It was undertaken within an industrial collaboration with the car manufacturer Renault on the first electric model of its brand (ZOE) and in cooperation with the composer and sound designer Andrea Cera. As electric vehicles are rather quiet but often move in noisy environments, the fundamental issue arises whether the silence of EVs is a blessing or a curse (Cocron et al. 2011), resulting in animated debates about questions like: 'Are vehicles driven in electric mode so quiet that they need acoustic warning signals?' (Sandberg et al. 2010). In fact, we are already facing national or international sound regulations about sound for EVs.[5]

The project was based on three main postulates: quietness potentially induces danger within the car's vicinity (pedestrians, cyclists, visually impaired persons, etc.); quietness deprives the driver of auditory feedback; and EV sound design introduces a new type of sound into the soundscape. The main goal of the project was to identify the properties an EV sound must have in order to fulfil elementary security rules without increasing environmental noise (Misdariis et al. 2012).

Method used

While some elements of the methodology used for the previous case study were adapted to the specificities of the current context, the approach differed significantly. The main reasons for this were the emerging nature of the topic and a very thin 'state-of-the-art', preventing us from working in a purely analytical way, as in case study 1. Also, the integration of the project in a large industrial process with limited control over the technical implementation forced us to adapt a methodology with regards to sound production tools and stakeholders' interaction.

The EV project was characterized by a certain level of *non-linearity* (several phases overlapping and interacting). It consisted of the following main steps:

- Multiple sources of inspiration and orientation
    - Designing *brief* sessions during which the industrial partner delivered global insights about the project. This allowed us to grasp the essence of the project, especially regarding its aesthetic and inspirational dimensions, for instance, that an electric engine should not sound like a classic internal combustion engine.
    - The *state-of-the-art* could be taken into account, in spite of a low number of inputs, at the beginning of the project. Seminal studies based on perceptual experiments and modelling were gathered and helped to formulate first ideas. For instance, 'engine', 'hum', 'music', 'whistle', 'beeps', 'horn', 'clicking', or 'exhaust' represent relevant EV sound categories in terms of perceptual object-sound associations (Wogalter et al. 2001). But, 'engine', 'hum', and 'white noise' are more acceptable than 'horn', 'siren', or 'whistle' (Nyeste et al. 2008).
    - A *context analysis* was executed on the basis of acoustic ecology principles (Schafer 1977) and auditory scene analysis rules (Bregman 1994). This led to the identification of spectro-temporal areas where an EV sound could purposefully inform without conflicting with other urban sounds or emitting too much acoustic energy, and define other basic design guidelines such as temporal invariance (with regards to urban background noise modulations).
    - *Inspiration from popular media* focused on the film industry and how sound designers of this field addressed the silent vehicle sonification issue. Despite some difficulties in transposing cinematographic concepts into true-to-life concepts, this approach started from the following hypothesis: acceptability is influenced by collective imagination and memory that are partially shaped by the film-making culture (on that latter point, see also Hug 2010). This could be put in the larger frame of cultural-interpretational exploration (see below).
- Multiple strands for conception and prototyping
    - *Sonic-functional prototyping*, focused on sound production, was executed using a custom-made real-time sound synthesis engine. The sound prototyping engine aimed at best simulating the embedded industrial sound device architecture provided by the client (wavetable synthesis, see Figure 44.2).

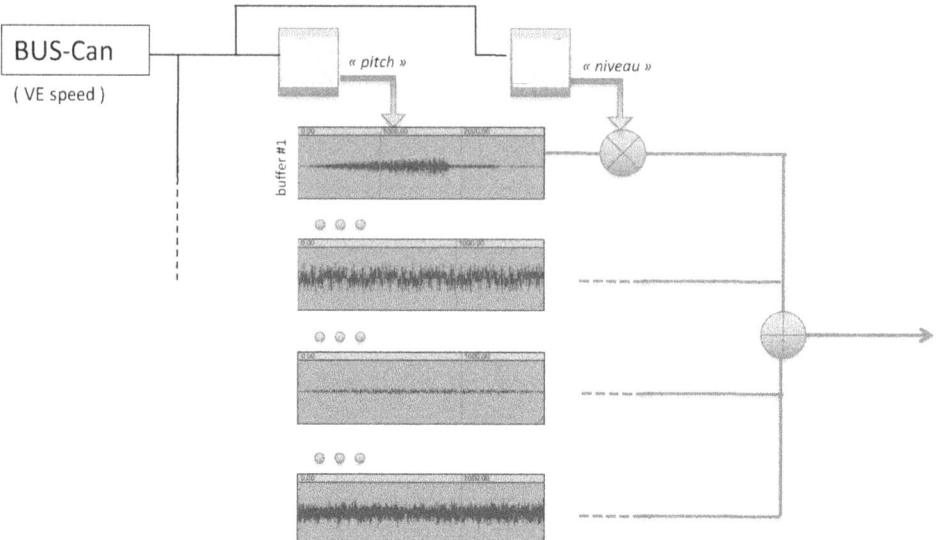

Figure 44.2 Functional scheme of the prototyped sound synthesis engine: a wavetable synthesis with four parallel buffers whose frequency and gain are driven by the vehicle's speed. Schematic block diagram made by Nicolas Misdariis.

- *Contextual-functional prototyping* concerned the test of the proposed solutions. Different contexts (concept cars or prototypes) were used to progressively evaluate the mapping between sound and object, and the acoustic issues emerging from the filtering effect of the physical construction of the engine compartment.
- *Socio-organizational prototyping* concerned the respective role of the stakeholders involved in the project and how they constituted operational entities serving the project goal. Three committees were formed and evolved during the project with regards to decision making (department heads), expertise (direct stakeholders), and techniques (specific crafts). Within these groups, collaborative conception methods were applied.

- Multiple processes of evaluation
  - *Acoustical:* measuring to verify the complementarity between content (background noise) and form (sound signature) as well as the saliency level of the designed sounds.
  - *Perceptual:* observing and experimenting to prove the ability of the sounds to inform about the presence of an EV. Experimental protocols delivered perceptual data that outlined significant differences between different sounds.
  - *Emotional:* interviewing and experimenting to examine evocations generated by EV sounds. Free verbalizations, as well as hedonic and emotional judgements allowed us to draw conclusions related to the coherence between final evocations and initial inspirations, associations between EV sound and visual appearance, or socioculturally motivated differences in subjective assessments.

Outcomes and insights

There were two main outcomes for this project: the design and production of the ZOE sound signature and, additionally, the production of *concept sounds* for different concept cars as orientation and sources of inspiration, being less fixed by technical constraints. The project can be considered as *project-grounded research* (where the project itself constitutes a proper field of research [Findeli 2015]) from which we learnt many things about the sound design discipline. By adapting Cross's three research loci concept (see above), we gained knowledge on:

- *People*: sound designers must show openness, self-effacement (the ability to put aside their own convictions), and pedagogy (the use of demonstrations to convince).
- *Process*: an unusual analysis approach can produce positive side effects by opening up creative opportunities independent from history or heritage. This relates to the challenge that trying to expand the present is oftentimes less innovative than starting from the ground up and trying to 'think out of the box'.
- *Products*: this concerns several issues, which are far from being resolved. For instance, how should we deal with the rather short or repeated sounds usually produced in sound design, the way they are diffused or listened to (lo-fi devices, noisy environments), and the relationships they maintain with the objects they are attached to (e.g. how a heavy moving object should sound like)? All of these reflections lead to a fundamental question, derived from Cross's approach: what does it mean to compose for sound design – is there a *designerly way of thinking* in sound design?

Regarding the relevance of scientific and artistic impulses, the phase of inspiration and orientation combined several approaches as *layers* running through the process in parallel. The integration of discussions with stakeholders adopted methods of qualitative research and creativity alike. The scientific contributions in this phase contained 'empirical' elements, based on studies and experiments, as well as 'theoretical' elements, based on theoretical principles or results from scientific work. In the creation and production phase, we could observe a strong prototype cycle oscillating between sound design, engineering, and empirical testing in an actual car. This included various configurations to involve the stakeholders. Here we also noticed a layering of parallel streams, integrating and interweaving methods and procedures from science and the arts alike. The final evaluation and decision-making phase involved more rigorous scientific procedures, comparable to those of the first case.

Case study 3: Explorative sound design – Sound of interactive commodities

Design issue

As opposed to the two previous case studies, this case addresses design for potential application scenarios rather than for an existing product. Approached from this open starting point, this project's goal was to address a challenge: sound design for interactive commodities faces the possibility to integrate a practically limitless range of computer-controlled electroacoustic sounds into physical objects. In such 'schizophonic' artefacts, sounds can originate both from the physical object itself or from an electroacoustic source. In the latter case, sounds can be anything between identifiable, vaguely reminding us of a material quality or entirely unfamiliar. Moreover, the cause of the sound can be a complex manipulation, just a button press, or an action initiated by the object's *intelligence* or *agency*. Thus, the design process for schizophonic interactive artefacts faces particular challenges regarding function, aesthetics, and semantics (Hug 2008) and has to deal with missing reference points, with a lack of comparable sonic examples or with non-applicable design paradigms. A suitable design method should support the creation of solutions without a priori knowledge. At the same time, it should contribute to a better understanding of interpretation of sounds for novel (interactive) applications (Hug 2017).

Method used

In order to address these challenges, this specific design method afforded the explorative creation and the systematic performative study of novel sounds for interaction in an ongoing reflective process related to interpretation and judgement of sounds.[6] This investigation took place in the sound creation process itself, using real-time sound making during the interaction, and a dialogue between spectators, users, and soundmakers. Moreover, the sound design and prototyping process supported aesthetic openness and avoided biases caused by following a specific auditory display design strategy.

The method for *drafting* or *sketching* sounds for and through interaction was based on the Wizard-of-Oz prototyping approach (Cross 1977), where a computational system is simulated by an invisible human, triggering events in real time, while another person uses the prototype. For this research, Foley-style sound making with MIDI-mapped multi-samples were used, exploiting the ventriloquism effect (Figure 44.3). This method is referred to as 'Electroacoustic Wizard-of-Oz Mockup' and described in more detail by Hug and Kemper (2014).

Participants were asked to formulate concepts and design hypotheses, and to conduct evaluations in a three-step iteration. The design process was implemented in a workshop setting, which allowed the production of a multitude of cases of 'plausible experiences' that could be analysed and discussed. Various fictional yet realistic design briefs were available as starting points. They covered applications related to social interaction in public space, medical self-monitoring, task-oriented activities at home or at work (team-building, time management), and navigation in space or eyes-free object manipulation. In the following, the resulting design process and the related steps, as followed by all projects, are discussed.[7]

Design brief, first scenario, and initial design approaches

After formulating an initial interaction scenario, interaction steps were analysed and sound categories devised. This involved brainstorming and free association methods, which were visualized in flowcharts or as sequences involving a task analysis combined with identifying

Figure 44.3 (a) Two 'wizards', performing their interaction mock-up. (b) Live try-out and exchange with some participants.

potentials for sound. Then, sounds that would match the initial characteristics were produced: sound libraries and sound recordings provided inspiring and concrete starting points. They were turned into sonic *moodboards* and served to derive semantic profiles; initial sound material was created based on reference sounds taken from the application context.

First sound design and evaluation

A first sound design draft was used to test initial design assumptions. The Wizard-of-Oz method was used to explore sound variations or various levels of effect parameters in test sessions or to explore the relationship between sounds, interaction gestures, and spatial movements. Semantic profiles were re-formulated and compared with semantic ratings by test users in order to identify suitable sounds. This stage often focused on reducing the broad range of possibilities as well as the design variables so that a simple test procedure which loosely followed methods from empirical research could be applied.

Expert review and revised design approach

An expert review was conducted with the workshop organizers, Simon Pfaff and Daniel Hug, both experienced sound designers in media and interactive product sound. Adopting the role of the clients, they questioned and challenged the proposed designs. This step often led to a shift of focus from a rather functional, often oversimplifying 'design to the test', to aesthetic, affective, or emotional aspects of the design. At this stage, a reduction of design ideas and conceptual focus on the core aspects or strengths took place.

Second sound design (iteration) and evaluation

Here, the designs were revised to optimize functionality as well as sonic elaboration. The sounds should be more distinctive while still adhering to the functional findings of the first evaluation. Design strategies were focused on fewer aspects (e.g. rhythm, distortion, musical patterns) with the aim of strengthening them. This also helped to integrate the individual sounds to a coherent design with a common identity. Moreover, the designs were oriented towards deriving mapping specifications, preparing for the transition from Wizard-of-Oz mock-ups to functional prototypes. Evaluations at this stage were usually user tests with a strong focus on usability and user experience. For that, 'situated lab settings' were staged, which allowed us to study the interactions in realistic, yet controllable environments.

Final revision, demonstration session, and conclusions

The process was concluded with a group presentation of the final design revisions, mimicking a client meeting. Here, reconstructing the design process including evaluations and insights from the situated lab tests became central, as establishing new design approaches also involved convincing clients of the approach by presenting data to back up the final results.

Outcomes and insights

This aggregate case study provides an example of an explorative, strongly design-led process, combining subjective artistic intuition and design methods with empirical methods and expert reviews. The process was in correspondence with a typical design process as described in the 'general framework' (see above).

First iterations focused on basic isolated functions and sound aesthetics. Together with attempts to visualize the interaction process and related sounds, this resulted in a reduction of design parameters and variables. At the same time, the first steps included intuitive, abductive explorations of design directions, dealing with aspects not covered in existing guidelines. The process also afforded 'unorthodox' design strategies beyond established design paradigms.

In later steps, the focus shifted to usability in the overall application context and refinement of the sound designs. Methodically relevant here is the notion of the situated lab for user tests. At the same time, interactive demo sessions with users could provoke unforeseen user behaviour leading to a sonic improvisation of the 'wizards', integrating sonic or musical improvisation and intuition into the decision-making process. Furthermore, the iterations were intertwined with expert reviews as a means to maintain quality expectations from a sound designer's perspective and to introduce a client perspective. This helped to expand the design beyond the test results by reintroducing artistic exploration and elaboration of sound designs.

Overall, even without formulating initial guidelines or empirical points of reference, the approaches all followed some kind of implicit or explicit reference points. This included 'filmic' listening, where interpretational references to mainstream audiovisual media were made, as well as design references to stereotypical auditory displays (computer startup sound, typical notification sounds). Other references for design and interpretation came from musical parameters, such as harmonic progression, consonances, or dissonances. Furthermore, many solutions worked with multimodal analogies, for example relating the object's visual quality to its sonic quality or integrating affective sound qualities with indexicality, for instance by using distortion to indicate that something is 'wrong'. The method, organized as an ongoing interpretational exchange, thus allowed interpretational patterns to be revealed, afforded play with familiarity of sounds, and helped to explore solutions beyond conventional design paradigms in auditory display.[8] Such interpretational patterns can support the creation of innovative, even surprising solutions which still offer a certain level of familiarity and orientation, thus opening up new aesthetic and functional directions for sound.

Throughout the process, we could see a strongly abductive, explorative approach ensuring a sufficiently broad spectrum of creative solutions. Scientific methods were implemented rather loosely, yet strategically, in terms of orienting the decision-making process and informing lightweight empirical evaluation methods. The goal of the process was not to create and sell a product at the end; hence an elaborate scientific evaluation in the final phase was not necessary. However, scientific methods were adopted in the large-scale study on design and interpretation of sound for interactive applications and thus addressed a higher-order research interest.

Discussion and conclusion

We have outlined above three incarnations of sound design processes incorporating scientific and artistic methods and procedures in varying configurations as they emerged from actual real-world design activities. In particular, the focus was on their relationship to specific design challenges, on the various emerging forms of knowing and making: between artistic inspiration and scientific perspiration, and between conceptual development and analytical insight and the related methods and tools.

These different case studies reveal some interesting analogies and differences. Each of them combines artistic skills and intuition with scientific knowledge and approaches, but the role played by each has a different impact on the whole process and therefore the outcomes. In some cases, we encounter the following of stringent and rigorous acoustic specifications as a basis for creative exploration; in other cases the exploration of sonic inspiration and creativity sets the stage for a refinement process informed by scientific methods. Moreover, they all concern the creation of purposeful sounds of artefacts, but they differ in the amount and type of technologies used to implement them, ranging from a mechanical watch to electro-acoustically and computationally augmented objects. Finally, some general methods and tools are applied and used in all cases (e.g. state-of-the-art formation, prototyping approaches, and evaluation procedures), regardless of the balance between artistic and scientific approaches. The underlying processes are in principle reproducible in other comparable design situations, but their practical application differs from one case to another, ranging from model-driven drafting tools to real-time prototyping setups or handmade ephemeral devices.

The following methodological components can be considered persistent throughout all cases:

- The role of *sonic references* is crucial in the sound design process, either as elements for analysing the existing state (case study 1) – acting as probes to investigate percepts, and as such, being a more or less limiting guide for design – or as inspirational actual sonic material to create novel sounds (case studies 2 and 3). Moreover, referent sounds may be derived from the envisaged application context directly or 'imported' from related domains (case study 2).
- There is a strong need for *quick-and-dirty*[9] sound prototyping methods and tools, whether they be sonic (case studies 1 and 2) or interaction-centred (case study 3). This will mitigate the challenge that sound is immaterial and consequently difficult to sketch or mock-up. Incidentally, this prototype-driven approach follows Crilly's observation that making – in the sense of 'thinking by doing' – tends to discourage fixation mechanisms.
- Evaluation procedures are inherent to the design process and constitute the difference, among other things, between a *designerly* and an 'artistic' way of composing sounds. Evaluations can be implemented following a more strictly scientific procedure aiming at 'objectivity', on the one hand, by means of acoustic and perceptual measurements (case studies 1 and 2) or, on the other hand, by more subjective protocols with a strong focus on ecological validity and pragmatic application, for instance, using expert knowledge or reviews, usability or user experience feedbacks (case study 3).
- Evaluation can also lead to creativity, or at least inspiration, in a duality between normativity and generativity. In fact, a scientific study might show some strange *outbreaks* in the data. The scientist would try to filter them out, while the designer might become curious and focus on such extremes on purpose, drawing inspiration

from them. Further, in the hands of a skilled artist, even seemingly restricting specifications and guidelines can stimulate creativity and lead to a large amount of possible solutions, provided that the artist obtains enough space to explore the possibilities without interventions from the outside, as the first case study shows nicely.
- Finally, sound design seems to be often a quest into developing the *schizophonic sound–artefact* relationship, whether the artefact be a mechanical (case study 1), an electro-acoustically equipped (case study 2), or a computationally augmented object (case study 3). In this quest, which stands at a very early pioneering stage, science and art have to work jointly in order to produce the best solutions in terms of relevance, usability, and pleasure.

This short comparative analysis can then lead to further discussion and address more formal issues regarding links and relationships between methods, tools, and creativity, and finally, on the basis of Cross's formulation, the relevance of a *sound-designerly* way of knowing, thinking, or acting.

Notes

1. This distinguishes the design process from an engineering approach which is often found in science-driven auditory display design processes where a relatively clear functional and formal-aesthetic goal is formulated as starting point.
2. The production of a 'systematic collection of accounts of successful design practices, design methods and their lessons' can be the basis of the data for a design science, which at the same time provides 'methods for validating designs' (Krippendorff 2005: 209).
3. For a biography of Sébastien Gaxie, see IRCAM 2013.
4. For example, 5h34min = 5 hours + 2 halves an hour + 4 minutes = H-H-H-H-H + M/H-M/H + M-M-M-M.
5. See, for instance, the American Pedestrian Safety Enhancement Act of 2010 or the regulation 540/2014/EC from the European Union which makes it obligatory to fit a sound device on all electric or hybrid vehicles by 2019.
6. In that case, we adopted 'designerly ways of knowing and acting' (Cross 2001).
7. For more details, see Hug and Pfaff (2019).
8. An in-depth discussion of these emerging points of reference for design and interpretation of interactive commodities was proposed by Hug (2017).
9. '*Quick and dirty*' is a conventional term in design/human–computer interaction (HCI) referring to early agile prototyping phases that occur during conceptual phases, (long) before functional prototyping. (See, for instance, Design Methods Finder n.d.)

References

Archer, B. (1979). 'Design as a Discipline'. *Design studies* 1(1): 17–20.
Blattner, M. M., D. A. Sumikawa, and R. M. Greenberg (1989). 'Earcons and Icons: Their Structure and Common Design Principles'. *Human–Computer Interaction* 4(1).

Bregman, A. S. (1994). *Auditory Scene Analysis: The Perceptual Organization of Sound*. Cambridge, MA: MIT Press.
Bronner, K. and R. Hirt (2009). *Audio-Branding: Brands, Sound and Communication*. Baden-Baden: Nomos.
Brown, T. (2008). 'Design Thinking'. *Harvard Business Review*, June.
Cocron, P., F. Bühler, T. Franke, I. Neumann, and J. F. Krems (2011). 'The Silence of Electric Vehicles: Blessing or Curse'. In *Proceedings of the 90th Annual Meeting of the Transportation Research Board*, Washington, DC, January.
Crilly, N. (2015). 'Fixation and Creativity in Concept Development: The Attitudes and Practices of Expert Designers'. *Design Studies* 38: 54–91.
Cross, N. (1977). *The Automated Architect*. London: Pion.
Cross, N. (2001). 'Designerly Ways of Knowing: Design Discipline versus Design Science'. *Design issues* 17(3): 49–55.
Cross, N. (2006). *Designerly Ways of Knowing*. London: Springer.
Design Methods Finder (n.d.). 'Quick and Dirty Prototyping'. Available online: https://www.designmethodsfinder.com/methods/quick-and-dirty-prototyping (accessed 9 July 2020).
Ekman, I. and M. Rinott (2010). 'Using Vocal Sketching for Designing Sonic Interactions'. In *Proceedings of the 8th ACM conference on Designing Interactive Systems*, August 2010.
Findeli, A. (2015). 'La recherche-projet en design et la question de la question de recherche: essai de clarification conceptuelle'. *Sciences du design* (1): 45–57.
Gaver, W. W. (1986). 'Auditory icons: Using Sound in Computer Interfaces'. *Human-Computer Interaction* 2(2): 167–177.
Hug, D., & Pfaff, S. (2019). Bringing Sound to Interaction Design. In M. Filimowicz (Ed.), *Foundations in Sound Design for Embedded Media* (1st ed., pp. 102–130). Routledge. https://doi.org/10.4324/9781315106359-5
Hug, D. (2008). 'Genie in a Bottle: Object-Sound Reconfigurations for Interactive Commodities'. In *Proceedings of Audio Mostly 2008 – 3rd Conference on Interaction with Sound*, Interactive Institute, Pitea, Sweden, 1–8.
Hug, D. (2010). 'Investigating Narrative and Performative Sound Design Strategies for Interactive Commodities'. In S. Ystad, M. Aramaki, R. Kronland-Martinet, and K. Jensen (eds), *Auditory Display*, CMMR 2009, ICAD 2009. Lecture Notes in Computer Science, vol. 5954, 12–40. Berlin: Springer. https://doi.org/10.1007/978-3-642-12439-6_2.
Hug, D. (2017). 'CLTKTY? CLACK! Exploring Design and Interpretation of Sound for Interactive Commodities'. PhD dissertation, University of Art & Design Linz.
Hug, D. and M. Kemper (2014). 'From Foley to Function: A Pedagogical Approach to Sound Design for Novel Interactions'. *Journal of Sonic Studies* 6(1): 1–23.
Hug, D. and N. Misdariis (2011). 'Towards a Conceptual Framework to Integrate Designerly and Scientific Sound Design Methods'. In *Proceedings of the 6th Audio Mostly Conference: A Conference on Interaction with Sound*, ACM.
IRCAM (2013). 'Sébastien Gaxie'. Available online: http://brahms.ircam.fr/sebastien-gaxie (accessed 9 July 2020).
IRCAM (2020). 'Perception and Sound Design'. Available online: https://www.ircam.fr/recherche/equipes-recherche/pds/ (accessed 9 July 2020).
Jansson, D. G. and S. M. Smith (1991). 'Design Fixation'. *Design Studies* 12(1): 3–11.
Kolko, J. (2010). 'Abductive Thinking and Sensemaking: The Drivers of Design Synthesis'. *Design Issues* 26(1): 15–28.

Krippendorff, K. (2005). *The Semantic Turn: A New Foundation for Design*. Boca Raton, FL: CRC Press.

LoBrutto, V. (1994). *Sound-on-Film: Interviews with Creators of Film Sound*. Westport, CT: Greenwood Publishing.

Mareis, C. (2016). *Theorien des Designs*. Hamburg: Junius.

Misdariis, N. and A. Cera (2017). 'Knowledge in Sound Design: The Silent Electric Vehicle—A Relevant Case Study'. In *Proceedings of the Conference on Design and Semantics of Form and Movement-Sense and Sensitivity, DeSForM 2017*, 185–195. London: InTech.

Misdariis, N., A. Cera, E. Levallois, and C. Locqueteau (2012). 'Do Electric Cars Have to Make Noise? An Emblematic Opportunity for Designing Sounds and Soundscapes'. In *Acoustics 2012*, Nantes, France, April 2012.

Murphy, E., A. Pirhonen, G. McAllister, and W. Yu (2006). 'A Semiotic Approach to the Design of Non-speech Sounds'. In D. McGookin and S. Brewster (eds), *International Workshop on Haptic and Audio Interaction Design (HAID)*, 121–132. Berlin: Springer.

Nyeste, P. and M. S. Wogalter (2008). 'On Adding Sound to Quiet Vehicles'. In *Proceedings of the Human Factors and Ergonomics Society Annual Meeting* 52 (21). Los Angeles: Sage.

Nykanen, A. (2008). 'Methods for Product Sound Design'. PhD dissertation, Luleå University of Technology.

Peeters, G., B. L. Giordano, P. Susini, N. Misdariis, and S. McAdams (2011). 'The Timbre Toolbox: Extracting Audio Descriptors from Musical Signals'. *Journal of the Acoustical Society of America* 130(5): 2902–2916.

Rittel, H. W. and M. M. Webber (1973). 'Dilemmas in a General Theory of Planning'. *Policy Sciences* 4(2): 155–169.

Rocchesso, D., G. Lemaitre, P. Susini, S. Ternstrm, and P. Boussard (2015). 'Sketching Sound with Voice and Gesture'. *Interactions* 22 (1): 38–41.

Rodriguez, W. (2003). 'Le design sonore, naissance d'une catagorie musicale'. MA thesis, EHESS.

Sandberg, U., L. Goubert, and P. Mioduszewski (2010). 'Are Vehicles Driven in Electric Mode so Quiet That They Need Acoustic Warning Signals?'. In *20th International Congress on Acoustics*, Sydney, Australia, 23–27 August 2010.

Schafer, R. M. (1977). *The Soundscape: Our Sonic Environment and the Tuning of the World*. 2nd edition. New York: Destiny Books.

Schuler, D. & Namioka, A. (1993). *Participatory design: Principles and practices*. Hillsdale, NJ: Erlbaum.

Spehr, G. (ed.) (2015). *Funktionale Klänge: hörbare Daten, klingende Geräte und gestaltete Hörerfahrungen*, vol. 2. Bielefeld: Transcript Verlag.

Strzalecki, A. (2000). 'Creativity in Design: General Model and Its Verification'. *Technological Forecasting and Social Change* 64(2–3): 241–260.

Susini, P., O. Houix, and N. Misdariis (2014). 'Sound Design: An Applied, Experimental Framework to Study the Perception of Everyday Sounds'. *New Soundtrack* 4(2): 103–121.

Wogalter, M. S., R. N. Ornan, R. W. Lim, and M. R. Chipley(2001). 'On the Risk of Quiet Vehicles to Pedestrians and Drivers'. In *Proceedings of the Human Factors and Ergonomics Society Annual Meeting* 45 (23). Los Angeles: Sage.

45

Listening to the 2001 Argentine Crisis: Soundscapes of Protest, Music, and Sound Art

Violeta Nigro Giunta

Introduction

> The special sense of a town is formed in part for its inhabitants – and perhaps even in the memory of the traveler who has stayed there – by the timbre and intervals with which its town-clocks begin to chime. The special sense of a city maybe no longer is given by tower-clocks and church-bells – by sound, that is, which tell time – but rather by those that tell of motion. The peculiar sounds of transit are the signature tunes of modern cities. These are sounds that remind us the city is a sort of machine. The diesel stammer of London taxis the wheeze of its buses. The clatter of the Melbourne tram. The two-stroke sputter of Rome. The note that sounds as the door shut on the Paris metro, and the flick, flick, flick of the handles. The many sirens of different cities.
>
> —Walter Benjamin (1985: 82)

In 2001, the city of Buenos Aires lived one of the biggest economic, political, and social crises in the country's recent history – the state was in turmoil and the citizenship no longer felt represented. The neoliberal model implemented in the 1990s had immersed the country in poverty. Social protest, which had been increasing exponentially during the 1990s, intensified, and by the year 2000 was led by the massive numbers of unemployed and by people below the poverty line.

As studied by sociologists during this period, the ways of protest took new and different forms (Giarracca 2001), which often involved finding innovative and creative ways of intervening in the public sphere and catching the attention of the press. A whole new vocabulary of protest specific to this time and place emerged: protests and demonstrations were joined by *piquetes* (roadblocks) that would often last up to 72 hours; *escraches*, a collective action meant to single out an institution or the house of a corrupt politician; the occupation of lands and auction houses to prevent the sale of land; and by 2001, the occupation and recovery of factories by their workers (*fábricas recuperadas*) and the *cacerolazo* – masses of people hitting casseroles and pans in protest against the political class.

Due to the economic cuts in education, by 2001 most of the art schools in Buenos Aires were mobilized and the Conservatory was no exception. The repetition of sounds that came from hitting the once-kitchen utensil could also be heard in musical manifestations of that year such as *VejacionesX8*, a protest organized by the students of the National Conservatory in which Erik Satie's *Véxations* was performed during an entire week. Erik Satie's 1893 piece, consisting of a 1-minute piano motive with the additional performing indication 'to be performed 840 times' has been subject to many aesthetic discussions and interpretations (see Giarracca et al. 2001; Nigro-Giunta 2014).[1] In Buenos Aires, they played the work eight times, gaining a stellar role as one of these new ways of protest (Giarracca 2001: 34) that characterized the period. *VejacionesX8* began on 29 September 2001 and lasted until 6 October, 7 days, 10,080 minutes, and 6,720 repetitions later. Over two hundred pianists (or whoever could more or less play the music) succeeded each other, playing the longest version of *Vexations* in the work's history. The manifesto behind the performance was written in pamphlets around the conservatory: 'If they continue to vex us, we will double the bet. The Student Centre of the National Conservatory, force in the fingers, the eyelids, and the chest.' The students took up the etymology of the French word *vexation*, the violation of one's dignity, as a symbol to their action. The protest finds its force in its duration: *VejacionesX8* proclaims itself as a 'gandhian' manifestation, in alignment with pacifist manifestations (such as fasting strikes) and the *piquetes*. But it also finds its force in repetition: getting a message through is sometimes a question of repeating it. Activism during that time became a part of everyday life to the point where artistic activity could not be disentangled from it.

The month of December 2001 saw massive manifestations, during which the *cacerolazos* became the iconic sound of the protest. In the sections that follow, I will analyse how the *cacerolazos,* sound manifestations that can be traced back to the Middle Ages, took on different meanings in different contexts. I will then zoom in, with a magnifying ear, on three very significant days: 19, 20, and 21 December 2001 in the city of Buenos Aires, to finally consider how these sounds later permeated into artistic musical practices.

To analyse this kaleidoscopic quality of sounds in protest, music, and sound art, I will draw from a multiplicity of sources, coupling historical and musical analysis: newspaper articles, television broadcasts, musical performances, and musical works. I will argue that the heterogeneous quality of the sources allows for a more comprehensive understanding of sound in its context. One the one hand, I will point at different uses of sound within certain historical contexts; on the other, the emergence of certain aesthetic practices will be considered as a direct consequence of how sound and politics came to be intertwined.

Listening back to noise

The *cacerolazo* as a way of protest has a long history. When analysing the *casserolades* in Quebec in 2012, Jonathan Sterne and Natalie Zemon Davis trace these manifestations of banging pots, pans, and kitchen utensils back to medieval *charivari*. In their origins, *charivari* – a 'noisy, masked demonstration to humiliate some wrongdoer in the community' – served to denounce sexual conducts that were deemed 'inconvenient' (Zemon Davis 1971: 24, [1975] 1992), to later become a way of expressing indignation towards authorities. These demonstrations called attention

> to a breach of community standards in the village or neighborhood. The English called it 'rough music,' and there were versions of it all over Europe and its colonies. Disguising themselves, young men would bang on pots and pans and ring cow bells in front of the house of, say, a widow or widower who was remarrying someone much too young. The youths were the voice of the community, given license by their elders to restore order. The charivari was an alternative to violent exclusion, instead shaming its target into compensation or reparation.
> (Sterne and Zemon Davis 2012)

The practice of *charivari* then evolved into a form of political protest, where the youths were joined by the elders, and targeted royal tax collectors that oppressed the families of peasants and artisans, for example. 'In the 20th century, rough music got less rough […] and the political charivari became a form of peaceful protest' (Sterne and Zemon Davis 2012).

In the case of Quebec 2012, the *casserolades* were a way to protest a law that banned large gatherings (Sterne and Zemon Davis 2012). Sterne characterizes them as quite 'festive', as a way for sound to become an 'expression of popular will', but also as being inclusive, since 'nothing is more simple than the rhythm of a protest, meaning that everyone can do music, and that this music is linked to larger social and collective meanings' (Sterne, Sklower, and Heuguet 2017: 181). In their analysis, these forms of aural protest serve as a critique of authority and power abuse.

From the medieval *charivari* to Quebec 2012, *casserolades* were heard during the celebration of Acadian independence in the 1950s, in Algeria by the OAS (Organisation Armée Secrète) on 23 September 1961, in Spain against the invasion of Iraq in 2003, in Iceland during the collapse of the banks in 2008, and opposing right-wing candidate François Fillon during the 2017 French presidential election. In Latin America, during the twentieth century, one of the most remembered *cacerolazos* was the one that took place on 1 December 1971 against Salvador Allende in Chile, also called 'the march of the empty casseroles' and organized by the right-wing women's movement 'Poder Femenino'. In Uruguay *cacerolazos* were organized against the dictatorial regime, with people hitting the pots from their balconies to avoid being detained by the military who had imposed a curfew. In Argentina's recent history, the groups that opposed the neoliberal government of the 1990s organized *cacerolazos*, and in 2008, *cacerolazos* were planned by conservative groups to oppose populist measures of the Kirchner government. They were also heard in Brazil, mostly by those opposed to the governments of Lula da Silva and Dilma Rousseff.

As a form of protest in Argentina as well as in the rest of Latin America, *cacerolazos* emerged from both the conservative and the progressive factions of society. They had multiple meanings, most often as a protest against the rise of food prices (as was the case in Chile and in the 1990s in Argentina). However, I would like to argue that the *cacerolazos* of 2001 had two particular meanings: they became a symbol of the citizenship no longer feeling represented by the political class on the one hand, and a way to defend democracy on the other. On 19 and 20 December 2001, people heard casseroles from their homes and went out with their own, deliberately defying the state of siege imposed by President Fernando de la Rúa (Granovsky 2001), thus opposing not only the government but also the possibility of a military intervention.[2] The protests of those days led to the resignation of the president, and became associated with the citizenship's rights to feel represented by the political class. This is symbolized in the *Monumento a la cacerola* (Monument to the Casserole) built in San Juan Province in 2002: a cooking pan that stands on a pedestal with the sign 'Civil servant, the casserole is watching you.' This is the meaning that the journalist Angela Dillon gave to the casseroles when writing on the August 2018 senate vote against the law that would have legalized abortion in Argentina:

> It's ten o'clock in the evening, the debate in the Senate is still ongoing, in the country, the noise of casseroles can be heard. It is the noise of the crisis of representation that this demand – that does not seem to be heard in the political agora – brings to light. It won't be without consequences. Because we won't return to clandestinely, abortion will be said out aloud, and maternity shall be wished for or shall not be at all. And without a doubt, the revolution is feminist.
>
> (Dillon 2018)

The sounds of the Argentine *cacerolazos* after those of 2001 can be thus considered as the sounds of participatory democracy (Kunreuther 2018: 24). In his analysis of the Argentine *cacerolazos* between 1982 and 2013, Tomás Gold characterized the 1990s as the moment during which the material metaphor of the 'empty casseroles' transformed into a sonic metaphor of a crisis of representation, through which the protesters asked to be 'heard' by their representatives (Gold 2018). In the next two sections, I will analyse the sounds of the 2001 *Argentinazo* and its aftermath.

Listening to the *Argentinazo* (December 2001)

During 2001, the government underwent increasing political and economic instability, leading to the resignation of two ministers for economy, after which President De la Rúa assigned Domingo Cavallo. During the mid-term elections in October that year, the president's party lost the majority in both chambers. On 1 December, Cavallo announced 'the corralito', which froze people's savings but also their salaries, and which was to last ninety days. These measures led to Cavallo being named the 'intellectual author' ('Dos días que cambiaron la Argentina' 2001) of what was to come, and many of the written sources refer to the events that began on that first day of December, mounting to 20 December as an 'acceleration of historical time'.

On 12 December, political dissent appealed to both eyes and ears: casseroles were heard in the city of Buenos Aires, and a coordinated power blackout[3] took place. A day later, the seventh national strike that year demanded Cavallo's resignation. The economic situation continued to deteriorate and on 19 December supermarkets and stores were plundered during lootings in the main cities. At 5.30 pm, the president addressed the nation through an official broadcast, minimizing social discomfort and declaring the nation in a state of siege (which he annulled on 21 December – his last decision as president), in the hopes of controlling the social riots. Many historians have attributed this speech as one of the detonators of the events that night. The state of siege was a reminder of the repressive politics of the 1976 to 1983 dictatorship, and to oppose it people took spontaneously to the street armed with casseroles and cooking utensils demanding for the resignation of both the minister for economy and the president. This protest lasted throughout the night and continued – in spite of the government's decision to repress protesters, and with the police also repressing the Mothers of *Plaza de Mayo* – through 20 December. At 7.00 pm that day, President De la Rúa presented his handwritten resignation to Congress and had to leave the house of government by helicopter, due to the crowds of people outside. On 2 January 2002, he was finally succeeded by Eduardo Duhalde, who called for national elections in March 2003, which resulted in the election of Nestor Kirchner on 25 May. The events of 19 and 20 December are known as 'The Argentinazo' (Rieznik 2014).

When reconstructing the soundscape of those hectic days in Buenos Aires, I follow R. Murray Schafer's classic definition of an acoustic environment as 'those sounds which are important either because of their individuality, their numerousness or their domination' (Schafer 1993: 9) with consideration to the specifics of an urban soundscape, studying the sound of cities as a fundamental part and not as a by-product (Wissmann 2016). But I especially consider contributions from anthropology to the study of sensory experiences (Porcello et al. 2010; Samuels et al. 2010), adhering to an integrated approach of the senses including language and discourse. In order to determine the way in which not only certain images but mostly certain sounds prevailed, my first sources are ear-witnesses through writings in the press and through texts by historians and sociologists written in retrospect. My second sources are audiovisual recordings of those days.

Sound signals. Those days were very loud. Most of the titles in the newspapers of the time allude to some kind of aural manifestation: 'The explosion of truth' (Kovaloff 2001), 'Screaming, crying, running, cars breaking, sirens', 'The police radio stunning from inside a patrol car', 'Aida cried. Her neighbors cried too' (Palacios 2001), 'De la Rúa between confusion and the denial of a social outburst' (Natanson 2001); and also many descriptions of 'screaming' and 'firing of rubber bullets.' The following quote from a press article resumes the importance of sound signals during these two days:

> The midday sun was burning. A few blocks away one can still hear a *cacerolazo* of some shopkeepers of the neighborhood. The sound of a siren triggers a grimace of panic in the face of a woman. It was enough for her to start running in a random direction. As if the only thing that was important was to be safe. Images such as these were present all over the city.
> (Himitian 2001)

The casseroles. The sound of wood on metal or metal on metal was almost omnipresent during these days, being part of what Tomás Gold called the 'action repertoire' of the crisis (Gold 2018: 455). A journalist writes that 'the sound was a blow into the ears of the government' ('Dos días que cambiaron la Argentina' 2001). The physical presence of the people in the street, occupying public spaces against the government's official stance, had its aural manifestation in the casseroles:

> The *clanc clanc clanc* began in a balcony, gained forces in the corners and exploded in Plaza de Mayo. Thousands of Argentine men and women, children and adults, were screaming and hitting casseroles. *Clanc, clanc, clanc,* throw out the bald one.[4] *Clanc, clanc, clanc,* throw them all out. *Clanc.* All of them, OK? Let them go to ... *clanc, clanc* and more *clanc*.
>
> (O'Donnell 2001)

As it reads from this quote, sound was moving through the city, giving its listener the possibility of reconstructing what was happening and where. A sociologist writes about that day, also reflecting on this idea of a signal, a sound-call, that, when heard, provoked a precise reaction:

> It is 22:41 of a hectic day. The sound of the TV yields to the sound of the street. 'This is why I have decided to declare state of siege in all the nation ...' The sound of metal is deafening. The insulation of the walls shatters. We must go out.
>
> (Benítez Larghi 2009)

The *cacerolazos* took place in the major cities of the country, especially in Buenos Aires. In those three days its streets served as arteries all leading towards Plaza de Mayo, the main city square and backdrop of the most significant events in Argentine history (in the square's surroundings is situated the Casa Rosada – the excecutive office of the President of Argentina). To provide an idea of how this sound evolved during time we can consider the following numbers: during the last thirteen days of December 2001 there were 859 *cacerolazos*, 706 during January 2002, 310 during February, and 'only' 139 in March. This decrease reflects the extreme quality of the sounds of those December days. Although the protests did not come to a halt, they mutated; they found an outlet in new nets of cooperation and solidarity that emerged, for example, through the consolidation of neighbourhood assemblies (Villalón 2007).

Silence. So, those days were not silent. However, we find silence metaphorically, as a lack of response from the authorities. But also when a journalist described the president sending his resignation 'in a sepulchral silence' (Obarrio 2001); as the 'calm before the storm': before a looting, a journalist stated that one 'could cut through the silence' (Palacios 2001); and from the fact that 'in the neighborhoods further away from the center of the city, it was not necessary to turn the TV on to know that the country had exploded. Shutters lowered, empty streets, and a deafening silence' (Himitian 2001).

Music. On 21 December, another journalist writes the following answer to a variation of Adorno's question formulated as thus: 'Can we talk about music in these days of sorrow?'

> Music is necessary, indispensable, as is all artistic expression, to help us live better. But today, it is not possible; there are more pressing needs. Other necessities. Today we need to find a way out of so much sorrow, so much confusion, and so much impossibility. Today there is no music, but reality. Today there is no music, but hope. Today, there is no music. Hopefully, tomorrow there will be.
>
> (Amiano 2001)

For the author, the lack of music and art is used as a metaphor to characterize those two days as a moment of exceptional violence and despair, and also as a symbol of hope for the future. Nevertheless, the national anthem was sung in protests in cities as well as rural areas (Giarracca 2001). Charly García's controversial rock cover of the patriotic march was often played, and it was also solemnly sung by Mirtha Legrand, a conservative TV talk-show host, and her guests on 19 December. The anthem, as a type of state music, had its unique political history during the nineteenth and twentieth centuries, undergoing several transformations and different uses (Buch 2013). Different governments and parties had encouraged its use as a patriotic symbol, but it had also been sung by social movements that opposed the state. In 2001, its symbolic quality proved to be as ambiguous as ever, being used by both protesters in the streets as well as conservative TV hosts, perhaps in both cases as a sonic embodiment of a nation that existed beyond its inept leadership. However, the last verse, 'O juremos con gloria a morir' (Or let's swear, in glory, to die!) took a bleak meaning considering that the repression by the police and security forces took the lives of thirty-nine people. Yet it could be argued, that it was a new chant that became the spontaneous anthem of this crisis, sang directly to the establishment: 'Qué se vayan todos, que no quede, ni uno solo' (Throw them all out! Leave no one!).

Hence, a complex soundscape can be reconstructed: singing in protest, hitting casseroles, feet marching and running, store-windows breaking, screaming, gunshots. But also phones ringing in the private offices of the politicians involved in decision making, the sound of a helicopter leaving the Casa Rosada with the soon to be ex-president. Sound was omnipresent, to occupy spaces, used to make oneself present. It was sound as an embodied manifestation that came to the front line when the state vanished, demanding a new government and a democratic transition.

Musical exegesis of the crisis

In the years before and especially after the 2001 crisis, and mainly in Buenos Aires, a new wave of sound art and site-specific sound installations emerged. These works referred to elements of national history and were politically engaged, mainly through the inclusion of urban sounds. Many artists collaborated and performed in 'recovered factories' (Figari 2007), factories that were expropriated by the workers who turned them into cooperatives, meaning that they all shared ownership. One of the biggest was Fábrica Ciudad Cultural, which started in 1999 at the IMPA (Industrias Metalúrgicas y Plásticas Argentina) in the city centre, and which held over seventy shows in 2000 (Friera 2001). Artists began to collaborate with factory workers, and factories became venues for cultural activities. Hosting these events meant that there were now new places for an alternative culture, simultaneously serving a political 'back-up plan': factories generated work but also culture (Benito 2010).

During the riots of December 2001, the ensemble La Bandina, made up of about twenty musicians playing wind instruments, guitar, and percussion, and directed by new music composer Marcelo Delgado, performed at La Fábrica. This ensemble was created in 1994 as the band of yet another collective, El Galpón de Catalinas, a theatre and circus group. Delgado arranged popular and classical musics for their show 'Fulgor argentino' in 1998 and 1999. From Igor Stravinsky's *L'histoire du soldat* to George Gershwin's *I love you Porgy*, from Astor Piazzolla's *Libertango* to Los Chalchaleros's *Samba de mi esperanza*, Delgado made his own versions using sounds and instrumental techniques from avant-garde and experimental music.[5]

On 21 December, La Bandina played at the IMPA. A music critic characterized this recovered factory as the ideal setting for the group:

> The sounds of this ensemble fit perfectly among machines and pieces of metal [...]. Towards the end of this particular concert and after playing a wedding song by Kusturika, all the musicians grabbed casseroles and began hitting them, proving that they can show a path of musical liberation and change.
>
> (Kohan 2001)

For many avant-garde musicians these spaces were what was left of experimental culture; they were part of the cultural resistance to a certain pre-existing model of culture (Moreno 2001).

Also in 2001, in another recovered factory, composers Juan Pampín and Nicolás Varchausky presented the site-specific work *La Estrella Federal*, made up of two electronic pieces: *UOM*, by Pampín, and *La Bonaerense/La Federal* by Varchausky. Varchausky's work uses materials hacked from police radios, as a way, according to the composer, to come to terms with his fear of a powerful and repressive entity such as the federal police. Of his work, Pampín stated:

> UOM is the acronym of the Argentine metal workers' union (Unión Obrera Metalúrgica), well known for the lack of representation of its corrupted leaders and their gangster-like approach to politics. The piece explores the sound of metal in an allegorical way, using digital samples deployed in space as a representation of the 'metallic' without mass, as the sonic essence of metal. The distance between what is represented and its representation, somewhat similar to the one between the metal workers and their union, constitutes the dialectic core of the work [...]. The text used for the piece is quoted from '¿Quién Mató a Rosendo?' (Who Killed Rosendo?), a book by writer Rodolfo Walsh, a central figure of Argentine culture, who disappeared during the 1976–83 military dictatorship. In his book Walsh investigated one of the darker chapters in the history of Argentine unions: the murder of UOM leader Rosendo García in 1966, perpetrated by gunmen of his own union.
>
> (Pampín 2001)

Both composers use reality as material for fiction (in the way Rodolfo Walsh's literature also conceives it) – art as a way to tell history. Other sound-art works that deal with history and politics are Pampín's *OID* (2003) and Carmen Baliero's *Oíd el ruido* (2015) – both critical deconstructions of the Argentine national anthem; *Tertulia*, by Nicolás Varchausky and Edgardo Molinari – an installation at the Recoleta Cemetery proposing a metaphorical conversation with the ancestry of the city and questioning the social construction of its heroes (Varchausky and Molinari 2004); Carmen Baliero's *Caleidoscopio/Bocinas* (2006, 2008) – a work for cars that took place in the city centres of Buenos Aires and Córdoba, using 'undesired' sounds of cars and car horns, while beginning with the performers whistling a military march (*The Battle of San Lorenzo*), that is meant to celebrate victory but sounds like a defeat when whistled; and *Sodot nayavesek* (see Figure 45.1), an electronic piece by Luciano Azzigotti in which he reversed the 2001 chant *Que se vayan todos* (Throw them all out), and recomposed it as a 'partisan' song for the artistic collective ETCETERA, thus making it the group's own 'Internationale'.

Figure 45.1 Score of Luciano Azzigotti's *International Errorista*.

After 2001, in an unforeseen manner, Argentina was in the centre of the world (Giunta 2009), and the Argentine crisis became a source material for artists, both locally and internationally. The artist Santiago Sierra, for example, conceived the collective work *The Displacement of a Cacerolada* (Sierra 2002): after recording manifestations in Buenos Aires, he asked people to play it through their speakers in London, Frankfurt, Geneva, Vienna, and New York. Finally, the collective BAS (Buenos Aires Sonora) put up a nine-channel sound installation in the central square of the city, the heart of the 2001 crisis and *cacerolazos*. In their work *Mayo, los sonidos de la plaza* (2003, 2006), they tell the history of the country up to 2001, using documentary materials from the radio and recorded political speeches, but also from fictional sources such as the explosions from the film *Apocalypse Now* (Francis Ford Coppola, 1979) and also an electroacoustic piece by an Argentine composer (see Figure 45.2). The idea was to tell history as it was heard from and by the central square, Plaza de Mayo (Liut 2008; Liut 2010).

Figure 45.2 Buenos Aires Sonora, *Mayo, los sonidos de la Plaza* (2003), press release.

When considering aural perception, not only as a physical phenomenon but also as part of the sensorial and as a cultural construction, in a work such as *Mayo, los sonidos de la plaza*, it is social relationships that are recalled through the experience of audition (Porcello et al. 2010; Samuels et al. 2010). In the finale of *Mayo, los sonidos de la plaza* we hear casseroles, steps of protesters, and the sound of a helicopter leaving the Casa Rosada.

Conclusion

I would like to conclude with a plea for interdisciplinarity as fundamental for the study of auditory culture. It is only through the analysis of heterogeneous sources that we can fully place sonic phenomena in the sociohistorical context where it takes on meaning and gains influence in society. In this case study, the omnipresence of the aural during the crisis and the unique and powerful sounds of those days were transformed into musical materials. In other words, the sonic quality of protests permeated into the musical works. At that moment in time, both music and sound, as part of larger cultural and social spheres, came to the foreground and channelled a battle cry to transform society.

Notes

1. Among its interpretations, it has been considered as a conceptual piece (not to be performed); an ironic message by Satie to one of his lovers; an exercise for pianists to play as many times as they wish (840 being just a way of saying 'as many times as necessary'); or a proper music work to be performed, in which 840 times is a specific performance instruction. This last interpretation was that of John Cage, who organized the first performance of the work, in 1972, also as a means to put forward his own musical philosophy around the importance of time and duration in music. When the motive is repeated 840 times, the piece usually lasts between 12 and 24 hours, although it can be done in less or more, depending on the speed.
2. During the twentieth century in Argentina, states of siege would often be followed by military coups.
3. An *apagón* (blackout) is a form a protest that was employed to reject increased prices of electricity, common during the late 1990s and early 2000s. It consisted of neighbours turning off all their lights at a certain day and time.
4. 'The bald one' refers to President Fernando de la Rúa.
5. As an example of how protesting was part of everyday life activities: one of the latest members to join La Bandina in 2001 was asked to do so by Delgado when they met in a roadblock.

References

Amiano, Daniel (2001). 'Música para días mejores'. *La Nación*, 21 December. Available online: https://www.lanacion.com.ar/360826-musica-para-dias-mejores (accessed 20 May 2019).

Benítez Larghi, Sebastián (2009). 'Una cultura trasnochada: Los usos culturales de los sectores movilizados de la clase media argentina a partir de diciembre de 2001'. In Ana Wortman (ed.), *Entre la política y la gestión de la cultura y el arte. Nuevos actores en la Argentina contemporánea*, 125–153. Buenos Aires: Eudeba.

Benito, Karina (2010). '"Piedra libre para todos los compañeros": análisis de la experiencia IMPA La Fábrica Ciudad Cultural'. *Nómadas* 32: 45–57.

Benjamin, Walter (1985). *One Way Street and Other Writings*. London: Verso.

Buch, Esteban (2013). *O juremos con gloria morir: Una historia del Himno Nacional Argentino, de la Asamblea del Año XIII a Charly García*. Buenos Aires: Eterna Cadencia Editora.

Dillon, Marta (2018). 'La revolución es feminista'. *Página/12*, 8 September. Available online: https://www.pagina12.com.ar/133995-la-revolucion-es-feminista (accessed 20 May 2019).

'Dos días que cambiaron la Argentina' (2001). *La Nación*, 23 December. Available online: https://www.lanacion.com.ar/361569-dos-dias-que-cambiaron-la-argentina (accessed 20 May 2019).

Gold, Tomás (2018). 'Conceptualización e historia de los cacerolazos en la Argentina reciente (1982-2013)'. *POSTData: Revista de Reflexión y Análisis Político* 23 (2): 453–489.

Figari, Carlos (2007). '"Ocupar, resistir, producir y educar": Fábricas y empresas recuperadas en la Ciudad de Buenos Aires'. *Labour Again Publications, International Institute of Social History (IISH)* 24: 1–24. Available online: http://www.iisg.nl/labouragain/documents/figari.pdf (accessed 20 May 2019).

Friera, Silvina (2001). 'La Fábrica, una experiencia inédita en el barrio de Almagro: Una auténtica cultura del trabajo'. *Página/12*, 2 January. Available online: https://www.pagina12.com.ar/2001/01-02/01-02-26/pag21.htm (accessed 20 May 2019).

Giarracca, Norma (2001). *La protesta social en la Argentina*. Buenos Aires: Alianza Editorial.

Giarracca, Norma, Karina Bidaseca, Pablo Lapegna, Daniela Mariotti, Cecilia Aramendy, Martín Lío, Celina Mingo Acuña, Elena Mingo Acuña, Flora Partenio, and Julia Sosa (2001). '"Vejaciones x 8": arte y protesta social en Buenos Aires'. *Informes de Coyuntura* 2. Buenos Aires: Instituto de Investigaciones Gino Germani, Facultad de Ciencias Sociales, UBA.

Giunta, Andrea (2009). *Poscrisis: Arte Argentino después de 2001*. Buenos Aires: Siglo XXI editores.

Guerriero, Leila (2001). 'Mundo Impa'. *La Nación*, 22 April. Available online: https://www.lanacion.com.ar/212782-mundo-impa (accessed 20 May 2019).

Granovsky, Martín (2001). 'El día (y la noche) del no va más'. *Página/12*, 20 December. Available online: https://www.pagina12.com.ar/2001/01-12/01-12-20/pag03.htm (accessed November 2019).

Himitián, Evangelina (2001). 'Escenas de una ciudad sobrepasada'. *La Nación*, 21 December. Available online: https://www.lanacion.com.ar/361017-escenas-de-una-ciudad-sobrepasada (accessed 20 May 2019).

Kohan, Pablo (2001). 'La Bandina: sonido moderno y atractivo con cacerolas propias'. *La Nación*, 23 December. Available online: https://www.lanacion.com.ar/361378-la-bandina-sonido-moderno-y-atractivo-con-cacerolas-propias (accessed 20 May 2019).

Kovaloff, Santiago (2001). 'El estallido de la verdad'. *La Nación*, 21 December. Available online: https://www.lanacion.com.ar/360926-el-estallido-de-la-verdad (accessed 20 May 2019).

Kunreuther, Laura (2018). 'Sounds of Democracy: Performance, Protest, and Political Subjectivity'. *Cultural Anthropology* 33 (1): 1–31. Available online: doi:https://doi.org/10.14506/ca33.1.01 (accessed 20 May 2019).

Liut, Martín (2008). 'De frontera y horizontes: música y arte sonoro'. *Clang*: 45–51.

Liut, Martín (2010). 'Creación artística y reflexión teórica en la Universidad Pública: La experiencia del grupo Buenos Aires Sonora'. *Revista LIS: Ciudad mediatizada* 5 (3): 163–170.

Moreno, Maria (2001). 'Los sonidos de la resistencia'. *Página/12*, 2 January. Available online: https://www.pagina12.com.ar/2001/suple/Las12/01-02/01-02-16/nota4.htm (accessed 20 May 2019).

Natanson, José (2001). 'De La Rúa entre el desconcierto y la negación del estallido social'. *Página/12*, 20 December. Available online: https://www.pagina12.com.ar/2001/01-12/01-12-20/pag06.htm (accessed 15 July 2020).

Nigro Giunta, Violeta (2014). 'Vexations: Les deux temps d'une œuvre'. *Marges: Revue d'art contemporain* 19: 61–73.

Obarrio, Mariano (2001). 'El fallido plan de De la Rúa para sobrevivir'. *La Nación*, 23 December. Available online: https://www.lanacion.com.ar/361550-el-fallido-plan-de-de-la-rua-para-sobrevivir (accessed 20 May 2019).

O'Donnell, Santiago (2001). 'El cacerolazo, la nueva forma de fiscalizar'. *La Nación*, 23 December. Available online: https://www.lanacion.com.ar/361507-el-cacerolazo-la-nueva-forma-de-fiscalizar (accessed 20 May 2019).

Palacios, Cynthia (2001). 'Imágenes de miedo, angustia y descontrol'. *La Nación*, 20 December. Available online: https://www.lanacion.com.ar/360734-imagenes-de-miedo-angustia-y-descontrol (accessed 20 May 2019).

Pampín, Juan (2001). 'UOM'. Available online: http://www.pampin.org/uom/index.htm (accessed 20 May 2019).

Porcello, Thomas, Louise Meintjes, Ana María Ochoa Gautier, and David W. Samuels (2010). 'The Reorganization of the Sensory World'. *Annual Review of Anthropology* 3: 51–66.

Rieznik, Pablo (2014). '1982–2002: Dos décadas de democracia y el Argentinazo'. *Razón y Revolución*. Available online: http://revistaryr.org.ar/index.php/RyR/article/view/250 (accessed 20 May 2019).

Samuels, David W., Louise Meintjes, Ana Maria Ochoa, and Thomas Porcello (2010). 'Soundscapes: Toward a Sounded Anthropology'. *Annual Review of Anthropology* 39 (1): 329–345.

Schafer, R. Murray (1993). *The Soundscape: Our Sonic Environment and the Tuning of the World*. Rochester, VT: Destiny Books.

Sierra, Santiago (2002). 'The Displacement of a Cacerolada'. Available online: http://www.santiago-sierra.com/200210_1024.php (accessed 20 May 2019).

Sterne, Jonathan and Natalie Zemon Davis (2012). 'Quebec's Manifs Casseroles are a Call for Order'. *Globe and Mail*, 31 May. Available online: https://www.theglobeandmail.com/opinion/quebecs-manifs-casseroles-are-a-call-for-order/article4217621/(accessed 20 May 2019).

Sterne, Jonathan, Jedediah Sklower, and Guillaume Heuguet (2017). 'Du charivari au big data: Les musiques populaires au prisme des sound studies'. *Volume !* 14 (1): 175–192. Available online: http://journals.openedition.org/volume/5437 (accessed 17 December 2018).

Varchausky, Nicolás and Eduardo Molinari (dir.) (2004). *Tertulia: Intervención en el Cementerio de la Recoleta*. Quilmes: Universidad Nacional de Quilmes Editorial.

Villalón, Roberta (2007). 'Neoliberalism, Corruption, and Legacies of Contention: Argentina's Social Movements, 1993–2006'. *Latin American Perspectives* 34: 139–156.

Wissmann, Torsten (2016). *Geographies of Urban Sound*. New York: Routledge.

Zemon Davis, Natalie (1971). 'The Reasons of Misrule: Youth Groups and Charivaris in Sixteenth-Century France'. *Past & Present* 50: 41–75.

Zemon Davis, Natalie ([1975] 1992). *Les Cultures du peuple: rituels, savoir et résistance au XVIe Siècle*. Trans. Marie Noelle Bourget. Paris: Aubier.

46
The Sound System of the State: Critical Listening as Performative Resistance
Tom Tlalim

What is heard and what goes unheard in contemporary sonic experience is subject to constant negotiation. Although high-powered industrial emissions overshadow fainter organic vibrations, meaningful signals are still frequently intercepted in spite of the noisy environment. Such sonic signifiers or 'cues' could be considered as the fundamental 'sonic blocks' of ideology.[1]

This chapter discusses the sonic methodology of critical listening as a means of interpreting these cues in their political context. This methodology can be used to reveal the ways in which sound operates as an ideological sphere. I will examine critical listening both as a method for analysing state sound systems and as a performative act of political resistance in its own right. The text draws on John L. Austin's influential theory of speech acts, outlining the role of the listening agency in setting the conditions for the failure or success of illocutionary acts. Critical listening is then conceptualized as a means of resistance that can challenge or subvert the ideological signification of state-produced sounds. Building upon this performative role, critical listening is theorized as a method which broadens our understanding of how ideological sound systems can be challenged and resisted.

The chapter also includes a case study of critical listening, based on my experience of listening to state-produced sounds in Tel Aviv during the 2014 Israel–Gaza conflict. The case explores Israel's Iron Dome missile defence system and its part in the state's self-inflicted soundscape of war. The case helps to support the underlying argument that reading and observing politics and ideology must also be supplemented by listening to the 'Sound System of the State' as one of the central tools of ideology.

The performativity of listening

When considering listening as a method of political critique, it is helpful to think of sound as a language and of listening as a performative act. In 1962, the philosopher and linguist, John L. Austin, first introduced his influential theory that utterances can be understood in terms of the rules governing their social use as 'speech acts' (e.g. promising, confirming, vowing, commanding, exclaiming, questioning, warning, etc.) (Austin 1975: 4). Austin argues, pragmatically, that a performative utterance is only deemed effective if the conditions for its success have been met. He divides speech acts into a three-stage framework, where 'locution' is the very act of uttering (and sounding); the 'illocutionary force' is the intent of the speaker and the contextual or social function of the uttered statement; and the 'perlocutionary effect' is the resulting act in the particular context in which the locution is made (99). In this sense, an utterance has a performative significance, since it operates in a particular contextual setting and has agency, in much the same way as a physical act. Utterances such as 'I do' in a marriage ceremony, or 'I commend this statement to the House' in a legislative assembly, such as the UK Houses of Parliament, are notable examples of how speech can usher in new realities and is considered as action in the eyes (and ears) of the law (6). But the discourse around speech acts often neglects to mention either the corresponding agency of the listener who confirms the performative function of the uttered words, or the capacity for different modes of listening to yield different realities.

Austin touches on listening when he considers the situations in which an utterance would fail because it is not accepted by the other party (Austin 1975: 27). Once again, he uses a matrimonial example: if someone says 'I divorce you' but the intended listener does not accept this statement, the speech act fails. A speech act is subject to particular conventions and contexts, and it fails to be performative if it is not accepted by the listening party. Speech acts thus require a listener; in the absence of a listening agency which registers and responds to it, a speech act will fail. Moreover, for a performative utterance to be effective, the listening agency must be aware of the specific meaning the sounds carry in the particular context in which they are uttered.

A stable operating relationship between sounding and listening is indeed crucial for ideology to work. Mladen Dolar offers a poignant if amusing anecdote as an illustration of this. In the midst of battle, an Italian officer shouts, 'Soldiers, attack!', three times in a loud, clear voice – yet none of his soldiers move. Following his third and loudest cry, a tiny voice rises from the trenches, commenting appreciatively, *che bella voce!* (what a beautiful voice!) (Dolar 2006: 3). For the command to be made manifest, the soldiers (the listening agents) have to be aware of the contextual significance of the officer's utterance in order to respond to its interpellation. In this case, the speech act failed because the contexts of the listeners and the speaker did not align. The listening agency did not register the logic or discursive meaning in the officer's performative order, only the phonic beauty of the calling voice.

This example illustrates how a change in the listening mode can subvert a performative act; it is the listener who *listens for*, selects, filters, identifies, and determines the utterance's capacity to *act*. In this case, the different mode of listening led to the officer's illocution becoming a musical perlocution instead of a military one. The listening agency altered the conditions and thus stripped the utterance from its ideological significance, illustrating how indispensable the listening party is to the performative sequence. Changing the listener, their attention or their mode of listening, altered the performative function of the uttered speech act. A similar albeit more conscious performative act of resistance takes place in critical listening.

Critical listening as resistance

Critical listening requires an awareness of *how* we listen and what we listen *for*. An apprehension of its performative agency enables listening to become an act of resistance since it breaks the chain between illocution and perlocution. Critical listening entails recognizing the performative meaning of sounds and considering the ideological significance embedded in them. It involves suspending any immediate response to the sound, in order to identify the cultural or political expectation it holds. By withstanding the automatic urge to heed the meaning of the call, the listener resists the 'hail' of ideology.

'Hailing', or 'interpellation', is the process by which a dominant ideology transforms individuals into subjects. Louis Althusser uses the example of the moment when a police officer shouts, 'Hey, you there!', and a startled individual turns round; the very act of turning transforms that individual into a subject as they identify themselves as the addressee (Althusser 2014: 191). They might not even have turned; their attention and recognition alone is sufficient as an act of self-production. This recognition happens at the point of listening. Critical listening becomes an act of resistance precisely at that level – where the listener acknowledges their position as the sound's addressee yet questions who the instigator is, and the purpose and consequence of their call, and then considers whether and how to respond. For Althusser, ideology is the 'imaginary relationship of individuals to their real conditions of existence' (256). This relationship is mediated in listening, through the knowledge, stories, or primed expectations listeners have towards the sounds they hear. In order to resist the call of ideology, critical listening requires some disengagement from the immediate meaning and affect that sounds can provoke. To adopt Roland Barthes's advice, critical thinking involves asking not only what signals mean but also what they tell us about their producers (Barthes 1991: 245). Listening can establish a critical relationship between the listener and the emitter if the former questions the message and, by doing so, interrogates the emitter. Thus, listening critically implicates the listener in the distribution of ideological sounds by producing a buffer within which the performative link between illocution and perlocution can be questioned.

Here, I would like to extend the notion of illocutionary acts beyond pure linguistics into 'sonic acts' which, much like speech acts, carry their own performative ideological meaning and 'speak' to the cultural context by which they are heard. Sounds such as car alarms, sirens, engine revs, or the sound design cues on a phone or game console, all transmit performative messages and have a similar function and effect to words. Sonic acts therefore warrant a similar treatment to speech acts, when they are listened to critically, within their political context.

There are therefore two stages to the critical listening method. First, an attempt is made by the listener to undo the causal link between sounds and their performative significance. This act suspends the affective impact of the sounds and questions their ideological meaning. One way to do this is by distinguishing between the sound's 'phonos' (material presence) and its 'logos' (the discursive signification it carries). This separation is achieved by withholding interpretation of the abstract meaning of sounds, delving instead into their concrete material properties so as to describe the sounds and chart their timbral, temporal, and spatial organization. Second, the illocutionary (social, political, contextual) meaning of the sounds is isolated from their locutionary (abstract) presence. This stage aims to unpick the performative significance of the sounds from their immediate affect. While listening, an attempt is made to listen for the intrinsic structures of the sonic event and thus to reveal the underlying ideological mechanisms at play. If we follow Austin's pragmatism, such an act of critical listening may lead to the failure of a sonic act since its immediate function as an ideological hail has been undermined by the very act of questioning.

The 'sound system of the state'

The case study below exemplifies the use of critical listening during a political conflict, drawing on the notion of the 'sound system of the state' (SSS) which has emerged from my analysis of the use of sound in conflicts in Palestine–Israel (Tlalim 2017). The SSS refers to the sonic aspect of the state's 'apparatus' – the discourses, legislation, emissions, and interceptions the state employs to assert its power and sovereignty, manage its flows,[2] and forge and propagate its identity, both internally and externally (Althusser 2006). The performative role played by sounds in constructing ideological spaces is central to this investigation. As Leonardo Cardoso notes in his introduction to *Hearing Like a State*, sound is a particularly 'tricky' medium for the state to grasp, due to its 'ontological fluidity, measurement complexity, and legal instability' (Cardoso 2019: 2). Yet, the power of language, the sound of the voice, of amplification and music, and the echoes of landscapes and architecture are all too great for the state to ignore.

The ideological use of sound is explored in Carolyn Birdsall's influential book *Nazi Soundscapes*. Birdsall's investigation rests on the underlying premise that the study of soundscapes can be particularly helpful in gaining insight into social organizations and the ways in which power relationships between authoritarian states and civilians unfold within public spaces (Birdsall 2012: 12). A very different relationship is expressed sonically within myriads of interactions in the contested borders of Palestine–Israel, where confrontations often take place outdoors, in and around border spaces. The gamut of noises produced by state apparatuses, a vast range from military emissions to festive sounds, have been studied widely by sound scholars. The military use of sonic tactics, for example, includes the sonic booms produced by fighter jets flying at supersonic speeds, drone sounds emitted by unmanned aerial vehicles (UAVs), and the deployment of sonic weaponry such as long-range acoustic devices (LRADs), sirens, megaphones, and other 'crowd control' devices (Goodman 2010: 14; Tlalim and Schuppli 2014; Cusick 2015: 379; Schuppli, Tlalim, and Hoare 2015).

The term sound system refers to the use of sonic techniques and technologies in social gatherings as a means for sharing knowledge, cohabiting, and directing communal gatherings. Sound, according to Julian Henriques, offers a dynamic model of thinking, where the traditional barrier between thinking and doing is crossed, and where embodied knowledge and gestural codes can be rehearsed, practiced, and exchanged (Henriques 2011: xviii, 3, 252). As group identity (national or otherwise) is often celebrated and expressed through sound, music, dance, and/or voice, informal groups frequently use sound amplification systems as part of a process of identity formation. These systems provide a peaceful means by which to differentiate and demarcate a shared space.

The SSS also encompasses more hostile or violent soundings produced by organizations or individuals who identify with or embody the state's ideology. Israeli settler groups, for example, frequently use sonic territorialization practices, such as song and dance, traditional herding calls, whistling, and other utterances, as well as sound amplification devices such as megaphones, to dominate spaces in contested areas of the West Bank. Such sounds are deployed to produce an exclusionary ideological space using minimal infrastructure. Many of the settlers' sonic tactics have been documented by videographers working with the B'Tselem Video Archive and are used as evidence of the tactical deployment as part of the Israeli civilian occupation of the West Bank. Some of these documentary videos are showcased in the performance piece *Archive,* on which I collaborated with choreographer Arkadi Zaides and B'Tselem (Zaides and Tlalim 2014; see also Abeliovich 2016; Segal, Weizman, and Tartakover 2003).

The methodology of critical listening proposed in this chapter can serve as a means of exposing the presence of ideological sonic cues within the varied soundscapes around the State and its borders.

Case study: The 2014 Israel–Gaza conflict

As discussed above, critical listening offers an analytical tool, a performative act, and a potentially powerful means of resistance. Listening to war, sounds are often heard without the corresponding visual image of their sources. The experience involves listening to acousmatic sounds as the vibrational forces of weaponry propagate through the air (Kane 2014). Martin Daughtry describes in his important study of a US soldier's experience of the 2003 Iraq War, that violence was often first encountered as sound, emanating from those epicentres of explosions into which the eye had as yet no access (Daughtry 2015: 272). Listening in the midst of battle is a hyper-charged form of listening, involving constant frantic auditioning, interpretation, and speculation about the origins and nature of the sounds, their sources, and their spatial location. Critical listening is therefore a particularly challenging methodology in the midst of conflict as it requires dissociation from the immediate sonic affects, and a focus on the messaging, patterns, and organizations of specific sounds.

The following case study offers a specific examination of the ideological role played by the sounds of the Iron Dome, which is an Israeli missile defence system that was used extensively during the 2014 Israel–Gaza conflict. The system was deemed by some military experts to be a political rather than a strategic weapon, and I argue here that the changing patterns of explosions emitted by the system, alongside the blare of sirens, produced a soundscape that reified the reality of war for Israelis in civilian areas, affecting their mood and morale throughout the conflict.

The Israel–Gaza conflict unfolded during the summer of 2014. During the fifty days of the conflict, the high and extremely asymmetric civilian death toll reflected the horrors of modern warfare (United Nations Human Rights Council [UNHCR] 2015a). The asymmetry was also reflected in the difference in costs of the opposing military systems, as Israeli Iron Dome missiles were estimated (by Israeli analysts) to cost up to a thousand times that of missiles deployed by Hamas (Azoulay 2014; Blay 2015). The United Nations Independent Commission of Inquiry on the 2014 Gaza conflict found that the scale of the devastation in Gaza was unprecedented, as 'Palestinians struggled to find ways to save their own lives and those of their families' under the intense Israeli bombardment (UNHCR 2015b). In Israel itself, there was a sense of panic among civilians, especially those living in the southern regions closest to Gaza, due to the constant threat of rocket and mortar attacks, with particular anxiety focused on the threat of assaults from tunnels penetrating into Israel. Residents of major Israeli cities experienced disruptions to their daily lives, with the regular wail of sirens announcing yet another emergency, forcing them to run for shelter, followed by the thuds of loud explosions, although a high percentage of Hamas rockets fired from Gaza were, in fact, intercepted by Israel's Iron Dome. Meanwhile, Israel retaliated with ground operations and intense aerial bombardment, reducing large areas of Gaza to dust. The region is still reeling from the intensity of that conflict as civilians were profoundly shaken by the events (UNHCR 2015b).

The immense destruction, suffering, and horror experienced by civilians, the different ways in which online and broadcast media were used, and the many violations of international humanitarian law comprise only a fraction of the aspects of this asymmetric conflict that call for further investigation. As the sheer volume of subjects to be interrogated greatly exceeds the scope of this chapter, I will focus mainly on the conflict's sonic dimensions, drawing on my experience working in Tel Aviv during the summer of 2014. I hope that my findings on critical listening during that period can help shed light on the sonic experience of the conflict from a civilian perspective.

In the early days of the war, I wrote the following account:

2.30 am: I am shaken from my sleep. 'Quick. There's a siren!' my wife whispers, and she gathers up our six-month-old baby, cautiously trying not to wake her. We grab our mobile phones and sprint to the *Mamad* or 'Sealed Room' – a reinforced nuclear, chemical, and biological security room, which has been a statutory requirement in all residential properties in Israel since the 1992 Gulf War [Weisenberg et al. 1993: 462; 'IDF Home Front Engineering

Advice' 2018]. We lock the shelter's vault-like door and shut the fortified metal window. It is an eerie feeling to shut ourselves in like this, in the dead of the night, behind thick walls of reinforced concrete and under an all-scrutinizing white neon light. We have not prepared ourselves for this. Most households would have installed some comfortable furniture in the room as well as food supplies, first-aid kits, emergency lamps, spare batteries, and other emergency provisions. As we are only visiting here for three months, our sealed room is empty and bare.

Before we have time to reflect on the situation or give rein to our anxiety, we hear four deep thuds. It is the first time I have ever heard such loud explosions. These are blasts that shake the room, setting off car and property alarms. Growing up here, we are used to the shrill of sirens that trigger well-rehearsed, embodied emergency routines. Our physical memory knows exactly what to do: grab essential items, run to the shelter, ensure everyone is in, seal the doors, switch the radio and mobiles on, and then wait for confirmation that it is safe to come out. But these blasts are new to us; they announce themselves very clearly. Sitting on the floor of the *Mamad*, my wife is breastfeeding our daughter as we try to keep calm. We wait for fifteen minutes. Nothing happens. How are we meant to know when it is safe to come out? I check my phone for news. Eventually, as we have heard the explosions, we decide that this specific attack has probably passed and we can emerge. Things seem quiet. No unusual signs anywhere. We go back to bed, distraught, lulling our baby back to sleep.

Inside the sealed room, the connection with events outside was primarily mediated through sound. The room was isolated and the thick concrete walls would muffle the sounds, providing some distance from the immediate impacts. Despite its eerie and claustrophobic atmosphere, the space was conducive to critical listening as it provided the distance required to evaluate and question the sonic patterns heard outside. As sirens and explosions would be heard several times a day, the wails and thuds became recognizable sonic cues. The traditional shelter routine would involve hearing the sirens, entering the shelter, then listening to the radio for updates. The sirens provided the cue to enter the shelter, but it was always far more difficult to ascertain when it was safe to venture out. Listening inside the shelter, the terrifyingly visceral explosions had the effect of punctuating the moment when an interception 'event' had occurred. They signalled that an attack was over and that it would soon be possible to emerge from the shelter. The thuds reified the moment of attack, rendering it audible.

Later, it became clear that the immense explosions we were hearing were not caused by rockets launched by Hamas but by Israeli interceptor missiles fired by the Iron Dome (Landau and Bermant 2014), a missile defence system developed by the Israeli defence manufacturer Rafael and US defence contractor Raytheon. The system is funded by an annual package from the US Congress; by 2018, it had received a cumulative US investment of about $6.5 billion and its operational costs are about $1 million a day (Shapir 2013; Bash and Cohen 2014; Hamblen 2014; Samaan 2015; Winer and Ari Gross 2018).

The loud thuds caused by the interceptors provided Israeli civilians with an awe-inspiring orchestration of power that boosted their confidence and had a positive effect on their morale: civilians would cheer and often film the interceptor missile launches, sharing their videos online. The noise of the explosions emanating from the skies dominated the soundscape of Israeli cities during the conflict (Samaan 2015; Wood 2016). According to military expert Yiftach Shapir, due to pressure from the mayors of Israeli cities most

Iron Dome batteries were stationed near city limits rather than next to strategic military infrastructure. Their audible and visual presence provided civilian populations with a sense of security. Shapir argues that this supports the view held by some military analysts that the political role of the Iron Dome was as important as its strategic one (Yehoshua 2011; Harkham 2012; Shapir 2013; Blay 2015; Richemond-Barak and Feinberg 2016).[3] One decorated Israeli missile expert caused much controversy by claiming that the Iron Dome was not intercepting missiles at all, but providing an 'audio-visual display that merely intercepts Israeli public opinion' (Broad 2013; Pedahtzur 2013; 'Israel Security Prize Laurate' 2014). The morale-boosting effect of the system helped secure popular support for continued operations in Gaza, as Emily Landau and Azriel Bermant explain in their analysis of the effects of missile defence systems:

> Additional benefits of missile defense systems relate to the public mood. Critics of the Iron Dome have overlooked the positive impact that successful missile defense has had on Israeli national morale, and its contribution to strengthening public resolve in a war situation. This is borne out by the very positive response of the Israeli public to the Iron Dome system's success in intercepting missiles from Gaza, both in 2012 and 2014.
>
> (Landau and Bermant 2014)

Critical listening in the sealed room

In late August 2014, the temporal and spatial relations between the sounding of the sirens and the sounds of the explosions underwent a noticeable change. Suddenly, although the loud blasts continued, the wailing of sirens was significantly reduced, disorienting civilians who were used to hearing them as an accompaniment to the explosions. In answer to complaints voiced in the media, the civil defence authorities explained that the reduction in siren soundings was made 'in order to prevent unnecessary anxiety among civilians' (Zeytun 2014). Of course the change caused some initial panic but also reduced the anxiety involved in running to a shelter in anticipation of the blasts; the panic was replaced by a strangely mundane experience of simply hearing the explosions and nothing more. The blasts would produce a momentary shock but then would be gone, causing less disruption and panic overall. Without the sirens, the explosions' emotional affect was somehow diminished. As civilians were no longer primed by the wail of sirens to seek shelter, once the explosions were over, life continued as normal.

The Israeli distribution of sirens soundings is centrally controlled through a system called 'Wall and Tower'. This system analyses the path of projectiles and then isolates the area where a missile is likely to hit, selecting its landing point from a system of 204 spatial 'polygons' into which the state is divided (Cohen 2014). It is a human decision whether or not to sound the sirens and across how many of the polygons surrounding the epicentre. The soundings are operated by soldiers in a military operations room and based on policy authorized by the head of operations of the Israeli Defense Force (IDF). The policy is often changed tactically during warfare, and decisions not to sound, which may have caused less anxiety, did sometimes lead to loss of life (Cohen 2014). The siren-distribution system is an emblematic example of the SSS. For example, the IDF stated directly that the decision to reduce siren soundings in August 2014 was aimed at reducing the levels of 'unnecessary' anxiety in the population and not due to a reduction in the quantity of missiles. The sound system was thus operationalized in order to change the civilian experience of the conflict, which was primarily aurally mediated. The ability to manipulate the public mood in such a way gave politicians and military planners important strategic advantages. Landau and Bermant, for example, discuss the strategic significance of the Iron Dome's effect on public morale:

> Public mood can translate into concrete strategic benefits [...] [T]he public's sense of protection by Iron Dome gave time and space for the government to make calculated decisions [...]. No serious military expert would claim that missile defense systems are able to provide hermetic protection, but missile defenses do create conditions for enhanced freedom of action for decision makers – defense systems ensure that they have time, and are not compelled to resort automatically to pre-emption and retaliation.
>
> (Landau and Bermant 2014)

For the Israeli government, the Iron Dome system provided a lever of control during the conflict as its effect of raising public morale allowed policymakers space and time to carry out ambitious ground operations. It can be inferred from Landau and Bermant's study, and from the IDF's statements, that the spatial distribution of siren soundings had a similar political effect. As the frequency of siren soundings affected the levels of anxiety in the population, the ability to influence the public mood by controlling the soundscape in this way provided the government with a second lever of control. When used in tandem, these two levers formed part of a wider political sound system in which different 'mixes' could create different sonic experiences that had implications for both military and political strategy.

As the voices of political pundits and military analysts dominated the media and were heard everywhere – in homes, in shops, and on public transport – during that period, alongside the pervasive beeping and buzzing of mobile alert apps such as 'Red Alert' and 'Home Front Command', the political soundscape of war was almost exclusively rendered audible through the state's own sound systems and media. These systems operated together as a 'heterogeneous ensemble',[4] producing the soundscape of conflict and war (Foucault and Gordon 1980; Cohen 2014; Hamblen 2014; Sales 2016). This soundscape gave Israelis in cities far from Gaza a palpable sense of being under attack. It carried a dual meaning: on the one hand, it had a materializing effect as it reified the population's anxiety of an imminent attack; on the other, it boosted their sense of confidence in the state's military apparatus (Chion, Gorbman, and Murch 1994; Landau and Bermant 2014). The soundscape of the Iron Dome thus created an orchestrated 'ecology of fear', directed at Israel's own population, while projecting a sense of complete protection. The Iron Dome system seemed to appeal to both the public's sense of fear and to its need to experience a feeling of power, security, and confidence (Davis 1999; Goodman 2010: 15).

Abigail Wood's insightful ethnographic work in Palestine–Israel provides a lucid reading of the sonic at work. In an article called 'The Siren's Song', she quotes Brian Massumi's proposition, that 'a history of modern nation-states […] could be written following the regular ebb and flow of fear rippling their surface, punctuated by outbreaks of outright hysteria' (Massumi 1993: viii quoted in Wood 2016). Wood adds:

> While the experiences of most civilians living in Israel's central regions during the 2012 and 2014 military operations were very far from the physical destruction that civilians in Gaza experienced at that time, the soundscape of the war touches on the ripples of fear that armed conflict causes in the stable surface of the state.
>
> (Wood 2016)

The Middle East's postcolonial history is fraught with conflict, and Israeli civilians' modes of listening are well trained to follow prescribed emergency guidance in response to set 'sonic cues'. These responses to actionable sounds such as sirens, explosions, red alert app sounds, coded slogans, and performative military-expert speak, have been rehearsed and re-performed repeatedly during every person's life in peacetime, instilling the habit of fear and institutionalizing trauma. The sonic techniques of the state exacerbate this trauma as each generation of civilians is trained to embody the emergency response to these cures following the state's ongoing emergency response training. The devastation of the 2014 conflict might not have been experienced first-hand by many Israelis living in central cities, but the collective embodied impulse to respond to the emergency was provoked by the state's own sonic apparatus which produced the bulk of war sounds in major cities through its various defence systems. In the 2014 Israel–Gaza conflict, the earwitnessed elements of the system described here included the Iron Dome missile interception system, the 'Wall and Tower' siren-distribution system, and a plethora of analysts' voices, mobile app signals, and other sonic cues. The Iron Dome, a most recent addition to the state's missile defence apparatus, added a terrifying sonic component to the soundscape of the conflict. The awe-inducing soundscape of explosions had a dualistic effect of on the one hand reifying civilians' fears from attack, and on the other signalling to civilians that they were absolutely secure. This dualism is an essential attribute of ideological interpellation.

Conclusion

Listening critically to the soundscape of war in 2014 revealed that many of the sounds heard in central cities – sirens, missile explosions, radio signals, mobile phone alerts, and the like – were produced by the state's own sonic apparatus. In this chapter I have referred to the sound systems and infrastructures producing these sounds as the 'sound system of the state' – a sonic interpellation machine that prompts civilians to respond affectively, either with ripples of doubt and fear or with surges of confidence and pride. These conflicting affects are emblematic of the dual nature of ideology which, according to Althusser, simultaneously attracts and repels its subjects. The dualism is also embodied in the very term 'subject' which connotes, on the one hand, 'a free subjectivity, a center of initiatives, author of and responsible for its actions', while on the other, it refers to 'a subjected being, who submits to a higher authority, [and] is therefore stripped of all freedom' (Althusser 2006: 108).

The methodology of critical listening mirrors this dualism by enabling an analysis of state-produced ideological sounds while at the same time constituting an act of performative resistance. Critical listening resists the performative power of the state's sonic apparatus by questioning and subverting its illocutionary instruction. It is at the point of listening where performative acts may fail or succeed, and the methodology of critical listening proposed in this chapter therefore prioritizes the act of listening as an act of resistance. The listening agency can make or break the causal link between performative sonic acts and their intended political consequences. Critical listening as a method separates the ideological content, meaning, and affect of state-produced sounds from their material properties, temporal organization, and acoustic qualities. As such, critical listening can be instrumental in unpacking the workings of ideology in its sonic form, and in interrogating the workings of the sound system of the state.

Notes

1. 'Sonorous or vocal components are very important: a wall of sound, or at least a wall with some sonic bricks in it […]. Radios and television sets are like sound walls around every household and mark territories (the neighbor complains when it gets too loud) […] [O]ne draws a circle, or better yet walks in a circle as in a children's dance, combining rhythmic vowels and consonants that correspond to the interior forces of creation […]. A mistake in speed, rhythm, or harmony would be catastrophic because it would bring back the forces of chaos, destroying both creator and creation' (Deleuze and Guattari 2008: 311).
2. The term flow (flux) is used here in reference to Deleuze and Guattari. who regarded social theory as a generalized theory of flows (economic, commercial, material, cultural), the decoding of which is the business of every society (Deleuze and Guattari 1977: 262; Smith 2011).
3. According to Richemond-Barak and Feinberg (2016): 'IDSs [Intelligent Defence Systems] neither qualify as weapons nor as military objectives under humanitarian law […]. An in-depth analysis of the little-known concept of civil defense shows that its rationale to afford absolute protection to those specifically assigned to protect the civilian population, even if they are members of the armed forces, is much better suited to IDSs and furthers the policy-oriented objective of incentivizing the use of IDSs.'
4. In a 1977 interview with Colin Gordon, Michel Foucault refers to the apparatus of the state – the system of relations between discourses, institutions, architecture, legislation, science, philosophy, and morality, both spoken and unspoken – using the term 'heterogeneous ensemble' (Foucault and Gordon 1980: 194). Foucault does not mean 'ensemble' in its specific musical sense, but rather refers to a system of relations between heterogeneous elements operating 'in-simul', in agreement or in concert. In this sense, it has a political meaning speaking of unity and coordinated organization. The valorization of 'simultaneity', 'synchronicity', 'harmony', or 'accord' in Western music traditions is perhaps precisely a reflection of how deeply embedded politics is in Western musical aesthetics.

References

Abeliovich, Ruthie (2016). 'Choreographing Violence: Arkadi Zaides's Archive'. *TDR/The Drama Review* 60 (1): 165–170.

Althusser, Louis (2006). 'Ideology and Ideological State Apparatuses (Notes towards an Investigation)'. *Anthropology of the State: A Reader* 9 (1): 86–98.

Althusser, Louis (2014). *On the Reproduction of Capitalism: Ideology and Ideological State Apparatuses*. London: Verso Books.

Austin, John Langshaw (1975). *How to Do Things with Words*. Oxford: Clarendon Press.

Azoulay, Yuval (2014). 'Secrets of the Iron Dome'. *Globes*, 10 July. Available online: https://www.globes.co.il/news/article.aspx?did=1000953639 (accessed 9 July 2020).

Barthes, Roland (1991). 'Listening'. In *The Responsibility of Forms: Critical Essays on Music, Art, and Representation*, 245–260. Trans. Richard Howard. Berkeley, CA: University of California Press.

Bash, Dana and Tom Cohen (2014). 'Congress Approves More Iron Dome Funding for Israel'. CNN, 27 August. Available online: https://www.cnn.com/2014/08/01/politics/congress-israel-iron-dome/index.html (accessed 9 July 2020).

Birdsall, Carolyn (2012). *Nazi Soundscapes: Sound, Technology and Urban Space in Germany, 1933–1945*. Amsterdam: Amsterdam University Press.

Blay, Gali (2015). 'Interview with Yiftah Shapir'. Unpublished.

Broad, William J. (2013). 'Weapons Experts Raise Doubts about Israel's Antimissile System'. *New York Times*, 20 March.

Cardoso, Leonardo (2019). 'Introduction: Hearing like a State'. *Sound Studies* 5 (1): 1–3. https://doi.org/10.1080/20551940.2018.1564461.

Chion, Michel, Claudia Gorbman, and Walter Murch (1994). *Audio-Vision: Sound on Screen*. New York: Columbia University Press.

Cohen, Gili (2014). 'How Israel's Sirens System Works'. *Haaretz*, 24 August. Available online: https://www.haaretz.co.il/news/politics/.premium-1.2413887 (accessed 9 July 2020).

Cusick, Suzanne G. (2015). 'Music as Torture / Music as Weapon'. In Michael Bull and Les Back (eds), *The Auditory Culture Reader*, 379–391. London: Bloomsbury.

Daughtry, J. Martin (2015). *Listening to War: Sound, Music, Trauma, and Survival in Wartime Iraq*. New York: Oxford University Press.

Davis, Mike (1999). *Ecology of Fear: Los Angeles and the Imagination of Disaster*. New York: Vintage Books.

Deleuze, Gilles and Félix Guattari (1977). *Anti-Oedipus: Capitalism and Schizophrenia*. New York: Viking Press.

Deleuze, Gilles and Félix Guattari (2008). *A Thousand Plateaus: Capitalism and Schizophrenia*. New edition, repr. London: Continuum.

Dolar, Mladen (2006). *A Voice and Nothing More*. Cambridge, MA: MIT Press.

Foucault, Michel and Colin Gordon (1980). *Power/Knowledge: Selected Interviews and Other Writings, 1972–1977*. New York: Pantheon Books.

Goodman, Steve (2010). *Sonic Warfare: Sound, Affect, and the Ecology of Fear*. Cambridge, MA: MIT Press.

Hamblen, Matt (2014). 'Red Alert App Warns of Imminent Missile Attacks in Israel'. Computerworld, 22 July. Available online: https://www.computerworld.com/

article/2490280/mobile-apps/red-alert-app-warns-of-imminent-missile-attacks-in-israel.html (accessed 9 July 2020).

Harkham, Ariel (2012). 'Trapped under the Iron Dome: Israel's Siege Mentality Represents a Fundamental Strategic Failure'. *Jerusalem Post*, 1 December. Available online: https://www.jpost.com/opinion/op-ed-contributors/trapped-under-the-iron-dome (accessed 9 July 2020).

Henriques, Julian (2011). *Sonic Bodies: Reggae Sound Systems, Performance Techniques, and Ways of Knowing*. New York: Continuum International.

'IDF Home Front Engineering Advice' (2018). IDF Home Front Command. Available online: http://www.oref.org.il/11146-he/Pakar.aspx (accessed 29 June 2018).

'Israel Security Prize Laurate: "The Iron Dome is a Bluff"' (2014). *Globes*, 13 July, sec. עמוד הבית. Available online: https://www.globes.co.il/news/article.aspx?did=1000953982 (accessed 9 July 2020).

Kane, Brian (2014). *Sound Unseen: Acousmatic Sound in Theory and Practice*. Oxford: Oxford University Press.

Landau, Emily B. and Azriel Bermant (2014). 'Iron Dome Protection: Missile Defense in Israel's Security Concept'. In Anat Kurz and Shlomo Brom (eds), *The Lessons of Operation Protective Edge*, 37–42. Tel Aviv: Institute for National Security Studies.

Massumi, Brian (1993). *Politics of Everyday Fear*. 1st edition. Minneapolis, MN: University of Minnesota Press.

Pedahtzur, Reuven (2013). 'How Many Rockets Has Iron Dome Really Intercepted?'. *Haaretz*, 9 March. Available online: http://www.haaretz.com/opinion/how-many-rockets-has-iron-dome-really-intercepted.premium-1.508277 (accessed 9 July 2020).

Richemond-Barak, Daphné and Ayal Feinberg (2016). 'The Irony of the Iron Dome: Intelligent Defense Systems, Law, and Security'. *7 Harvard National Security Journal* 469. Available online: https://ssrn.com/abstract=2685858 (accessed 9 July 2020).

Sales, Ben (2016). 'IDF Releases Smartphone App that Warns of Attacks'. *Times of Israel*, 22 February. Available online: http://www.timesofisrael.com/idf-releases-smartphone-app-that-warns-of-attacks/ (accessed 9 July 2020).

Samaan, Jean-Loup (2015). 'Another Brick in the Wall: The Israeli Experience in Missile Defence'. Carlisle, PA: Army War College Carlisle Barracks PA Strategic Studies Institute. http://www.dtic.mil/docs/citations/ADA615822 (accessed 9 July 2020).

Schuppli, Susan, Tom Tlalim, and Natasha Hoare (2015). 'Susan Schuppli, Tom Tlalim and Natasha Hoare In Conversation'. In Defne Ayas, Adam Kleinman, and centrum voor hedendaagse kunst Witte de With (eds), *Art in the Age of …*, 150–155. Rotterdam: Witte de With Center for Contemporary Art.

Segal, Rafi, Eyal Weizman, and David Tartakover (eds) (2003). *A Civilian Occupation: The Politics of Israeli Architecture*. Rev. edition. London: Verso.

Shapir, Yiftah S. (2013). 'Lessons from the Iron Dome'. *Military and Strategic Affairs: Institute for National Security Studies (INSS)* 5 (1): 81–94.

Smith, Daniel W. (2011). 'Flow, Code and Stock: A Note on Deleuze's Political Philosophy'. *Deleuze Studies* 5 (suppl.): 36–55. https://doi.org/10.3366/dls.2011.0036.

Tlalim, Tom (2017). 'The Sound System of the State: Sonic Strategies for Political Critique at the Borders of Palestine-Israel'. London: Goldsmiths, University of London.

Tlalim, Tom and Susan Schuppli (2014). 'Casino Luxembourg · Casino Channel · Susan Schuppli & Tom Tlalim'. Casino Channel, 17 May. Available online: https://www.

casino-luxembourg.lu/en/Casino-Channel/Susan-Schuppli-Tom-Tlalim (accessed 9 July 2020).

United Nations Human Rights Council (UNHCR) (2015a). 'Report of the Detailed Findings of the Independent Commission of Inquiry Established Pursuant to Human Rights Council Resolution S-21/1'. The United Nations Independent Commission of Inquiry on the 2014 Gaza Conflict A/HRC/29/52.

United Nations Human Rights Council (UNHCR) (2015b). 'The United Nations Independent Commission of Inquiry on the 2014 Gaza Conflict'. Available online: https://www.ohchr.org/EN/HRBodies/HRC/CoIGazaConflict/Pages/CommissionOfInquiry.aspx (accessed 9 July 2020).

Weisenberg, Matisyohu, Joseph Schwarzwald, Mark Waysman, Zahava Solomon, and Avigdor Klingman (1993). 'Coping of School-Age Children in the Sealed Room during Scud Missile Bombardment and Postwar Stress Reactions'. *Journal of Consulting and Clinical Psychology* 61 (3): 462–467.

Winer, Stuart and Judah Ari Gross (2018). 'Defense Minister Welcomes "Record" $705 Million US Funding for Missile defence'. *Times of Israel*, 26 March. Available online: https://www.timesofisrael.com/defense-minister-welcomes-record-705-million-us-funding-for-missile-defense/ (accessed 9 July 2020).

Wood, Abigail (2016). 'The Siren's Song: Sound, Conflict, and the Politics of Public Space in Tel Aviv'. *AJS Perspectives: The Magazine of the Association for Jewish Studies*, 24 July. Available online: http://perspectives.ajsnet.org/sound-issue/sirens-song-sound-conflict-politics-public-space-tel-aviv/ (accessed 9 July 2020).

Yehoshua, Yossi (2011). 'Municipalities 'battle' over Iron Dome'. *Ynetnews*, 21 August. Available online: https://www.ynetnews.com/articles/0,7340,L-4111602,00.html (accessed 9 July 2020).

Zaides, Arkadi and Tom Tlalim (2014). 'Works'. Available online: http://arkadizaides.com (accessed 9 July 2020).

Zeytun, Yoav (2014). 'Why Were Loud Blasts Heard Without Sirens?'. ynet, 19 August. Available online: https://www.ynet.co.il/articles/0,7340,L-4560603,00.html (accessed 9 July 2020).

47
Sonifications Sometimes Behave So Strangely
Paul Vickers

The comprehension of phenomena by analysing and exploring data collected for the purpose is an old and established practice. Statistical methods have become quite sophisticated and are the bedrock of much modern scientific enquiry. Ever since William Playfair introduced the line, area, and bar chart (1786) and the pie chart and circle graph (1801) to the world, the field of information visualization research has refined and extended his ideas and has developed rules and heuristics for the visual representation of data. In all of this, it is not evident that the ontological nature of vision has been taken into account. And why would it be? Phenomenologists and anthropologists have presented varied and competing theories as to how we perceive the world visually, but it seems that much of that can be bracketed when it comes to choosing how to lay out a plot or a chart.

Sonification is a family of representational techniques that use non-speech audio to communicate data and data relations (think Geiger counter for data). With its recent use in the discovery of gravitational waves, sonification has begun to gain some cultural traction, but for the most part it lacks the ubiquity and acceptance of its graphical cousin, information visualization. The term 'sonification' was adopted to describe the use of non-speech sound for communicating data and data relations, and when Greg Kramer established the International Community for Auditory Display and its associated conference series, the International Conference on Auditory Display in 1992, the emergent field of sonification research put down roots.

The idea of sonification at first seems so simple: take some data values and use them to control the properties of an acoustic signal such that listening to the signal reveals something about the data or the data relations that are driving it. Tools such as the Sonification Sandbox (Walker and Cothran 2003) make this process very easy, generating auditory graphs that step through tabular data with each value altering the pitch of a chosen tone.

Following the emergence of affordable digital audio processing hardware in the 1980s and 1990s researchers began to investigate the possibilities afforded by the auditory modality for data and information analysis and exploration. As they began to explore more deeply the use of sound as a complement to (and in some limited cases, a replacement for) visual display techniques, it became evident quite early on that unlike visualizations, and to borrow from Diana Deutsch, sonifications 'sometimes behave so strangely' (Deutsch, Lapidis, and Henthorn 2011). There was something about the auditory representation of data that meant issues of ontology and phenomenology kept raising their (often unwelcome) heads. Unlike graphs, which do not immediately come across as paintings or pieces of visual art, sonifications kept raising questions of their relationship to music and the sonic arts. From an engineer's, computer scientist's, or even psychologist's point of view, all of whom in the early days of the field were trying to find good ways to map data to sound without any composerly intent, sonification is not music. And yet, as Deutsch (Deutsch, Lapidis, and Henthorn 2011) rediscovered – Pierre Schaeffer arguably being the first to document the phenomenon with his account of the *sillon fermé* (Schaeffer 1967) – the mind, regardless of our volition, sometimes adopts a musical orientation to listening (Vickers, Hogg, and Worrall 2017).

Sonification, it goes, 'is not visualization for the ears, it follows completely different rules' (Kosara 2009). At one level this is perfectly obvious and self-evidently true for vision is (primarily) spatial and hearing is (primarily) temporal. A graph persists over time, the whole can be seen at a single glance, and it may be compared side by side with another graph. But the physical phenomenon of sound exists only in its production. To experience an entire sound requires it be listened to as it unfolds over time, and comparing one sound with the memory of another is fraught with difficulty.

Further, Cartesian dualism holds that perception involves an outside that we see, hear, feel, smell, and taste which we then internally interpret by cognition to form an understanding of the world. This fits very well with a bottom-up account of sensory processing. But in recent years there has been a shift in understanding of perception, from the Cartesian dualism of body and mind to an embodied phenomenological account which involves the 'whole organism in its environmental setting' (Ingold 2000: 258), an understanding which has been embraced by the third wave of human-computer interaction (HCI) research.

Information visualization has gradually accreted conventions for the visual layout of data. Guided by writers such as Jacques Bertin (1981) and Edward Tufte (2001) standardized techniques and aesthetic heuristics have been adopted. In contrast, since the inaugural International Conference on Auditory Display in 1992 the question of how best to specify the data-to-sound mappings remains, to a large extent, an open one in sonification research.

Certain physical properties of sound are well understood thanks to the extensive body of psychoacoustic literature. Equal loudness contours, the relationship between perceived pitch and loudness, and so on are well documented and can be factored into sonification designs. Rules for some types of sonification have been proposed, such as John Flowers's (2005) heuristics for successful auditory graph design, with pitch being used as the main carrier of data values. But, as Bruce Walker's programme of work demonstrated (Walker, Kramer, and Lane 2000), there is no universal property obtaining to the polarity of data-to-pitch mappings; some data are better understood where a rise in value corresponds to a rise in pitch, while others seem to work better the other way around. A partial explanation for this might be that we associate sounds with real world events. While we see objects, we do not hear them, rather we hear the sounds they make, that is we hear events (Rosenblum 2004). Further, the sounds objects and events produce give us knowledge about the objects' size, density, and type. Low-frequency sounds typically belong to heavy, dense objects so an increase in weight might be sensibly sonified with an inverse pitch mapping. On the other hand, physical height conceptually works the other way around, so the greater the height, the higher the pitch of the sonification will be.

Psychoacoustics is based largely on a laboratory-based bottom-up information processing model in which raw sounds are given meaning by attending to their context, what has been heard most recently, prior listener training, experience, and so forth. In this model the physical properties of sounds are decoded, then cognition is employed to classify the sounds according to their form, organization, rhythm, and so on. Finally, at the top level the listener applies social and cultural filters to attribute aesthetic value, meaning, and any referential properties (Clarke 2005: 11–14). As the sensory interrelatedness of perception and our interactions with the environment lead us to needing to embrace an embodied account of perception, we discover that sonification becomes much more complex than we first thought. As John Neuhoff (2004) realized, we need to discuss real-world psychoacoustics in terms of ecology and embodied experience. Al Bregman's magisterial work *Auditory Scene Analysis* (1990) serves as a stepping stone between this bottom-up information processing Cartesian dualistic approach to perception and the rich embodied experience it is being seen as by many today.

Sonification listening may be said to be an embodied, interactional, and practically situated activity. Interaction can be with the sonification tool itself, as in the case of interactive sonification (see Weinberg and Thatcher 2006), but also with the environment and space in which the listening takes place. Sonification is a lot more interesting than lab-based stimulus-response tests. Within information visualization there are some established aesthetic principles which, if followed, are deemed to lead to more successful representations. That is, representations that the intended user is able to read and understand without confusion or ambiguity. For example, consider graph layout aesthetics, such as the goal of minimizing the number of edges that cross each other in order to reduce the visual complexity. At this point it is not yet clear what an aesthetics of sonification entails or even if such a thing exists. Music philosophy has several competing aesthetic accounts but, as has been pointed out repeatedly elsewhere, sonification is not music, that is, it is typically not designed with composerly intent or with the goal of producing a musical aesthetic experience. Indeed, if one looks at sonification through the various lenses of music philosophy it appears to inhabit the (musically) contradictory position of referential formalism. It is referential because its very purpose is to point the listener to something beyond itself (the data) yet also formal because the meaning of the sonification lies within its syntactic and organizational structures.

If the view is taken that aesthetics deals with sensory perception (Barrass and Vickers 2011; Vickers, Hogg, and Worrall 2017) – and this appears to be the reason why graph aesthetics have been developed – then a way to approach the question of sonification aesthetics is to come at it pragmatically in terms of how we might design sonifications that are, as Stephen Roddy (2015) puts it, 'communicatively effective'.

How do we choose the mapping?

How, then, do we choose the mapping? How does the translation of data into sound affect the data and our understanding of it and how do we come to decide to translate those data through particular sounds and not through others (which might influence how we attribute meaning to the data)? An ungenerous answer to the question (from looking at many of the sonifications put forward over the last quarter of a century) is that a great deal of thought was not always given to this aspect. This is, of course, unfair, and belies much serious consideration, but there is a sense in which much early sonification work was motivated by the novelty of simply being able to map data to sound. Questions of aesthetics were usually limited to whether or not the sonification sounded pleasant and there also appears to have been an underlying assumption that sonifications should be easy to use, that is, easy for the listener to understand that information being communicated (more on this later).

More recently, there has been a deeper interrogation of how we listen to sonifications, what role the aesthetic plays in the experience and the nature of the relationship between sonification and the sonic arts (including music). This has been informed largely by the aesthetic turns in the field of HCI which moved from the functional approaches of traditional HCI through considerations of user experience informed by a pragmatist aesthetics (Barrass and Vickers 2011) to today's third-wave which deals with the phenomenological nature of embodied perception and interaction, for which Richard Shusterman (1999) coined the term 'somaesthetics'. Stephen Barrass and I put forth the case for sonification to consider these pragmatist experiential ideas in thinking about sonification aesthetics (Barrass and Vickers 2011), and Bennett Hogg, David Worrall, and I took this further by directly addressing the question of embodied perception in sonification design (Vickers, Hogg, and Worrall 2017). This was motivated by questions around the nature of sonification listening, the directness of a sonification, and the prior listening experiences of the sonification user. The question now becomes 'how might we in future decide on the mapping?' Such an enquiry affords the opportunity to consider the factors involved in sonification as an embodied and interactional listening experience. Just as no 'widely accepted model of an aesthetic interaction' exists (Lenz, Hassenzahl, and Diefenbach 2017: 81) so is there no current definition of an aesthetics of sonification. However, as we move from the very functional view of early sonification research to considerations of the somaesthetic issues, then three factors become very important in the design of sonifications: directness, space, and listening, and I will address these below.

Directness

The choice of sound depends, in large part, on the type of sonification approach adopted. Sonification approaches span a continuum from the very direct, indexical processes involved in audification to the conventional representations (in semiotic terms) used in parameter mapping sonifications which can be very indirect and highly metaphorical. In audification the dataset defines the sonification as it involves transposing the frequencies of a time series dataset into the human audible range, together with any necessary filtering to remove unwanted linear distortions and occasional dynamic range compression to flatten out large variations in sound level. Because the data itself is transposed such that each data value effectively becomes an individual sample in a digital audio signal, the resultant auditory stream is very direct and tightly coupled to the dataset. The choice of what sound to use then becomes one of what filtering and scaling to apply to the signal in order to best make the audification 'readable' and fit for purpose (for a fuller treatment of audification, see Dombois and Eckel 2011).

When it comes to sonifications in which there is no inherent link between the data and the chosen sounds, the directness of the representation is determined by the mapping strategy chosen by the sonification designer. Perhaps the most direct sonifications that use the data to drive the parameters of an audio signal are auditory graphs. They are so called because just like a visual graph maps one dimension (typically time) of the data to the abscissa and the values of the data to the ordinate, an auditory graph represents the abscissa by elapsed time and the data values by some change in the audio signal. The simplest way to effect this is to control the frequency of a sinusoidal oscillator with the data values. A high value gives a high pitch, a lower value a lower pitch. As each data value is plotted the pitch of the signal rises and falls accordingly. Historically, pitch has been most often chosen in auditory graphing and parameter mapping strategies alike. For auditory graphs it is a simple but effective mapping. For parameter mapping sonifications pitch seems to have been chosen as often for its ease of implementation as for any other reason.

Directness is a multivalent term in sonification as different writers have used the word to express different ideas about the relationship between sound and data. For example, Bovermann, Tünnermann, and Hermann (2010) use directness as a measure of the responsiveness of an auditory display, such that user interactions lead to quick changes in output. By contrast, and taking a steer from semiotics, Bennett Hogg and I viewed directness as the conceptual distance between the data and its mapping, that is, a measure of the arbitrariness of the data-to-sound mapping (Vickers and Hogg 2006). For example, a symbolic mapping involving sonic metaphors that stand for features of the data – for example, the use of real-world sounds such as bird song and frog croaks to represent features of network traffic (Vickers, Laing, and Fairfax 2017; Debashi and Vickers 2018) – is an arbitrary mapping in the sense that the sounds chosen bear no direct relationship to those data or phenomena represented. Contrast this with an audification in which the sound generated is directly caused by the scaling of the data. There arises, then, a question as to what sort of mapping is best (indexical or symbolic), a question which, at this point in time, remains unanswered. A representational view of sonification holds that the data being referenced should somehow be a part of how the sonification is properly experienced so that the sonification is experienced in terms of the data it represents (Vickers, Hogg, and Worrall 2017: 96). If Deniz Peters is correct in his assertion that 'an essential part of our listening experience draws on what our own body suggests might have gone into the making of [a] sound' (Peters 2012: 22) then the directness of a sonification's mappings ought to play a very important role in how successful the sonification is at communicating its underlying data. On the face of it, the mappings from data to sound should be as direct as possible with the implication that the more symbolic a mapping is, the less successful it might be. To this end, Robert Höldrich and I have begun work to explore how to implement good direct mapping strategies which we call 'Direct Sonification' (see Vickers and Höldrich 2019; Vickers 2020). But this view does not account for the occasions when a symbolic mapping might be considered by the listener to be direct. For example, if one wished to sonify the comings and goings of worker bees in a hive over the course of a week, sensors could be added to register each time a bee arrives and leaves, and this data could be mapped to a buzzing sound that mimics that of a bee in flight. This is not a direct mapping in the sense that the data themselves are not the cause of the sound (in the way that they are in audifications or in the direct sonification mentioned above); because the data are generated by the activity of bees, and the sounds are of bees, one could argue that the data have become part of the sonification experience and are thus an authentic representation.

The idea that the more (causally) direct a mapping is, the less conceptual distance there is between the data and the sonic parameters, the more likely a sonification is to be successful is an attractive one. The more complex and richer the mapping, the greater the possibility that artefacts of the sonic rendering will be mistaken for properties of the data. For example, the use of tonal musical frameworks and rhythms could lead to expectations and understandings on the part of the listener that are based in the listener's prior experience rather than pointing to characteristics of the data. Perhaps a particular chord sequence is generated by a particular combination of data, a sequence that calls to the listener's mind a meaning that is not intended and which leads to incorrect inferences being drawn. This is one of the reasons why Hogg, Worrall, and I began a programme of work to explore how accounting for the subject position in sonification design might lead to clearer, less ambiguous renderings (Vickers, Hogg, and Worrall 2017).

So far, we have considered the translation between data and sound only as a one-way activity, but we do need to consider the effect the rendering might have on the data. Of course, the objection is immediately raised that such an effect is impossible; how can any sonification affect the data it represents? It cannot, in any real sense alter the data values, or the underlying phenomenon from which the data were measured. The user can, of course, on listening to a sonification, choose to change the phenomenon or system which was being sonified. For example, if I am sonifying my heart and respiratory rates during exercise, the feedback might cause me to increase or decrease my activity which will, in turn, lead to changes in my heart beat and breathing. But here the sonification is a messenger, not an actor. Alternatively, and this is perhaps the more interesting consideration, the sonification might influence the way we interpret the data, leading us to change the way we perceive it, a sort of auditory version of seeing something in a new light: it causes us to appreciate the data, or the phenomenon from which it was measured, anew. The phenomenon hasn't actually changed, but it certainly appears different than before.

Space and listening

The act of listening to a sonification is always situated within a space. Sonifications can be designed for monophonic, stereophonic, or multichannel sound, or three-dimensional playback. If headphones are used then virtual listening spaces and ambiences can be created using combinations of convolution reverberation, binaural recording and reproduction techniques, ambisonics, head-related transfer functions (HRTFs), surround sound, and so forth. To create multichannel or three-dimensional sound fields without headphones requires multi-loudspeaker arrays, or sophisticated equipment such as Sonible's IKO, an icosahedral loudspeaker that employs beamforming and ambisonics to create a three-dimensional sound image (Sonible Gmbh n.d.).

In the early days the majority of sonifications were designed for stereo playback either with headphones or the small loudspeakers commonly used with desktop computers. The focus here was on producing the data-to-sound mappings with little regard given to the listening experience. Headphones provide convenient isolation to reduce the effect of environmental noise during listening tests and also allow experiments to be conducted with multiple participants in a single laboratory. Experimental hypotheses revolved around whether the use of sound (either on its own or in conjunction with a visual display) improved participants' ability to construct knowledge about the data. Even when spatial audio reproduction systems were used, the focus was largely on whether spatial audio could be used to communicate information rather than on the listening experience as an interactional embodied activity.

When we consider the subject position and think about designing for embodied experience, we begin to realize that the sonification designer's past experiences, listening skills, and frames of reference could be very different from those of the intended listener. As Karin Bijsterveld observes, sonification designers tend to have 'trained ears' (Bijsterveld 2019: 104), and it is not always going to be the case that the intended listener will have developed their listening skills to the same extent. In the case where the listener and the sonification designer are not the same person, such as when designers and domain experts come together to collaborate on producing sonifications for the domain experts it is entirely possible that what the designers are able to infer from the sonification is not the same as the listeners whose data is being sonified.

Not only does the mapping itself affect how we perceive and experience a sonification, the spatial aspects of the presentation also play a role. Gerriet Sharma's (2016) concept of the 'shared perceptual space' provides a framework for exploring the sculptural aspects of spatial audio and how to approach the perceptual issues that arise during spatial audio production (Wendt et al. 2017). The shared perceptual space is the space 'within which the perceptions of composers, scientists and audience intersect in respect of three-dimensional sound objects' (Sharma 2016: 3). With it, Sharma discovered that he could construct generalized descriptions of sound objects and that the 'collisions of perceptions gradually informed the ensuing compositional process and led to an expanded understanding and a different practice of artistic work with these phenomena' (3). The idea of 'situated perspective' (Harrison, Tatar, and Sengers 2007) has gained traction in the wider field of HCI, but sonification research has not yet caught up. Even if a sonification is to be designed for stereo headphone presentation, it would still be instructive to consider the situated perspective of the listening and to use concepts such as Sharma's to explore how better to design and construct sonifications. When moving to more ambitious spatialized presentations we can ask questions such as what is the impact of spatial attributes (foreground/background, inside/outside, high/low, 2D/3D, direction) on perception of spatial sound-textures produced by mapped data? How can an understanding of shared perception inform and improve sonification design?

Listener experience

All of the above inexorably draws us to consideration of the listener, both in terms of the embodied experience that occurs during listening, as well as the listener's past experience, skill, and knowledge. The subject position is the stance a listener adopts towards the objects of perception (Clarke 2005). Designing for the subject position is about careful direction of the listener towards what the sonification designer desires to reveal about the underlying data. That is, the 'aesthetic enters at the point of constructing the subject-position such that [...] something in the aesthetic of the sound has to match the phenomenon being revealed' (Vickers, Hogg, and Worrall 2017: 105). This, coupled with knowledge gained from understanding the shared perceptual space, lets us focus on the embodied interactional experience of sonification listening.

However, in our endeavours to address the complexities of embodied listening experience it is easy to fail to deal with listener skill. It has often been assumed that sonification should be designed so as to be as easy as possible to listen to, to require as little training as possible to use. Sometimes this is because the experiments to evaluate the usefulness of a sonification are designed to be run over short periods with large groups of listeners who are typically not domain experts (undergraduates are often recruited as participants for this purpose). Other times it may be motivated by the fact that sonification still often fails to be treated as a serious field of scientific research and enquiry, and so designers have felt that sonifications that are not simple to use will be quickly dismissed. However, it has long been accepted in other fields that sound-based exploratory tools require skill to be used well. In the hands of an adept physician, a stethoscope can be used to diagnose heart conditions; sonar operators need to be trained to use their equipment to be able to distinguish between different underwater objects and structures; and a skilled mechanic can often troubleshoot a car engine by listening to the sounds it makes (Bijsterveld 2019: 2). So why should we insist that sonifications require little skill to use? If we are using sonification to explore complex data then there is every reason to expect that the subtle differences in the sounds produced will require a degree of training to detect. Complex tools require training and skill to use well and if we are to go beyond the very simple sonifications (that are also often not very interesting) the issue of listener training needs to be tackled. Of course, someone joining the navy as a sonar operator would have the expectation of receiving training on how to listen to sonar signals. A climate scientist interested in modelling the effects of pollution on global temperatures, on the other hand, might not reasonably have the expectation that they will need to develop analytical listening skills in order to do their job. But, if sonification users can be trained to listen more analytically than they might be used to, can we choose richer, more subtle, data-to-sound mappings that allow deeper and more valuable sonic exploration of data than has been hitherto accomplished? It will be necessary, then, to determine how 'ordinary' users can be trained to listen in a skilful manner and, hence, to use sonifications more effectively. It will be interesting to discover what the practical limitations and the implications of such training for sonification design are.

In the early days, it was largely sufficient to show that sonification *could* be done, and some preliminary heuristics on how to map certain types of data to sound were produced. The underpinning theory was drawn from music philosophic accounts of listening (particularly those of Pierre Schaeffer, Michel Chion, and R. Murray Schafer) and from psychoacoustics. More recently, the role of aesthetics has become a branch of sonification research in its own right as researchers have started to tackle the rich issues associated with sonification listening as an embodied and interactional experience. It is hoped that this recent programme of research, with a particular focus on sonification directness, listener skill, and the space(s) in which sonification listening takes place will yield valuable insights into how to successfully map rich and complex (and increasingly 'big') data to sound.

References

Barrass, Stephen and Paul Vickers (2011). 'Sonification Design and Aesthetics'. In Thomas Hermann, Andrew D. Hunt, and John Neuhoff (eds), *The Sonification Handbook*, 145–172. Berlin: Logos Verlag.

Bertin, Jacques (1981). *Graphics and Graphic Information Processing*. Berlin: Walter de Gruyter.

Bijsterveld, Karin (2019). *Sonic Skills: Listening for Knowledge in Science, Medicine and Engineering (1920s–Present)*. London: Palgrave Macmillan.

Bovermann, Till, René Tünnermann, and Thomas Hermann (2010). 'Auditory Augmentation'. *International Journal of Ambient Computing and Intelligence* 2 (2): 27–41.

Bregman, Albert S. (1990). *Auditory Scene Analysis: The Perceptual Organization of Sound*. Cambridge, MA: MIT Press.

Clarke, Eric F. (2005). *Ways of Listening: An Ecological Approach to the Perception of Musical Meaning*. Oxford: Oxford University Press.

Debashi, Mohamed and Paul Vickers (2018). 'Sonification of Network Traffic Flow for Monitoring and Situational Awareness'. *PLoS One* 13 (4). https://doi.org/10.1371/journal.pone.0195948.

Deutsch, Diana, Rachael Lapidis, and Trevor Henthorn (2011). 'Illusory Transformation from Speech to Song'. *Journal of the Acoustical Society of America* 129 (4): 2245–2252.

Dombois, Florian and Gerhard Eckel (2011). 'Audification'. In Thomas Hermann, Andrew D. Hunt, and John Neuhoff (eds), *The Sonification Handbook*, 301–324. Berlin: Logos Verlag.

Flowers, John H. (2005). 'Thirteen Years of Reflection on Auditory Graphing: Promises, Pitfalls, and Potential New Directions'. In Eoin Brazil (ed.), *Proceedings of the 2005 International Conference on Auditory Display (ICAD 2005)*, 406–409.

Goudarzi, Visda, Katharina Vogt, and Robert Höldrich (2015). 'Observations on an Interdisciplinary Design Process Using a Sonification Framework'. In Katharina Vogt, Areti Andreopoulou, and Visda Goudarzi (eds), *Proceedings of the 2015 International Conference on Auditory Display (ICAD 2015)*, 81–85. Graz: Institute of Electronic Music and Acoustics (IEM), University of Music and Performing Arts Graz (KUG).

Harrison, Steve, Deborah Tatar, and Phoebe Sengers (2007). 'The Three Paradigms of HCI'. In *Alt. Chi. Session at the SIGCHI Conference on Human Factors in Computing Systems*, San Jose, CA, 1–18.

Ingold, Tim (2000). *The Perception of the Environment: Essays on Livelihood, Dwelling and Skill*. London: Routledge.

Kosara, Robert (2009). 'New Sister Site: EagerEars.org'. [Blog] *EagerEyes*, 1 April. Available online: https://eagereyes.org/blog/2009/new-sister-site-eagerears (accessed 9 July 2020).

Lenz, Eva, Marc Hassenzahl, and Sarah Diefenbach (2017). 'Aesthetic Interaction as Fit Between Interaction Attributes and Experiential Qualities'. *New Ideas in Psychology* 47: 80–90. https://doi.org/10.1016/j.newideapsych.2017.03.010.

Neuhoff, John G. (ed.) (2004). *Ecological Psychoacoustics*. Amsterdam: Elsevier.

Peters, Deniz (2012). 'Touch: Real, Apparent, and Absent – On Bodily Expression in Electronic Music'. In Deniz Peters, Gerhard Eckel, and Andreas Dorschel (eds), *Bodily Expression in Electronic Music: Perspectives on Reclaiming Performativity*, 17–34. London: Routledge.

Roddy, Stephen (2015). 'Embodied Sonification'. PhD thesis, Trinity College Dublin.

Rosenblum, Lawrence D. (2004). 'Perceiving Articulatory Events: Lessons for an Ecological Psychoacoustics'. In John G. Neuhoff (ed.), *Ecological Psychoacoustics*, 219–248. London: Elsevier.

Schaeffer, Pierre (1967). *Traité Des Objets Musicaux*. Rev. edition. Paris: Seuil.

Sharma, Gerriet K. (2016). 'Composing with Sculptural Sound Phenomena in Computer Music'. PhD thesis, Kunst Uni Graz, Graz, Austria.

Shusterman, Richard (1999). 'Somaesthetics: A Disciplinary Proposal'. *Journal of Aesthetics and Art Criticism* 57 (3): 299–313. http://doi.org/10.2307/432196.

Sonible Gmbh (n.d.). 'IKO'. Available online: http://iko.sonible.com/ (accessed 9 July 2020).

Tufte, Edward R. (2001). *The Visual Display of Quantitative Information*. 2nd edition. Cheshire, CT: Graphics Press.

Vickers, Paul (2020). 'Direct Segmented Sonification'. Available online: https://paulvickers.github.io/DSSon/ (accessed 9 July 2020).

Vickers, Paul and Bennett Hogg (2006). 'Sonification Abstraite/Sonification Concrète: An "Aesthetic Perspective Space" for Classifying Auditory Displays in the Ars Musica Domain'. In *Proceedings of the 2006 International Conference on Auditory Display (ICAD 2006)*, 210–216.

Vickers, Paul and Robert Höldrich (2019). 'Direct Segmented Sonification of Characteristic Features of the Data Domain'. In *Proceedings of the 2019 International Conference on Auditory Display (ICAD 2019)*, 244–253.

Vickers, Paul, Bennett Hogg, and David Worrall (2017). 'Aesthetics of Sonification: Taking the Subject-Position'. In Clemens Wöllner (ed.), *Body, Sound and Space in Music and Beyond: Multimodal Explorations*, 89–109. SEMPRE Studies in the Psychology of Music. London: Routledge.

Vickers, Paul, Christopher Laing, and Tom Fairfax (2017). 'Sonification of a Network's Self-Organized Criticality for Real-Time Situational Awareness'. *Displays* 47 (April): 12–24. https://doi.org/10.1016/j.displa.2016.05.002.

Walker, Bruce N. and Joshua T. Cothran (2003). 'Sonification Sandbox: A Graphical Toolkit for Auditory Graphs'. In *Proceedings of the 2003 International Conference on Auditory Display (ICAD 2003)*, 161–163.

Walker, Bruce N., Gregory Kramer, and David M. Lane (2000). 'Psychophysical Scaling of Sonification Mappings'. In Perry R. Cook (ed.), *Proceedings of the 2000 International Conference on Auditory Display (ICAD 2000)*, 99–104. Atlanta, GA: International Community for Auditory Display.

Weinberg, Gil and Travis Thatcher (2006). 'Interactive Sonification: Aesthetics, Functionality and Performance'. *Leonardo Music Journal* 16: 9–12.

Wendt, Florian, Gerriet K. Sharma, Matthias Frank, Franz Zotter, and Robert Höldrich (2017). 'Perception of Spatial Sound Phenomena Created by the Icosahedral Loudspeaker'. *Computer Music Journal* 41 (1): 76–88. https://doi.org/10.1162/COMJ_a_00396.

48

The Conflicting Sounds of Urban Regeneration in Liverpool

Jacqueline Waldock

How can we approach sonic field work in urban areas? In this chapter I discuss a co-production approach that I have employed in a number of urban sound projects. I will address why we should use a co-production approach to urban sound studies and how this approach can be applied in practice.

Late one evening at a conference I was introduced to an academic who asked me about the methodological approach I was taking in my research. As I was a relatively new PhD student, I timidly responded that my methodology involved partnering with local residents to record, listen, and analyse sounds in a form of co-production. This was met with scorn: 'Co-production, of course everyone is doing co-production, what a load of meaningless jargon.' Sadly, I have not had the opportunity to meet this academic since, however his response always stuck with me. In the years since, the term 'co-production' has in many ways reached semantic saturation in the academic world. In this chapter I will argue that co-production is an important methodological approach in understanding sonic communities and analysing conflicting urban spaces within the sound environment.

Co-production allows us to draw upon specific experiential expertise situated within the sonic terrain by those who live, work, and/or utilize that area. Co-produced research does not simply consult local residents about their sound environment; instead it is a partnership with them that values the expertise that they have, such as a specific experience and understanding of place. In the case of this methodology their expertise is coined from the everydayness of their experience, the layers of quotidian ritual, and personal connections. Engaging with this experiential expertise enables us as researchers to be better placed to decipher the construction of sonic space and its significance in forming sense of place.

Co-production approaches commonly engage with groups of people. This allows for multiple voices to be heard, compared, and shared. Tripta Chandola, a New Delhi based ethnographer, in her study 'Listening into Others: Moralising the Soundscapes in Dehli', questioned 'whose experience of soundscapes – that is, *listenings* – are given preference and whose *listening* is not?' (Chandola 2012: 397, emphasis in the original). A co-production methodology allows for a broader spectrum of listening experiences to be explored. This is particularly poignant in urban areas. The highly populated nature of the city environment often means that sonic terrains intersect and overlap with each other; that differing, and sometimes conflicting, experiences of sound spaces crossover and rally against one another.

I developed a co-production approach, that I have utilized in several urban sound projects, called the 'trinitarian methodology'. This methodology partnered with the residents to record their sound environments. They were considered as artists, academics, and activists in producing their own sound recordings: artists, through making sound recordings, capturing, editing, and presenting their sounds; academics, through listening to and commenting upon the sounds that they had recorded, giving a unique insight into their relationship with the sonic events; and activists, through sharing the sounds with one another and with the wider public. This has led to the residents leading radio shows on BBC radio Merseyside, participating in the curation of sound exhibitions, and creating public sound sculptures.

Each resident was given a digital sound recorder and asked to record sounds in their everyday lives. They also had the freedom to delete sounds or edit recordings as they wished. Alongside the recordings, they were sometimes prompted by questions such as: What did you record? Why did you record it? This led to residents recording sounds and then recording their thoughts about the sounds on the next track. Usually these descriptions included an interpretation of the sounds: 'This is my door, it squeaks a little.' Sometimes the recordings were accompanied by an emotional description: 'It makes me feel safe when I hear it click, it tells me that I can relax and not worry.' These reflections allowed me not only to be guided by the sound itself but also to provide a context and insight into how the recording resident was listening to them.

One of the early applications of this methodology was in Toxteth, an inner-city area of Liverpool facing social and political conflict. There were rows of houses in what is collectively known as the Welsh Streets that were under a Compulsory Purchase Order (CPO).[1] In this case, the CPO was applied to the homes in order to demolish them. However, some residents resisted the council's desire to knock down the houses and formed a resistance that staved off the council's plans for over a decade. The streets were redeveloped rather than demolished in 2017/2018 with many of the past residents applying to move back to their homes. Prior to 2017 the resistant residents of these streets formed the focus of my project. They recorded their sound environments during this period of change, they listened to the sounds and they commentated upon their own recordings (Waldock 2015). This resulted in 40 hours of recording over a nine-month period. Sounds ranged from kettles brewing and doors locking to community meetings, house fires, and the boarding up of homes. The recordings gave me a collection that extended beyond the public streets. The residents captured their personal spaces, collective experiences, and the sounds that seeped from one space to another. Although the sounds were recorded and collected in a small parcel of streets, they highlight the issues surrounding the regeneration process: how the changes, that regeneration dictates, transform the sonic space and create acoustic conflicts. The sounds that were recorded in the Welsh Streets lay bare the way in which regeneration can irreparably change a sound community.

The streets themselves are layered with history: industrial, musical, and personal. They mark a significant point in Liverpool's industrial history (Roberts 1986), the houses primarily being built to support Welsh families moving to find work at Liverpool's docks. The streets are also the birth place of the Beatles' drummer Ringo Starr (Madryn Street), and form part of the pilgrimage route of fans to the city. The admiral pub that featured on the cover of his first solo album is also part of the Welsh streets. The significance of this space to fans can be seen by the graffiti on the door of Starr's home, the boarded-up house on Madryn Street which has become a site where fans prize the opportunity to leave their mark.

Many of the original residents in the street grew up there; their families and grandparents had also lived within a few streets of one another. One ex-resident described to me how her parents had lived just one road over from where she grew up because their house was bombed in the war and when she got married she had moved only a few houses down. The layers of history in the streets, the length of time that many of the houses had lay empty, and the dividing battle ground of the residents has led to the houses being seen in several conflicting ways: an eye sore, a site of pilgrimage (to visit Ringo Starr's birthplace), a sub-par housing crisis, a home, a family, a community. These perspectives sit at odds with each other socially and sonically. There were distinct power relationships at play between the residents and the council. In *The Sonic Color Line* Jennifer Stoever writes about the interplay of these power relationships with sound and within listening: 'Although often deemed an unmediated physical act, listening is an interpretive, socially constructed practice conditioned by historically contingent and culturally specific value systems riven with power relations' (Stoever 2016: 14). My project sought not only to capture the changing sounds of the streets but also the specific value of the sounds to those who listen to them. What became clear in people's reflections upon the sound environment was that the transformation which regeneration had inflicted on the area had led to a change in the accepted sonic order, and that, without doubt, the listening to certain sounds had been utterly transformed. The complexities of a divided community, the political rifts, and economic tensions surrounding the streets had created a group of residents who felt let down, and who struggled to exert agency in the decision-making process. When I first met them their sentiments were very much like 'nothing about us without us'. They were in the middle of a long battle and had become disillusioned with academics, politicians, and charities speaking for them instead of with them and running projects that appeared to make no significant difference. Distinctively the processes employed through this project enabled me to gain insight into how people's listenings were dismissed in favour of 'progress'.

The project gave a considerable amount of freedom to those recording.[2] This freedom enabled a breadth of sounds to be captured but it also meant that some recordings were hindered by wind, muffled microphones, batteries running out, and many other technical issues. It took time for people to become confident with the technology. One resident would only record with the help of another, and spoke to the recorder throughout as if commentating her world personally to me, often starting recordings with 'Hiya Jacky, its me.' She used the digital recorder much like an answering machine, leaving me messages and sounds to listen to. The partnering residents of the Welsh Streets, despite being part of the same project approached recording in different ways. These approaches predominantly fell into two groups, those who recorded every sound, and those who recorded a specific sound.

One of the residents had left the digital recorder on the window sill and then later listened back to what had been captured. At first, she was surprised by the amount of bird song that she could hear; she commented on how the birds could be heard so clearly. However, she went on to note that the recording lacked human sounds. There were no neighbours talking in the streets or children chatting on their way to school. The joy of hearing the birds quickly passed into a melancholy for the sounds of a community which she missed. The sound recorder that was left on the window sill acted as a tool for a later, more concentrated listening to the sonic space. The recorders that were left in situ captured many of the sounds that surrounded that particular home; however, when listening back to these recordings people commonly talked about sounds that they hadn't heard on the recording. When listening back they were struck by the sounds that weren't captured, sounds that had disappeared from their sonic space without them noticing. Most of these sounds will be considered as background, everyday city noises: people getting into their cars, people walking to the shops or the bus stop, parents out with their prams. It is only in the removal of these conventional sounds that their significance in regular community life is felt and appreciated. If I had not partnered with local residents and utilized their expertise, I would not have been able to understand the importance of these missing sounds.

The second type of recording occurred when partnering residents recorded a sound they had previously heard. Here, the recording became a reference for a previous listening experience and it was often commented on in conjunction with the sound. This happened because the listener had already heard and considered the sounds, and therefore was able to comment immediately. This approach is only possible when sounds are repeated. What often occurred was that tracks would contain a description of a sound that had occurred but not the sound itself. As I was actually interested in exploring both the existence of sounds in the environment and reflections of the listeners to those sounds, the description and reference to sounds that had not been recorded was something I had to accept as part of the variation in using a co-production approach. However, the advantage of having multiple residents recording the space at the same time meant that sometimes people spoke about sounds that other people had recorded. This highlighted the importance of sharing and collectively listening to the recordings made.

In both approaches, whether it was a general capturing of sound or a specific capturing of a particular moment, the recorded sound would be listened to for an additional time with me either in the home of the person recording or in the studio. This listening often highlighted their sonic memories of the place distilled into their sonic present (Jarviluoma 2009; Waldock 2016). In some cases the recordings produced an anamnesis (Augoyard and Torgue 2006: 21) in the listener, reviving the home that no longer exists.

The Welsh Streets project was the first time I employed the trinitarian methodology. I have since utilized the approach in several other urban settings. However, the first application was significant as it highlighted some of the problematics of implementing this co-production approach.

Inscribing conflict

All of the residents who partnered in recording were women, and over the time they recorded they gained a confidence in using their recorders and being unapologetic in their recording presence. The Canadian sound artist and researcher Andra McCartney has drawn attention to 'the way a woman's movements through public space are marked and regulated' (McCartney and Gabriele 2001). She tells of the experience of her research assistant who refused to turn off her recording device when asked by a passing male cyclist.

> At the moment Gabriele met this passing cyclist, he asked her to stop recording, to stop inscribing her sonic presence in this context. Her refusal and the agency she insists on by continuing to record are important parts of the sound walk recording experience, and the sense of entitlement access to such technology can provide.
>
> (McCartney and Gabriele 2001)

Conflict occurs in the inscribing of sonic presence. Because of this many residents were nervous at first to record outside of their home or yards. To them using the technology in public drew attention to their presence in what they considered to be an unstable space. However, as they grew in confidence they felt increasingly comfortable with the agency that the technology allowed them, particularly the ability to etch their existence into the public space. The recordings of the actual streets and pathways call attention to the power of the residents in inscribing their presence onto the very streets. This was a marker of the agency that they have in that space.

The trinitarian methodology by its nature requires partners to record sound themselves in places of political and social conflict. However, asking people to step out and inscribe their sonic presence in a public space comes with added tensions and risks. It is important to acknowledge this and to accept that, sometimes, the instruction to record in a public realm is too great an ask. In projects where I have employed this methodology, it has often been necessary to give people time to become comfortable with the recorders and to walk alongside them when they first record publicly.

Recording conflict

The method of recording each other allowed me to hear multiple recordings of the same sound and references to the same sounds by different people. One resident recorded the fire that had started in the empty houses adjacent to them whilst another recorded themselves talking about the same fire. This allowed me to understand how people's lives intersect with one another and how significant sonic communal touchpoints develop, not just in spaces such as the paths that connected the houses or the communal stairs in the flats, but sounds that seep through the bounds of domestic spaces and become part of the quotidian ritual of the residents' lives.

The residual community became more connected by uniting against the homes being demolished; people who were acquaintances had become friends through the neighbourhood meetings. As the houses had emptied the area had become a target of anti-social behaviour (thirteen incidents in one month in 2015), arson, and drug taking. Streets that had been designated as play streets – spaces closed to traffic so children can play – became considered 'dangerous', both within and outside the community. The streets that had been boarded up ceased to be familiar. Many residents noted that they avoided walking through them, that going to the shops or visiting friends took longer as they didn't feel safe walking along the empty houses. This heightened sense of fear also changed people's relationships with their neighbours: they used to listen out for people coming home, be concerned for those who were late or if they hadn't seen them for a few days. This sense of communal care challenges one of the key discussions of urban soundscapes, that of 'neighbour noise'. Within the sounds captured we had several references to neighbour sounds such as, 'I always hear her radio at tea time'; 'I can hear him singing on the stairs'; 'I know he is home when I hear his door bang.' Quotes such as these are often accompanied by a barrage of negativity, the familiar tale of noise pollution by people who selfishly invade other's spaces, breaking into someone else's private cocoon. However, this was not the case here as these sounds were not recorded as examples of noise pollution or annoyance; they were recorded as a sign of life, a sign of community, a sign that life still happened, a sign of resistance. There is increasingly an assumption that 'our mutual harmony and peacefulness as a society is predicated on an increasing amount of aural segregation' (Fluegge 2011). What became evident from my projects was that aural desegregation in the streets strengthened communal ties.

Stoever sought to challenge the idea that 'one has to listen similarly to power valuing the same sounds in the same way and reproducing only certain sounds that the listening ear deems appropriate, pleasurable, and respectful' (Stoever 2016: 20). One of the selling points of the new houses that the council offered to people was that they were well insulated, with double glazing and small private gardens. Despite these benefits some of those who moved felt isolated, unable to hear those around them and being in fear of others not hearing them. An elderly lady who had moved to her new home worried if anyone would notice if she fell down the stairs or hear her if she cried for help. She exclaimed that 'in the old houses you knew your neighbours, they would look out for you.' To the council, neighbour sounds are a problem to be tackled and deemed inappropriate, but in the desire to reduce 'noise pollution' there is an element of isolating people from their community. However, this didn't imply that the residents accepted all sounds into their private sphere: there were sounds that they found problematic such as the sound of fire engines at night treating the arson attacks that intermittently occurred; or the sound of workmen cutting off utilities, emptying the homes, and boarding them up. Sociologists Hugh Pickering and Tom Rice write:

> In a social context, noise is used as a descriptor for deviant sonic behavior, highlighting societal norms and delineating what is acceptable and what is not. A neighbor's dog barking late at night is 'noise,' despite the fact that it is not opposed to 'music' or 'signal' but because

it occurs outside of the accepted sonic order. Its designation as 'noise' implies that the night is a quiet time and denotes an expectation that dogs be kept under control.

(Pickering and Rice 2017: 2)

By employing a co-production approach through the trinitarian methodology, I was able to reflect on a number of residents' experiences of the local environment and better understand the norms around acceptable community sounds. In this particular project neighbour noise was not seen as 'outside of the accepted sonic order'; rather it became a significant symbol of a sense of place that had been lost and was now yearned for.

Conflict in listening

It is common for people to not like the sound of their own voice, much like seeing a photograph of oneself; there is a disconnection between how we hear ourselves and how others do. However, it was not their own voices that caused this unwillingness, it was emotional distress. Some recordings were heavily emotionally charged, for example two residents recording another's moving day (Waldock 2016). This became, unsurprisingly, too difficult for them to listen back to. Another resident recorded their neighbours flat being boarded up and the resident who had actually lived there found it too emotional to hear it again. Since the recording of the house being boarded was made, it has been played in several places. It is the sound of metal being drilled on to stone; it is loud and it is uncomfortable, but to those who lived there it holds an extra layer of significance: it is the breaking of a community and the stealing of a home.

This sound was new to the area, brought in by the council as emptied houses need to be kept from vandals. The sound was unanimously the most hated sound in the area, partly because of what it represented and partly because of how it interacted with the space. Boarding up a home is not a quick process nor is it a quiet one. The drills need to be able to bore into the stone masonry, or brick of the houses in order to fix steel sheets in place. The sheets sometimes have to be adapted to the size of the window and are cut on site with electric saws. To cover every window and door of a three-story house front and back takes approximately 5 hours, and a row of houses thus took days. The workmen would wear muffs protecting their ears from the noisy drills and saws, but the residents had no protection or warning for these infiltrating sounds. There is no other sound that so represented the relationship between the council and the residents at that time. Anthropologist Pauline Destree writes:

> If […] we can consider noisy tenants as the 'abject subject' that fails to comply to the auditory norms of tenancy, as the 'haunting specter' of the normative subject, then we can understand noise events not only as invoking this specter but as harmful penetrations of this 'polluting outside' into the subject, producing it partly from its own defiling substance.
>
> (Destree 2013: 17)

The sonic penetration of the drilling into the houses nearby was a marker of power – power over the homes that were being boarded up, power over the personal lives of the neighbours who heard it.

Despite never being considered as a sonic weapon by the council it acts very much as a tool to subdue the remaining residents, to hinder community, and to disrupt daily life. I received a call early one morning to tell me that the home of one of the recording partners was being boarded up and they wanted to capture the sound but they needed my help. I sat with them in their home hearing the workmen board up their neighbours' homes. It was a warm sunny day but the sound of drilling made it difficult to have any windows open and hard to hold conversations in the house. On the street it was impossible to speak to one another or move without stopping for more than a few moments through the experience of physical pain from the noise. The men carrying out the work were friendly, non-inflammatory, even apologetic; however, the sound created by the boarding up was damaging, physically, socially, and mentally. It is clear from the recordings and analysis by the residents that this sound stopped people from congregating; it disabled people from directly communicating with one another and hindered people from gathering. It sonically stamped 'we will win, we have the power and we have the control' into every home.

The methodology that sought to co-produce with the residents – as artists, activists, and academics – was both broadened and strengthened by its partnering in recording, listening, and analysing. However, this also came with some complications and risk. Throughout the project, digital recorders were lost, some residents never made any recordings, and some sounds were erased before they could be analysed or shared. The project was also intensive in meeting with and encouraging those who partnered. It enabled an understanding of how urban regeneration transforms not only the sounds that occur in the streets but also shifts the sonic order and the listening of the community that lives there. I heard how the changes in housing and development changed the community and enabled them to accept and rarefy sounds that prior to the houses being bought for demolish were commonplace. These insights provoke a discussion about counter-cultural approaches to sound and how sound is a tool for exercising power. This was seen most prominently in the council's exertion of acoustic control by boarding up the homes and the way that sound was understood. However, it could also be seen in the way the residents, as a small but significant resistance, inscribed their own presence into their recordings and, most crucially, their space.

Notes

1. A CPO is an order that give the local council authority to purchase your home.
2. They were given instructions on how to use the digital recorder and advice on various ethical protocols.

References

Augoyard, Jean-François and Henry Torgue (2006). *Sonic Experience: A Guide to Everyday Sounds*. Trans. Andra McCartney and David Paquette. Montreal: McGill-Queen's University Press.

Chandola, Tripta (2012). 'Listening into Others: Moralising the Soundscapes in Delhi'. *International Development Planning Review* 34 (4): 391–408.

Destrée, Pauline (2013). '"Dirty Dirt" and Sonic Relationality: The Politics of Noise in a London Estate Community'. *Ethnographic Encounters* 3 (2).

Fluegge, Elen (2011). 'The Consideration of Personal Sound Space'. *Journal of Sonic Studies* 1. Available online: https://www.researchcatalogue.net/view/223095/223096 (accessed 9 July 2020).

Jarlivouma, Helmi, Meri Kytö, Barry Truax, Heikki Uimonen, and Noora Vikman (eds) (2009). *Acoustic Environments in Change and the Five Village Soundscapes*. Tampere: Tampere University of Advanced Studies (TAMK).

McCartney, Andra and Gabriele, Sandra (2001). 'Soundwalking at Night'. *The Night and the City Conference*, McGill University, Montreal, Quebec, 15–18 March.

Pickering, Hugh and Tim Rice (2017). 'Noise as "Sound Out of Place": Investigating the Links between Mary Douglas' Work on Dirt and Sound Studies Research'. *Journal of Sonic Studies* 14. Available online: https://www.researchcatalogue.net/view/374514/374515 (accessed 9 July 2020).

Roberts, Thomas A. (1986). 'The Welsh Influence on the Building Industry in Victorian Liverpool'. In M. Doughty (ed.), *Building the Industrial City*, 106–149. Leicester: Leicester University Press.

Stoever, Jennifer (2016). *The Sonic Color Line: Race & the Cultural Politics of Listening*. New York: New York University Press.

Waldock, Jacqueline (2015). 'Hearing Urban Change'. In Les Black and Michael Bull (eds), *Auditory Cultural Reader*. 2nd rev. edition. London: Bloomsbury.

Waldock, Jacqueline (2016). 'Crossing the Boundaries: Community Composition and Sensory Ethnography'. *Senses and Society* 11 (1): 60–67.

49

Ethnographies Sounded on What? Methodologies, Sounds and Experiences in Cairo

Vincent Battesti

The last decade has seen an increase in academic literature in the humanities and social sciences on the sonic dimension of our world. This chapter asks 'what' sounds are of interest to ethnography, 'why', and consequently 'how?'

This chapter, like the rest of this Handbook, is dedicated to the thorny issue of methodologies. From a heuristic point of view, a method (i.e. 'how') is only worthwhile if it is based on the answers given to the first two questions, 'what' and 'why'. It is important to draw clear distinctions between these three questions, although they often overlap in the complex and varied context of ethnography (at least for ethnographies that are not fettered upstream of the fieldwork by theory and programmatic approach).

Therefore, it goes without saying that I cannot suggest a 'good' and sound methodology to be used in doing ethnographic research, as it depends on the purposes and the context of the fieldwork; and also, as I think of the senses as interconnected and perceive the environment as a whole, an exclusive sonic modality or category will be avoided.

I will give a quick and partial overview of ethnographies that are concerned with sound in order to present my own questions in a more precise and informed way. The journey I propose here does not quite illustrate a categorization that I would come to terms with, it is too incoherent as such, but it is rather a way of bringing about the different methodologies that are, or could be, used in sonic ethnography (Figure 49.1).

Ethnomusicologists and their music

Anthropology has inherited a great deal from a long tradition of Western musicology. Scholars of this field forged a range of tools and concepts to grapple with music. Music is a highly diverse social practice understood as organized and performed sounds which

Figure 49.1 This chapter is dedicated to Ahmed Wahdan, my master of the Cairo night, who died in 2018, far too early. Giza, Cairo, Egypt, 30 November 2016, 11.00 p.m. Photograph by Vincent Battesti.

usually express or communicate ideas or feelings. But music also represents a minute part of the audible world for human beings, whatever the time or place. Ethnographers have every right to be interested in music as a part of social life, but the entirety of social life, in its audible dimension, exceeds that of music. Ethnomusicologists today are scarcely interested in working on music extracted from its performance conditions, its social and cultural context of real practice. From this perspective two attitudes prevail towards music, which are not necessarily in conflict: to deepen the relationship between the production of music and its environment; and to focus on an aural ethnography that investigates the ways music is used, diffused, and plays a role at the heart of a society.

A renewed approach towards music phenomenon among ethnomusicologists, coupled with a greater awareness of sound, has emphasized the social context of music production (for instance, see Christine Guillebaud [2008] on the Kerala travelling musicians in India), as well as the natural environment, which refers to the work of Steven Feld (1982), who described the highly developed practices of listening, hearing, and sounding that characterized Kaluli engagement with their rainforest environment in New Guinea (and especially through ethno-ornithology). Methodologically, long-term participant observation (usual amongst ethnologists and anthropologists) is used together with field recordings for this approach. Steven Feld's work was a turning point for ethnomusicology. Along with classical ethnography methodology, Feld explained,

> [I used] extensive playback of the recordings as soon as I made them [with stereo Nagra and AKG studio microphones] and over long periods of time as a key methodology. I felt that

not just being with people in sound but listening and talking about the recordings was an important way to gain a sense of how to be an ethnographic listener.

(Feld and Brenneis 2004: 465)

I should mention also other prominent ethnographers, such as Ellen B. Basso (1985) working on the Brazilian Kalapalo Amerindians, or Marina Roseman (1991) on the Senoi of the Malaysian rainforest, connecting music and medicine, and Nicole Revel (1992) on the Palawan (in the Philippines) on songs and birds, etc. Again, among Suyá in Amazonia, Anthony Seeger (1987) reveals the centrality of the music phenomenon in a society and how Suy singing creates euphoria out of silence, a village community out of a collection of houses, a socialized adult out of a boy, and contributes to the formation of ideas about time, space, and social identity.

Ethnomusicologists are not the only ones to practice field recording for the purpose of analysing sonic data, but they are the ones who have the expertise and social legitimacy to edit music – often marketed as world music and sold to the public. Many specialized labels have been created to distribute these albums. Before that, field recordings were used to analyse musical systems. Due to the existence of different schools of thought, no universal or objective analytical system has been established (using the Western system of notation or one fitting the local system, for example, is a dilemma); see, for instance, the criticism of Alan Lomax's method of cantometrics (1968) – intended for universal comparison – by Steven Feld (1984), who rather insists on the local context of performance.

Music phenomena beyond ethnomusicology

Working out methodologies deployed for the second option – which does not focus on 'the study of people making music' (Titon 1992: xiv) but on people playing back and listening to music – is uneasy as such works suffer from a common flaw when it comes to writing, the usual lack of an explicit description of the methodology used: it is assumed (and not necessarily erroneously) that the method, if not described, is the 'regular' one, that of a participating and floating observation and semi-structured along a majority of non-structured interviews conducted with an indefinite number of informants over a long period of time during the fieldwork, with, in general, a command of the language used by the social group in question (Figure 49.2). The intention here is obviously not to lay the blame on the habits of a discipline; in fact, most of the time, I do not do otherwise and often end up omitting the 'materials and methods' part of my research. Ethnology and anthropology do not apply the most prominent norm for the organizational structure of a scientific journal article, IMRaD (Introduction, Methods, Results, and Discussion), for valid reasons that will not be discussed here. Regarding the fields we are talking about, for

Figure 49.2 Workshop open on the street, al-Gamaliyya, Cairo, Egypt, 28 November 2016, 3.30 p.m. Photograph by Vincent Battesti.

instance, Nahid Siamdoust in 'Tehran's Soundscape as a Contested Public Sphere' (2015) deals with music (and not sound despite using the notion of soundscape): not the making of music but the music consumption diffused in the public space through a small shop, a taxi's window, or new media technologies.

Most of the work on music, beyond ethnomusicology, uses classical field data collection tools in which we cannot discern a methodology specific to the sound dimension, because it is not necessarily the sound materials and sensory perceptions that are important to the study. For example, the work of anthropologist Nicolas Puig (2010) on Egyptian wedding musicians in Cairo deals with the social group of these relegated musicians and their performance in street weddings.

Forgetting music for other sonic phenomena

But what if we forget about music as a practice and keep it only as one tiny part of the myriad of sonic experiences that inform everyday life? What if we are interested in the broader spectrum of sounds that bathe our lives in order to expand beyond the notion of 'world*views*' and address sensoria? Canonical ethnographic methodologies do not yet exist, because these academic works that aim to cover the whole sonic dimension of a social group remain scarce.

However, some works focus, in a very relevant way, on specific sonic phenomena (other than music) or sonic devices. For instance, Patrick Eisenlohr (2018) provides an account of the sonic dimensions of Islam in Mauritius, exploring how the voice, as a site of divine manifestation, becomes refracted in media practices that have become integral parts of religious traditions. For once, we have an explicit, though classic, account of the methodology for addressing 'the sonic incitement of sensations' in Mauritian Islam:

> In addition to recording these interviews and semi structured and open-ended conversations, many of which took place in several sessions, I audio-recorded the performances I attended and video-recorded some of them. I also collected a corpus of cassette and CD *naʿt* recordings that were sold or otherwise distributed in Mauritius. And I participated in the regular social life of several of my interlocutors and made visits to the homes of many others far beyond the context of *naʿt* performances and other events connected to the genre.
>
> (Eisenlohr 2018: 19–20)[1]

Eisenlohr's audio recordings (which were either recorded by the author or acquired on the market) do not seem to have been processed in any way other than by listening to them with care through a linguistic analysis (and trying too to understand the desired qualities of the voice). Charles Hirschkind (2006), before him, conducted a similar approach in Cairo by focusing on the sermon tapes that circulate in the city and more importantly are played in shops, taxis, etc. to offer an understanding of the ethics of listening and an analysis of rhetorical styles. These are studies that focus on particular sonic phenomena. The methodologies remain fairly classic, while at times resorting to sound recording and listening to their own recordings and marketed tapes. To my knowledge, sounds are not analysed from an acoustic angle either but, rather, the moral, political, and competitive dimensions of public space are analysed from the perspective of the humanities. In a way, sociologist Iman Farag's work on the wide public debate in the press over the governmental decision to replace the thousands of amplified muezzins by an automatic, synchronous, and centralized broadcasting of recorded *adhān*, call to prayer, is not far from this kind of research (Farag 2009).

Voice, both as a sonic phenomenon and a sonic device, has been an issue for sonic researchers, beyond calling in linguistic expertise. Olivier Féraud (2010) (discussed below in more detail), for instance, produced a noteworthy work on the local economy of sounds, the voice on the one hand and firecrackers on the other hand, in a popular neighbourhood of Naples, Italy. Jean-Jacques Luthi (1985) inventoried and transcribed and translated the calls from street vendors in Cairo, Egypt. Noha Gamal Said (2014) did the same in Cairo recently, extending the description beyond street vendors to all professions making calls in public space. She used a classical methodology of observation in this work, illustrating it with sonograms of the street (actually, just time and decibels).

Spectrograms, the visual representation of the spectrum of sound frequencies and volume changing through time, are notably used by Julien Meyer (2015) in his complete worldwide study of human whistled languages to demonstrate that whistled speech 'is adapted to the structure of each language, to specific traditional rural activities such as hunting or shepherding, and to specific ecological milieus'. Paul-Louis Colon (2013) has produced a precise ethnography of professional and amateur ornithologists and their listening skills when observing birds in their environment. In this context as well, the sonic phenomenon studied is narrow, the song of birds and its approach by humans, and the methodology classical, but it allows him to access relevant parts of the mutual dependencies of the senses and their ambiguities in our sensory models. The same can be said of Nicolas Puig's (2017) work on sound techniques (amplification, electro-acoustic effects, etc.) of urban rituals in Cairo, or Vincent Andrisani's (2010) on the 'electrified soundscapes' of Havana, Cuba, and Vancouver, Canada, two distinct environments with differences in the social uses of technology.

Another device that has interested researchers working on sound is the Walkman or the mobile phone, which allow an urban, mobile person to build their own sound/musical permeable bubble and to create a unique relationship with the city: for instance, the work of Jean-Paul Thibaud (1994), Michael Bull (2004), and Anthony Pecqueux (2009a, b). Bull and Pecqueux acknowledged the difficulty of such an ethnography of a sonic practice. Bull's methodology consisted primarily of in-depth, qualitative interviews with over one hundred personal stereo users living in large British cities. Pecqueux deployed a sophisticated methodology to circumvent the problem, based on the 'commented walk method' (Thibaud 2001), which I discuss more fully below. The solution chosen in order to follow the daily musical journeys made by individuals belonging to a previously constituted sample is described as the 'method of (post-)commented journeys': on a settled sample, the observer conducts a micro-ethnography throughout the journey (as in a close tailing, he remains about three metres behind the subject) mainly by taking notes and collecting in situ and regular verbalizations of the respondents during 'interview breaks' about their listening and their journey. These interruptions are infrequent, in order to minimize disruption to the continuity of the route. In situ, the gestures are observed, then confirmed by the respondent during the interview break; the music listened to is known, listed; the things seen, heard, and felt are both observed by the researcher and revealed by the respondent (Pecqueux 2009b: 57–58). While the observations made were able to highlight most of the ordinary, routine reactions to sudden noise in the urban environment, verbalizations were essential to account for the more complex interactions, and in particular to highlight the operations of auditory selection of relevant sounds to be heard (Pecqueux 2009a).

We note, at this stage, that holistic approaches to all sound phenomena remain insufficiently addressed. Sound has remained unexplored, because it has not been thought of, or forgotten and excluded within the construction of modern humanities where sight and visuals have until recently clearly prevailed (Figure 49.3).

No discipline for sound studies

If we deviate from ethnology and disciplinary approaches in general, suddenly, ethnologists are not alone anymore. Today, a huge panel of researchers offer their multidisciplinary approaches of the sonic dimension of our world, with urban planning, architectural, psychological, geographical, historical, sociological (and so on) backgrounds. This multidisciplinary galaxy focused on the sonic object is often labelled as sound studies. Their

Figure 49.3 In the street of Gamaliyya, Cairo, Egypt, 28 November 2016, 3.00 p.m. Photograph by Vincent Battesti.

approach is radically different to disciplinary ones as they usually start with a question requiring them to find their methodology rather than the other way around.

Jean-François Augoyard is a leading figure in the pioneering spirit of research in the field of sound. Summarizing the work of this sociologist (and of the multidisciplinary lab, the CRESSON, which he created) would be a difficult task: he has produced considerable innovative work on the sonic issues of the urban world. He inscribed his work within an architectural and urban planning school problematic. Inspired by micro-sociology, he developed theories based on the difference between the space designed by urban planners and the space experienced by city dwellers, and worked out notions and methods on the everyday sonic phenomena. 'In the research, when you're stuck, you're forced to innovate' (Augoyard interviewed by Sevin and Voilmy 2009). First, he fully reversed the canonical method of sociological inquiries in his field by letting people tell him what they wanted to about urban spaces rather than by asking people to answer a questionnaire: 'I'm not asking you anything, I'll come back in 15 days, in 15 days you'll tell me where you walked [oral account or diary]. What interests me most is how you walk. How is the city for you?' (Sevin and Voilmy 2009).

Augoyard coined an innovative methodology he called 'reactivated listening' (*écoute réactivée*). It is a semi-in situ technique to collect data (even if Augoyard insists on its innovative in situ feature in the field of urban development). 'In short, it is about collecting the reactions of residents or users who are made to hear the sounds of their own environment' (Augoyard 2001). A sound recordist (possibly the researcher or a technician) records various auditory scenes of the urban space under investigation and then makes a montage. 'The objective is not to render with absolute fidelity the sound experience of others – which is impossible – but to awaken it' (Augoyard 2001). Two types of soundtracks proposed for reactivated listening are distinguished. 'The first is the local track, i.e. the one that evokes an identifiable concrete place. The second, typologies of objects, spaces, or situations, is composed of various forms chosen for their evocative capacity; such as a typical southern market, a Lyon street, a Haussmannian building courtyard, a household vacuum cleaner' (Augoyard 2001). 'The important thing is what people would tell me in retrospect about the journey or the sound sequences […]. It is not possible to render exactly one lived experience, but with this method, we get closer to it because the respondents reactivate all their experience' (Augoyard interviewed by Sevin and Voilmy 2009). Augoyard fully developed the possibilities of this survey methodology in 'Entretien sur écoute réactivée' (2001). His school has developed, along with interesting notions and tools such as the 'sonic effects' (*effets sonores*) (Augoyard and Torgue 2006) – which can be seen as the coupling of hard-acoustic with phenomenology – other methodologies that address a range of urban sonic issues, such as the commented-walk method (Thibaud 2001). In short, the commented-walk is an urban walk sound-recorded by the investigator and performed alongside a respondent describing (while walking) the sound as well as their impressions in situ.

Acknowledging that contemporary epistemology constantly affirms the impossibility of a dominant position of the researcher vis-à-vis his object of study, on the one hand, and certain that we can overcome a long tradition of Western philosophy that tends to oppose the sensitive and the intelligible (in other words, certain that the sensible can be a driving force for speech and the grounds for verbalization of local ambiances), on the other hand, Jean-Paul Thibaud considers, with reference to phenomenology (Maurice Merleau-Ponty) and to the ecology of perception (James J. Gibson), that it is now illusory to dissociate perception from movement (personally, I am receptive to this idea of 'movement as the foundations of perception'). In concrete terms, multiple variations of this technique exist and were already pointed out by Thibaud (2001). Sound studies practitioners took up the idea and adapted it in their research (sometimes under the label 'soundwalk'). Thibaud presented this commented-walk method as an 'open method' (Thibaud 2001: 98), but there is a definite general trend towards closure: closing the method and granting privilege to an objectivist approach to data on phenomenology.

In this objectivist logic, Catherine Semidor (2006, 2007) suggests that methods of 'soundwalks' can do without informants: she records herself and analyses herself in her urban journeys. Her soundwalks, inspired by *The Image of the City* by Kevin Lynch (1960), are

> sound recordings made with a binaural system (SEB) and a DAT recorder, on walks along a route with different urban forms. The different data (commented analysis of each track, recordings, photos, map, etc.) extracted from the soundwalks are examined in order to gather information about the relationship between the urban characteristics, urban activities and the sound environment. The acoustical images (time versus frequency graphs) of the sound signals can be seen as a representation of the soundscapes. The purpose of this approach is to enable us to evaluate what is pleasant and relevant in an urban sound environment in accordance with activities in the area.
>
> (Semidor 2006)

Like many in this field of urban sound studies, Semidor also supports the usefulness of this soundwalk method 'to help urban planners and other town designers to improve the outside acoustics in cities' (Semidor 2006).

I will not review here all research work committed to improving (sonic) urban planning using the 'soundwalk' as a method, but I will mention two or three articles that give an overview of the state-of-the-art of the method and offer to further improve, not only its technical efficiency but also its reproducibility, starting with Jin Yong Jeon, Joo Young Hong, and Pyoung Jik Lee's paper, to which reference can be made for a review of these approaches (Jeon, Hong, and Lee 2013).

> Soundwalks were developed starting in the 1970s by several pioneers in soundscape research [They] have been conducted individually as well as in groups, [but] even when soundwalks are performed individually, researchers accompany participants to mark evaluation locations and audio-visual scenes on maps. [...] Recent studies have mainly focused on urban contexts including urban streets, residential areas, parks, and urban squares [and] several [of them] planned soundwalk routes to include various types of urban spaces and elements that contribute to the urban environment.
>
> (Jeon, Hong, and Lee 2013: 803–804)

Jeon, Hong, and Lee's study 'proposes a soundwalk procedure that is applicable to the evaluation of urban soundscapes' in order to narrow the variability of factors that may influence the results of soundwalks (Jeon, Hong, and Lee 2013: 804).

This whole article deals with acoustic studies or sound studies, not ethnographies. It means that the sonic dimension revealed has a poor connection with the culture of the studied social group. Moreover, these soundwalk methods are used in urban and architecture studies to assess the soundscape qualities of the environment. This is not the usual programme of ethnographies by anthropologists, more interested in understanding how local people of the targeted social group deal with the sonic dimension in their everyday lives. The nuance may appear slight, but it explains why a soundwalk, sometimes determined by the researcher, based on a listening at 'evaluation positions' and survey using a questionnaire to fill out (when respondents noticed any positive or negative characteristics of the urban soundscape during that walk), remains disconnected from their mundane experience of the sonic environment. To complete the survey, structured or semi-structured interviews concerning soundscape elements such as soundmarks, keynote sounds, preferred sound sources, contexts affecting soundscape perception, expectations, and impressions of sites may be administered to participants (inspired of course by Raymond Murray Schafer's [(1977) 1994] work). Along the same lines, we can refer to the 'Positive Soundscape Project' of the interdisciplinary team of William J. Davies et al. (2013).

In the same vein as the methods developed for urban sound engineering, Julien Tardieu and colleagues (2015) proposes a method to improve the method used by others, much as Jeon, Hong, and Lee (2013) do for soundwalks. This time, the focus is on *ex situ* (laboratory) listening tests to facilitate and improve or make less 'subjective' the choice of sound samples that are played back to subjects. The samples mentioned are the soundtracks proposed for reactivated listening by Augoyard (2001), to name but one. 'A typical first step of most studies involves collecting soundscape samples, which has one major limitation: it is a selection from the scientist's own representation of the urban soundscape being studied' (Tardieu et al. 2015: 1). The tool is being improved, so that the protocol is worthwhile in all situations, a tool detached from sociocultural contexts and their in-depth studies. Three questions/situations are offered to online respondents: 'imagine what you hear in the city in this situation x'. This is followed by an analysis that groups words according to semantic categories ('deduced from the semantic links between the forms, grouped in two macro categories: sound sources and human activities') set ex post by the researchers. 'These different distributions of sound sources and human activities can be used to guide future soundscape recordings in the field [...]. These results can also be used to guide the selection of soundscapes samples either from a database [...] or from a soundscape design tool' (5). To summarize, it is about making a sound recording (real or mounted) that most closely resembles what should sound like an urban sound environment for respondents. I am not further developing an exhaustive inventory of this type of work here, as the degree of technical refinement and, above all, the objectives of which are of no real interest to the questions raised by ethnological and anthropological research.

Psycholinguists finally took over Schafer's work by revisiting the notions of sound categories (Schafer forged and used etic categories whilst ethnologists are interested in emic categories). These categorization issues are of great interest to ethnologists, and particularly to ethno-ecologists (part of my own work pertains to ethnoecology, which is interested in 'ecology, perceived, experienced, practiced, by others'). Danièle Dubois, Catherine Guastavino, and Manon Raimbault (2006) propose to 'bridge the gap between individual sensory experiences and sociological representations of soundscapes', using the psycholinguistic analysis which mediates between individual experiences and collective representations shared in language and elaborated as knowledge. For that purpose, these psycholinguists apply an ex situ methodology: they give subjects pre-recorded sounds to hear.

> In the following experiments, the acoustic stimuli were acoustic recordings of complex and meaningful everyday sounds and environments, rather than *a priori* parameterised stimuli as the ones involved in studying perceptual categories. We first present results of experiments using recordings of isolated domestic sounds and recordings of actual urban soundscapes. We then discuss the findings in terms of cognitive representations of environmental sounds.
> (Dubois, Guastavino, and Raimbault 2006: 867)

This cognitive (but non-cognitivist) approach to urban soundscapes uses verbal data to access everyday life auditory categories. Puig and I didn't know about this study when we published our paper presenting our methodology 'Mics in the ear' (see below). My 'sound postcards experiment' differs, as I was offering informants pieces of the recorded urban sonic environment to listen to. The different approach here has noteworthy results which suggest 'that the meaning attributed to sounds act as a determinant for sound quality evaluations' (Dubois, Guastavino, and Raimbault 2006: 865). Heuristically speaking, the proposal deserves to be deepened and above all tested in real, ethnographic fieldwork. In any case, this highlights the value of interdisciplinary avenues.

Figure 49.4 Part of a loud sound system unpacked for a birth celebration (subu'), Bashtīl, Cairo, Egypt, 18 November 2016, 4.30 p.m. Photograph by Vincent Battesti.

My criticism of those sound studies, as an anthropologist concerns the failure to take into account the social and cultural dimension of experience. Many of these studies may not mention the social space in which they are conducted (Figure 49.4): urban informants are subjects of experience, in Taiwan, Los Angeles, Berlin, or Nairobi, it is the universality of their psychological processes towards the sonic world that is the focus of the methodologies deployed.

The global approach of the World Soundscape Project

This chapter does not follow a chronological logic. I proceed by discussing the World Soundscape Project and Raymond Murray Schafer's work dealing with 'soundscapes' (Schafer [1977] 1994). Its history is well known, but I cannot afford not to mention some of these fundamentals in order to position our own approach as ethnographers. These works inevitably seduce us with their holistic approach to the sound environment. The concept of 'soundscape' is intended to conceive of the sound environment by integrating the listener's relationship (conscious or not) and the symbolic significance of sounds, especially those of nature. Musician and environmentalist, Schafer's pioneering work – contemporaneous with Steven Feld's – has had a considerable impact yet contains a range of ethnocentric assumptions: natural sounds, for example, are presumed to be the object of a harmonious orchestration ('the harmony of the world') that is superior to industrial sounds. His musical analysis of a landscape structured by tonic and dynamic perspectives (keynote sounds, signals sounds, soundmarks, etc.) drove him to propose 'sound design', understood as the way 'to improve the acoustic environment'. His melioristic perspective aims to reduce 'unhealthy' sounds or attempts to preserve others that are deemed characteristic of a place or community (the famous Vancouver foghorn, for instance, see Schafer and World Soundscape Project 1978). 'Acoustic ecology' (Truax 1978) is a direct result of Murray Schafer's World Soundscape Project.

An educational ambition to raise public awareness of the sound environment also led Schafer to propose the first of the soundwalks. 'Schafer's method is often still used today and typically involves a group of people being led around an area or along a route. Silence is maintained for the duration – which may be as long as an hour – and impressions of the entire soundscape discussed only at the end' (Davies et al. 2013: 226). This is an experiment that voluntarily places the subjects in a state of heightened sensibility to their sound environment. In many respects, the work initiated in Vancouver by Schafer in the early days of sound studies promotes a positive approach in the analysis of our sound environments and analytical tools aimed at an unrecognized cultural universality.

Schafer and sound studies seem to represent a mix of disciplines: history, psychology, acoustic, ecology, etc. My own approach is equally interdisciplinary; therefore, it is not an issue I want to criticize. I believe cross-disciplinary analysis opens up new science frontiers. In a chapter on ethnographic methodologies, however, I am hard put to acknowledge sound studies to be ethnographic. These approaches are undoubtedly stimulating, and in recent years there have been many attempts to address the issue of sound, especially in urban contexts, to overcome the work on 'noise' when they were just broaching it in terms of one-dimensional sound maps; that of decibels.

Methodologies for sonic ethnographies?

Anthropologists have the advantage over other researchers in that, through long-term participant in situ observation, they have first-hand experience of the sensory lives of the peoples they study. Participant observation also requires long-term training or apprenticeship, which they should not play down or forget. The use of ethnographic observation permits, on the one hand, the acquisition of background knowledge essential for an understanding and analysis of the practices of daily life of a social group. On the other hand, it represents a commitment by the researcher, an embodied apprenticeship of the sensitivities at work, to the practice of decentring their sensory universe to learn that of others, a new balance of the senses, the always arbitrary (I mean social and cultural) local division of the senses and their practices, the meaning and emphasis attached to each of the modalities of perception (Howes 1991: 3) and so forth.

With their fieldwork-proven methodology, do ethnologists explore all dimensions of existence? Ethnographic works that focus clearly on – or seriously consider – the sonic dimensions of existence with ad hoc methodologies are (still) rare. This does not mean that no one is working on it. Many researchers are committed to an anthropology of the senses – or a sensory anthropology as promoted by Sarah Pink (2009). For the time being, it is the plurality of approaches that dominates this aspect of anthropology in fieldwork: cognitivist approaches, use of empathy and the researcher's body, field recordings.

I will not dwell at length on the many articles and books with *pro-domo* pleas that state the importance of the ethnographies of the sensible: the methodologies invented and/or used to ethnograph the sound dimension of existence are what interests me here.

Some anthropologists argue in favour of a quite radical shift in anthropology: a cognitivist perspective (for instance, Wathelet and Candau 2013). This represents a return to the laboratory to conduct experiments and collect data. I regard this, perhaps mistakenly, as the consequence of a crisis of confidence in ethnographic data and/or an overconfidence in the supposedly 'objective' and often quantitative data provided by a laboratory experiment governed by positive science. David Howes's criticism of Sarah Pink's position in a debate in the journal *Social Anthropology/Anthropologie sociale* seems appropriate here:

> By stating that anthropologists should look to neurology for 'essential understandings of sensory perception and experience', Pink makes it clear that any indigenous ideas about perception must play second fiddle to the 'essential' pronouncements of the neurologists. While I think that dialogue between anthropologists and neurologists can be informative for both sides (indeed, anthropologists might be able to tell neurologists something about how culture tunes the neurons), it is important to keep in mind that neuroscience is itself a product of culture in its particular research aims, methods and interpretations, and therefore cannot provide an a-cultural, a-historical paradigm for understanding cultural phenomena.
>
> (David Howes in Pink and Howes 2010: 335)

Even advocates of cognitive approaches such as Dubois, Guastavino, and Raimbault cool their ardour stating that 'the recent cognitivist approach to audition has remained within the same experimental paradigm, focusing mainly on low-level perceptual features rather than semantic features resulting from identification and categorisation processes' (Dubois, Guastavino, and Raimbault 2006).

Sarah Pink, in attempting to reformulate the anthropology of senses (founded by David Howes), argues for a sensuous ethnography. In the great debates that anthropology has been experiencing in recent years, this is one of the issues up in the air, and many people are trying to operationalize it (but, let us be clear, not necessarily on sound issues). For instance, Paul Stoller speaks highly of fieldwork through sensuous experience and the 'importance of understanding the "sensuous epistemologies" of many non-Western societies so that we can better understand the societies themselves and what their epistemologies have to teach us about human experience in general' (Stoller 1997).

Indeed, original approaches have been proposed to deepen the classic ethnographic methodology: among others, through empathy (Sayeux 2010) and the body (Jackson 2006). In the field on music and dance, a fieldworker's body can very voluntarily become a tool of knowledge (this is always the case in the field, but not always in a conscious and claimed fashion). Evangelos Chrysagis and Panas Karampampas emphasize

> the sheer physicality of the ethnographic encounter and the forms of sociality that gradually emerge between self and other. Researchers' immersion in sonic events and the flow of movement induces bodily responses that render fieldwork an intensely visceral experience. By employing their bodies as tools of research, ethnographers find themselves in spaces of sonic and kinetic intimacy and reciprocity with their informants, which articulate what Rouch called 'shared anthropology' (*anthropologie partagée*) (2003).
>
> (Chrysagis and Karampampas 2017: 3)

It is the case, for instance, of Phil Jackson, who offers a 'sensual ethnography' of the clubbing world:

> I started with the notion of 'presence' specific to the methodology of participant observation, but I modified this methodology a little bit so that it sounds more like a 'participant sensation'. The first step was to let my feelings run free in order to situate myself in the social space under investigation. So, I danced, flirted, used drugs and alcohol, spent nights with friends and strangers. For the first three months, I didn't ask any questions; I just learned to practice clubbing.
>
> (Jackson 2006: 95–96)

The basis of Jackson's 'sensual anthropology' represents: 'the ability to feel, share and understand with empathy the sensations and feelings of the people around us without imposing a pre-existing theoretical framework' (Jackson 2006: 96). The word is out: empathy. I mentioned above the usual difficult issue of reflexivity that ethnographers encounter when collecting data during fieldwork, the real 'black box' of the discipline (Battesti and Puig 2006). Pink expresses the same assessment:

> I was often disappointed to find how little other ethnographers (whose work demonstrates so well the significance of the senses in culture and society) have written about the processes through which they came to these understandings. In this vein, I would urge contemporary ethnographers of the senses to be more explicit about the ways of experiencing and knowing that become central to their ethnographies, […] to acknowledge the processes through which their sensory knowing has become academic knowledge.
>
> (Pink 2009: 2)

I have previously tried to explain the general use of empathy, always in use, consciously or not (except for hardliner ethno-methodologists) (Battesti 2006: 168). Similar to Phil Jackson, Anne-Sophie Sayeux made use of her body experience to understand how one is feeling about electronic music, developing the idea of the 'body-ear' (*corps-oreille*) (Sayeux 2010: 230).

If I have advocated to take into account the use of empathy, use we make anyway, and finally frame it methodologically, I retain a certain mistrust on its heuristic qualities if empathy is to become the phenomenological alpha and omega of a fieldwork methodology, on the sensory as on other fields. As Howes pointed out, 'by universalising the subjective sensations of the individual, phenomenology ignores the extent to which perception is a cultural construct' (Howes in Pink and Howes 2010: 335).

Whatever the methodology used in ethnographies, the aim is always the same: approaching as much as possible an adequate description of how people live, experience, practice, in our case, the sonic dimension of (a part of) their lives. Agata Stanisz, through 'audioethnography', opens up a new perspective to do fieldwork. She puts into practice audioethnography with truck drivers working for Western European freight companies who inhabit the cabs of their tractor units. It is field recording as a tool, her goal being to describe through sound the daily life of this occupational group: 'In the case of my studies, doing ethnographic research through sound means listening, recording, editing registered sounds and, with their help, developing acoustic representations of not so much the drivers' community, but rather their activities, events which they co-created and took part in' (Stanisz 2017b: 58–59). In my opinion, this method is particularly well suited to research on an occupational group.

> Field recording must be understood as one of the research methods of sonic ethnography and a way to create alternative representations of fieldwork knowledge. Application of field recording in ethnographic research makes it possible [to] track down social patterns, connections between collective emotions, auditory practices and social structures, which usually remain hidden and faded out in textual representations. Doing field recordings also allows the researcher to experience anthropological fieldwork in a more sensuous and immersive way. Field recording as an ethnographic practice focuses on extra-musical and extra-verbal sounds. It is a sort of deep listening: listening with particular attention.
> (Stanisz 2017a: 1)

'What about ethnography in and through sound?', the question that Steven Feld has constantly repeated (Feld and Brenneis 2004). Feld not only worked to understand the sound and aural universe of the Kaluli, but used and edited his field recordings of Kaluli daily life on LPs (and actually on radio programmes on National Public Radio [NPR], using multitrack recording and *musique concrète* compositional techniques) to continue his ethnological reflection, documentation, and demonstration, since his first LP *Music of the Kaluli* in 1982. Inspired by Jean Rouch, Feld introduced what he called 'dialogic editing', playback and feedback with the local community. The continuators of his approach – 'to have the sound raise the question about the indexicality of voice and place, to provoke you to hear sound making as place making' (Feld in Feld and Brenneis 2004: 465) – can be found in quite different fields of ethnography. Nadja Monnet and Maribel Tovar, for instance, depict an ethnological portrait of an urban space, the central square of Barcelona (the square of Catalonia), through photographs and sound recordings. Photographs and sound ambiances have been 'truly converted into working tools and sources of reflection, which have enabled us to build the ethnography of this square' (Monnet and Tovar 2006: 5). This is done through a distancing from the sound material itself: it is no longer the devotion to the most faithful sound recording of the reality, nor a calculation of the best sampling to submit it to a panel of experimental subjects: 'The photographic and sound registers that we have created have not been considered as mimetic productions of the reality, but rather as traces thereof. Like ethnographic text, they are constructed objects' (5).

Olivier Féraud's ethnography on the sonic environments in a popular neighbourhood of Naples expresses a clear statement:

> Field recording was a central practice during the case study, the whole methodology developed around it. Field recording is not only justified by the fact that we are interested in sound, it is not only a 'sampling' tool. It constitutes the methodological pivot of sonic anthropology [...] because it allows, by considering the positioning of the sound recordist, to reflect the relationships that are established in a sonic dimension between individuals in their environment.
>
> (Féraud 2010: 243–244)

It is to his credit that he dedicated an entire chapter of his PhD thesis to his methodology. Along with participant observation, he placed, for his own work documentation, his microphones at particular 'points of listening' (as we have particular 'points of view', see Chion 1985), the actors' point of auditory reception: from the window of a building where women hear the call of the street vendor, for example. For the collection of discourses, which is the third level of his ethnographic methodology, he used field recordings and played them back to people from the very same neighbourhood or from other neighbourhoods, allowing for socially situated feedback and to develop his research problematic. He also used the commented-walk methodology together with 'situated interviews', interviews in a context of listening with the inhabitants.

In those cases, working on sound with sound, a challenging but effective project, makes perfect sense. It should be mentioned that there are cases of ethnographies whose fieldwork dictates the default method: Mohamad Hafeda used the pretext of exploring sonic material to examine notions of division, connection, and negotiation between and within contested sectarian communities in Beirut, and especially on the borderline between two adjacent neighbourhoods divided along the Sunni–Shiite sectarian. He collected the street sounds from inside taxis he used to get around and from walking journeys in the two neighbourhoods, including voices and sounds overheard from inside shops as well as those collected outside and on pavements and streets. 'This research method responds to the previously found site sensitivities defined by the prohibition of photography and the difficulty of site accessibility due to security and surveillance performed by formal political/militia groups in both neighborhoods' (Hafeda 2011: 32).

Tentative ethnographic research in Cairo for a comprehensive sonic dimension

For the methodological overview of possible approaches to the sonic dimensions of existence, this chapter does not aim to list exhaustively ethnographies on sonic dimensions but, rather, to display some notes on part of my readings that can contribute to explain the methodologies I chose and tested (while sometimes they were then unknown to me) and their a-historic phylogeny.

Perhaps I need to clarify why it is necessary to use alternative methodologies to the canons of ethnography to examine this relationship to the sonic dimensions of existence, beyond that congenital deafness of the discipline. In our field, in a word: verbalization. For those who are not familiar with the working methods of fieldwork anthropologists, let us keep in mind that the customary method is, in addition to – or through – participant observation, interviewing (clearly distinguished from the questionnaire). The purpose of the experimental procedure I undertook in Cairo (Figure 49.5) was to obtain verbal descriptions of the acoustic experiences of city dwellers of diverse social and residential backgrounds. This endeavour poses a particular methodological challenge in the field of sensory studies. An ethnography of local acoustic ambiances, received and produced in Cairo, becomes possible when these are treated as 'social productions' (Battesti 2009). Research in sensory studies and sound studies has demonstrated the need for precise observations of this neglected dimension of our relationship – which is always first sensory – to our social and ecological environment. While an analytical grid has been proposed (Battesti 2009, 2013) for understanding the different ways of relating to this sound matter – deliberately repeating and modifying the framework proposed by Steven Feld (1984) – no satisfactory tools have yet been devised or perfected that would allow us to overcome the obstacle that the weak verbalization that most of our sensory activities represent. Without a discourse to build on, ethnographers find themselves at a loss. Even keeping my distance from linguistic determinism (the Sapir-Whorf hypothesis) which assumes that 'we conceive a universe that the language has already patterned' (Benveniste 1966: 6), the fact remains that I have to consider semantic distinctions and inbuilt ontologies within the language: obviously, language affects the ways we conceptualize our world, our worldview, or world hearing as it may be. We need verbalization.

Figure 49.5 Promenade on a bridge over the Nile, Downtown, Cairo, Egypt, 3 November 2016, 5.00 p.m. Photograph by Vincent Battesti.

Studies on sound perception suggest that this mechanism most often operates almost unconsciously. This is widely verified in the sensory domain: the retroactive loops between perception and sensation, very fast (almost synchronous), are most often unconscious, probably avoiding a cognitive overload. Is it the unconscious nature of the mechanism of sound perceptions that explains why they are weakly able to be voiced? This may not always be the case. At any rate, in Cairo, the few words spontaneously used to talk about this sonic dimension give it its evanescent quality: sonic ambiances are a central dimension of urban life – they teach us about long-term and situated participant observation, a 'thickly-textured thorough', in the few but central words of David Howes. To say more than 'the ambiance is good' (*al-gaw halū*) is rare and difficult for urban actors in public spaces; the judgements tend to be summed up in mere polarized hedonistic terms. The difficulty is real to verbalize. 'The atmosphere is good. There you go.' This relationship between urbanites and the sonic dimensions of their city, one of the objectives of this research, is largely unspeakable for the actors themselves, which is not the least of the difficulties of this ethnographic study.

How to work on the sound dimension? How to design the 'satisfactory tool' to make people talk? These 'how' questions depend of course, once again, on the 'what' and the 'why'. The 'what' and 'why' said in few words is the everyday relationship Cairo people have with an essential but scholarly underestimated sonic dimension of the city, varying with the neighbourhoods, received and produced, highlighting also the relationships between the inhabitants of Cairo. The overall sonic dimension of the city – that was thick and present to the outsiders we were – rather than the effects of a sonic device or a special situation: the everyday life. I, and when the ethnologist Nicolas Puig fortunately supported me in this project, we aimed to get as close as possible to the local daily experience. I confessed at the beginning of this chapter that I had unfortunately not established the complete methodological state of the art in the field of sound studies, a quick overview in ethnology had not revealed many things to me. So, everything was to be invented. Let's evoke two of these trial and error experiments with ethnographic methods.

The aural postcard experiment

To bypass the apparent difficulty of verbalization, I first tried the 'aural postcard experiment', a kind of 'reactivated listening' (Augoyard 2001). I used sound ambiances that I had recorded in different neighbourhoods and situations. The informants had to comment on the recordings while listening to them on headphones. The experience showed that the inhabitants of Cairo are able to determine – if not always specific to the neighbourhood – the type (and therefore a categorization) of neighbourhoods offered for listening based on the sonic ambiances I had recorded for playback. However, to objectify their answer, I made them justify themselves, the subjects of the experiment highlighted some characteristic sounds, some 'soundmarks'. Listening to a lively evening in the city centre, towards Azbakiyya, it was the density of repeated hits from the horn of car traffic or the cry of newspaper merchants ('*al-Ahram, al-Akhbar, al-Gumhuriyya!*'); listening to a shopping street (*sūq*) in the popular Fatimid district of Darb al-Ahmar, it was the rubbing in the dust of the *šibšib* (sandals) and the familiar address of the merchants or greetings between acquaintances (which sign territories where mutual acquaintance prevails, unlike the places of anonymous walks downtown). Some sounds can clearly signal an urban space: the cry of merchants, for example. Nonetheless, this 'aural postcard experiment' remained steeped in my culture, as I choose what to record selecting the orientation of the mic (and therefore the 'frame'), I choose the sounding object or situation recorded, I choose when to press the 'record' and 'stop' buttons of my minidisc and then my Zoom H4n (a digital portable recording device), the day and the moment, and so on. This means that I was creating a kind of sound montage of the sounds I was giving to hear, a montage embedded in my culture. And the difficulty of capturing the other's intimate experience of a sonic city persists: it has to be seized in situ and by the people concerned.

The mics in the ears experiment

To get at least a set of local descriptors and local, emic categorizations of this sonic material in which Cairo is immersed, a terminological survey seems unavoidable. It was the first purpose of this procedure, the 'mics in the ears', Nicolas Puig and I set up. We opted for field recording made in situ by the people of Cairo (Battesti and Puig 2016). So, we opted for informants walking in the street through their neighbourhood with a sound recorder. Our approach, nonetheless, was radically different from the soundwalks previously mentioned and different too from the commented-walk method (Thibaud 2001). We asked local people we chose to perform one of their very daily walks, interacting as usual with their environment, but equipped with binaural mics: they were to forget they were wearing mics and not to focus on the recording aspect but on their usual business. A commented-walk method – an informant walking and the investigator recording the sounds and the informant's impressions – would have been too artificial and would have missed the very everyday interactions of the informant with their socioecological environment. Binaural mics in an informant's ears record the most possible intimate exposure to sound ambiance during a routine alone trip. We used stereo binaural microphones/earphones (Roland), small enough to fit inside the ears (like intra-aural headphones), along with a digital recorder and a GPS device. We asked the respondents to take a daily walk route in their neighbourhood – during their commute from home to work, or while shopping in local stores, etc. – for a duration of 20–30 minutes, alone (without us anthropologists) with this non-intrusive equipment. This method does not record a 'soundscape'. This technique is a unique and personal experiment and gives unique and personal results, specific to an informant and a space-time: not an objective or unbiased recording, as even the shape of the informant's head plays a role (the shape of the nose and ears, or the use or not of a veil by women, etc.), the informant's attitude in public space, whether they bend their head, greet people, turn their head to talk or to react to sound, smell, contact, and visual stimuli – especially because while walking, their environment is changing. But still, only sounds reaching the informant's ears are recorded, not what this informant is listening to.

The soundtrack was then replayed on the headphones by the informant, with us, and the binaural's 3D sound reproduction (and its spatial of psychoacoustic quality) used to trigger off verbalization, the precious data all anthropologists are looking for. 'Re-immersed', informants are brought back into their own actions, movements, and displacements in their neighbourhood. They relive their routine trip in the city, and we record their comments: 'Please, tell us what do you hear?' Within this approach, what participants perceived as negative or positive acoustic elements is not an end in itself (to build a better city, etc.). What motivates anthropologists performing such an ethnography is not noise mapping, or 'to enhance and design urban soundscapes from urban planning levels' (Jeon, Hong, and Lee 2013: 806) – actually a kind of hygienist paradigm or agenda – but to increase their knowledge of how social groups or a society works, deals with its environment, or participates in it.

The GPS device that the informant was equipped with on their walk helped us to check the route the informant took afterward, and accompanying the transcribed comments, it helped us to spot the threshold effects, when 'entering' a *ḥāra* (sub-neighbourhood, alley), entering a shopping street, home, a new ambiance, etc. After listening to the recording, we had a further brief talk with the informant to deepen some topics.[2]

Conclusion

To conclude, we have explored other possibilities in our tentative ethnographic methodologies, such as work with blind people in Cairo, which took place while Florian Grond and Piet Devos (2016) were exploring the same path (but with a slightly different purpose) in Canada. Furthermore, other researchers have highlighted the heuristics of sensory impairment (Keating and Hadder 2010). I would like to emphasize the necessity to not forget that working on the sonic dimension of the world and of our existence is an arbitrary cut-out: in Western sensorium, we tie 'sonic' to 'auditory', but the 'auditory' modality is surely embedded in multisensory practices and competences. To go back to an uneasy multisensory ethnography that explores the multiplicity of sensorialities, as Puig did along with his team in Beirut (Kassatly et al. 2016), seems to me unavoidable.

Notes

1. The word *na't* is a Urdu term referring to poetry in praise of the prophet Muhammad, which is usually said in Arab-speaking contexts as *madḥī* or *madḥī nabawī*.
2. This chapter devoted to methodologies is not the place to provide our results – they are stimulating, by the way. The methodology and initial results (Battesti and Puig 2016) and then the main part of the analysis have been published along with the raw and analysed data (Battesti and Puig 2020).

References

Andrisani, Vincent (2010). 'Relocating the Ear: A Cross-Cultural Exploration of the Electrified Soundscape'. *Canadian Acoustics / Acoustique canadienne* 38 (3): 102–103.

Augoyard, Jean-François (2001). 'Entretien sur écoute réactivée'. In Michèle Grosjean and Jean-Paul Thibaud (eds), *L'espace urbain en méthodes*, 127–152. Marseille: Éditions Parenthèses.

Augoyard, Jean-François and Henry Torgue (2006). *Sonic Experience: A Guide to Everyday Sounds*. Montreal: Mc-Gill-Queen's University Press.

Basso, Ellen B. (1985). *A Musical View of the Universe: Kalapalo Myth and Ritual Performances*. Philadelphia, PA: University of Pennsylvania Press.

Battesti, Vincent (2009). 'Ambiances sonores du Caire: proposer une anthropologie des environnements sonores'. *Les Cahiers du GERHICO* (13): 35–49. Available online: http://hal.archives-ouvertes.fr/halshs-00341934 (accessed 14 July 2020).

Battesti, Vincent (2013). '"L'ambiance est bonne" ou l'évanescent rapport aux paysages sonores au Caire: Invitation à une écoute participante et proposition d'une grille d'analyse'. In Joël Candau and Marie-Barbara Le Gonidec (eds), *Paysages sensoriels: Essai d'anthropologie de la construction et de la perception de l'environnement sonore*, 70–95. Paris: éditions du CTHS. Available online: http://hal.archives-ouvertes.fr/hal-00842075 (accessed 14 July 2020).

Battesti, Vincent (2006). '"Pourquoi j'irais voir d'en haut ce que je connais déjà d'en bas ?" Comprendre l'usage des espaces dans l'oasis de Siwa'. In Vincent Battesti and Nicolas Puig (eds), *Terrains d'Égypte, anthropologies contemporaines*, 139–179. Le Caire: CEDEJ. Available online: http://hal.archives-ouvertes.fr/halshs-00004050 (accessed 14 July 2020).

Battesti, Vincent and Nicolas Puig (2016). '"The Sound of Society": A Method for Investigating Sound Perception in Cairo'. *The Senses & Society* 11 (3): 298–319. Available online: https://hal.archives-ouvertes.fr/hal-01380972 (accessed 14 July 2020).

Battesti, Vincent and Nicolas Puig (2006). 'Terrains d'Égypte: Introduction'. In Vincent Battesti and Nicolas Puig (eds), *Terrains d'Égypte, anthropologies contemporaines*, 11–22. Le Caire: CEDEJ. Available online: http://hal.archives-ouvertes.fr/halshs-00126186 (accessed 14 July 2020).

Battesti, Vincent and Nicolas Puig (2020). 'Towards a Sonic Ecology of Urban Life: Ethnography of Sound Perceptions in Cairo'. *The Senses & Society* 15 (2): 170–191. Available online: https://hal.archives-ouvertes.fr/hal-02890453 (accessed 14 July 2020).

Benveniste, Émile (1966). *Problèmes de linguistique générale*, 1. Paris: Gallimard, Tel.

Bull, Michael (2004). 'Thinking about Sound, Proximity, and Distance in Western Experience: The Case of Odysseus's Walkman'. In Veit Erlmann (ed.), *Hearing Cultures, Essays on Sound, Listening and Modernity*, 173–190. London: Berg Publishers, Wenner-Gren International Symposium Series.

Chion, Michel (1985). *Le son au cinéma*. Paris: Cahiers du cinéma, Éditions de l'Étoile.

Chrysagis, Evangelos and Panas Karampampas (2017). 'Introduction: Collaborative Intimacies'. In Evangelos Chrysagis and Panas Karampampas (eds), *Collaborative Intimacies in Music and Dance, Anthropologies of Sound and Movement*, 1–24. New York: Berghahn.

Colon, Paul-Louis (2013). 'Pourquoi est-il si difficile d'apprendre à écouter les oiseaux ?'. In Marie-Luce Gélard (ed.), *Corps sensibles, Usages et langages des sens*, 181–207. Nancy: Éditions universitaires de Lorraine, Presses universitaires de Nancy (PUN).

Davies, William J. et al. (2013), 'Perception of Soundscapes: An Interdisciplinary Approach'. *Applied Acoustics* 74 (2): 224–231. https://doi.org/10.1016/j.apacoust.2012.05.010

Dubois, Danièle, Catherine Guastavino, and Manon Raimbault (2006). 'A Cognitive Approach to Urban Soundscapes: Using Verbal Data to Access Everyday Life Auditory Categories'. *Acta Acustica united with Acustica* 92 (6): 865–874. Available online: https://www.researchgate.net/publication/200045136 (accessed 14 July 2020).

Eisenlohr, Patrick (2018). *Sounding Islam: Voice, Media, and Sonic Atmospheres in an Indian Ocean World*. Oakland, CA: University of California Press. Available online: https://www.oapen.org/download?type=document&docid=1000231 (accessed 14 July 2020).

Farag, Iman (2009). 'Querelle de minarets en Égypte, Le débat public sur l'appel à la prière'. *Revue des mondes musulmans et de la Méditerranée* (125): 47–66. Available online: http://remmm.revues.org/index6170.html (accessed 14 July 2020).

Feld, Steven (1982). *Sound and Sentiment: Birds, Weeping, Poetics, and Song in Kaluli Expression*. Publications of the American Folklore Society New series, 5. Philadelphia, PA: University of Pennsylvania Press.

Feld, Steven (1984). 'Sound Structure as Social Structure'. *Ethnomusicology* 28 (3): 383–409.

Feld, Steven and Donald Brenneis (2004). 'Doing Anthropology in Sound'. *American Ethnologist* 31 (4): 461–474.

Féraud, Olivier (2010). 'Voix publiques: Environnements sonores, représentations et usages d'habitation dans un quartier populaire de Naples'. PhD dissertation, École des hautes études en sciences sociales (EHESS), LAHIC-IIAC - Laboratoire d'Anthropologie et d'Histoire de l'Institution de la Culture/Equipe IIAC, Paris. Available online: http://tel.archives-ouvertes.fr/tel-00462396 (accessed 14 July 2020).

Grond, Florian and Piet Devos (2016). 'Sonic Boundary Objects: Negotiating Disability, Technology and Simulation'. *Digital Creativity* 27 (4): 334–346.

Guillebaud, Christine (2008). *Le chant des serpents musiciens itinérants du Kerala*. Paris: CNRS, Collection Monde indien Sciences sociales XVe-XXIe siècle.

Hafeda, Mohamad (2011). 'This Is How Stories of Conflict Circulate and Resonate'. In Hiba Bou Akar and Mohamad Hafeda (eds), *Narrating Beirut from its Borderlines*, 18–57. [Berlin]: Heinrich Böll Foundation, Middle East. Available online: https://lb.boell.org/en/2012/02/13/narrating-beirut-its-borderlines (accessed 14 July 2020).

Hirschkind, Charles (2006). *The Ethical Soundscape: Cassette Sermons and Islamic Counterpublics*. New York: Columbia University Press.

Howes, David (ed.) (1991). *The Varieties of Sensory Experience: a Sourcebook in the Anthropology of the Senses*, Toronto, Buffalo, London: University of Toronto Press, Anthropological horizons.

Jackson, Phil (2006). 'Des corps "ecstatiques" dans un monde séculier: Une ethnographie sensuelle du clubbing'. *Anthropologie et sociétés* 30 (3): 93–107. https://doi.org/10.7202/014927ar

Jeon, Jin Young, Joo Young Hong, and Pyoung Jik Lee (2013). 'Soundwalk Approach to Identify Urban Soundscapes Individually'. *Journal of the Acoustical Society of America* 134 (1): 803–812. Available online: https://www.researchgate.net/publication/249996426 (accessed 14 July 2020).

Kassatly, Hoda, Nicolas Puig, and Michel Tabet (2016). 'Le marché de Sabra à Beyrouth par l'image et le son: Retour sur une enquête intensive'. *Revue européennes des migrations internationales* 32 (3–4): 37–68. Available online: http://remi.revues.org/8179 (accessed 14 July 2020).

Keating, Elizabeth and R. Neill Hadder (2010), 'Sensory Impairment'. *Annual Review of Anthropology*, 39 (1): 115–29. http://doi.org/10.1146/annurev.anthro.012809.105026

Lomax, Alan (ed.) (1968). *Folk Song Style and Culture*. Washington DC: American Association for the Advancement of Science, Publication AAAS.

Luthi, Jean-Jacques (1985). 'Les cris du Caire'. *Le Caire, Mille et une villes*, 79–83. Paris: Autrement.

Lynch, Kevin (1960). *The Image of the City*. Cambridge, MA: MITMIT Press.

Meyer, Julien (2015). *Whistled languages, A Worldwide Inquiry on Human Whistled Speech*. New York: Springer Berlin Heidelberg.

Monnet, Nadja and Maribel Tovar (2006). 'Essai d'ethnographie visuelle ou comment cerner le pouls de la place de Catalogne au travers de photographies et d'ambiances sonores'. In *Colloque 25 ans du Bilan du Film Ethnographique*, 15. Paris.

Pecqueux, Anthony (2009a). 'Les ajustements auditifs des auditeurs-baladeurs: Instabilités sensorielles entre écoute de la musique et de l'espace sonore urbain'. *ethnographiques.org* 19. Available online: https://www.ethnographiques.org/2009/Pecqueux (accessed 14 July 2020).

Pecqueux, Anthony (2009b). 'Embarqués dans la ville et la musique: Les déplacements préoccupés des auditeurs-baladeurs'. *Réseaux* 156 (4): 49–80. Available online: https://www.cairn.info/revue-reseaux-2009-4-page-49.htm (accessed 14 July 2020).

Pink, Sarah (2009). *Doing Sensory Ethnography*. London: Sage.

Pink, Sarah and David Howes (2010). 'The Future of Sensory Anthropology/the Anthropology of the Senses'. *Social Anthropology/Anthropologie Sociale* 18 (3): 331–340. Available online: https://onlinelibrary.wiley.com/toc/14698676/2010/18/3#heading-level-1-2 (accessed 14 July 2020).

Puig, Nicolas (2010). *Farah, Musiciens de noces et scènes urbaines au Caire*. Arles: Actes Sud-Sindbad, La bibliothèque arabe, Collection Hommes et sociétés.

Puig, Nicolas (2017). 'La ville amplifiée, Synthétiseurs, sonorisation et effets électro-acoustiques dans les rituels urbains au Caire'. *Techniques & Culture* (suppl. 67). Available online: http://tc.revues.org/8504 (accessed 14 July 2020).

Revel, Nicole (1992). *Fleurs de paroles. Histoire naturelle palawan. Tome III: Chants d'amour/chants d'oiseaux*. Ethnosciences, 7 [Paris]: Peeters Publishers, Société d'études linguistiques et anthropologiques de France (SELAF).

Roseman, Marina (1991). *Healing Sounds from the Malaysian Rainforest: Temiar Music and Medicine*. Berkeley, CA: University of California Press.

Rouch, Jean (2003). *Ciné-ethnography*. Visible evidence, 13. Minneapolis, MN: University of Minnesota Press.

Said, Noha Gamal (2014). 'Les crieurs publics: Un dispositif sonore dans les quartiers populaires du Caire'. In Claire Guiu, Guillaume Faburel, Marie-Madeleine Mervant-Roux, Henry Torgue, and Philippe Woloszyn (eds), *Soundspaces: espaces, expériences et politiques du sonore*, 173–181. Rennes: Presses universitaires de Rennes. Available online: https://hal.archives-ouvertes.fr/hal-01098615 (accessed 14 July 2020).

Sayeux, Anne-Sophie (2010). 'Le corps-oreille, Une approche anthropologique sensuelle des musiques électroniques'. *Communications* 86 (1): 229–246. Available online: http://www.persee.fr/doc/comm_0588-8018_2010_num_86_1_2546 (accessed 14 July 2020).

Schafer, R. Murray ([1977] 1994). *The Soundscape: Our Sonic Environment and the Tuning of the World*. Rochester, VT: Destiny Books.

Schafer, R. Murray and World Soundscape Project (1978). *The Vancouver Soundscape*. Music of the environment series, 2. Burnaby, BC: World Soundscape Project, Sonic Research Studio, Department of Communication, Simon Fraser University.

Seeger, Anthony (1987). *Why Suyá Sing: A Musical Anthropology of an Amazonian People*. Cambridge: Cambridge University Press.
Semidor, Catherine (2007). 'Le paysage sonore de la rue comme élément d'identité urbaine'. *Flux* 2006/4–2007/1 (66–67): 120–126. Available online: https://www.cairn.info/revue-flux1-2006-4-page-120.htm (accessed 14 July 2020).
Semidor, Catherine (2006). 'Listening to a City With the Soundwalk Method'. *Acta Acustica united with Acustica* 92 (6): 959–964.
Sevin, Jean-Christophe and Dimitri Voilmy (2009). 'Une pensée de la modalité: Entretien avec Jean-François Augoyard'. *ethnographiques.org* 19. Available online: https://www.ethnographiques.org/2009/Augoyard-Sevin-Voilmy (accessed 14 July 2020).
Siamdoust, Nahid (2015). 'Tehran's Soundscape as a Contested Public Sphere: Blurring the Lines between Public and Private'. *Orient-Institut Studies* 3 (2015). Available online: http://www.perspectivia.net/publikationen/orient-institut-studies/3-2015/siamdoust_soundscape (accessed 14 July 2020).
Stanisz, Agata (2017a). '*Field Recording* jako metoda etnografii poprzez dźwięk'. *Przegląd Kulturoznawczy* 31 (1): 1–19.
Stanisz, Agata (2017b). 'Tractor Unit Acoustemology, Sounds of a Dwelling on the Road'. *Mobile Culture Studies: The Journal* 3: 53–76.
Stoller, Paul (1997). *Sensuous Scholarship*. Philadelphia, PA: University of Pennsylvania Press.
Tardieu, Julien, Cynthia Magnen, Marie-Mandarine Colle-Quesada, Nathalie Spanghero-Gaillard, and Pascal Gaillard (2015). 'A Method to Collect Representative Samples of Urban Soundscapes'. In *Science and Technology for a Quiet Europe (Euronoise 2015)*, Maastricht, Netherlands, 31 May–3 June 2015, p. 6. Available online: https://hal.archives-ouvertes.fr/hal-01572478 (accessed 14 July 2020).
Thibaud, Jean-Paul (1994). 'Composer l'espace: les territoires du pas chanté'. In Michel Bassand and Jean-Philippe Leresche (eds), *Les faces cachées de l'urbain*, 183–195. Bern: Éditions Peter Lang. Available online: https://hal.archives-ouvertes.fr/halshs-00379493 (accessed 14 July 2020).
Thibaud, Jean-Paul (2001) 'La méthode des parcours commentés'. In Michèle Grosjean and Jean-Paul Thibaud (eds), *L'espace urbain en méthodes*, 79–99. Marseille: Éditions Parenthèses, Eupalinos. Available online: http://doc.cresson.grenoble.archi.fr/index.php?lvl=notice_display&id=2111 (accessed 14 July 2020).
Titon, Jeff Todd (ed.) (1992). *Worlds of Music: An Introduction to the Music of the World's Peoples*. New York: Schirmer Books, Simon & Schuster Macmillan, Prentice Hall International. 2nd edition. Available online: https://archive.org/details/worldsofmusicint00titon (accessed 14 July 2020).
Truax, Barry (ed.) (1978). *Handbook for Acoustic Ecology*. Vancouver: Simon Fraser University, and ARC Publications, 1st edition, The World Soundscape Project, The music of the Environment series.
Wathelet, Olivier and Joël Candau (2013). 'Considérations méthodologiques en anthropologie sensorielle: pour une ethnographie cognitive des perceptions'. In Joël Candau and Marie-Barbara Le Gonidec (eds), *Paysages sensoriels: Essai d'anthropologie de la construction et de la perception de l'environnement sonore*, 213–239. Paris: éditions du CTHS, Orientations et Méthodes.

50
Podcast Preservation and the Noise of Saved Sounds

Jeremy Wade Morris

Most popular press histories of podcasting reference 2004 as the year the format came into its own. This was the year Ben Hammersely inadvertently coined the term, and a year before the *New Oxford American Dictionary* voted it as the 'word of the year' for 2005. Although many academics studying podcasting might dispute this origin point – there were audioblogging services in the early 2000s and other podcast-like web audio experiments as far back as the early 1990s (Sterne et al. 2008; Bottomley 2020) – the coining of the term in 2004 seems to overshadow the lineage of these prototypical podcasts.

Imagine my excitement, then, when I stumbled across what I thought was a rare and early podcast from July 1970. I found the audio file in a large-scale database/collection of podcasts my colleagues and I have been building since 2014. Called PodcastRE – short for Podcast Research and found at http://podcastre.org – the database aims to provide a searchable, researchable, and preservable collection of podcasts for the study and exploration of audio culture. We were analysing the date ranges of the nearly 2.5 million episodes our database houses, and the 1970 data point stuck out. What is more, there are nine other podcasts that seem to have started that same month. This early flurry seems to have stopped immediately after 1970, though, and is followed by a lengthy hiatus; the database shows no further podcasts until 1977, when, in March, there's evidence of another feed going live. There's another one in 1979 and another in 1985, and the ensuing decades show the publication of close to another few dozen podcasts, all before Hammersely had even coined the term.

Of course, after examining the files more closely, our initial excitement over our momentous discovery of these historical podcasts was quickly dashed. Though the metadata clearly lists the publication dates, the 1970 'podcasts' in question appear to be from Earwolf productions, the comedy podcasting network formed in 2010 by Scott Aukerman and Jeff Ullrich. Playing the audio files reveals that they are all 30-second advertisements encouraging listeners to sign up for Stitcher Premium – a podcast distribution service – in order to access archives of popular Earwolf shows such as *Cracked Movie Club*, *James Bonding*, and *Never Not Funny*. The other 'early' shows – those from 1977, 1979, and through the 1980s and 1990s – appear to come from various BBC Radio 4 shows, such as *Desert Island Discs*, *The Reith Lectures*, or *In Our Time with Melvin Bragg*. Unlike the Earwolf/Stitcher advertisements, which could only have come after 2010, the BBC podcasts are rebroadcasts of old interviews and recorded conversations that presumably aired on radio shows during those dates but were made available as podcasts over the last decade. Unless we stretch our definition of podcasting far enough to include general public radio broadcasts in the 1970s, then unfortunately the hunt for the earliest ever podcast is still on.

These are just two of many oddities we've come across building and researching our database of podcasts, and they speak to much bigger questions about audio preservation and sonic methodologies more broadly. Many of the challenges we have faced are common to all kinds of digital data collection and preservation, as Lisa Gitelman so accurately describes in her discovery of the 'Internet of 1854' (Gitelman 2006: 123). However, audio files in general, and podcasts in particular, present a very particular and acute set of challenges that undermine our assumptions about studying and saving sonic culture. Accordingly, this chapter looks at the difficulties and opportunities podcasts represent for studying the longer history of audio, and at some of the sonic methodologies podcasts make possible. Wrapped in RSS feeds, XML files, and ID3tags, podcasts and other forms of digital audio create a series of traces that offer new possibilities for the automated collection of audio files and their accompanying metadata. New sonic analysis tools (frequency analysis, waveform display, audio fingerprinting, etc.) also bring new digital means of analysing and studying sound. But audio still remains mystifyingly difficult to preserve in ways that also make it searchable, analysable, and useable. Qualitative audio analysis, for example, still has to be done in real time (or longer) and quantitative methods still largely rely on analysing audio after it has been transcribed into text, often stripping digital audio of its most important component: the sound itself. Inconsistent metadata standards across podcast producers and the major distribution platforms as well as changing technologies for delivering podcasts stifle both the ability to study podcasts and initiatives to preserve them.

Using the case of the PodcastRE database, I argue that the development of new sonic methodologies is crucial for researchers to capitalize on the wealth of insight and information available in the booming audio culture emerging around podcasting. There are certainly advantages to the affordances provided by digital formats (i.e. metadata, automated collection, elaborate tools for the visualization of sound waves, etc.) but I argue we still need to foreground sound itself in our methods and audio collections. Developing sonic methodologies in conjunction with more traditional/typical digital methodologies for sound and audio archives will expand the utility of sound databases and will also help alleviate the problems that accompany the primacy of the visual (Sterne 2003; Hilmes 2005) afforded to most archives. Approaching sound archives and objects sonically, instead of just visually or textually, offers new perspectives and paths for research, since, as Tara Rodgers notes, 'vibrations – including that specific class of audible vibrations experienced as sound – present alternative ways of apprehending reality that can point to political sensibilities that emphasize complexity, interconnection and interdependence rather than modes of distancing and control' (Rodgers 2018: 234). Placing sound at the forefront, not just of media artefacts we need to save but as a method for analysing that which we have saved, gives us a new mode for approaching archives and databases, as well as the sonic ephemera that lies within them.

Building a database for sonic study

Podcasting represents a significant addition to the collective repository of sound and audio culture. While music recordings, radio broadcasts, oral histories, and soundscapes have been the primary forms of audio in archives, collections, and libraries that scholars and everyday users have turned to in their attempts to explore sounds of the past (and present), the proliferation of technologies allowing everyday users to make, record, and distribute audio via podcasts has resulted in a proliferation of sonic artefacts. In 2015 there were over 525,000 podcast feeds and close to 18.5 million individual episodes in over 100 languages on the iTunes store alone (Locker 2018). Factoring in the last four years of exponential growth in feeds and episodes, as well as numbers from other aggregation and distribution sites, such as Podbean, Anchor, Soundcloud, Google Podcasts, and Spotify, pushes those numbers up even further. Podcasting has grown steadily over the last fifteen years, and while it may not generate the total number of listeners that traditional broadcasting does, nor does it draw the influx of 'users' that most social media companies hope to garner, there now exists a substantial body of sound recordings that are publicly and widely available, with a healthy number of listeners listening to them: 37 per cent of Americans listen to podcasts monthly, according to Edison Research (2020). This explosion in audio activity has led to a general consensus in the popular press in the last few years that we are currently experiencing a 'Golden Age of Podcasts' (Blattberg 2014; Roose 2014; Sillesen 2014); a moment where the choice for quality digital audio abounds, and where new voices and new listeners connect daily through earbuds, car stereos, or office computers.

We know from earlier media 'golden ages', however, that excitement over new technologies, styles, and techniques does not necessarily translate to long-term preservation and care of the media content produced during those 'golden ages'. A huge amount of US silent films, early television broadcasts and radio recordings, for example, have been lost or destroyed. Although podcasting is far newer than these legacy media and although podcasts are currently relatively ubiquitous and available, this alone does not ensure they will always be so. Podcasts may not face the same preservation risks as, say, decaying old radio tape reels, celluloid film stock, or misplaced transcription discs, but they face their own kinds of vulnerability (podcast feeds go dead, audio files get improperly stored or migrated, once 'free' episodes get put behind paywalls, etc.). Given podcasting's ability to allow for near-instant and constant commentary on current social, cultural, and political events, the sounds emerging from the format provide a wealth of insight for cultural critics (Berry 2006, 2016; Hilmes 2013; McHugh 2016). Because today's podcasts are part of the format's infant years, they are also inherently historical; their content, like other media we use to shape our understanding of history, gives a glimpse into current trends, ideas, and ideologies and their format tells industrial and political economic stories about the evolution of audio technology.

Despite the potential the sounds of podcasting represent for researchers of media and culture, there are few institutional repositories or collections that enable their study and analysis. The Internet Archive has an impressively robust collection of podcast episodes, though the user-generated nature of the collection means the interface is not necessarily built with media researchers in mind, with limited details about the sounds and shows and ability to search within specific shows or episodes. There are a few library-based initiatives to archive podcasts and present them in a researchable collection – the Library of Congress, for example, is earmarking a small subset of shows for long-term preservation – but these are developing slowly and haphazardly. There are, of course, large personal or community collections of podcasts available via the internet, though these tend to be show-specific and highly fan or listener driven and are often hard to locate.

Even if there were large collections of podcasts available, though, the tools for analysing sound remain difficult to use or highly expensive. Whereas there are a host of accessible services that cater to textual and visual data (Word Clouds, N-Gram, Google Image Search, etc.), there has not been similar innovation of sonic analysis tools. There are a number of good transcription tools for helping to turn audio into text, and tools for mining the metadata that comes via podcasts and their RSS feeds, but these generally limit users to studying textual or visual representations of metadata.

The PodcastRE database aims to address some of these needs and gaps, though the very process of building the database highlights some of the difficulties audio in general and podcasting specifically poses for saving and studying sonic culture. The database began as a small and simple affair in 2014 – a personal collection of podcasts in an iTunes database – but has grown into a much more complex and robust podcast preservation effort that now holds over 100 terabytes of data, across a dozen hard drives and relies on dozens of python scripts to fetch and sort podcasts on a daily basis. The collection now indexes over 2.5 million podcasts from nearly 15,000 feeds, along with metadata and audio files for the majority of those episodes.

I have detailed elsewhere some of the challenges my team and I have faced in both defining the very object we are trying to collect and the larger curatorial questions of what to save and what to ignore (Morris, Hansen, and Hoyt 2019). These are, of course, questions that face anyone who endeavours to create a collection, database, or archive, particularly with digital artefacts. As web objects, podcasts are sonic media and digital artefacts. They exist online as sound files that can be downloaded to some kind of physical form (stored on a computer, mobile device, or hard drive), but they also exist as a part of a wider assemblage of artefacts that often include a website, show notes, production files, thumbnail images, and the context that comes with how they are presented in the various podcast aggregator apps and software they appear. If one's goal is to archive and preserve podcasts, then this goal is immediately complicated by the question of what, exactly, is a podcast. Is it simply the audio file? Is it the audio file plus its accompanying metadata, which provide contextual clues about the object? Is it the audio file, the metadata, and the supplemental website the producers built to host the show? The list of possible questions goes on.

These seem like relatively simple questions, but each decision affects how a researcher or archivist will approach the process of preservation, which in turn affects the kinds of sonic methodologies that can be developed. Building a database of audio files is one task; building a database that somehow preserves those audio files as well as all the features of the websites that helped deliver those files is another matter. Each of these options will also give researchers different kinds of objects and data to incorporate into their methodologies. Stepanyan et al. (2012), looking at the blurry boundaries of blogs, raise similar questions (i.e. should we preserve just the text of the entries? The comments? The internal dashboard? etc.). They conclude that any preservation effort requires identifying the 'significant properties' of an object, though debate over what counts as 'significant' for any resource or artefact will likely persist among the various stakeholders and communities who want to access and use the preserved resources.

Beyond the difficulties of accounting for the 'significant' elements of any digital object, any effort to archive or database podcasts is also challenged by the fact that, as natively digital web objects, podcasts suffer from the dynamic and ever-changing nature of the web itself (Brügger 2009). As Mél Hogan argues, 'The web constantly overwrites itself, but unlike the palimpsest, past iterations are cached in layers rather than made visible underneath current iterations, if at all retrievable' (Hogan 2015: 20). Not only do podcast feeds suffer from the same kinds of link rot that other webpages do, the various platforms that host podcasts also go through changes and iterations that can affect the ability to locate a particular file or show. Several of the podcasts in our databases have multiple entries, each with a different RSS feed, indicating either personal or institutional changes in terms of who is producing or hosting the podcast. The hit Australian Podcast *Science Vs.*, for example, has had three different RSS feeds over the course of its lifespan, each indicating different hosting locations and production arrangements (i.e. the move from the public Australian Broadcasting Corporation to the podcasting network Gimlet Media):

1 http://feeds.soundcloud.com/users/soundcloud:users:154052752/sounds.rss
2 http://www.abc.net.au/radio/programs/sciencevs/feed/8604304/podcast.xml
3 http://feeds.gimletmedia.com/sciencevs.

There are also hundreds of shows in our database that have changed their title, yet kept the same RSS feed, whether that be for clearer marketing purposes (e.g. *Sleep With Me*'s change in subtitle from *Sleep With Me: Helps You Fall Asleep Via Silly Boring Bedtime Stories* to *Sleep With Me: The Podcast That Puts You To Sleep*) or more fundamental changes to the name and content (e.g. *The Ethnic Revolution*'s switch to *MK on the Mic* while maintaining the same RSS feed URL). These slight variations in title matter little for the everyday listener. Presumably, fans who enjoyed *The Ethnic Revolution* would also enjoy the show that replaced it in the RSS feed given that it is from the same producer and covers the same subject. However, for researchers interested in tracking podcasts over time, these minor changes in title or RSS feed location create issues for preservation and documentation. This becomes even more true as podcast producers and distributors shift to models that rely more heavily on streaming than on downloading, thus making the original audio files or a fixed URL even harder to obtain.

Further, the reality that not every text, film, radio broadcast, tweet, etc. can be saved means that choices need to be made and that someone – be it an organization, an individual, a state or federal institution, etc. – needs to decide what should be saved and what criteria will govern the choices for what is saved and what is not. Archivists and librarians are well versed in the strategies for dealing with these kinds of questions, even if their answers still involve debates over 'significance', resources, formats, availability, space, rationale, and need. Media scholars who build or acquire their own personal collections of media usually do so with much narrower goals in mind: to facilitate research on a particular set of texts, films, radio broadcasts, tweets, etc. and sometimes saving them to return to the data in question. This is not to say media scholars are ignorant of the above questions and issues facing librarians and archivists, but that often, media collections are a means to an end, rather than the end in and of itself. We are, in other words, accidental archivists (Gold 2008); our research data can become accidental archives.

PodcastRE has tried to address the issues of 'defining the object' by bringing in the audio files of various podcasts, but also by saving the web links and links to extraneous material associated with the preserved shows. It may not capture the entirety of the podcast experience for users (i.e. how podcasts appear in podcasting apps and aggregators), nor does it preserve the many companion materials that go along with a show (though it does link to the original sites so long as they are maintained), but it brings in a significant amount of contextual data around the various entities in the database. We have also developed a curation strategy that lets us capture the most popular and discussed podcasts through the automated collection of top-ranked podcasts in iTunes across various regions and we actively seek out more amateur, marginalized, and vulnerable content through our extended network of researchers and through the general 'submit a podcast' link feature on the site.

As with other digital humanities projects (see, for example, projects described in Burdick et al. 2012; Schreibman et al. 2016; and Berry and Fagerjord 2017), PodcastRE allows for basic keyword searches across a number of fields (e.g. show and episode title, show and episode description, producer, etc.) as well as more advanced, faceted searches that allow users to specify one or multiple keywords over a specific date range or across a specific attribute (e.g. author, description, podcast source). We have also created some visualization tools that allow for graphic representations of data. For example, users can create a word cloud of a particular keyword search, with the surrounding words representing the words that most frequently show up alongside that keyword in descriptions. Users can also graph the occurrence of keywords in the dataset over time and compare multiple keywords.

While these tools provide a more robust research interface for studying podcasts than many of the currently available options, they are still highly visual and textual: they rely on text from metadata, transcripts, and XML files in order to present their results. As far as sonic methodologies go, they help us study sound, but they don't use sound to study. Take a transcription for example. Transcription errors aside, transcriptions generally provide an excellent ability to search through an audio file much quicker than sitting through the entire audio file. So, if one was to search for the keyword 'Shenzhen' in the *This American*

Life episode 'Mr. Daisey and the Apple Factory', they could quickly find multiple points when the narrator, Mike Daisey, references the Chinese town of Shenzhen, which he visited in order to see where many Apple products are made in the city's teeming factories. But relying solely on the transcript strips the audio of some of its most essential elements: pace, tone, pauses, sentiment, and emotion. For example, in the follow-up episode, 'Retraction', where *This American Life* host Ira Glass is pressing Daisey on the reasons why he fabricated elements of the story about his visits to the factories in Schengen and why he lied to *This American Life* producers, the transcript utterly fails to communicate Glass's frustration, Daisey's floundering, and the extremely long, awkward, and telling pause between that takes place between question and answer. An automated transcript might simply look like this:

> Ira Glass: And, and at that point you could have come back to us and said 'oh no no no I didn't meet these workers, you know, this is just something I inserted in the monologue based on things I had read and things I had heard in Hong Kong' um, but instead you lied further […]. Why not just tell us what really happened at that point?
> Mike Daisey: I think I was terrified.
> Ira Glass: Of what?
> Mike Daisey: That […] I think I was terrified that if I untied these things, that the work, that I know is really good, and tells a story, that does these really great things for making people care, that it would come apart in a way where, where it would ruin everything.

A more accurate transcription taking into account the non-verbal utterances and emotional cues might instead look like this:

> Ira Glass: And, and at that point you could have come back to us and said 'oh no no no I didn't meet these workers, you know, this is just something I inserted in the monologue based on things I had read and things I had heard in Hong Kong' um, but instead you lied further […]. Why not just tell us what really happened at that point?
> [*long pause*]
> [*long pause*]
> Mike Daisey: I think I was terrified. [*breathing, uneasily, exasperated*]
> Ira Glass: Of what?
> [*long pause*]
> [*long pause*]
> Mike Daisey: – That—
> [*long pause*]
> […]
> […]
> Mike Daisey: I think I was terrified that if I untied these things, that the work, that I know is really good, and tells a story, that does these really great things for making people care, that it would come apart in a way where, where it would ruin everything.

While automated transcriptions enable new ways to search through large quantities of audio files, they also impose a particular understanding of sound, one that centres the dialogue or the transcribe-able content at the expense of the sound's greater meaning, significance, or impact, as well as any of the important sound cues and sonic design.

Relying on standard textual and visual tools also means relying heavily on the metadata that podcasts offer. Many digital humanities projects involve the collection and ordering of a significant amount of metadata – information about information – around particular objects. I have detailed elsewhere some of the challenges we have faced with PodcastRE in terms of the various kinds of metadata that accompany the audio files for any podcast (Morris, Hansen, and Hoyt 2019). Podcasts are troublesome objects in some sense because there are multiple sources for metadata. There's the information that comes embedded in the files themselves, usually in the form of ID3 tags (e.g. title, author, date, duration, genre). Each podcast also has metadata in the RSS feed, which usually includes fields for the show title, URL, description, and creator. Since most podcasts contain many separate episodes, the RSS feed has data at the show level as well as at the episode level. Podcasts, then, are a combination of file and feed metadata; sometimes these data match, other times they do not. Additionally, with the exception of some automated fields, this metadata needs to be intentionally produced by the show's creator or the platform that hosts these shows. RSS and ID3 tags may have certain customs (i.e. the producer's name should go in the 'author' field), there's no guarantee that these fields will be used or interpreted the same by various producers.

Returning to the 'early' podcasts at the start of this chapter, for example, is instructive. The Earwolf shows from '1970' are dated as such because in iTunes and other feed aggregators, the oldest shows appear first in the feed so that if users subscribe to a new show, they can begin with its earliest episode. The Earwolf advertisements are meant to drive users to the paid, premium Stitcher service, so dating the shows '1970' allows Earwolf to ensure the advertisements for Stitcher will always be the first thing users hear when they subscribe to new Earwolf shows. In other words, Earwolf engineered the metadata to support the commercial aims of their podcast network responsible for producing the content.

The BBC podcasts, on the other hand, seem to present a more honest conundrum. The interviews and shows they were re-airing originally aired during the dates they listed, 1977, 1979, etc. As a national public broadcaster, the BBC holds huge archives and by rereleasing these archives in new formats (iPlayer, Podcasts, etc.), they meet their public mission to make culture and history available for its listener-public. The RSS standard, however, does not offer the nuance between an 'original air date' versus an 'uploaded as podcast' date. 'Publication Date' in this case, can be either the date the recording originally aired, or the date the BBC exported the recording as an mp3 and uploaded it to iTunes, and this decision is wholly at the whim of the producer in question.

So, while RSS is a technical standard – and as a standard it technically shapes and defines what information is included and excluded in the description of audio files such as podcasts – this does not necessarily mean there is a standard way that podcast producers encode the details of their feeds. It is also a standard that originated before podcasts. It was originally a standard for blogging and a way to push textual content from websites to feed readers such as Google Reader or NetNewsWire. It was only later that users and programmers figured out how to enclose audio files within RSS feeds. Although there have been additions and builds to the capabilities of RSS through the various podcast XML namespaces Apple, Google, and others have developed (Morris, Hansen, and Hoyt 2019), it is hardly a standard designed with the needs of today's librarians or preservationists in mind. The development of RSS was driven by the needs of a relatively small and likely homogenous group of bloggers and tech enthusiasts in the late 1990s and early 2000s. It was not meant for, nor did it envision, the potential needs of the various communities making and studying podcasts nearly two decades later. But as with other standards, its impact persists in its continued use; it continues to be a major infrastructural component of the podcasting ecosystem.

Listening to audio databases

As useful as transcriptions, advanced keyword searches, word clouds, and other kinds of textual and metadata tools may be, they do not capture the musical or sonic aspects of podcasts, which may contain data or experiences of data that scholars might find useful. As Tanya Clement notes, most audio analysis tools 'are designed to leverage human-generated transcripts and metadata and do not provide any means for analyzing the audio itself' (Clement et al. 2014: 2). Clement's research, along with a number of other 'sonically-attuned digital humanists' (Clement 2016a, b; Foka and Arvidsson 2016; MacArthur 2016; Mustazza, forthcoming) represent some of the few concerted efforts to build usable, accessible tools for audio analysis, even if they require a bit of learning and a healthy technological infrastructure to support their implementation. The overall lack of tools, however, has meant that 'humanists have few opportunities to use advanced technologies for analyzing large, messy sound archives' (Clement and McLaughlin 2016), as well as the specific cultural markers one might find in those archives, be it a database of podcasts or a collection of old poetry recordings.

Rather than textual tools that seek to turn audio into text or that visualize metadata, there is an emerging set of visual tools that focus more centrally on audio as an object to be mined in its own right. At the most basic level, most audio editing software programs provide visuals of soundwaves and, with practice, one can learn to read these visuals for various sonic characteristics. For example, examining the 'thickness' of soundwaves can be an indication of the relative level of production levels that go into a podcast. The two soundwaves in Figure 50.1 show different levels of polish and finish. The first is from the *StartUp* podcast – a highly edited and professionally produced podcast from Gimlet Media, that takes care to record with good microphones and employs mixers and masterers to polish the finished audio file. The resulting audio file is highly compressed and normalized and thus looks 'thicker'. The second sound wave shows a more independent and lo-fi podcast – an interview podcast that is recorded much more informally in coffee shops and restaurants with presumably one microphone on the table in between the host and the guest(s). The finished file is sparingly edited and mastered upon export, giving it a much 'thinner' resulting sound wave. An analysis such as this across a much larger number of podcasts or segments of shows might give an indication as to the level of professional versus amateur podcasts in a certain corpus or genre or category, and whether or not the

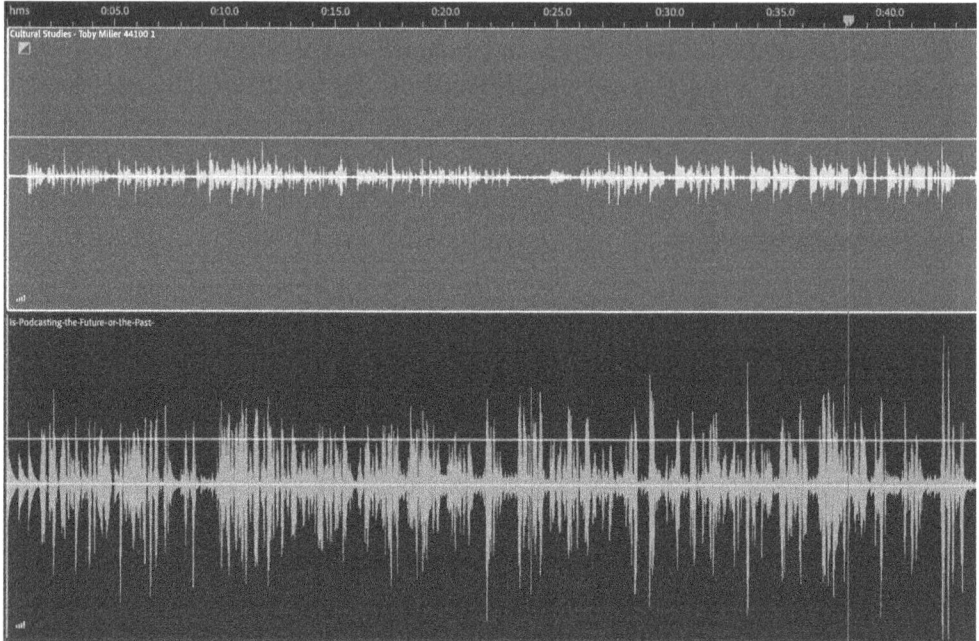

Figure 50.1 Two sound waves from different podcasts, indicating different levels of production, editing, and mastering for each.

rise in professionally produced podcasts has crowded out shows with lower production values. While this is still a more visual analysis than a sonic one, unlike transcripts, it focuses researchers on the sounds embedded in the audio file, not just the speech.

Similarly, many audio editing software programs also provide spectral and frequency analysis of sound files, meaning that researchers can look for similar patterns, pitches, and noises that recur throughout sound files. Patrick Sullivan (2017) uses such tools to examine the sound effects library used in Hanna-Barbera's animated productions between 1957 and 1985, using a process he calls 'distant listening' to explore the 'rich texture of the sonic register of the studio's output' and to question the 'deep formal rhythm' that structures the sounds of the early cartoons. Given the difficulty in 'transcribing' cartoon sound effects, or in creating useful metadata visualizations, Sullivan's project shows how visual representations of sound can turn us towards specific sonic elements of a particular sample of audio files. A similar type of analysis might be employed to analyse, for example, the formal elements of podcasts; looking at spectrograms could help researchers identify different sections of a podcast such as the introductions or advert breaks, or even to compare the sonic characteristics of various podcast networks (i.e. Gimlet, Audible, Earwolf, etc.). When Neil Verma (2018) discusses Audible's 'house sound', for example, there might be ways to track this quantitatively as well as qualitatively.

New audio frequency analysis tools are also emerging that make the study of pitch and frequency more user-friendly and accessible. Tools such as Gentle and Drift allow users to create 'aligned' transcripts of audio files – a transcript where the words dynamically link to the audio in the file – and track frequency changes in the voice over the duration of the audio clip (MacArthur et al. 2018). Looking at the pitch and frequency characteristics of 'poet voice' across a series of poetry recordings, Marit MacArthur and her colleagues argue that their method 'in some sense, […] slows down speech by giving us new ways (new to literary study, anyway) to think about our perceptions of it' (MacArthur, Zellou, and Miller 2018: 69). Although Drift has largely been used for poetry and other short recordings, frequency analysis could easily apply to a number of different kinds of audio files (for instance, podcasts). For example, frequency analysis may provide a first-pass look at a large swath of audio files to help direct attention to the kinds of speakers hosting podcasts. Generally, adult male voices tend to have lower frequencies that feature a fundamental frequency range of about 85 to 180 hertz, while adult female voices fall in the 165 to 255 hertz range. Although the human voice is incredibly fluid and measures like this would obviously miss nuances of voices that fall along a broader spectrum of gender, tools such as Drift might allow us to test assumptions about how many podcasts within a certain subset are hosted by men or by women, or how many different speakers appear in a particular show. Frequency analysis is not, in and of itself, sufficient for answering these questions, but its mode of distant listening can direct closer listening to specific shows, episodes, or moments. Mertens, Hoyt, and Morris (forthcoming) use a similar strategy to measure the variation in pitch among different categories of podcasts and find that, for example, sports and talk podcasts feature more frequency variation than traditional news-based podcasts. Mertens argues that news podcasts often employ what has been called an 'NPR voice' – the concerted use of looser language, generous pauses, and controlled but emphatic inflection to create a professional but personal vocal performance – and this has led to a disadvantaging of speakers with a wider natural frequency range who do not conform to these sonic ideals.

There are also 'audio fingerprinting' technologies which allow for the identification and comparison of sounds across a larger subset of audio. Unfortunately, most of these tools are proprietary and highly cost-prohibitive for researchers, like those employed by private companies such as Shazam or Gracenote. Open source or more cost-efficient versions of such software, for instance AcoustID, however, could help researchers find trends across shows and episodes by listening for similar sonic cues, sound effects, or music samples and by using these results to help support their aesthetic or industrial analyses of the audio in question.

These more sound-focused tools are, admittedly, still nascent and they often require technical resources or specialized knowledge of particular software programs, making their widespread adoption in media and cultural studies, or other disciplines, not currently feasible. They do, however, speak to new ways of privileging sound in our methods. Sonic methodologies should, by their very nature, require researchers to engage with the sound itself and make use of the affordances of audio to provide a different kind of analysis than might otherwise be possible. This is not to suggest that these tools should replace close listening; in fact, in most cases, they work better as supplements to close listening. But sonic methodologies should open up the kinds of questions we can ask about audio, and thus about the role of sound in exploring and understanding media and culture.

Sonic methodologies from sound collections

The 'early' Earwolf and BBC podcasts from the 1970s are now, like most media, historical documents. They can tell us about particular moments in time and give us clues to what might have been happening culturally, industrially, or technologically at those moments. The Earwolf advertisements, for example, point us to stories about the rise of podcast networks, the tug of war between free digital content and monetization through paywalls and premium services, as well as the role comedy podcasts have played in raising the profile of podcasting more generally. The BBC podcasts are doubly historical; they trace a moment in the mid-2010s when it became possible to share culture from the 1970s in a different format.

These examples also remind us that any archive or database for the study of large-scale artefacts or phenomenon is always as much an exploration of the archive/database in question as it is of the artefacts/phenomenon. As Hogan argues, 'The archive is less about reality, and more about contrived conjecture' (Hogan 2015: 13) since the archive must always be as much a site of inquiry itself as it is a platform for primary research (Stoler 2009). The process of building sound archives and collections is fraught with decisions that will eventually affect the way users and researchers make use of the sounds within that collection. This is partly why Hogan argues that 'in a technological landscape with ubiquitous recording and disseminating, the crux of the archive could become about what is forgotten, erased, thrown out, deleted and never there' (Hogan 2015: 14). A 30-second advertisement from Earwolf directing users to sign up for their premium service is, by most accounts, highly forgettable and deletable. But its metadata aberration, its presence in our database remains.

Ultimately, the conclusions we can draw from PodcastRE are limited by its inclusions and exclusions. In other words, the work researchers are able to do with the database depend on what the automated scripts and researchers working on it have decided to collect and not collect and on what the programmers and developers have been able to build (or not) as features for search and analysis. While we've made significant efforts to capture a representative sample of podcasting culture, and to provide useful tools for their analysis, there will always be omissions. Many of the earliest podcasts from the mid-2000s, for example, are already lost or unfindable. Their producers did not bother to save them, or did not have the knowledge, foresight, or the resources to do so. They may not have realized that just by the sheer fact they were taking part in a format's infancy, they were also, inadvertently, making history. Similarly, today's podcasts are at risk of becoming inaccessible to researchers less because of issues of ageing formats or obsolescence – though these are certainly issues – and more because the sheer ubiquity and availability of podcasts lull us into thinking that our access to this booming sonic culture will always be thus. As Hogan argues, there is a lack of archival understanding about what makes digital content valuable 'not because digital content is without worth, but because we still do not know how to collectively assign it to content outside of a scarcity/capitalist model, or how best to organize large amounts of data within a framework that is about more than the moment of search' (Hogan 2015: 13). It is difficult to create a sense of urgency around preserving something that is seen as both mundane and ubiquitous, and how to allow for the analysis of such artefacts that go beyond present-day issues and concerns.

All histories of digital objects are, to some extent, marked by their absences – by what cannot be captured in a dynamic and often-changing environment of code, objects, pages, sites, and spheres (Brügger 2009). Audio is especially hard to save and study because there are so few tools to enable these kinds of activities, but also because we rarely think of what sound might add to our research and about how it might document affect, power, and the intimacy of human voices and sounds differently than other media. We sorely need more collections that house and feature audio, and that make audio not just playable but researchable and analysable. For sonic methodologies to develop, we need sound collections that enable new ways to explore, work, and play with audio and new ways to focus on which voices are being amplified through new media, and which others remain silenced. Developing more robust sonic analysis tools helps place sound at the forefront of our methods and opens up new affordances through which to study modalities of culture that are much more difficult to grasp using traditional textual and visual methodologies.

References

Berry, D. M. and A. Fagerjord (2017). *Digital Humanities: Knowledge and Critique in a Digital Age*. Cambridge: Polity. Available online: https://search.library.wisc.edu/catalog/9912336945502121 (accessed 22 August 2018).

Berry, R. (2006). 'Will the iPod Kill the Radio Star? Profiling Podcasting as Radio'. *Convergence: The International Journal of Research into New Media Technologies* 12(2): 143–162. https://doi.org/10.1177/1354856506066522.

Berry, R. (2016). 'Podcasting: Considering the Evolution of the Medium and Its Association with the Word "Radio"'. *Radio Journal: International Studies in Broadcast & Audio Media* 14(1): 7–22. https://doi.org/10.1386/rjao.14.1.7_1.

Blattberg, E. (2014). 'The Podcast Enters a New Golden Age'. Digiday, 19 November. Available online: https://digiday.com/media/nielsenes-rise-podcast/ (accessed 12 February 2015).

Bottomley, A. (2020). *Sound Streams: A Cultural History of Radio-Internet Convergence*. Ann Arbor, WI: University of Michigan Press.

Brügger, N. (2009). 'Website History and the Website as an Object of Study'. *New Media & Society* 11(1–2): 115–132. https://doi.org/10.1177/1461444808099574.

Burdick, A., J. Drucker, P. Lunefeld, T. Presner, and J. Schnapp (2012). *Digital Humanities*. Cambridge, MA: MIT Press. Available online: https://search.library.wisc.edu/catalog/9910275767102121 (accessed 22 August 2018).

Clement, T. E. (2016a). 'Towards a Rationale of Audio-Text'. *Digital Humanities Quarterly* 10 (3). Available online: http://www.digitalhumanities.org/dhq/vol/10/3/000254/000254.html (accessed 18 July 2020).

Clement, T. E. (2016b). 'When Texts of Study Are Audio Files: Digital Tools for Sound Studies in DH'. In S. Schreibman, R. Siemens, and J. Unsworth (eds), *A New Companion to Digital Humanities*, 348–357. Chichester: John Wiley.

Clement, T. E. and S. McLaughlin (2016). 'Measured Applause: Toward a Cultural Analysis of Audio Collections'. *Journal of Cultural Analytics* 1(May 23). https://doi.org/10.22148/16.002.

Clement, T. E., D. Tcheng, L. Auvil, and T. Borries (2014). 'High Performance Sound Technologies for Access and Scholarship (HiPSTAS) in the Digital Humanities'. *Proceedings of the American Society for Information Science and Technology* 51(1): 1–10. https://doi.org/10.1002/meet.2014.14505101042.

Edison Research (2020). 'The Infinite Dial 2020'. [Blog] *The Infinite Dial*, 19 March. Available online: https://www.edisonresearch.com/the-infinite-dial-2020/ (accessed 18 July 2020).

Foka, A. and V. Arvidsson (2016). 'Experiential Analogies: A Sonic Digital Ekphrasis as a Digital Humanities Project'. *Digital Humanities Quarterly* 10 (2). Available online: http://www.digitalhumanities.org/dhq/vol/10/2/000246/000246.html (accessed 18 July 2020).

Gitelman, L. (2006). *Always Already New: Media, History and the Data of Culture*. Cambridge, MA: MIT Press.

Gold, D. (2008). 'The Accidental Archivist: Embracng Chance and Confusion in Historical Scholarship'. In G. E. Kirsch and L. Rohan (eds), *Beyond the Archives: Research as a Lived Process*, 13–19. Carbondale, IL: Southern Illinois University Press.

Hammersley, B. (2004). 'Audible Revolution'. *Guardian*, 12 February. Available online: https://www.theguardian.com/media/2004/feb/12/broadcasting.digitalmedia (accessed 13 June 2018).

Hilmes, M. (2005). 'Is There a Field Called Sound Culture Studies? And Does It Matter?'. *American Quarterly* 57(1): 249–259. https://doi.org/10.1353/aq.2005.0006.

Hilmes, M. (2013). 'The New Materiality of Radio: Sound on Screens'. In J. Loviglio and M. Hilmes (eds), *Radio's New Wave: Global Sound in the Digital Era*, 43–61. New York: Routledge.

Hogan, M. (2015). 'The Archive as Dumpster'. *Pivot* 4(1): 7–38.

Locker, M. (2018). 'Apple's Podcasts Just Topped 50 Billion All-Time Downloads and Streams'. Fast Company, 25 April. Available online: https://www.fastcompany.com/40563318/apples-podcasts-just-topped-50-billion-all-time-downloads-and-streams (accessed 27 May 2018).

MacArthur, M. (2016). 'Introducing Simple Open-Source Tools for Performative Speech Analysis: Gentle and Drift'. *Jacket2* June (6). Available online: http://jacket2.org/commentary/introducing-simple-open-source-tools-performative-speech-analysis-gentle-and-drift (accessed).

MacArthur, M., G. Zellou, and L. Miller (2018). 'Beyond Poet Voice: Sampling the (Non-) Performance Styles of 100 American Poets'. *Journal of Cultural Analytics*. https://doi.org/10.22148/16.022.

McHugh, S. (2016). 'How Podcasting Is Changing the Audio Storytelling Genre'. *Radio Journal: International Studies in Broadcast & Audio Media* 14(1): 65–82.

Mertens, J., E. Hoyt, and J. W. Morris (forthcoming). 'Drifting Voices: Studying Emotion and Pitch in Podcasting with Digital Tools'. In J. Morris and E. Hoyt (eds), *Saving New Sounds: Podcast Preservation and Historiography*. Ann Arbor: University of Michigan Press.

Morris, J. W., S. Hansen, and E. Hoyt (2019). 'The PodcastRE Project: Curating and Preserving Podcasts (and Their Data)'. *Journal of Radio & Audio Media* 26 (1): 8–20. https://doi.org/10.1080/19376529.2019.1559550.

Rodgers, T. (2018). 'Approaching Sound'. In J. Sayers (ed.), *The Routledge Companion to Media Studies and Digital Humanities*, 233–242. New York: Routledge.

Roose, K. (2014). 'What's Behind the Great Podcast Renaissance?'. *New York Magazine*, 30 October. Available online: https://nymag.com/intelligencer/2014/10/whats-behind-the-great-podcast-renaissance.html (accessed 10 July 2020).

Schreibman, S., R. Siemens, and J. Unsworth (eds). (2016). *A New Companion to Digital Humanities*. Chichester: John Wiley & Sons. Available online: https://search.library.wisc.edu/catalog/9912220876302121. (accessed 22 August 2018)

Sillesen, L. B. (2014). 'Is This the Golden Age of Podcasts?'. *Columbia Journalism Review*, 24 November. Available online: https://archives.cjr.org/behind_the_news/is_this_the_golden_age_of_podc_1.php?page=all#sthash.BZ13l2b8.dpuf (accessed 12 February 2015).

Stepanyan, K., G. Gkotsis, H. Kalb, Y. Kim, A. Cristea, M. Joy, S. Ross (2012). 'Blogs as Objects of Preservation: Advancing the Discussion on Significant Properties'. In *Proceedings of the 9th International Conference on Preservation of Digital Objects (iPress 2012)*, 218–224. Toronto, ON: Digital Curation Insititute. Available online: https://www.researchgate.net/publication/233935527_Blogs_as_Objects_of_Preservation_Advancing_the_Discussion_on_Significant_Properties (accessed 13 June 2018).

Sterne, J. (2003). *The Audible Past: Cultural Origins of Sound Reproduction*. Durham, NC: Duke University Press.

Sterne, J., J. W. Morris, M. Baker, and A. Moscote Freire (2008). 'The Politics of Podcasting'. *Fibreculture* (13). Available online: http://thirteen.fibreculturejournal.org/fcj-087-the-politics-of-podcasting/ (accessed 10 July 2020).

Stoler, A. L. (2009). *Along the Archival Grain: Epistemic Anxieties and Colonial Common Sense*. Princeton, NJ: Princeton University Press.

Sullivan, P. (2017). 'Hanna-Barbera's Cacophony: Distant Listening and TV Sound'. Presented at the *Great Lakes Association for Sound Studies 2017 Fall Conference*, University of Chicago, Chicago, IL.

Verma, N. (2018). 'Speaking to You Wherever You Are: Streaming, Podcasting, and Audible's House Sound'. Presented at the *Society for Cinema and Media Studies*, Toronto, ON.

51

The Earview as a Border Epistemology: An Analytical and Pedagogical Proposition for Design

Pedro J. S. Vieira de Oliveira

One of the main ideas I deal with in my research is that design, both as a field of study and a set of material practices, privileges modern, colonial, and colonizing histories and stories of listening. This privileging, embedded in the affordances of designed artefacts, validates and perpetuates the hierarchization of bodies as well as the relegation of certain bodies to a state of otherness within and beyond the auditory space. I use the term 'auditory space' here to refer to the sets of actions, performances, materials, and means which shape and are shaped by experiences of bodily engagement with vibrations. That I use the term 'auditory' rather than 'acoustic' is not at all arbitrary; rather, I understand that the spaces of negotiation amongst bodies and phenomena imply different (and uneven by design) agencies and performances. The design of an auditory space implies delegating, assigning, and codifying of different agencies over sonic phenomena. I understand the auditory space to be a transient space, constituted by different engagements and levels of said engagements among vibrations – not constrained by those perceived by the ear – bodies – not predicated by the ability to hear – and their performances within said space.

Design is to be understood as an inherent condition of being human and as such interacting – in different forms and with different agencies – with the materiality of lived and living phenomena. In that sense my thinking is much aligned with scholar Anne-Marie Willis, who argues for an understanding of design as an ontological force. For her, design inevitably engages with four different aspects of mattering, that is, (1) the object itself; (2) the process through which this material object comes into being; (3) the agency of those implicated in this endeavour; and (4) the consequences of said endeavour to society (Willis 2006). 'There is never a beginning or end of design', Willis contends, because 'once the comfortable fiction of an originary human agent evaporates, the inscriptive power of the designed is revealed and stands naked' (Willis 2006: 95). Thus design is a set of ontological actions which may, but not necessarily must, include designing as it is usually understood; as such, this ontological force is inextricably implicated on configuring the world, and this configuring extends well beyond and much before the industrial production of artefacts. In other words, design is implicated in what informs our human condition as such, and cannot be anything but product and producer of contingent material practices. Concomitantly, design acts on the tension between different stories of listening, of bodily and material engagements with sound; it shapes and produces the auditory space and its power relationships. This entanglement is what I have named the *earview*.[1]

The coalescing of a visual marker – view – with an apparatus that performs hearing – the ear takes inspiration from the theory of the gaze, much discussed both in race studies and feminist scholarship.[2] The gaze is produced not only by patriarchal but also by racist ideas and ideals towards othered bodies; in an analysis of 'human zoos' – from the early twentieth century to 2005 Augsburg, Germany – scholar Obioma Nnaemeka describes how the separation between colonizers and colonized – slave owners and African enslaved peoples – was deliberately fostered by the European subject so as to produce enough distance for establishing a 'viewpoint' of the black body as alien, deprived of humanity. 'The Europeans', she writes, 'are not interested in what the Africans "think" (they are not credited with the capacity to think/reason) but "on what they do" and "how they look"' (Nnaemeka 2005: 95). The European gaze directly evokes the coloniality of looking, the notion that certain bodies are made to be looked at from a distance that vouches for their dehumanization.

The idea of the gaze makes it clear that who looks matters. For the earview, it matters who listens, and who defines what and how is to be listened to. While seemingly trivial, the coalescing of these ideas into a concept such as this is a useful analytical tool for design, because it re-centres the field's concern towards not only what design designs but also the scope of actions, engagements, and performances that emerge from the act of designing, including those that shape and form the auditory space. Instead of simply transposing the gaze, then, I use the term earview to demonstrate the narratives conveyed and sustained by an engagement with and around designed objects and systems – that is, their sonic affordances. The earview understands sonic affordances as a process of listening forward as well as backward, projected to the listened phenomena and towards the enunciator of the listening act. This multidirectional process highlights the co-constitutive aspect of sonic affordances – and thus configures the earview as a plural entity, composed of phenomena at the borders of what can be understood as 'objective' and 'subjective' – though never regarding them as fundamentally disconnected or separated from one another. I began a rough outline of the earview in a past publication (Vieira de Oliveira 2016); yet this idea deserves to be further unravelled to encompass what design designs back into the auditory space, and how it does so.

The earview can be thought of as the amalgamation and negotiation of three aspects: first, a rupture with Eurocentric notions of listening as the idealized and quasi-spiritual counterpoint to seeing, albeit not eschewing subjectivity in favour of a modern/enlightened rationalization of listening but instead accounting for other forms of 'objectivity' that break from modern and colonial histories of audition. Secondly, a bodily engagement with the sonic affordances of designed artefacts, systems, procedures, and spaces – and the narratives they convey. Third, personal, idiosyncratic, and situated forms of storytelling that emerge from experiences of listening, never divorcing them from their role in co-constituting reality. These aspects are not separate but rather always take place interrelated to one another; therefore, in outlining them in text, overlaps may occur.

Alternative modernities yield alternative histories of audition

Jonathan Sterne has famously argued that the canonical and almost dogmatic separation of the aural from the visual, ear from eye, as well as their dichotomous, binary oppositional framework configures a form of ideology that constrains studies into aural culture to the position of the 'other' of the visual. Sterne calls this framework the 'audiovisual litany'; for him, the litany relies on physical and psychological oppositions to build 'a cultural theory of the senses' (Sterne 2003: 15). Within the 'audiovisual litany', the separation between the ear and the eye as sensory dichotomies flattens the understanding of listening and hearing as being practices of a similar nature – an assumption that, he maintains, is erroneous at best. For him, listening is 'a directed, learned activity […] not simply reducible to hearing' (19). From the moment the ear is trained into rationalizing acoustic phenomena, listening becomes a technique of hearing, a 'set of practical orientations' (93) that manage and instrumentalize the auditory space.

While for Sterne listening should be considered as much a fundamental constitutive of modernity as vision, his consideration relies on the discourse of a technicized, rational objectivity. His focus on a single history of listening in modernity is a constraint of his own research and properly acknowledged as such – by focusing on what he names the 'hearing elites', who are able to produce enough historical (i.e. technical-scientific) documentation to sustain his research (Sterne 2003: 28). In that sense, Sterne inevitably sanctions certain practices of listening – and the bodies that convey them – in detriment of other bodies that inevitably fall outside these (white, male-centric, Anglo/European) elites; in understanding that such technological 'audile techniques' are enacted solely within the confines of the colonial/modern framework, Sterne limits the very notion of what audile techniques might in fact be to the privatization and capitalization of the acoustic space qua narratives of development and progress.

Colombian sound scholar Ana María Ochoa Gautier provides an insight on alternative histories of audition happening within alternative modernities. Focusing on nineteenth-century, colonial Colombia, she thoroughly engages with entangled yet oftentimes conflicting listening modes across different modernities. For her, aurality 'is central to the constitution of ideas about Latin American nature and culture [and] also imply different ecologies of acoustics' (Gautier 2014: 75). In one of her case studies, she compares different historical testimonies of the sounds produced by the *bogas* – boat rowers that transported contraband along the Magdalena River. For European 'explorers'[3] such as Alexander von Humboldt, colonial/modern audile techniques were used to privilege sounds understood as 'from the nature' and in turn dismiss the conversations from his (racialized) subalterns. However, he would waive such auditory techniques when certain conversations could potentially offer him insights into 'traditional knowledge' (68). In writing about the bogas, he would describe their sounds as 'barbarous, lustful, ululating and angry shouting, which is sometimes like a lament and sometimes joyful' (von Humboldt quoted in Gautier 2014: 32) or resemble them to the howling of dogs. Similarly, other colonial travellers were seemingly 'shocked' by the bogas' 'lack of civilization' and 'blasphemous' practices of praying simultaneously to Catholic saints, Afro-Caribbean entities, and 'many more of their own invention' (Cochrane quoted in Gautier 2014: 38) in a mishmash of Spanish, Latin, 'Lengua Franca' and other indigenous languages (Holton quoted in Gautier 2014: 39).

Often engaging in such a syncretism that is as religious as it is linguistic, the sounds of the bogas, Ochoa Gautier argues, are 'a remix practice that also involve[s] vocables and the acoustic incorporation of the sounds of natural entities around them, all in the rhythmic regularity of vocalization for labor that involve[s] repetitive movement' (Gautier 2014: 41). In understanding the bogas' chantings and vocalizations as structured, learned techniques of (aural) navigation and communication with both human and nonhuman entities, Ochoa Gautier subverts the dehumanizing logic of the colonial European ear to reposition the bogas as the centre of an 'audile technique' that has in 'envoicing [...] multiplicity', that is, mimicking and speaking different languages, both human and nonhuman, a powerful mode of 'transformational [...] becoming' (65). In other words, she envisions this form of nonhuman embodiment to be a conscious, objective technique, a mode of social/ interspecies communication, which understands every human and animal language and voice to be a potentiality in and by itself rather than the reductionist, racialized view of a 'lower condition of animality' (63). It is a form of redistribution of acoustic perception to different corporealities and materialities of sound which defy and often negate the hegemonic, colonial/modern ear.

Sonic affordances

Listening is a subjective, yet cogent mode of attending to the world. A methodological proposition such as Lydia French's 'differential listening' helps re-frame listening outside of its constraining, rational logics of 'technique'. Her writing calls for distinct accounts of listening – those that move beyond instrumental hearing and as such refrain from a direct connection with the supposed empirical materialism of auditory phenomena. Drawing tangentially from Chela Sandoval's 'differential oppositional consciousness', French argues for a positioned and situated research endeavour that is able to encompass alternative, plural modes of listening; these modes reconfigure the scope of what we acknowledge as 'real', for they are a form of a 'border epistemology that understands temporal movement differently' (French 2014), one which displaces memory and affect by making use of sound reproduction technologies. In other words, she calls for a decolonizing mode of listening that embraces the ambiguity of sound reproduction, and its capability of disconnecting time and event, thus positioned at the constant becoming of the present. A subjective and interiorized method of listening – clairaudience – is a method for breaking with the 'interior/exterior' separation fostered by traditional histories of listening constrained to the artefacts they enact and are enacted from – in other words, Sterne's 'audile techniques'.

The methodology proposed by French acknowledges sound reproduction technologies to possess the ability to cast memories that are intrinsically bound to their political and cultural loci. More than that, she argues that from the moment these technologies actualize said memories in the now of the listening experience, they also displace them in time. Past and future are reconstructed from the present by clairaudient hearings of othered voices or, in her words, by 'alternative histories [which] live in the transitive play between sounds clairaudiently remembered and those reproduced through sonic media' (French 2014). In applying a differential listening approach, one is able to pinpoint differences and dissonances between one's own positionality and that which is actualized in the materiality of listening. This mode of listening is thus constantly reconfiguring sonic reality as a space inherently plural and ambiguous, one which accommodates both dominant practices – albeit stripping them down from their dominative aspect – and their 'outcast' counterparts, equating both to the same discursive and epistemological relevance. French's subjectivities possess, perform, and negotiate distinct agencies within sonic phenomena rather than conforming to a place of silent (or silenced) otherness.

In regarding Sterne's histories of listening-for as being histories of highly specialized techniques, French directly interrogates contemporary practices of 'pure listening' or 'better hearing' by technological delegation. These practices, she maintains, seek to 'render the machine's noise as ancillary and exterior to faithful sonic reproduction' (French 2014) with the desire to accommodate to normative soundscapes, and thus authorizes certain practices in detriment of others. For her, the search for familiarity (an expanded notion of fidelity) in listening is also reproduced in social relations that emerge out of these practices, as a way to cast out the 'peripheral noise' of otherness. She contends that Sterne's history of listening in modernity is but a history among many, and a practice that is oppositional and transformative should listen-for non-universalizing ways of hearing that challenges dominant discourses of what is said to be heard. Thus, a 'differential listening' practice seeks to identify these prevalent discourses of listening and dismantle them; or, in her words, to '[denaturalise] the historical trajectory of the ideological premises on which fidelity is founded, suggesting that there may exist alternative histories of audition in the age of sonic reproduction' (French 2014).

Border thinking

To think of sound and listening beyond the framework of coloniality is an exercise in border thinking. As both a place from where one inquires as it is a form of counter-hegemonic epistemology, border thinking relies on the idea of the borderlands as a transitional space with no fixed point of origin. Walter Mignolo proposes border thinking as 'not a matter of "thinking the other" but rather, the "other thinking"' (Mignolo quoted in Kalantidou and Fry 2014: 173). For him, it is a way to 'think otherwise' that does not necessarily connect or refer back to Greek, French, British, or German ontologies and epistemologies; rather, it refers to the 'colonial wounds' that emerge from the erased subjectivities of subaltern bodies and the knowledge that was left in the aftermath of colonization (Mignolo 2000: 66–69).

The image of the border, which Mignolo draws his concept from, has its genesis in a direct encounter with the borderlands as a transient and contradictory space. Latina feminist writer Gloria Anzaldúa writes from the perspective of a border subject herself. First published in 1987, *Borderlands/La Frontera: The New Mestiza* is an active call for the decolonization of knowledge, spirit, and the self, exactly by seeing them as inextricably related, an entanglement of bodily knowledge, political identity, and ancestral reconciliation. For Anzaldúa, the border is not only the metaphor of how knowledge is positioned but also its very physical location. As a Tejana – that is, a person whose ancestry comes from Mexican instead of US-American Texas – Anzaldúa did not move over borders but had the borders moved around her (Anzaldúa 2007: 28; Mignolo 2000: 72). In articulating her own multiple identities – queer, indigenous, Mexican, US American, Chicana – her thinking is inevitably displaced into multiple perceptions of reality, 'forced to live in the interface [...] to become adept at switching modes' (Anzaldúa 2007: 59). Anzaldúa reminds us that a 'border culture' determines the places that 'distinguish us from them' (25); there is little we can do but to turn this border culture on its head, and use the border as the locus of our enunciation. If the modern/colonial reality categorizes her very existence into a dichotomous narrative, her method for emancipation is to nurture a tolerance for the ambiguous and the contradictory (101). Border subjects 'cambian en punto de referencia [...] locating her/himself in this border lugar [the Borderlands she speaks of and from], tearing apart and then rebuilding the place itself' (Anzaldúa 2015: 49).[4] They are thus forced to think of the world through dichotomous philosophies, not opposing them directly nor dwelling on either side, but rather navigating with and within their very differences (Mignolo 2000: 85).

Thinking from the border is a not a replacement of the dominant discourses of enunciation with subaltern ones, for this would subscribe to the same logic that governs epistemological colonialism (Kalantidou and Fry 2014). Instead, it is a promotion of alternatives that must be situated and ever 'universally marginal, fragmentary and unachieved' (Mignolo 2000: 68). To be able to think beyond the subject–object dualism, to act on this 'saber', is what Anzaldúa calls a 'conocimiento' (Anzaldúa 2015: 119) – or in Mignolo's words, border thinking. Moreover, border thinking assumes the locus of thinking – where one thinks from – as an important point of departure; a place of ontological design that is never fixed (Kalantidou and Fry 2014: 6). The ability to balance epistemologies by mapping them to their ontological loci allows one to listen to the cracks and pops that lie between and underneath them, and, from there, extract new ways of knowing and sensing the world. A border enunciation experiences reality through multiple perspectives, shifting from one to the other while never exactly fully departing from them (Anzaldúa 2015: 127).

Mignolo proposes then, based on Maghreb philosopher Abdelkebir Khatibi's 'an other thinking' (*pensée-autre*), a double form of critique (akin to, for example, W. E. B. DuBois's 'double consciousness') seeking to emancipate its perspective from both Western and non-Western thought, belonging instead to the border itself. It is exactly this negotiation of different forms, that is, how they cannot be compared to one another nor cancel each other out but rather be seen as juxtapositions – what Mignolo (2000: 84) calls an 'irreducible difference' that paves the way for a decolonization of Western logocentrism. Similarly, Anzaldúa reminds us that 'it's not enough to denounce the culture's old account – you must provide narratives that embody alternative potentials [...] we need a more expansive conocimiento. The new stories must partially come from outside the system of ruling powers' (Anzaldúa 2015: 140).

The earview understands the locus of auditory enunciation as a site of confrontational orientations; on the one hand, the predominance of an auditory culture embedded in material and discursive practices of designing, and on the other hand, practices of misusing, insolent listening, and relocating auditory practices that enunciate different earviews, which complicate the boundaries of any form of normative listening practice. To understand how design articulates the earview becomes, then, a question of understanding orientation; of attuning listening to certain things rather than others (Ahmed 2010). In casting attention to how these agencies, performativities and things may connect, they are regarded as orbiting around a same 'gravitational centre', which might not be observable in the first place but reveals its connections when it is 'pulled up'. As feminist scholar Sara Ahmed contends, 'we touch things and are touched by things [...] bodies as well as objects take shape through being orientated toward each other, an orientation that may be experienced as the cohabitation or sharing of space' (Ahmed 2010: 245). Alas, listening is a bodily engagement with vibration – a haptic phenomenon (Kassabian 2013). Inferring the entanglement of certain bodies, discourses, and objects by drawing them closer, or taking one element of these sets as a 'nodal point', reveals the underlying structures and systems that make these agents come into existence and come together as a normative earview. The observation or a normative earview, in turn, creates the possibility of other, displaced earviews. It is by shifting these entanglements from background to foreground and vice versa, picking up loose threads and intervening in them, that we begin to gain an understanding of the world from different orientations (Ahmed 2010); other sets of temporary connections may emerge, revealing otherwise silenced earviews.

The earview as a border epistemology: Yarn Sessions

The earview was developed not only as an analytical tool but also a pedagogical framework. My design practice makes extensive use of open-ended, horizontal, and collective narratives to make sense of political problems, often starting from a clearly defined, tangible phenomenon. From 2014 to 2016 I worked closely with the question of racialized police violence in Brazil, in particular violence that made use of, or instrumentalized sounds and music – from the outlawing of jukeboxes in Rio to the deployment of sound bombs in São Paulo, to name a few. To understand the accountability and implications of design as that which configures material practices around these phenomena I created a semi-fictional story, narrated predominantly through soundscape compositions and field recordings (Vieira de Oliveira 2019). In setting the stage for a discussion around sensitive topics to happen through sound, and through storytelling, participants of the projects were encouraged to fabulate, situate themselves, and try to make sense of what they were listening to based mostly on what they could fathom from listening.

This hybrid format of design project, workshop, and storytelling session was named 'Yarn Sessions' and developed together with Luiza Prado de O. Martins. The name comes from the verb 'to yarn', which is a way of telling a story without necessarily achieving an endpoint but also from its Portuguese translation 'tricotar', which in Brazil can also mean a long conversation session (Vieira de Oliveira and Prado de O. Martins 2019). In total, five sessions of the same project were held in four different countries (Brazil, Germany, Switzerland, and Croatia) and welcomed participants from design, arts, architecture, cultural studies, and science and technology studies. The sessions were always free of charge and promoted as a designerly inquiry on sound and violence – which prompted participants to come with already a curiosity and/or prior knowledge on the subject. Sessions lasted from 3 to 6 hours in length, and merged conversation with more focused tasks.

In these sessions, participants were asked to take a position in which they deal with their own accountability as privileged researchers, trying not to speak for others, and not to research from the position of others; rather, they were encouraged to stand together and in phase with the stories, anecdotes, sonic fictions (Eshun 1998; Schulze 2013; Vieira de Oliveira 2016), and struggles for which the Yarn Sessions were interlocutors. The proposal for the earview suggests a form of border thinking as the locus of enunciation for decolonizing practices. Border epistemologies highlight the controversial, dichotomous, and conflicting ontological mechanisms imposed by coloniality; however, instead of negating these conflicts, border thinking embraces them as the point of departure of a research inquiry. From the moment we understand that the borders of the auditory space are enforced to cross certain bodies rather than the opposite, and that normative modes of listening reenact this delimitation of borders through design, we require border thinking to allow us to confront these borders.

Together in these sessions we attended to the materiality of sound, albeit never divorced from its cultural implications – they knew, from the beginning, that these sounds were coming from stories and histories of police brutality. The sounds, however, seldom conveyed violence, relying instead on crafting an open scenario from which fabulations could emerge. Participants attempted to plot their own point of listening, and to create a brief cultural history of the objects, voices, and spaces present in the soundscapes. Some of them developed short theatre scenes to subjectively convey ideas of rhythm and consonance; others crafted telltales of revolution and civil disobedience. Attuning to the multiple layers of the earview encouraged them to think of the recording apparatus, of the software used to craft the soundscapes and collages, and also the decisions taken by the recordist to focus on certain sounds rather than others. In the end, most of the sessions ended up yielding a broad ecology of how listening is able to traverse multiple layers of meaning which are not disconnected from one another but instead craft together a limited glimpse of a cultural, social, and political composition which is conveyed by – but never limited to – sounds themselves.

Conclusion

In my work I take inspiration from other histories and stories of listening as part of a proposal for delinking from the Eurocentrism implicit and complicit in designerly language. I constantly negotiate my relationship of subject and object within this study, my own shifting perspectives and performed identities. The colonized body, Anzaldúa contends, is a shapeshifter in nature, living and speaking from the borders of different languages, identities, and knowledges. Thus I attempt to make these shapeshifting processes as visible as possible, synthesizing the 'dualities, contradictions, and perspectives from these different selves and worlds'; as a border subject, I also cross 'other mundos' and 'speak in tongues' (Anzaldúa 2015: 3), thus inhabiting a space of transitions and intersectionalities.

Earlier on I defined the earview as 'stories that orbit around design language but that start, evolve, and end at the ear [which help to] craft narratives that theorize and produce new knowledge through listening practices' (Vieira de Oliveira 2016: 51). Hence, I already positioned the earview as a rough form of a locus of auditory enunciation – a proposition which I expanded here. In rough terms, the earview should be understood as a form of storytelling, not constrained to fabulation and speculation only, but also assuming that these modern/colonial techniques of hearing, as well as the differential listening modes discussed above also weave a specific narrative with a specific point of listening. What I argue is that design privileges certain narratives constrained by designed sonic affordances, enabling the modern/colonial earview and erasing or obliterating others.

Normative listening practices trace the borders of the auditory space within which we determine the scope of sonic reality. Privileging thinking through sound at the border opens up for understanding the earview falling outside the sonic affordances of design as being a proper facultad (Anzaldúa 2007), enabling counter-hegemonic techniques with which to locate, interrogate, and negotiate the borders of the auditory space. Thinking through sound comes to the aid of the work of design by exposing the normative earview, while concomitantly making ground for other histories of listening – both those beyond rational-scientific techniques of hearing as well as those who emerge as practices of resistance. Thus, in my work I propose a radical shift in the locus of enunciation to that of the border between conflicting earviews. Only so we can unwrap and scrutinize the agencies and accountabilities of design in enforcing a segregation amongst divergent earviews, the domestication of difference towards consensual pluralism, and the privileging of hegemonic, violent narratives of sound and listening.

Notes

1. My idea for the earview was inspired by cultural theorist Kodwo Eshun's 'rearview hearing', albeit not with the same meaning as he conveys in his book *More Brilliant Than The Sun: Adventures in Sonic Fiction* (1998). With rearview hearing, Eshun was more concerned with the insistence on cultural theory and musicology to 'hear backwards', for instance, by calling beat-making hardware a 'drum machine' instead of a 'rhythm synthesizer' (Eshun 1998: 78–79, 186).
2. See, for example, Mulvey 1975.
3. I deliberately use the word 'explorer' with quotes here; 'exploration' of what was perceived to be 'untamed land' by European travellers laid grounds for the understanding of indigenous peoples and cultures to be in a state of 'primitiveness' which needed to be properly catalogued, described, and sanctioned by the curious 'explorer'.
4. Anzaldúa writes in 'Spanglish', switching back and forth between Spanish and English arbitrarily and often multiple times within the same sentence. This deliberate attitude highlights her own border thinking, a stream-of-consciousness which switches between languages at will and which she does not intend to tame but rather to embrace and emphasize in and through her writing. Hence, in her work – and in particular *Light in the Dark/Luz en lo Oscuro* (2015) – Spanish words and terms are seldom italicized, and translations to English are only provided when done by Anzaldúa herself. I deliberately chose to keep these faithful to the original text.

References

Ahmed, S. (2010). 'Orientations Matter'. In D. H. Coole and S. Frost(eds), *New Materialisms: Ontology, Agency, and* Politics, 234–258. Durham, NC: Duke University Press.

Anzaldúa, G. (2007). *Borderlands/La Frontera: The New Mestiza*. 3rd edition. San Francisco: Aunt Lute Books.

Anzaldúa, G. (2015). *Light in the Dark/Luz En Lo Oscuro: Rewriting Identity, Spirituality, Reality*. Durham, NC: Duke University Press.

Eshun, K. (1998). *More Brilliant Than the Sun: Adventures in Sonic Fiction*. London: Quartet Books.

French, L. (2014). 'Chican@ Literature of Differential Listening'. *Interference Journal* 4. Available online: http://www.interferencejournal.org/chican-literature-of-differential-listening/ (accessed 18 July 2020).

Gautier, A. M. O. (2014). *Aurality: Listening and Knowledge in Nineteenth-Century Colombia*. Durham, NC: Duke University Press.

Kalantidou, E. and T. Fry (eds) (2014). *Design in the Borderlands*. 1st edition. New York: Routledge.

Kassabian, A. (2013). *Ubiquitous Listening: Affect, Attention, and Distributed Subjectivity*. 1st edition. Berkeley, CA: University of California Press.

Mignolo, W. D. (2000). *Local Histories/Global Designs*. Princeton, NJ: Princeton University Press.

Mulvey, L. (1975). 'Visual Pleasure and Narrative Cinema'. *Screen* 16: 6–18. https://doi.org/10.1093/screen/16.3.6.

Nnaemeka, O. (2005). 'Bodies that Don't Matter: Black Bodies and the European Gaze'. In M. Eggers, G. Kilomba, P.Pesche, and S. Arndt (eds), *Mythen, Masken Und Subjekte: Kritische Weissseinsforschung in Deutschland*, 90–104. Münster: Unrast.

Schulze, H. (2013). 'Adventures in Sonic Fiction: A Heuristic for Sound Studies'. *Journal of Sonic Studies* 4.

Sterne, J. (2003). *The Audible Past: Cultural Origins of Sound Reproduction*. Durham, NC: Duke University Press.

Vieira de Oliveira, P. J. S. (2016). 'Design at the Earview: Decolonizing Speculative Design through Sonic Fiction'. *Design Issues* 32: 43–52.

Vieira de Oliveira, P. J. S., and L. Prado de O. Martins(2019). 'Designer/Shapeshifter: A Decolonising Redirection for Speculative and Critical Design'. In T. Fisher and L. Gamman (eds), *Tricky Design: The Ethics of Things*, 103–114. London: Bloomsbury.

Vieira de Oliveira, Pedro (2019) 'Algerinha Vive'. Soundcloud. Available online: https://soundcloud.com/ppedrooliveiraa/algerinha-vive-mixtape (accessed 18 July 2020).

Willis, A.-M. (2006). 'Ontological Designing'. *Design Philosophy Papers* 4: 69–92. https://doi.org/10.2752/144871306X13966268131514.

52
Hacking Composition: Dialogues with Musical Machines
Ezra J. Teboul

Electronic music was invented before it was composed (Teboul 2017). Hacking as a mode of invention is intimately related to the continual development of the genre. Yet, there are few critical accounts of hacking as *equivalent* to composition, and the implications thereof. In this chapter I will present examples which illustrate hacking as a shape-shifting method of instrument design, and discuss its privileged ideological and material connection with experimental music. How are material decisions equivalent to musical decisions and what does looking at hacking as a process reveal about both?

Electromechanical hacks

Experimental turntablist Christian Marclay and his 1985 *Record Without A Cover* offers a starting point. Each vinyl copy of this record is exposed to the everyday dirt and micro-violences of record selling, owning, and playing. They evolve independently to reward each owner with a progressively more unique version, augmented by the pop and crunch of each new damage. They slowly lose material, via new scratches, and gain material, from dust and particulates (Thompson 2017: 66). A new conception of additive and subtractive synthesis as negotiated by the unique material realities of these objects and listeners.

Marclay brings attention to the temporality of the vinyl record as medium of music. *All* vinyl slowly warps and decays, even when it is sold *with* a cover. The very action of a stylus reading the groove scrapes away some of any disc's plastic with every revolution. But more than simply acknowledging that, he takes one aspect of what makes the vinyl record special as a medium to fix sound – that it is a slab of plastic slowly spinning to its own noisy death – and turns that into the focus of the work.

This is done by taking away the sleeve traditionally associated with the medium, but also by the way in which traditional musical material and sound effects are complemented by sections of layered recordings of silent parts of other records and *their* vinyl noise

(Thompson 2017: 65). More than a variant of John Cage's silent classic 4'33", the interplay of the generative crackle and various levels of sonic to musical material makes the piece an exposition of what Marclay was interested in and had access to, seen through the eye of the vinyl record as a temporarily living medium. In other words, *Record Without a Cover* is the record of a dialogue between the materiality of this specific sound technology, its potential as a musically generative device, and Marclay's ideas as an artist. Through the process of care (or lack thereof) he invites the costumer of these records and their respective environment into this dialogue.

Record Without A Cover is a significant example of hacking in a musical context because it is uniquely effective in conveying both Marclay's expertise in creating poetically generative experiences while creating the opportunity of a rewarding audience participation.

'Hackers create the possibility of new things entering the world. Not always great things, or even good things, but new things' (Wark 2004: [004]). Marclay's piece creates a new musical work out of an old technology: by challenging the traditional mode of consumption of what was then the principal medium for music playing, he crystallizes both his artistic vision and the limitations of our understanding of the medium, while offering room for personal salvation from our blindness. Fittingly, here, that salvation can be purchased: what better way to acknowledge the dual nature of the record as both an artefact of art and a commodity? The technical bar for innovation is also rendered fairly accessible: all that was required here was to sell the *Record*, without a cover.

Record is actively revealed by the electromechanical process of playing the physical object, which involves the use of the record cartridge, a preamplifier, amplifier, and speakers. The process that *makes* this piece is purely mechanical, but it is mostly imperceptible to us as more than a slab of plastic until its enactment. In this sense, the hack implemented in the slow personalization of each copy of the record via wear and tear is disconnected from our experience of the hack. Therefore, *Record* helps us imagine the breadth of possible mediums and experiences in hacking composition: what other *stuff* can we engage with as composers, and in what ways have these engagements already happened? How can informed technical experimentation, whether we call it hacking, tinkering, bricolage, do it yourself, or plain making, reveal both something about the 'music *implicit* in technology' (Collins 2007: 47, emphasis in the original) and our relationship to it?

Electromagnetic hacks

Christina Kubisch's *Electrical Walks* series, begun in 2003 and sporadically reiterated in different cities to this day, consists of a walk through urban spaces with the help of custom built headphones. These convert the electromagnetic energy into sound waves (Cox 2006). Kubisch provides a map of interesting sounding locations she has identified in advance. It is then up to the participant to explore this somewhat indeterminate experience and form their own composition, as they explore the space. Here, the medium is the electromagnetic spectrum, but like *Record* the piece is also participative. By combining the familiar device of headphones with the more unusual property of inductors to transduce electromagnetic waves into electrical signals and framing the resulting device as the enabler of a new form of soundwalk, Kubisch's system is an effective techno-artistic development.

The hack is once more only perceptible as it is enacted, using a static device from components built by others to mediate an active experience. However, unlike *Record*, *Electrical Walks* reveal almost more about the technology that surrounds it than that which makes it. Kubisch's work moves us towards the electromagnetic spectrum as a medium, reminding us that all electronic devices which have come to define modern human life, as well as some natural geomagnetic events such as storms, emit electromagnetic waves.

And yet the agency of the listener is not just enabled through the act of walking. *Electrical Walks* teaches us about the music implicit in those waves that happen to occur at audio frequencies, and therefore, we also engage with the limits of our perception. Electromagnetic waves occur at various frequencies below and over the audible frequency range, and they may be converted to sound waves by Kubisch's headphones; regardless we may not hear them without additional signal processing. If *Record* does in a way produce a record of dust and action as recorded through scratches, that record is somewhat obscured and unrecoverable: most pops on a record are indistinguishable from another. In *Walks* what the hack says about us is clear: we as a species make a lot of electromagnetic waves that our natural bodies can't acknowledge directly. This hack is an extension of perception that enables us to engage with those signals, those devices, and with the unique boundaries of each person's hearing.

Electrical Walks is somewhat less accessible than *Record*: custom headphones need to be manufactured, and their use requires walking, if not repeated walking. Where *Record* connected the large-scale concepts of manufacturing and consumption with the personal lives of record buyers, *Electrical Walks* connects us with our local technological reality through this inhuman medium of electromagnetic fields. Furthermore, where *Record*'s process was almost visible and permanent, the underlying mechanics of *Electrical Walks* is somewhat more ethereal. With each medium come nuances in clarity and scale between the hack (technical decisions) and its results (musical consequences).

Record and *Electrical Walks* have introduced two mediums with which to engage through musical hacking: the electromechanical and the electromagnetic. They have helped me identify three key questions: How can hacking help reveal Collins's 'music *implicit* in technology?' What can that say of our relation to those devices and associated technologies? Since a hack is so deeply related to the material nature of the system at hand, what is the unique connection between the technical act enabled by hacking and its musical consequences?

Electrical and cultural hacks

The no-input mixer technique is an interesting case where the concept of the recording studio as instrument is reduced to its most literal and compact form. By plugging in outputs for channels of an audio mixer back into its inputs, one can create feedback loops. Most channels on consumer grade mixers include equalization or gain stages which can produce harmonic and inharmonic distortions when overdriven: by playing with these, one can tune these feedback loops to specific pitches and timbres which heavily depend on the characteristics of that particular mixer. Adding additional signal processing devices in the feedback loop can lead to complex timbral, melodic, and rhythmic structures (Thompson 2017: 148). The composer and performer David Tudor experimented with variations on this arrangement as early as 1967 to 1968 (Nakai 2016: 137, 267). The behaviour of the system as revelatory of the components placed in the feedback loop is the focus of the 1978 piece *Star Networks At The Singing Point* by Ralph Jones (2004), while Marko Ciciliani developed a composition and performance technique of no-input mixing through the 1990s and early 2000s (see *Mask*, composed in 2002).[1] Japanese performer and composer Toshimaru Nakamura, who has released *No Input Mixing Board* records since 2000, comments on the system's behaviour:

> You can't totally control no-input music because it's all about feedback. Things like turning the tuning knob, even by one millimeter, make a big difference to the sound … It's very hard to control it. The slightest thing can change the sound. It's unpredictable and uncontrollable. Which makes it challenging. But, in a sense, it's because of the challenges that I play it. I'm not interested in playing music that has no risk.
>
> (Nakamura, 'No-input, Sachiko M and Toshimaru Nakamura', [YouTube Video, 2003], quoted in Sanfilippo and Valle 2015: 15)

In this case, the musical work operates at a purely electrical level. There is no audience participation, but the structure of what an audience hears is directly related to the structure of the technical systems and its settings at a given time. Each loop has resonant frequencies produced by the system. In a sense, this is one of the most direct modes of sonifying the 'preferred' signals of a particular electrical system.

This system complicates the boundary between hack and adjustment or tweak. Some no-input mixer setups do not modify the device itself. However, by simply transgressing on the previously accepted use of the mixer as a tool, no-input mixing turns this tool into a rich and rewarding instrument. Nakamura and other no-input practitioners have effectively made a new system out of existing technology. Therefore, although the purely electrical medium of operation of the no-input mixer makes it a somewhat more confined practice in technical terms, the cultural transgression it represents suggests that the physical black box of the mixer does not need to be broken for a hack to occur; that perhaps some valuable hacks can in some specific technological circumstances be cultural. This is historically consonant with the history of technical and artistic experimentation with electronics: the post-Second World War boom of amateur radio was built on a series of cultural evolutions just as much as technical ones, and our modern society is effectively a techno-culture (Haring 2007).

Here this techno-cultural hack enables us to learn about the mixer as a system through the sounds it produces, as negotiated by Nakamura's, Tudor's, or Ciciliani's experienced hands. The dialogue between the mixer made 'unpredictable and uncontrollable' and its carer make the performance and recordings. Once again, the technologically charged act of plugging the mixer into itself allows an audience to perceive a negotiation between human and nonhumans through a poetic, rather than didactic, experience. The music produced could not be more implicit in the technology: it is literally the produce of its machine noise building onto itself. The relationship to its user is exposed in real time as we hear the player attempts to push the mixer in specific directions and listen to the mixer push back with its own suggestions.

Hacking the music out of technology: Scale, control, and autonomy with/in electronics

The complication of instrumental expertise brought on by semi-autonomous musical instruments is a central theme of do-it-yourself (DIY) systems and musical hacking. Revealing Collins's 'music *implicit* in technology' involves relinquishing some control to the system itself at the moment of performance (bringing attention to the *asynchronous* nature of expertise in some electronic music works). As such, Nakamura's expertise is not in playing famous tunes on his no-input mixing board, although one could technically modify the system to play feedback-based covers. Neither is it in his ability to play his own past work. Rather, it is in setting up and continually reorienting his system in ways compelling to himself and his audience. In that sense, part of his knowledge is expressed prior to a performance, in his effective setting up of a no-input mixer. For these musicians working with hacked instruments, there is a decoupling of expertise and performance. Quoting David Tudor:

> Electronic components & circuitry, observed as individual & unique rather than as servomechanisms, more & more reveal their personalities, directly related to the particular musician involved with them. The deeper this process of observation, the more the components seem to require & suggest their own musical ideas, arriving at that point of discovery, always incredible, where music is revealed from 'inside,' rather than from 'outside.'
> (Tudor 1976 quoted in Nakai 2016: 276)

Nakamura, Kubisch, and Marclay all pay attention to the individuality of components in order to let them develop and reveal their musical potential. These follow directly from their materiality, but are inextricably linked to the way these materialities were perceived by these artists. They show that hacking as the basis for musical work is not restricted to the acoustic, the electronic, the mechanical, or the magnetic but, rather, is defined by the messy and real-life technologies we may have access to and all the baggage they carry along because of that.

The composer Nicolas Collins, who worked with Tudor for several years as part of the latter's *Composers Inside Electronics* ensemble, developed *The Royal Touch* in 2014. *Touch* takes any circuit board (for Collins, it is sometimes a dead channel strip from a Neotek mixer taken out of its enclosure) and uses it as a complex collection of electronic components. Resistors, inductors, and capacitors do not need power to act as such, but transistors and integrated circuits act as unpredictable combinations of diodes when

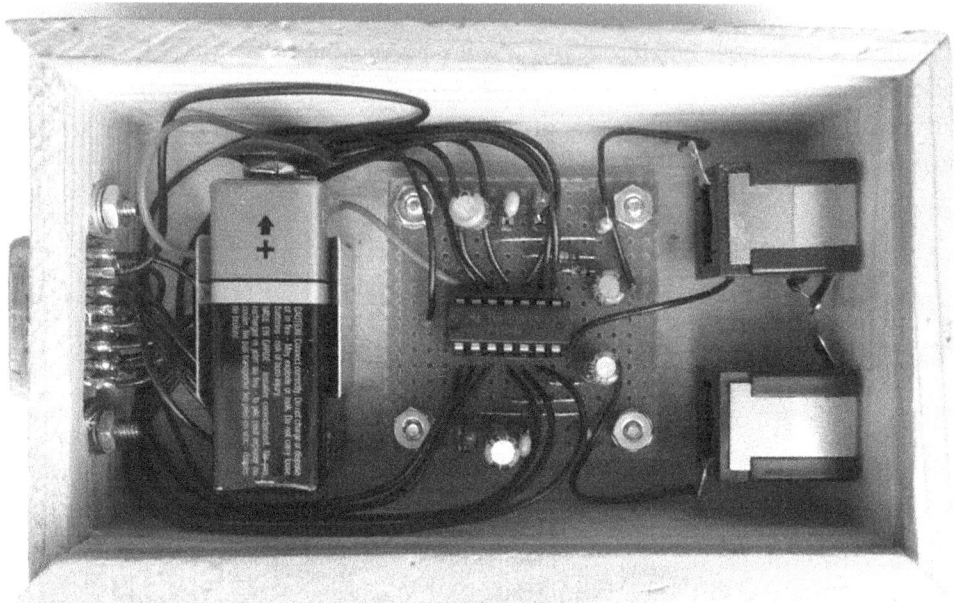

Figure 52.1 The inside of the oscillator box, containing one Hex Schmitt Trigger, 74C14 chip. The capacitors surrounding it set the range of frequencies produced by each of the six square wave oscillators, and the multipin connector on the left links to the ribbon connector used to 'probe' the dead circuit.

unpowered. A device containing six square wave oscillators based on the 74C14 CMOS chip (Figure 52.1) connects to the dead board by a ribbon connector terminated with lead fishing weights. The weights link the oscillator to a part of the circuit, and the erratic nature of the components in the path set the frequency of the oscillator within the range defined by a capacitor in the oscillator box (Figure 52.2). As the fishing weights are slowly nudged across the board, they come in and out of contact with various components, and the frequencies of each oscillator cut out or evolve accordingly. The final timbrally important processing step is a diode-based mixing arrangement which takes the already harmonically rich square waves and implements ring modulation style mixing for three oscillators at a time.[2]

The circuit board and the way in which this piece is performed – slowly moving lead beads – implies a degree of indeterminacy. In *The Royal Touch* each movement of a fishing weight also means a new variation on the actual circuit producing sound. Once again, the boundary between tweak and hack is unclear, although the identity and performability of the system is quite clear and straightforward. Like Nakamura, Collins explores the board for 'good' sounds, stays as still as possible when he has found some, or makes minuscule movements in search of variation, and sometimes waits for the circuit to charge and discharge back to silence or a less interesting steady state. The result is a chaotic, unpredictable exploration of a physical space mediated by a revelatory circuit. *The Royal Touch* operates an electrification of exploration at a smaller scale from Kubisch's

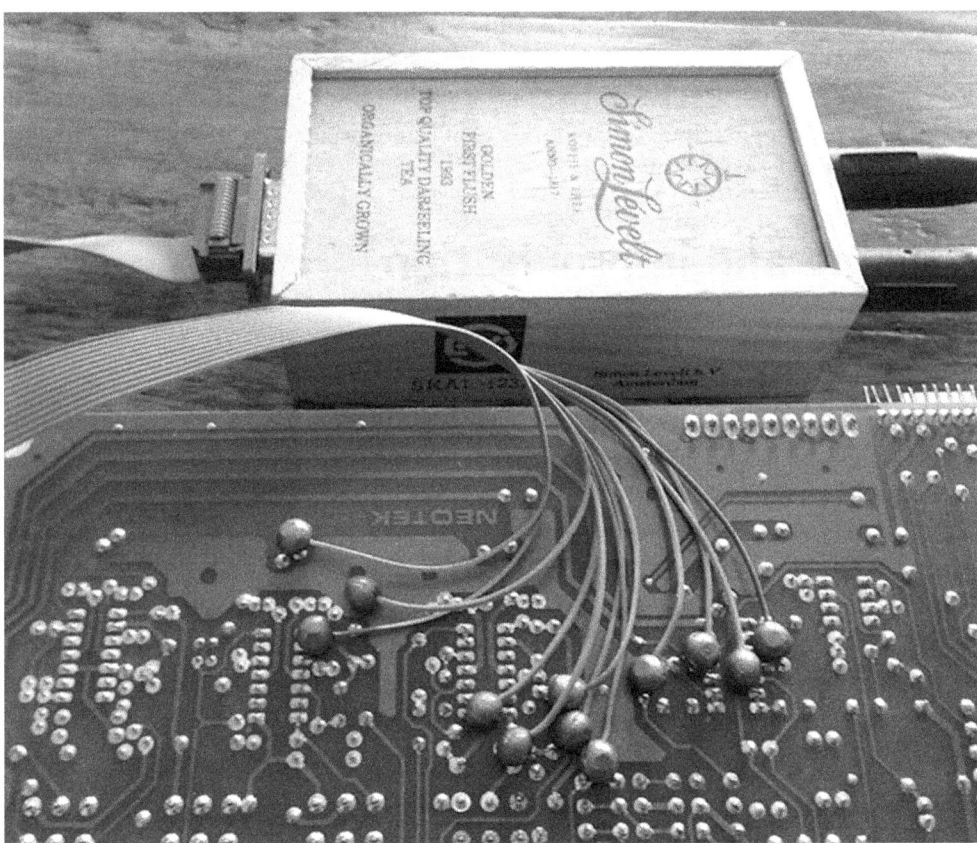

Figure 52.2 The oscillator box connected to the mixer circuit components using a ribbon connector attached to fishing weights for *The Royal Touch* set-up.

Electrical Walks, but it also offers a different relationship to the probed circuits. Indeed, where Kubisch's headphones are non-destructive, almost furtive modes of listening, Collins's circuit sees that target board as effectively *dead* until it is activated by his own oscillators. The systems are still all components, voltages, and fields, but, paraphrasing Tudor, the particular musician involved with them relates to them in dramatically different ways. Hacking is a multifaceted and ideologically malleable practice, and this is also true in musical contexts.

In his book *Handmade Electronic Music: The Art of Hardware Hacking*, Collins presents a series of rules meant to facilitate hardware hacking in a musical context.

> **Rule #6: Many hacks are like butterflies: beautiful but short-lived.**
> Many hacks you perform, especially early in your career, may destroy the circuit eventually. Accept this. If it sounds great, record it as soon as possible, and make note of what you've done to the circuit so you can try to recreate it later.
>
> (Collins 2006: 7)

No Input Mixing Board and *The Royal Touch* are both ways to create a performance system robust enough to turn hacked instruments' inherent unpredictability into a recognizable practice. They are indirect rebuttals to *Rule #6*. Nakamura does it by becoming extensively familiar with the recurring behaviours of his system and knowing how to work with unusual developments. Collins does so by developing a similar familiarity and complementing it with a novel patching method, through the use of fishing weights.

Hacking and improvisation

Although there are plenty of hacked instruments that offer no particular affinity with improvisatory contexts, the significant number of systems that are 'allowed to speak for themselves' means hacking has a privileged relationship to the generative and the improvised. Here, Derek Bailey's concept of 'mutual subversion' in free improvisation can be connected to the 'tacit knowledge' theorized by science scholar Harry Collins (Bailey 1993: 95–96; Collins 2001). Because of the relative 'user-unfriendliness'[3] of working with an instrument that can never be fully predictable, hacking systems like no-input mixers or dead circuit boards act as viable improvising partners which challenge and subvert their human equivalents. In response, the human player will, if not become better at controlling systems, better at intuiting the complex nature of the system, internalizing its behaviour – developing a tacit, embodied knowledge of it. Quoting Bailey:

> One of the basic characteristics of his [*sic*] improvising, detectable in everything he plays, will be how he harnesses the instrumental impulse. Or how he reacts against it. And this makes the stimulus and the recipient of this impulse, the instrument, the most important of his musical resources.
>
> (Bailey 1993: 97)

The Royal Touch setup is built with an early digital logic chip, a 'Hex Inverting Schmitt Trigger'. Inverting Schmitt triggers look at an incoming voltage, producing a low output if that input is above a threshold, and a high output if the input is under that threshold (Schmitt 1938). The 74C14 chip was developed in a family of low-power logic chip intended for the development and manufacture of digital systems (Lancaster 1988: 255). A circuit similar to the one used for *Royal Touch* is described in *Handmade*, it is the first circuit presented there: 'In the contrarian spirit of hacking', Collins writes, 'the first circuit we build from scratch is based on the misuse of an Integrated Circuit (IC) never intended for making sound' (Collins 2006: 111). More than simply fitting into a narrative of creative misuse, this fact helps clarify another nuance in the medium associated with hacking in and out of musical contexts. As evidenced by Wark's and Jordan's discussion of the topic, hacking is often meant to imply software hacking, even when evidence of hardware hacking is well known (Jordan 2008: 123). In a musical context, however, hacking holds no allegiance to vibratory mediums (as proven by the previous examples, which hacked electromechanical, electromagnetic, and purely electrical vibratory schemes), let alone hardware or software. In fact, the latter distinction breaks down quickly when considering that because of Collins's successful book, a wide number of hacked instruments function with digital logic chips used as signal generators and processors (Teboul 2015: 42). A number of toys prized by hackers as the basis for circuit bending, such as the Speak & Spell™, are based on digital signal processing or memory chips (129). The recent resurgence of modular synthesizers is correspondingly hybrid (Paradiso 2017), and follows a pragmatic approach to getting music done. Once again, this makes sense historically: early computer music experiments in academia and industry made materiality a central part of the process because of their non-real time nature and the physically demanding infrastructure necessary to turn code into recordings (Kahn 2012). Outside of academia, using early personal microcomputers such as the KIM-1 meant building custom analogue to digital converters, interfaces, network protocols, etc. (Perkis and Bischoff 2007). In musical hacking, there is room for hardware and software, if only because our fingers can only press buttons and turn knobs and our ears can only hear fluctuations in air pressure. Going further, I would suggest that in fact, because of the relatively low stakes of electronic instrument design, hacking in a musical context is uniquely privileged to blend analogue and digital technologies in an earnestly ruthless fashion. It would almost be a shame not to.

Bonnie Jones is a musician and educator whose practice often focuses on digital delay pedals. These are opened to reveal their circuit board, plugged into themselves (like a no-input mixer), flipped around, and powered on. Jones then takes ⅛" audio connectors and uses them to probe various copper traces and connectors looking for sounds. Digital delay guitar pedals are not meant to have their output plugged into their input, yet Jones has developed this setup in the same way that Nakamura has turned the mixer into a feedback instrument with lifetimes of possibilities. Jones states:

> I like when an instrument is indeterminate but there is also a desire to be able to play something – so yes while the pedals have a certain level of indeterminacy – playing them for over a decade has really allowed me to understand many of the ways to produce specific sounds. As for complexity – I use tools for the sounds they make vs the nature of their construction or code. I like getting to know an electronic instrument, and I appreciate when that instrument has some surprises or enables me to create and discover sounds that I wouldn't expect – but I wouldn't care if something was complex if I didn't like the way it sounded.
>
> (Jones interviewed in Teboul 2015: 188–189)

Considering the highly unusual mode of operation Jones puts these delays through, a traditional understanding of what happens at the circuit level during her performances seems, in this case, almost superfluous, coincidental at best. In that sense, Jones is closer to circuit bending values tacit, embodied knowledge such as the one she describes above to explicitly technical expertise. Early circuit bending practitioner Qubais Reed Ghazala states of the practice:

> Circuit bending's chance approach is an act of clear illogic. As opposed to fuzzy logic, a seeking of norm within chaos, clear illogic seeks chaos within the norm. It is through this chaos, a powerful creative force, that the instruments are allowed to behave beyond the theoretical intentions (and limitations) of the designer.
>
> (Ghazala 2004: 100)

Jones's practice goes beyond repurposing digital delay pedals, also incorporating live instruments, found sound, text art, etc. Hacking comes in many flavours and can fit within a larger grouping of methods, as simply one way to approach materiality and meaning. Jones's use of digital delay pedals clearly gives these devices an opportunity to build off of themselves. Paraphrasing Tudor once more, they reveal as much about themselves as what Jones may want an audience to hear.

With Jones and each previous artist, a local, specialized, and personal understanding of audio engineering emerged. Up to this point every example has foregrounded the materiality of hardware, but this perspective of hacking as a method of composition can also be helpful to understand mostly digital and software-based projects.

George Lewis is a musician, technologist, and professor whose 1993 work *Voyager* is both the name of an improvising computer music program and of a record made with the program along with various human players. The system is innovative in that it could both produce music independently and respond to musical input. The underlying mechanisms were programmed using predetermined scales, phrases, and progressions, coded in the programming language FORTH. The software was built between 1986 and 1988 at STEIM, the Studio for Electro Instrumental Music in the Netherlands (Lewis 2000: 34). Lewis had a history of building software mostly on his own, occasionally with advice on test drafts from various programmers and musicians (Born 1995: 191).

> Voyager functions as an extreme example of a 'player' program, where the computer system does not function as an instrument to be controlled by a performer […] I conceive a performance of Voyager as multiple parallel streams of music generation, emanating from both the computers and the humans – a nonhierarchical, improvisational, subject-subject model of discourse, rather than a stimulus/response setup.
>
> (Lewis 2000: 34)

The non-hierarchical nature of this custom, shape-shifting system of humans and machines clearly illustrate the 'dialogical' nature of both designing and playing along *Voyager*. This is evident in Lewis's commentary on FORTH:

> Seemingly anti-authoritarian in nature, during the early 1980s Forth appealed to a community of composers who wanted an environment in which a momentary inspiration could quickly lead to its sonic realization – a dialogic creative process, emblematic of an improvisor's way of working.
>
> (Lewis 2000: 34)

Musical hacking as a dialogue that mimics the dynamics of non-hierarchical improvisation is therefore just as possible in software as it is in hardware. The notion of agency is central to Lewis's scholarship (Lewis 1996). Here I use it to complicate Tudor's notion of 'letting the components speak for themselves'. Although *Voyager* speaks for itself, it only has agency over the phrases, scales, and other melodic and rhythmic structures programmed by Lewis in the software. This is apparent in *Voyager*'s reliance on MIDI to enact its musicality, which signals Lewis's relative disinterest in timbre as opposed to note and melody-scale events. The non-features are almost as significant a statement as the features themselves.

Hacking: For whom?

These examples have spanned the following mediums: the electromechanical, the electroacoustic, the electronic, digital hardware, and software. Real time, techno-cultural, circuit bending: these are just some of the flavours of hacking where experimental music meets experimental technology. This is not an exhaustive selection, but a map presenting some of the principal spaces of musical hacking's multidimensional affordances. In offering this, I have introduced hacking in a musical context as a mutually subversive dialogue between material realities and musical ideas, which often include a large number of people and objects across time and space. I have presented the resulting system-compositions as materialized compromises of these negotiated and nested agencies enacted through various intermediaries, across time and space. These compromises can be partially read and reconstructed by inspection of the artefacts left behind, a process facilitated by open sourcing thorough textual or technical documentation and which can serve as the basis for new works.[4] In each case, hacking's engagement with the materiality of each medium has allowed the specificities of each device and the users relationship to them to be partially revealed through sound.

The humanity of technologies, then, is clear. Not because machines act like us, but because we made machines to act in ways we imagined as potentially interesting: the arbitrariness of some decisions and standards is visible after close examination of hacked musical systems. This doesn't value machine–improviser relationships less than human–human ones, as a dialogue with machines can be conceived of as a dialogue with nested and crystallized past human and material agencies. However, as with all methodologies, in music or otherwise, we should consider the ethics of these techno-social affordances. Quoting Tara Rodgers:

> As soundmakers and sound students […] we should be attuned to how historically and culturally specific metaphors and descriptive language frame our knowledge of sound, as well as to sound's potential for complex communications of its own kind. Instruments and interfaces are often where these two trajectories join forces: technological designs crystallize sound knowledge into material forms that, in turn, generate more sounds. Knowing an instrument's history and interrogating the logic of its design can be a productive starting point for creative interruption and innovation […]. Sound is both a carrier of cultural knowledge and an expressive medium modulated by individual and collaborative creativity.
> (Rodgers 2018: 239–240)

I will conclude this chapter by emphasizing the power dynamic inherent to hacking's mutually subversive relationship and how this may shape the creation of new technological systems and musical works. I want to be critical of the optimistic presentations of the practice, in and out of music, as a 'democratizing' force, rather than as an ideologically malleable tool with undeniable techno-cultural potential. The instances of hacking described above helped musicians engage with their audiences and the increasingly technological in-between that connects them, on their own terms, but the degree of self-awareness varied. Jones and Lewis frame their technical and artistic work and perspective in the context of an inescapably political existence. Paraphrasing Rodgers, building the critical tools for connecting technological systems with the labour necessary to their construction offers new grounds for artistic creation, which hacking and tinkering are uniquely fit to explore. It also presents the possibility of progress in adequately crediting and acknowledging the various humans and nonhumans involved in these generative processes, poetically, through works of art, and explicitly, through written or oral research.[5] In framing electronic music composition as a dialogue with machines, I hoped to hint at the ethical nature of picking who we wish to listen to and who we hope to let speak through or with technology. If, as Wark suggests, hacking makes the new out of the old, then by better understanding hacking we can better appreciate its inherently utopian aspects. If these are realized through music, then the act of the hack comes with the responsibility to perhaps make our musical systems and compositions embody the values we wish to see in our fields and beyond. To summarize: I have presented hacking as a compositional method with the potential to make new works out of existing materials and ideologies at various scales, where the decision to implement modes of control over technology connects those materials and ideologies. In situating ourselves within this question of control between sound and its sources, we reiterate the connection between tools, art making, and cultural critique, and reify listening and responding as tactics for actively participating in the making of our actually existing experimentalisms (Piekut 2011: 195).

Notes

1. Email exchange with Marko Ciciliani, 31 May 2018. Additional no-input pieces by Ciciliani are included on *81 Matters in Elemental Order* (Evil Rabbit Records 2008).
2. Email exchange with Nicolas Collins, 21 May 2018.
3. For a longer discussion of electronic instruments, 'user-unfriendliness' and Anthony Dunne's concept of 'post-optimality' in the context of audio technologies, see Teboul 2018.
4. This is, in part, the work of media archeology (see Huhtamo and Parikka 2012).
5. See Haring 2007; Rodgers 2015; Vagnerova 2017.

References

Bailey, Derek (1993). *Improvisation: Its Nature and Practice in Music*. New York: Da Capo.
Born, Georgina (1995). *Rationalizing Culture: IRCAM, Boulez, and the Institutionalization of the Musical Avant-Garde*. Berkeley, CA: University of California Press.
Collins, Harry M. (2001). 'Tacit Knowledge, Trust and the Q of Sapphire'. *Social Studies of Science* 31 (1): 71–85.
Collins, Nicolas (2006). *Handmade Electronic Music: The Art of Hardware Hacking*. New York: Routledge.

Collins, Nicolas (2007). 'Live Electronic Music'. In *The Cambridge Companion to Electronic Music*, 38–54. Cambridge: Cambridge University Press.

Cox, Christoph (2006). *Invisible Cities: An interview with Christina Kubisch*. Available online: http://cabinetmagazine.org/issues/21/cox.php (accessed 27 May 2018).

Dunne, Anthony (2008). *Hertzian Tales: Electronic Products, Aesthetic Experience, and Critical Design*. Cambridge, MA: MIT Press.

Ghazala, Qubais Reed (2004). 'The Folk Music of Chance Electronics: Circuit-Bending the Modern Coconut'. *Leonardo Music Journal*: 97–104.

Haring, Kristen (2007). *Ham Radio's Technical Culture*. Cambridge, MA: MIT Press.

Huhtamo, Erkki and Jussi Parikka (eds) (2012). *Media Archaeology: Approaches, Applications, and Implications*. Berkeley, CA: University of California Press.

Jones, Ralph (2004). 'Composer's Notebook: Star Networks at the Singing Point'. *Leonardo Music Journal* 14: 81–82.

Jordan, Tim (2008). *Hacking: Digital Media and Technological Determinism*. New York: Polity.

Kahn, Douglas (2012). 'James Tenney at Bell Labs'. in *Mainframe Experimentalism: Early Computing and the Foundations of the Digital Arts*, 131–146. Berkeley, CA: University of California Press.

Lancaster, Don (1988). *CMOS Cookbook*. 2nd edition. New York: SAMS.

Lewis, George E. (1996). 'Improvised Music after 1950: Afrological and Eurological Perspectives'. *Black Music Research Journal*: 91–122.

Lewis, George E. (2000). 'Too many notes: Computers, complexity and culture in Voyager.' *Leonardo Music Journal* 10: 33–39.

Nakai, You (2016). 'On the Instrumental Natures of David Tudor's Music'. PhD dissertation, New York University.

Paradiso, Joe (2017). 'The Modular Explosion: Deja Vu or Something New?'. Presented at the *Voltage Connect Conference*, Berklee College of Music, Boston, MA, 10–11 March 2017.

Perkis, Tim and John Bischoff (2007). *The League of Automatic Music Composers 1978–1983*. Liner Notes. Brooklyn, NY: New World Records.

Piekut, Benjamin (2011). *Experimentalism Otherwise: The New York Avant-Garde and Its Limits*. Berkeley, CA: University of California Press.

Rodgers, Tara (2015). 'Cultivating Activist Lives in Sound'. *Leonardo Music Journal* 25: 79–83.

Rodgers, Tara (2018). 'Approaching Sound'. In Jentery Sayers (ed.), *The Routledge Companion to Media Studies and Digital Humanities*, 233–242. New York: Routledge.

Sanfilippo, Dario and Andrea Valle (2015). 'Towards a Typology of Feedback Systems'. In *Proceedings of the International Computer Music Conference*, 30–37.

Schmitt, Otto H. (1938). 'A Thermionic Trigger'. *Journal of Scientific Instruments* 15 (1): 24–26.

Teboul, Ezra J. (2015). 'Silicon Luthiers: Contemporary Practices in Electronic Music Hardware'. MA Thesis, Dartmouth College.

Teboul, Ezra J. (2017). 'The Transgressive Practices of Silicon Luthiers'. In *Guide to Unconventional Computing for Music*, 85–120. London: Springer.

Teboul, Ezra J. (2018). 'Electronic Music Hardware and Open Design Methodologies for Post-Optimal Objects'. In *Making Things and Drawing Boundaries: Experiments in the Digital Humanities*, 177–184. Minneapolis, MN: University of Minnesota Press.

Thompson, Marie (2017). *Beyond Unwanted Sound: Noise, Affect and Aesthetic Moralism.* New York: Bloomsbury.

Tudor, David (1976). 'The View from Inside'. Box 19, David Tudor Papers, Getty Research Institute.

Vágnerová, Lucie (2017). '"Nimble Fingers" in Electronic Music: Rethinking Sound through Neo-colonial Labour'. *Organised Sound* 22 (2): 250–258.

Wark, McKenzie (2004). *A Hacker Manifesto.* Cambridge, MA: Harvard University Press.

Index

acoustics 203–4, 206, 209, 660
   electro-acoustics 205, 206
   psycho-acoustics 205
acoustic environments in change 602
acoustic profiling 159–62, 165
   management 217
   Tag 535
Adorno, Theodor W. 10–12, 18, 24, 17, 373, 710
aerial projection 584, 588, 592
affect 116–17, 120–4, 128–31, 135
affective tonality 672–3, 680, 683
affordance 644
against method 62, 65, 285–7, 293, 294n16
ambiance 671–74, 678–82, 684, 722
Amsterdam 557–8, 562–3, 569
animal communication 75
   monitoring 7
   hearing 81
Anzaldúa, Gloria 800–1, 803, 804n4
archive 779–81, 784, 786–7, 790
   colonial 533–534
   internet archive 782
   Library of Congress 782
Argentinazo 708–9
artistic research 327–8, 363
   method 426
   atmosphere 423
   atmospheric 671–4, 680–4
attunement 322
audibility, thresholds 651–652
audiograph 159–60, 165
audioposition 156, 160, 162, 165
audio
   diagramming 511–25
   file 660
   procedural 398, 402
   technology 481, 483–4, 491
auditory experience, (on-site) 558, 564–5, 567, 573–5
   diagramming 514–15
   graph 733–4, 737
   perception 187–92, 195–7
Augoyard, Jean Francios 761
auralization 567, 569–70
auraltypical 604
aural postcard 772
auto-ethnography 119, 121–2, 129, 352
Azzigotti, Luciano 713

Barad, Karen 2, 10–13, 175–6, 180
Benjamin Walter 21
Bennett, Jane 171–2, 173, 174, 179
Big Ben Silent Minute 96, 99–100, 105, 111
binaural 773
Böhme, Gernot 317–19, 321–22
border thinking 799–802, 804n4
Bowie, David 663
Brady, Erica 11
Buenos Aires Sonora (BAS) 714

Cacerolazo 705–710
Cage, John 27, 173–4, 176
Canetti, Elias 601
Cairo (Egypt) 770
Cartesian dualism 734–5
cartography 118–19, 131–3, 135
Cave, Nick 664
Chiari, Giuseppe 305, 308
city maker 557–8, 560, 569, 575–6
city user 557–60, 562, 574–6
clairaudience 648
climate change 449–51, 453–4, 457–9, 465
cohesion, cohesive 660, 664, 666, 668
Collectif Rendez-Vous 602
Collins, Nicolas 811
community engagement 452–3, 456
competition 643, 637
compositional process 391
conflict 745–7, 749, 751
co-production 745–6, 748, 751

contamination 671, 674, 676, 679–80
Corbin, Alain 647, 651, 653
Cox, Christoff 370
crowds 636–637, 643
Cusak, Peter 5, 461, 582, 590

Darlington Quiet Town Experiment 94, 96, 99–100, 106–11
de Certeau, Michel 299–300
De Luca, Erik 461
Delgado, Marcelo 711, 712, 715
delocalization 399, 403–4
Derrida, Jacques 9, 13
design process 686–8, 690, 696, 698–700
diffraction 10
digitization 532–3
directness 736–8, 741
Dunn, David 461
Dreyfus, Laurence 361, 362

ear piece 668
earwitness 650
earview (defined) 796
earworms 163–164
ecoacoustics 614–15, 711
echolocation 80
ecosystem 451–2, 455–8
environment 612–14, 616–700, 712
    monitoring 87
    impact assessment 223, 589
    strategic assessment 225
Eidsheim, Nina Sun 172–3, 174, 179–80
embodied experience 734–6, 739–41
Eshun, Kodwo 668
ethnography 118, 122–4, 126, 131, 135, 612–13, 616–71, 674, 680–83, 711, 756
    multi-sited 123, 125
ethnology 756
ethnomusicology 755
eurocentrism 2, 457, 803
experiential 745
experiment, experimental, experimentation 5, 11, 41, 43, 47, 53, 55, 79, 85, 87, 90, 94, 96, 99, 100, 106–8, 111, 113, 117, 123, 125, 131–3, 143, 145, 157, 163, 166, 167, 170, 179, 182, 191–2, 197, 199, 203, 205–9, 213, 215, 221, 229, 250, 256, 264, 267, 273–5, 278, 283–4, 286, 289–94, 304–5, 308, 312–13, 315–7, 326–7, 330, 334, 336, 339, 343, 357–359, 363–7, 387, 400–1, 407, 418, 420–1, 423, 427, 433, 457, 479, 481, 483–4, 486, 490–1, 493, 499, 507, 559, 577, 580–1, 586, 597, 600, 603, 610, 612, 631, 654, 656, 671, 672, 673, 686, 688–91, 693–5, 703, 712, 739, 740, 764, 766, 767, 769, 771–3, 779, 807–8, 815, 817–19
evaluation 687, 689, 690, 692, 694–9, 700
exhibition 567, 569, 571

Feld, Steven 756
feminist critique 553–554
Ferrara, Lawrence 287, 288, 294
field recording 119, 130, 132–3, 450–3, 455, 482, 484–7, 492, 493, 582
fieldwork 766
film (sound making) 688, 693
fixation 685, 686, 687, 688, 700
flâneur 297–8, 311, 301–2
foon/phon 500, 502, 503–7, 508 (fn), 509–10

Hamdan, Lawrence Abu 176–7, 178–9, 591
Harraway, Donna 13
health impact assessment 226
heuristics 660, 664–665
hearing
    imaginative 517
    way of hearing 97, 99–100, 103, 111
Howes, David 4, 767
hydrophones 460–2, 470–1, 476

ideology 719, 720, 721, 722, 723, 724, 728, 729
illocution 719, 720, 721, 722, 729
IMPA 711, 712
improvisation 325, 328–9, 330–6
    critical studies in 326, 328–9
information visualization 733–5
interdisciplinarity 3, 651–4
internet humour 546
interview (go-along) 564, 565–7
infrastructure 174–5, 176–7, 180
installation 450–4
instrument 203–4, 206–7
    musical 204–5, 207, 211
    scientific 205
    builders 204, 207
    digital 207

interaction design 423
interpretation 358–66
Iron Dome 719, 724, 725, 726, 727, 728
Israel–Gaza Conflict 719, 723, 724, 728

Jacob, Wendy 461
Jeck, Philip 461
Jones, Bonnie 815

Kahn, Douglas 177–8, 179–80
Klangumweld 513
Knowledge
    elicitation (tacit and explicit) 558, 561, 564, 575, 576
    production 532–4
Kubisch, Kristina 5, 174–5, 176, 178–9, 808
Kunzra, Hari (*White Tears*) 155–8

La Bandina 711–12, 715
Labelle, Brandon 469, 471
Lane, Cathy 592
laser-Doppler vibrometry 84
Lewis, George 816
librettization 163
listening 177–9, 208–9, 673–5, 681, 683–4, 745–8, 750–2
    acousmatic listening 489
    active 405
    attentive listening 469, 482–4, 488, 491
    body 661
    close 481, 493, 535–7
    critical listening 719, 721–6, 729
    cultures 205
    directing 403
    habits 203, 207
    new forms of urban 559
    peripheral 404
    together 565, 575–6
    ubiquitous 659
    underwater 469–78
Lockwood, Annea 461
loudness 262

magnetic tape 660
Manning, Erin 286, 287, 290, 294
Marclay, Christian 807
maps, mapping 118–21, 131, 132, 612–15, 710 735–8, 739–41

counter-mapping 591
distortion 587
inside and outside 589
live 594
noise map 583
Marconi, Gugliemo 157–8
materialism 170, 175
    materialism new 171
    materialism, sonic 170, 173, 176–7, 178–9, 180–1
materiality 171–3, 175–7, 179, 201–3, 207, 211, 358–61, 363–4
mediation 173, 176, 178, 180
media urbanism 120–1
metadata 779–80, 782, 784, 786–8, 790
metaphor 61–2, 67–71
method, as invention 57–8
    live 117, 128–9, 132–3, 135
    as *meta-hodos* 63–67
    triangulation 118, 135
micro CT-scanning 85
microphone 773
modelling 219
Mowitt, John 173–5, 176, 178
music 202–3, 205–7, 210, 755
    aesthetics 207
    culture 204, 205–6
    circulation 210
    industry 203
    technology/instrument 204–5, 207
    performance 206, 210
    mobile 659
    metaphor 207
    notation 358–65
    pitch 207
    musician 204, 207

Nakamura, Toshimara 810
neoliberalism 543–4, 550, 554
    and feminism 543, 548–9, 553
Neuhaus, Max 315, 316–17, 322
no-input mixer 810
noise 94, 99–103, 217
    city noise 647–51
    control 219
    cries of Paris 648
    of machines 481
    noise pollution 482, 492

pink 401–3
tolerance 651–2
traffic 482, 487

Oliveros, Pauline 3, 473, 668
ontological design 795
oscillations 661

Pardoen, Mylène/Projet Bretez 653, 655
participatory-collaborative process 557–8, 574, 576, 612–13, 616, 712
perception 169, 172–3, 175–6, 179
performativity 719–23, 728–9
phenomenology 169–70, 178
    spatial 193–5
phonautograph 648, 660
phonograph 78
Picker, John 647, 650
Piper, Adrian 369, 373, 377
podagogy 142–3
podcasting 141–2, 779–80, 782, 784, 786, 790
Polli, Andrea 461
pollution 633–634
print textualized orality 161–2
practice 201–10
prototyping 686–8, 693–4, 696, 700
psychoacoustics 253, 262
Pynchon, Thomas 164–165

quantification 544–545, 548–551
quiet 94–5, 99–103, 106–8, 111

Raaijmakers, Dick 283, 284, 289, 292n1, 294
recording 746–9, 751–2, 756
reflexivity 341–3, 348, 351, 353
reflective practice 344, 352
reification 172, 177–8
Rheinberger, Hans-Jorg 286, 287, 289, 290, 292n3, 292n5, 293n7, 295
risk 671–5, 677, 679–80
RSS 780, 782–3, 786–8

San Cipriano Picentino, Italy 582
Satie, Erik 706, 715
Schafer, Murray R 3, 28, 253, 512, 600–5, 608, 648, 650
schizophonic artifacts 696, 701

Schweighauser, Philipp (*The Noises of American Literature*) 159–60
scratch orchestra 305–7
sense/sensory 671–3, 677, 680–4
    anthropology 767
    sensory corpus 662, 665, 668
    felt 662, 665, 667
sensuous sociology 117, 120
Serres, Michel 664
Shepherd, Nan 590
Shiomi, Mieko 181
silence 59, 66–7, 671–81
situationism 304, 308, 311
skills 201, 203, 207, 739–41
Smith, Bruce R. 647
Smith, Mark M. 98–9, 647, 653–4
social constructivism 201–4, 208
socioecological environment 773
sonagraph 79
sonification 453–4, 733
    aesthetics 734, 735, 736, 740–1
    listening 735, 739–41
sonogram (spectral analysis) 637, 639–43
song learning 79
sonic
    acts 721–2, 729
    agency 5
    encounter 97, 103, 105–6, 108, 111
    experience 659–62, 664–5, 667
    density 634, 639
    extended environment 404–5
    material 663, 667
    past (reconstruction of) 653–4
    percepts 663
    persona 665–6, 668
    representation 400, 405
    rupture 315–6, 321
    situation 665–7
    writing 662, 664–8
sonic methodology
    and philosophy 187–99
sonic research, types of 647–51
sound
    sound awareness 557–61, 567, 575–6
    archive 533–4, 536
    ambient 481, 487
    and color 190, 195–7

design 397–401, 405
everyday 483, 489, 491
installations 229, 315–22
as metaphor 61–2, 70
milieu 634
philosophy of 187–98
sculpture 451, 453
and space 187, 192–5
technology 147, 151
and time 187, 195–7
wanted 228
writing 151–2
underwater 459–67
sound art 325, 333–4, 370, 372, 705, 711
ecological 449–58
sound artists 558, 562, 566–7
sounding Dartmoor 604
soundscape 155–7, 159, 254, 255, 258, 260, 450–6, 612, 614–16, 711
composition 486
environmental soundscape 481, 488
everyday soundscape 489, 492
lo-fi soundscape 482
hi-fi soundscape 482, 492
planning 227
urban soundscapes 485
sound sources 187, 189–91
sound studies 343–5, 760
soundwalk 119, 121, 129, 132, 135, 259, 261, 298–300, 762
augmented 567, 569, 570
commented 564–565
self-guided 565, 567
Sound System of the State 719, 721–23, 725, 727–29
space 735, 736, 739–41
public 304–5, 308–11, 426, 633–4
space-use-sound model 560, 566, 571, 574
urban 745–6, 748, 750, 752
spectrograph 584
speech acts 719–22
spotify 549–52
Steingo, Gavin 2–3, 20
Sterne, Jonathan 18, 99, 156, 647
Stoever, Jenny-Lynn 747, 750
Stravinsky, Igor 358–62, 364
stridulation 82

subject position 738–40
syrinx 82
Szendy, Peter 62, 66, 73, 351, 355, 504, 507, 510

technologies (sound) 201–9
telephone 68–70
theory, grounded 260
Thompson, Emily 99
Thompson, Marie 101, 104–5
Toop, David 648
transduction 179–81, 661
transportation (bus) 634–7
Trower, Shelly 170–1, 173, 175, 179
Thulin, Samuel 581
tympanum 60–1, 65
Tyndall, John 584
two-voice phenomenon 83

urban 298–301, 303–5, 308–10
planning 217
urban square, redesign of 558, 560, 565, 566–7, 569, 570, 571, 574
urban sound 118–21, 135

Varchausky, Nicolás 712
verbalization 771
véxations 706, 715
vibration 170–1, 172–3, 175–6
vinyl records 383–95
visual perception 187, 192–5
vulnerability 671, 679–80

Waldock, Jacqueline 586
walking 297–304, 308, 310–12, 612–13, 618–19
Wang, Hong Kai 593
Watson, Chris 461
Werktreue 346
Westerkamp, Hildegard 3, 600–2, 609–10, 668
White, Hayden 9
wilderness 610–14, 616, 620, 700–11
Wizard-of-Oz (design method) 696–8
workshop 561, 564, 566–8, 570, 574
World Soundscape Project 3, 486, 585–6, 600–4, 608
World Trade Center 588

Zukofsky, Louis 169–70, 171, 173, 178–9, 181